T0183420

Graduate Texts in Mathematics

Series Editors

Sheldon Axler, *San Francisco State University*
Kenneth Ribet, *University of California, Berkeley*

Advisory Board

Alejandro Adem, *University of British Columbia*
David Eisenbud, *University of California, Berkeley & MSRI*
Brian C. Hall, *University of Notre Dame*
Patricia Hersh, *University of Oregon*
J. F. Jardine, *University of Western Ontario*
Jeffrey C. Lagarias, *University of Michigan*
Eugenia Malinnikova, *Stanford University*
Ken Ono, *University of Virginia*
Jeremy Quastel, *University of Toronto*
Barry Simon, *California Institute of Technology*
Ravi Vakil, *Stanford University*
Steven H. Weintraub, *Lehigh University*
Melanie Matchett Wood, *University of California, Berkeley*

Graduate Texts in Mathematics bridge the gap between passive study and creative understanding, offering graduate-level introductions to advanced topics in mathematics. The volumes are carefully written as teaching aids and highlight characteristic features of the theory. Although these books are frequently used as textbooks in graduate courses, they are also suitable for individual study.

More information about this series at http://www.springer.com/series/136

Konrad Schmüdgen

An Invitation to Unbounded Representations of *-Algebras on Hilbert Space

 Springer

Konrad Schmüdgen
Fakultät für Mathematik und Informatik
Universität Leipzig
Leipzig, Germany

ISSN 0072-5285 ISSN 2197-5612 (electronic)
Graduate Texts in Mathematics
ISBN 978-3-030-46368-7 ISBN 978-3-030-46366-3 (eBook)
https://doi.org/10.1007/978-3-030-46366-3

Mathematics Subject Classification: 47L60, 16G99, 16W10, 81S05

© The Editor(s) (if applicable) and The Author(s), under exclusive license to Springer Nature Switzerland AG 2020
This work is subject to copyright. All rights are solely and exclusively licensed by the Publisher, whether the whole or part of the material is concerned, specifically the rights of translation, reprinting, reuse of illustrations, recitation, broadcasting, reproduction on microfilms or in any other physical way, and transmission or information storage and retrieval, electronic adaptation, computer software, or by similar or dissimilar methodology now known or hereafter developed.
The use of general descriptive names, registered names, trademarks, service marks, etc. in this publication does not imply, even in the absence of a specific statement, that such names are exempt from the relevant protective laws and regulations and therefore free for general use.
The publisher, the authors and the editors are safe to assume that the advice and information in this book are believed to be true and accurate at the date of publication. Neither the publisher nor the authors or the editors give a warranty, expressed or implied, with respect to the material contained herein or for any errors or omissions that may have been made. The publisher remains neutral with regard to jurisdictional claims in published maps and institutional affiliations.

This Springer imprint is published by the registered company Springer Nature Switzerland AG
The registered company address is: Gewerbestrasse 11, 6330 Cham, Switzerland

Preface and Overview

Everything Should Be Made as Simple as Possible, But Not Simpler[1]

The purpose of this book is to give an introduction to the unbounded representation theory of *-algebras on Hilbert space. As the title indicates, the book should be considered as an invitation to this subject rather than a monograph or a comprehensive presentation.

Let us briefly explain the two main concepts explored in this book.

A complex *-algebra A is a complex algebra with an involution, denoted by $a \mapsto a^+$. An involution is an antilinear mapping of A into itself which is antimultiplicative (that is, $(ab)^+ = b^+ a^+$) and involutive (that is, $(a^+)^+ = a$). The complex conjugation of functions and the Hilbert space adjoint of operators are standard examples of involutions.

Just as rings are studied in terms of their modules in algebra, it is natural to investigate *-representations of *-algebras. Let $\mathcal{D}$ be a complex inner product space, that is, $\mathcal{D}$ is a complex vector space equipped with an inner product $\langle \cdot, \cdot \rangle$, and let $\mathcal{H}$ be the corresponding Hilbert space completion. A *-representation of a *-algebra A on $\mathcal{D}$ is an algebra homomorphism π of A into the algebra of linear operators on $\mathcal{D}$ such that

$$\langle \pi(a)\varphi, \psi \rangle = \langle \varphi, \pi(a^+)\psi \rangle, \quad \varphi, \psi \in \mathcal{D}, \tag{1}$$

for all $a \in$ A. In general, the operators $\pi(a)$ are *unbounded*. Equation (1) is crucial, because it translates algebraic properties of elements of A into operator-theoretic properties of their images under π. For instance, if $a \in$ A is hermitian (that is, $a^+ = a$), then the operator $\pi(a)$ is symmetric, or if a is normal (that is, $a^+ a = aa^+$), then $\pi(a)$ is formally normal (that is, $\|\pi(a)\varphi\| = \|\pi(a^+)\varphi\|$, $\varphi \in \mathcal{D}$). Since the closure of the symmetric operator $\pi(a)$ for $a = a^+$ on the Hilbert space $\mathcal{H}$ is not necessarily self-adjoint, we are confronted with all the difficulties of unbounded operator theory.

---

[1] Attributed to Albert Einstein.

In quantum mechanics the canonical commutation relation

$$PQ - QP = -i\hbar I \qquad (2)$$

plays a fundamental role. Here P is the momentum operator, Q is the position operator, and $\hbar = \frac{h}{2\pi}$ is the reduced Planck's constant. Historically, relation (2) is attributed to Max Born (1925)[2]. It implies Werner Heisenberg's uncertainty principle [Hg27]. Born and Jordan [BJ26] found a representation of (2) by infinite matrices. Schrödinger [Schr26] discovered that the commutation relation (2) can be represented by the *unbounded* operators P and Q, given by

$$(Q\varphi)(x) = x\varphi(x) \text{ and } (P\varphi)(x) = -i\hbar\frac{d\varphi}{dx}, \qquad (3)$$

acting on the Hilbert space $L^2(\mathbb{R})$. It was shown later by Wielandt [Wie49] and Wintner [Wi47] that (2) cannot be realized by bounded operators. For the *mathematical* treatment of the canonical commutation relation (2), there is no loss of generality in setting $\hbar = 1$, upon replacing P by $\hbar^{-1}P$.

The unital $*$-algebra W with hermitian generators p and q satisfying the relation $pq - qp = -i \cdot 1$ is called the *Weyl algebra*. Since relation (2) cannot hold for bounded operators, W has no $*$-representation by bounded operators, but it has many unbounded $*$-representations. Among them there is one distinguished "well-behaved" representation, the Schrödinger representation π_S, or its unitarily equivalent version, the Bargmann–Fock representation. The $*$-representation π_S acts on the Schwartz space $\mathcal{S}(\mathbb{R})$, considered as a subspace of the Hilbert space $L^2(\mathbb{R})$, by

$$\pi_S(p)\varphi = P\varphi \text{ and } \pi_S(q)\varphi = Q\varphi, \ \varphi \in \mathcal{S}(\mathbb{R}),$$

where P and Q are given by (3) with $\hbar = 1$. The Weyl algebra has a rich algebraic structure and an interesting representation theory. This $*$-algebra will be our main guiding example through the whole book; it is treated in detail in Chap. 8.

Aims of the Book

For decades, operator theory on Hilbert space and operator algebras have provided powerful methods for quantum theory and mathematical physics. Among the many books on these topics, two can be recognized as standard textbooks for graduate students and researchers. These are the four volumes [RS72]–[RS78] by

---

[2]In a letter to Pauli [Pa79, pp. 236–241], dated September 18, 1925, Heisenberg called the commutation relation (2) "eine sehr gescheite Idee von Born" ("a very clever idea of Born"). In the literature the relation (2) was first formulated by Born and Jordan [BJ26] and by Dirac [D25].

Reed–Simon covering operator theory and the two volumes [BR87]–[BR97] by Bratteli-Robinson for C^*- and W^*-algebras. The present book might be considered a supplement covering unbounded representations of general $*$-algebras.

The aims and features of this book are the following:

- *The main aim is to provide a careful and rigorous treatment of the basic concepts and results of unbounded representation theory on Hilbert space.*

Our emphasis is on representations of important nonnormed $*$-algebras. In general, representations of $*$-algebras on Hilbert space act by unbounded operators. It is well known that algebraic operations involving unbounded operators are delicate matters, so it is not surprising that unbounded representations lead to new and unexpected difficulties and pathologies. Some of these are collected in Sect 4.7. In fact, these phenomena already occur for very simple algebras such as the Weyl algebra or polynomial algebras.

Compared to bounded Hilbert space representations, many results and developments require additional assumptions, concepts, and technical arguments. We point out possible pathologies and propose concepts to circumvent them.

- *In the exposition and presentation we try to minimize the use of technicalities and generalities.*

So we treat the representation theory of the Weyl algebra only in dimension one; positivity only for functionals rather than complete positivity of mappings; decomposition theory only for functionals and not for representations; and we avoid details from the theory of quantum groups. Some results with long and technically involved proofs, such as the trace representation Theorem 3.26 and the integrability Theorems 9.49 and 9.50 for Lie algebra representations, are stated without proofs. (The reader can find these topics and complete proofs in the author's monograph [Sch90].) We hope to fulfill Einstein's motto stated above in this manner, at least to some extent.

- *The choice of topics illustrates the broad scope and the usefulness of unbounded representations.*

There are various fields in mathematics and mathematical physics where representations of general $*$-algebras on Hilbert space appear. The canonical commutation relation of quantum mechanics was already mentioned and is only one example. Quantum algebras and noncompact quantum groups can be represented by unbounded operators. Unitary representations of Lie groups lead to in general unbounded representations of enveloping algebras. Representations of polynomial algebras play a crucial role in the operator approach to the classical multi-dimensional moment problem. Noncommutative moment problems are closely related to Hilbert space representations. Properties of states on general $*$-algebras are important in noncommutative probability theory. Dynamical systems appear in the representation theory of operator relations. Noncomutative real algebraic geometry asks when elements, which are positive operators in certain representations, are sums of hermitian squares, possibly with denominators. These topics will appear in this book; for most of them we provide introductions to these subjects. Some of them are treated in great detail, while others are only touched upon.

- *Our aim is to present fundamental general concepts and their applications and basic methods for constructing representations.*

The GNS construction is a powerful tool that is useful to reformulate or to solve problems by means of Hilbert space operators. We carry out this construction in detail and apply it to the study of positive functionals on $*$-algebras. Further, we develop general methods for the construction of classes of representations such as induced representations, operator relations, and well-behaved representations. Representations on rigged modules or Hilbert C^*-modules is a new topic which belongs to this list as well. Throughout, our main focus is on basic ideas, concepts, examples, and results.

- *For some selected topics self-contained and deeper presentations are given.*

This concerns the representation theory of the Weyl algebra and the theory of infinitesimal representations of enveloping algebras. Both topics are extensively developed including a number of advanced and deep results. Also, Archimedean quadratic modules and the corresponding C^*-algebras are explored in detail.

Brief Description of the Contents

Chapter 1 should be considered as a prologue to this book. We give a brief and informal introduction into the algebraic approach to quantum theories thereby provided some physical motivation for the study of general $*$-representations and states of $*$-algebras.

Chapter 2 deals with the algebraic structure of general involutive algebras. Basic constructions (tensor products, crossed products, matrix algebras), examples (semigroup $*$-algebras, $*$-algebras defined by relations), and concepts (characters, positive functionals, quadratic modules) are introduced and investigated.

Chapter 3 gives a short digression into O^*-algebras. These are $*$-algebras of linear operators on an invariant dense domain of a Hilbert space. The involution is the restriction of the Hilbert space adjoint to the domain. We treat three special topics (graph topology, bounded commutants, and trace functionals) that are used later in the study of representations.

With Chap. 4 we enter the main topic of this book: $*$-representations on Hilbert space. We develop basic concepts (closed, biclosed, self-adjoint, essentially self-adjoint representations), in analogy to single operator theory, and standard notions on representations (invariant subspaces, irreducible representations). The heart of this chapter is the GNS construction which associates a $*$-representation with each positive functional. It is probably the most important and useful technical tool in Hilbert space representation theory.

Chapter 5 is devoted to a detailed study of positive linear functionals on *-algebras. The GNS representation allows one to explore the interplay between properties of Hilbert space representations and positive functionals. Ordering, orthogonality, transition probability, and a Radon–Nikodym theorem for positive functionals are treated in this manner. Choquet's theory is applied to obtain extremal decompositions of states. Quadratic modules defined by representations are introduced.

Chapters 6–9 are devoted to the representation theories of some important special classes of *-algebras.

Chapter 6 deals with tensor algebras and free *-algebras. Positive functionals are approximated by vector functionals of finite-dimensional representations and faithful representations are constructed. We define topological tensor algebras such as the field algebra of quantum field theory and develop continuous representations.

Chapter 7 is about "well-behaved" representations and states of commutative *-algebras. We characterize these representations by a number of conditions and express well-behaved representations of finitely generated *-algebras in terms of spectral measures.

Chapters 8 and 9 are two core chapters that stand almost entirely by themselves.

Chapter 8 gives an extensive treatment of Hilbert space representations of the canonical commutation relation (2) and the Weyl algebra. After collecting algebraic properties of this algebra we treat the Bargmann–Fock representation and the corresponding uniqueness theorem. Then the Schrödinger representation is studied and the Stone–von Neumann uniqueness theorem is proved. The Bargmann transform establishes the unitary equivalence of both representations. Kato's theorem on the characterization of Schrödinger pairs in terms of resolvents is derived. Further, the Heisenberg uncertainty principle and the Groenewold-van Hove "no-go" theorem for quantization are developed in detail.

Chapter 9 is about infinitesimal representations of universal enveloping algebras of finite-dimensional Lie algebras. Each unitary representation of a Lie group yields a *-representation of the corresponding enveloping algebra. Basic properties of these representations (C^∞-vectors, Gårding domains, graph topologies, essential self-adjointness of symmetric elements) are studied in detail and elliptic regularity theory is used to prove a number of advanced results.

Analytic vectors, first for single operators and then for representations, are investigated. They play a crucial role for the integrability theorems of Lie algebra representations due to Nelson and to Flato, Simon, Snellman, and Sternheimer. These results are presented without proof, but with references. Finally, we discuss K-finite vectors for unitary representations of $SL(2, \mathbb{R})$ and the oscillator representation.

Chapter 10 is concerned with Archimedean quadratic modules and the associated *-algebras of bounded elements. Two abstract Stellensätze give a glimpse into noncommutative real algebraic geometry. As an application we derive a strict Positivstellensatz for the Weyl algebra. Finally, a theorem about the closedness of the cone of finite sums of hermitian squares in certain *-algebras is proved and some applications are obtained.

Chapter 11 examines the operator relation $XX^* = F(X^*X)$, where F is Borel function on $[0, +\infty)$ and X is a densely defined closed operator on a Hilbert space. The representation theory of this relation is closely linked to properties of the dynamical system defined by the function F. For instance, finite-dimensional irreducible representations correspond to cycles of the dynamical system. The hermitian q-plane and the q-oscillator algebra are treated as important examples.

Chapter 12 presents an introduction to unbounded induced representations of $*$-algebras. For group graded $*$-algebras there exists a canonical conditional expectation which allows one to define induced representations. We develop this theory for representations which are induced from characters of commutative subalgebras. The Bargmann–Fock representation of the Weyl algebra is obtained in this manner.

An important topic of advanced Hilbert space representation theory is to describe classes of "well-behaved" representations of general $*$-algebras. In Chap. 13 we propose some general methods (group graded $*$-algebras, fraction algebras, compatible pairs) and apply them to the Weyl algebra and to enveloping algebras.

Chapter 14 provides a brief introduction to $*$-representations on rigged modules and Hilbert C^*-modules. This is a new subject of theoretical importance. A rigged space is a right or left module equipped with an algebra-valued sesquilinear mapping which is compatible with the module action. First we explore $*$-representations of $*$-algebras on rigged modules purely algebraically. If the riggings are positive semi-definite (in particular, in the case of Hilbert C^*-modules), induced representations on "ordinary" Hilbert spaces can be defined and imprimitivity bimodules yield equivalences between $*$-representations of the corresponding $*$-algebras.

Guide to Instructors and Readers

Various courses and advanced seminars can be built on this book. All of them should probably start with some basics on $*$-algebras (Sects. 2.1 and 2.2), positive functionals and states (Sect. 2.4), and $*$-representations (Sect. 4.1).

One possibility is a graduate course on unbounded representation theory. The basics should be followed with important notions and tools such as irreducibility (Sect. 4.3), GNS representations (Sect. 4.4), and bounded commutants (Sects. 3.2 and 5.1). Then there are many ways to continue. One way is to treat representations of special classes of $*$-algebras such as tensor algebras (Chap. 6), commutative algebras (Chap. 7), or the Weyl algebra (Chap. 8). One may also continue with a detailed study of states (with material taken from Chap. 5) or by developing general methods such as induced representations (Chap. 12), operator relations (Chap. 11), and fraction algebras (Sects. 13.2 and 13.3).

Another possible course for graduate students of mathematics and theoretical physics is on representations of the canonical commutation relation and the Weyl algebra. Such a course could be based entirely on Chap. 8. Here, after considering some basics and algebraic properties of the Weyl algebra, the Bargmann–Fock and

Schrödinger representations, the Fock space, the Bargmann–Segal transform, the Stone–von Neumann uniqueness theorem should be developed and continued until the sections on the Heisenberg uncertainty principle and the Groenewald–von Hove "no-go" theorem.

Chapter 9, which treats integrable representations of enveloping algebras, could be used in general or advanced courses or as a reference for researchers. Material from this chapter, for example, the "elementary" parts from Sect. 9.2 on infinitesimal representations, C^∞-vectors and Garding domains, can be integrated into any general course on infinite-dimensional unitary representation theory of Lie groups. More complex material such as elliptic elements or analytic vectors (see e.g. Sects. 9.4 and 9.6) would suit an advanced course. Because Chap. 9 contains a number of strong results on infinitesimal representations, their domains, and commutation properties, it might be also useful as a reference for researchers.

Apart from basic concepts and facts, most chapters are more or less self-contained and could be studied independently of each other. Special topics can be easily included into courses, treated in seminars or read on their own. Examples are the noncommutative Positivstellensätze (Chap. 10) or operator relations and dynamical systems (Chap. 11).

Each chapter is followed by a number of exercises. They vary in difficulty and serve for different purposes. Most of them are examples or counter-examples illustrating the theory. Some are slight variations of results stated in the text, while others contain additional new results or facts that are of interest in themselves.

Prerequisites

The main prerequisite for this book is a good working knowledge of unbounded Hilbert space operators such as adjoint operators, symmetric operators, self-adjoint operators, and the spectral theorem. The corresponding chapters of the author's Graduate Text [Sch12] contain more material than really needed. The reader should be also familiar with elementary techniques of algebra, analysis, and bounded operator algebras. Chapter 9 assumes a familiarity with the theory of Lie groups and Lie algebras. In three appendices, we have collected some basics on unbounded operators, C^*-algebras and their representations, and locally convex spaces and separation of convex sets. In addition, we have often restated facts and notions at the places where they are most relevant.

For parts of the book or for single results, additional facts from other mathematical fields are required, which emphasize the interplay with these fields. There we have given links to the corresponding literature. In most cases these results are not needed elsewhere in the book, so the unfamiliar reader may skip these places.

Leipzig, Germany   Konrad Schmüdgen
March 2020

Acknowledgements I am grateful to Prof. J. Cimprič and Prof. V. L. Ostrovskyi for careful reading of some chapters and for many useful comments. Also, I would like to thank Dr. R. Lodh from Springer-Verlag for his indispensable help getting this book published.

Contents

General Notation

Throughout the book, we use the following notational conventions:

The involution of an abstract $*$-algebra is denoted by $a \mapsto a^+$.

The symbol a^* is only used for the *adjoint* of a Hilbert space operator a.

The symbol $\mathbb{K}$ denotes either the real field $\mathbb{R}$ or the complex field $\mathbb{C}$.

All algebras or vector spaces are either over $\mathbb{R}$ or $\mathbb{C}$.

All inner products of complex inner product spaces or Hilbert spaces are linear in the first and conjugate linear in the second variables.

Unless stated explicitly otherwise, all inner products and Hilbert spaces are over the complex field.

We denote

- abstract $*$-algebras by sanserif letters such as A, B, F, M, W, X,
- unit elements of a unital $\mathbb{K}$-algebras by 1 and write $\alpha \cdot 1$ by α for $\alpha \in \mathbb{K}$,
- O^*-algebras by script letters such as $\mathcal{A}$, $\mathcal{B}$,
- Hilbert spaces by $\mathcal{H}$, $\mathcal{H}_0$, $\mathcal{G}$, $\mathcal{K}$,
- inner products by angle brackets $\langle \cdot, \cdot \rangle$, $\langle \cdot, \cdot \rangle_1$
- dense domains or inner product spaces by $\mathcal{D}$, $\mathcal{D}(T)$,
- representations by π, π_f, ρ,
- Hilbert space vectors by φ, ψ, η, ξ.

$\mathbb{N}_0$	Set of nonnegative integers,
$\mathbb{N}$	Set of positive integers,
$\mathbb{Z}$	Set of integers,
$\mathbb{R}$	Set of real numbers,
$\mathbb{R}_+$	Set of nonnegative real numbers,
$\mathbb{C}$	Set of complex numbers,
$\mathbb{T}$	Set of complex numbers of modulus one.

$\mathbb{C}_d[\underline{x}] := \mathbb{C}[x_1, \ldots, x_d]$, $\mathbb{R}_d[\underline{x}] := \mathbb{R}[x_1, \ldots, x_d]$.

For a Hilbert space $\mathcal{H}$, we denote by

- $\mathbf{B}(\mathcal{H})$ the bounded operators on $\mathcal{H}$,
- $\mathbf{B}_1(\mathcal{H})$ the trace class operators on $\mathcal{H}$,
- $\mathrm{Tr}\,t$ the trace of a trace class operator t,
- $\mathbf{B}_1(\mathcal{H})_+$ the positive trace class operators on $\mathcal{H}$,
- $\mathbf{B}_2(\mathcal{H})$ the Hilbert-Schmidt operators on $\mathcal{H}$.

For a $*$-algebra $\mathbf{A}$ we denote by

- $\mathbf{A}^1$ the unitization of $\mathbf{A}$,
- $\mathbf{A}_{\mathrm{her}}$ the hermitian part of $\mathbf{A}$,
- $\mathcal{P}(\mathbf{A})^*$ the positive linear functionals on $\mathbf{A}$,
- $\mathcal{P}_e(\mathbf{A})^*$ the extendable positive linear functionals on $\mathbf{A}$,
- $\mathcal{S}(\mathbf{A})$ the states of $\mathbf{A}$,
- $\hat{\mathbf{A}}$ the hermitian characters of $\mathbf{A}$, if $\mathbf{A}$ is commutative and unital,
- 1 its unit element, if $\mathbf{A}$ is unital.

$C_c(\mathcal{X})$ Compactly supported continuous functions on a topological space $\mathcal{X}$.
$C_0(\mathcal{X})$ Continuous functions on a locally compact space $\mathcal{X}$ that vanish at infinity.
$L^2(M)$ L^2-space with respect to the Lebesgue measure if M is a Borel set of $\mathbb{R}^d$.
$\mathcal{F}$ Fourier transform $\mathcal{F}(f)(x) = (2\pi)^{-d/2} \int_{\mathbb{R}^d} e^{-\mathrm{i}(x,y)} f(y)\, dy$.

Chapter 1
Prologue: The Algebraic Approach to Quantum Theories

Let us begin by recalling some well-known concepts from **quantum mechanics**. For details, the reader can consult one of the standard textbooks such as [SN17] or [Ha13].

The mathematical formulation of quantum mechanics is based on a complex Hilbert space $\mathcal{H}$, which is called the *state space*. The two fundamental objects of a quantum theory, observables and states, are described by the following postulates.

(QM1) *Each **observable** is a self-adjoint operator on the Hilbert space $\mathcal{H}$.*

(QM2) *Each pure **state** is given by the unit ray $[\varphi] := \{\lambda\varphi : \lambda \in \mathbb{T}\}$ of a unit vector $\varphi \in \mathcal{H}$.*

In general, not all self-adjoint operators on $\mathcal{H}$ are physical observables and not all unit vectors of $\mathcal{H}$ correspond to physical states. In the subsequent informal discussion we will ignore this distinction and consider all unit rays as states and all bounded self-adjoint operators on $\mathcal{H}$ as observables.

That each observable A is a self-adjoint operator by axiom (QM1) has important consequences. Then the spectral theorem applies, and there exists a unique projection-valued measure $E_A(\cdot)$, called the spectral measure of A, on the Borel σ-algebra of $\mathbb{R}$ such that

$$A = \int_{\mathbb{R}} \lambda \, dE_A(\lambda).$$

This spectral measure E_A is a fundamental mathematical object in operator theory and in quantum mechanics as well. All properties of the self-adjoint operator and the observable A are encoded in E_A. First we note that the support of the spectral measure E_A coincides with the spectrum of the operator A.

© The Editor(s) (if applicable) and The Author(s), under exclusive license to Springer Nature Switzerland AG 2020
K. Schmüdgen, *An Invitation to Unbounded Representations of *-Algebras on Hilbert Space*, Graduate Texts in Mathematics 285,
https://doi.org/10.1007/978-3-030-46366-3_1

The probabilistic interpretation of quantum mechanics and the measurement theory of observables are essentially based on spectral measures. To explain this, we consider a unit vector $\varphi \in \mathcal{H}$. It is clear that

$$\mu_{[\varphi]}(\cdot) := \langle E_A(\cdot)\varphi, \varphi \rangle$$

defines a probability measure $\mu_{[\varphi]}$ on $\mathbb{R}$ which depends only on the unit ray $[\varphi]$. The probabilistic interpretation says that $\mu_{[\varphi]}(M)$ is the probability that the measurement outcome of the observable A in the state $[\varphi]$ lies in the Borel set M of $\mathbb{R}$. Two observables A_1 and A_2 are simultaneously measurable if and only if their spectral measures E_{A_1} and E_{A_2} commute.

Now let φ be a unit vector of the domain of A. Then the number

$$\langle A\varphi, \varphi \rangle = \int_{\mathbb{R}} \lambda \, d\mu_{[\varphi]}(\lambda) = \int_{\mathbb{R}} \lambda \, d\langle E_A(\lambda)\varphi, \varphi \rangle$$

is interpreted as the *expectation value* and $(\Delta_{[\varphi]}A)^2 := \|A\varphi\|^2 - \langle A\varphi, \varphi \rangle^2$ as the *variance* of the observable A in the state $[\varphi]$. Finally, the spectral measure allows one to define a function $F(A)$ of an observable A by $F(A) = \int F(\lambda)dE_A(\lambda)$ for any Borel function F on the spectrum of A.

Let $[\varphi]$ and $[\psi]$ be states of $\mathcal{H}$. Then the number

$$P([\varphi], [\psi]) := |\langle \varphi, \psi \rangle|^2$$

depends only on the unit rays, and it is called the *transition probability* between the states $[\varphi]$ and $[\psi]$.

A *symmetry* of the quantum system is a bijection of the set of states $[\varphi]$ which preserves the transition probabilities between states. By Wigner's theorem (see, e.g., [Em72]), each symmetry θ is implemented by a unitary or an antiunitary operator U of the Hilbert space $\mathcal{H}$, that is, $\theta([\varphi]) = U[\varphi]U^{-1}$ for all states $[\varphi]$. (An antiunitary operator is an operator U on $\mathcal{H}$ such that $U(\alpha\varphi + \beta\psi) = \overline{\alpha}U\varphi + \overline{\beta}U\psi$ and $\langle U\varphi, U\psi \rangle = \overline{\langle \varphi, \psi \rangle}$ for $\varphi, \psi \in \mathcal{H}$ and $\alpha, \beta \in \mathbb{C}$.)

Let U be a unitary or an antiunitary operator on $\mathcal{H}$. If A is an observable, then the operator $\theta(A) := UAU^{-1}$ is self-adjoint and hence an observable. For arbitrary $A \in \mathbf{B}(\mathcal{H})$, we set $\theta_0(A) = UAU^{-1}$ if U is unitary and $\theta_1(A) = UA^*U^{-1}$ if U is antiunitary. For self-adjoint operators A, both $\theta_0(A)$ and $\theta_1(A)$ coincide with $\theta(A)$. Then, θ_0 is a $*$-automorphism and θ_1 is a $*$-antiautomorphism of the C^*-algebra $\mathbf{B}(\mathcal{H})$ of bounded operators on $\mathcal{H}$.

There are also mixed states and states given by density matrices. Assume for a moment that the observables are bounded operators. Then each positive trace class operator t on $\mathcal{H}$ of trace one defines also a state. The corresponding probability measure is $\mu_t(\cdot) := \mathrm{Tr}\, t E_A(\cdot)$, and the expectation value is $\mathrm{Tr}\, tA$.

That was the classic approach to quantum mechanics. Let us explain now the **algebraic approach**, in which the main objects of study of this book appear.

Here the *observable algebra* is the central object of the theory. This is an abstract complex unital algebra A equipped with an algebra involution $a \mapsto a^+$, that is, A is a complex unital $*$-algebra. The key postulates in this approach are the following:

(A1) *Each* **observable** *is a hermitian element* $a = a^+$ *of the $*$-algebra* A.
(A2) *Each* **state** *is a linear functional* f *on* A *such that* $f(a^+ a) \geq 0$ *for* $a \in$ A *and* $f(1) = 1$.

If f is a state and a is an observable of A, then the real number $f(a)$ is considered as the *expectation value* and the nonnegative number $\Delta_f(a)^2 := f(a^2) - f(a)^2$ as the *variance* of a in the state f.

Let us motivate this definition of a state. Elements of the form $a^+ a$ are always hermitian, and they should be positive, because Hilbert space operators of the form $A^* A$ are positive. Then the condition $f(a^+ a) \geq 0$ says that the expectation value of the "positive" observable $a^+ a$ is nonnegative. A functional f with this property is called *positive*. The requirement $f(1) = 1$ is a normalization condition for the trivial observable $1 \in$ A.

Since A is a $*$-algebra, one can form algebraic operations (linear combinations, products, adjoints) of elements of A. It is easily verified that the product of two hermitian elements is hermitian if and only if the elements commute. Hence the product of two observables can be only an observable if they commute in the algebra A.

To remedy this failure it is convenient to consider the *Jordan product*

$$a \circ b := \frac{1}{2}(ab + ba)$$

of elements $a, b \in$ A. Obviously, if the elements a and b are hermitian, so is $a \circ b$. Clearly, $a \circ b = \frac{1}{2}((a + b)^2 - a^2 - b^2)$. Therefore, if we agree that real linear combinations and squares of observables are also observables, then the Jordan product $a \circ b$ of observables $a, b \in$ A is again an observable. Note that the Jordan product "$\circ$" is distributive and commutative, but it is not associative in general.

Before we continue our discussion we introduce a few more mathematical notions. Let θ be a linear map of A into another $*$-algebra B such that $\theta(a^+) = \theta(a)^+$ for $a \in$ A. Then θ is called a $*$-*antihomomorphism* if $\theta(ab) = \theta(b)\theta(a)$ for $a, b \in$ A and a *Jordan homomorphism* if $\theta(a \circ b) = \theta(a) \circ \theta(b)$ for $a, b \in$ A. In this case, if A $=$ B and θ is bijective, then θ is said to be a $*$-*antiautomorphism* and a Jordan automorphism of A, respectively. Clearly, $*$-homomorphisms and $*$-antihomomorphisms are Jordan homomorphisms.

Roughly speaking, a symmetry of a physical system should be a bijection that preserves the main structures of the system. In the case of pure states on a Hilbert space, the transition probability of states was chosen as the relevant concept. In the algebraic approach, it is natural to require that symmetries preserve the Jordan product. Thus, we define a *symmetry* to be a *Jordan automorphism* of the $*$-algebra A. Then any symmetry θ preserves observables, and the map $f \mapsto f \circ \theta$ preserves states. Various symmetry concepts for C^*-algebras are treated and discussed in [Ln17, Chap. 5], [Em72, Sect. 2.2.a], [Mo13, Sect. 12.1], and [K65].

In particular, *-automorphisms and *-antiautomorphisms of A are symmetries. We say that a group G acts as a *symmetry group* on the observable algebra A if we have a homomorphism $g \mapsto \theta_g$ of G into the group of *-automorphisms of A.

We collect the main concepts introduced so far in the following table:

	Quantum mechanics	Algebraic approach
	State Hilbert space $\mathcal{H}$	Observable algebra A
Observable	Self-adjoint operator on $\mathcal{H}$	Hermitian element of A
State	Unit ray $[\varphi]$ of $\varphi \in \mathcal{H}$, $\|\varphi\| = 1$	Positive functional f with $f(1) = 1$
Symmetry	Unitary or antiunitary operator on $\mathcal{H}$	Jordan automorphism of A

It should be emphasized that for the study of quantum theories usually specific sets of further axioms and topics are added. Important examples are the Gårding–Wightman axioms and the Haag–Kastler axioms in algebraic quantum field theory [Hg55] and the KMS states in quantum statistical mechanics [BR97].

Next we discuss the role of representations of the observable algebra. To avoid technical difficulties, let us assume throughout the following discussion that the observable algebra A is a unital C^*-algebra. Recall that a *-representation of A is a *-homomorphism ρ of A into the *-algebra $\mathbf{B}(\mathcal{H})$ of bounded operators of some Hilbert space $\mathcal{H}$. Then the image of each abstract observable $a \in$ A is a bounded self-adjoint operator $\rho(a)$, hence an observable on the Hilbert space $\mathcal{H}$, and each unit vector $\varphi \in \mathcal{H}$ defines a state $f_{\rho,\varphi}(\cdot) := \langle \rho(\cdot)\varphi, \varphi \rangle$ on A. These states $f_{\rho,\varphi}$ are called the vector states of the representation ρ. Conversely, if f is a state on A, then the GNS construction provides a *-representation ρ_f of A on a Hilbert space $\mathcal{H}$ such that $f(\cdot) = \langle \rho_f(\cdot)\varphi_f, \varphi_f \rangle$ for some unit vector $\varphi_f \in \mathcal{H}$. Thus, the abstract state f on A gives a concrete state $[\varphi_f]$ on the Hilbert space $\mathcal{H}$.

Further, two *-representations of A are *physically equivalent* if and only if each vector state of one is a weak limit of convex combinations of vector states of the other, or equivalently, if the kernels of both representations coincide [Em72, Theorem II.1.7]. It is obvious that unitarily equivalent representations are physically equivalent, but the converse is not true.

Let us turn to symmetries. Suppose ρ is a *-representation of A on a Hilbert space $\mathcal{H}$. A *-automorphism θ of A is called *unitarily implemented* in the representation ρ if there exists a unitary operator U on $\mathcal{H}$ such that $\rho(\theta(a)) = U\rho(a)U^{-1}$ for $a \in$ A. Likewise, an action $g \mapsto \theta_g$ of a group G on A is said to be *unitarily implemented* in the representation ρ if there is a homomorphism $g \mapsto U(g)$ of G into the group of unitaries on $\mathcal{H}$, called then a unitary representation of G on $\mathcal{H}$, such that

$$\rho(\theta_g(a)) = U(g)\rho(a)U(g^{-1}) \quad \text{for} \ a \in \mathsf{A}, \ g \in G. \tag{1.1}$$

It can be shown that (1.1) holds, for instance, for the GNS representation associated with any state which is invariant under θ_g. In important cases, G is a Lie group; then appropriate continuity assumptions on θ_g and $U(g)$ have to be added.

According to a result of Kadison [K65], [BR97, Proposition 3.2.2], any Jordan homomorphism into $\mathbf{B}(\mathcal{H})$ can be decomposed into a sum of a $*$-homomorphism and a $*$-antihomomorphism. More precisely, if $\rho : \mathsf{A} \mapsto \mathsf{B}$ is a Jordan homomorphism of A on a C^*-subalgebra B of $\mathbf{B}(\mathcal{H})$, then there is a projection $P \in \mathsf{B}' \cap \mathsf{B}''$ such that $a \mapsto \rho(a)P$ is a $*$-homomorphism and $a \mapsto \rho(a)(I - P)$ is a $*$-antihomomorphism of A into $\mathbf{B}(\mathcal{H})$. Here B' and B'' denote the commutant and bicommutant of B, respectively. In particular, if the von Neumann algebra B'' is a factor, then $P = 0$ or $P = I$, so ρ is a $*$-homomorphism or a $*$-antihomomorphism.

Any $*$-representation ρ of the observable algebra allows one to pass from the fixed abstract observable algebra A to the observable algebra $\rho(\mathsf{A})$ of operators acting on a Hilbert space. There the power of operator theory on Hilbert spaces can be used to study the quantum system. The flexibility of choosing the $*$-representation has a number of advantages. First, various realizations of unitarily equivalent representations may provide new methods and structural insight. For instance, the Schrödinger representation and the Bargmann–Fock representation of the Weyl algebra are unitarily equivalent, but their realizations on $L^2(\mathbb{R}^d)$ and on the Fock space, respectively, lead to different approaches for the study of the canonical commutation relations. Second, unitarily or physically inequivalent realizations of quantum systems can be treated by means of the same abstract observable algebra. Here the canonical commutation relations for infinitely many degrees of freedom form an interesting example. There exist unitarily inequivalent irreducible representations which are physically equivalent [BR97, Em72]. Third, let $g \mapsto \theta_g$ be an action of a Lie group G as $*$-automorphisms of A. In "good" cases there exists a $*$-representation ρ of A such that this action is implemented by a unitary representation $g \mapsto U(g)$ of G, as in formula (1.1). Then the representation theory of Lie groups on Hilbert space can be used to study the $*$-automorphism group.

The preceding was a brief sketch of some basic general concepts and ideas of quantum mechanics and the algebraic approach to quantum theories.

In the case of general $\star$-algebras a number of additional technical problems appear in the study of $*$-representations and states. For instance, it may happen that the image of a hermitian element under a $*$-representation has no self-adjoint extension, so it cannot be considered as an observable on the representation Hilbert space. An aim of this book is to lay down a rigorous mathematical foundation of the theory of representations and states of general $*$-algebras.

The pioneering work for the algebraic approach goes back to Neumann [vN32], Segal [Se47a], and others. Modern treatments of this approach and various sets of axioms can be found in the books of Emch [Em72], Moretti [Mo13] and Landsman [Ln17]; see also [K65]. Standard references are [Hg92, Ak09] for algebraic quantum field theory and [BR87, BR97] for quantum statistical mechanics.

Chapter 2
*-Algebras

The aim of this chapter is to develop algebraic properties and structures of *-algebras and of positive functionals and states. Also, we introduce a number of basic concepts, notations, and facts that will be used later in this book.

Section 2.1 contains basic definitions and examples of *-algebras. In Sect. 2.2, we treat some general constructions of *-algebras (tensor products, matrix algebras, crossed products, group graded algebras). Positivity in *-algebras is expressed in terms of quadratic modules; they are introduced in Sect. 2.3 and studied later in Sect. 5.7 and Chap. 10.

Sections 2.4–2.8 deal with positive linear functionals. In Sect. 2.4, we develop basic facts on positive functionals and states on complex *-algebras. Positive functionals on real *-algebras are briefly considered in Sect. 2.5. In Sect. 2.6 we study characters of general algebras and prove the Gleason–Kahane–Zelazko characterization of characters (Theorem 2.56). Section 2.7 is about hermitian characters and pure states of commutative *-algebras (Theorem 2.63). In Sect. 2.8, we give a short digression into hermitian and symmetric *-algebras.

Throughout this chapter, A is an **algebra** over the field $\mathbb{K}$, where $\mathbb{K}$ is $\mathbb{R}$ or $\mathbb{C}$.

2.1 *-Algebras: Definitions and Examples

The following definitions introduce the first main notions which this book is about.

Definition 2.1 An *algebra* over $\mathbb{K}$ is a vector space A over $\mathbb{K}$, equipped with a mapping $(a, b) \mapsto ab$ of $A \times A$ into A, such that for $a, b, c \in A$ and $\alpha \in \mathbb{K}$:

$$a(bc) = (ab)c, \quad (\alpha a)b = \alpha(ab) = a(\alpha b), \quad a(b + c) = ab + ac, \quad (b + c)a = ba + ca.$$

© The Editor(s) (if applicable) and The Author(s), under exclusive license
to Springer Nature Switzerland AG 2020
K. Schmüdgen, *An Invitation to Unbounded Representations of *-Algebras on Hilbert Space*, Graduate Texts in Mathematics 285,
https://doi.org/10.1007/978-3-030-46366-3_2

The element ab is called the *product* of a and b; we also write $a \cdot b$ for ab.

An algebra A is called *unital* if it has a unit element $1 \in \mathsf{A}$, that is, $1a = a1 = a$ for all $a \in \mathsf{A}$. An algebra A is *commutative* if $ab = ba$ for $a, b \in \mathsf{A}$.

Definition 2.2 An *algebra involution*, briefly an *involution*, of an algebra A over $\mathbb{K}$ is a mapping $a \mapsto a^+$ from A into A such that for $a, b \in \mathsf{A}$ and $\alpha, \beta \in \mathbb{K}$:

$$(\alpha a + \beta b)^+ = \overline{\alpha} \, a^+ + \overline{\beta} \, b^+, \quad (ab)^+ = b^+ a^+, \quad (a^+)^+ = a. \qquad (2.1)$$

An algebra (over $\mathbb{K}$) equipped with an involution is called a *$*$-algebra* (over $\mathbb{K}$).

Example 2.3 Let $d \in \mathbb{N}$. The polynomial algebra $\mathbb{K}_d[\underline{x}] := \mathbb{K}[x_1, \ldots, x_d]$ is a unital $*$-algebra with involution defined by

$$f^+(x) := \sum_{\alpha} \overline{a}_{\alpha} x^{\alpha} \quad \text{for} \quad f(x) = \sum_{\alpha} a_{\alpha} x^{\alpha} \in \mathbb{K}_d[\underline{x}],$$

where we set $x^{\alpha} := x_1^{\alpha_1} \cdots x_d^{\alpha_d}$ for $\alpha = (\alpha_1, \ldots, \alpha_d) \in \mathbb{N}_0^d$ and $x_j^0 := 1$. Note that the involution on $\mathbb{R}_d[\underline{x}]$ is just the identity mapping. $\qquad \bigcirc$

Let A be a $*$-algebra over $\mathbb{K}$. It is easily verified that if A has a unit element 1 and $a \in \mathsf{A}$ is invertible in A, then $1^+ = 1$ and $(a^{-1})^+ = (a^+)^{-1}$.

Definition 2.4 An element $a \in \mathsf{A}$ is called *hermitian* if $a = a^+$ and *skew-hermitian* if $a^+ = -a$.

The *hermitian part* A_{her} and the *skew-hermitian part* A_{sher} of A are

$$\mathsf{A}_{\text{her}} := \{a \in \mathsf{A} : a^+ = a\}, \quad \mathsf{A}_{\text{sher}} := \{a \in \mathsf{A} : a^+ = -a\}. \qquad (2.2)$$

Clearly, both parts are real vector spaces, A_{her} is invariant under the Jordan product $a \circ b := \frac{1}{2}(ab + ba)$, and A_{sher} is invariant under the commutator $[a, b] := ab - ba$. Further, $\mathsf{A} = \mathsf{A}_{\text{her}} + \mathsf{A}_{\text{sher}}$ and each $a \in \mathsf{A}$ can be uniquely written as

$$a = a_{\text{h}} + a_{\text{sh}}, \quad \text{where} \quad a_{\text{h}} \in \mathsf{A}_{\text{her}}, \ a_{\text{sh}} \in \mathsf{A}_{\text{sher}}. \qquad (2.3)$$

Indeed, for $a_{\text{h}} := \frac{1}{2}(a^+ + a)$ and $a_{\text{sh}} := \frac{1}{2}(a - a^+)$ we have (2.3). Conversely, if $\tilde{a}_{\text{h}} \in \mathsf{A}_{\text{her}}$ and $\tilde{a}_{\text{sh}} \in \mathsf{A}_{\text{sher}}$ satisfy $a = \tilde{a}_{\text{h}} + \tilde{a}_{\text{sh}}$, then $a^+ = \tilde{a}_{\text{h}} - \tilde{a}_{\text{sh}}$ and hence $\tilde{a}_{\text{h}} = a_{\text{h}}$ and $\tilde{a}_{\text{sh}} = a_{\text{sh}}$.

Now suppose $\mathbb{K} = \mathbb{C}$. Then, obviously, $\mathsf{A}_{\text{sher}} = \mathrm{i}\, \mathsf{A}_{\text{her}}$, so that $\mathsf{A} = \mathsf{A}_{\text{her}} + \mathrm{i}\mathsf{A}_{\text{her}}$. Therefore, by (2.3), each element $a \in \mathsf{A}$ can be uniquely represented in the form

$$a = a_1 + \mathrm{i}a_2, \quad \text{where} \quad a_1, a_2 \in \mathsf{A}_{\text{her}}, \qquad (2.4)$$

and we have $a_1 = \mathrm{Re}\, a := \frac{1}{2}(a^+ + a)$ and $a_2 = \mathrm{Im}\, a := \frac{1}{2}(a^+ - a)$.

If A is a commutative real algebra, the identity map is obviously an involution.

There exist algebras A which admit no algebra involution and others which have infinitely many involutions making A into a $*$-algebra; see, e.g., [CV59].

Example 2.5 (*An algebra which has no involution*)
Let A be the $\mathbb{K}$-algebra of 2×2 matrices $(a_{kl})^2_{k,l=1}$, with $a_{kl} \in \mathbb{K}$, $a_{21} = a_{22} = 0$. Clearly, the algebra A is isomorphic to the vector space $\mathbb{K}^2$ with multiplication

$$(x_1, x_2)(y_1, y_2) = (x_1 y_1, x_1 y_2). \tag{2.5}$$

The algebra A has no involution such that A becomes a *-algebra.

Indeed, assume to the contrary that $a \mapsto a^+$ is an algebra involution of A. Set $x := (1, 0)$ and $y := (0, 1)$. We have $x^2 = x$, $y^2 = 0$, $xy = y$, $yx = 0$ by (2.5). Then $(x^+)^2 = x^+$ and $(y^+)^2 = 0$. By (2.5), these equations imply $x^+ = (1, x_2)$ and $y^+ = (0, y_2)$. Then $0 = (yx)^+ = x^+ y^+ = (1, x_2)(0, y_2) = (0, y_2) = y^+$ and hence $0 = (y^+)^+ = y$, a contradiction.   ○

We develop different involutions in Example 2.15 below using the next lemma.

Lemma 2.6 *Suppose A is an algebra. If $\varphi : a \mapsto a^+$ is an algebra involution and θ is an algebra automorphism of A such that*

$$(\theta \circ \varphi) \circ (\theta \circ \varphi) = \mathrm{Id}, \quad \text{that is,} \quad \theta(\theta(a^+)^+) = a \quad \text{for } a \in \mathsf{A}, \tag{2.6}$$

then $\psi := \theta \circ \varphi$ is also an algebra involution of A.

Conversely, if φ and ψ are algebra involutions of A, then $\theta := \psi \circ \varphi$ is an automorphism of the algebra A such that $\psi = \theta \circ \varphi$ and condition (2.6) holds.

The proof of this lemma is given by simple algebraic manipulations based on (2.1). Equation (2.6) is equivalent to the last condition in (2.1). We omit the details; see Exercise 1.

Next let us introduce some standard notions.

A map θ of a *-algebra A into another *-algebra B is called a *-*homomorphism* if θ is an algebra homomorphism such that $\theta(a^+) = \theta(a)^+$ for $a \in \mathsf{A}$. A *-*isomorphism* is a bijective *-homomorphism of A and B; in this case, A and B are said to be *-*isomorphic*. A *-*automorphism* of A is a *-isomorphism of A on itself. A *-*ideal* of A is a two-sided ideal of A which is invariant under the involution.

Next we consider two useful general constructions.

Unitization of a *-algebra

For many considerations it is necessary that the *-algebra possesses a unit element. If a *-algebra has no unit, it can be embedded into a unital *-algebra by *adjoining a unit*. Let A be a *-algebra. It is easy to check that the $\mathbb{K}$-vector space $\mathsf{B} := \mathsf{A} \oplus \mathbb{K}$ is a unital *-algebra with multiplication and involution defined by

$$(a, \alpha)(b, \beta) := (ab + \alpha b + \beta a, \alpha \beta) \quad \text{and} \quad (a, \alpha)^+ := (a^+, \overline{\alpha}) \tag{2.7}$$

for $a, b \in \mathsf{A}$ and $\alpha, \beta \in \mathbb{K}$. Obviously, $1 := (0, 1)$ is the unit element of B. By identifying a and $(a, 0)$, the *-algebra A becomes a *-subalgebra of B. For notational simplicity we write $a + \alpha$ instead of (a, α). Note that if A has a unit element, this element is no longer a unit element of the larger *-algebra B.

If A is not unital, we denote the unital *-algebra $B = A \oplus \mathbb{K}$ by A^1. If A is unital, we set $A^1 := A$.

Definition 2.7 The unital *-algebra A^1 is called the *unitization* of the *-algebra A.

For real *-algebras we may have $\mathrm{Lin}\, A_{\mathrm{her}} \neq A$, as the following example shows.

Example 2.8 On the vector space $A := \mathbb{R}$ we define a product by $x \cdot y := 0$ and an involution by $x^+ := -x$. Then A is a real *-algebra and $A_{\mathrm{her}} = \{0\} \neq A$. For the unitization A^1 we have $(A^1)_{\mathrm{her}} = \{(0, \alpha) : \alpha \in \mathbb{R}\}$ by (2.7). Hence the linear span of $(A^1)_{\mathrm{her}}$ is different from A^1.   ○

Complexification of a real *-algebra

Suppose A is a *real* *-algebra. Let $A_{\mathbb{C}}$ be the Cartesian product $A \times A$. It is not difficult to verify that $A_{\mathbb{C}}$ becomes a *complex* *-algebra with addition, multiplication by complex scalars, multiplication, and involution defined by

$$(a, b) + (c, d) = (a + c, b + d), \quad (\alpha + i\beta)(a, b) = (\alpha a - \beta b, \alpha b + \beta a),$$
$$(a, b)(c, d) = (ac - bd, bc + ad), \quad (a, b)^+ := (a^+, -b^+),$$

where $a, b, c, d \in A$ and $\alpha, \beta \in \mathbb{R}$. The map $a \mapsto (a, 0)$ is a *-isomorphism of A on a real *-subalgebra of $A_{\mathbb{C}}$. We identify $a \in A$ with $(a, 0) \in A_{\mathbb{C}}$. Then A becomes a real *-subalgebra of $A_{\mathbb{C}}$, and we have $(a, b) = a + ib$ for $a, b \in A$.

Definition 2.9 The complex *-algebra $A_{\mathbb{C}}$ is called the *complexification* of the real *-algebra A.

We define $\theta(a + ib) = a - ib$ for $a, b \in A$. Then we have

$$\theta(\alpha a + \beta b) = \overline{\alpha}\, \theta(a) + \overline{\beta}\, \theta(b), \quad \theta(x^+) = \theta(x)^+, \tag{2.8}$$
$$\theta(xy) = \theta(x)\theta(y), \quad (\theta \circ \theta)(x) = x \tag{2.9}$$

for $\alpha, \beta \in \mathbb{C}$, $a, b \in A$, $x, y \in A_{\mathbb{C}}$, and $A = \{x \in A_{\mathbb{C}} : \theta(x) = x\}$. Conversely, if B is a complex *-algebra and $\theta : B \mapsto B$ is a map satisfying (2.8) and (2.9), then $A := \{x \in B : \theta(x) = x\}$ is a real *-algebra and B is the complexification of A.

Now let A be a *commutative* real algebra. Then A is a real *-algebra with the identity map as involution and we have $(a + ib)^+ = a - ib$ in $A_{\mathbb{C}}$, where $a, b \in A$. Hence A is the hermitian part $(A_{\mathbb{C}})_{\mathrm{her}}$ of its complexification $A_{\mathbb{C}}$. For instance, if $A = \mathbb{R}[x_1, \ldots, x_d]$, we obtain $A_{\mathbb{C}} = \mathbb{C}[x_1, \ldots, x_d]$.

Now we turn to examples of *-algebras. Large classes of examples of *-algebras are defined by means of generators and defining relations.

1. *-Algebras defined by relations

Let $\mathbb{K}\langle x_1, \ldots, x_m \rangle$ denote the free unital $\mathbb{K}$-algebra with generators $x_1, \ldots, x_m$. The elements of this algebra can be considered as noncommutative polynomials f in $x_1, \ldots, x_m$; for instance, $f(x_1, x_2) = 5x_1 x_2^7 x_1^3 - 3x_1 x_2 + x_2 x_1 + 1$.

Let $n + k \in \mathbb{N}$, where $k, n \in \mathbb{N}_0$. The algebra $\mathbb{K}\langle x_1, \ldots, x_n, y_1, \ldots, y_{2k}\rangle$ has an involution determined by $(x_j)^+ = x_j$ for $j = 1, \ldots, n$ and $(y_l)^+ = y_{l+k}$ for $l = 1, \ldots, k$; the corresponding ∗-algebra is denoted by

$$\mathbb{K}\langle x_1, \ldots, x_n, y_1, \ldots, y_{2k} \mid (x_j)^+ = x_j, j = 1, \ldots, n; (y_l)^+ = y_{l+k}, l = 1, \ldots, k\rangle. \tag{2.10}$$

(If $n = 0$ or $k = 0$, we interpret (2.10) by omitting the corresponding variables.)

Now let $f_1, g_1 \ldots, f_r, g_r$ be elements of the ∗-algebra (2.10) and let J be the ∗-ideal of this ∗-algebra generated by the elements $f_1 - g_1, \ldots, f_r - g_r$. We write

$$\mathbb{K}\langle x_1, \ldots, x_n, y_1, \ldots, y_{2k} \mid (x_j)^+ = x_j, j = 1, \ldots, n; (y_l)^+ = y_{l+k}, l = 1, \ldots, k;$$
$$f_1 = g_1, \ldots, f_r = g_r\rangle \tag{2.11}$$

for the quotient ∗-algebra of (2.10) by the ∗-ideal J. Thus, (2.11) is the unital ∗-algebra with generators $(x_1)^+ = x_1, \ldots, (x_n)^+ = x_n, (y_1)^+ = y_{k+1}, \ldots, (y_k)^+ = y_{2k}$ and defining relations $f_1 = g_1, \ldots, f_r = g_r$.

Example 2.10 (*Weyl algebra* $\mathsf{W}(d)$)
For $d \in \mathbb{N}$, the d-dimensional *Weyl algebra* $\mathsf{W}(d)$ is the complex unital ∗-algebra

$$\mathsf{W}(d) := \mathbb{C}\langle p_1, \ldots, p_d, q_1, \ldots, q_d \mid (p_k)^+ = p_k, (q_k)^+ = q_k, p_k q_k - q_k p_k = -\mathrm{i};$$
$$p_j p_l = p_l p_j, q_j q_l = q_l q_j, p_j q_l = q_l p_j, k, j, l = 1, \ldots, d, j \neq l\rangle,$$

where i is the complex unit. The one-dimensional Weyl algebra or *CCR-algebra* is

$$\mathsf{W} := \mathbb{C}\langle p, q \mid p^+ = p, q^+ = q, \; pq - qp = -\mathrm{i}\rangle. \tag{2.12}$$

For elements p, q of a complex unital algebra, $a := \frac{1}{\sqrt{2}}(q + \mathrm{i}p)$, $a^+ := \frac{1}{\sqrt{2}}(q - \mathrm{i}p)$ satisfy $aa^+ - a^+a = 1$ if and only if $pq - qp = -\mathrm{i}$. From this fact it follows that the map $\frac{1}{\sqrt{2}}(q + \mathrm{i}p) \mapsto a$ extends to a ∗-isomorphism of W on the ∗-algebra $\mathbb{C}\langle a, b \mid a^+ = b, ab - ba = 1\rangle$. We shall write this ∗-algebra as

$$\mathbb{C}\langle a, a^+ \mid aa^+ - a^+a = 1\rangle. \tag{2.13}$$

Thus, (2.12) and (2.13) are ∗-isomorphic versions of the Weyl algebra; see Sect. 8.1. Chapter 8 is devoted to the study of representations of the Weyl algebra. ○

As angle brackets $\langle \cdot \rangle$ denote *free* algebras, squared brackets $[\cdot]$ always refer to *commutative* polynomial algebras. In particular, $\mathbb{C}_d[\underline{x}] := \mathbb{C}[x_1, \ldots, x_d]$ and $\mathbb{R}_d[\underline{x}] := \mathbb{R}[x_1, \ldots, x_d]$ are commutative ∗-algebras of polynomials with involution $(x_j)^+ = x_j$, $j = 1, \ldots, d$. Commutative algebras with relations are defined similarly as above and are self-explanatory. For instance, $\mathbb{C}[x, y \mid x^+x + y^+y = 1]$ denotes the commutative ∗-algebra of polynomials in x, x^+, y, y^+ satisfying the equation $x^+x + y^+y = 1$ of the unit sphere in $\mathbb{C}^2$.

Another source of important *-algebras are *-semigroups and groups.

2. Semigroup *-algebras and group *-algebras

By a *semigroup* we mean a nonempty set S with an associative binary operation $(a, b) \mapsto a \cdot b$. We consider this operation as multiplication and call $a \cdot b$ the product of a and b. We say an element $e \in S$ is a *unit* of S if $e \cdot a = a \cdot e = a$ for $a \in S$.

Definition 2.11 A *semigroup with involution*, briefly a *-semigroup, is a semigroup S with a mapping $s \mapsto s^+$ of S into itself, called *involution*, satisfying

$$(s \cdot t)^+ = t^+ \cdot s^+ \text{ and } (s^+)^+ = s \quad \text{for all } s, t \in S.$$

If S is an abelian semigroup, then the identity map of S is an involution. If S is a group, then the map $s \mapsto s^+ := s^{-1}$ is an involution.

Suppose S is a *-semigroup. Let $\mathbb{K}[S]$ be the vector space of all sums $\sum_{s \in S} \alpha_s s$, where $\alpha_s \in \mathbb{K}$ and only finitely many numbers α_s are nonzero. Then $\mathbb{K}[S]$ becomes a *-algebra over $\mathbb{K}$ with product and involution defined by

$$\left(\sum_{s \in S} \alpha_s s \right) \cdot \left(\sum_{t \in S} \beta_t t \right) := \sum_{s,t \in S} \alpha_s \beta_t \, s \cdot t,$$
$$\left(\sum_{s \in S} \alpha_s s \right)^+ := \sum_{s \in S} \overline{\alpha}_s \, s^+.$$

Definition 2.12 The *-algebra $\mathbb{K}[S]$ is called the *semigroup *-algebra of S*. If S is a group (with involution $s^+ = s^{-1}$), we say $\mathbb{K}[S]$ is the *group algebra of S*.

Example 2.13 Let S be the additive semigroup $\mathbb{N}_0^d$ with identity involution. Then the map $(n_1, \ldots, n_d) \mapsto x_1^{n_1} \cdots x_d^{n_d}$ extends to a *-isomorphism of $\mathbb{K}[\mathbb{N}_0^d]$ on the polynomial *-algebra $\mathbb{K}_d[\underline{x}] := \mathbb{K}[x_1, \ldots, x_d]$.   ○

Example 2.14 For the group $\mathbb{Z}$ with involution $n^+ = -n$, there is a *-isomorphism $n \mapsto z$ of the group algebra $\mathbb{C}[\mathbb{Z}]$ on the *-algebra $\mathbb{C}[z, \overline{z} \,|\, z\overline{z} = \overline{z}z = 1]$ of polynomials in $z, \overline{z} \in \mathbb{T}$.   ○

2.2 Constructions with *-Algebras

In this section, we consider four general constructions and structures of *-algebras.

1. Tensor product of *-algebras

Let $\mathsf{A}_1, \ldots, \mathsf{A}_n$ be *-algebras over $\mathbb{K}$. The $\mathbb{K}$-tensor product $\mathsf{A}_1 \otimes \cdots \otimes \mathsf{A}_n$ of vector spaces $\mathsf{A}_1, \ldots, \mathsf{A}_n$ becomes a *-algebra with the product and involution defined on elementary tensors by

$$(a_1 \otimes \cdots \otimes a_n)(b_1 \otimes \cdots \otimes b_n) = a_1 b_1 \otimes \cdots \otimes a_n b_n, \tag{2.14}$$
$$(a_1 \otimes \cdots \otimes a_n)^+ = (a_1)^+ \otimes \cdots \otimes (a_n)^+, \tag{2.15}$$

where $a_1, b_1 \in \mathsf{A}_1, \ldots, a_n, b_n \in \mathsf{A}_n$. This ∗-algebra is called the *tensor product* ∗-*algebra* and denoted also by $\mathsf{A}_1 \otimes \cdots \otimes \mathsf{A}_n$ if no confusion can arise.

Suppose that $\mathsf{A}_1, \ldots, \mathsf{A}_n$ are unital. Then there is an injective ∗-homomorphism $\theta_j : \mathsf{A}_j \mapsto \mathsf{A}_1 \otimes \cdots \otimes \mathsf{A}_n$ defined by $\theta_j(a) = 1 \otimes \cdots \otimes 1 \otimes a \otimes 1 \otimes \cdots \otimes 1$, where a stands at the jth position. If we identify a and $\theta_j(a)$, then A_j becomes a ∗-subalgebra of $\mathsf{A}_1 \otimes \cdots \otimes \mathsf{A}_n$.

There is another algebra involution that makes the algebra $\mathsf{A} \otimes \mathsf{A}$ (with product (2.14)) into a ∗-algebra; it is defined by $(a_1 \otimes a_2)^+ = (a_2)^+ \otimes (a_1)^+$, $a_1, a_2 \in \mathsf{A}$.

2. Matrices over A

Let $n \in \mathbb{N}$. The set $M_n(\mathsf{A})$ of $n \times n$-matrices with entries in A is a ∗-algebra with the "usual" algebraic operations: For $a = (a_{ij})_{i,j=1}^n$, $b = (b_{ij})_{i,j=1}^n$ in $M_n(\mathsf{A})$ and $\lambda \in \mathbb{K}$, the (i, j)-entries of $a + b$, λa, ab and a^+ are defined by

$$(a+b)_{ij} = a_{ij} + b_{ij}, \quad (\lambda a)_{ij} = \lambda a_{ij}, \quad (ab)_{ij} = \sum_{k=1}^n a_{ik} b_{kj}, \quad (a^+)_{ij} = (a_{ji})^+.$$

Suppose A is unital. For $i, j \in \{1, \ldots, n\}$, let $e_{ij} \in M_n(\mathsf{A})$ be the matrix with 1 as (i, j)-entry and zeros elsewhere. These *matrix units* e_{ij} satisfy the relations

$$e_{ij} e_{kl} = \delta_{jk} e_{il} \quad \text{and} \quad e_{ij}^+ = e_{ji} \quad \text{for} \quad i, j, k, l = 1, \ldots, n.$$

There is a ∗-isomorphism of the ∗-algebras $M_n(\mathsf{A})$ and $\mathsf{A} \otimes M_n(\mathbb{K})$ given by

$$(a_{ij})_{i,j=1}^n \mapsto \sum_{i,j=1}^n a_{ij} \otimes e_{ij}.$$

Example 2.15 (*Involutions on $M_n(\mathbb{R})$*)
Let $\mathsf{A} = \mathbb{R}$. Then, by the preceding, $M_n(\mathbb{R})$ is a real ∗-algebra with involution $\varphi : a \mapsto a^+ := a^T$, where a^T is the transposed matrix. Let $b \in M_n(\mathbb{R})$ be invertible. Then $\theta_b(a) = bab^{-1}$, $a \in M_n(\mathbb{R})$, defines an algebra automorphism θ_b of $M_n(\mathbb{R})$.

Statement 1: *Suppose $b^T = b$ or $b^T = -b$. Then the map $\psi := \theta_b \circ \varphi$, that is, $\psi : a \mapsto a^{+b} := \theta_b(\varphi(a)) = ba^T b^{-1}$, is an algebra involution of $M_n(\mathbb{R})$.*

Proof For $a \in M_N(\mathbb{R})$, we compute

$$\theta(\theta(a^+)^+) = \theta((ba^T b^{-1})^T) = \theta((\pm b^{-1})a(\pm b)) = \theta(b^{-1}ab) = bb^{-1}abb^{-1} = a.$$

Therefore, condition (2.6) holds, so the assertion follows from Lemma 2.6. □

For instance, we have

$$\begin{pmatrix} a_{11} & a_{12} \\ a_{21} & a_{22} \end{pmatrix}^{+b} = \begin{pmatrix} a_{22} & \tau a_{21} \\ \tau a_{12} & a_{11} \end{pmatrix} \quad \text{for} \quad b = \begin{pmatrix} 0 & \tau \\ 1 & 0 \end{pmatrix}, \quad \tau = \pm 1.$$

Statement 2: *Each algebra involution ψ of $M_n(\mathbb{R})$ is of the form $\psi = \theta_b \circ \varphi$ for some invertible matrix $b \in M_n(\mathbb{R})$ such that $b^T = b$ or $b^T = -b$.*

Proof Let $\psi : a \mapsto a^\dagger$ be an algebra involution of $M_n(\mathbb{R})$. Then $\theta := \varphi \circ \psi$ is an automorphism of $M_n(\mathbb{R})$. It is well known that all automorphisms of $M_n(\mathbb{R})$ are inner, so $\theta = \theta_{b^{-1}}$ for some invertible $b \in M_n(\mathbb{R})$. Thus, $\theta(a) = (a^\dagger)^+ = b^{-1}ab$. Then, $a^\dagger = (b^{-1}ab)^+ = b^+ a^+ (b^+)^{-1}$ and hence

$$a = (a^\dagger)^\dagger = b^+ [b^+ a^+ (b^+)^{-1}]^+ (b^+)^{-1} = b^+ [b^{-1}ab](b^+)^{-1} = b^+ b^{-1} ab(b^+)^{-1},$$

so $(b(b^+)^{-1})a = a(b(b^+)^{-1})$ for all $a \in M_n(\mathbb{R})$. This means that $b(b^+)^{-1}$ belongs to the center of $M_n(\mathbb{R})$. Hence $b(b^+)^{-1} = \lambda I$ for some $\lambda \in \mathbb{R}$. Thus, $b = \lambda b^+$ and $b^+ = (\lambda b^+)^+ = \lambda b = \lambda^2 b^+$. Hence $\lambda = \pm 1$ and $b^+ = b^T = \pm b$. Therefore, we have $\psi(a) = a^\dagger = b^+ a^+ (b^+)^{-1} = ba^T b^{-1}$, that is, $\psi = \theta_b \circ \varphi$.   □

Let $b \in M_n(\mathbb{R})$ be a diagonal matrix with nonzero diagonal entries $b_1, \ldots, b_n$. Then $(e_{ij})^{+_b} e_{ij} = be_{ji}b^{-1}e_{ij} = b_j b_i^{-1} e_{jj}$. In particular, $(e_{jj})^{+_b} e_{jj} = e_{jj}$.

Assume further that $b_i = -b_j \neq 0$. Then we have

$$(e_{ij})^{+_b} e_{ij} + (e_{jj})^{+_b} e_{jj} = 0, \quad \text{while} \quad e_{ij}, e_{jj} \neq 0.$$

Note that, in contrast, for the standard involution $a \mapsto a^+ := a^T$ on $M_n(\mathbb{R})$, any equation $\sum_i (a_i)^+ a_i = 0$ implies that $a_i = 0$ for all i.   ○

3. Crossed product *-algebras

Let A be a *-algebra and G a group. Suppose $g \mapsto \theta_g$ is a homomorphism of G into the group of *-automorphisms of A. Then $\theta_e = \text{Id}$.

We define a *-algebra $A \times_\theta G$. As a vector space it is the $\mathbb{K}$-tensor product $A \otimes \mathbb{K}[G]$, or equivalently, the vector space of A-valued functions on G with finite support. By straightforward computations it follows that the vector space $A \otimes \mathbb{K}[G]$ becomes a *-algebra over $\mathbb{K}$ with product and involution on A determined by

$$(a \otimes g)(b \otimes h) = a\theta_g(b) \otimes gh \quad \text{and} \quad (a \otimes g)^+ = \theta_{g^{-1}}(a^+) \otimes g^{-1}, \qquad (2.16)$$

where $a, b \in A$ and $g, h \in G$.

Definition 2.16 This *-algebra is called the *crossed product *-algebra* of A and G and denoted by $A \times_\theta G$.

If the action of G is trivial, that is, $\theta_g = \text{Id}$ for all $g \in G$, it follows from (2.16) that the crossed product *-algebra $A \times_\theta G$ is the tensor product *-algebra $A \otimes \mathbb{K}[G]$.

From now on we suppose that A is unital. Let us identify b with $b \otimes e$ and g with $1 \otimes g$. Then A and $\mathbb{K}[G]$ are *-subalgebras of $A \times_\theta G$, and the crossed product *-algebra $A \times_\theta G$ is just the *-algebra generated by the two *-subalgebras A and $\mathbb{K}[G]$ with cross-commutation relations

$$gb = \theta_g(b)g \quad \text{for } b \in \mathsf{A}, \ g \in G. \tag{2.17}$$

Suppose now that G is a finite group of n elements. We define a linear mapping $\varphi : \mathsf{A} \times_\theta G \mapsto M_n(\mathsf{A}) \cong \mathsf{A} \otimes M_n(\mathbb{K})$ by

$$\varphi : a \otimes g \mapsto \sum_{h \in G} \theta_h(a) \otimes e_{h,hg}. \tag{2.18}$$

(We consider $e_{h,hg}$ as matrix unit after enumerating the elements of G.)

Lemma 2.17 *Then φ is an injective $*$-homomorphism of $\mathsf{A} \times_\theta G$ into $M_n(\mathsf{A})$.*

Proof Suppose $\varphi(x) = 0$ for $x = \sum_i a_i \otimes g_i \in \mathsf{A} \times_\theta G$. Then, for the matrix entries e_{h,hg_i} with $h = e$, we get $\theta_e(a_i) = a_i = 0$, so $x = 0$. Thus, φ is injective.

Now let $a \otimes g, \ b \otimes k \in \mathsf{A} \times_\theta G$. Then

$$\varphi(a \otimes g)\varphi(b \otimes k) = \left(\sum_{h \in G} \theta_h(a) \otimes e_{h,hg} \right) \left(\sum_{l \in G} \theta_l(b) \otimes e_{l,lk} \right)$$

$$= \sum_{h,l \in G} \delta_{hg,l}\, \theta_h(a)\theta_l(b) \otimes e_{h,lk} = \sum_{h \in G} \theta_h(a)\theta_{hg}(b) \otimes e_{h,hgk}$$

$$= \sum_{h \in G} \theta_h(a\theta_g(b)) \otimes e_{h,hgk} = \varphi(a\theta_g(b) \otimes gk) = \varphi((a \otimes g)(b \otimes k)).$$

Similarly, $(\varphi(a \otimes g))^+ = \varphi((a \otimes g)^+)$. Therefore, φ is a $*$-homomorphism. $\qquad\square$

Lemma 2.17 says that $\mathsf{A} \times_\theta G$ can be considered as a $*$-subalgebra of $M_n(\mathsf{A})$ via the embedding φ. We illustrate this with a simple example.

Example 2.18 Suppose σ is a $*$ automorphism of A such that $\sigma^n = \text{Id}$. For instance, if $\mathsf{A} = \mathbb{K}[x_1, \dots, x_n]$ and $\varepsilon_1, \dots, \varepsilon_n \in \{-1, 1\}$ satisfy $\varepsilon_1 \cdots \varepsilon_n = 1$, we may take the $*$-automorphism σ defined by

$$\sigma(f)(x_1, \dots, x_n) = f(\varepsilon_1 x_n, \varepsilon_2 x_1, \dots, \varepsilon_n x_{n-1}).$$

Let $G \cong \{0, \dots, n-1\}$ be the cyclic group of order n. There is a homomorphism θ of G into the group of $*$-automorphisms of A given by $k \mapsto \sigma^k$. Then

$$\varphi\left(\sum_{k=0}^{n-1} a_k \otimes k \right) = \begin{pmatrix} a_0 & a_1 & \dots & a_{n-1} \\ \sigma(a_{n-1}) & \sigma(a_0) & \dots & \sigma(a_{n-2}) \\ \vdots & \vdots & \ddots & \vdots \\ \sigma^{n-1}(a_1) & \sigma^{n-1}(a_2) & \dots & \sigma^{n-1}(a_0) \end{pmatrix}. \tag{2.19}$$

That is, $\mathsf{A} \times_\theta G$ is ($*$-isomorphic to) the $*$-algebra of matrices (2.19), where $a_0, \dots, a_{n-1}$ are arbitrary elements of A. $\qquad\bigcirc$

4. Group graded *-algebras

Another important class of *-algebras is introduced in the following definition.

Definition 2.19 Let G be a (discrete) group. A *G-graded *-algebra* is a *-algebra
A which is a direct sum $A = \bigoplus_{g \in G} A_g$ of vector spaces A_g, $g \in G$, such that

$$A_g \cdot A_h \subseteq A_{g \cdot h} \quad \text{and} \quad (A_g)^+ \subseteq A_{g^{-1}} \quad \text{for } g, h \in G. \tag{2.20}$$

From the conditions in (2.20) we conclude that a G-grading of a *-algebra A is
completely determined if we know the components for a set of algebra generators of
A. It is convenient to describe G-gradings in this manner.

The following is probably the simplest example of a G-graded *-algebra.

Example 2.20 (*Super *-algebra*)
Let G be the group $\mathbb{Z}/2 = \{0, 1\}$. Then each G-graded *-algebra A is a direct sum
$A = A_0 \oplus A_1$ of vector spaces A_0, A_1 such that

$$A_0 \cdot A_0 \subseteq A_0, \ A_0 \cdot A_1 + A_1 \cdot A_0 \subseteq A_1, \ A_1 \cdot A_1 \subseteq A_0, \ (A_0)^+ = A_0, \ (A_1)^+ = A_1.$$

Such a *-algebra is usually called a *super *-algebra*.   ◯

Example 2.21 Let J be an index set and let $F = \mathbb{C}\langle z_i, w_i; i \in J | (z_i)^+ = w_i \rangle$ be the
free unital *-algebra with generators z_i, w_i, $i \in J$, and involution given by $(z_i)^+ =
w_i$, $i \in J$. Then F is a $\mathbb{Z}$-graded *-algebra with $\mathbb{Z}$-grading determined by $z_i \in F_1$,
$i \in J$. The vector space F_n is the span of all finite products of $k \in \mathbb{N}_0$ factors of
elements w_i and $(n - k) \in \mathbb{N}_0$ factors of elements z_j.   ◯

The proof of the following simple lemma is straightforward and will be omitted.

Lemma 2.22 *If* $A = \bigoplus_{g \in G} A_g$ *is a G-graded *-algebra and* J *is a two-sided *-
ideal of* A *generated by subsets of* A_g, $g \in G$, *then the quotient *-algebra* A/J *is
also G-graded.*

Proofs of the existence of gradings are usually given by the following pattern:
First we define a G-grading on the free *-algebra (as done with $G = \mathbb{Z}$ in Example
2.21). If the polynomials of the defining relations belong to single components of
this grading, Lemma 2.22 applies and gives the grading of the *-algebra.

Example 2.23 (*Weyl algebra* $A = \mathbb{C}\langle a, a^+ | aa^+ - a^+a = 1 \rangle$)
Let $F = \mathbb{C}\langle a, a^+ \rangle$ be the free unital *-algebra with single generator a equipped with
$\mathbb{Z}$-grading given by $a \in F_1$. Then we have $a^+ \in F_{-1}$, so the defining polynomial
$aa^+ - a^+a - 1$ belongs to the component F_0 and Lemma 2.22 applies. Hence A is
a $\mathbb{Z}$-graded *-algebra with grading determined by $a \in A_1$.

Set $N := a^+a$. It is not difficult to verify that A_0 is the polynomial algebra $\mathbb{C}[N]$
and $A_n = a^n \mathbb{C}[N]$, $A_{-n} = (a^+)^n \mathbb{C}[N]$ for $n \in \mathbb{N}_0$.   ◯

Example 2.24 (*Enveloping algebra of the Virasoro algebra*)
Let A be the unital ∗-algebra with generators z and x_n, $n \in \mathbb{Z}$, relations

$$[x_k, x_n] = (n-k)x_{k+n} + \delta_{k+n,0}\frac{1}{12}(n^3-n)z, \quad [x_n, z] = 0, \quad k, n \in \mathbb{Z}, \quad (2.21)$$

and involution determined by $(x_n)^+ = x_{-n}$ and $z^+ = z$, where $[x, y] := xy - yx$. The Lie algebra with generators z and x_n, $n \in \mathbb{Z}$, and Lie brackets (2.21) is the *Virasoro algebra* and the algebra A is called its enveloping algebra. From Lemma 2.22 it follows that A is a $\mathbb{Z}$-graded ∗-algebra with $x_n \in A_n$ and $z \in A_0$. ○

In the literature, a G-graded algebra $A = \oplus_{g \in G} A_g$ is often defined by requiring that A_{gh} is *equal* to the span of $A_g \cdot A_h$ for $g, h \in G$, see, e.g., [Mk99]. We do not assume this, because it does not hold for our standard example. For the Weyl algebra A (Example 2.23), we have $A_0 = \mathbb{C}[N]$, $A_1 = aA_0$, $A_{-1} = a^+A_0 = A_0a^+$. Hence the linear span of $A_{-1} \cdot A_1$ is equal to $N \cdot \mathbb{C}[N]$ and different from A_0.

2.3 Quadratic Modules

Throughout this section, A denotes a ∗-**algebra** over $\mathbb{K}$.

In commutative real algebraic geometry [Ms08], positivity notions are usually described in terms of quadratic modules. Their counterpart for general ∗-algebras is introduced in the following definition.

Definition 2.25 A nonempty subset Q of A_{her} is a *pre-quadratic module* of A if

$$a + b \in Q \quad \text{and} \quad \lambda a \in Q \quad \text{for} \quad a, b \in Q, \lambda \geq 0, \quad (2.22)$$

$$x^+ax \in Q \quad \text{for} \quad u \in Q, x \in A. \quad (2.23)$$

Any element a^+a, where $a \in A$, is called a *hermitian square* of A.

A quadratic module Q of a unital ∗-algebra such that $1 \in Q$ is called a *quadratic module*.

Example 2.26 Let $A = \mathbb{C}[x]$. The set of finite sums of elements x^2p^+p, where $p \in \mathbb{C}[x]$, is a pre-quadratic module of A which is not a quadratic module. ○

It is easily checked that the set

$$\sum A^2 := \left\{ \sum_{k=1}^n (a_k)^+a_k : a_k \in A, n \in \mathbb{N} \right\} \quad (2.24)$$

of all finite sums of hermitian squares is a pre-quadratic module of A.

If A is unital, it follows from condition (2.23), with $a = 1$, that $\sum A^2$ is the *smallest* quadratic module of A.

Example 2.27 Suppose A is unital and X is a subset of A_{her}. Then the set of all finite sums of elements a^+a and a^+xa, where $a \in A$ and $x \in X$, is the smallest quadratic module of A that contains X.   ○

The next lemma collects a number of useful *polarization identities*.

Lemma 2.28 *If A is a complex ∗-algebra and $a, x, y \in A$, we have*

$$4\, yax = \sum_{k=0}^{3} i^k \, (x + i^k y^+)^+ a \, (x + i^k y^+), \tag{2.25}$$

$$4\, xay = \sum_{k=0}^{3} i^{-k} \, (x^+ + i^k y)^+ a \, (x^+ + i^k y), \tag{2.26}$$

$$4\, yx = \sum_{k=0}^{3} i^k \, (x + i^k y^+)^+ (x + i^k y^+), \tag{2.27}$$

$$2\, (x^+ ay + y^+ ax) = (x + y)^+ a(x + y) - (x - y)^+ a(x - y), \tag{2.28}$$

$$2\, (x^+ y + y^+ x) = (x + y)^+ (x + y) - (x - y)^+ (x - y). \tag{2.29}$$

If A is a complex unital ∗-algebra and $a, x, y \in A$, then

$$4\, ax = \sum_{k=0}^{3} i^k \, (x + i^k 1)^+ a \, (x + i^k 1), \tag{2.30}$$

$$4\, xa = \sum_{k=0}^{3} i^k \, (x + i^k 1) \, a \, (x + i^k 1)^+. \tag{2.31}$$

Proof The equalities (2.25) and (2.28) are proved by direct computations of the corresponding right-hand sides. Setting $y = 1$ in (2.25) gives (2.30). The identities (2.26) and (2.31) follow from (2.25) and (2.30) by taking the adjoints and replacing a, x, y by a^+, x^+, y^+, respectively. By applying (2.25) and (2.28) to the unitization A^1 of A and setting $a = 1$, we obtain (2.27) and (2.29).   □

For any complex ∗-algebra A, it follows easily from (2.27) and (2.29) that

$$A^2 = \sum A^2 - \sum A^2 + i \sum A^2 - i \sum A^2, \tag{2.32}$$

$$A^2 \cap A_{her} = \sum A^2 - \sum A^2.$$

Here A^2 denotes the $\mathbb{C}$-linear span of products ab for $a, b \in A$. If Q is a quadratic module of a complex *unital* ∗-algebra A, then $Q \supseteq \sum A^2$ and $A^2 = A$, so that

$$A = Q - Q + i\, Q - i\, Q, \quad A_{her} = Q - Q.$$

Proposition 2.29 *Let Q be a pre-quadratic module of a complex $*$-algebra A and set $Q^0 := Q \cap (-Q)$. Then $\mathcal{I}_Q := Q^0 + i \, Q^0$ is a two-sided $*$-ideal of A; it is called the* support ideal *of Q.*

Proof It is clear that $\mathcal{I}_Q$ is a $*$-invariant complex vector space and $Q^0 - Q^0 \subseteq Q^0$. For $c \in Q^0$ and $x, y \in A$, it follows from (2.23) that $(x^+ + i^k y)^+ c (x^+ + i^k y) \in Q^0$. Therefore, by (2.26), $4xcy \in (Q^0 - Q^0 + iQ^0 - iQ^0) \subseteq Q^0 + i \, Q^0 = \mathcal{I}_Q$. Thus, $A \cdot Q^0 \cdot A \subseteq \mathcal{I}_Q$ and hence $A \cdot (Q^0 + iQ^0) \cdot A = A \cdot \mathcal{I}_Q \cdot A \subseteq \mathcal{I}_Q$, that is, $\mathcal{I}_Q$ is a two-sided ideal of A. $\qquad\square$

Important examples of quadratic modules are defined by $*$-representations in Sect. 5.7. We close this section with two examples from the commutative case.

Example 2.30 (*Motzkin polynomial $p(x_1, x_2) := x_1^2 x_2^2 (x_1^2 + x_2^2 - 3) + 1$*) Let $A := \mathbb{C}[x_1, x_2]$. We show that $p(x_1, x_2)$ is nonnegative on $\mathbb{R}^2$, but $p \notin \sum A^2$.

For $(x_1, x_2) \in \mathbb{R}^2$, the arithmetic–geometric mean inequality yields

$$x_1^4 x_2^2 + x_1^2 x_2^4 + 1 \geq 3 \sqrt[3]{x_1^4 x_2^2 \cdot x_1^2 x_2^4 \cdot 1} = 3x_1^2 x_2^2,$$

which in turn implies $p(x_1, x_2) \geq 0$.

We prove that $p \notin \sum \mathbb{C}[x_1, x_2]^2$. Assume to the contrary that $p = \sum_j q_j^+ q_j$, where $q_j \in \mathbb{C}[x_1, x_2]$. Then

$$p(x_1, x_2) = \sum_j |q_j(x_1, x_2)|^2, \quad (x_1, x_2) \in \mathbb{R}^2. \tag{2.33}$$

Comparing the monomials of highest degrees on both sides we conclude that $\deg(q_j) \leq 3$. Since $p(0, x_2) = p(x_1, 0) = 1$, it follows from (2.33) that the polynomials $q_j(0, x_2)$ and $q_j(x_1, 0)$ are constant. Hence each q_j is of the form $\lambda_j + x_1 x_2 r_j$, with $\lambda_j \in \mathbb{C}$ and linear $r_j \in \mathbb{C}[x_1, x_2]$. Then, comparing the coefficients of $x_1^2 x_2^2$ in (2.33) yields $\sum_j |r_j(0, 0)|^2 = -3$, which is the desired contradiction. $\qquad\bigcirc$

Example 2.31 (*Real algebraic geometry*) Let $\mathbb{K}$ be either $\mathbb{R}$ or $\mathbb{C}$ and let A be the commutative $*$-algebra $\mathbb{K}[x_1, \dots, x_d]$. Then $A_{her} = \mathbb{R}[x_1, \dots, x_d]$. For any closed subset K of $\mathbb{R}^d$ we define

$$A(K)_+ := \{ p \in \mathbb{R}[x_1, \dots, x_d] : p(t) \geq 0 \ \text{ for all } \ t \in K \}.$$

Obviously, $A(K)_+$ is a quadratic module which is invariant under multiplication, that is, $p, q \in A(K)_+$ implies $p \cdot q \in A(K)_+$. In real algebraic geometry, a quadratic module which has this property is called a *pre-ordering*.

In particular, $A_+ := A(\mathbb{R}^d)_+$ is the set of polynomials of $\mathbb{R}[x_1, \dots, x_d]$ that are nonnegative on $\mathbb{R}^d$. Clearly, $\sum A^2 \subseteq A_+$. If $d = 1$, it follows from the fundamental theorem of algebra that $\sum A^2 = A_+$. For $d \geq 2$, we have $\sum A^2 \neq A_+$. Indeed, if p is the polynomial from Example 2.30, then $q(x_1, \dots, x_d) := p(x_1, x_2)$ belongs to A_+, but q is not in $\sum A^2$. $\qquad\bigcirc$

2.4 Positive Functionals and States on Complex *-Algebras

Throughout this section, A denotes a **complex** *-algebra.

Definition 2.32 A linear functional $f : \mathsf{A} \mapsto \mathbb{C}$ is called *positive* if

$$f(a^+a) \geq 0 \quad \text{for} \quad a \in \mathsf{A}$$

and *hermitian* if

$$f(a^+) = \overline{f(a)} \quad \text{for} \quad a \in \mathsf{A}.$$

A *state* on a unital *-algebra is a positive linear functional f such that $f(1) = 1$. The set of positive linear functionals on A is denoted by $\mathcal{P}(\mathsf{A})^*$.

Clearly, the positive linear functionals on A are precisely the linear functionals that are nonnegative on the pre-quadratic module $\sum \mathsf{A}^2$ defined by (2.24).

The following fundamental inequality (2.34) is the *Cauchy–Schwarz inequality*.

Proposition 2.33 *Let f be a positive linear functional on* A. *Then, for $a, b \in \mathsf{A}$,*

$$|f(b^+a)|^2 \leq f(a^+a)f(b^+b), \tag{2.34}$$

$$f(a^+b) = \overline{f(b^+a)}. \tag{2.35}$$

If A *is unital, then f is hermitian.*

Proof For $\alpha, \beta \in \mathbb{C}$ we compute

$$f((\alpha a + \beta b)^+(\alpha a + \beta b))$$
$$= \overline{\alpha}\alpha f(a^+a) + \overline{\alpha}\beta f(a^+b) + \alpha\overline{\beta} f(b^+a) + \overline{\beta}\beta f(b^+b) \geq 0. \tag{2.36}$$

Hence $\overline{\alpha}\beta f(a^+b) + \alpha\overline{\beta} f(b^+a)$ is real, so its imaginary part vanishes. We set first $\overline{\alpha}\beta = 1$ and then $\overline{\alpha}\beta = \mathrm{i}$. This gives $\mathrm{Im}\, f(a^+b) = -\mathrm{Im}\, f(b^+a)$ and $\mathrm{Re}\, f(a^+b) = \mathrm{Re}\, f(b^+a)$, respectively. Both equalities together imply (2.35).

The expression in (2.36) is a positive semi-definite quadratic form, so its discriminant has to be nonnegative. By (2.35), this yields $f(a^+a)f(b^+b) - |f(b^+a)|^2 \geq 0$, which proves (2.34).

Now suppose A is unital. Then we set $b = 1$ in (2.35) and obtain $f(a^+) = \overline{f(a)}$. Thus f is hermitian. $\qquad\square$

Definition 2.34 A positive linear functional f on A is called *extendable* if it can be extended to a positive linear functional on its unitization A^1. We denote the set of extendable positive functionals on A by $\mathcal{P}_e(\mathsf{A})^*$.

If A is unital, $\mathsf{A} = \mathsf{A}^1$ and all positive functionals on A are trivially extendable.

Proposition 2.35 *Suppose f is a positive linear functional on A. Then f is extendable if and only if f is hermitian and there exists a constant $c \geq 0$ such that*

$$|f(a)|^2 \leq cf(a^+a) \quad \text{for} \quad a \in A. \tag{2.37}$$

If A is unital, then $f(1)$ is the smallest constant $c \geq 0$ such that (2.37) holds.

Proof First assume that f is extendable. Let f_1 be a positive linear functional on A^1 which extends f. Then, by Proposition (2.33), f_1 is hermitian, so is f, and (2.37) holds with $c := f_1(1)$ by the Cauchy–Schwarz inequality (2.34), applied to f_1 with $b = 1$. Therefore, both conditions on f are satisfied.

Conversely, first suppose A is unital. Then (2.34) implies (2.37) with $c = f(1)$. If (2.37) holds for some $c \geq 0$, setting $a = 1$ yields $f(1)^2 \leq cf(1)$. Hence $f(1) \leq c$. Thus, $f(1)$ is indeed the smallest $c \geq 0$ for which (2.37) is fulfilled.

Now we suppose that A is not unital and both conditions are satisfied. We can assume that $f \not\equiv 0$. Clearly, there exists a smallest number $c \geq 0$ for which (2.37) holds. Then $c > 0$, since $c = 0$ would imply $f \equiv 0$ by (2.37). We fix a real number $C \geq c$ and define a linear functional f_1 on A^1 by

$$f_1(a + \alpha) := f(a) + \alpha C \quad \text{for} \quad a \in A, \ \alpha \in \mathbb{C}. \tag{2.38}$$

Since f is hermitian, $f(a^+) = \overline{f(a)}$. Therefore, using (2.38) and (2.37) we obtain

$$\begin{aligned}
f_1((a + \alpha)^+(a + \alpha)) &= f(a^+a) + \overline{\alpha}f(a) + \alpha f(a^+) + |\alpha|^2 C \\
&= f(a^1a) + \overline{\alpha}f(a) + \alpha\,\overline{f(a)} + |\alpha|^2 C \\
&\geq f(a^+a) - 2|\alpha|\,|f(a)| + |\alpha|^2 c \\
&= f(a^+a) - c^{-1}|f(a)|^2 + \left(|\alpha|c^{1/2} - |f(a)|c^{-1/2}\right)^2 \geq 0.
\end{aligned}$$

Hence f_1 is a positive functional on A^1. By (2.38), f_1 is an extension of f. □

The last assertions of Propositions 2.33 and 2.35 suggest how to define a state in the nonunital case and they show that for unital ∗-algebras the following definition of a state coincides with the one given in Definition 2.32.

Definition 2.36 A positive functional f on a ∗-algebra A is called a *state* if f is hermitian and 1 is the smallest number $c \geq 0$ such that $|f(a)|^2 \leq cf(a^+a)$ for all $a \in A$. The set of states of A is denoted by $\mathcal{S}(A)$.

Corollary 2.37 *Any state on a ∗-algebra A has an (obviously unique) extension to a state on its unitization A^1.*

Proof Suppose A is not unital and f is a state on A. Since (2.37) holds with $c = 1$, it follows from (2.38) and the proof of Proposition 2.35 that $f_1(1) = 1$ defines a state on A^1. □

Corollary 2.38 *Let f be a positive linear functional on a $*$-algebra A and $x \in \mathsf{A}$. Then $f_x(\cdot) := f(x^+ \cdot x)$ is an extendable positive linear functional.*

Proof Let $a \in \mathsf{A}$. Since $f_x(a^+a) = f((ax)^+ax) \geq 0$, f_x is a positive functional on A. Using (2.35) and (2.34) we obtain

$$f_x(a^+) = f((ax)^+x) = \overline{f(x^+(ax))} = \overline{f_x(a)}\,,$$
$$|f_x(a)|^2 = |f(x^+(ax))|^2 \leq f(x^+x)f((ax)^+ax) = f(x^+x)f_x(a^+a).$$

Hence f_x is extendable by Proposition 2.35. □

Example 2.39 (*A positive functional that is not extendable and not hermitian*)
Let A be the $*$-algebra $x\mathbb{C}[x]$ with involution given by $x^+ := x$. Fix $\alpha \in \mathbb{C}$, $\alpha \neq 0$. We define a linear functional f on A by $f(xp(x)) := \alpha p(0)$, $p \in \mathbb{C}[x]$. Since $f((xp)^+xp) = 0$ for $p \in \mathbb{C}[x]$, f is a positive functional on A. But condition (2.37) does not hold, so f is not extendable. Clearly, f is not hermitian if $\alpha \notin \mathbb{R}$. ○

Example 2.40 (*Positive semi-definite functions on $*$-semigroups*)
Let S be a $*$-semigroup. Clearly, a linear functional f on $\mathbb{C}[S]$ is uniquely determined by its values $f(s)$ at $s \in S$. Hence linear functionals on $\mathbb{C}[S]$ correspond to functions on S. By definition, a linear functional f on $\mathbb{C}[S]$ is positive if and only if

$$f\left(\left(\sum_s \alpha_s s\right)^+ \left(\sum_t \alpha_t t\right)\right) = \sum_{s,t} \overline{\alpha_s}\alpha_t f(s^+t) \geq 0 \tag{2.39}$$

for all $\sum_s \alpha_s s \in \mathbb{C}[S]$. ○

Definition 2.41 A function f on a $*$-semigroup S is called *positive semi-definite*, or of *positive type*, if

$$\sum_{j,k=1}^{n} \overline{\alpha_j}\alpha_k f(s_j^+ s_k) \geq 0 \tag{2.40}$$

for arbitrary elements $s_1, \ldots, s_n \in S$ and numbers $\alpha_1, \ldots, \alpha_n \in \mathbb{C}$, $n \in \mathbb{N}$.

Comparing (2.39) and (2.40) we see that a linear functional on $\mathbb{C}[S]$ is positive if and only if its restriction to S is a positive semi-definite function. ○

In the rest of this section, we assume that f is a **state** on a **unital** $*$-algebra A.

Definition 2.42 The *standard deviation* $\Delta_f(a)$ of the state f in $a \in \mathsf{A}$ is defined by

$$\Delta_f(a) := \left[f\left((a - f(a)1)^+(a - f(a)1)\right)\right]^{1/2} = \left(f(a^+a) - |f(a)|^2\right)^{1/2}. \tag{2.41}$$

The second equality in (2.41) follows by a simple computation using the fact that the state f is hermitian. Note that we have $f(a^+a) - |f(a)|^2 \geq 0$ by the Cauchy–Schwarz inequality, since $f(1) = 1$.

Lemma 2.43 *For $a, b \in A$ we have*

$$|f(ba) - f(b)f(a)| \leq \Delta_f(a)\,\Delta_f(b^+). \tag{2.42}$$

Proof Using the equations $f(1) = 1$ and (2.35) and the inequality (2.34) we derive

$$\begin{aligned}
\left|f(ba) - f(b)f(a)\right| &= \left|f\big((b - f(b)1)(a - f(a)1)\big)\right| \\
&= \left|f\big((b^+ - f(b^+)1)^+(a - f(a)1)\big)\right| \\
&\leq \left[f\big((a - f(a)1)^+(a - f(a)1)\big)\right]^{1/2}\left[f\big((b^+ - f(b^+)1)^+(b^+ - f(b^+)1)\big)\right]^{1/2} \\
&= \Delta_f(a)\,\Delta_f(b^+).
\end{aligned}$$

$\square$

The following inequality (2.43) is the algebraic version of the *uncertainty relation* studied in Sect. 8.7.

Proposition 2.44 *For each state f on A and hermitian elements $a, b \in A$,*

$$\Delta_f(a)\Delta_f(b) \geq \frac{1}{2}\,|f(ab - ba)|. \tag{2.43}$$

Proof Since $a = a^+$, $b = b^+$, it follows from (2.42) and its counterpart obtained by interchanging a and b that

$$|f(ab) - f(ba)| \leq |f(ab) - f(a)f(b)| + |f(ba) - f(b)f(a)| \leq 2\Delta_f(a)\Delta_f(b).$$

This proves (2.43).

$\square$

Example 2.45 (*Noncommutative probability space*)
A (noncommutative) *probability space* is a pair (A, f) of a unital ∗-algebra A and a fixed state f of A. Any hermitian element $a \in A$ is considered as a random variable, and the number $f(a)$ is interpreted as the expectation value of f at a.

The classical example is the following: Let $(X, \mathfrak{A})$ be a measurable space and μ a probability measure on $(X, \mathfrak{A})$. Then $f(a) = \int a\,d\mu$ defines a state f on the ∗-algebra $A := L^\infty(X, \mathfrak{A})$ and the pair (A, f) is a probability space. In this case, $f(a)$ and $\Delta_f(a)$ are the "usual" expectation value and standard deviation, respectively, of a random variable $a = a^+ \in A$. $\bigcirc$

2.5 Positive Functionals on Real ∗-Algebras

In this section, A denotes a **real** ∗-algebra.

The first two notions of the following definition are verbatim the same as for complex ∗-algebras; see Definition 2.32.

Definition 2.46 A linear functional $f : A \mapsto \mathbb{R}$ is called
- *positive* if $f(a^+a) \geq 0$ for $a \in A$,
- *hermitian* if $f(a^+) = f(a)$ for $a \in A$.

If A is unital, we say that f is a *state* on A if f is positive, hermitian, and $f(1) = 1$.

The counterpart of Proposition 2.33 for real *-algebras is the following.

Proposition 2.47 *Let $f : A \mapsto \mathbb{R}$ be a positive linear functional and $a, b \in A$. Then*

$$f(b^+a + a^+b)^2 \leq 4f(a^+a)f(b^+b). \tag{2.44}$$

In particular, if b^+a is hermitian, then

$$f(b^+a)^2 \leq f(a^+a)f(b^+b), \tag{2.45}$$

and if A is unital, then $f(a + a^+)^2 \leq 4f(1)f(a^+a)$.

Proof As in the proof of Proposition 2.33, see (2.36), we obtain for *real* α, β,

$$f((\alpha a + \beta b)^+(\alpha a + \beta b)) = \alpha^2 f(a^+a) + \alpha\beta[f(a^+b) + f(b^+a)] + \beta^2 f(b^+b) \geq 0.$$

Hence the expression in the middle is a positive semi-definite real quadratic form. Therefore, its discriminant $f(a^+a)f(b^+b) - (\frac{1}{2}[f(b^+a + a^+b)])^2$ is nonnegative, which implies (2.44).

If b^+a is hermitian, $b^+a = (b^+a)^+ = a^+b$, so (2.44) gives (2.45). Setting $b = 1$ in (2.44) yields the last inequality. □

Let $f_{\mathbb{R}} : A \mapsto \mathbb{R}$ be an $\mathbb{R}$-linear functional on the real *-algebra A. Clearly, $f_{\mathbb{R}}$ has a unique extension to a $\mathbb{C}$-linear functional $f_{\mathbb{C}} : A_{\mathbb{C}} \mapsto \mathbb{C}$ on the complexification $A_{\mathbb{C}}$ (Definition 2.9) of A and $f_{\mathbb{C}}$ is given by $f_{\mathbb{C}}(a + ib) = f_{\mathbb{R}}(a) + if_{\mathbb{R}}(b), a, b \in A$. If $f_{\mathbb{C}}$ is a positive functional on A, then $f_{\mathbb{C}}$ is not necessarily positive on $A_{\mathbb{C}}$, as shown in Example 2.50 below. The following simple lemma clarifies when $f_{\mathbb{C}}$ is positive.

Lemma 2.48 *$f_{\mathbb{C}}$ is a positive functional on $A_{\mathbb{C}}$ if and only if $f_{\mathbb{R}}$ is a positive functional on A and $f_{\mathbb{R}}(a^+b) = f_{\mathbb{R}}(b^+a)$ for all $a, b \in A$.*

Suppose that A is unital. Then $f_{\mathbb{C}}$ is positive on $A_{\mathbb{C}}$ if and only if $f_{\mathbb{R}}$ is positive and hermitian on A. Further, $f_{\mathbb{C}}$ is a state on $A_{\mathbb{C}}$ if and only if $f_{\mathbb{R}}$ is a state on A.

Proof Recall that the elements of $A_{\mathbb{C}}$ are $a + ib$, where $a, b \in A$, and in the *-algebra $A_{\mathbb{C}}$ we have $(a + ib)^+(a + ib) = a^+a + b^+b + i(a^+b - b^+a)$. Hence $f_{\mathbb{C}}$ is positive on $A_{\mathbb{C}}$ if and only if $f_{\mathbb{R}}$ is positive on A and $f_{\mathbb{R}}(a^+b - b^+a) = 0$, that is, $f_{\mathbb{R}}(a^+b) = f_{\mathbb{R}}(b^+a)$ for $a, b \in A$.

If A is unital, we can set $b = 1$ and the latter holds if and only if the functional $f_{\mathbb{R}}$ is hermitian on A. Therefore, by the definition of a state on a real *-algebra (Definition 2.46), $f_{\mathbb{C}}$ is a state if and only if $f_{\mathbb{R}}$ is so. □

The following examples show a number of peculiarities that can occur for positive linear functionals on real *-algebras.

Example 2.49 Let A be the complex *-algebra $\mathbb{C}$ considered as a *real* *-algebra. Then $f(x + iy) := x + y, x, y \in \mathbb{R}$, defines an $\mathbb{R}$-linear (!) functional f on A. This functional is positive (by $f((x + iy)^+(x + iy)) = x^2 + y^2$), but it is not hermitian. Set $a = 1, b = 1 + i$. Then $|f(b^+a)|^2 = 4$ and $f(a^+a)f(b^+b) = 2$, so the inequality (2.45) does not hold in general if the element b^+a is not hermitian.   ○

Example 2.50 (*Example 2.8 continued*)
Let A^1 be the commutative real unital *-algebra from Example 2.8. From the definitions (2.7) of the algebraic operations of the unitization it follows that the *-algebra A^1 is just the vector space $\mathbb{R}^2$ with multiplication and involution

$$(x_1, x_2)(y_1, y_2) = (x_2y_1 + x_1y_2, x_2y_2), \quad (x_1, x_2)^+ = (-x_1, x_2). \tag{2.46}$$

The unit element is $(0, 1)$. Let $t = (t_1, t_2) \in \mathbb{R}^2$. We define a linear functional f_t on A^1 by $f_t(x) = t_1x_1 + t_2x_2$ for $x = (x_1, x_2) \in \mathbb{R}^2$. Since $(x_1, x_2)^+(x_1, x_2) = (0, x_2^2)$ by (2.46), f_t is a positive linear functional if and only if $t_2 \geq 0$. Clearly, f_t is a hermitian functional if and only if $t_1 = 0$.

Now suppose $t_1 \neq 0$ and $t_2 \geq 0$. Then f_t is a positive functional on A^1 that is *not* hermitian. Hence, by Lemma 2.48, the positive functional f_t on the real *-algebra A^1 has *no extension* to a positive linear functional on the complexification $(A^1)_\mathbb{C}$.

For $x = (x_1, 0) \neq 0$, we have $f_t(x) = t_1x_1 \neq 0$ and $f_t(x^+x) = 0$. Hence the Cauchy–Schwarz inequality $|f(a)|^2 \leq f(1)f(a^+a)$ for *complex* unital *-algebras does not hold for real *-algebras. Note that the element $b^+a + a^+b$ on the left-hand side of (2.44) is hermitian. But if the functional f on A is positive and hermitian, then (2.44) gives the "usual" Cauchy–Schwarz inequality (2.45).   ○

2.6 Characters of Unital Algebras

In this section, A is a (not necessarily commutative) **unital** algebra over $\mathbb{K}$.

Definition 2.51 A *character*, or a *multiplicative linear functional*, of A is a nonzero algebra homomorphism $\chi : A \mapsto \mathbb{K}$, that is, χ is a linear functional on A such that $\chi \not\equiv 0$ and

$$\chi(ab) = \chi(a)\chi(b) \quad \text{for } a, b \in A. \tag{2.47}$$

If χ is a character, then $\chi(1) = 1$. (Indeed, since χ is nonzero, we can find $a \in A$ such that $\chi(a) \neq 0$; then $\chi(a)\chi(1) = \chi(a)$, so that $\chi(1) = 1$.)

Example 2.52 Let A be the algebra of rational functions in one variable x. If χ is a character of A, then $1 = \chi(1) = \chi(\chi(x) - x)\chi((\chi(x) - x)^{-1}) = 0$, a contradiction. This shows that the algebra A has no character.   ○

Proposition 2.53 *Let χ be a linear functional on A such that $\chi(1) = 1$. Then the following are equivalent:*

(i) $\chi(a) = 0$ *implies* $\chi(a^2) = 0$ *for* $a \in$ A.
(ii) $\chi(a^2) = \chi(a)^2$ *for* $a \in$ A.
(iii) $\chi(a) = 0$ *implies* $\chi(ab) = 0$ *for* $a, b \in$ A.
(iv) χ *is a character of* A.

Proof (i)$\to$(ii): Let $a \in$ A. Since $\chi(1) = 1$, $\chi(a - \chi(a)1) = 0$. Hence, by (i),

$$0 = \chi\big((a - \chi(a)1)^2\big) = \chi\big(a^2 - 2a\chi(a) + \chi(a)^2 1\big) = \chi(a^2) - \chi(a)^2.$$

(ii)$\to$(iii): Applying (ii) with $a = u + v$ we obtain

$$\big(\chi(u) + \chi(v)\big)^2 = \big(\chi(u + v)\big)^2 = \chi\big((u + v)^2\big) = \chi(u^2) + \chi(v^2) + \chi(uv + vu).$$

Combined with $\chi(u^2) = \chi(u)^2$ and $\chi(v^2) = \chi(v)^2$ again by (ii), the latter implies

$$\chi(uv + vu) = 2\chi(u)\chi(v) \quad \text{for } u, v \in \text{A}. \tag{2.48}$$

Now suppose $\chi(a) = 0$. Then, (2.48) applied with u, v replaced by a, b, yields

$$\chi(ab + ba) = 0. \tag{2.49}$$

Applying first (ii) twice, then (2.48) with $u = a$, $v = bab$, (2.49), and finally $\chi(a) = 0$ we compute

$$\big(\chi(ab - ba)\big)^2 = \chi\big((ab - ba)^2\big) = \chi\big(2abab + 2baba - (ab + ba)^2\big)$$
$$= 2\chi\big(a(bab) + (bab)a\big) - \big(\chi(ab + ba)\big)^2 = 4\chi(a)\chi(bab) + 0 = 0,$$

so that $\chi(ab - ba) = 0$. Adding the latter equation and (2.49) we get $\chi(ab) = 0$.

(iii)$\to$(iv): Since $\chi(a - \chi(a)1) = 0$, it follows from (iii) that

$$0 = \chi((a - \chi(a)1)b) = \chi(ab) - \chi(a)\chi(b).$$

Hence $\chi(ab) = \chi(a)\chi(b)$, so χ is a character.
(iv)$\to$(i) is trivial. $\square$

From now on, A is a **complex unital** algebra. Let A^{-1} denote the set of elements $a \in$ A which are invertible in A, that is, there exists an element $b \in$ A such that $ab = ba = 1$. Then b is uniquely determined by a and denoted by a^{-1}.

Definition 2.54 For $a \in$ A, the set of numbers $\lambda \in \mathbb{C}$ for which the element $a - \lambda \cdot 1$ is not in A^{-1} is called the *spectrum* of a and denoted by $\sigma_\text{A}(a)$.

Example 2.55 For $\text{A} = \mathbb{C}[x]$, the spectrum of each nonconstant polynomial is $\mathbb{C}$. By contrast, for the algebra of rational functions of x the spectrum of each nonconstant function is empty. $\bigcirc$

Simple properties of the spectrum are listed in Exercise 13. The spectrum is a fundamental tool in the theory of Banach algebras and C^*-algebras. However, the main classes of $*$-algebras treated in this book (polynomial algebras, Weyl algebra, enveloping algebras) do not contain "enough" inverses, so the concept of the spectrum is not useful for their study. In fact, for these $*$-algebras the spectrum of all nonconstant elements is $\mathbb{C}$.

The following result is the *Gleason–Kahane–Zelazko theorem.*

Theorem 2.56 *Suppose* A *is a unital complex algebra for which the spectrum of every element of* A *is bounded. If* χ *is a linear functional on* A *such that* $\chi(1) = 1$, *then the following statements are equivalent:*

(i) $\chi(a) \in \sigma_A(a)$ *for* $a \in A$.
(ii) $\chi(a) \neq 0$ *for* $a \in A^{-1}$.
(iii) χ *is a character of* A.

Proof (i)→(ii): Suppose $a \in A^{-1}$. If we would have $\chi(a) = 0$, then $0 \in \sigma_A(a)$ by (i), so $a \notin A^{-1}$, which is a contradiction.

(ii)→(i): Since $\chi(a - \chi(a)1) = 0$, $(a - \chi(a)1) \notin A^{-1}$ by (ii), so $\chi(a) \in \sigma_A(a)$.

(iii)→(ii): From $1 = \chi(1) = \chi(aa^{-1}) = \chi(a)\chi(a^{-1})$ we get $\chi(a) \neq 0$.

(ii)→(iii): Let $a \in A$. Suppose that $\chi(a) = 0$.

Let $n \subset \mathbb{N}$, $n \geq 2$. Consider the polynomial $p(\lambda) := \chi((\lambda - a)^n)$. Let $\lambda_1, \dots, \lambda_n$ be its (complex!) roots. Since $0 = p(\lambda_i) = \chi((\lambda_i - a)^n)$, we have $(\lambda_i - a)^n \notin A^{-1}$ by (ii), hence $(\lambda_i - a) \notin A^{-1}$, so that $\lambda_i \in \sigma_A(a)$. Further,

$$p(\lambda) = \prod_{i=1}^{n}(\lambda - \lambda_i) = \lambda^n - n\chi(a)\lambda^{n-1} + \binom{n}{2}\chi(a^2)\lambda^{n-2} + \dots \qquad (2.50)$$

Comparing the coefficients of λ^{n-1} and λ^{n-2} in (2.50) yields $\sum_i \lambda_i = n\chi(a) = 0$ and $\sum_{i<j} \lambda_i \lambda_j = \binom{n}{2}\chi(a^2)$. Therefore,

$$0 = \left(\sum_i \lambda_i\right)^2 = \sum_i \lambda_i^2 + 2\sum_{i<j} \lambda_i \lambda_j = \sum_i \lambda_i^2 + n(n-1)\chi(a^2),$$

which implies that

$$n(n-1)|\chi(a^2)| = \left|-\sum_{i=1}^{n} \lambda_i^2\right| \leq \sum_{i=1}^{n} |\lambda_i|^2.$$

Hence for at least one i we have $(n-1)|\chi(a^2)| \leq |\lambda_i|^2$. Since $n \geq 2$ is arbitrary and λ_i belongs to the bounded set $\sigma_A(a)$ (by assumption), it follows that $\chi(a^2) = 0$. Therefore, by the implication (i)→(iv) of Proposition 2.53, χ is a character. $\square$

Recall that a *Banach algebra* A is an algebra which is also a complete normed space with norm satisfying $\|ab\| \leq \|a\| \|b\|$ for $a, b \in A$. By a *unital Banach algebra* we mean a Banach algebra with unit element 1 such that $\|1\| = 1$.

Suppose A is a complex unital Banach algebra. Then $|\lambda| \leq \|a\|$ for $a \in A$ and $\lambda \in \sigma_A(a)$; see Exercise 12. Hence $\sigma_A(a)$ is bounded. Therefore, the assertion of Theorem 2.56 holds for each complex unital Banach algebra A.

The following is another important fact about characters of Banach algebras.

Theorem 2.57 *Each character χ on a complex unital Banach algebra is continuous and has norm one.*

Proof We prove that $|\chi(a)| \leq \|a\|$ for all $a \in A$. Assume to the contrary that there exists an $a \in A$ such that $|\chi(a)| > \|a\|$. By scaling we can assume that $\|a\| < 1$ and $\chi(a) = 1$. Since $\|a^n\| \leq \|a\|^n$ and $\|a\| < 1$, the series $\sum_{n=1}^{\infty} a^n$ converges in A and defines an element $b \in A$. Then $b = ab + a$. Hence $\chi(b) = \chi(a)\chi(b) + \chi(a) = \chi(b) + 1$, a contradiction. Thus, $|\chi(a)| \leq \|a\|$ for $a \in A$, so that $\|\chi\| \leq 1$.

Since $\chi(1) = 1$ and $\|1\| = 1$, we have $\|\chi\| \geq 1$. Therefore, $\|\chi\| = 1$.   □

2.7 Hermitian Characters of Unital *-Algebras

Throughout this section, A is a **unital** *-algebra over $\mathbb{K} = \mathbb{C}$ or $\mathbb{K} = \mathbb{R}$.

Definition 2.58 A *hermitian character* of A is a character of the algebra A (Definition 2.51) that is a hermitian functional. The set of hermitian characters of A is denoted by $\hat{A}$.

A hermitian character of A is a nonzero *-algebra homomorphism $\chi : A \to \mathbb{K}$. This means that χ is a nonzero linear functional on A such that

$$\chi(ab) = \chi(a)\chi(b) \quad \text{and} \quad \chi(a^+) = \overline{\chi(a)} \quad \text{for} \ a, b \in A. \tag{2.51}$$

Each hermitian character χ is a state on A. Indeed, $\chi(1) = 1$, and from (2.51) we obtain $\chi(a^+a) = \chi(a^+)\chi(a) = \overline{\chi(a)}\,\chi(a) \geq 0$ for $a \in A$.

Obviously, a character on A is a hermitian character if and only if it is a state.

Characters on commutative C^*-algebras are always hermitian [Dv96, Theorem I.3.1]. For general commutative *-algebras this is not true. For instance, $\chi(p) := p(\mathrm{i})$, $p \in \mathbb{C}[x]$, defines a character χ on $A = \mathbb{C}[x]$ which is not hermitian.

The following proposition provides characterizations of characters among states in terms of the standard deviation Δ_f (Definition 2.42).

Proposition 2.59 *Suppose f is a state on A.*

(i) *Let $a \in A$. If $\Delta_f(a) = 0$, then $f(ba) = f(b)f(a)$ for all $b \in A$.*
(ii) *f is a character if and only if $\Delta_f(a) = 0$ for all $a \in A$.*
(iii) *If A is a complex *-algebra and $\Delta_f(a^2) = 0$ for all $a \in A_{\mathrm{her}}$, then $f \in \hat{A}$.*

Proof (i): is an immediate consequence of the inequality (2.42).

(ii): Using (2.51) we easily derive that $\Delta_f(a) = 0$ if f is a character. The converse implication follows at once from (i).

(iii): Let $c \in A_{\text{her}}$. By assumption, $\Delta_f((c \pm 1)^2) = 0$. Therefore, by (i), applied with $b = c, a = (c \pm 1)^2$, we compute

$$4f(c^2) = f\big(c(c+1)^2\big) - f\big(c(c-1)^2\big) = f(c)f((c+1)^2) - f(c)f((c-1)^2)$$
$$= f(c)f\big((c+1)^2 - (c-1)^2\big) = 4f(c)^2.$$

Hence $\Delta_f(c) = 0$. Then $f(bc) = f(b)f(c)$ for $b \in A$ by (i). Since $\mathbb{K} = \mathbb{C}$, we have $A = A_{\text{her}} + iA_{\text{her}}$. Therefore, $f(bc) = f(b)f(c)$ for all $b, c \in A$. Thus, f is a character. By assumption, f is a state. Hence f is hermitian, so $f \in \hat{A}$. $\square$

Definition 2.60 A state on a unital *-algebra A is called *pure* if it is an extreme point of the convex set $\mathcal{S}(A)$ of states on A.

For states f, g on A, $\lambda \in [0, 1]$, and $a \in A$, a straightforward computation yields the following very useful identity:

$$\Delta_{\lambda f+(1-\lambda)g}(a)^2 = \lambda\Delta_f(a)^2 + (1-\lambda)\Delta_g(a)^2 + \lambda(1-\lambda)\big|(f-g)(a)\big|^2. \qquad (2.52)$$

Corollary 2.61 *Each hermitian character on A is a pure state.*

Proof Let $h \in \hat{A}$. Suppose $h = \lambda f + (1 - \lambda)g$, where f, g are states and $\lambda \in (0, 1)$. Let $a \in A$. Since h is a character, $\Delta_h(a) = 0$ by Proposition 2.59(ii). Thus, since $\Delta_f(a) \geq 0, \Delta_g(a) \geq 0$, and $\lambda(1 - \lambda) > 0$, the identity (2.52) gives $(f - g)(a) = 0$. Hence $f = g$ and therefore $f = g = h$. $\square$

A natural question is the following:

*When are pure states on commutative unital *-algebras characters?*

In general this is not true: As shown in Example 7.5 below, there exists a pure state on the *-algebra $\mathbb{C}[x_1, x_2]$ that is not a character. Positive results concerning this question are given in Theorem 2.63 and Proposition 2.72.

We begin with a simple technical lemma.

Lemma 2.62 *Let C be a subset of A such that $C \supseteq \sum A^2$. Suppose f is a state on A which is an extreme point of the convex set of all C-positive states. Let $g \in \mathcal{P}(A)^*$. If g and $f - g$ are C-positive, then $g = g(1)f$.*

Proof Since the assertion is trivial for $g = 0$, we can assume that $g \neq 0$. Then we have $g(1) > 0$ by the Cauchy-Schwarz inequality.

Suppose $g(1) = 1$. Then $(f-g)(1) = 0$. Since $C \supseteq \sum A^2$, $f - g \in \mathcal{P}(A)^*$. Hence $f = g$ again by the Cauchy-Schwarz inequality, so the assertion is also true.

Now assume $0 < g(1) < 1$. Then $f_1 := g(1)^{-1}g$ and $f_2 := (1-g(1))^{-1}(f - g)$ are C-positive states such that $f = g(1)f_1 + (1-g(1))f_2$. Since f is an extreme point of the C-positive states, we obtain $f_1 = f$. This gives $g = g(1)f$. $\square$

Theorem 2.63 *Let f be a state on a commutative (!) complex unital ∗-algebra A. Suppose Q is a quadratic module of A such that for any $x \in \mathsf{A}$ and $a \in \mathsf{A}_{\mathrm{her}}$,*

$$x^+x - 1 \in Q \quad \text{and} \quad x^+ax \in Q \quad \text{imply} \quad a \in Q. \tag{2.53}$$

If f is Q-positive and an extreme point of the convex set of Q-positive states, then f is a hermitian character and hence a pure state.

Proof Fix $b \in \mathsf{A}_{\mathrm{her}}$. Clearly, $S_b = \{(1 + b^2)a : a \in \mathsf{A}_{\mathrm{her}}\}$ is a linear subspace of $\mathsf{A}_{\mathrm{her}}$. First we show that there is a well-defined linear functional g on S_b given by

$$g((1 + b^2)a) = f(a), \quad a \in \mathsf{A}_{\mathrm{her}}. \tag{2.54}$$

Suppose $(1 + b^2)a = (1 + b^2)a'$ with $a, a' \in \mathsf{A}_{\mathrm{her}}$. Then $(b - i)^+(b - i) - 1 = b^2 \in Q$ and $0 = (1 + b^2)(a - a') = (b + i)(a - a')(b - i) \in Q$, so $a - a' \in Q$ by condition (2.53). Hence $f(a - a') \geq 0$, since f is Q-positive. Interchanging the role of a, a' yields $f(a' - a) \geq 0$. Thus, $f(a) = f(a')$, so g is well defined.

Suppose $(1 + b^2)a \in Q$, where $a \in \mathsf{A}_{\mathrm{her}}$. Then, since $(1 + b^2)a = (b - i)^+ a(b - i)$, condition (2.53), applied with $x = b - i$, yields $a \in Q$. Hence, since f is Q-positive, g is nonnegative on $Q \cap S_b$ by (2.54). Further, if $c \in \mathsf{A}_{\mathrm{her}}$, then $(1 + b^2)(1 + c)^2 \in Q$ and $(1 + b^2)(1 + c)^2 - 4c = b^2(1 + c)^2 + (1 - c)^2 \in Q$. This shows that S_b is Q-cofinal in $\mathsf{A}_{\mathrm{her}}$. Therefore, by Proposition C.4, g admits an extension to a Q-positive $\mathbb{R}$-linear functional on $\mathsf{A}_{\mathrm{her}}$ and hence to a $\mathbb{C}$-linear Q-positive functional, denoted again g, on A. In particular, $g \in \mathcal{P}(\mathsf{A})^*$.

Let $c \in Q$. Then $b^2c = bcb \in Q$, so $g(b^2c) \geq 0$ and $g(c) \leq g((1 + b^2)c) = f(c)$. Thus, $f - g$ is Q-positive. By assumption, f is an extreme point of the Q-positive states, so Lemma 2.62 implies $g = g(1)f$. Note that $g(1) > 0$, since otherwise $g = 0$ and hence $f(1) = 0$ by (2.54), a contradiction. Then, by (2.54),

$$f(a) = g((1 + b^2)a) = g(1)^{-1}g(a) \quad \text{for} \quad a \in \mathsf{A}_{\mathrm{her}}. \tag{2.55}$$

Set $a = 1 + b^2$ and $a = (1 + b^2)^2$ in (2.55). Since $g(1 + b^2) = f(1) = 1$, we obtain

$$f(1 + b^2) = g(1)^{-1}g(1 + b^2) = g(1)^{-1},$$
$$f((1 + b^2)^2) = g(1)^{-1}g((1 + b^2)^2) = g(1)^{-2}g(1 + b^2) = g(1)^{-2}.$$

Hence $\Delta_f(1 + b^2) = 0$ and therefore $f(x(1 + b^2)) = f(x)f(1 + b^2)$, $x \in \mathsf{A}$, by Proposition 2.59(i). Because A is the linear span of elements $1 + b^2$, where $b \in \mathsf{A}_{\mathrm{her}}$, f is a character. Since f is a state, it is hermitian. By Corollary 2.61, f is pure. □

Corollary 2.64 *Let f be a state on a commutative complex unital ∗-algebra A and*

$$\mathsf{A}_+ := \{a \in \mathsf{A}_{\mathrm{her}} : \chi(a) \geq 0 \text{ for } \chi \in \hat{\mathsf{A}}\}. \tag{2.56}$$

*If f is A_+-positive and an extreme point of the set of A_+-positive states on A, then
f is a hermitian character and a pure state.*

Proof By Theorem 2.63 it suffices to check that the quadratic module A_+ satisfies
condition (2.53). Indeed, suppose $x^+x - 1 \in A_+$ and $x^+ax \in A_+$. Then, for $\chi \in \hat{A}$,
$\chi(x^+x - 1) = \chi(x^+x) - 1 \geq 0$ and $\chi(x^+ax) = \chi(x^+x)\chi(a) \geq 0$, which implies
that $\chi(a) \geq 0$. Thus, $a \in A_+$ and condition (2.53) holds.   □

Extreme points of the convex set of Q-positive states play a crucial role in the
decomposition theory of positive linear functionals (see Theorem 5.35). Obviously,
each Q-positive pure state is an extreme point of the set of Q-positive states.

Now we turn to the description and existence of hermitian characters.

First let A be a real unital subalgebra of the algebra $C(\mathcal{X}; \mathbb{R})$ of continuous real-
valued functions on a topological Hausdorff space $\mathcal{X}$, equipped with the identity
involution. For $x \in \mathcal{X}$, let ev_x denote the evaluation functional at x, that is, $\mathrm{ev}_x(f) = f(x)$ for $f \in A$. Obviously, $\mathrm{ev}_x \in \hat{A}$. It is natural to ask when the functionals ev_x,
$x \in \mathcal{X}$, exhaust the set $\hat{A}$. A simple fact is the following.

Lemma 2.65 *Suppose $\mathcal{X}$ is a compact topological Hausdorff space. For each char-
acter χ on $A = C(\mathcal{X}; \mathbb{R})$ there exists a unique point $x \in \mathcal{X}$ such that $\chi = \mathrm{ev}_x$.*

Proof We consider the ideal $J := \{f \in A : \chi(f) = 0\}$. First we show that there
exists an $x \in \mathcal{X}$ such that $f(x) = 0$ for all $f \in J$. Assume the contrary. Then for
each $x \in \mathcal{X}$ there exists a function $\varphi_x \in J$ such that $\varphi_x(x) \neq 0$. Clearly, the sets
$O_x = \{y \in \mathcal{X} : \varphi_x(y) \neq 0\}$ are open and cover the compact set $\mathcal{X}$, since
$x \in O_x$. Hence there exists a finite subcover, say $O_{x_1} \cup \cdots \cup O_{x_k} = \mathcal{X}$. The function $g := \varphi_{x_1}^2 + \cdots + \varphi_{x_k}^2$ is in J (since $\varphi_{x_j} \in J$) and satisfies $g(x) > 0$ on $\mathcal{X}$. Hence
$g^{-1} \in A = C(\mathcal{X}; \mathbb{R})$. Since $g \in J$, $1 = gg^{-1} \in J$ and therefore $J = A$. Thus, $\chi = 0$,
a contradiction.

We fix an $x \in \mathcal{X}$ at which all functions of J vanish. Let $f \in A$. Since $\chi(1) = 1$,
$(f - \chi(f)1) \in J$ and hence $(f - \chi(f)1)(x) = f(x) - \chi(f) = 0$, that is, we have
$\chi(f) = f(x) = \mathrm{ev}_x(f)$. Since the functions of $C(\mathcal{X}; \mathbb{R})$ separate the points of $\mathcal{X}$,
x is uniquely determined by χ.   □

For arbitrary topological spaces $\mathcal{X}$ the assertion of Lemma 2.65 does not hold.
The point evaluations on $C(\mathcal{X}; \mathbb{R})$ exhaust the characters if and only if the space
$\mathcal{X}$ is *realcompact*; see [GJ60, Chap. 8] for the definition and basic results. Large
classes of spaces (such as metrizable spaces, σ-compact spaces) have this property.
A topological space that is *not* realcompact is sketched in the next example.

Example 2.66 Let $\mathcal{X}$ be the space of all ordinals less than the first uncountable
ordinal, equipped with the order topology. Each function $f \in C(\mathcal{X}; \mathbb{R})$ is a constant
$\chi(f)$ on a set $\{\alpha \in \mathcal{X} : \alpha \geq \beta\}$ for some $\beta \in \mathcal{X}$ depending on f. Then $\chi(\cdot)$ is a
character on A which is not of the form ev_x for some point $x \in \mathcal{X}$. (Full details and
proofs concerning this example can be found in [GJ60, pp. 72–75].)   ○

For a commutative real algebra A with identity involution, the set $\hat{A}$ of hermitian characters is just the set of nonzero algebra homomorphisms of A into $\mathbb{R}$.

Next we describe the set $\hat{A}$ for a *finitely generated commutative* unital $\mathbb{K}$-algebra A. First we fix a set $\{f_1, \ldots, f_d\}$ of hermitian generators of A. Then there exists a unique surjective unital *-homomorphism $\theta : \mathbb{K}_d[\underline{x}] \mapsto A$ such that $\theta(x_j) = f_j$ for $j = 1, \ldots, d$. The kernel $\mathcal{J}$ of θ is a *-ideal of $\mathbb{K}_d[\underline{x}]$ and A is *-isomorphic to the quotient *-algebra $\mathbb{K}_d[\underline{x}]/\mathcal{J}$, that is, $A \cong \mathbb{K}_d[\underline{x}]/\mathcal{J}$.

Let $\chi \in \hat{A}$. Since f_j are hermitian generators of A, $\chi(f_j) \in \mathbb{R}$ and χ is uniquely determined by the point $x_\chi := (\chi(f_1), \ldots, \chi(f_d)) \in \mathbb{R}^d$. For $p \in \mathbb{K}_d[\underline{x}]$, we obtain

$$\chi(\theta(p)) = \chi(p(f_1, \ldots, f_d)) = p(\chi(f_1), \ldots, \chi(f_d)) = p(x_\chi). \qquad (2.57)$$

For $p \in \mathcal{J}$, we have $\theta(p) = 0$ and hence $p(x_\chi) = 0$ by (2.57), so x_χ is in the zero set $\mathcal{Z}(\mathcal{J})$ of $\mathcal{J}$. Conversely, suppose $x \in \mathbb{R}^d$ is in $\mathcal{Z}(\mathcal{J})$. Then, since $A \cong \mathbb{K}_d[\underline{x}]/\mathcal{J}$, there is a well-defined (!) hermitian character χ on A given by $\chi(\theta(p)) = p(x)$ for $p \in \mathbb{K}_d[\underline{x}]$. That is, by (2.57), $\chi(f) = p(x_\chi)$, $\theta(p) = f$, is well defined on A. We identify χ with x_χ and write $f(x_\chi) := \chi(f)$ for $f \in A$. Then, by the preceding, $\hat{A}$ becomes the *real algebraic set*

$$\hat{A} = \mathcal{Z}(\mathcal{J}) := \{x \in \mathbb{R}^d : f(x) = 0 \text{ for } f \in \mathcal{J}\}. \qquad (2.58)$$

Since $\mathcal{Z}(\mathcal{J})$ is closed in $\mathbb{R}^d$, $\hat{A}$ is a locally compact Hausdorff space in the induced topology of $\mathbb{R}^d$ and the elements of A can be considered as continuous functions on this space. In the special case $A = \mathbb{K}_d[\underline{x}]$ we can take $f_1 = x_1, \ldots, f_d = x_d$ and obtain $\hat{A} = \mathbb{R}^d$.

Before we continue let us introduce another general concept. Note that locally convex topologies are defined in Appendix C.

Definition 2.67 A *topological *-algebra* is a *-algebra A, together with a locally convex topology, such that the involution $a \mapsto a^+$ is continuous and the multiplication maps $a \mapsto ab$ and $a \mapsto ba$ of A are continuous for each element $b \in A$.

The continuity of the maps $a \mapsto ab$ and $a \mapsto ba$ as in the preceding definition is called the *separate continuity* of the multiplication. Since the involution is continuous, it suffices to require the continuity of one of these mappings. The separate continuity is weaker than the *joint continuity* which means that the map $(a, b) \mapsto ab$ of $A \times A$ into A is continuous. It follows from [Sh71, Theorem III.5.1] that the multiplication in Frechet topological *-algebras is always jointly continuous.

The Arens algebra is a commutative unital Frechet *-algebra without characters.

Example 2.68 (*Arens algebra $L^\omega(0, 1)$*)
Let $\| \cdot \|_p$ denote the norm of $L^p(0, 1)$ with respect to the Lebesgue measure on $[0, 1]$. Let $p \in (1, +\infty)$. For $f, g \in L^p(0, 1)$, we have by the Hölder inequality,

$$\|fg\|_p \leq \|f\|_{2p} \|g\|_{2p}. \qquad (2.59)$$

The vector space $L^\omega(0, 1) := \cap_{1 < p < \infty} L^p(0, 1)$, equipped with the locally convex topology defined by the norms $\| \cdot \|_p$, $p \in (1, \infty)$, is a Frechet space.

From (2.59) it follows that for $f, g \in L^\omega(0, 1)$ the pointwise product fg is also in $L^\omega(0, 1)$ and the multiplication is jointly continuous in the topology of $L^\omega(0, 1)$. Further, $L^\omega(0, 1)$ is a complex unital *-algebra with involution $f^+(t) := \overline{f(t)}$ and we have $\|f^+\|_p = \|f\|_p$ for $f \in L^\omega(0, 1)$. Thus,

$$L^\omega(0, 1) = \cap_{1 < p < \infty} L^p(0, 1)$$

is a commutative Frechet *-algebra, called the *Arens algebra*.

We prove that the *-algebra $L^\omega(0, 1)$ has no character. Assume to the contrary, it has a character χ. Its restriction to $C([0, 1]; \mathbb{R})$ is a character, so by Lemma 2.65 there exists an $x_0 \in [0, 1]$ such that $\chi(f) = f(x_0)$ for $f \in C([0, 1]; \mathbb{R})$. Define

$$g(x) = \log|x - x_0|, \quad h(x) = (\log|x - x_0|)^{-1} \text{ for } x \in [0, 1], x \neq x_0, \quad h(x_0) = 0.$$

Then $h \in C([0, 1]; \mathbb{R})$, $g \in L^\omega(0, 1)$, and $gh = 1$ in $L^\omega(0, 1)$, so we obtain

$$1 = \chi(1) = \chi(gh) = \chi(g)\chi(h) = \chi(g)h(x_0) = 0,$$

a contradiction.

But $L^\omega(0, 1)$ has plenty of positive functionals. For any $\varphi \in L^\infty(0, 1)$ there is a positive linear functional defined by $L_\varphi(f) = \int_0^1 f(x)|\varphi(x)|dx$, $f \in L^\omega(0, 1)$. $\quad\bigcirc$

2.8 Hermitian and Symmetric *-Algebras

In this section, A is a **complex unital** *-algebra.

Definition 2.69 A complex unital *-algebra A is called
- *hermitian* if $1 + a^2$ is invertible in A for each hermitian element $a = a^+ \in A$,
- *symmetric* if $1 + a^+a$ is invertible in A for each element $a \in A$.

Obviously, each symmetric *-algebra is hermitian. The converse is not true, as Example 2.73 below shows.

Example 2.70 Obviously, $C(\mathbb{R}^d)$, $C^\infty(\mathbb{R}^d)$ and the *-algebra of rational functions on $\mathbb{R}^d$ are symmetric. The polynomial *-algebra $\mathbb{C}_d[\underline{x}]$ is not hermitian. $\quad\bigcirc$

The next proposition gives characterizations of these notions in terms of spectra.

Proposition 2.71 (i) A *is hermitian if and only if* $\sigma_A(a) \subseteq \mathbb{R}$ *for each hermitian element* $a \in A$.
(ii) A *is symmetric if and only if* $\sigma_A(a^+a) \subseteq [0, +\infty)$ *for each* $a \in A$.

Proof (i): Suppose $\sigma_A(a) \subseteq \mathbb{R}$ for $a = a^+ \in A$. Then, since $\pm i \notin \sigma_A(a)$, $a - i$ and $a + i$ are invertible, and so is $1 + a^2 = (a - i)(a + i)$. Thus A is hermitian. Conversely, suppose A is hermitian. Then $1 + a^2$ is invertible for $a = a^+ \in A$. From $a(1 + a^2) = (1 + a^2)a$ we obtain $(1 + a^2)^{-1}a = a(1 + a^2)^{-1}$. Using this relation and the identity $1 + a^2 = (a - i)(a + i)$ we derive $(a \pm i)(1 + a^2)^{-1}$ is the inverse of $a \mp i$. Hence $i, -i \notin \sigma_A(a)$. Upon scaling by a factor it follows that all nonreal numbers are not in $\sigma_A(a)$. Thus, $\sigma_A(a) \subseteq \mathbb{R}$.

(ii): Let $a \in A$ and suppose that $\sigma_A(a^+a) \subseteq [0, +\infty)$. Then $-1 \notin \sigma_A(a^+a)$, hence $1 + a^+a$ is invertible. Hence A is symmetric.

Conversely, assume that A is symmetric. Let $a \in A$. Since A is also hermitian, $\sigma_A(a^+a) \subseteq \mathbb{R}$ by (i). If $\lambda > 0$, then $-\lambda - a^+a = -\lambda(1 + (\lambda^{-1/2}a)^+(\lambda^{-1/2}a))$ is invertible, because A is symmetric. Therefore, $-\lambda \notin \sigma_A(a^+a)$. This proves that $\sigma_A(a^+a) \subseteq [0, +\infty)$. $\square$

Let $f, g \in \mathcal{P}(A)^*$. We write $f \leq g$ if $f(a^+a) \leq g(a^+a)$ for all $a \in A$. Recall that f_x denotes the positive functional defined by $f_x(\cdot) := f(x^+ \cdot x)$ for $x \in A$.

Proposition 2.72 *Suppose A is a commutative hermitian complex unital $*$-algebra. Then each pure state of A is a hermitian character.*

Proof Let $a \in A_{\text{her}}$. Then $b := (1 + a^2)^{-1} \in A_{\text{her}}$ and

$$1 = f(1)^2 = f(b(1 + a^2))^2 \leq f((1 + a^2)^2)f(b^2)$$

by the Cauchy-Schwarz inequality. Hence $f(b^2) > 0$.

From the relations

$$1 - (ab)^2 = (1 + a^4 + a^2)b^2 \in \sum A^2 \quad \text{and} \quad 1 - b^2 = (a^4 + 2a^2)b^2 \in \sum A^2$$

we obtain $f_{ab} \leq f$ and $f_b \leq f$. Therefore, since f is pure, it follows from Lemma 2.62, applied with $C = \sum A^2$, that $f_{ab} = f_{ab}(1)f$ and $f_b = f_b(1)f$. Using essentially that A is commutative and these two equalities we derive

$$f(a^2b^2x) = f_b(a^2x) = f_b(1)f(a^2x) = f(b^2)f(a^2x),$$
$$f(a^2b^2x) = f_{ab}(1)f(x) = f_b(a^2)f(x) = f_b(1)f(a^2)f(x) = f(b^2)f(a^2)f(x)$$

for $x \in A$. As noted above, $f(b^2) \neq 0$. Hence $f(a^2x) = f(a^2)f(x)$ for $x \in A$. Since squares of hermitian elements span A_{her} (by the identity $(a + 1)^2 - (a - 1)^2 = 4a$) and hence A, we conclude that $f(cx) = f(c)f(x)$ for all $c, x \in A$, that is, f is a character. As a state, f is hermitian. $\square$

Example 2.73 (*A commutative hermitian $*$-algebra which is not symmetric*)
Let $\mathbb{C}(x)[y]$ denote the polynomials in y over the field $\mathbb{C}(x)$ of rational functions in x. Then, $\mathbb{C}(x)[y]$ is a complex unital $*$-algebra, with involution determined by $x^+ = x$, $y^+ = y$. Let A be the quotient $*$-algebra of $\mathbb{C}(x)[y]$ by the $*$-ideal generated by the polynomial $1 + x^2 + y^2$.

Statement 1: A *is a field.*

Proof This follows from the fact that the polynomial $1 + x^2 + y^2$ in y is irreducible over $\mathbb{C}(x)$. We give a direct proof. We denote the images of x, y in the quotient algebra A also by x, y. Clearly, each element of A is of the form $f = a + by$, where $y^2 = -1 - x^2$ and $a, b \in \mathbb{C}(x)$. Suppose $f \neq 0$. Then $a \neq 0$ or $b \neq 0$.

Suppose $a \neq 0$. First we show that $a^2 + (1 + x^2)b^2 \neq 0$. Assume the contrary. Then, after clearing denominators and canceling common factors, we can assume that $a^2 = -(1 + x^2)b^2$, where $a, b \in \mathbb{C}[x]$ have no common divisor. But then $x + i$ is a divisor of a, so $(x + i)^2$ is a factor of a^2. This implies that $x + i$ is a factor of b, a contradiction.

Let c be the inverse of $a^2 + (1 + x^2)b^2 \neq 0$ in the field $\mathbb{C}(x)$. Then we compute

$$f(ac - bcy) = (a + by)(ac - bcy) = (a^2 - b^2y^2)c = (a^2 + (1 + x^2)b^2)c = 1.$$

Now let $a = 0, b \neq 0$. Then $f(-b^{-1}(1 + x^2)^{-1}y) = -byb^{-1}(1 + x^2)^{-1}y = 1$. Thus, in both cases, $f \neq 0$ is invertible in A. Hence A is a field. $\square$

Statement 2: A *is hermitian, but not symmetric.*

Proof Let $f = f^+ \in A$. Then $1 \pm if \neq 0$. (Otherwise, $1 = \mp if$ and therefore $1 = \pm if$. Adding both equations yields $2 = 0$, a contradiction.) Hence we have $1 + f^2 = (f - i)(f + i) \neq 0$, so $1 + f^2$ is invertible in the field A. This proves that A is hermitian.

For $f := x + iy$, we have $1 + f^+f = 1 + (x + iy)^+(x - iy) = 1 + x^2 + y^2 = 0$ in A, so $1 + f^+f$ is not invertible in A. Therefore, A is not symmetric. $\square\bigcirc$

A Banach *-algebra A is a complex Banach algebra which is also a *-algebra. The Shirali–Ford theorem ([SF70], see [DB86, Theorem 33.2] for a simplified proof), states that each hermitian Banach *-algebra is symmetric, thus solving an outstanding conjecture of I. Kaplansky. Even more, in any hermitian unital Banach *-algebra, $1 + (a_1)^+a_1 + \cdots + (a_n)^+a_n$ is invertible for arbitrary elements $a_1, \ldots, a_n \in A$ and $n \in \mathbb{N}$. Such *-algebras are called *completely symmetric.*

Group algebras of locally compact groups form an important class of Banach *-algebras. Group algebras of abelian or compact groups are symmetric. In general, group algebras are not symmetric. It is an interesting question of whether the group algebra of a locally compact group is symmetric; see, e.g., [Pl01, Chap. 12] for a detailed treatment.

2.9 Exercises

1. Carry out the proof of Lemma 2.6.
2. Suppose A is a commutative real *-algebra with the identity map as involution. Let $A_\mathbb{C}$ be its complexification (Definition 2.9). Show that $\sum A^2 = \sum(A_\mathbb{C})^2$.

Show that each $\mathbb{R}$-linear positive functional on A has a unique extension to a $\mathbb{C}$-linear positive functional on $\mathsf{A}_{\mathbb{C}}$.

3. Consider the real $\ast$-algebra $M_n(\mathbb{R})$, with involution $a \mapsto a^+ := a^T$. Suppose $b, c \in M_n(\mathbb{R})$ are invertible, $b^T = b$ or $b^T = -b$, and $c^T = c$ or $c^T = -c$. Show that the corresponding $\ast$-algebras $(M_n(\mathbb{R}), +_b)$ and $(M_n(\mathbb{R}), +_c)$ from Example 2.15 are $\ast$-isomorphic if and only if there exists an invertible matrix $u \in M_n(\mathbb{R})$ such that $c = u^T b u$ or $c = -u^T b u$.

4. Extend the considerations of Example 2.15 to the matrix $\ast$-algebra $M_n(\mathbb{C})$ with involution defined by $(a^+)_{ij} := \overline{a}_{ji}$. Prove that each algebra involution of the complex algebra $M_n(\mathbb{C})$ is of the form $a \mapsto a^{+b} := ba^+b^{-1}$ for some invertible hermitian matrix $b \in M_n(\mathbb{C})$. (Note that $b^+ = b$ in contrast to $M_n(\mathbb{R})$.)

5. (*Quaternions*) Let $\mathbb{H}$ denote the algebra of real quaternions, that is, $\mathbb{H}$ is the real unital algebra with generators i, j and defining relations $i^2 = j^2 = 0$, $ij = -ji$. Set $k := ij$. Each element $x \in \mathbb{H}$ is of the form $x = x_0 + ix_1 + jx_2 + kx_3$ with $x_0, x_1, x_2, x_3 \in \mathbb{R}$ uniquely determined. Define $x^+ = x_0 - ix_1 - jx_2 - kx_3$.

   a. Show that the map $x \mapsto x^+$ is an algebra involution of $\mathbb{H}$, so $\mathbb{H}$ becomes a real $\ast$-algebra.
   b. Prove $x^+x = xx^+ = x_0^2 + x_1^2 + x_2^2 + x_3^2$ for $x = x_0 + ix_1 + jx_2 + kx_3 \in \mathbb{H}$.
   c. Show that $\|x\| := \sqrt{x^+x}$, $x \in \mathbb{H}$, defines a norm on the real vector space $\mathbb{H}$ such that $(\mathbb{H}, \|\cdot\|)$ is a real C^*-algebra according to Definition B.3 and $\|xy\| = \|x\|\,\|y\|$ for $x, y \in \mathbb{H}$.
   d. Show that any $x \in \mathbb{H}$, $x \neq 0$, is invertible in $\mathbb{H}$ with $x^{-1} = \|x\|^{-2}x^+$.
   e. Let $b \in \mathbb{H}$, $b \neq 0$, be such that $b^+ = b$ or $b^+ = -b$. Prove that the map $x \mapsto x^{+b} := bx^+b^{-1}$ is an algebra involution of $\mathbb{H}$.
   f. Show that any involution of the real algebra $\mathbb{H}$ is of the form described in e.

   Hint for e. and f.: Mimic the proofs given in Example 2.15.

6. Show that the real $\ast$-algebras $M_2(\mathbb{R})$ and $\mathbb{H}$ (see Exercise 5) are not $\ast$-isomorphic, but their complexifications are $\ast$-isomorphic (as complex $\ast$-algebras) to $M_2(\mathbb{C})$, equipped with the involution $(a^+)_{ij} := \overline{a}_{ji}$.

7. Let $\mathcal{X}$ be a topological space and σ a homeomorphism of $\mathcal{X}$ such that $\sigma^2 = \mathrm{Id}$.

   a. Show that the algebra $C(\mathcal{X})$ is a $\ast$-algebra with involution defined by $f^+(x) := \overline{f(\sigma(x))}$.
   b. Let $\mathcal{X} = [0, 1]$ and $\sigma(x) = 1 - x$, $x \in \mathcal{X}$. Describe the elements of $\mathsf{A}_{\mathrm{her}}$ and $\sum \mathsf{A}^2$ for the $\ast$-algebra A from a. in this case.

8. Prove that the enveloping algebra of the Virasoro algebra from Example 2.24 is $\mathbb{Z}$-graded.

9. Let A be a complex $\ast$-algebra. Show that each linear functional f on A can be uniquely represented as $f = f_1 + if_2$, where f_1, f_2 are hermitian functionals.

10. Let A be a $\ast$-algebra. Show that $Q := \{a \in \mathsf{A}_{\mathrm{her}} : f(a) \geq 0, f \in \mathcal{P}(\mathsf{A})^*\}$ is a pre-quadratic module of A.

11. Let $\mathsf{A} = x\mathbb{C}[x]$. Describe the smallest pre-quadratic module of A containing x.

12. Let $(A, \| \cdot \|)$ be a complex unital Banach algebra. Prove that $|\lambda| \leq \|a\|$ for each element $a \in A$ and $\lambda \in \sigma_A(a)$.

13. Let A be a unital complex algebra and let $a, b \in A$. Prove the following:

    a. $\sigma_A(ab) \cup \{0\} = \sigma_A(ba) \cup \{0\}$.
    b. $\sigma_A(p(a)) = \{p(\lambda) : \lambda \in \sigma_A(a)\}$ for any nonconstant polynomial $p \in \mathbb{C}[x]$.
    c. If A is a $*$-algebra, then $\lambda \in \sigma_A(a)$ if and only if $\overline{\lambda} \in \sigma_A(a^+)$.

14. Let $A = C(\mathcal{X})$, where $\mathcal{X}$ is a topological space, and $f \in A$. Prove that $\sigma_A(f)$ is the closure of the set $f(\mathcal{X})$ in $\mathbb{C}$. Give an example for which $\sigma_A(f) \neq f(\mathcal{X})$.

15. Determine the hermitian characters of the real $*$-algebra $\mathbb{R} \oplus \mathbb{R}$.

16. Determine the hermitian characters of $A = \mathbb{C}$, first considered as a complex $*$-algebra and then as a $*$-real $*$-algebra.

17. Describe the semigroup $*$-algebras of the following $*$-semigroups as polynomial $*$-algebras and determine the corresponding sets of hermitian characters.

    a. $\mathbb{Z}^n$, $n^+ = -n$.
    b. $\mathbb{N}_0^{2d}$, $(m, n)^+ = (n, m)$ for $n, m \in \mathbb{N}_0^d$.
    c. $\mathbb{N}_0 \times \mathbb{Z}$, $(n, m)^+ = (n, -m)$.

18. Determine the hermitian characters of the following commutative $*$-algebras:

    a. $\mathbb{C}[x, y \mid x^+ x = x x^+ = 1, y = y^+]$.
    b. $\mathbb{R}[x, y \mid x = x^+, y = y^+, x^2 = y^2 = 1]$.
    c. $\mathbb{C}[x \mid x^+ x = 0]$ and $\mathbb{R}[x \mid x = x^+, x^2 = 0]$.

19. Let $A = \mathbb{C}[x, x^{-1}]$ be the algebra of complex Laurent polynomials. Show that A becomes a $*$-algebra B with involution $x^+ = x$ and a $*$-algebra C with involution $x^+ = x^{-1}$. Determine the hermitian characters of B and C.

20. Suppose f is a state on a commutative complex unital $*$-algebra A. Show that $A_0 := \{a \in A : \Delta_f(a) = 0\}$ is a unital $*$-subalgebra of A and the restriction of f to A_0 is a hermitian character of A_0. Show that A_0 is the largest $*$-subalgebra which has this property.

21. Let A be the vector space $\mathbb{C} \oplus \mathbb{C}$, with the multiplication $(a, b)(c, d) = (ac, bd)$ and involution $(a, b)^+ = (\overline{b}, \overline{a})$.

    a. Show that A is a commutative unital complex $*$-algebra that is not hermitian.
    b. Show that $\sum A^2 = A_{her} = \{(a, \overline{a}) : a \in \mathbb{C}\}$.
    c. Show that A has no nonzero positive functional.

22. Let A be the $*$-algebra $\mathbb{C}[x]$, with the norm topology of $\|p\| = \int_0^1 |p(x)|\, dx$.

    a. Show that A is a topological $*$-algebra.
    b. Show that the multiplication of A is not jointly continuous.
    c. Show that the multiplication of A cannot be extended by continuity to the completion of $(A, \| \cdot \|)$ such that the completion becomes an algebra.

2.10 Notes

A standard reference for involutions on rings and algebras is [KMRT98]. A standard text book for harmonic analysis on *-semigroups is [BCR84]. A large number of examples of group graded *-algebras can be found in [SS13]. The Gleason–Kahane–Zelazko Theorem 2.56 was proved in [Gl67] and in [KZ68]. The Arens algebra $L^\omega(0, 1)$ was discovered by R. Arens [Ar46]. Theorem 2.63, Exercise 20 and the proof of Corollary 2.61 are due to Schötz [Sz19, Sz18].

The symmetry condition appeared already in the classical paper [GN43]. Symmetric Banach *-algebras were first studied by Raikov [Ra46]. Hermitian Banach *-algebras were introduced by Rickart [Ri47] and Kaplansky [Kp47]. For general *-algebras these notions were studied by Wichmann [Wi74, Wi76, Wi78] who obtained a number of interesting results. Example 2.73 is taken from [Wi74]. Detailed presentations of symmetric and hermitian *-algebras are given in [DB86, Chap. 6] and [Pl01, Sect. 9.8].

Topological *-algebras are not the subject of this book; we refer to [Sch90, In98, AIT02] for topological algebras of unbounded operators, to [Fr05] for locally multiplicatively convex algebras, to [Pl01] for Banach *-algebras, to [Di77b, KR83, DB86, Dv96] for C^*-algebras, and to [Gol82, Li03] for real C^*-algebras.

Chapter 3
O^*-Algebras

An O^*-algebra is a $*$-algebra of linear operators acting on an invariant dense domain $\mathcal{D}$ of a Hilbert space. The involution is the restriction of the Hilbert space adjoint to $\mathcal{D}$. In this chapter, we treat a few selected topics on O^*-algebras that are needed later for the study of $*$-representations. These are the graph topology and the closure of an O^*-algebra in Sect. 3.1, weak and strong bounded commutants in Sect. 3.2, and functionals defined by positive trace class operators in Sect. 3.3. In Sect. 3.4, we prove and apply a useful technical result, the so-called Mittag-Leffler lemma.

On a first reading, the reader may skip this chapter and pass directly to the treatment of $*$-representations which begins in Chap. 4; if notions or facts from this chapter are needed, one can easily return to the corresponding places.

In this chapter, all algebras and vector spaces are over the **complex field**.

3.1 O^*-Algebras and Their Graph Topologies

Suppose $(\mathcal{D}, \langle \cdot, \cdot \rangle)$ is a complex inner product space, that is, $\mathcal{D}$ is a complex vector space with a $\mathbb{C}$-valued inner product[1] $\langle \cdot, \cdot \rangle$. Let $\mathcal{H}$ denote the Hilbert space completion of $(\mathcal{D}, \langle \cdot, \cdot \rangle)$ and $L(\mathcal{D})$ the algebra of linear operators mapping $\mathcal{D}$ into itself.

Definition 3.1 $\mathcal{L}^+(\mathcal{D})$ denotes the set

$$\big\{ a \in L(\mathcal{D}) : \text{There exists a } b \in L(\mathcal{D}) \text{ such that } \langle a\varphi, \psi \rangle = \langle \varphi, b\psi \rangle \text{ for } \varphi, \psi \in \mathcal{D} \big\}.$$

---

[1] In the literature, an "inner product" is often referred to as a "scalar product" and an "inner product space" as a "pre-Hilbert space." Occasionally, "complex inner product spaces" are called "unitary spaces."

© The Editor(s) (if applicable) and The Author(s), under exclusive license
to Springer Nature Switzerland AG 2020
K. Schmüdgen, *An Invitation to Unbounded Representations of *-Algebras
on Hilbert Space*, Graduate Texts in Mathematics 285,
https://doi.org/10.1007/978-3-030-46366-3_3

Clearly, the operator b is uniquely determined by a and denoted by a^+. Note that $b = a^+$ coincides with the restriction to $\mathcal{D}$ of the adjoint operator a^* on $\mathcal{H}$. Hence each operator $a \in \mathcal{L}^+(\mathcal{D})$ has a densely defined adjoint, so a is closable on $\mathcal{H}$.

By a slight reformulation, $\mathcal{L}^+(\mathcal{D})$ is the set of linear operators a on the Hilbert space $\mathcal{H}$ with domain $\mathcal{D}$ such that

$$a\mathcal{D} \subseteq \mathcal{D}, \quad \mathcal{D} \subseteq \mathcal{D}(a^*), \quad \text{and} \quad a^*\mathcal{D} \subseteq \mathcal{D}.$$

Lemma 3.2 $\mathcal{L}^+(\mathcal{D})$ *is a unital $*$-algebra with the addition, scalar multiplication and product of linear operators and the involution $a \mapsto a^+$. The unit element is the identity map $I = I_{\mathcal{D}}$.*

Proof The axioms of a $*$-algebra are proved by straightforward verifications. As a sample, we show that $ab \in \mathcal{L}^+(\mathcal{D})$ and $(ab)^+ = b^+a^+$ for $a, b \in \mathcal{L}^+(\mathcal{D})$. For vectors $\varphi, \psi \in \mathcal{D}$ we have $b\varphi \in \mathcal{D}$ and $a^+\psi \in \mathcal{D}$, so that

$$\langle ab\varphi, \psi \rangle = \langle b\varphi, a^+\psi \rangle = \langle \varphi, b^+a^+\psi \rangle.$$

Since $b^+a^+ \in L(\mathcal{D})$, this implies that $ab \in \mathcal{L}^+(\mathcal{D})$ and $(ab)^+ = b^+a^+$. $\qquad\square$

Definition 3.3 A $*$-subalgebra $\mathcal{A}$ of the $*$-algebra $\mathcal{L}^+(\mathcal{D})$ is called an O^*-*algebra* on the domain $\mathcal{D}(\mathcal{A}) := \mathcal{D}$. If $\mathcal{A}$ contains the identity map I, we say that $\mathcal{A}$ is a *unital O^*-algebra.*

If $\mathcal{A}$ is an O^*-algebra, then $\mathcal{A} + \mathbb{C} \cdot I$ is unital O^*-algebra which is $*$-isomorphic to the unitization $\mathcal{A}^1$ of $\mathcal{A}$.

It is straightforward to check that the set

$$\mathcal{A}_+ := \left\{ a = a^+ \in \mathcal{A} : \langle a\varphi, \varphi \rangle \geq 0 \text{ for } \varphi \in \mathcal{D}(\mathcal{A}) \right\} \tag{3.1}$$

of all positive operators of an O^*-algebra $\mathcal{A}$ is a *pre-quadratic module* of $\mathcal{A}$.

Let $\mathcal{A}$ be an O^*-algebra. In general, the operators of $\mathcal{A}$ are not continuous in the Hilbert space norm. But there is a natural locally convex topology on $\mathcal{D}(\mathcal{A})$, the graph topology, in which all operators of $\mathcal{A}$ are continuous. For $a \in \mathcal{A}$ let $\|\cdot\|_a$ denote the seminorm on $\mathcal{D}(\mathcal{A})$ defined by $\|\varphi\|_a := \|a\varphi\|$, $\varphi \in \mathcal{D}(\mathcal{A})$.

We refer to Appendix C for some basics on locally convex topologies.

Definition 3.4 The *graph topology* $t_{\mathcal{A}}$ is the locally convex topology on the vector space $\mathcal{D}(\mathcal{A})$ defined by the family of seminorms $\{\|\cdot\| + \|\cdot\|_a : a \in \mathcal{A}\}$.

A base of neighborhoods of $\varphi \in \mathcal{D}(\mathcal{A})$ in the topology $t_{\mathcal{A}}$ is given by the sets

$$\mathcal{U}_{\varepsilon, a_1, a_2, \ldots, a_n}(\varphi) := \{\psi \in \mathcal{D}(\mathcal{A}) : \|\varphi - \psi\| + \|a_k(\varphi - \psi)\| \leq \varepsilon, k = 1, \ldots, n\},$$

where $\varepsilon > 0$ and $a_1, \ldots, a_n \in \mathcal{A}, n \in \mathbb{N}$.

A net $(\varphi)_{i \in I}$ of vectors $\varphi_i \in \mathcal{D}(\mathcal{A})$ converges to a vector $\varphi \in \mathcal{D}(\mathcal{A})$ in the topology $t_{\mathcal{A}}$ if and only if $\lim_i \varphi_i = \varphi$ and $\lim_i a\varphi_i = a\varphi$ in $\mathcal{H}$ for all $a \in \mathcal{A}$.

From the relation $\|a \cdot \| + \|a \cdot \|_b = \| \cdot \|_a + \| \cdot \|_{ba}, a, b \in \mathcal{A}$, it follows that each $a \in \mathcal{A}$ is a continuous linear mapping of the locally convex space $\mathcal{D}(\mathcal{A})[t_{\mathcal{A}}]$.

If all operators $a \in \mathcal{A}$ are bounded on $\mathcal{D}(\mathcal{A})$, then the graph topology is given by the Hilbert space norm.

Lemma 3.5 *If $\mathcal{A}$ is a unital O^*-algebra, the family of seminorms $\{\| \cdot \|_a : a \in \mathcal{A}\}$ is directed and defines the graph topology $t_{\mathcal{A}}$. Given $a_1, \ldots, a_n \in \mathcal{A}$, setting $b = I + a_1^+ a_1 + \cdots + a_n^+ a_n$, we have $\| \cdot \| \leq \| \cdot \|_b$ and $\| \cdot \|_{a_k} \leq \| \cdot \|_b, k = 1, \ldots, n$.*

Proof For $\varphi \in \mathcal{D}(\mathcal{A})$ we derive

$$\|\varphi\|_b^2 = \|(I + a_1^+ a_1 + \cdots + a_n^+ a_n)\varphi\|^2$$
$$= \|(a_1^+ a_1 + \cdots + a_n^+ a_n)\varphi\|^2 + \|\varphi\|^2 + 2 \operatorname{Re} \langle (a_1^+ a_1 + \cdots + a_n^+ a_n)\varphi, \varphi \rangle$$
$$\geq \|\varphi\|^2 + 2 \|\varphi\|_{a_1}^2 + \cdots + 2 \|\varphi\|_{a_n}^2,$$

which implies the assertion. $\square$

Lemma 3.6 *Let $\mathcal{A}$ be a unital O^*-algebra. Suppose that $(a_n)_{n \in \mathbb{N}_0}$ is a sequence of operators $a_n \in \mathcal{A}$, where $a_0 = I$, such that the locally convex topology t on $\mathcal{D}(\mathcal{A})$ defined by the seminorms $\{\| \cdot \|_{a_n} : n \in \mathbb{N}_0\}$ is complete. Then t is equal to the graph topology $t_{\mathcal{A}}$.*

Proof Obviously, t is weaker than $t_{\mathcal{A}}$. Since t is defined by countably many seminorms, t is metrizable, so $\mathcal{D}(\mathcal{A})[t]$ is a Frechet space. Let $a \in \mathcal{A}$. The topology t is stronger than the Hilbert space topology on $\mathcal{D}(\mathcal{A})$, because $a_0 = I$. Since a is closable as a Hilbert space operator, a is a closed linear operator of the Frechet space $\mathcal{D}(\mathcal{A})[t]$ into $\mathcal{H}$. Therefore, by the closed graph theorem, $a : \mathcal{D}(\mathcal{A})[t] \mapsto \mathcal{H}$ is continuous, so the seminorm $\| \cdot \|_a$ is t-continuous. Thus, $t_{\mathcal{A}}$ is weaker than t. $\square$

Definition 3.7 An O^*-algebra $\mathcal{A}$ is said to be *closed* if $\mathcal{D}(\mathcal{A}) = \cap_{a \in \mathcal{A}} \mathcal{D}(\overline{a})$.

For an O^*-algebra $\mathcal{A}$ we define

$$\mathcal{D}(\overline{\mathcal{A}}) = \bigcap_{a \in \mathcal{A}} \mathcal{D}(\overline{a}), \quad \overline{\mathcal{A}} := \{j(a) : a \in \mathcal{A}\}, \quad \text{where } j(a) := \overline{a}\lceil \mathcal{D}(\overline{\mathcal{A}}). \quad (3.2)$$

Proposition 3.8 *Suppose $\mathcal{A}$ is an O^*-algebra. Then $\overline{\mathcal{A}}$ is a closed O^*-algebra with domain $\mathcal{D}(\overline{\mathcal{A}})$ and the map $a \mapsto j(a)$ is a $*$-isomorphism of $\mathcal{A}$ and $\overline{\mathcal{A}}$. Moreover, $\mathcal{D}(\mathcal{A})$ is dense in the locally convex space $\mathcal{D}(\overline{\mathcal{A}})[t_{\overline{\mathcal{A}}}]$.*

Proof Set $\mathcal{B} := \mathcal{A} + \mathbb{C} \cdot I$. Then $\mathcal{D}(\overline{\mathcal{B}}) = \mathcal{D}(\overline{\mathcal{A}})$ and $t_{\mathcal{B}} = t_{\mathcal{A}}$. Hence, upon replacing $\mathcal{A}$ by $\mathcal{B}$, we can assume that $\mathcal{A}$ is a unital O^*-algebra.

Suppose that $\varphi, \psi \in \mathcal{D}(\overline{\mathcal{A}})$ and $a, b \in \mathcal{A}$. Because the family of seminorms $\{\| \cdot \|_x : x \in \mathcal{A}\}$ is directed by Lemma 3.5, there exists an operator $c \in \mathcal{A}$ such that

$$\|a\eta\| + \|b\eta\| + \|ab\eta\| \leq \|c\eta\| \quad \text{for } \eta \in \mathcal{D}(\mathcal{A}). \tag{3.3}$$

Since $\varphi \in \mathcal{D}(\overline{c})$, there is a sequence $(\varphi_n)_{n \in \mathbb{N}}$ of vectors $\varphi_n \in \mathcal{D}(\mathcal{A})$ such that $\lim_n \varphi_n = \varphi$ and $\lim_n c\varphi_n = \overline{c}\,\varphi$. Applying (3.3) to $\eta = \varphi_n - \varphi_k$ we conclude that $(a\varphi_n)_{n \in \mathbb{N}}$, $(b\varphi_n)_{n \in \mathbb{N}}$, and $(ab\varphi_n)_{n \in \mathbb{N}}$ are Cauchy sequences, hence they converge in $\mathcal{H}$. The operators b, ab, and a are closable. Hence it follows that

$$\overline{b}\,\varphi = \lim_n b\varphi_n, \quad \overline{ab}\,\varphi = \lim_n ab\varphi_n, \quad \overline{a}\,\varphi = \lim_n a\varphi_n. \tag{3.4}$$

Since a is closable, the first two relations of (3.4) imply $\overline{b}\,\varphi \in \mathcal{D}(\overline{a})$ and $\overline{a}\,\overline{b}\,\varphi = \overline{ab}\,\varphi$. Since $a \in \mathcal{A}$ was arbitrary, $\overline{b}\,\varphi \in \cap_{a \in \mathcal{A}} \mathcal{D}(\overline{a}) = \mathcal{D}(\overline{\mathcal{A}})$. Hence $j(b)\varphi = \overline{b}\,\varphi$ and $j(a)j(b)\varphi = \overline{a}\,\overline{b}\,\varphi = \overline{ab}\,\varphi = j(ab)\varphi$. Because $a + b$ is closable, from the first and third relation of (3.4) we obtain $j(a+b)\varphi = (\overline{a+b})\varphi = (j(a) + j(b))\varphi$. Obviously, $j(\lambda a) = \lambda j(a)$ for $\lambda \in \mathbb{K}$.

Next we prove that $j(a^+) = j(a)^+$ for $a \in \mathcal{A}$. Since $\psi \in \mathcal{D}(\overline{a^+})$, we can find a sequence (ψ_n) of vectors $\psi_n \in \mathcal{D}(\mathcal{A})$ such that $\psi = \lim_n \psi_n$ and $j(a^+)\psi = \overline{a^+}\,\psi = \lim_n a^+\psi_n$. Then

$$\langle j(a)\varphi, \psi \rangle = \lim_n \langle a\varphi_n, \psi_n \rangle = \lim_n \langle \varphi_n, a^+\psi_n \rangle = \langle \varphi, j(a^+)\psi \rangle \tag{3.5}$$

for $\varphi, \psi \in \mathcal{D}(\overline{\mathcal{A}})$. As shown above, each operator $j(b)$ maps $\mathcal{D}(\overline{\mathcal{A}})$ into itself. Hence $j(a)$ and $j(a^+)$ leave $\mathcal{D}(\overline{\mathcal{A}})$ invariant, so (3.5) implies $j(a) \in \mathcal{L}^+(\mathcal{D}(\overline{\mathcal{A}}))$ and $j(a)^+ = j(a^+)$. Putting the preceding together we have proved that j is a $*$-isomorphism of $\mathcal{A}$ on the O^*-algebra $j(\mathcal{A}) = \overline{\mathcal{A}}$ on $\mathcal{D}(\overline{\mathcal{A}})$.

Since $a \subseteq j(a) \subseteq \overline{a}$ for $a \in \mathcal{A}$, we have $\mathcal{D}(\overline{\mathcal{A}}) = \cap_{a \in \mathcal{A}} \mathcal{D}(\overline{j(a)})$, that is, the O^*-algebra $j(\mathcal{A}) = \overline{\mathcal{A}}$ is closed.

Suppose that $\varphi \in \mathcal{D}(\overline{\mathcal{A}})$. Let $\varepsilon > 0$ and $a \in \mathcal{A}$. Since $\varphi \in \mathcal{D}(\overline{a})$, we can find a vector $\varphi_{a,\varepsilon} \in \mathcal{D}(\mathcal{A})$ such that $\|\overline{a}\,(\varphi - \varphi_{a,\varepsilon})\| \equiv \|\varphi - \varphi_{a,\varepsilon}\|_{j(a)} < \varepsilon$. Since the family of seminorms $\{\|\cdot\|_{j(a)} : a \in \mathcal{A}\}$ is directed by Lemma 3.5, this means that $\mathcal{D}(\mathcal{A})$ is dense in the locally convex space $\mathcal{D}(\overline{\mathcal{A}})[\mathsf{t}_{\overline{\mathcal{A}}}]$. $\qquad\square$

Proposition 3.9 *An O^*-algebra $\mathcal{A}$ on $\mathcal{D}$ is closed if and only if the locally convex space $\mathcal{D}(\mathcal{A})[\mathsf{t}_{\mathcal{A}}]$ is complete.*

Proof First suppose $\mathcal{A}$ is closed, that is, $\mathcal{D}(\mathcal{A}) = \cap_{a \in \mathcal{A}} \mathcal{D}(\overline{a})$. Let $(\varphi_i)_{i \in J}$ be a Cauchy net in the locally convex space $\mathcal{D}(\mathcal{A})[\mathsf{t}_{\mathcal{A}}]$. Fix $a \in \mathcal{A}$. Then $(\varphi_i)_{i \in J}$ and $(a\varphi_i)_{i \in J}$ are Cauchy nets in the underlying Hilbert space $\mathcal{H}$. Hence there exist vectors φ and φ_a of $\mathcal{H}$ such that $\lim_i \varphi_i = \varphi$ and $\lim_i a\varphi_i = \varphi_a$ in $\mathcal{H}$. Since a is closable, the latter implies that $\varphi \in \mathcal{D}(\overline{a})$ and $\varphi_a = \overline{a}\,\varphi$. Therefore, it follows that $\varphi \in \cap_{a \in \mathcal{A}} \mathcal{D}(\overline{a}) = \mathcal{D}(\mathcal{A})$ and hence $\overline{a}\varphi = a\varphi$. From the relations $\lim_i \varphi_i = \varphi$ and $\lim_i a\varphi_i = a\varphi$ for all $\in \mathcal{A}$ we conclude that $\lim_i \varphi_i = \varphi$ in $\mathcal{D}(\mathcal{A})[\mathsf{t}_{\mathcal{A}}]$. This proves that the locally convex space $\mathcal{D}(\mathcal{A})[\mathsf{t}_{\mathcal{A}}]$ is complete.

Now assume that $\mathcal{D}(\mathcal{A})[\mathsf{t}_{\mathcal{A}}]$ is complete. By Proposition 3.8, $\mathcal{D}(\mathcal{A})$ is dense in the locally convex space $\mathcal{D}(\overline{\mathcal{A}})[\mathsf{t}_{\overline{\mathcal{A}}}]$. Since $\mathsf{t}_{\mathcal{A}}$ is the induced topology of $\mathsf{t}_{\overline{\mathcal{A}}}$ and $\mathcal{D}(\mathcal{A})[\mathsf{t}_{\mathcal{A}}]$ is complete, this implies $\mathcal{D}(\mathcal{A}) = \mathcal{D}(\overline{\mathcal{A}})$. $\qquad\square$

From Propositions 3.8 and 3.9, $\mathcal{D}(\overline{\mathcal{A}})[t_{\overline{\mathcal{A}}}]$ is the completion of the locally convex space $\mathcal{D}(\mathcal{A})[t_\mathcal{A}]$ and $\overline{\mathcal{A}}$ is the *smallest* closed O^*-algebra which extends the O^*-algebra $\mathcal{A}$. This justifies the following definition.

Definition 3.10 The O^*-algebra $\overline{\mathcal{A}}$ is called the *closure* of the O^*-algebra $\mathcal{A}$.

We illustrate the preceding with an important example.

Example 3.11 (*Differential operators with polynomial coefficients*)
Let $\mathcal{D}$ be the Schwartz space $\mathcal{S}(\mathbb{R}^d)$ of rapidly decreasing C^∞-functions on $\mathbb{R}^d$ considered as a dense domain of the Hilbert space $\mathcal{H} = L^2(\mathbb{R}^d)$. Let $\mathcal{A}$ be the set of operators a acting on $\mathcal{D}$ of the form

$$a = \sum_{|\alpha| \le n} f_\alpha(x) D^\alpha, \quad \text{where } f_\alpha \in \mathbb{C}_d[\underline{x}], \ D^\alpha := \left(\frac{\partial}{\partial x_1}\right)^{\alpha_1} \cdots \left(\frac{\partial}{\partial x_d}\right)^{\alpha_d},$$

and $n \in \mathbb{N}$. Here $\alpha = (\alpha_1, \dots, \alpha_d) \in \mathbb{N}_0^d$ and $|\alpha| := \alpha_1 + \cdots + \alpha_d$. Clearly, a leaves $\mathcal{D}$ invariant. It is easily checked that $\mathcal{A}$ is a unital O^*-algebra on $\mathcal{D}$.

The assignment $p_j \mapsto -i\frac{\partial}{\partial x_j}, q_j \mapsto x_j, j = 1, \dots, d$, defines a $*$-homomorphism π of the Weyl algebra $W(d)$ (see Example 2.10) on the O^*-algebra $\mathcal{A}$. (In order to prove this it suffices to verify that the operators $-i\frac{\partial}{\partial x_j}, x_j$ satisfy the defining relations of $W(d)$, which is easily done.) Further, if $a \in \mathcal{A}$ as above acts as the zero operator on $\mathcal{S}(\mathbb{R}^d)$, then all coefficients f_α are zero. Hence π is injective, so π is a $*$-isomorphism of $W(d)$ and the O^*-algebra $\mathcal{A}$. In the terminology of Definition 4.2 below, this means that π is a faithful $*$-representation of the Weyl algebra $W(d)$ on the domain $\mathcal{S}(\mathbb{R}^d)$. In fact, π is the d-dimensional *Schrödinger representation*.

The Schwartz space $\mathcal{S}(\mathbb{R}^d)$ is a Frechet space with respect to the locally convex topology t given by the countable family of seminorms

$$q_m(\varphi) = \sup_{|\alpha| \le m} \sup_{x \in \mathbb{R}^d} \left|(1 + x_1^2 + \cdots + x_d^2)^m (D^\alpha \varphi)(x)\right|, \quad m \in \mathbb{N}_0.$$

It is well known (see, e.g., [RS72, Appendix to V.3]) that this topology t is also generated by the family of seminorms $\{\|\cdot\|_{c^n} : n \in \mathbb{N}_0\}$, where

$$c := x_1^2 + \cdots + x_d^2 - \frac{\partial^2}{\partial x_1^2} - \cdots - \frac{\partial^2}{\partial x_d^2} \in \mathcal{A}. \tag{3.6}$$

Hence, by Lemma 3.6, $t = t_\mathcal{A}$. That is, *the graph topology of the O^*-algebra $\mathcal{A}$ coincides with the "natural" locally convex topology of the Schwartz space $\mathcal{S}(\mathbb{R}^d)$.* In particular, the O^*-algebra $\mathcal{A}$ is closed, because $\mathcal{S}(\mathbb{R}^d)$ is complete. ◯

3.2 Bounded Commutants

In this section we define and study basic notions of bounded commutants of O^*-algebras. They will serve later for different purposes: While the strong commutant and the symmetrized commutant are used to characterize irreducible representations (Proposition 4.26), the weak commutant appears in the study of orderings of positive linear functionals and pure states (Corollary 5.4).

Throughout this section, we assume that $\mathcal{A}$ is an O^*-algebra on the dense domain $\mathcal{D}(\mathcal{A})$ of the Hilbert space $\mathcal{H}$. Define

$$\mathcal{D}(\mathcal{A}^*) = \bigcap_{a \in \mathcal{A}} \mathcal{D}(a^*).$$

Definition 3.12 An O^*-algebra $\mathcal{A}$ is called *self-adjoint* if $\mathcal{D}(\mathcal{A}) = \mathcal{D}(\mathcal{A}^*)$.

Definition 3.13 The *weak commutant* $\mathcal{A}'_w$ is defined by

$$\mathcal{A}'_w = \big\{T \in \mathbf{B}(\mathcal{H}) : \langle Ta\varphi, \psi \rangle = \langle T\varphi, a^+\psi \rangle \text{ for } a \in \mathcal{A}, \ \varphi, \psi \in \mathcal{D}(\mathcal{A})\big\},$$

the *strongly commutant* $\mathcal{A}'_s$ by

$$\mathcal{A}'_s = \big\{T \in \mathbf{B}(\mathcal{H}) : T\mathcal{D}(\mathcal{A}) \subseteq \mathcal{D}(\mathcal{A}) \text{ and } Ta\varphi = aT\varphi \text{ for } a \in \mathcal{A}, \ \varphi \in \mathcal{D}(\mathcal{A})\big\},$$

and the *symmetrized commutant* $\mathcal{A}'_{sym}$ by

$$\mathcal{A}'_{sym} = \big\{T \in \mathbf{B}(\mathcal{H}) : T\overline{a} \subseteq \overline{a}T \text{ and } T^*\overline{a} \subseteq \overline{a}T^* \text{ for } a \in \mathcal{A}\big\}.$$

It is convenient to have weak and strong commutants also for single operators.

Definition 3.14 The *strong commutant* $\{a\}'_s$ of a linear operator a on $\mathcal{H}$ is

$$\{a\}'_s := \big\{T \in \mathbf{B}(\mathcal{H}) : T\mathcal{D}(a) \subseteq \mathcal{D}(a) \ \ Ta\varphi = aT\varphi \text{ for } \varphi \in \mathcal{D}(a)\big\}.$$

If a is a symmetric operator, its *weak commutant* $\{a\}'_w$ is

$$\{a\}'_w = \big\{T \in \mathbf{B}(\mathcal{H}) : \langle Ta\varphi, \psi \rangle = \langle T\varphi, a\psi \rangle \text{ for } \varphi, \psi \in \mathcal{D}(a)\big\}.$$

For linear operators a and b on $\mathcal{H}$ and $T \in \mathbf{B}(\mathcal{H})$ the relation $Ta \subseteq bT$ means that T maps the domain $\mathcal{D}(a)$ into $\mathcal{D}(b)$ and $Ta\varphi = bT\varphi$ for all $\varphi \in \mathcal{D}(a)$.

It is not difficult to verify that the preceding definitions can be rewritten as

$$\{a\}'_s := \big\{T \in \mathbf{B}(\mathcal{H}) : Ta \subseteq aT\big\}, \quad \{a\}'_w = \big\{T \in \mathbf{B}(\mathcal{H}) : Ta \subseteq a^*T\big\}.$$

For the second equality we have to assume that $\mathcal{D}(a)$ is dense in $\mathcal{H}$ in order to ensure that the adjoint operator a^* is defined.

If the operator a is self-adjoint, we clearly have $\{a\}'_s = \{a\}'_w$ and we write $\{a\}'$ for $\{a\}'_s = \{a\}'_w$.

For the following results we need the definition and some basic facts of von Neumann algebras (see Definition B.2 and Appendix B).

Proposition 3.15 *The weak commutant $\mathcal{A}'_w$ is a $*$-invariant linear subspace of $\mathbf{B}(\mathcal{H})$ which is closed in the weak operator topology of $\mathbf{B}(\mathcal{H})$ and spanned by its positive elements. We have $\mathcal{A}'_w = (\overline{\mathcal{A}})'_w$,*

$$\mathcal{A}'_w = \{T \in \mathbf{B}(\mathcal{H}) : T\mathcal{D}(\mathcal{A}) \subseteq \mathcal{D}(\mathcal{A}^*),\ Ta\varphi = (a^+)^*T\varphi \text{ for } a \in \mathcal{A},\ \varphi \in \mathcal{D}(\mathcal{A})\},$$
$$\mathcal{A}'_s = \{T \in \mathcal{A}'_w : T\mathcal{D}(\mathcal{A}) \subseteq \mathcal{D}(\mathcal{A})\}.$$

If $\mathcal{A}$ is self-adjoint, then $\mathcal{A}'_w = \mathcal{A}'_s$. If $\mathcal{A}'_w = \mathcal{A}'_s$, this set is a von Neumann algebra which is denoted by $\mathcal{A}'$.

Proof From its definition it follows at once that $\mathcal{A}'_w$ is a linear subspace of $\mathbf{B}(\mathcal{H})$ that is closed in the weak operator topology. Suppose $T \in \mathcal{A}'_w$. Then

$$\langle T^*a\varphi, \psi \rangle = \langle a\varphi, T\psi \rangle = \langle \varphi, Ta^+\psi \rangle = \langle T^*\varphi, a^+\psi \rangle$$

for $a \in \mathcal{A}$ and $\varphi, \psi \in \mathcal{D}(\mathcal{A})$, where the second equality holds by the assumption $T \in \mathcal{A}'_w$. Therefore, $T^* \in \mathcal{A}'_w$. Thus $\mathcal{A}'_w$ is $*$-invariant.

Since obviously $I \in \mathcal{A}'_w$, the positive operators $T + \|T\| \cdot I$ and $\|T\| \cdot I - T$ are in $\mathcal{A}'_w$ for any $T = T^* \in \mathcal{A}'_w$. Hence $\mathcal{A}'_w$ is the span of its positive elements.

We show that $\mathcal{A}'_w = (\overline{\mathcal{A}})'_w$. Clearly, $(\overline{\mathcal{A}})'_w \subseteq \mathcal{A}'_w$. Suppose $T \in \mathcal{A}'_w$. Further, let $a \in \mathcal{A}$ and $\varphi, \psi \in \mathcal{D}(\overline{\mathcal{A}})$. Then there are nets (φ_i) and (ψ_j) from $\mathcal{D}(\mathcal{A})$ such that $\varphi = \lim_i \varphi_i$ and $\psi = \lim_j \psi_j$ in the graph topology $t_{\overline{\mathcal{A}}}$. Passing to the limit in $\langle Ta\varphi_i, \psi_j \rangle = \langle T\varphi_i, a^+\psi_j \rangle$ yields

$$\langle Tj(a)\varphi, \psi \rangle = \langle T\varphi, j(a^+)\psi \rangle = \langle T\varphi, j(a)^+\psi \rangle.$$

This proves that $T \in (\overline{\mathcal{A}})'_w$. Thus $\mathcal{A}'_w = (\overline{\mathcal{A}})'_w$.

Next we prove the formula for $\mathcal{A}'_w$. Suppose $T \in \mathcal{A}'_w$ and $\varphi \in \mathcal{D}(\mathcal{A})$. Let $a \in \mathcal{A}$. From the relation $\langle Ta\varphi, \psi \rangle = \langle T\varphi, a^+\psi \rangle$ for all $\psi \in \mathcal{D}(\mathcal{A})$ it follows that $T\varphi \in \mathcal{D}((a^+)^*)$ and $Ta\varphi = (a^+)^*T\varphi$. Since $a \in \mathcal{A}$ is arbitrary, $T\varphi \in \mathcal{D}(\mathcal{A}^*)$. Thus, $T\mathcal{D}(\mathcal{A}) \subseteq \mathcal{D}(\mathcal{A}^*)$, which proves one inclusion. The proof of the reverse inclusion is straightforward.

Since $(a^+)^*\psi = a^{++}\psi = a\psi$ for $\psi \in \mathcal{D}(\mathcal{A})$, the formula for $\mathcal{A}'_w$ implies the formula for $\mathcal{A}'_s$ and also the equality $\mathcal{A}'_w = \mathcal{A}'_s$ when $\mathcal{A}$ is self-adjoint.

Finally, suppose $\mathcal{A}'_w = \mathcal{A}'_s$. Since $\mathcal{A}'_s$ is obviously an algebra, $I \in \mathcal{A}'_s$, and $\mathcal{A}'_w$ is $*$-invariant and weak operator closed, $\mathcal{A}'_w = \mathcal{A}'_s$ is a von Neumann algebra. $\qquad \square$

The next lemma contains some technical facts on commutants of single operators.

Lemma 3.16 (i) *If a is a closable operator on $\mathcal{H}$, then the closure of $\{a\}'_s$ in the weak operator topology of $\mathbf{B}(\mathcal{H})$ is contained in $\{\overline{a}\}'_s$.*

(ii) *If the operator a is closed, then $\{a\}'_s$ is closed in the weak operator topology.*

(iii) *If a is closable symmetric operator, then $\{a\}'_w = \{\overline{a}\}'_w$.*

(iv) *If a is a self-adjoint operator and $E_a(\lambda)$, $\lambda \in \mathbb{R}$, are its spectral projections, then $\{a\}'_s = \{a\}'_w = \big\{ E_a(\lambda) : \lambda \in \mathbb{R} \big\}'$.*

Proof (i): First we prove that $\{a\}'_s \subseteq \{\overline{a}\}'_s$. Fix $T \in \{a\}'_s$. Let $\varphi \subseteq \mathcal{D}(\overline{a})$. Then there exists a sequence $(\varphi_n)_{n \in \mathbb{N}}$ of vectors $\varphi_n \in \mathcal{D}(a)$ such that $\varphi = \lim_n \varphi_n$ and $\overline{a}\,\varphi = \lim_n a\varphi_n$. Since $T\varphi = \lim_n \varphi_n$ and $T\overline{a}\,\varphi = \lim_n Ta\varphi_n = \lim_n aT\varphi_n$, it follows that $T\varphi \in \mathcal{D}(\overline{a})$ and $T\overline{a}\,\varphi = \overline{a}\,T\varphi$. That is, $T \in \{\overline{a}\}'_s$, which proves that $\{a\}'_s \subseteq \{\overline{a}\}'_s$.

To complete the proof it suffices to show that $\{\overline{a}\}'_s$ is weak operator closed in $\mathbf{B}(\mathcal{H})$. For convex subsets of $\mathbf{B}(\mathcal{H})$ the weak operator closure coincides with its strong operator closure (see, e.g., [KR83, Theorem 5.1.2]), so it suffices to prove that $\{\overline{a}\}'_s$ is strong operator closed. Let T be an operator of the strong operator closure of $\{\overline{a}\}'_s$. Then there is a net $(T_i)_{i \in}$ of operators $T_i \in \{\overline{a}\}'_s$ such that $T\psi = \lim_i T_i\psi$ for all $\psi \in \mathcal{H}$. Let $\varphi \in \mathcal{D}(\overline{a})$. Since the operator $\overline{a}$ is closed, we conclude from $T\varphi = \lim_i T_i\varphi$ and $T\overline{a}\,\varphi = \lim_i T_i\overline{a}\,\varphi = \lim_i \overline{a}\,T_i\varphi$ that $T\varphi \in \mathcal{D}(\overline{a})$ and $T\overline{a}\,\varphi = \overline{a}T\varphi$. Thus, $T \in \{\overline{a}\}'_s$ and $\{\overline{a}\}'_s$ is strong operator closed.

(ii): Follows immediately from (i).

(iii): is proved by the same reasoning as used in the proof of Proposition 3.15 to derive the equality $\mathcal{A}'_w = (\overline{\mathcal{A}})'_w$.

(iv): Since a is self-adjoint, $\{a\}'_s = \{a\}'_w$. The equality $\{a\}'_s = \{E_a(\lambda) : \lambda \in \mathbb{R}\}'$ is a well-known fact in unbounded operator theory [Sch12, Proposition 5.15].   □

A densely defined closed operator a on a Hilbert space $\mathcal{H}$ is called *affiliated* with a von Neumann algebra $\mathcal{N}$ on $\mathcal{H}$ if $Ta \subseteq aT$ for all $T \in \mathcal{N}'$; see Appendix B.

The next proposition collects basic properties of the strong commutant $\mathcal{A}'_s$ and the symmetrized commutant $\mathcal{A}'_{sym}$ for an O^*-algebra $\mathcal{A}$ on $\mathcal{H}$.

Proposition 3.17 (i) *$\mathcal{A}'_s$ is a subalgebra of $\mathbf{B}(\mathcal{H})$ such that $\mathcal{A}'_s \subseteq (\overline{\mathcal{A}})'_s$.*

(ii) *If the O^*-algebra $\mathcal{A}$ is closed, then $\mathcal{A}'_s$ is closed in the weak operator topology of $\mathbf{B}(\mathcal{H})$ and we have $\mathcal{A}'_{sym} = \mathcal{A}'_s \cap (\mathcal{A}'_s)^*$.*

(iii) *If the O^*-algebra $\mathcal{A}$ is self-adjoint, then $\mathcal{A}'_s = \mathcal{A}'_w = \mathcal{A}'_{sym}$ and this set is a von Neumann algebra on $\mathcal{H}$ which is denoted by $\mathcal{A}'$.*

(iv) *$\mathcal{A}'_{sym}$ is a von Neumann algebra on $\mathcal{H}$ and $\mathcal{A}'_{sym} = (\overline{\mathcal{A}})'_{sym}$. For each $a \in \mathcal{A}$, the closed operator $\overline{a}$ is affiliated with $(\mathcal{A}'_{sym})'$ and $(\mathcal{A}'_{sym})'$ is the smallest von Neumann algebra on $\mathcal{H}$ which has this property.*

Proof (i): From its definition it is clear that $\mathcal{A}'_s$ is a subalgebra of $\mathbf{B}(\mathcal{H})$. Lemma 3.16(i) implies that $\mathcal{A}' \subseteq (\overline{\mathcal{A}})'_s$

(ii): Let $T \in \mathcal{A}'_s \cap (\mathcal{A}'_s)^*$. Then, for $a \in \mathcal{A}$, T and T^* are in $\{a\}'_s$ by definition and hence in $\{\overline{a}\}'_s$ by Lemma 3.16(i). Therefore, $T \in \mathcal{A}'_{sym}$. Conversely, let $T \in \mathcal{A}'_{sym}$.

Then $T \in \{\overline{a}\}'_s$ and $T^* \in \{\overline{a}\}'_s$ for each $a \in \mathcal{A}$. Since $\mathcal{A}$ is closed by assumption, $\mathcal{D}(\mathcal{A}) = \cap_{a \in \mathcal{A}} \mathcal{D}(\overline{a})$, so the latter implies that T and T^* are in $\mathcal{A}'_s$. Hence $T \in \mathcal{A}'_s \cap (\mathcal{A}'_s)^*$, so we have shown that $\mathcal{A}'_{\text{sym}} = \mathcal{A}'_s \cap (\mathcal{A}'_s)^*$.

Let T be in the weak operator closure of $\mathcal{A}'_s$ in $\mathbf{B}(\mathcal{H})$. Since $\mathcal{A}'_s \subseteq \{a\}'_s$ for $a \in \mathcal{A}$, we get $T \in \{\overline{a}\}'_s$ by Lemma 3.16(i). Thus, $T\mathcal{D}(\mathcal{A}) \subseteq \cap_{a \in \mathcal{A}} \mathcal{D}(\overline{a}) = \mathcal{D}(\mathcal{A})$ and hence $T \in \mathcal{A}'_s$. This proves that $\mathcal{A}'_s$ is closed in the weak operator topology.

(iii): By Proposition 3.15, $\mathcal{A}'_w$ is $*$-invariant and $\mathcal{A}'_s = \mathcal{A}'_w$ is a von Neumann algebra, because $\mathcal{A}$ is self-adjoint. Hence (ii) implies $\mathcal{A}'_s = \mathcal{A}'_w = \mathcal{A}'_{\text{sym}}$.

(iv): Since $\mathcal{A}'_{\text{sym}} = \cap_{a \in \mathcal{A}} (\{\overline{a}\}'_s \cap (\{\overline{a}\}'_s)^*)$ by definition, $\mathcal{A}'_{\text{sym}}$ is closed in the weak operator topology by Lemma 3.16(ii). Obviously, $I \in \mathcal{A}'_{\text{sym}}$ and $\mathcal{A}'_{\text{sym}}$ is a $*$-algebra. Hence $\mathcal{A}'_{\text{sym}}$ is a von Neumann algebra. Since $\mathcal{A}'_{\text{sym}} = (\mathcal{A}'_{\text{sym}})''$, it is clear from the definition of $\mathcal{A}'_{\text{sym}}$ that each operator $\overline{a}$, where $a \in \mathcal{A}$, is affiliated with $(\mathcal{A}'_{\text{sym}})'$.

Suppose $\mathcal{N}$ is another von Neumann algebra on $\mathcal{H}$ which has this property. Fix $x \in \mathcal{N}'$. Let $a \in \mathcal{A}$. Since $\overline{a}$ is affiliated with $\mathcal{N}$ and $x^* \in \mathcal{N}'$, we have $x\overline{a} \subseteq \overline{a}x$ and $x^*\overline{a} \subseteq \overline{a}x^*$, so $x \in \mathcal{A}'_{\text{sym}}$. Thus $\mathcal{N}' \subseteq \mathcal{A}'_{\text{sym}}$ and hence $(\mathcal{A}'_{\text{sym}})' \subseteq \mathcal{N}'' = \mathcal{N}$. $\qquad \square$

Remark 3.18 While the weak commutant and symmetrized commutant of an O^*-algebra and its closure coincide, this is not true in general for the strong commutant. A simple example is the following: Consider the O^*-algebra $\mathcal{A} = \mathbb{C}[x]$ on the dense domain $\mathbb{C}[x]$ of $L^2(0,1)$, where the polynomials act as multiplication operators. Then $\mathcal{D}(\overline{\mathcal{A}}) = L^2(0,1)$ and the multiplication operators of characteristic functions of intervals $[a,b] \subseteq (0,1)$ are in $(\overline{\mathcal{A}})'_s$. But they are not in $\mathcal{A}'_s$, because they do not leave the domain $\mathbb{C}[x]$ invariant. $\qquad \bigcirc$

The following simple lemma describes weak and strong commutants of an O^*-algebra in terms of commutants of algebra generators.

Lemma 3.19 *Let $\mathcal{A}$ be an O^*-algebra and let $\mathcal{B}$ be a subset of $\mathcal{A}$ which generates $\mathcal{A}$ as an algebra.*

(i) $\mathcal{A}'_s = \cap_{b \in \mathcal{B}} \{b\}'_s$.

(ii) *If $\mathcal{B} \subseteq \mathcal{A}_{\text{her}}$, then $\mathcal{A}'_w = \cap_{b \in \mathcal{B}} \{b\}'_w$.*

Proof (i) is obvious. We prove (ii). It is trivial that $\mathcal{A}'_w \subseteq \cap_{b \in \mathcal{B}} \{b\}'_w$. Conversely, suppose $T \in \cap_{b \in \mathcal{B}} \{b\}'_w$. Let $\mathcal{A}_0$ denote the set of operators $a \in \mathcal{A}$ for which $\langle Ta\varphi, \psi \rangle = \langle T\varphi, a^+\psi \rangle$ for all $\varphi, \psi \in \mathcal{D}(\mathcal{A})$. Clearly, $\mathcal{A}_0$ is a vector space. Let $a_1, a_2 \in \mathcal{A}_0$. For $\varphi, \psi \in \mathcal{D}(\mathcal{A})$, we have $a_2\varphi, a_1^+\psi \in \mathcal{D}(\mathcal{A})$ and therefore

$$\langle Ta_1a_2\varphi, \psi \rangle = \langle Ta_2\varphi, a_1^+\psi \rangle = \langle T\varphi, a_2^+a_1^+\psi \rangle = \langle T\varphi, (a_1a_2)^+\psi \rangle.$$

This shows that $a_1a_2 \in \mathcal{A}_0$. Thus $\mathcal{A}_0$ is a subalgebra of $\mathcal{A}$. Since $\mathcal{B} \subseteq \mathcal{A}_0$ by assumption, $\mathcal{A}_0 = \mathcal{A}$, which proves that $T \in \mathcal{A}'_w$. $\qquad \square$

In the next example we describe the commutants $\mathcal{A}'_s$ and $\mathcal{A}'_w$ explicitly, and we shall see that $\mathcal{A}'_s$ is not $*$-invariant and $\mathcal{A}'_w$ is not an algebra. This example requires some facts on Toeplitz operators; all of them can be found in [H67].

Example 3.20 Let S denote the shift operator on the Hardy space $\mathcal{H} = H^2(\mathbb{T})$. Since $(I - S)\mathcal{H}$ is dense in $\mathcal{H}$ as easily checked, $A := \mathrm{i}(I + S)(I - S)^{-1}$ is a densely defined closed symmetric operator on $\mathcal{H}$ with deficiency indices $(0, 1)$. Further, for $\xi \in L^\infty(\mathbb{T})$ let T_ξ denote the Toeplitz operator on $H^2(\mathbb{T})$ with symbol ξ, that is, $T_\xi \varphi = P(\xi \cdot \varphi)$ for $\varphi \in H^2(\mathbb{T})$, where P is the projection of $L^2(\mathbb{T})$ on $H^2(\mathbb{T})$.

Statement 1: $\{A\}'_w = \{T_\xi : \xi \in L^\infty(\mathbb{T})\}$ and $\{A\}'_s = \{T_\xi : \xi \in H^\infty(\mathbb{T})\}$.

Proof Suppose $T \in \mathbf{B}(\mathcal{H})$. Then $T \in \{A\}'_w$ if and only if $\langle TA\varphi, \psi \rangle = \langle T\varphi, A\psi \rangle$, or equivalently, $\langle T(A + \mathrm{i})\varphi, \psi \rangle = \langle T\varphi, (A - \mathrm{i})\psi \rangle$ for $\varphi, \psi \in \mathcal{D}(A)$. Writing $\varphi = (I - S)\eta$ and $\psi = (I - S)\zeta$ with $\eta, \zeta \in \mathcal{H}$, we have $(A + \mathrm{i})\varphi = 2\mathrm{i}\,\eta$ and $(A - \mathrm{i})\psi = 2\mathrm{i}\,\zeta$. Inserting this it follows easily that the latter is equivalent to the equation $T = S^* T S$. But this holds if and only if T is a Toeplitz operator with symbol $\xi \in L^\infty(\mathbb{T})$; see e.g., [H67, Nr. 194]. This proves the first equality.

We show the second equality. It is easily verified that $T_\xi \in \{A\}'_s$ for $\xi \in H^\infty(\mathbb{T})$. Conversely, let $T \in \{A\}'_s$. Then $T \in \{A\}'_w$ and hence $T = T_\xi$ with $\xi \in L^\infty(\mathbb{T})$. Since $T_\xi \in \{A\}'_s$, we have $T_\xi(A - \mathrm{i})\mathcal{D}(A) \equiv T_\xi S\mathcal{H} \subseteq (A - \mathrm{i})\mathcal{D}(A) \equiv S\mathcal{H}$. The relation $T_\xi S\mathcal{H} = S\mathcal{H}$ implies that the negative Fourier coefficients of ξ are zero. Therefore, $\xi \in H^\infty(\mathbb{T})$.   □

Since A has finite deficiency indices, $\mathcal{D}^\infty(A) := \cap_{n\in\mathbb{N}}\mathcal{D}(A^n)$ is dense in $\mathcal{H}$ and a core for A by Proposition 3.31 proved in Sect. 3.4. Then the symmetric operator $a := A \lceil \mathcal{D}^\infty(A)$ is in $\mathcal{L}^+(\mathcal{D}^\infty(A))$ and $\bar{a} = A$. Let $\mathcal{A}$ denote the unital O^*-algebra on $\mathcal{D}^\infty(A)$ generated by a.

Statement 2: $\mathcal{A}'_w = \{T_\xi : \xi \in L^\infty(\mathbb{T})\}$ and $\mathcal{A}'_s = \{T_\xi : \xi \in H^\infty(\mathbb{T})\}$.

Proof From Lemmas 3.19(ii) and 3.16(iii) and $\bar{a} = A$ we get $\mathcal{A}'_w = \{a\}'_w = \{A\}'_w$. If $T \in \{A\}'_s$, then T maps $\mathcal{D}^\infty(A)$ into itself and hence $T \in \{a\}'_s = \mathcal{A}'_s$. Conversely, let $T \in \mathcal{A}'_s$. Then $T \in \{a\}'_s$. Hence, since $\bar{a} = A$, $T \in \{A\}'_s$ by Lemma 3.16(i). Thus, $\mathcal{A}'_s = \{A\}'_s$. Now the assertions follow from Statement 1.   □

If $\xi \in L^\infty(\mathbb{T})$ and both ξ and $\bar{\xi}$ are not in $H^\infty(\mathbb{T})$, then $T_\xi \in \mathcal{A}'_w$, but $(T_\xi)^2$ is not in $\mathcal{A}'_w$ by Statement 2, because $(T_\xi)^2$ is not a Toeplitz operator [H67, Exercise 195]. This shows that the *weak commutant $\mathcal{A}'_w$ is not an algebra*.

Suppose $T = T^* \in \mathcal{A}'_s$. Then $T = T_\eta$ with $\eta \in H^\infty(\mathbb{T})$ by Statement 2. The relation $T_\eta = (T_\eta)^* = T_{\bar{\eta}}$ implies $\eta = \bar{\eta} \in H^\infty(\mathbb{T})$. Hence η is constant [H67, Exercise 26], so $T = \lambda \cdot I$ for some $\lambda \in \mathbb{R}$. In particular, this implies that the *strong commutant $\mathcal{A}'_s$ is not $*$-invariant*.   ○

3.3 Trace Functionals on O^*-Algebras

Throughout this section, $\mathcal{A}$ is a **closed unital** O^*-**algebra** on the Hilbert space $\mathcal{H}$.

Our aim is to study functionals of the form $f_t(a) = \operatorname{Tr} at$, $a \in \mathcal{A}$, for some positive trace class operator t. These functionals are an important source of positive functionals and of $*$-representations of $*$-algebras; see, e.g., Example 4.10 below.

First we recall some facts on trace class operators. For notational simplicity we assume that the Hilbert space $\mathcal{H}$ has infinite dimension. We denote the trace class operators on $\mathcal{H}$ by $\mathbf{B}_1(\mathcal{H})$, the positive trace class operators by $\mathbf{B}_1(\mathcal{H})_+$, and the operator $\langle \cdot, \varphi \rangle \psi$, where $\varphi, \psi \in \mathcal{H}$, by $\varphi \otimes \psi$.

By an *absolutely convergent series* on $\mathcal{H}$ we mean a series $\sum_{n=1}^\infty \varphi_n \otimes \psi_n$, where $\varphi_n, \psi_n \in \mathcal{H}$, such that $\sum_n \|\varphi_n\| \, \|\psi_n\| < \infty$. Since $\|\varphi \otimes \psi_n\| = \|\varphi_n\| \, \|\psi_n\|$, such a series converges in the operator norm and defines a bounded operator t on $\mathcal{H}$. We say that t is represented by this series and write $t = \sum_n \varphi_n \otimes \psi_n$.

The following lemma reviews some basic results on trace class operators; see, e.g., [BS87, Chap. 11] for proofs of these facts.

Lemma 3.21 (i) *If $\sum_{n=1}^\infty \varphi_n \otimes \psi_n$ is an absolutely convergent series on $\mathcal{H}$, then the operator $t := \sum_n \varphi_n \otimes \psi_n$ belongs to $\mathbf{B}_1(\mathcal{H})$ and*

$$\operatorname{Tr} t = \sum_{n=1}^\infty \langle \psi_n, \varphi_n \rangle. \tag{3.7}$$

Each operator $t \in \mathbf{B}_1(\mathcal{H})$ is of this form $t = \sum_{n=1}^\infty \varphi_n \otimes \psi_n$.

(ii) *Suppose $t \in \mathbf{B}_1(\mathcal{H})_+$. There exist a sequence $(\lambda_n)_{n \in \mathbb{N}}$ of numbers $\lambda_n \geq 0$ and an orthonormal sequence $(\varphi_n)_{n \in \mathbb{N}}$ of $\mathcal{H}$ such that $\sum_{n=1}^\infty \lambda_n < \infty$ and t is represented by the absolutely convergent series $\sum_n \varphi_n \otimes (\lambda_n \varphi_n)$:*

$$t = \sum_{n=1}^\infty \varphi_n \otimes (\lambda_n \varphi_n). \tag{3.8}$$

Now we begin the study of *trace functionals* $f_t(\cdot) = \operatorname{Tr} \cdot t$ on O^*-algebras.

Definition 3.22 Set $f_t(a) := \operatorname{Tr} at$ for $a \in \mathcal{A}$ and $t \in \mathbf{B}_1(\mathcal{A})_+$, where

$$\mathbf{B}_1(\mathcal{A})_+ := \left\{ t \in \mathbf{B}_1(\mathcal{H})_+ : t\mathcal{H} \subseteq \mathcal{D}(\mathcal{A}), \ \overline{atb} \in \mathbf{B}_1(\mathcal{H}) \ \text{for all} \ a, b \in \mathcal{A} \right\}.$$

Note that since $t\mathcal{H} \subseteq \mathcal{D}(\mathcal{A})$ for $t \in \mathbf{B}_1(\mathcal{A})_+$, the operator at is defined on the whole Hilbert space $\mathcal{H}$.

Remark 3.23 Let $\{a_i : i \in I\}$ be a subset of $\mathcal{A}$ such that the graph topology $t_\mathcal{A}$ is generated by the seminorms $\{\| \cdot \|_{a_i} : i \in I\}$. The proof of Proposition 3.24 below shows that in the definition of $\mathbf{B}_1(\mathcal{A})_+$ the condition "$\overline{atb} \in \mathbf{B}_1(\mathcal{H})$ for $a, b \in \mathcal{A}$" can be replaced by the weaker requirement "$\overline{a_i t a_i^+} \in \mathbf{B}_1(\mathcal{H})$ for $i \in I$". $\bigcirc$

Let $(\varphi_n)_{n\in\mathbb{N}}$ be a sequence of vectors $\varphi_n \in \mathcal{D}(\mathcal{A})$. We shall say that the series $\sum_{n=1}^{\infty} \varphi_n \otimes \varphi_n$ is *absolutely $\mathcal{A}$-convergent* if

$$\sum_{n=1}^{\infty} \|a\varphi_n\|^2 \equiv \sum_{n=1}^{\infty} \|\varphi_n\|_a^2 < \infty \quad \text{for} \quad a \in \mathcal{A}.$$

Since the family of seminorms $\{\|\cdot\|_a : a \in \mathcal{A}\}$ is directed (Lemma 3.5), this implies that $\sum_{n=1}^{\infty} \|a\varphi_n\| \, \|b\varphi_n\| < \infty$ for all $a, b \in \mathcal{A}$.

The following proposition collects basic properties of trace functionals f_t.

Proposition 3.24 (i) *If $t \in \mathbf{B}_1(\mathcal{A})_+$ and $t = \sum_{n=1}^{\infty} \varphi_n \otimes (\lambda_n \varphi_n)$ is a representation (3.8) of t, then the series $\sum_{n=1}^{\infty} (\lambda_n^{1/2} \varphi_n) \otimes (\lambda_n^{1/2} \varphi_n)$ is absolutely $\mathcal{A}$-convergent and represents the operator t.*

(ii) *If $\sum_{n=1}^{\infty} \psi_n \otimes \psi_n$ is absolutely $\mathcal{A}$-convergent, then $t := \sum_{n=1}^{\infty} \psi_n \otimes \psi_n$ is in $\mathbf{B}_1(\mathcal{A})_+$ and for $a, b \in \mathcal{A}$ we have*

$$f_t(ab) \equiv \text{Tr } abt = \text{Tr } \overline{tab} = \text{Tr } \overline{bta} = \sum_{n=1}^{\infty} \langle ab\psi_n, \psi_n \rangle. \tag{3.9}$$

(iii) *Suppose $t \in \mathbf{B}_1(\mathcal{A})_+$ and $t^{1/2}\mathcal{H} \subseteq \mathcal{D}(\mathcal{A})$. Then, $t^{1/2}at^{1/2} \in \mathbf{B}_1(\mathcal{H})$ and*

$$f_t(a) = \text{Tr } t^{1/2}at^{1/2} \quad \text{for} \quad a \in \mathcal{A}. \tag{3.10}$$

(iv) *For $t \in \mathbf{B}_1(\mathcal{A})_+$, the linear functional $f_t(\cdot) \equiv \text{Tr } \cdot t$ is $\mathcal{A}_+$-positive, that is, $f_t(c) \geq 0$ for $c \in \mathcal{A}_+ := \{a = a^+ \in \mathcal{A} : \langle a\varphi, \varphi \rangle \geq 0, \ \varphi \in \mathcal{D}(\mathcal{A})\}$.*

Proof (i): Let $a \in \mathcal{A}$. Since $t \in \mathbf{B}_1(\mathcal{A})_+$, we have $s := \overline{ata^+} \in \mathbf{B}_1(\mathcal{H})$ and $t \geq 0$. The latter implies that $\langle s\varphi, \varphi \rangle = \langle ta^+\varphi, a^+\varphi \rangle \geq 0$ for $\varphi \in \mathcal{D}(\mathcal{A})$. Hence $s \geq 0$. We apply the Gram–Schmidt procedure to the sequence $(a\varphi_n)_{n\in\mathbb{N}}$ and construct an orthonormal sequence $(\psi_k)_{k\in\mathbb{N}}$ of vectors $\psi_k \in \mathcal{D}(\mathcal{A})$ such that $\mathcal{G} := \text{Lin } \{\psi_k : k \in \mathbb{N}\} \supseteq \text{Lin } \{a\varphi_n : n \in \mathbb{N}\}$. Then

$$\sum_{k=1}^{\infty} \|s^{1/2}\psi_k\|^2 = \sum_{k=1}^{\infty} \langle s\psi_k, \psi_k \rangle = \sum_{k=1}^{\infty} \langle ata^+\psi_k, \psi_k \rangle$$

$$= \sum_{k=1}^{\infty} \langle ta^+\psi_k, a^+\psi_k \rangle = \sum_{k=1}^{\infty} \sum_{n=1}^{\infty} \lambda_n \langle a^+\psi_k, \varphi_n \rangle \langle \varphi_n, a^+\psi_k \rangle$$

$$= \sum_{k=1}^{\infty} \sum_{n=1}^{\infty} |\langle \psi_k, a(\lambda_n^{1/2}\varphi_n) \rangle|^2 = \sum_{n=1}^{\infty} \sum_{k=1}^{\infty} |\langle \psi_k, a(\lambda_n^{1/2}\varphi_n) \rangle|^2$$

$$= \sum_{n=1}^{\infty} \|a(\lambda_n^{1/2}\varphi_n)\|^2, \tag{3.11}$$

where the last equality follows from Parseval's identity, applied to the orthonormal basis $\{\psi_k : k \in \mathbb{N}\}$ of the Hilbert space $\overline{\mathcal{G}}$. Since $s \in \mathbf{B}_1(\mathcal{H})$, its square root $s^{1/2}$ is a Hilbert–Schmidt operator and therefore $\sum_k \|s^{1/2}\psi_k\|^2 < \infty$ (see [BS87]). Thus, $\sum_n \|a(\lambda_n^{1/2}\varphi_n)\|^2 < \infty$ by (3.11). Since $a \in \mathcal{A}$ was arbitrary, this proves that the series $\sum_n (\lambda_n^{1/2}\varphi_n) \otimes (\lambda_n^{1/2}\varphi_n)$ is absolutely $\mathcal{A}$-convergent. By construction this series represents the operator t.

(ii): In particular, $\sum_n \psi_n \otimes \psi_n$ is an absolutely convergent series on $\mathcal{H}$, so by Lemma 3.21(i) it represents an operator $t \in \mathbf{B}_1(\mathcal{H})$. Clearly, $t \geq 0$, since

$$\langle t\varphi, \varphi \rangle = \sum_{n=1}^{\infty} \langle \varphi, \psi_n \rangle \langle \psi_n, \varphi \rangle = \sum_{n=1}^{\infty} |\langle \varphi, \psi_n \rangle|^2 \geq 0, \quad \varphi \in \mathcal{H}.$$

Next we prove that $t\mathcal{H} \subseteq \mathcal{D}(\mathcal{A})$. Suppose $\varphi \in \mathcal{H}$. Let $a \in \mathcal{A}$. Since $I \in \mathcal{A}$, we have $\sum_n \|\psi_n\| \, \|a\psi_n\| < \infty$. Hence the series $\sum_n \langle \varphi, \psi_n \rangle \psi_n$ converges in the Hilbert space $(\mathcal{D}(\overline{a}), \langle \cdot, \cdot \rangle_{\overline{a}} := \langle \overline{a} \cdot, \overline{a} \cdot \rangle + \langle \cdot, \cdot \rangle)$. It converges to the vector $t\varphi$ in $\mathcal{H}$. Thus, $t\varphi \in \mathcal{D}(\overline{a})$. Because the O^*-algebra $\mathcal{A}$ is closed, $\mathcal{D}(\mathcal{A}) = \cap_{a \in \mathcal{A}} \mathcal{D}(\overline{a})$. Therefore, $t\varphi \in \mathcal{D}(\mathcal{A})$ and $t\mathcal{H} \subseteq \mathcal{D}(\mathcal{A})$.

Now let $a, b \in \mathcal{A}$. Since the series $\sum_n \psi_n \otimes \psi_n$ is absolutely $\mathcal{A}$-convergent, $\sum_n a^+\psi_n \otimes b\psi_n$ is an absolutely convergent series on $\mathcal{H}$, hence its represents an operator $t_{a,b} \in \mathbf{B}_1(\mathcal{H})$ by Lemma 3.21(i). For $\xi, \eta \in \mathcal{D}(\mathcal{A})$,

$$\langle t_{a,b}\xi, \eta \rangle = \sum_{n=1}^{\infty} \langle \xi, a^+\psi_n \rangle \langle b\psi_n, \eta \rangle = \sum_{n=1}^{\infty} \langle a\xi, \psi_n \rangle \langle \psi_n, b\eta \rangle = \langle ta\xi, b^+\eta \rangle.$$

$$(3.12)$$

Since $t\mathcal{H} \subseteq \mathcal{D}(\mathcal{A})$ as shown in the preceding paragraph, it follows from (3.12) that $\langle t_{a,b}\xi, \eta \rangle = \langle bta\xi, \eta \rangle$ for $\xi, \eta \in \mathcal{D}(\mathcal{A})$. Therefore, $bta = t_{a,b}\lceil \mathcal{D}(\mathcal{A})$ and hence $\overline{bta} = t_{a,b} \in \mathbf{B}_1(\mathcal{H})$. Putting these facts together we have proved that $t \in \mathbf{B}_1(\mathcal{A})_+$.

Setting $a = I$, we get $\overline{bt} = bt \in \mathbf{B}_1(\mathcal{H})$. By the preceding we have shown that

$$\overline{bta} = t_{a,b} = \sum_{n=1}^{\infty} a^+\psi_n \otimes b\psi_n,$$

$$abt = t_{I,ab} = \sum_{n=1}^{\infty} \psi_n \otimes ab\psi_n,$$

$$\overline{tab} = t_{ab,I} = \sum_{n=1}^{\infty} (ab)^+\psi_n \otimes \psi_n.$$

We apply formula (3.7) to these three series and derive

$$f_t(ab) \equiv \text{Tr } abt = \sum_{n=1}^{\infty} \langle ab\psi_n, \psi_n \rangle = \sum_{n=1}^{\infty} \langle b\psi_n, a^+\psi_n \rangle = \text{Tr } \overline{bta}$$

$$= \sum_{n=1}^{\infty} \langle \psi_n, (ab)^+\psi_n \rangle = \text{Tr } \overline{tab}.$$

(iii): It suffices to prove the assertion for $a = c^2$, where $c = c^+ \in \mathcal{A}$, because these elements span $\mathcal{A}$. By the assumption $t^{1/2}\mathcal{H} \subseteq \mathcal{D}(\mathcal{A}), s := t^{1/2}c^2t^{1/2}$ is defined on $\mathcal{H}$. Therefore, since s is symmetric, s is bounded. Let $t = \sum_{n=1}^{\infty} \varphi_n \otimes (\lambda_n\varphi_n)$ be a representation (3.8) of $t \in \mathbf{B}_1(\mathcal{A})_+$. By adding vectors ψ_k if necessary, we obtain an orthonormal basis $\{\psi_k, \varphi_n\}$ of $\mathcal{H}$. Clearly, $t\varphi_n = \lambda_n\varphi_n$ implies $t^{1/2}\varphi_n = \lambda_n^{1/2}\varphi_n$. Since ψ_k is orthogonal to $\overline{t\mathcal{H}} = \overline{t^{1/2}\mathcal{H}}$, we have $t^{1/2}\psi_k = 0$ and hence $s\psi_k = 0$. Using these facts we derive

$$\sum_{k} \langle s\psi_k, \psi_k \rangle + \sum_{n=1}^{\infty} \langle s\varphi_n, \varphi_n \rangle = \sum_{n=1}^{\infty} \langle s\varphi_n, \varphi_n \rangle$$

$$= \sum_{n=1}^{\infty} \langle c^2 t^{1/2}\varphi_n, t^{1/2}\varphi_n \rangle = \sum_{n=1}^{\infty} \langle c^2(\lambda_n^{1/2}\varphi_n), (\lambda_n^{1/2}\varphi_n) \rangle. \tag{3.13}$$

Since $\sum_n (\lambda_n^{1/2}\varphi_n) \otimes (\lambda_n^{1/2}\varphi_n)$ is absolutely $\mathcal{A}$-convergent, the sum in (3.13) is finite. Therefore, since $s = t^{1/2}c^2t^{1/2} \geq 0$ and $\{\psi_k, \varphi_n\}$ is an orthonormal basis, this implies that s is trace class [BS87] and $\text{Tr } s = \sum_n \langle c^2(\lambda_n^{1/2}\varphi_n), (\lambda_n^{1/2}\varphi_n) \rangle$. By (3.9), applied with $\psi_n = \lambda_n^{1/2}\varphi_n$, $a = b = c$, this yields $\text{Tr } s \equiv \text{Tr } t^{1/2}c^2t^{1/2} = f_t(c^2)$. This proves (3.10).

(iv): By (i) and (ii), formula (3.9) holds. If $c \in \mathcal{A}_+$, then we have $\langle c\varphi, \varphi \rangle \geq 0$ for $\varphi \in \mathcal{D}(\mathcal{A})$. Therefore, setting $a = c, b = I$ in (3.9), we obtain $f_t(c) \geq 0$. $\square$

Corollary 3.25 *For a linear functional f on $\mathcal{A}$ the following are equivalent:*

(i) *There is an operator $t \in \mathbf{B}_1(\mathcal{A})_+$ such that $f(a) = f_t(a) \equiv \text{Tr } at$ for $a \in \mathcal{A}$.*
(ii) *There exists a sequence $(\varphi_n)_{n\in\mathbb{N}}$ of vectors $\varphi_n \in \mathcal{D}(\mathcal{A})$ such that*

$$f(a) = \sum_{n=1}^{\infty} \langle a\varphi_n, \varphi_n \rangle \text{ for } a \in \mathcal{A}. \tag{3.14}$$

Proof (i)→(ii): Apply formula (3.9) with $b = I$.

(ii)→(i): By (3.14), $f(a^+a) = \sum_n \langle a^+a\varphi_n, \varphi_n \rangle = \sum_n \|a\varphi_n\|^2 < \infty$ for $a \in \mathcal{A}$, so $\sum_n \varphi_n \otimes \varphi_n$ is absolutely $\mathcal{A}$-convergent. By Proposition 3.24(ii), it represents an operator $t \in \mathbf{B}_1(\mathcal{A})_+$ and formula (3.9) yields $f = f_t$. $\square$

A *Frechet–Montel space* (see, e.g., [Sh71, IV, 5.8.]) is a locally convex space which is a Frechet space (i.e., a complete metrizable space) and has the Montel property (i.e., each bounded subset is relatively compact). For a Frechet space the Montel property means that each bounded sequence has a convergent subsequence.

The following theorem, due to the author [Sch78], shows that for a large class of O^*-algebras *all* $\mathcal{A}_+$-positive linear functionals are of the form f_t with $t \in \mathbf{B}_1(\mathcal{A})_+$.

Theorem 3.26 *Suppose $\mathcal{A}$ is a unital O^*-algebra such that the locally convex space $\mathcal{D}(\mathcal{A})[t_\mathcal{A}]$ is a Frechet–Montel space. Then, for each $\mathcal{A}_+$–positive linear functional f on $\mathcal{A}$, there exists an operator $t \in \mathbf{B}_1(\mathcal{A})_+$ such that*

$$f(a) = \mathrm{Tr}\, at \quad \text{for} \quad a \in \mathcal{A}. \tag{3.15}$$

Proof The proof is long and technical. Complete proofs of this theorem are given in the original papers [Sch78, Sch79] and also in [Sch90, Theorem 5.3.8]. $\square$

For a closable operator a we denote by $\mathcal{H}_a$ the Hilbert space $\mathcal{D}(\overline{a})$ with inner product $\langle \cdot, \cdot \rangle_{\overline{a}} := \langle \overline{a}\, \cdot, \overline{a}\, \cdot \rangle + \langle \cdot, \cdot \rangle$.

Corollary 3.27 *Suppose $\mathcal{D}(\mathcal{A})[t_\mathcal{A}]$ is a Frechet space. Assume that there exists an operator $c \in \mathcal{A}$ such that the embedding of the Hilbert space $\mathcal{H}_c$ into the Hilbert space $\mathcal{H}$ is compact. Then each $\mathcal{A}_+$–positive linear functional on $\mathcal{A}$ is of the form (3.15) for some operator $t \in \mathbf{B}_1(\mathcal{A})_+$.*

Proof Since the graph topology $t_\mathcal{A}$ is metrizable, there is a sequence $(a_k)_{k \in \mathbb{N}}$ of elements $a_k \in \mathcal{A}$ such that the family of seminorms $\{\| \cdot \|_{a_k} : k \subset \mathbb{N}\}$ defines the topology $t_\mathcal{A}$. Let $(\varphi_n)_{n \in \mathbb{N}}$ be a bounded sequence of $\mathcal{D}(\mathcal{A})[t_\mathcal{A}]$. Fix $k \in \mathbb{N}$. Then $\sup_{n \in \mathbb{N}} (\|ca_k \varphi_n\| + \|a_k \varphi_n\|) < \infty$, so the sequence $(a_k \varphi_n)_{n \in \mathbb{N}}$ is bounded in the Hilbert space $\mathcal{H}_c$. Hence, by the compactness of the embedding of $\mathcal{H}_c$ into $\mathcal{H}$, there is a subsequence $(\varphi_{n(k)_j})_{j \in \mathbb{N}}$ such that the sequence $(a_k \varphi_{n(k)_j})_{j \in \mathbb{N}}$ converges in $\mathcal{H}$. By a diagonal procedure we choose a subsequence $(\varphi_{n_j})_{j \in \mathbb{N}}$ (independent of k) such that $(a_k \varphi_{n_j})_{j \in \mathbb{N}}$ converges in $\mathcal{H}$ for each $k \in \mathbb{N}$. Then $(\varphi_{n_j})_{j \in \mathbb{N}}$ is a Cauchy sequence in the graph topology. Since $\mathcal{D}(\mathcal{A})[t_\mathcal{A}]$ is complete, the subsequence $(\varphi_{n_j})_{j \in \mathbb{N}}$ converges in the locally convex space $\mathcal{D}(\mathcal{A})[t_\mathcal{A}]$. This proves that $\mathcal{D}(\mathcal{A})[t_\mathcal{A}]$ has the Montel property, so the assertion follows from Theorem 3.26. $\square$

Functionals (3.15) are extensively studied in [Sch90, Chap. 5]. In the preceding we have presented a simplified approach to some of the main results on this topic.

Let us add a few remarks.

Remark 3.28 1. Let c be a densely defined closable operator on $\mathcal{H}$. It is easily verified that the embedding of the Hilbert space $\mathcal{H}_c$ into $\mathcal{H}$ is compact if and only if the (bounded) operator $(I + \overline{c}^*\overline{c})^{-1}$ on $\mathcal{H}$ is compact.

2. There exists an O^*-algebra $\mathcal{A}$ such that $\mathcal{D}(\mathcal{A})[t_\mathcal{A}]$ is a Frechet–Montel space, but there is no operator $c \in \mathcal{A}$ such that the embedding of $\mathcal{H}_c$ into $\mathcal{H}$ is compact. In fact, Theorem 3.26 is stronger than Corollary 3.27 and its proof is much more involved.

3. If $\mathcal{D}(\mathcal{A})[t_\mathcal{A}]$ is a Frechet space, then by Theorem 3.26 the Montel property is *sufficient* for representing all $\mathcal{A}_+$-positive linear functionals as f_t, $t \in \mathbf{B}(\mathcal{A})_+$. The following result [Sch90, Proposition 5.5.1] shows that if the O^*-algebra is "large" enough, the Montel property of the graph topology is also *necessary*:

Suppose $\mathcal{A}$ is a unital O^*-algebra such that $\mathcal{D}(\mathcal{A})[t_\mathcal{A}]$ is a Frechet space and $\mathcal{A}$ contains all operators $\varphi \otimes \psi$, where $\varphi, \psi \in \mathcal{D}(\mathcal{A})$. If each $\mathcal{A}_+$-positive linear functional on $\mathcal{A}$ is of the form $f_t = \mathrm{Tr} \cdot t, t \in \mathbf{B}(\mathcal{A})_+$, then the locally convex space $\mathcal{D}(\mathcal{A})[t_\mathcal{A}]$ has the Montel property.

4. The problem of representing functionals in the form f_t, $t \in \mathbf{B}(\mathcal{A})_+$, is called the *quantum moment problem* ; see [Wo70, Sch91]. The trace can be viewed as a noncommutative integral and the operator t as a counterpart of a Radon measure. For the quantum moment problem the Weyl algebra $\mathsf{W}(d)$ plays a similar role as the polynomial algebra $\mathbb{R}[x_1, \ldots, x_d]$ does for the classical moment problem.   ○

Example 3.29 (*Example* 3.11 *continued*)
Let $\mathcal{A}$ be the O^*-algebra on $\mathcal{D}(\mathcal{A}) = \mathcal{S}(\mathbb{R}^d)$ from Example 3.11. Then *all $\mathcal{A}_+$-positive linear functionals on $\mathcal{A}$ are of the form $f_t = \mathrm{Tr} \cdot t$, where $t \in \mathbf{B}(\mathcal{A})_+$.*

This follows from Theorem 3.26, since the graph topology $t_\mathcal{A}$ is the usual topology of the Schwartz space, which is known to be a Frechet–Montel space [Tr67].

It can be also derived from Corollary 3.27. Indeed, let c be the operator (3.6). Then $\overline{c}$ is a self-adjoint operator with discrete spectrum of eigenvalues tending to infinity. Hence $(I + \overline{c}^* \overline{c})^{-1}$ is compact, so the embedding of $\mathcal{H}_c$ into $\mathcal{H}$ is compact by Remark 3.28.1 and the assumptions of Corollary 3.27 are satisfied.   ○

3.4   The Mittag-Leffler Lemma

The following result is usually called the *Mittag-Leffler lemma*. It is a useful tool to prove that (under certain assumptions) intersections of countably many dense domains of operators are dense.

Proposition 3.30 *Suppose $(E_n, \| \cdot \|_n), n \in \mathbb{N}_0$, is a sequence of Banach spaces such that E_{n+1} is a dense subspace of $(E_n, \| \cdot \|_n)$ and the embedding of E_{n+1} into E_n is continuous for $n \in \mathbb{N}_0$. Then $E_\infty := \bigcap_{n=0}^{\infty} E_n$ is dense in each space $(E_k, \| \cdot \|_k)$ for $k \in \mathbb{N}_0$.*

Proof There is no loss of generality to assume that $k = 0$. Since the embedding of E_{n+1} in E_n is continuous, there is a constant $a_n > 0$ such that $\| \cdot \|_n \leq a_n \| \cdot \|_{n+1}$ on E_{n+1}. Upon rescaling the norms we can assume that $a_n = 1$.

Fix $x \in E_0$. Let $\varepsilon > 0$. Since E_{n+1} is dense in E_n, we can define inductively a sequence $(x_n)_{n \in \mathbb{N}_0}$ such that $x_0 = x$, $x_n \in E_n$, and $\| x_{n+1} - x_n \|_{n+1} \leq \varepsilon 2^{-(n+1)}$ for $n \in \mathbb{N}_0$. Then

$$\|x_{m+n+r} - x_{m+n}\|_m = \left\| \sum_{j=1}^{r} (x_{m+n+j} - x_{m+n+j-1}) \right\|_m$$

$$\leq \sum_{j=1}^{r} \|x_{m+n+j} - x_{m+n+j-1}\|_m \leq \sum_{j=1}^{r} \|x_{m+n+j} - x_{m+n+j-1}\|_{m+n+j}$$

$$\leq \sum_{j=1}^{r} \varepsilon \, 2^{-m-n-j} \leq \varepsilon \, 2^{-n} \tag{3.16}$$

for $m, n \in \mathbb{N}_0$ and $r \in \mathbb{N}$. From (3.16) it follows that for each $m \in \mathbb{N}$ the sequence $X_m := (x_{m+n})_{n \in \mathbb{N}}$ is a Cauchy sequence in the Banach space E_m. Let y denote the limit of X_1 in E_1. Since $\|y - x_{m+n}\|_1 \leq \|y - x_{m+n}\|_m$, y is also the limit of the sequence X_m in E_m for each $m \in \mathbb{N}_0$. Therefore, $y \in E_\infty$.

Setting $m = n = 0$ and letting $r \to \infty$ in (3.16) we get $\|y - x_0\| = \|y - x\| \leq \varepsilon$. This proves that E_∞ is dense in E_0. $\qquad\square$

The next proposition is an application of the Mittag-Leffler lemma.

First we recall the *Cayley transform* u of a densely defined closed symmetric operator a on a Hilbert space $\mathcal{H}$; see, e.g., [Sch12, Sect. 13.1]. The operator u is defined by $u(a + \mathrm{i}I)\varphi = (a - \mathrm{i}I)\varphi$, $\varphi \in \mathcal{D}(a)$. The operator u is a partial isometry with initial space $(a + \mathrm{i}I)\mathcal{D}(a) = \mathcal{H} \ominus \mathcal{H}_+$ and final space $(a - \mathrm{i}I)\mathcal{D}(a) = \mathcal{H} \ominus \mathcal{H}_-$, where $\mathcal{H}_\pm := \mathcal{N}(a^* \mp \mathrm{i}I)$. The cardinal numbers $\dim \mathcal{H}_\pm$ are the *deficiency indices* of a. Then we have $\mathcal{D}(a) = (I - u)(\mathcal{H} \ominus \mathcal{H}_+)$ and

$$(a + \mathrm{i})(I - u)\varphi = 2\mathrm{i}\,\varphi \quad \text{for} \quad \varphi \in \mathcal{H} \ominus \mathcal{H}_+. \tag{3.17}$$

Proposition 3.31 *Suppose a is a densely defined closed symmetric operator on $\mathcal{H}$ such that at least one of its deficiency indices is finite. Then $\mathcal{D}^\infty(a) := \bigcap_{n=1}^{\infty} \mathcal{D}(a^n)$ is a core for each power a^k, $k \in \mathbb{N}_0$. In particular, $\mathcal{D}^\infty(a)$ is dense in $\mathcal{H}$.*

Proof Upon replacing a by $-a$ we can assume that $\mathcal{H}_+$ is finite-dimensional. Let u be the Cayley transform of a and let q_n denote the projection of $\mathcal{H}$ on the finite-dimensional, hence closed, subspace $\mathcal{H}_n := \mathcal{H}_+ + \cdots + (u^*)^{n-1}\mathcal{H}_+$, $n \in \mathbb{N}$.

First we show by induction that for $n \in \mathbb{N}$ we have

$$\mathcal{H}_{n+1} = (I - u^*)\mathcal{H}_n + \mathcal{H}_+. \tag{3.18}$$

Since $\mathcal{H}_1 = \mathcal{H}_+$, this is clear for $n = 1$. Assume that (3.18) holds for n. Then, since obviously $\mathcal{H}_{k+1} = u^*\mathcal{H}_k + \mathcal{H}_+$ by definition, we derive

$$\mathcal{H}_{n+2} = u^*\mathcal{H}_{n+1} + \mathcal{H}_+ = u^*[(1 - u^*)\mathcal{H}_n + \mathcal{H}_+] + \mathcal{H}_+$$
$$= (1 - u^*)[u^*\mathcal{H}_n + \mathcal{H}_+] + \mathcal{H}_+ = (1 - u^*)\mathcal{H}_{n+1} + \mathcal{H}_+,$$

which proves (3.18) for $n + 1$.

Our next aim is to prove, again by induction on n, that

$$\mathcal{D}(a^n) = (I - u)^n (I - q_n)\mathcal{H} \quad \text{for} \quad n \in \mathbb{N}. \tag{3.19}$$

Since $\mathcal{D}(a) = (I - u)(\mathcal{H} \ominus \mathcal{H}_+) = (I - u)(I - q_1)\mathcal{H}$, (3.19) holds for $n = 1$. Assume now that (3.19) is satisfied for n.

Let $\varphi \in \mathcal{D}(a^{n+1})$. Then $\varphi \in \mathcal{D}(a^n)$, so $\varphi = (I - u)^n \psi$ for some $\psi \in (I - q_n)\mathcal{H}$ by the induction hypothesis. Further, $(a + iI)^n \varphi = (2i)^n \psi \in \mathcal{D}(a)$ by (3.17) implies that $\psi = (I - u)\zeta$ with $\zeta \in (I - q_1)\mathcal{H}$. Since $\zeta \perp \mathcal{H}_+$ and $\psi \perp q_n\mathcal{H}$, that is, $\psi = (I - u)\zeta \perp \mathcal{H}_+, u^*\mathcal{H}_+, \dots, (u^*)^{n-1}\mathcal{H}_+$, it follows $\zeta \perp \mathcal{H}_+, \dots, (u^*)^n\mathcal{H}_+$, that is, $\zeta \perp q_{n+1}\mathcal{H}$. Thus, $\varphi = (I - u)^n \psi = (I - u)^{n+1}\zeta \in (I - u)^{n+1}(I - q_{n+1})\mathcal{H}$. This proves that $\mathcal{D}(a^{n+1}) \subseteq (I - u)^{n+1}(I - q_{n+1})\mathcal{H}$. The converse inclusion is easily verified. This completes the proof of (3.19).

We want to apply Proposition 3.30 with the normed space $E_n := (\mathcal{D}(a^n), \|\cdot\|_n)$, where $\|\cdot\|_n := \|(a + iI)^n \cdot \|$. Then $\mathcal{D}(a^n) = (I - u)^n (I - q_n)\mathcal{H}$ by (3.19) and

$$\|(I - u)^n (I - q_n)\psi\|_n = \|(a + iI)^n (I - u)^n (I - q_n)\psi\| = \|(2i)^n (I - q_n)\psi\|$$

by (3.17) for $\psi \in \mathcal{H}$. This implies that E_n is complete and hence a Banach space. Since the operator a is symmetric, we have $\|(a + i)^n \cdot \| \leq \|(a + i)^{n+1} \cdot \|$, so that $\|\cdot\|_n \leq \|\cdot\|_{n+1}$ on E_{n+1}. Hence the embedding of E_{n+1} into E_n is continuous.

Next we prove that for each $n \in \mathbb{N}_0$ the subspace E_{n+1} is dense in the Banach space E_n. Note that E_n is a Hilbert space with respect to the inner product given by $\langle \cdot, \cdot \rangle_n := \langle (a + iI)^n \cdot, (a + iI)^n \cdot \rangle$. Therefore, it is sufficient to prove that the orthogonal complement of E_{n+1} in $(E_n, \langle \cdot, \cdot \rangle_n)$ is $\{0\}$. Suppose that $\varphi \in E_n$ is orthogonal to all vectors $\eta \in E_{n+1}$ in $(E_n, \langle \cdot, \cdot \rangle_n)$. By (3.19), we can write $\varphi = (I - u)^n \psi$ and $\eta = (I - u)^{n+1}(I - q_{n+1})\zeta$ with $\psi \in (I - q_n)\mathcal{H}$, $\zeta \in \mathcal{H}$. Then, using again (3.17), we deduce

$$0 = \langle \varphi, \eta \rangle_n = \langle (a + iI)^n \varphi, (a + iI)^n \eta \rangle = 4^n \langle \psi, (I - u)(I - q_{n+1})\zeta \rangle$$
$$= 4^n \langle (I - u^*)\psi, (I - q_{n+1})\zeta \rangle$$

for all $\zeta \in \mathcal{H}$. Hence $(I - u^*)\psi \in q_{n+1}\mathcal{H} = \mathcal{H}_{n+1}$ and from (3.18) it follows that $(I - u^*)\psi = (I - u^*)\xi + \xi_+$ with $\xi \in \mathcal{H}_n = q_n\mathcal{H}$ and $\xi_+ \in \mathcal{H}_+$. Thus,

$$\langle \psi - \xi, (I - u)(I - q_1)\delta \rangle = \langle (I - u^*)(\psi - \xi), (I - q_1)\delta \rangle = \langle \xi_+, (I - q_1)\delta \rangle = 0$$

for all $\delta \in \mathcal{H}$. Therefore, $\psi - \xi \perp (I - u)(I - q_1)\mathcal{H} = \mathcal{D}(a)$. Since $\mathcal{D}(a)$ is assumed to be dense in $\mathcal{H}$, we get $\psi = \xi$. By construction, $\psi \in (I - q_n)\mathcal{H}$ and $\xi \in q_n\mathcal{H}$. Therefore, $\psi = 0$, hence $\varphi = 0$, which proves that E_{n+1} is dense in E_n.

By the preceding, the sequence $(E_n)_{n \in \mathbb{N}_0}$ of Banach spaces satisfies the assumptions of Proposition 3.30. Therefore, $\mathcal{D}^\infty(a) = \cap_n E_n$ is dense in each Banach space $E_k = (\mathcal{D}(a^k), \|(a + iI)^k \cdot \|)$, $k \in \mathbb{N}_0$. This implies that $\mathcal{D}^\infty(a)$ is a core for each operator a^k. In the special case $k = 0$ this means that $\mathcal{D}^\infty(a)$ is dense in $\mathcal{H}$. $\qquad\square$

3.5 Exercises

1. Suppose there exists an element $a \in \mathcal{L}^+(\mathcal{D})$ which is a closed operator on the underlying Hilbert space $\mathcal{H}$. Prove that $\mathcal{D} = \mathcal{H}$ and $\mathcal{L}^+(\mathcal{D}) = \mathbf{B}(\mathcal{H})$.

2. Let $\mathcal{J} \subseteq \mathbb{R}$ be an interval and let $(h_n)_{n \in \mathbb{N}}$ be a sequence of (finite) real Borel functions on $\mathcal{J}$ such that $h_1(x) = 1$ and $h_n(x)^2 \leq h_{n+1}(x)$ for $x \in \mathcal{J}$, $n \in \mathbb{N}$. Define $\mathcal{D} = \{\varphi \in L^2(\mathcal{J}) : h_n \cdot \varphi \in L^2(\mathcal{J}), n \in \mathbb{N}\}$ and let h_n act on $\mathcal{D}$ as a multiplication operator in the Hilbert space $L^2(\mathcal{J})$.

   a. Show that each operator h_n belongs to $\mathcal{L}^+(\mathcal{D})$.
   b. Show that the O^*-algebra $\mathcal{A}$ on $\mathcal{D}$ generated by h_n, $n \in \mathbb{N}$, is self-adjoint.
   c. Determine the commutants $\mathcal{A}'_s$, $\mathcal{A}'_w$, $\mathcal{A}'_{sym}$.

3. Let $t = \varphi \otimes \psi$, where $\varphi, \psi \in \mathcal{H}$. When is $t \in \mathbf{B}(\mathcal{A})_+$? What is f_t in this case?

4. Let $t = \sum_{k=1}^n \varphi_k \otimes \psi_k$, where $\varphi_k, \psi_k \in \mathcal{H}$ for $k = 1, \ldots, n$, $n \in \mathbb{N}$. When is t in $\mathbf{B}_1(\mathcal{A})_+$? What is f_t in this case?

5. Let $\mathcal{A}$ denote the O^*-algebra from Example 3.29 for $d = 1$. Find $t_1, t_2 \in \mathbf{B}_1(\mathcal{A})_+$ such that $t_1 \neq t_2$ and $f_{t_1}(a) = f_{t_2}(a)$ for all $a \in \mathcal{A}$.
   Hint: Look for rank one and rank two operators.

3.6 Notes

This chapter covers only a very small portion of the theory of O^*-algebras. The monograph [Sch90] gives an extensive treatment with many references and historical comments. A nice presentation is given in A. Inoue's Lecture Notes [In98]. For the theory of partial $*$-algebras of unbounded operators we refer to [AIT02]. Proposition 3.31 was obtained in [Sch83a].

Chapter 4
*-Representations

In this chapter we begin the study of *-representations which is the main topic of this book. In Sect. 4.1, we define *-representations and introduce basic notions such as adjoint, closed, biclosed, and self-adjoint representations. In Sect. 4.2, we express the domain of the adjoint representation in terms of generators and construct a natural maximal representation (Theorem 4.18). Section 4.3 deals with invariant subspaces and reducing subspaces. It is shown that in this respect pathological behavior can be avoided by adding self-adjointness assumptions (Theorem 4.29).

In Sects. 4.4 and 4.5, we develop the most important technical tool in Hilbert space representation theory, the GNS construction. It associates a *-representation with any positive linear functional (Theorems 4.38 and 4.41). The *-radical is defined and characterized in terms of positive functionals. In Sect. 4.6, we use the GNS representation to represent positive semi-definite functions on groups in terms of unitary group representations (Theorem 4.56). The GNS construction is crucial for a deeper analysis of positive functionals which is carried out in the next chapter.

In Sect. 4.7, we briefly discuss some technical difficulties and pathologies that can occur for unbounded *-representations, compared to the bounded case.

Throughout this chapter, A denotes a (not necessarily unital) **complex** *-algebra and all inner product spaces and Hilbert spaces are **complex** (except for Remark 4.13 in Sect. 4.1). Further, we shall use some definitions and facts on O^*-algebras from Chap. 3 and on unbounded operators from Appendix A.

4.1 Basic Concepts on *-Representations

Let $(\mathcal{D}, \langle \cdot, \cdot \rangle)$ be a complex inner product space and let $(\mathcal{H}, \langle \cdot, \cdot \rangle)$ be its Hilbert space completion. We denote by $L(\mathcal{D})$ the algebra of linear operators $a : \mathcal{D} \mapsto \mathcal{D}$.

© The Editor(s) (if applicable) and The Author(s), under exclusive license
to Springer Nature Switzerland AG 2020
K. Schmüdgen, *An Invitation to Unbounded Representations of *-Algebras
on Hilbert Space*, Graduate Texts in Mathematics 285,
https://doi.org/10.1007/978-3-030-46366-3_4

Definition 4.1 A *representation* of A on $\mathcal{D}$ is an algebra homomorphism π of A into $L(\mathcal{D})$ such that $\pi(a)$ is a closable operator on $\mathcal{H}$ for each $a \in$ A. The inner product space $\mathcal{D}$ is called the *domain* of π and denoted by $\mathcal{D}(\pi)$, and we write $\mathcal{H}(\pi) := \mathcal{H}$.

If the domain $\mathcal{D}(\pi)$ is equal to the Hilbert space $\mathcal{H}(\pi)$, then each operator $\pi(a)$ is closed on $\mathcal{H}(\pi)$ and hence bounded by the closed graph theorem, that is, π is a representation of A by *bounded operators* on $\mathcal{H}(\pi)$.

We introduce three standard notions. Let π_1 and π_2 be representations of A. We say that π_1 is a *restriction* of π_2, or equivalently, π_2 is an *extension* of π_1, and write $\pi_1 \subseteq \pi_2$ if $\mathcal{D}(\pi_1) \subseteq \mathcal{D}(\pi_2)$ and $\pi_1(a) = \pi_2(a) \lceil \mathcal{D}(\pi_1)$ for all $a \in$ A.

Next, π_1 and π_2 are called *unitarily equivalent* if there exists a unitary operator U of $\mathcal{H}(\pi_2)$ on $\mathcal{H}(\pi_1)$ such that $U\mathcal{D}(\pi_2) = \mathcal{D}(\pi_1)$ and $\pi_1(a)\varphi = U\pi_2(a)U^{-1}\varphi$ for $a \in$ A and $\varphi \in \mathcal{D}(\pi_1)$; then we write $\pi_1 = U\pi_2 U^{-1}$.

Finally, we define the direct sum of representations. Suppose $\{\pi_i : i \in I\}$ is a family of representations π_i of A. Let $\mathcal{D}(\pi)$ denote the set of vectors $\varphi = (\varphi_i)_{i \in I}$ of the direct sum Hilbert space $\mathcal{H}(\pi) := \oplus_{i \in I} \mathcal{H}(\pi_i)$ for which $\varphi_i \in \mathcal{D}(\pi_i)$ for $i \in I$ and $\pi(a)\varphi := (\pi_i(a)\varphi_i)$ is in $\mathcal{H}(\pi)$ for all $a \in$ A. It is easily checked that π is a representation of A on the dense linear subspace $\mathcal{D}(\pi)$ of the Hilbert space $\mathcal{H}(\pi)$. We write $\pi := \oplus_{i \in I} \pi_i$ and call π the *direct sum* of the family $\{\pi_i : i \in I\}$.

In this book representations appear only as auxiliary objects to study adjoints of ∗-representations. Our main objects are *∗-representations* of ∗-algebras:

Definition 4.2 A *∗-representation* of A on $\mathcal{D}$ is an algebra homomorphism π of A into $L(\mathcal{D})$ such that for all $a \in$ A,

$$\langle \pi(a)\varphi, \psi \rangle = \langle \varphi, \pi(a^+)\psi \rangle, \quad \varphi, \psi \in \mathcal{D}. \tag{4.1}$$

By definition, the domain $\mathcal{D}(\pi(a)^*)$ of the adjoint operator $\pi(a)^*$ consists of all vectors $\psi \in \mathcal{H}$ for which there exists an $\eta \in \mathcal{H}$ such that $\langle \pi(a)\varphi, \psi \rangle = \langle \varphi, \eta \rangle$ for all $\varphi \in \mathcal{D}$; in this case $\eta = \pi(a)^*\psi$. Hence Eq. (4.1) says that $\mathcal{D} \subseteq \mathcal{D}(\pi(a)^*)$ and $\pi(a)^*\psi = \pi(a^+)\psi$ for $\psi \in \mathcal{D}$. That is, by (4.1), the operator $\pi(a^+)$ is the restriction to $\mathcal{D}$ of the adjoint operator $\pi(a)^*$ of $\pi(a)$ on the Hilbert space $\mathcal{H}$. If $a = a^+$, then (4.1) means that $\pi(a)$ is a symmetric operator on $\mathcal{D}$.

In general, the domain $\mathcal{D}(\pi(a)^*)$ of the adjoint operator is much larger than $\mathcal{D}$ and the adjoint of the symmetric operator $\pi(a)$ for $a = a^+$ is no longer symmetric. Since the distinction between $\pi(a^+)$ and $\pi(a)^*$ is crucial in unbounded operator theory, we denote throughout this book the algebra involution always by $a \mapsto a^+$ and the adjoint of an operator T by T^*.

As noted above, $\mathcal{D} \subseteq \mathcal{D}(\pi(a)^*)$ by (4.1). Hence the adjoint of $\pi(a)$ is densely defined, so the operator $\pi(a)$ is closable. Therefore, *each ∗-representation is a representation*.

The ∗-algebra $\mathcal{L}^+(\mathcal{D})$ was introduced in Definition 3.1 and Lemma 3.2. If π is an algebra homomorphism of A into $L(\mathcal{D})$, then condition (4.1) holds for $a \in$ A if and only if

$$\pi(a) \in \mathcal{L}^+(\mathcal{D}) \quad \text{and} \quad \pi(a^+) = \pi(a)^+ \quad \text{for} \quad a \in \text{A}. \tag{4.2}$$

Therefore, by a slight reformulation of Definition 4.2, a *∗-representation of* A *on* $\mathcal{D}$ *is a ∗-homomorphism of the ∗-algebra* A *in the ∗-algebra* $\mathcal{L}^+(\mathcal{D})$.

The counterpart of ∗-representations in single operator theory is densely defined symmetric operators. For such operators, closures, adjoints, and notions such as closedness, self-adjointness, and essential self-adjointness are useful. Self-adjoint and essentially self-adjoint representations will be important in this book. Before we define similar concepts for ∗-representations, we prove three technical lemmas.

Let π be a representation of A. First we define the adjoint representation π^* :

$$\mathcal{D}(\pi^*) := \bigcap_{a \in A} \mathcal{D}(\pi(a)^*) \quad \text{and} \quad \pi^*(a) := \pi(a^+)^* {\restriction} \mathcal{D}(\pi^*), \quad a \in A. \qquad (4.3)$$

Let $\mathcal{H}(\pi^*)$ denote the closure of $\mathcal{D}(\pi^*)$ in $\mathcal{H}(\pi)$. If π is a ∗-representation, then $\mathcal{D}(\pi) \subseteq \mathcal{D}(\pi^*)$ and hence $\mathcal{H}(\pi) = \mathcal{H}(\pi^*)$. Since $\pi^*(a^+) \subseteq \pi(a)^*$ by definition,

$$\langle \pi(a)\psi, \varphi \rangle = \langle \psi, \pi^*(a^+)\varphi \rangle \quad \text{for} \quad a \in A, \ \psi \in \mathcal{D}(\pi), \ \varphi \in \mathcal{D}(\pi^*). \qquad (4.4)$$

Lemma 4.3 *Let π be a representation of* A. *Then:*

(i) π^* *is a representation of* A *on* $\mathcal{D}(\pi^*)$, *called the* adjoint representation *to* π.
(ii) π *is a ∗-representation of* A *if and only if* $\pi \subseteq \pi^*$.
(iii) *Suppose* $\mathcal{H}(\pi^*) = \mathcal{H}(\pi)$. *Then we have* $\pi \subseteq \pi^{**}$, $\pi^* = \pi^{***}$, $\pi^{**} = \pi^{****}$, *where* $\pi^{**} := (\pi^*)^*$, $\pi^{***} := (\pi^{**})^*$, $\pi^{****} := (\pi^{***})^*$.

Proof (i): Suppose $a, b \in A$ and $\alpha, \beta \in \mathbb{C}$. Let $\varphi \in \mathcal{D}(\pi^*)$ and $\psi \in \mathcal{D}(\pi)$. Using (4.3) and (4.4) we derive

$$\langle \pi^*(\alpha a + \beta b)\varphi, \psi \rangle = \langle \pi(\overline{\alpha} a^+ + \overline{\beta} b^+)^*\varphi, \psi \rangle$$
$$= \langle \varphi, \pi(\overline{\alpha} a^+ + \overline{\beta} b^+)\psi \rangle = \alpha\langle \varphi, \pi(a^+)\psi \rangle + \beta\langle \varphi, \pi(b^+)\psi \rangle$$
$$= \langle (\alpha\pi(a^+)^* + \beta\pi(b^+)^*)\varphi, \psi \rangle = \langle (\alpha\pi^*(a) + \beta\pi^*(b))\varphi, \psi \rangle.$$

Since $\mathcal{D}(\pi)$ is dense in $\mathcal{H}(\pi)$, this implies $\pi^*(\alpha a + \beta b) = \alpha\pi^*(a) + \beta\pi^*(b)$. Further, again by (4.3) and (4.4), we obtain

$$\langle \pi(a^+)\psi, \pi^*(b)\varphi \rangle = \langle \pi(a^+)\psi, \pi(b^+)^*\varphi \rangle = \langle \pi(b^+)\pi(a^+)\psi, \varphi \rangle$$
$$= \langle \pi((ab)^+)\psi, \varphi \rangle = \langle \psi, \pi((ab)^+)^*\varphi \rangle = \langle \psi, \pi^*(ab)\varphi \rangle$$

and conclude that $\pi^*(b)\varphi \in \mathcal{D}(\pi(a^+)^*)$ and $\pi(a^+)^*\pi^*(b)\varphi = \pi^*(ab)\varphi$. Since $a \in A$ was arbitrary, $\pi^*(b)\varphi \in \mathcal{D}(\pi^*)$ and hence $\pi^*(a)\pi^*(b)\varphi = \pi^*(ab)\varphi$ by (4.3). Since $\pi^*(a) \subseteq \pi(a^+)^*$ by definition and the operator $\pi(a^+)^*$ on $\mathcal{H}(\pi)$ is closed, $\pi^*(a)$ is closable on $\mathcal{H}(\pi^*)$. This proves that π^* is a representation of A on $\mathcal{D}(\pi^*)$.

(ii): Comparing (4.1) with (4.3) resp. (4.4) it follows that π is a ∗-representation if and only if $\mathcal{D}(\pi) \subseteq \mathcal{D}(\pi^*)$ and $\pi(a) \subseteq \pi(a^+)^*$ for $a \in A$, that is, $\pi \subseteq \pi^*$.

(iii): Since $\mathcal{H}(\pi^*) = \mathcal{H}(\pi)$, we conclude from (4.4) that $\mathcal{D}(\pi) \subseteq \mathcal{D}(\pi^*(a^+)^*)$ and
$\pi(a) \subseteq \pi^*(a^+)^*$ for all $a \in \mathsf{A}$, so that $\mathcal{D}(\pi) \subseteq \mathcal{D}((\pi^*)^*) = \mathcal{D}(\pi^{**})$ and $\pi(a) =$
$\pi^*(a^+)^* \lceil \mathcal{D}(\pi) = \pi^{**}(a) \lceil \mathcal{D}(\pi)$. This shows that $\pi \subseteq \pi^{**}$.

By (i), π^{**} is a representation and so is π^{***}. We prove that $\pi^* = \pi^{***}$. Since
$\pi \subseteq \pi^{**}$ and hence $\mathcal{H}(\pi) = \mathcal{H}(\pi^{**})$, it follows at once from the definition
of the adjoint representation that $\pi^{***} \subseteq \pi^*$. (This follows also from Lemma
4.4(ii) below, applied with $\rho = \pi^{**}$.) Further, since $\mathcal{H}(\pi^*) = \mathcal{H}(\pi) = \mathcal{H}(\pi^{**})$
by assumption, we can replace π by π^* in the inclusion $\pi \subseteq \pi^{**}$ and get
$\pi^* \subseteq \pi^{***}$. Thus, $\pi^* = \pi^{***}$.

Finally, replacing π by π^* in the equality $\pi^* = \pi^{***}$ we obtain $\pi^{**} = \pi^{****}$. $\square$

Lemma 4.4 *Let π and ρ be representations of A such that $\pi \subseteq \rho$. Let $P_{\mathcal{H}(\pi)}$ denote
the projection of the Hilbert space $\mathcal{H}(\rho)$ on $\mathcal{H}(\pi)$. Then:*

(i) $P_{\mathcal{H}(\pi)}\rho^*(a) \subseteq \pi^*(a)P_{\mathcal{H}(\pi)}$ *for $a \in \mathsf{A}$.*
(ii) *If $\mathcal{H}(\pi) = \mathcal{H}(\rho)$, then $\rho^* \subseteq \pi^*$.*
(iii) *If $\mathcal{H}(\rho^*) = \mathcal{H}(\rho)$, then $\pi^{**} \subseteq \rho^{**}$.*

Proof (i): We abbreviate $P := P_{\mathcal{H}(\pi)}$ and fix $\psi \in \mathcal{D}(\rho^*)$. Let $\varphi \in \mathcal{D}(\pi)$ and $a \in \mathsf{A}$.
Using the assumption $\pi \subseteq \rho$ we obtain

$$\langle \pi(a^+)\varphi, P\psi \rangle = \langle P\pi(a^+)\varphi, \psi \rangle = \langle \pi(a^+)\varphi, \psi \rangle = \langle \rho(a^+)\varphi, \psi \rangle$$
$$= \langle \varphi, \rho(a^+)^*\psi \rangle = \langle \varphi, \rho^*(a)\psi \rangle = \langle P\varphi, \rho^*(a)\psi \rangle = \langle \varphi, P\rho^*(a)\psi \rangle.$$

From this equality it follows that $P\psi \in \mathcal{D}(\pi(a^+)^*)$ and $\pi(a^+)^* P\psi = P\rho^*(a)\psi$.
Hence $P\psi \in \cap_{b \in \mathsf{A}} \mathcal{D}(\pi(b)^*) = \mathcal{D}(\pi^*)$ and $\pi^*(a)P\psi = P\rho^*(a)\psi$ for $\psi \in \mathcal{D}(\rho^*)$.
This proves that $P\rho^*(a) \subseteq \pi^*(a)P$.

(ii) follows at once from (i), since $P = I$ by the assumption $\mathcal{H}(\pi) = \mathcal{H}(\rho)$.

(iii): Let $\xi \in \mathcal{D}(\pi^{**})$ and $\psi \in \mathcal{D}(\rho^*)$. Since $\mathcal{H}(\pi^{**}) \subseteq \mathcal{H}(\pi^*) \subseteq \mathcal{H}(\pi)$ by definition, we have $P\xi = \xi$ and $P\pi^{**}(a)\xi = \pi^{**}(a)\xi$. From (i) we get $P\psi \in \mathcal{D}(\pi^*)$
and $P\rho^*(a^+)\psi = \pi^*(a^+)P\psi$ for $a \in \mathsf{A}$. Using these facts we derive

$$\langle \rho^*(a^+)\psi, \xi \rangle = \langle \rho^*(a^+)\psi, P\xi \rangle = \langle P\rho^*(a^+)\psi, \xi \rangle = \langle \pi^*(a^+)P\psi, \xi \rangle \quad (4.5)$$
$$= \langle \psi, P\pi^*(a^+)^*\xi \rangle = \langle \psi, P\pi^{**}(a)\xi \rangle = \langle \psi, \pi^{**}(a)\xi \rangle. \quad (4.6)$$

Since $\mathcal{H}(\rho) = \mathcal{H}(\rho^*)$ by assumption and $\mathcal{H}(\pi^{**}) \subseteq \mathcal{H}(\pi) \subseteq \mathcal{H}(\rho)$ by $\pi \subseteq \rho$,
both vectors ξ and $\pi^{**}(a)\xi$ belong to the Hilbert space $\mathcal{H}(\rho^*)$. Hence we con-
clude from (4.5)–(4.6) that $\xi \in \mathcal{D}(\rho^*(a^+)^*)$ and $\rho^*(a^+)^*\xi = \pi^{**}(a)\xi$ for all
$a \in \mathsf{A}$. These relations imply that $\xi \in \cap_{b \in \mathsf{A}} \mathcal{D}(\rho^*(b)^*) = \mathcal{D}(\rho^{**})$ and $\rho^{**}(a)\xi =$
$\rho^*(a^+)^*\xi = \pi^{**}(a)\xi$. This proves that $\pi^{**} \subseteq \rho^{**}$. $\square$

For a *-representation π of A, we define

$$\mathcal{D}(\overline{\pi}) := \bigcap_{a \in \mathsf{A}} \mathcal{D}(\overline{\pi(a)}) \quad \text{and} \quad \overline{\pi}(a) := \overline{\pi(a)} \lceil \mathcal{D}(\overline{\pi}), \quad a \in \mathsf{A}, \quad (4.7)$$

$$\mathcal{D}(\pi^{**}) := \bigcap_{a \in \mathsf{A}} \mathcal{D}(\pi^*(a)^*) \quad \text{and} \quad \pi^{**}(a) := \pi^*(a^+)^* \lceil \mathcal{D}(\pi^{**}), \quad a \in \mathsf{A}. \quad (4.8)$$

Lemma 4.5 *Let* π *be a* ∗*-representation of* A. *Then* $\overline{\pi}$ *and* π^{**} *are also* ∗*-representations of* A, *called the* closure *and* biclosure *of* π, *respectively, and*

$$\pi \subseteq \overline{\pi} \subseteq \pi^{**} \subseteq \pi^*, \tag{4.9}$$

$$\mathcal{D}(\pi^{**}) = \bigcap_{a \in A} \mathcal{D}(\overline{\pi^{**}(a)}). \tag{4.10}$$

Proof First we show that $\overline{\pi}$ is a ∗-representation. Comparing equations (4.7) and (3.2) it follows that $\mathcal{D}(\overline{\pi})$ is the domain $\mathcal{D}(\overline{\pi(A)})$ of the closure $\overline{\pi(A)}$ of the O^*-algebra $\pi(A)$ and $j(\pi(a)) = \overline{\pi}(a)$ for $a \in A$. Then $\overline{\pi}$ is the composition of the ∗-homomorphism $\pi : A \mapsto \pi(A)$ and the ∗-isomorphism $j : \pi(A) \mapsto \overline{\pi(A)}$ (by Proposition 3.8), so $\overline{\pi}$ is a ∗-homomorphism and hence a ∗-representation of A.

We verify (4.10). Let $\mathcal{D}_0$ denote the set on the right-hand side of (4.10). Trivially, $\mathcal{D}(\pi^{**}) \subseteq \mathcal{D}_0$. By definition, $\pi^{**}(a) \subseteq \pi^*(a^+)^*$ and hence $\overline{\pi^{**}(a)} \subseteq \pi^*(a^+)^*$ for $a \in A$. Taking the intersections of the domains over $a \in A$ yields $\mathcal{D}_0 \subseteq \mathcal{D}(\pi^{**})$. Thus, $\mathcal{D}(\pi^{**}) = \mathcal{D}_0$, which proves (4.10).

Since π is a ∗-representation, it follows from Lemma 4.3(ii) that $\pi \subseteq \pi^*$. Applying Lemma 4.4(iii) to $\pi \subset \pi^*$ gives $\pi^{**} \subseteq \pi^{***}$. Therefore, π^{**} is a ∗-representation again by Lemma 4.3(ii).

Further, since $\pi^* = \pi^{***}$ by Lemma 4.3(iii) and $\pi^{**} \subseteq \pi^{***}$, we have $\pi^{**} \subseteq \pi^*$, which is the third inclusion of (4.9). The first inclusion $\pi \subseteq \overline{\pi}$ is obvious.

From the relation $\pi \subseteq \pi^{**}$ (by Lemma 4.3(iii)) it follows that $\overline{\pi}(a) \subseteq \pi^{**}(a)$ for $a \in A$. Hence, by (4.7) and (4.10), $\mathcal{D}(\overline{\pi}) \subseteq \mathcal{D}(\pi^{**})$ and $\overline{\pi} \subseteq \pi^{**}$, which proves the middle inclusion of (4.9). □

The following definition collects a number of basic concepts of ∗-representations.

Definition 4.6 A ∗-representation π of A is called
- *closed* if $\pi = \overline{\pi}$, or equivalently, if $\mathcal{D}(\pi) = \mathcal{D}(\overline{\pi})$,
- *biclosed* if $\pi = \pi^{**}$, or equivalently, if $\mathcal{D}(\pi) = \mathcal{D}(\pi^{**})$,
- *self-adjoint* if $\pi = \pi^*$, or equivalently, if $\mathcal{D}(\pi) = \mathcal{D}(\pi^*)$,
- *essentially self-adjoint* if π^* is self-adjoint, that is, if $\pi^* = \pi^{**}$, or equivalently, if $\mathcal{D}(\pi^{**}) = \mathcal{D}(\pi^*)$.

It should be noted that this definition of *essential self-adjointness* differs from the one given in [Sch90], where a ∗-representation π was called essentially self-adjoint if $\overline{\pi}$ is self-adjoint, or equivalently, if $\overline{\pi} = \pi^*$.

The adjoint π^*, the closure $\overline{\pi}$, and the biclosure π^{**} of a ∗-representation π and their domains been defined by formulas (4.3), (4.7), and (4.8), respectively. By comparing the corresponding domains we obtain necessary and sufficient criteria when the conditions in Definition 4.6 are fulfilled. Note that the adjoint π^* of a ∗-representation π is not necessarily a ∗-representation just as the adjoint of a symmetric operator is not symmetric in general.

Though these notions are in close analogy with the corresponding concepts for single symmetric operators, there is at least one important difference: It may happen that $\overline{\pi} \neq \pi^{**}$, that is, $\overline{\pi}$ is closed, but not biclosed; see Example 4.20 below.

Let π be a *-representation. The *-representations $\overline{\pi}$ and π^{**} are closed (by (4.7) and (4.10)) and π^{**} is biclosed (by Lemma 4.3(iii)). It is easily seen that $(\overline{\pi})^* = \pi^*$. The following two definitions introduce further useful notions.

Definition 4.7 The *graph topology* of π is the locally convex topology t_π on the domain $\mathcal{D}(\pi)$ defined by the family of norms $\{ \| \cdot \| + \| \pi(a) \cdot \| : a \in \mathsf{A} \}$.

Thus, a *-representation π is a *-homomorphism of A on the O^*-algebra $\pi(\mathsf{A})$, and the graph topology t_π coincides with the graph topology $t_{\pi(\mathsf{A})}$ of the O^*-algebra $\pi(\mathsf{A})$ from Definition 3.4. By comparing the corresponding definitions we see that a *-representation π of A is closed resp. self-adjoint if and only if the O^*-algebra $\pi(\mathsf{A})$ is closed resp. self-adjoint according to Definition 3.7 resp. 3.12. Further, since $t_\pi = t_{\pi(\mathsf{A})}$, it follows from Proposition 3.9 that a *-representation π is closed if and only if the locally convex space $\mathcal{D}(\pi)[t_\pi]$ is complete.

If all operators $\pi(a)$, $a \in \mathsf{A}$, are bounded (in particular, if $\mathcal{D}(\pi) = \mathcal{H}(\pi)$), then the graph topology t_π is just the topology given by the Hilbert space norm.

Definition 4.8 A *-representation π of A is called
• *faithful* if $\pi(a) = 0$ for $a \in \mathsf{A}$ implies $a = 0$,
• *nondegenerate* if $\pi(\mathsf{A})\mathcal{D}(\pi):=\mathrm{Lin}\,\{\pi(a)\varphi : a \in \mathsf{A}, \varphi \in \mathcal{D}(\pi)\}$ is dense in $\mathcal{H}(\pi)$.

The nondegeneracy excludes trivial cases such as $\pi(a) = 0$ for all $a \in \mathsf{A}$. In the unital case, it is equivalent to $\pi(1) = I$, as shown in Lemma 4.9(iii).

In [Sch90], *-algebras are always unital and the requirement $\pi(1) = I$ is part of the definition of a *-representation.

*Throughout this book, all *-representations π of unital *-algebras are nondegenerate and satisfy $\pi(1) = I$.*

Lemma 4.9 *Suppose π is a *-representation of a unital *-algebra A. For $a \in \mathsf{A}$ and $\psi \in \mathcal{D}(\pi_0) := \pi(\mathsf{A})\mathcal{D}(\pi)$, we define $\pi_0(a)\psi = \pi(a)\psi$. Then:*

(i) *π_0 is a nondegenerate *-representation of A.*
(ii) *For $a \in \mathsf{A}$, we have $\pi(a) = 0$ if and only if $\pi_0(a) = 0$.*
(iii) *π is nongenerate if and only if $\pi(1) = I_{\mathcal{D}(\pi)}$. In this case, $\pi_0 = \pi$.*

Proof (i): Since π is a *-representation, $\mathcal{D}(\pi_0)$ is invariant under $\pi_0(a)$ and π_0 is again a *-representation. Since $\pi_0(1)\pi(a)\varphi = \pi(1)\pi(a)\varphi = \pi(1 \cdot a)\varphi = \pi(a)\varphi$ for $\varphi \in \mathcal{D}(\pi)$, we have $\pi_0(\mathsf{A})\mathcal{D}(\pi_0) = \mathcal{D}(\pi_0)$, so π_0 is nondegenerate.

(ii): Since $\pi_0 \subseteq \pi$, $\pi(a) = 0$ implies $\pi_0(a) = 0$. Conversely, suppose $\pi(a) \neq 0$. Then $\pi(a)\varphi \neq 0$ for some $\varphi \in \mathcal{D}(\pi)$, so $\pi_0(a)\pi(1)\varphi = \pi(a)\pi(1)\varphi = \pi(a)\varphi \neq 0$.

(iii): If $\pi(1) = I$, then trivially $\pi(\mathsf{A})\mathcal{D}(\pi) = \mathcal{D}(\pi)$ and $\pi_0 = \pi$ is nondegenerate. Conversely, suppose π is nondegenerate. Clearly, $\pi(1)\pi(a)\varphi = \pi(a)\varphi$ for $a \in \mathsf{A}$, $\varphi \in \mathcal{D}(\pi)$, so $\pi(1)\psi = \psi$ for $\psi \in \pi(\mathsf{A})\mathcal{D}(\pi)$. Since $\pi(1)$ is closable and $\pi(\mathsf{A})\mathcal{D}(\pi)$ is dense in $\mathcal{H}(\pi)$ (because π is nondegenerate), we conclude that $\pi(1)\psi = \psi$ for $\psi \in \mathcal{D}(\pi)$, that is, $\pi(1) = I$. $\square$

In the next example we develop an important general construction. It will be used in Example 4.48 to describe the GNS representations of trace functionals.

Example 4.10 (∗-*Representations on Hilbert–Schmidt operators*)
First we recall some facts on Hilbert–Schmidt operators (see [BS87, Chap. 11, Sect. 3]) on a Hilbert space $\mathcal{H}$. An operator $x \in \mathbf{B}(\mathcal{H})$ is called *Hilbert–Schmidt* if for one (and then for any) orthonormal basis $\{\varphi_i : i \in J\}$ of the Hilbert space $\mathcal{H}$,

$$\sum_i \|x\varphi_i\|^2 < \infty. \tag{4.11}$$

The set of Hilbert–Schmidt operators on $\mathcal{H}$ is denoted by $\mathbf{B}_2(\mathcal{H})$. It is a two-sided ∗-ideal of $\mathbf{B}(\mathcal{H})$. Each Hilbert–Schmidt operator is compact and the product of two Hilbert–Schmidt operators is trace class. The complex vector space $\mathbf{B}_2(\mathcal{H})$ is a Hilbert space with inner product given by

$$\langle x, y \rangle_{\mathrm{HS}} = \operatorname{Tr} y^* x, \quad x, y \in \mathbf{B}_2(\mathcal{H}).$$

Now we suppose that π is a ∗-representation of $\mathbf{A}$ and define

$$\mathcal{D}(\pi_{\mathrm{HS}}) := \big\{ x \in \mathbf{B}_2(\mathcal{H}(\pi)) : x\mathcal{H}(\pi) \subseteq \mathcal{D}(\pi), \ \pi(a) \cdot x \in \mathbf{B}_2(\mathcal{H}(\pi)) \text{ for } a \in \mathbf{A} \big\},$$
$$\pi_{\mathrm{HS}}(a)x := \pi(a) \cdot x \quad \text{for } a \in \mathbf{A}, \ x \in \mathcal{D}(\pi_{\mathrm{HS}}),$$

where "$\pi(a) \cdot x$" means the product of the operators $\pi(a)$ and x.

We abbreviate $\mathcal{H} := \mathcal{H}(\pi)$. Each finite rank operator $x = \sum_{k=1}^n \varphi_k \otimes \psi_k$, with $\varphi_k \in \mathcal{H}$, $\psi_k \in \mathcal{D}(\pi)$, is in $\mathcal{D}(\pi_{\mathrm{HS}})$ and we have $\pi_{\mathrm{HS}}(a)x = \sum_{k=1}^n \varphi_k \otimes \pi(a)\psi_k$. Clearly, the set of such finite rank operators x is dense in the Hilbert space $\mathbf{B}_2(\mathcal{H})$.

Statement 1: π_{HS} *is* ∗-*representation of* $\mathbf{A}$ *on the Hilbert space* $(\mathbf{B}_2(\mathcal{H}), \langle \cdot, \cdot \rangle_{\mathrm{HS}})$.

Proof From the definition of $\mathcal{D}(\pi_{\mathrm{HS}})$ it is clear that the operator $\pi(a) \cdot x$ is again in $\mathcal{D}(\pi_{\mathrm{HS}})$ and the map $a \mapsto \pi_{\mathrm{HS}}(a)$ is an algebra homomorphism of $\mathbf{A}$ into $L(\mathcal{D}(\pi_{\mathrm{HS}}))$. It remains to show that π_{HS} preserves the involution.

Let $x, y \in \mathcal{D}(\pi_{\mathrm{HS}})$ and $a \in \mathbf{A}$. We fix an orthonormal basis $\{\varphi_i : i \in J\}$ of the Hilbert space $\mathcal{H}$. Recall that $\operatorname{Tr} t = \sum_i \langle t\varphi_i, \varphi_i \rangle$ for each trace class operator t on $\mathcal{H}$. Therefore, since $x\varphi_i, y\varphi_i \in \mathcal{D}(\pi)$ by the definition of $\mathcal{D}(\pi_{\mathrm{HS}})$, we derive

$$\langle \pi_{\mathrm{HS}}(a)x, y \rangle_{\mathrm{HS}} = \operatorname{Tr} y^*(\pi_{\mathrm{HS}}(a)x) = \sum_{i \in J} \langle (y^*\pi_{\mathrm{HS}}(a)x)\varphi_i, \varphi_i \rangle$$
$$= \sum_{i \in J} \langle \pi_{\mathrm{HS}}(a)x\varphi_i, y\varphi_i \rangle = \sum_{i \in J} \langle x\varphi_i, \pi_{\mathrm{HS}}(a^+)y\varphi_i \rangle$$
$$= \sum_{i \in J} \langle (\pi_{\mathrm{HS}}(a^+)y)^*x\varphi_i, \varphi_i \rangle = \operatorname{Tr} (\pi_{\mathrm{HS}}(a^+)y)^*x = \langle x, \pi_{\mathrm{HS}}(a^+)y \rangle_{\mathrm{HS}},$$

which proves that π_{HS} is a ∗-representation of $\mathbf{A}$. $\square$

Statement 2: *If* π *is self-adjoint, so is* π_{HS}.

Proof Let $y \in \mathbf{B}_2(\mathcal{H})$ and suppose $y \in \mathcal{D}((\pi_{HS})^*)$. Let $\varphi \in \mathcal{H}$ and $\psi \in \mathcal{D}(\pi)$. Then $x = \varphi \otimes \psi \in \mathcal{D}(\pi_{HS})$. For $a \in \mathsf{A}$, we compute

$$\langle \pi_{HS}(a)x, y \rangle_{HS} = \langle \varphi \otimes \pi(a)\psi, y \rangle_{HS} = \text{Tr } y^*(\varphi \otimes \pi(a)\psi)$$
$$= \text{Tr } (\varphi \otimes y^*\pi(a)\psi) = \langle y^*\pi(a)\psi, \varphi \rangle = \langle \pi(a)\psi, y\varphi \rangle,$$
$$\langle x, (\pi_{HS})^*(a^+)y \rangle_{HS} = \text{Tr } [(\pi_{HS})^*(a^+)y]^*(\varphi \otimes \psi) = \text{Tr } \varphi \otimes [(\pi_{HS})^*(a^+)y]^*\psi$$
$$= \langle [(\pi_{HS})^*(a^+)y]^*\psi, \varphi \rangle = \langle \psi, [(\pi_{HS})^*(a^+)y]\varphi \rangle.$$

Therefore, since $\langle \pi_{HS}(a)x, y \rangle_{HS} = \langle x, (\pi_{HS})^*(a^+)y \rangle_{HS}$, these equalities imply that $\langle \pi(a)\psi, y\varphi \rangle = \langle \psi, [(\pi_{HS})^*(a^+)y]\varphi \rangle$ for $\psi \in \mathcal{D}(\pi)$ and $a \in \mathsf{A}$. Hence, because π is self-adjoint, $y\varphi \in \mathcal{D}(\pi)$ and $\pi(a^+)y\varphi = [(\pi_{HS})^*(a^+)y]\varphi$ for all $\varphi \in \mathcal{H}$. Therefore, since $[(\pi_{HS})^*(a^+)y] \in \mathbf{B}_2(\mathcal{H})$, we have $\pi(a^+) \cdot y \in \mathbf{B}_2(\mathcal{H})$. It follows from the definition of the domain $\mathcal{D}(\pi_{HS})$ that $y \in \mathcal{D}(\pi_{HS})$. Thus, we have proved that $\mathcal{D}((\pi_{HS})^*) \subseteq \mathcal{D}(\pi_{HS})$. Hence the *-representation π_{HS} is self-adjoint. $\qquad \square$

There is a unitary operator U of the Hilbert space $\mathbf{B}_2(\mathcal{H})$ of Hilbert–Schmidt operators on the tensor product $\overline{\mathcal{H}} \otimes \mathcal{H}$ of Hilbert spaces which maps each rank one operator $\varphi \otimes \psi$ of $\mathbf{B}_2(\mathcal{H})$ on the vector $\varphi \otimes \psi$ of $\overline{\mathcal{H}} \otimes \mathcal{H}$ [KR83, Proposition 2.6.9]. Here $\overline{\mathcal{H}}$ denotes the conjugate Hilbert space to $\mathcal{H}$, that is, the scalar multiplication and the inner product of $\mathcal{H}$ are replaced by the corresponding complex conjugate numbers. Using this unitary U the *-representation π_{HS} can be also realized on the Hilbert space $\overline{\mathcal{H}} \otimes \mathcal{H}$, which is sometimes more convenient to deal with. $\qquad \bigcirc$

Next we derive a simple but useful technical result.

Lemma 4.11 *Let $a = a^+ \in \mathsf{A}$ and $\lambda_\pm \in \mathbb{C}$, $\text{Im } \lambda_+ > 0$, $\text{Im } \lambda_- < 0$. Suppose A is unital and $a - \lambda_\pm$ is invertible in A with inverse denoted by $x_\pm$. If π is a nondegenerate *-representation of A, then $\overline{\pi(a)}$ is self-adjoint, $\overline{\pi(x_\pm)}$ is bounded, and $\overline{\pi(x_\pm)} = (\overline{\pi(a)} - \lambda_\pm I)^{-1}$.*

Proof Set $\mathcal{D} := \mathcal{D}(\pi)$. Since π is nondegenerate, $\pi(1) = I$ and therefore

$$\mathcal{D} \supseteq (\pi(a) - \lambda_\pm I)\mathcal{D} = \pi(a - \lambda_\pm)\mathcal{D} \supseteq \pi(a - \lambda_\pm)\pi(x_\pm)\mathcal{D} = \pi(1)\mathcal{D} = \mathcal{D},$$

so $(\pi(a) - \lambda_\pm I)\mathcal{D} = \mathcal{D}$ is dense and $\overline{\pi(a)}$ is self-adjoint by Proposition A.1.

Let $\varphi \in \mathcal{D}$ and set $\psi_\pm := (\pi(a) - \lambda_\pm I)\varphi$. Then $\psi_\pm = (\overline{\pi(a)} - \lambda_\pm I)\varphi$ and hence $\varphi = (\overline{\pi(a)} - \lambda_\pm I)^{-1}\psi_\pm$. Since $\pi(a)$ is symmetric, we have

$$\|\psi_\pm\| = \|(\pi(a) - \lambda_\pm I)\varphi\| \geq |\text{Im } \lambda_\pm| \, \|\varphi\|. \tag{4.12}$$

From $\pi(x_\pm)\psi_\pm = \pi(x_\pm(a - \lambda_\pm))\varphi = \pi(1)\varphi = \varphi = (\overline{\pi(a)} - \lambda_\pm I)^{-1}\psi_\pm$ and (4.12) we conclude that

$$\|\pi(x_\pm)\psi_\pm\| = \|\varphi\| \leq |\text{Im } \lambda_\pm|^{-1}\|\psi_\pm\|,$$

so $\pi(x_\pm)$ is bounded on $(\pi(a) - \lambda_\pm I)\mathcal{D} = \mathcal{D}$. Hence $\overline{\pi(x_\pm)} = (\overline{\pi(a)} - \lambda_\pm I)^{-1}$ is a bounded operator defined on the whole Hilbert space $\mathcal{H}(\pi)$. □

We state an immediate consequence of Lemma 4.11 and Proposition 2.71(i).

Corollary 4.12 *If π is a nondegenerate ∗-representation of a hermitian ∗-algebra, then $\overline{\pi(a)}$ is self-adjoint for all hermitian elements $a \in \mathsf{A}$.*

Remark 4.13 In this book, ∗-representations are treated only for complex ∗-algebras on complex inner product spaces. In this remark we discuss the case of real ∗-algebras.

Let us suppose that A is a *real* ∗-algebra and $\mathcal{D}$ is a *real* inner product space. As in the complex case, a ∗-representation π of the real ∗-algebra A on $\mathcal{D}$ is defined by Definition 4.2.

Let $\mathsf{A}_\mathbb{C}$ denote the complexification of A according to Definition 2.9. Further, if $\mathcal{K}$ denotes the Hilbert space completion of $\mathcal{D}$, let $\mathcal{K}_\mathbb{C}$ be the complexification of the real Hilbert space $\mathcal{K}$ (see Appendix A). Then $\mathcal{D}_\mathbb{C} := \mathcal{D} + i\mathcal{D}$ is a complex inner product space and a dense subspace of $\mathcal{K}_\mathbb{C}$. For $a, b \in \mathsf{A}$ and $\varphi, \psi \in \mathcal{D}$, we define

$$\pi_\mathbb{C}(a + ib)(\varphi + i\psi) := \pi(a)\varphi - \pi(b)\psi + i(\pi(a)\psi + \pi(b)\varphi).$$

By lengthy, but straightforward computations it is shown that $\pi_\mathbb{C}$ is a ∗-representation of the *complex* ∗-algebra $\mathsf{A}_\mathbb{C}$ on the *complex* inner product space $\mathcal{D}_\mathbb{C}$.

Since $\pi(a)\varphi = \pi_\mathbb{C}(a)(\varphi + i0)$, the ∗-representation π of the real ∗-algebra A can be recovered from the ∗-representation $\pi_\mathbb{C}$ of its complexification $\mathsf{A}_\mathbb{C}$. Thus, by the close interplay between operators T on $\mathcal{K}$ and its extensions $T_\mathbb{C}$ to the complexification $\mathcal{K}_\mathbb{C}$ (see Appendix A and [MV97]), the study of the real ∗-representation π can be reduced to that of the complex ∗-representation $\pi_\mathbb{C}$.

Further, as discussed in Sect. 2.5 (see Lemma 2.48), each positive *and* hermitian functional on A can be uniquely extended to a positive functional on the complexification $\mathsf{A}_\mathbb{C}$. In particular, states on A extend to states on $\mathsf{A}_\mathbb{C}$. This allows one to apply all results concerning states on complex ∗-algebras to the real case. ○

4.2 Domains of Representations in Terms of Generators

Throughout this section, π denotes a ∗-**representation** of the ∗-algebra A.

First we describe the domain $\mathcal{D}(\pi^*)$ in terms of algebra generators.

Lemma 4.14 *Suppose $\{a_j : j \in I\}$ is a subset of A such that A is the linear span of products $\{a_{j_1} \cdots a_{j_r} : (j_1, \ldots, j_r) \in J\}$, where J is an index set. Then*

$$\mathcal{D}(\pi^*) = \bigcap_{(j_1,\ldots,j_r) \in J} \mathcal{D}\big(\pi(a_{j_r})^* \cdots \pi(a_{j_1})^*\big). \tag{4.13}$$

Proof Let $\mathcal{D}_0$ denote the set on the right-hand side of (4.13).

Suppose $\varphi \in \mathcal{D}(\pi^*)$ and fix $(j_1, \ldots, j_r) \in J$. Then $\varphi \in \mathcal{D}(\pi(a_{j_1})^*)$ by the definition of $\mathcal{D}(\pi^*)$ and $\pi^*(a_{j_1}^+)\varphi = \pi(a_{j_1})^*\varphi$ is again in $\mathcal{D}(\pi^*)$ by Proposition 4.5 and hence in $\mathcal{D}(\pi(a_{j_2})^*)$. Continuing this reasoning we get $\varphi \in \mathcal{D}(\pi(a_{j_r})^* \cdots \pi(a_{j_1})^*)$. Thus $\varphi \in \mathcal{D}_0$. This proves that $\mathcal{D}(\pi^*) \subseteq \mathcal{D}_0$.

Let $a \in \mathsf{A}$. By the assumption a is a finite sum $\sum \lambda_{(j_1,\ldots,j_r)} a_{j_1} \cdots a_{j_r}$ with complex coefficients $\lambda_{(j_1,\ldots,j_r)}$. Then $\pi(a)^* \supseteq \sum \overline{\lambda_{(j_1,\ldots,j_r)}} \, \pi(a_{j_r})^* \cdots \pi(a_{j_1})^*$. This in turn implies $\mathcal{D}(\pi^*) \supseteq \mathcal{D}_0$.   □

Corollary 4.15 *Suppose there are elements* $a_1, \ldots, a_d \in \mathsf{A}$ *such that* A *is the span of* $\{a_1^{n_1} \cdots a_d^{n_d} : (n_1, \ldots, n_d) \in \mathbb{N}_0^d\}$. *Then*

$$\mathcal{D}(\pi^*) = \bigcap_{(n_1,\ldots,n_r)\in\mathbb{N}_0^d} \mathcal{D}\big((\pi(a_d)^*)^{n_d} \cdots (\pi(a_1)^*)^{n_1}\big).$$

Proof Apply Lemma 4.14 to the set $\{a_1 \cdots a_1 a_2 \cdots a_2 \cdots a_d \cdots a_d\}$.   □

Corollary 4.16 *Suppose* B *is a subset of* A *which generates* A *as an algebra. Let* π_1 *and* π_2 *be* ∗-*representations of* A *acting on the same Hilbert space* $\mathcal{H}(\pi_1) = \mathcal{H}(\pi_2)$. *If* $\overline{\pi_1(b)} = \overline{\pi_2(b)}$ *for all* $b \in \mathsf{B}$, *then* $\pi_1^* = \pi_2^*$.

Proof Clearly, $\overline{\pi_1(b)} = \overline{\pi_2(b)}$ implies $\pi_1(b)^* = \pi_2(b)^*$ for $b \in \mathsf{B}$. Therefore, by Lemma 4.14, $\mathcal{D}(\pi_1^*) = \mathcal{D}(\pi_2^*)$ and hence

$$\pi_1^*(b^+) = \pi_1(b)^* \lceil \mathcal{D}(\pi_1^*) = \pi_2(b)^* \lceil \mathcal{D}(\pi_2^*) = \pi_2^*(b^+).$$

Since π_1^* and π_2^* are algebra homomorphisms and $\{b^+ : b \in \mathsf{B}\}$ generates the algebra A as well, it follows that $\pi_1^*(a) = \pi_2^*(a)$ for all $a \in \mathsf{A}$. Thus $\pi_1^* = \pi_2^*$.   □

The next corollary is another application of Lemma 4.14. The main assertion therein follows also from Lemma 3.19(ii).

Corollary 4.17 *Suppose* B *is a subset of* $\mathsf{A}_{\mathrm{her}}$ *such that* B *generates the* ∗-*algebra* A. *If* π *is a* ∗-*representation of* A, *then* $\pi(\mathsf{A})'_w = \cap_{b\in\mathsf{B}} \{\pi(b)\}'_w = \cap_{b\in\mathsf{B}} \{\overline{\pi(b)}\}'_w$.

Proof By Lemma 3.16(iii), we have $\{\pi(b)\}'_w = \{\overline{\pi(b)}\}'_w$ for $b \in \mathsf{B}$. This gives the second equality. If $T \in \pi(\mathsf{A})'_w$, then obviously $T \in \{\pi(b)\}'_w$ for $b \in \mathsf{B}$.

Conversely, let $T \in \cap_{b\in\mathsf{B}} \{\pi(b)\}'_w$. Then $T\pi(b) \subseteq \pi(b)^*T$ for $b \in \mathsf{B}$. Using this fact it follows by induction on n that $T\varphi \in \mathcal{D}(\pi(b_1)^* \cdots \pi(b_n)^*)$ and

$$T\pi(b_1) \cdots \pi(b_n)\varphi = \pi(b_1)^* \cdots \pi(b_n)^* T\varphi \tag{4.14}$$

for $b_1, \ldots, b_n \in \mathsf{B}, n \in \mathbb{N}$, and $\varphi \in \mathcal{D}(\pi)$. Since the ∗-algebra A is generated by B, A is the span of products $b_1 \cdots b_n$, where $b_j \in \mathsf{B}$. Therefore, by Lemma 4.14, T maps $\mathcal{D}(\pi)$ into $\mathcal{D}(\pi^*)$. Because π and π^* are representations, (4.14) yields

$$T\pi(b_1 \cdots b_n)\varphi = T\pi(b_1) \cdots \pi(b_n)\varphi = \pi(b_1)^* \cdots \pi(b_n)^* T\varphi$$
$$= \pi^*(b_1) \cdots \pi^*(b_n) T\varphi = \pi^*(b_1 \cdots b_n) T\varphi.$$

Hence $T\pi(a)\varphi = \pi^*(a)T\varphi$ for $a \in A$, so $T \in \pi(A)'_w$; see, e.g., Proposition 3.15. $\square$

Let π be a $*$-representation of A and let B be a set of algebra $\underline{\text{generators of A}}$. We are looking for closed $*$-representations ρ such that $\pi \subseteq \rho$ and $\overline{\pi(b)} = \overline{\rho(b)}$ for $b \in B$. (Definition 9.47 below deals with such a situation.) In general there are many representations having this property (see Example 4.20), but there is always a *largest* such $*$-representation, as Theorem 4.18(ii) shows.

Often $*$-representations are described in terms of the actions of generators, and one easily finds an invariant domain $\mathcal{D}$ for these operators. (For instance, if the generators act as weighted shifts on an orthonormal basis, the span of basis vectors is invariant.) Then we obtain a $*$-representation π of A on $\mathcal{D}$. Theorem 4.18 provides a "natural" extension of π to a $*$-representation ρ of A that has nice properties.

Theorem 4.18 *Suppose π is a $*$-representation of A and B is a subset of A which generates A as an algebra. We define $\rho := \pi^* \lceil \mathcal{D}(\rho)$, where*

$$\mathcal{D}(\rho) = \bigcap_{k \in \mathbb{N}} \bigcap_{b_1, \ldots, b_k \in B} \mathcal{D}(\overline{\pi(b_k)} \cdots \overline{\pi(b_1)}). \tag{4.15}$$

(i) *ρ is a closed $*$-representation of A on the domain $\mathcal{D}(\rho)$ and*

$$\rho(b_k \cdots b_1)\varphi = \overline{\pi(b_k)} \cdots \overline{\pi(b_1)}\,\varphi \quad \text{for } b_1, \ldots, b_k \in B, \ \varphi \in \mathcal{D}(\rho). \tag{4.16}$$

Moreover, we have $\pi \subseteq \rho$ and $\rho^* = \pi^*$.

(ii) *ρ is the largest among all closed $*$-representations ρ_0 acting on $\mathcal{H}(\rho_0) = \mathcal{H}(\pi)$ and satisfying $\pi \subseteq \rho_0$ and $\overline{\pi(b)} = \overline{\rho_0(b)}$ for all $b \in B$.*

(iii) *If $\overline{\pi(b)} = \pi(b^+)^*$ for $b \in B$, then $\rho = \pi^*$ and π is essentially self-adjoint.*

Proof (i): For $b_1, \ldots, b_n \in B$, we have $\overline{\pi(b_k)} \cdots \overline{\pi(b_1)} \subseteq \pi(b_k^+)^* \cdots \pi(b_1^+)^*$. Therefore, since B^+ also generates A, (4.13) implies that $\mathcal{D}(\rho) \subseteq \mathcal{D}(\pi^*)$. From its definition it is obvious that $\mathcal{D}(\rho)$ is invariant under each operator $\overline{\pi(b)}, b \in B$. We have $\overline{\pi(b)} \subseteq \pi(b^+)^*$ and hence $\overline{\pi(b)} \lceil \mathcal{D}(\rho) = \pi^*(b) \lceil \mathcal{D}(\rho) = \rho(b)$. Thus $\mathcal{D}(\rho)$ is invariant under $\pi^*(b)$. Since π^* is an algebra homomorphism and B generates A, $\mathcal{D}(\rho)$ is invariant under $\pi^*(a)$ for all $a \in A$ and ρ is an algebra homomorphism of A into $L(\mathcal{D}(\rho))$.

Next we show that ρ preserves the involution. It suffices to prove this for elements of B. Let $\psi \in \mathcal{D}(\rho)$. As noted above, $\mathcal{D}(\rho) \subseteq \mathcal{D}(\pi^*)$, so that

$$\langle \pi(b)\eta, \psi \rangle = \langle \eta, \pi^*(b^+)\psi \rangle, \quad \eta \in \mathcal{D}(\pi). \tag{4.17}$$

Since $\rho(b)\varphi = \overline{\pi(b)}\,\varphi$ and $\rho(b^+)\varphi = \pi^*(b^+)\varphi$ for $\varphi \in \mathcal{D}(\rho)$, (4.17) implies that

$$\langle \rho(b)\,\varphi, \psi \rangle = \langle \overline{\pi(b)}\,\varphi, \psi \rangle = \langle \varphi, \pi^*(b^+)\psi \rangle = \langle \varphi, \rho(b^+)\psi \rangle.$$

Putting the preceding together we have proved that ρ is a $*$-representation of **A**. Obviously, $\pi \subseteq \rho$. The equality $\rho(b) = \overline{\pi(b)} \lceil \mathcal{D}(\rho)$ mentioned above gives $\overline{\rho(b)} = \overline{\pi(b)}$ for $b \in$ **B**. Hence $\rho^* = \pi^*$ by Corollary 4.16.

Now we prove that ρ is closed. Let $\varphi \in \mathcal{D}(\overline{\rho})$. Since the O^*-algebra $\overline{\rho(\mathbf{A})}$ is the closure of $\rho(\mathbf{A})$, by Proposition 3.8 there is a net (φ_i) of vectors $\varphi_i \in \mathcal{D}(\rho)$ such that $\varphi = \lim_i \varphi_i$ in the graph topology of $\rho(\mathbf{A})$. To prove that ρ is closed we have to show that $\varphi \in \mathcal{D}(\rho)$. For this it suffices to prove that

$$\varphi \in \mathcal{D}\big(\overline{\pi(b_k)} \cdots \overline{\pi(b_1)}\big) \quad \text{and} \quad \overline{\rho}(b_k \cdots b_1)\varphi = \overline{\pi(b_k)} \cdots \overline{\pi(b_1)}\,\varphi \qquad (4.18)$$

for $b_1, \ldots, b_k \in$ **B**. We proceed by induction on k. For $k = 1$ this has been already noted above. Assume that (4.18) is true for fixed k. Let $b_1, \ldots, b_{k+1} \in$ **B**. Using that $\overline{\rho}$ is a representation and the relation $\overline{\rho}(b_{k+1}) \subseteq \overline{\pi(b_{k+1})}$ we obtain

$$\overline{\rho}(b_{k+1} \cdots b_1)\varphi_i = \overline{\rho}(b_{k+1})\,\overline{\rho}(b_k \cdots b_1)\varphi_i = \overline{\pi(b_{k+1})}\,\overline{\rho}(b_k \cdots b_1)\varphi_i. \quad (4.19)$$

Since $\varphi = \lim_i \varphi_i$ in the graph topology, $\overline{\rho}(b_{k+1} \cdots b_1)\varphi = \lim_i \overline{\rho}(b_{k+1} \cdots b_1)\varphi_i$ and $\overline{\rho}(b_k \cdots b_1)\varphi = \lim_i \overline{\rho}(b_k \cdots b_1)\varphi_i$. Therefore, it follows from (4.19) that $\overline{\rho}(b_k \cdots b_1)\varphi \in \mathcal{D}\big(\overline{(\pi(b_{k+1})}\big)$ and $\overline{\rho}(b_{k+1} \cdots b_1)\varphi = \overline{\pi(b_{k+1})}\,\overline{\rho}(b_k \cdots b_1)\varphi$. Inserting the induction hypothesis (4.18) into the latter we obtain the assertion (4.18) in the case $k + 1$. This completes the induction proof. Thus $\rho = \overline{\rho}$. Using that $\overline{\rho} = \rho$ the second equality of (4.18) yields (4.16).

(ii): Suppose ρ_0 is another $*$-representation of **A** on $\mathcal{H}(\rho_0) = \mathcal{H}(\pi)$ such that $\pi \subseteq \rho_0$ and $\overline{\rho_0(b)} = \overline{\pi(b)}$ for $b \in$ **B**. Let $\varphi \in \mathcal{D}(\rho_0)$. Then, for $b_1, \ldots, b_k \in$ B,

$$\rho_0(b_k \cdots b_1)\,\varphi = \overline{\rho_0(b_k)} \cdots \overline{\rho_0(b_1)}\,\varphi = \overline{\pi(b_k)} \cdots 4\overline{\pi(b_1)}\,\varphi$$

and $\varphi \in \mathcal{D}\big(\overline{\pi(b_1)} \cdots \overline{\pi(b_k)}\big)$. Therefore, it follows from (4.15) and (4.16) that we have $\varphi \in \mathcal{D}(\rho)$ and $\rho_0(b_k \cdots b_1)\varphi = \rho(b_k \cdots b_1)\varphi$. This implies that $\rho_0 \subseteq \rho$.

(iii): Suppose that $\overline{\pi(b)} = \pi(b^+)^*$ for $b \in$ **B**. From the definition (4.15) of $\mathcal{D}(\rho)$ and the formula for $\mathcal{D}(\pi^*)$ in Lemma 4.14 it follows that $\mathcal{D}(\rho) = \mathcal{D}(\pi^*)$. Hence $\rho = \pi^*$, since $\rho \subseteq \pi^*$ by definition. Thus π^* is a $*$-representation. Therefore, $\pi^* \subseteq (\pi^*)^*$. But $\pi \subseteq \pi^*$ implies $(\pi^*)^* \subseteq \pi^*$, so we get $\pi^* = (\pi^*)^*$, that is, $\pi^* = \rho$ is self-adjoint. $\square$

The next proposition clarifies the notions from Definition 4.6 in the simplest case of the polynomial algebra $\mathbf{A} = \mathbb{C}[x]$ in a single variable. Among others, it shows that the biclosure π^{**} depend only on the closed symmetric operator $\overline{\pi(x)}$. This is one reason why the biclosedness of representations is more useful than closedness.

In the proofs of Proposition 4.19 and Example 4.20 we use some technical results on symmetric operators and their adjoints (see, e.g., [Sch12] and Appendix A). Recall that $\mathcal{D}^\infty(T) := \bigcap_{n=1}^\infty \mathcal{D}(T^n)$ for an operator T.

Proposition 4.19 *Suppose π is a $*$-representation of* $\mathbf{A} = \mathbb{C}[x]$. *Then:*

(i) $\mathcal{D}(\overline{\pi}) = \cap_{n \in \mathbb{N}} \mathcal{D}\big(\overline{\pi(x^n)}\big)$ and $\overline{\pi}(x) = \overline{\pi(x)} \restriction \mathcal{D}(\overline{\pi})$.

(ii) $\mathcal{D}(\pi^*) = \mathcal{D}^\infty(\pi(x)^*)$ and $\pi^*(x) = \pi(x)^* \restriction \mathcal{D}(\pi^*)$.

(iii) $\mathcal{D}(\pi^{**}) = \mathcal{D}^\infty\big(\overline{\pi(x)}\big)$ and $\pi^{**}(x) = \overline{\pi(x)} \restriction \mathcal{D}(\pi^{**})$.

(iv) π is biclosed if and only if $\mathcal{D}(\pi) = \mathcal{D}^\infty\big(\overline{\pi(x)}\big)$.

(v) π is essentially self-adjoint if and only if $\pi(x)$ is essentially self-adjoint.

(vi) π is self-adjoint if and only if the operator $\pi(x)$ is essentially self-adjoint and
$\mathcal{D}(\pi) = \mathcal{D}^\infty\big(\overline{\pi(x)}\big)$.

(vii) $\overline{\pi}$ is self-adjoint if and only if $\pi(x)^n$ is essentially self-adjoint for all $n \in \mathbb{N}$.

Proof (i): If $p \in \mathbb{C}[x]$ is a polynomial of degree n, the domain of the closure of $\pi(p(x)) = p(\pi(x))$ is equal to the domain of the closure of $\pi(x^n) = \pi(x)^n$. This yields the first formula. The second formula only restates the definition of $\overline{\pi}(x)$.

(ii): Follows at once from Corollary 4.15 and the definition of $\pi^*(x)$.

(iii): Obviously, $\overline{\pi(x)} \subseteq \pi^{**}(x)$. First we prove that $\overline{\pi(x)} = \pi^{**}(x)$. Assume to the contrary that $\overline{\pi(x)} \subsetneq \pi^{**}(x)$. If T is a closed symmetric operator, then the map $\varphi \mapsto (T + \mathrm{i}I)\varphi$ is a bijection of $\mathcal{D}(T)$ on the closed subspace $\mathrm{ran}\,(T + \mathrm{i}I)$. Therefore, $\mathrm{ran}\,(\overline{\pi(x)} + \mathrm{i}I)$ and $\mathrm{ran}\,(\pi^{**}(x) + \mathrm{i}I)$ are closed subspaces of $\mathcal{H}(\pi)$ such that $\mathrm{ran}\,(\overline{\pi(x)} + \mathrm{i}I) \subsetneq \mathrm{ran}\,(\pi^{**}(x) + \mathrm{i}I)$, so that

$$\ker(\pi^{**}(x)^* - \mathrm{i}I) = (\mathrm{ran}\,(\pi^{**}(x) + \mathrm{i}I))^\perp$$
$$\subsetneq (\mathrm{ran}\,(\overline{\pi(x)} + \mathrm{i}I))^\perp = \ker(\pi(x)^* - \mathrm{i}I).$$

Hence there exists a vector $\xi \in \ker(\pi(x)^* - \mathrm{i}I)$ such that $\xi \notin \ker(\pi^{**}(x)^* - \mathrm{i}I)$. We show that $\xi \notin \mathcal{D}(\pi^{**}(x)^*)$. Otherwise, for $\eta \in \mathcal{D}(\pi) \subseteq \mathcal{D}(\pi^{**})$,

$$\langle \pi^{**}(x)^*\xi, \eta \rangle = \langle \xi, \pi^{**}(x)\eta \rangle = \langle \xi, \pi(x)\eta \rangle = \langle \pi(x)^*\xi, \eta \rangle = \langle \mathrm{i}\xi, \eta \rangle,$$

so that $\pi^{**}(x)^*\xi = \mathrm{i}\xi$, which contradicts the choice of ξ. Since $\xi \notin \mathcal{D}(\pi^{**}(x)^*)$, we have $\xi \notin \mathcal{D}(\pi^{***})$. But $\xi \in \mathcal{D}^\infty(\pi(x)^*) = \mathcal{D}(\pi^*)$ by (ii). This contradicts the relation $\pi^* = \pi^{***}$ and completes the proof of $\overline{\pi(x)} = \pi^{**}(x)$.

Let ρ be the $*$-representation of A defined by $\rho(x) = \overline{\pi(x)} \restriction \mathcal{D}(\rho)$ on the domain $\mathcal{D}(\rho) := \mathcal{D}^\infty(\overline{\pi(x)})$. Obviously, $\pi(x) \subseteq \rho(x) \subseteq \overline{\pi(x)}$, so that $\overline{\pi(x)} = \overline{\rho(x)}$. Therefore, $\pi^* = \rho^*$ by Corollary 4.16 and hence $\pi^{**} = \rho^{**}$. Using the equality $\overline{\pi(x)} = \pi^{**}(x)$ proved above, we derive

$$\mathcal{D}(\rho^{**}) = \mathcal{D}(\pi^{**}) \subseteq \cap_{n \in \mathbb{N}} \mathcal{D}\big(\overline{\pi^{**}(x)^n}\big) \subseteq \cap_{n \in \mathbb{N}} \mathcal{D}\big(\overline{\pi^{**}(x)}^n\big)$$
$$= \mathcal{D}^\infty\big(\overline{\pi^{**}(x)}\big) = \mathcal{D}^\infty\big(\overline{\pi(x)}\big) = \mathcal{D}(\rho).$$

Since $\rho \subseteq \rho^{**}$, this implies $\rho = \rho^{**}$. Thus, $\rho = \pi^{**}$, which is the assertion.

(iv): Follows from (iii) and the relation $\pi \subseteq \pi^{**}$.

(v): Suppose $\pi(x)$ is essentially self-adjoint. Then, since $\pi^*(x) \subseteq \pi(x)^* = \overline{\pi(x)}$, the operator $\pi^*(x)$ is symmetric, so π^* is a $*$-representation. Therefore, $\pi^* = \pi^{**}$ by Lemma 4.3(ii), that is, π is essentially self-adjoint.

Conversely, suppose that $\pi(x)$ is not essentially self-adjoint. Then there exists a vector $\xi \neq 0$ in a deficiency space of $\pi(x)$ [Sch12, Proposition 3.8]. Since $\xi \in \mathcal{D}^\infty(\pi(x)^*) = \mathcal{D}(\pi^*), \pi^*(x)$ is not symmetric, so π^* is not a ∗-representation. Hence $\pi^* \neq \pi^{**}$, which means that π is not essentially self-adjoint.

(vi): Clearly, π is self-adjoint (i.e., $\pi = \pi^*$) if and only if π is biclosed (i.e., $\pi = \pi^{**}$) and π is essentially self-adjoint (i.e., $\pi^* = \pi^{**}$). Using this fact the assertion of (vi) follows easily by combining (ii), (iii), and (v).

(vii): First suppose $\overline{\pi}$ is self-adjoint. Note that $\overline{\pi(a)} = \overline{\pi}(a)$ for $a \in \mathsf{A}$. Hence, by (vi), $\mathcal{D}(\overline{\pi}) = \mathcal{D}^\infty(\overline{\pi(x)})$ and $\overline{\pi(x)}$ is self-adjoint. From Proposition 3.31 (or from spectral theory) it follows that $\mathcal{D}^\infty(\overline{\pi(x)})$ is a core for $\overline{\pi(x)}^n$. Obviously, $\mathcal{D}(\overline{\pi}) \subseteq \mathcal{D}(\overline{\pi(x)^n})$. Combining these facts it follows that the closure of $\pi(x)^n$ is the self-adjoint operator $\overline{\pi(x)}^n$, so $\pi(x)^n$ is essentially self-adjoint.

Conversely, assume $\pi(x)^n$ is essentially self-adjoint for all $n \in \mathbb{N}$. Then, since $\pi(x^n) = \pi(x)^n \subseteq (\pi(x)^*)^n \subseteq (\pi(x)^n)^* = \pi(x^n)^*$, we obtain $\overline{\pi(x^n)} = (\pi(x)^*)^n = \pi(x^n)^*$ for $n \in \mathbb{N}$. Hence $\mathcal{D}(\overline{\pi}) = \mathcal{D}(\pi^*)$ by (i) and (ii). Since $\overline{\pi} \subseteq \pi^*$, we get $\overline{\pi} = \pi^* = (\overline{\pi})^*$, that is, $\overline{\pi}$ is self-adjoint. $\square$

The preceding proposition is nicely illustrated by the next example. The proofs of some assertions are only sketched; the reader may fill the details as an exercise.

Example 4.20 Let us consider the self-adjoint operator $T = -i\frac{d}{dx}$ with domain

$$\mathcal{D}(T) = \{\varphi \in H^1(0, 1) : \varphi(0) = \varphi(1)\}$$

in the Hilbert space $L^2(0, 1)$. We define a ∗-representation π_∞ of $\mathsf{A} = \mathbb{C}[x]$ on

$$\mathcal{D}(\pi_\infty) := \mathcal{D}^\infty(T) = \{\varphi \in C^\infty([0, 1]) : \varphi^{(k)}(0) = \varphi^{(k)}(1) \text{ for } k \in \mathbb{N}_0\}$$

by $\pi_\infty(p) = p(T)\lceil\mathcal{D}(\pi_\infty), p \in \mathbb{C}[x]$. By Proposition 4.19(vi), π_∞ is *self-adjoint*.

Let $\pi_n, n \in \mathbb{N}_0$, denote the restriction of π_∞ to the invariant linear subspace

$$\mathcal{D}(\pi_n) = \{\varphi \in \mathcal{D}(\pi_\infty) : \varphi^{(k)}(0) = 0 \text{ for } k \geq n\}.$$

Obviously, $\pi_n \subseteq \pi_m$ if $n \leq m$. The linear functional $\varphi \mapsto \varphi^{(k)}(0)$ on $\mathcal{D}(\pi_\infty)$ is continuous in the graph norm $\|T^{k+1} \cdot \| + \| \cdot \|$ of the operator T^{k+1} and hence in the graph topology t_{π_∞}. This implies that each ∗-representation π_n is *closed*.

Clearly, $\mathcal{D}(\pi_0) = \mathcal{D}^\infty(\overline{\pi_0(x)})$ and $\pi_0(x)$ has deficiency indices $(1, 1)$. Therefore, by Proposition 4.19,(iv), π_0 is *biclosed*, but not essentially self-adjoint.

It can be shown that, given $c_0, \dots, c_n \in \mathbb{C}$, there exists a function $\varphi \in \mathcal{D}(\pi_{n+1})$ such that $\varphi^{(k)}(0) = c_k$ for $k = 0, \dots, n$. The following assertions are derived from this fact. We omit the details and state only the results. For this suppose $n \in \mathbb{N}_0$.

First, we have $\mathcal{D}(\pi_n) \neq \mathcal{D}(\pi_m)$ and hence $\pi_n \neq \pi_m$ if $n \neq m, n, m \in \mathbb{N}_0$.

Next, for $k = 0, \dots, n$, the symmetric operator $\pi_n(x^k) = T^k\lceil\mathcal{D}(\pi_n)$ is essentially self-adjoint, or equivalently, $\mathcal{D}(\pi_n)$ is a core for T^k. Further, $\mathcal{D}(\pi_n)$ is not a core for T^j when $j > n$. That is, $\overline{\pi_n(x^k)} = T^k$ if and only if $k = 0, \dots, n$.

Finally, $(\pi_{n+1})^* = \pi_\infty$. Hence $(\pi_{n+1})^{**} = (\pi_\infty)^* = \pi_\infty = (\pi_{n+1})^*$, so π_{n+1} is *essentially self-adjoint*, and $(\pi_{n+1})^{**} = \pi_\infty \neq \pi_{n+1}$, so π_{n+1} is *not biclosed* for $n \in \mathbb{N}_0$. Recall that each $*$-representation π_{n+1} is closed, as noted above.

Now we set $\pi := \pi_1$ and $\mathsf{B} := \{x\}$ in Theorem 4.18. Then, each $*$-representation π_n, $n \in \mathbb{N} \cup \{\infty\}$, satisfies $\overline{\pi_n(b)} = \overline{\pi(b)}$ for $b \in \mathsf{B}$, and $\rho = \pi_\infty$ is the *largest* $*$-representation which has this property.

However, if we take $\pi := \pi_1$ and $\mathsf{B} = \{x, x^2\}$, then $\rho = \pi_1$ is the largest $*$-representation ρ such that $\overline{\rho(b)} = \overline{\pi(b)}$ for all $b \in \mathsf{B}$. This shows that the $*$-representation ρ in Theorem 4.18 depends also on the set B of generators and that ρ is not necessarily biclosed. $\bigcirc$

Example 4.21 (*Algebras of measurable functions*)
Let A be a unital $*$-algebra of measurable functions on $\mathbb{R}^d$ with pointwise algebraic operations and involution $f^+(t) := \overline{f(t)}$. Suppose that $\mathsf{A} \subseteq L^2_{\mathrm{loc}}(\mathbb{R}^d)$. Let μ be a Radon measure on $\mathbb{R}^d$. Define $\pi(f)\varphi = f \cdot \varphi$ for $f \in \mathsf{A}$ and φ in the domain

$$\mathcal{D}(\pi) = \{\varphi \in L^2(\mathbb{R}^d; \mu) : f \cdot \varphi \in L^2(\mathbb{R}^d; \mu) \text{ for } f \in \mathsf{A}\}.$$

Since $\mathsf{A} \subseteq L^2_{\mathrm{loc}}(\mathbb{R}^d)$, $\mathcal{D}(\pi)$ contains $C_c(\mathbb{R}^d)$. Hence $\mathcal{D}(\pi)$ is a dense linear subspace of $\mathcal{H}(\pi) = L^2(\mathbb{R}^d; \mu)$. It is easily checked that π is a $*$-representation of A.

We prove that π is *self-adjoint*. Let $\psi \in \mathcal{D}(\pi^*)$. Fix $f \in \mathsf{A}$ and set $\eta := \pi(f)^*\psi$. Then, for $\varphi \in \mathcal{D}(\pi)$ we derive

$$\langle f \cdot \varphi, \psi \rangle = \langle \pi(f)\varphi, \psi \rangle = \langle \varphi, \pi(f)^*\psi \rangle = \langle \varphi, \eta \rangle.$$

Thus, $\int \varphi \overline{(\overline{f} \cdot \psi - \eta)} \, d\mu = 0$ for $\varphi \in \mathcal{D}(\pi)$. Hence $\overline{f} \cdot \psi = \eta$, since $\mathcal{D}(\pi)$ is dense. In particular, $\overline{f} \cdot \psi = \eta \in L^2(\mathbb{R}^d; \mu)$. Therefore, $\psi \in \mathcal{D}(\pi)$ and π is self-adjoint.

An important special case is the $*$-algebra $\mathsf{A} = \mathbb{C}[x_1, \ldots, x_d]$. If μ is the Lebesgue measure on $\mathbb{R}^d$, then the $*$-representation π defined above is faithful. $\bigcirc$

4.3 Invariant Subspaces and Reducing Subspaces

In this section, π is a $*$-**representation** of A.

Definition 4.22 Let $\mathcal{E}$ be a linear subspace of $\mathcal{D}(\pi)$. We shall say that $\mathcal{E}$ is
- *invariant* under π if $\pi(a)\mathcal{E} \subseteq \mathcal{E}$ for all $a \in \mathsf{A}$,
- a *core* for π if it is a core for all operators $\pi(a)$, i.e., $\overline{\pi(a)\lceil\mathcal{E}} = \overline{\pi(a)}$ for $a \in \mathsf{A}$.

A closed linear subspace $\mathcal{K}$ of $\mathcal{H}(\pi)$ is called *invariant* under π if there exists a linear subspace $\mathcal{E}$ of $\mathcal{D}(\pi) \cap \mathcal{K}$ which is invariant under π and dense in $\mathcal{K}$.

Let $\mathcal{E}$ be an invariant linear subspace of $\mathcal{D}(\pi)$. The map $a \mapsto \pi_\mathcal{E}(a) := \pi(a)\lceil\mathcal{E}$ defines a $*$-representation $\pi_\mathcal{E}$ of A with domain $\mathcal{D}(\pi_\mathcal{E}) = \mathcal{E}$ in the Hilbert space $\overline{\mathcal{E}}$. Such a $*$-representation $\pi_\mathcal{E}$ is called a *subrepresentation* of π. Obviously, π itself and $\pi\lceil\{0\}$ are subrepresentations of π; they are the *trivial subrepresentations* of π.

As in the case of single operators, a *-representation can be restored from a core and a core is often easier to describe than the full domain. Clearly, $\mathcal{E}$ is a core for π if and only if $\mathcal{E}$ is dense in $\mathcal{D}(\pi)$ with respect to the graph topology t_π. Hence, if $\mathcal{E}$ is a core for π, then $\pi_{\mathcal{E}}$ has the same closure, biclosure, and adjoint as π.

Suppose now that $\mathcal{K}$ is a closed linear subspace of $\mathcal{H}(\pi)$ which is invariant under π according to the second part of Definition 4.22. Let $\mathcal{E}(\mathcal{K})$ denote the set of vectors $\varphi \in \mathcal{D}(\pi) \cap \mathcal{K}$ such that $\pi(a)\varphi \in \mathcal{K}$ for all $a \in \mathsf{A}$. Then $\mathcal{E}(\mathcal{K})$ is the largest linear subspace of $\mathcal{D}(\pi) \cap \mathcal{K}$ which is invariant in the sense of the first part of Definition 4.22. We denote the corresponding *-representation $\pi_{\mathcal{E}(\mathcal{K})} = \pi \lceil \mathcal{E}(\mathcal{K})$ also by $\pi_{\mathcal{K}}$.

Let $\mathcal{K}$ be a closed subspace of $\mathcal{H}(\pi)$ which is invariant under π. Then, in contrast to bounded *-representations acting on the whole Hilbert space, the orthogonal complement $\mathcal{K}^\perp$ is in general not invariant and π does not split into a direct sum with respect to the decomposition $\mathcal{H}(\pi) = \mathcal{K} \oplus \mathcal{K}^\perp$; see Example 4.33 below. To describe when the latter happens we introduce the following notions.

Definition 4.23 A linear subspace $\mathcal{E}$ of $\mathcal{D}(\pi)$ (resp. a closed linear subspace $\mathcal{K}$ of $\mathcal{H}(\pi)$) is called *reducing* for π if there exist *-representations π_1 and π_2 of π such that $\pi = \pi_1 \oplus \pi_2$ and $\mathcal{E} = \mathcal{D}(\pi_1)$ (resp. $\mathcal{K} = \mathcal{H}(\pi_1)$).

A *-representation π is said to be *irreducible*, or *indecomposable*, if $\{0\}$ and $\mathcal{H}(\pi)$ are the only closed linear subspaces of $\mathcal{H}(\pi)$ that are reducing for π.

Remark 4.24 In algebra there are the following (in general different) notions for modules over a ring: A module is called *simple*, or *irreducible*, if it has no proper submodule, *indecomposable* if it is not a direct sum of two proper submodules, and a *Schur module* if its endomorphism ring is a division algebra. For bounded *-representations acting on the whole Hilbert space the counterparts of these three notions are equivalent (see Exercise 6), but for unbounded *-representations they are not! To circumvent pathological behavior we have defined irreducibility as in Definition 4.23. This fits to the corresponding notion for single operators (see Appendix A), and it leads to the expected results: The Schrödinger representation of the Weyl algebra becomes irreducible (Example 4.32) and Proposition 9.29 holds.   ○

We mention some simple consequences of Definition 4.23.

Suppose that $\mathcal{E}$ is a linear subspace of $\mathcal{D}(\pi)$ which is reducing for π. Clearly, $\mathcal{E}$ is invariant under π and we have $\mathcal{D}(\pi) \cap \overline{\mathcal{E}} = \mathcal{E}$ and $\pi_{\mathcal{E}} = \pi_{\overline{\mathcal{E}}}$.

Let $\mathcal{K}$ be a closed linear subspace of $\mathcal{H}(\pi)$, and let $P_{\mathcal{K}}$ denote the orthogonal projection on $\mathcal{K}$. If $\mathcal{K}$ is reducing for π, then $\mathcal{K}$ is invariant under π, $\mathcal{K}^\perp$ is also reducing for π and we have $\mathcal{D}(\pi) \cap \mathcal{K} = P_{\mathcal{K}}\mathcal{D}(\pi) = \mathcal{D}(\pi_{\mathcal{K}})$.

Some further slight reformulations are given in the following two propositions.

Proposition 4.25 *For a closed subspace $\mathcal{K}$ of $\mathcal{H}(\pi)$ the following are equivalent:*

(i) *$\mathcal{K}$ is a reducing subspace for π.*
(ii) *$P_{\mathcal{K}} \in \pi(\mathsf{A})'_{\mathrm{s}}$.*
(iii) *The linear subspaces $\mathcal{D}(\pi) \cap \mathcal{K}$ and $\mathcal{D}(\pi) \cap \mathcal{K}^\perp$ of $\mathcal{D}(\pi)$ are invariant under π and $P_{\mathcal{K}}\mathcal{D}(\pi) \subseteq \mathcal{D}(\pi)$.*

If one of these conditions is satisfied, then $\pi = \pi_{\mathcal{K}} \oplus \pi_{\mathcal{K}^\perp}$, *with* $\pi_{\mathcal{K}} = \pi \lceil P_{\mathcal{K}} \mathcal{D}(\pi)$ *and* $\pi_{\mathcal{K}^\perp} = \pi \lceil P_{\mathcal{K}^\perp} \mathcal{D}(\pi)$.

Proof The proof is given by straightforward verifications. We sketch the proof of (iii)→(i). Since $P_{\mathcal{K}} \mathcal{D}(\pi) \subseteq \mathcal{D}(\pi)$ and $\mathcal{D}(\pi) \cap \mathcal{K}$ is invariant, $P_{\mathcal{K}} \mathcal{D}(\pi) = \mathcal{D}(P_{\mathcal{K}})$. Similarly, $P_{\mathcal{K}^\perp} \mathcal{D}(\pi) = \mathcal{D}(P_{\mathcal{K}^\perp})$. These relations imply that $\pi = \pi_{\mathcal{K}} \oplus \pi_{\mathcal{K}^\perp}$ and $\mathcal{K} = \mathcal{H}(\pi_{\mathcal{K}})$. $\square$

Proposition 4.26 *The following statements are equivalent:*

(i) π *is irreducible.*
(ii) *If* $\pi = \pi_1 \oplus \pi_2$ *is a direct sum decomposition, then* $\mathcal{H}(\pi_1 = \{0\}$ *or* $\mathcal{H}(\pi_2)=\{0\}$.
(iii) *The only projections in the strong commutant* $\pi(\mathsf{A})'_s$ *are* 0 *and* I.

If π *is closed, then* π *is irreducible if and only if* $\pi(\mathsf{A})'_{\mathrm{sym}} = \mathbb{C} \cdot I$.
If π *is self-adjoint, then* π *is irreducible if and only if* $\pi(\mathsf{A})'_w = \mathbb{C} \cdot I$.

Proof The equivalence of conditions (i)–(iii) follows at once from Proposition 4.25.

If π is closed, then $\pi(\mathsf{A})$ is a closed O^*-algebra, so by Proposition 3.17(ii), $\pi(\mathsf{A})'_{\mathrm{sym}} = \pi(\mathsf{A})'_s \cap (\pi(\mathsf{A})'_s)^*$ and this set is a von Neumann algebra. If π is self-adjoint, then $\pi(\mathsf{A})$ is a self-adjoint O^*-algebra and therefore $\pi(\mathsf{A})'_w = \pi(\mathsf{A})'_s$ is a von Neumann algebra by Proposition 3.15. Hence, in these cases, (iii) is equivalent to $\pi(\mathsf{A})'_{\mathrm{sym}} = \mathbb{C} \cdot I$ and $\pi(\mathsf{A})'_w = \mathbb{C} \cdot I$, respectively. $\square$

Corollary 4.27 *If the closure* $\overline{\pi}$ *of* π *is irreducible, so is* π.

Proof Since $\pi(\mathsf{A})'_s \subseteq \overline{\pi}(\mathsf{A})'_s$ by Proposition 3.17(i), this follows from Proposition 4.26,(iii)↔(i). $\square$

The next proposition contains the technical ingredients for Theorem 4.29, which is the main result of this section.

Proposition 4.28 *Let* $\mathcal{E}$ *be a linear subspace of* $\mathcal{D}(\pi)$.

(i) *Suppose* $\mathcal{E}$ *is invariant under* π. *If* $\pi_{\mathcal{E}} := \pi \lceil \mathcal{E}$ *is self-adjoint, then* $P_{\overline{\mathcal{E}}}$ *is in* $\pi(\mathsf{A})'_s$, $\overline{\mathcal{E}}$ *is reducing for* π, *and we have* $\pi_{\mathcal{E}} = \pi_{\overline{\mathcal{E}}}$ *and* $\mathcal{E} = P_{\overline{\mathcal{E}}} \mathcal{D}(\pi)$.
(i)′ *Suppose* $\mathcal{E}$ *is invariant under* π. *If* $\pi_{\mathcal{E}} := \pi \lceil \mathcal{E}$ *is essentially self-adjoint, then* $P_{\overline{\mathcal{E}}}$ *is in* $\pi^{**}(\mathsf{A})'_s$, $\overline{\mathcal{E}}$ *is reducing for* π^{**}, *and* $\mathcal{D}((\pi_{\mathcal{E}})^{**}) = P_{\overline{\mathcal{E}}} \mathcal{D}(\pi^{**})$.
(ii) *Suppose* $\overline{\mathcal{E}}$ *is reducing for* π. *If* π *is self-adjoint (resp. essentially self-adjoint), then* $\pi_{\overline{\mathcal{E}}}$ *is self-adjoint (resp. essentially self-adjoint).*

Proof In the proofs of (i) and (i)′ we abbreviate $P := P_{\overline{\mathcal{E}}}$ and $\rho := \pi_{\mathcal{E}}$.

(i): The crucial step in this proof is the relation $P\pi^*(a) \subseteq \rho^*(a)P$ which holds by Lemma 4.4(i). Note that $\rho \subseteq \pi$, $\pi \subseteq \pi^*$ (because π is a $*$-representation), and $\rho = \rho^*$ (since ρ is self-adjoint). Using these facts we derive for $\varphi \in \mathcal{D}(\pi), a \in \mathsf{A}$,

$$P\pi(a)\varphi = P\pi^*(a)\varphi = \rho^*(a)P\varphi = \rho(a)P\varphi = \pi(a)P\varphi,$$

that is, $P \in \pi(\mathsf{A})'_s$. Therefore, $\overline{\mathcal{E}}$ is reducing for π and $\rho \subseteq \pi_{\overline{\mathcal{E}}} = \pi \lceil P\mathcal{D}(\pi)$ by Proposition 4.25. Since ρ is a self-adjoint representation on $\overline{\mathcal{E}}$, it follows that $\rho = \pi_{\overline{\mathcal{E}}}$, so $\mathcal{E} = \mathcal{D}(\rho) = \mathcal{D}(\pi_{\overline{\mathcal{E}}}) = P\mathcal{D}(\pi)$.

(i)′: By Lemma 4.4(iii), the inclusion $\rho \subseteq \pi$ implies $\rho^{**} \subseteq \pi^{**}$. Since ρ is essentially self-adjoint, $\rho^* = \rho^{**}$ and hence $\rho^{**} = \rho^{***}$, which means that ρ^{**} is self-adjoint. Thus, (i) applies to ρ^{**} and π^{**}. Therefore, $P \in \pi^{**}(\mathsf{A})_s'$, $\overline{\mathcal{E}}$ is reducing for π^{**}, and $\mathcal{D}(\rho^{**}) = P\mathcal{D}(\pi^{**})$.

(ii): Since $\overline{\mathcal{E}}$ is reducing, there is a direct sum decomposition $\pi = \pi_{\overline{\mathcal{E}}} \oplus \pi_0$. It is straightforward to verify that $\pi^* = (\pi_{\overline{\mathcal{E}}})^* \oplus (\pi_0)^*$ and $\pi^{**} = (\pi_{\overline{\mathcal{E}}})^{**} \oplus (\pi_0)^{**}$. If π is self-adjoint (resp. essentially self-adjoint), then $\pi = \pi^*$ (resp. $\pi^* = \pi^{**}$) and therefore $\pi_{\overline{\mathcal{E}}} = (\pi_{\overline{\mathcal{E}}})^*$ (resp. $(\pi_{\overline{\mathcal{E}}})^* = (\pi_{\overline{\mathcal{E}}})^{**}$). This shows that $\pi_{\overline{\mathcal{E}}}$ is self-adjoint (resp. essentially self-adjoint).   □

Theorem 4.29 (i) *If ρ is a self-adjoint subrepresentation of π, then the projection $P_{\mathcal{H}(\rho)}$ of $\mathcal{H}(\pi)$ on $\mathcal{H}(\rho)$ is in $\pi(\mathsf{A})_s'$ and $\rho = \pi \upharpoonright P_{\mathcal{H}(\rho)}\mathcal{D}(\pi)$.*
(i)′ *If ρ is an essentially self-adjoint subrepresentation of π, then the projection $P_{\mathcal{H}(\rho)}$ of $\mathcal{H}(\pi)$ on $\mathcal{H}(\rho)$ is in $\pi^{**}(\mathsf{A})_s'$ and $\rho^{**} = \pi^{**} \upharpoonright P_{\mathcal{H}(\rho)}\mathcal{D}(\pi^{**})$.*
(ii) *If π is self-adjoint and P is a projection in $\pi(\mathsf{A})_s'$, then $\pi \upharpoonright P\mathcal{D}(\pi)$ is a self-adjoint subrepresentation of π.*

Proof (i): Apply Proposition 4.28(i) to $\mathcal{E} := \mathcal{D}(\rho)$.

(i)′: Apply Proposition 4.28(i)′ to $\mathcal{E} := \mathcal{D}(\rho)$.

(ii): Apply Propositions 4.25 and 4.28(ii) to $\overline{\mathcal{E}} := P\mathcal{H}(\pi)$.   □

The next two corollaries follow easily from Theorem 4.29 or Proposition 4.28. They illustrate how self-adjointness assumptions rule out pathological behavior.

Corollary 4.30 *A self-adjoint representation π is irreducible if and only if the only self-adjoint subrepresentations of π are the trivial subrepresentations π and $\pi \upharpoonright \{0\}$.*

Corollary 4.31 *Suppose π is a self-adjoint representation of A and ρ is a ∗-representation of A acting on a possibly larger Hilbert space such that $\pi \subseteq \rho$. Then there is a ∗-representation π_0 of A on the Hilbert space $\mathcal{H}(\rho) \ominus \mathcal{H}(\pi)$ such that $\rho = \pi \oplus \pi_0$. In particular, if $\mathcal{H}(\pi) = \mathcal{H}(\rho)$, then we have $\pi = \rho$.*

Let us discuss all this for our main guiding example in this book.

Example 4.32 (*Schrödinger representation of the Weyl algebra*)
Let W be the Weyl algebra $\mathbb{C}\langle p, q \, | \, p = p^+, q = q^+, pq - qp = -\mathrm{i}\rangle$. The self-adjoint operators $P := -\mathrm{i}\frac{d}{dx}$ and $Q := x$ on $L^2(\mathbb{R})$ leave the Schwartz space

$$\mathcal{S}(\mathbb{R}) = \big\{\varphi \in C^\infty(\mathbb{R}) : \sup\{|x^k \varphi^{(n)}(x)| : x \in \mathbb{R}\} < \infty \ \text{ for } \ k, n \in \mathbb{N}_0\big\}$$

invariant and satisfy the relation $PQ\varphi - QP\varphi = -\mathrm{i}\varphi, \varphi \in \mathcal{S}(\mathbb{R})$. Hence there is a ∗-representation π_S, called the *Schrödinger representation*, of W with domain $\mathcal{D}(\pi_S) := \mathcal{S}(\mathbb{R})$ on $L^2(\mathbb{R})$ given by $\pi_S(p) = P\upharpoonright \mathcal{D}(\pi_S), \pi_S(q) = Q\upharpoonright \mathcal{D}(\pi_S)$. This representation of the Weyl algebra will play a crucial role in Chap. 8.

The image $\pi_S(\mathsf{W})$ of W is just the O^*-algebra $\mathcal{A}$ for $d = 1$ from Example 3.11. As noted therein, the graph topology of $\pi_S(\mathsf{W}) = \mathcal{A}$ is the usual Frechet topology of the Schwartz space $\mathcal{S}(\mathbb{R})$. Hence π_S is closed.

Statement: π_S *is irreducible and self-adjoint.*

Proof Clearly, $\mathcal{S}(\mathbb{R})$ is a core for Q. Hence $\pi(W)'_s \subseteq \{\pi_S(q)\}'_s \subseteq \{Q\}'_s$ by Lemma 3.16. Let $T \in \pi(W)'_s$. Then T commutes with Q, hence with all functions of Q, and so with $L^\infty(\mathbb{R})$ acting by multiplication on $L^2(\mathbb{R})$. Since $L^\infty(\mathbb{R})' = L^\infty(\mathbb{R})$, T is a multiplication operator by some function $f \in L^\infty(\mathbb{R})$. Because T leaves $\mathcal{S}(\mathbb{R})$ invariant, $f \in C^\infty(\mathbb{R})$. Since T commutes with $\pi_S(p)$, f must be constant. Thus, $\pi(W)'_s = \mathbb{C} \cdot I$ and π_S is irreducible by Proposition 4.26.

Since $\mathcal{S}(\mathbb{R})$ is a core for the self-adjoint operators Q and P, $\pi_S(q)^* = Q$ and $\pi_S(p)^* = P$. Thus, $\mathcal{D}((\pi_S)^*) \subseteq \mathcal{D}(P) \cap \mathcal{D}(Q)$. Clearly, $\mathcal{S}(\mathbb{R})$ is the largest linear subspace of $L^2(\mathbb{R})$ which is invariant under P and Q. Hence it follows from Theorem 4.18 and formula (4.15), applied with $\mathsf{B} = \{p, q\}$, that $\mathcal{D}((\pi_S)^*) = \mathcal{D}(\rho) = \mathcal{S}(\mathbb{R})$. Therefore, since $\mathcal{S}(\mathbb{R}) = \mathcal{D}(\pi_S)$, π_S is self-adjoint. $\qquad\square$

Another proof of the self-adjointness goes as follows. The Hermite functions $H_n, n \in \mathbb{N}_0$, (see Sect. 8.4) form an orthonormal basis of the Hilbert space $L^2(\mathbb{R})$, and they are eigenfunctions of the operator $T := \pi_s(q^2 + p^2) = x^2 - \frac{d^2}{dx^2}$. Hence $T^n = \pi_S((q^2 + p^2)^n)$ is essentially self-adjoint for $n \in \mathbb{N}_0$. Therefore,

$$\mathcal{D}((\pi_S)^*) \subseteq \bigcap_{n=0}^\infty \mathcal{D}((T^n)^*) = \bigcap_{n=0}^\infty \mathcal{D}(\overline{T^n}).$$

But $\bigcap_{n=0}^\infty \mathcal{D}(\overline{T^n}) = \mathcal{S}(\mathbb{R})$; see, e.g., [RS72, Theorem V.13]), so $\mathcal{D}((\pi_S)^*) \subseteq \mathcal{D}(\pi_S)$. Hence π_S is self-adjoint. $\qquad\bigcirc$

Example 4.33 (*An invariant subspace which is not reducing*)
Let π be the Schrödinger representation π_S of the Weyl algebra W from Example 4.32. The closed linear subspaces $\mathcal{K} := \{f \in L^2(\mathbb{R}) : f(t) = 0 \text{ a.e. on } (0, +\infty)\}$ and $\mathcal{K}^\perp$ of $\mathcal{H}(\pi)$ are invariant under π, but they are not reducing. In fact,

$$\mathcal{D}(\pi_\mathcal{K} \oplus \pi_{\mathcal{K}^\perp}) = \{f \in \mathcal{D}(\pi) : f^{(n)}(0) = 0 \text{ for } n \in \mathbb{N}_0\} \neq \mathcal{D}(\pi).$$

Moreover, since π is self-adjoint and irreducible, $\pi(W)'_w = \pi(W)'_s = \mathbb{C} \cdot I$. Hence the projection $P_\mathcal{K}$ is not in $\pi(W)'_w$. $\qquad\bigcirc$

4.4 The GNS Construction

Suppose π is a $*$-representation of the $*$-algebra A and $\varphi \in \mathcal{D}(\pi)$. Then

$$f_{\pi,\varphi}(a) := \langle \pi(a)\varphi, \varphi \rangle, \quad a \in \mathsf{A}, \tag{4.20}$$

defines a positive linear functional $f_{\pi,\varphi}$ on A. Indeed, for $a \in \mathsf{A}$, we have

$$f_{\pi,\varphi}(a^+a) = \langle \pi(a^+a)\varphi, \varphi \rangle = \langle \pi(a)\varphi, \pi(a)\varphi \rangle \geq 0.$$

Any positive functional of the form (4.20) is called a *vector functional* of π. In this section we will show (by Corollary 4.39 below) that a positive linear functional f is a vector functional of some *-representation if and only if f is extendable. In particular, *each* positive functional on a *unital* *-algebra is of the form (4.20).

Definition 4.34 Let π be a *-representation of A. We say that π is
• *algebraically cyclic* if there exists a vector $\varphi \in \mathcal{D}(\pi)$ such that $\mathcal{D}(\pi) = \pi(A)\varphi$; in this case, φ is called an *algebraically cyclic vector*,
• *cyclic* if the there exists a vector $\varphi \in \mathcal{D}(\pi)$, called then a *cyclic vector* for π, such that $\pi(A)\varphi$ is dense in $\mathcal{D}(\pi)$ with respect to the graph topology t_π.

Obviously, cyclic *-representations are nondegenerate.
Before we develop the GNS construction we prove a preliminary result.

Lemma 4.35 *Suppose f is a positive linear functional on a *-algebra A. Then*

$$\mathcal{N}_f := \{x \in A : f(x^+x) = 0\} = \{x \in A : f(ax) = 0 \text{ for all } a \in A\} \qquad (4.21)$$

is a left ideal of the algebra A. There exists an inner product $\langle \cdot, \cdot \rangle$ on the quotient space $\mathcal{D}_f := A/\mathcal{N}_f$ defined by

$$\langle x + \mathcal{N}_f, y + \mathcal{N}_f \rangle = f(y^+x), \quad x, y \in A. \qquad (4.22)$$

Proof First we prove the second equality in (4.21). If $f(ax) = 0$ for all $a \in A$, setting $a = x^+$ yields $f(x^+x) = 0$. Conversely, suppose that $f(x^+x) = 0$. Let $a \in A$. Then, by the Cauchy–Schwarz inequality (2.34),

$$|f(ax)|^2 \le f(x^+x)f(aa^+) = 0,$$

so that $f(ax) = 0$.
From the second description in (4.21) it follows at once that $\mathcal{N}_f$ is a left ideal.
We define a sesquilinear form $\langle \cdot, \cdot \rangle'$ on A by $\langle x, y \rangle' = f(y^+x)$ for $x, y \in A$. Let $u, v \in \mathcal{N}_f$. Using that $f(v^+x) = \overline{f(x^+v)}$ (by (2.35)) and (4.21) we derive

$$\langle x + u, y + v \rangle' = \langle x, y \rangle' + \langle u, y + v \rangle' + \langle x, v \rangle'$$
$$= \langle x, y \rangle' + f((y+v)^+u) + f(v^+x) = \langle x, y \rangle' + 0 + \overline{f(x^+v)} = \langle x, y \rangle'.$$

Therefore, Eq. (4.22) defines unambiguously a sesquilinear form on the quotient space $\mathcal{D}_f = A/\mathcal{N}_f$. Since f is positive, (2.35) holds by Lemma 2.33, so the form $\langle \cdot, \cdot \rangle$ is positive semi-definite and hermitian. By (4.21), $\langle x, x \rangle = f(x^+x) = 0$ if and only if $x \in \mathcal{N}_f$. Thus $\langle \cdot, \cdot \rangle$ is positive definite and hence an inner product on $\mathcal{D}_f$. $\square$

Proposition 4.36 *Suppose f is a positive linear functional on* **A**. *There exists a* *-representation π_f *on the complex inner product space* $\mathcal{D}_f = \mathbf{A}/\mathcal{N}_f$, *with inner product (4.22), such that for $a, x, y \in \mathbf{A}$,*

$$\pi_f(a)(x + \mathcal{N}_f) = ax + \mathcal{N}_f, \tag{4.23}$$

$$f(y^+ax) = \langle \pi_f(a)(x + \mathcal{N}_f), y + \mathcal{N}_f \rangle. \tag{4.24}$$

Proof Since $\mathcal{N}_f$ is a left ideal by Lemma 4.21, we have $a(x + u) \in ax + \mathcal{N}_f$ for $a, x \in \mathbf{A}$ and $u \in \mathcal{N}_f$. Hence Eq. (4.23) defines unambiguously a linear operator $\pi_f(a)$ on the vector space $\mathcal{D}_f$. Obviously, $a \mapsto \pi_f(a)$ is an algebra homomorphism of **A** into $L(\mathcal{D}_f)$. For $a, x, y \in \mathbf{A}$ we have

$$\langle \pi_f(a)(x + \mathcal{N}_f), y + \mathcal{N}_f \rangle = \langle ax + \mathcal{N}_f, y + \mathcal{N}_f \rangle = f(y^+ax)$$
$$= f((a^+y)^+x) = \langle x + \mathcal{N}_f, a^+y + \mathcal{N}_f \rangle = \langle x + \mathcal{N}_f, \pi_f(a^+)(y + \mathcal{N}_f) \rangle.$$

This shows that π_f is a *-representation of **A** on the complex inner product space $\mathcal{D}_f$ and that Eq. (4.24) is satisfied. $\qquad\square$

From the view point of representation theory, the following assertion is the main reason for requiring condition (2.23) in the definition of a pre-quadratic module.

Corollary 4.37 *Let Q be a pre-quadratic module and f a positive linear functional on* **A**. *If f is Q-positive, so is the *-representation π_f, that is, $\pi_f(a) \geq 0$ for $a \in Q$.*

Proof Let $a \in Q$ and $x \in \mathbf{A}$. Since $x^+ax \in Q$ by (2.23) and f is Q-positive, $f(x^+ax) \geq 0$. Hence, by (4.24), the symmetric operator $\pi_f(a)$ is positive. $\qquad\square$

The next theorem is the major result of this chapter.

Theorem 4.38 *Suppose* **A** *is a unital *-algebra and f is a positive linear functional on* **A**. *Then the *-representation π_f on the complex inner product space $\mathcal{D}_f$ defined by (4.23) is algebraically cyclic with algebraically cyclic vector $\varphi_f := 1 + \mathcal{N}_f$ and*

$$f(y^+ax) = \langle \pi_f(a)\pi_f(x)\varphi_f, \pi_f(y)\varphi_f \rangle, \quad a, x, y \in \mathbf{A}. \tag{4.25}$$

In particular,

$$f(a) = \langle \pi_f(a)\varphi_f, \varphi_f \rangle, \quad a \in \mathbf{A}. \tag{4.26}$$

*If ρ is another algebraically cyclic *-representation of* **A** *with algebraically cyclic vector ψ such that $f(a) = \langle \rho(a)\psi, \psi \rangle$ for $a \in \mathbf{A}$, then there is a unitary operator U of $\mathcal{H}(\pi_f)$ on $\mathcal{H}(\rho)$ such that $U\mathcal{D}(\pi_f) = \mathcal{D}(\rho)$, $U\varphi_f = \psi$ and*

$$\rho = U\pi_f U^{-1}.$$

Proof Let π_f be the *-representation from Proposition 4.36. Since $z + \mathcal{N}_f = \pi_f(z)(1 + \mathcal{N}_f) = \pi_f(z)\varphi_f$ for $z = x, y$ by (4.23), (4.24) gives (4.25). Setting $x = y = 1$ in (4.25) we obtain (4.26). Since $\mathcal{D}_f = \{x + \mathcal{N}_f : x \in \mathsf{A}\} = \pi_f(\mathsf{A})\varphi_f$ by (4.23), φ_f is algebraically cyclic for π_f.

Next we prove the uniqueness assertion. Let $x \in \mathsf{A}$. Applying (4.25) with $x = y$, $a = 1$ and the assumption on ρ we derive

$$\|\pi_f(x)\varphi_f\|^2 = f(x^+x) = \langle \rho(x^+x)\psi, \psi \rangle = \langle \rho(x)\psi, \rho(x)\psi \rangle = \|\rho(x)\psi\|^2.$$

Hence there exists a well-defined isometric linear map U of $\mathcal{D}_f = \pi_f(\mathsf{A})\varphi_f$ on $\mathcal{D}(\rho) = \rho(\mathsf{A})\psi$ given by $U\pi_f(x)\varphi_f = \rho(x)\psi$, $x \in \mathsf{A}$. It extends by continuity to a unitary operator, denoted again by U, of the corresponding Hilbert spaces $\mathcal{H}(\pi_f)$ and $\mathcal{H}(\rho)$. Setting $x = 1$ we get $U\varphi_f = U\pi_f(1)\varphi_f = \rho(1)\psi = \psi$. For $a, x \in \mathsf{A}$, we have $U^{-1}\rho(x)\psi = \pi_f(x)\varphi_f$ and

$$U\pi_f(a)U^{-1}\rho(x)\psi = U\pi_f(a)\pi_f(x)\varphi_f = U\pi_f(ax)\varphi_f = \rho(ax)\psi = \rho(a)\rho(x)\psi.$$

This proves that $U\pi_f U^{-1} = \rho$. $\square$

Corollary 4.39 *A positive linear functional f on a *-algebra A is extendable if and only if there exists a *-representation π of A and a vector $\varphi \in \mathcal{D}(\pi)$ such that*

$$f(a) = \langle \pi(a)\varphi, \varphi \rangle \quad \text{for} \quad a \in \mathsf{A}. \tag{4.27}$$

Proof First suppose f is extendable. Then f extends to a positive linear functional f_1 of the unitization A^1. If π denotes the restriction of the *-representation π_{f_1} of A^1 to A and $\varphi := \varphi_{f_1}$, then (4.27) holds by (4.26).

Conversely, suppose (4.27) is satisfied. Then, for $a \in \mathsf{A}$,

$$f(a^+) = \langle \pi(a^+)\varphi, \varphi \rangle = \langle \varphi, \pi(a)\varphi \rangle = \overline{f(a)},$$
$$|f(a)|^2 = |\langle \pi(a)\varphi, \varphi \rangle|^2 \le \|\varphi\|^2 \|\pi(a)\varphi\|^2 = \|\varphi\|^2 \langle \pi(a^+a)\varphi, \varphi \rangle = \|\varphi\|^2 f(a^+a).$$

This shows that f is hermitian and condition (2.37) is satisfied. Therefore, f is extendable by Proposition 2.35. $\square$

Example 4.40 (*Example* 2.39 *continued*)
Let f be the non-extendable positive functional on the *-algebra $\mathsf{A} = x\mathbb{C}[x]$ defined by $f(xp(x)) = p(0)$, $p \in \mathbb{C}[x]$. Since $f((xp(x))^+xp(x)) = 0$ for $p \in \mathbb{C}[x]$, we have $\mathsf{A} = \mathcal{N}_f$, so that $\mathcal{D}_f = \{0\}$ and $\pi_f(a) = 0$ for all $a \in \mathsf{A}$. $\bigcirc$

The following is the counterpart of Theorem 4.38 for closed *-representations.

Theorem 4.41 *Suppose f is a positive linear functional on a unital *-algebra A. Then the closure $\overline{\pi}_f$ of the *-representation π_f defined by (4.23) is a closed cyclic *-representation with cyclic vector $\varphi_f := 1 + \mathcal{N}_f$ and*

$$f(a) = \langle \overline{\pi}_f(a)\varphi_f, \varphi_f \rangle, \quad a \in \mathsf{A}.$$

If ρ is a cyclic closed $*$-representation of A with cyclic vector ψ such that $f(a) = \langle \rho(a)\psi, \psi \rangle$ for $a \in \mathsf{A}$, then there is a unitary operator U of $\mathcal{H}(\pi_f)$ on $\mathcal{H}(\rho)$ such that $U\mathcal{D}(\overline{\pi}_f) = \mathcal{D}(\rho)$, $U\varphi_f = \psi$, and $\rho = U\overline{\pi}_f U^{-1}$.

Proof The first group of assertions follows at once from Theorem 4.38.

We prove the uniqueness. Let ρ_0 denote the restriction of ρ to $\mathcal{D}(\rho_0) := \rho(\mathsf{A})\psi$. Then ψ is algebraically cyclic for ρ_0 and $f(\cdot) = \langle \rho_0(\cdot)\psi, \psi \rangle$, so by the uniqueness assertion of Theorem 4.38 there is a unitary U of $\mathcal{H}(\pi_f)$ on $\mathcal{H}(\rho_0)$ such that $U\mathcal{D}(\pi_f) = \mathcal{D}(\rho_0)$, $U\varphi_f = \psi$, $\rho_0 = U\pi_f U^{-1}$. Taking the closures by using that ψ is cyclic for ρ we get $U\mathcal{D}(\overline{\pi}_f) = \mathcal{D}(\rho)$ and $\rho = U\overline{\pi}_f U^{-1}$. $\qquad\square$

Definition 4.42 The $*$-representation π_f with algebraically cyclic vector φ_f defined by (4.23) and satisfying (4.26), likewise its closure $\overline{\pi}_f$, is called the *GNS representation* associated with the positive functional f on the unital $*$-algebra A.

The notations $\pi_f, \overline{\pi}_f, \varphi_f$ will be kept throughout this book.

Theorems 4.38 and 4.41 are key results of the theory of $*$-representations on Hilbert spaces. The significance of these results and of the preceding considerations lies in the ingenious construction, discovered by I.M. Gelfand and M.A. Naimark and studied by I. Segal, of an inner product and a Hilbert space action of the $*$-algebra A derived from the positive functional. This method is usually called the *Gelfand–Naimark–Segal construction* or briefly the *GNS construction*. It is an important technical tool in various parts of mathematics such as group representation theory, operator algebras, reproducing kernels, moment problems, and others.

Definition 4.43 The $*$-*radical*, or the *reducing ideal*, Rad A of a $*$-algebra A is the intersection of kernels $\mathcal{N}(\pi) := \{a \in \mathsf{A} : \pi(a) = 0\}$ of all $*$-representations π of A. A $*$-algebra A is called $*$-*semisimple*, or *reduced*, if Rad $\mathsf{A} = \{0\}$.

That is, a $*$-algebra A is $*$-semisimple if and only if their $*$-representations separate the points of A. Clearly, Rad A is a two-sided $*$-ideal of A and the quotient $*$-algebra $\mathsf{A}/(\text{Rad } \mathsf{A})$ is always $*$-semisimple.

Proposition 4.44 *For any $*$-algebra A we have*

$$\text{Rad } \mathsf{A} = \{x \in \mathsf{A} : f(x^+ x) = 0 \ \text{ for all } \ f \in \mathcal{P}_e(\mathsf{A})^*\}$$
$$= \{x \in \mathsf{A} : f(ax) = 0 \ \text{ for all } \ a \in \mathsf{A}, f \in \mathcal{P}_e(\mathsf{A})^*\},$$

where $\mathcal{P}_e(\mathsf{A})^*$ denotes the set of all extendable positive linear functionals on A.

Proof By Corollary 4.39, the functionals f of $\mathcal{P}_e(\mathsf{A})^*$ are precisely the vector functionals $f(\cdot) = \langle \pi(\cdot)\varphi, \varphi \rangle$ of $*$-representations. Then $f(x^+ x) = \|\pi(x)\varphi\|^2$, which yields the first equality. The second equality follows from (4.21). $\qquad\square$

Corollary 4.45 *Let $a_1, \ldots, a_n$ be elements of* A *such that* $\sum_{k=1}^{n}(a_k)^+a_k = 0$. *Then* $a_1, \ldots, a_n \in$ Rad A. *In particular, if* A *admits a faithful *-representation, then we obtain* $a_1 = \cdots = a_n = 0$.

Proof Let $f \in \mathcal{P}_e(\mathsf{A})^*$. Then we have $f((a_k)^+a_k) \geq 0$. Therefore, the equation $0 = f\left(\sum_k (a_k)^+a_k\right) = \sum_k f((a_k)^+a_k)$ implies that $f((a_k)^+a_k) = 0$ for each k. Hence $a_k \in$ Rad A by Proposition 4.44. $\qquad\square$

Example 4.46 Let A be the *-algebra of rational functions in one variable with complex conjugation as involution. Then A has no nonzero positive linear functional and hence Rad A = A.

Indeed, assume to the contrary that there is a positive functional $f \neq 0$ on A. Then $\pi_f(\mathsf{A})$ is a complex division algebra *and* an O^*-algebra, so that $\pi_f(\mathsf{A}) = \mathbb{C} \cdot I$ by [Sch90, Proposition 2.1.12]. Since $f \neq 0$, $\mathcal{H}(\pi_f) \neq \{0\}$. Hence $\pi_f(a) = \lambda(a)I$, $a \in \mathsf{A}$, defines a hermitian character λ of A. This is a contradiction, since A has no character, as noted in Example 2.52. $\bigcirc$

4.5   Examples of GNS Representations

The GNS representation is a useful tool for the operator-theoretic approach to the multi-dimensional moment problem. This is elaborated in detail in [Sch17, Sect. 12.5]. We do not repeat this here. The following example relates the determinacy of the one-dimensional Hamburger moment problem to Hilbert space representations.

Example 4.47 (*One-dimensional Hamburger moment problem*)
Let $s = (s_n)_{n \in \mathbb{N}_0}$ be a real sequence. The Riesz functional of s is the linear functional f_s on $\mathsf{A} := \mathbb{C}[x]$ defined by $f_s(x^n) = s_n, n \in \mathbb{N}_0$. Suppose f_s is a positive functional on A; see also Example 2.40. Then, by a classical result on the Hamburger moment problem [Sch17, Theorem 3.8], s is a moment sequence, that is, there exists a Radon measure μ on $\mathbb{R}$ such that $f_s(p) = \int p(x)d\mu(x)$ for $p \in \mathsf{A}$. To avoid trivial cases assume that μ has infinite support. Then $\mathcal{N}_{f_s} = \{0\}$ and the GNS representation π_{f_s} is faithful and acts on the complex inner product space $\mathcal{D}(\pi_{f_s}) = \mathbb{C}[x]$, with inner product $\langle q_1, q_2 \rangle = \int q_1 \overline{q_2}\, d\mu, q_1, q_2 \in \mathbb{C}[x]$, by $\pi_{f_s}(p)q = p \cdot q, p \in \mathsf{A}, q \in \mathcal{D}(\pi_{f_s})$.

Statement: *The GNS representation π_{f_s} is essentially self-adjoint if and only if the moment sequence s is determinate.*

Proof A basic result on the Hamburger moment problem [Sch17, Theorem 6.10] states that s is determinate if and only if the operator $\pi_{f_s}(x)$ is essentially self-adjoint. By Proposition 4.19(v), this holds if and only if the *-representation π_{f_s} of $\mathsf{A} = \mathbb{C}[x]$ is essentially self-adjoint. $\square$

By Proposition 4.19(vii), the closure $\overline{\pi}_{f_s}$ is self-adjoint if and only if all powers of $\pi_{f_s}(x)$ are essentially self-adjoint. This is a strong condition, and there are moment sequences s such that π_{f_s} is essentially self-adjoint, but $\overline{\pi}_{f_s}$ is not self-adjoint. $\bigcirc$

Now we return to the setup at the beginning of Sect. 4.4 and consider a $*$-representation π of a unital $*$-algebra A and a vector $\varphi \in \mathcal{D}(\pi)$. We want to describe the GNS representations $\pi_f, \overline{\pi}_f$ of the positive functional $f(\cdot) := \langle \pi(\cdot)\varphi, \varphi \rangle$. Let ρ_φ denote the restriction of π to the domain $\mathcal{D}(\rho_\varphi) := \pi(A)\varphi$. Then ρ is an algebraically cyclic $*$-representation of A with algebraically cyclic vector φ and $f(\cdot) = \langle \rho_\varphi(\cdot)\varphi, \varphi \rangle$. Hence, by the uniqueness assertions of Theorems 4.38 and 4.41, ρ_φ (resp. $\overline{\rho}_\varphi$) is unitarily equivalent to the GNS representation π_f (resp. $\overline{\pi}_f$), with a unitary U defined by $U(\pi_f(a)\varphi_f) = \rho_\varphi(a)\varphi$, $a \in A$, and satisfying $\rho_\varphi = U\pi_f U^{-1}$ and $\overline{\rho}_\varphi = U\overline{\pi}_f U^{-1}$. In general, it can be difficult to determine the domains $\mathcal{D}(\rho_\varphi)$ and $\mathcal{D}(\overline{\rho}_\varphi)$ explicitly. We illustrate this with two examples.

Example 4.48 (*Trace functionals; see Sect. 3.3*)
Let ρ be a closed $*$-representation of a unital $*$-algebra A and t a trace class operator of $\mathbf{B}_1(\rho(A))_+$; see Definition 3.22. There is a positive functional f_t on A defined by

$$f_t(a) = \operatorname{Tr} \rho(a)t, \quad a \in A.$$

Let $t = \sum_n \varphi_n \otimes (\lambda_n \varphi_n)$ be a representation (3.8) of the operator t. We denote by $\mathcal{H}_\infty$ the direct sum $\oplus_{n=1}^\infty \mathcal{H}(\rho)$ of countably many copies of the Hilbert space $\mathcal{H}(\rho)$. Then, by Proposition 3.24(i), $(\rho(b)\lambda_n^{1/2}\varphi_n)_{n \in \mathbb{N}} \in \mathcal{H}_\infty$ for all $b \in A$. Hence

$$\rho_\Phi(a)(\rho(b)\lambda_n^{1/2}\varphi_n)_{n \in \mathbb{N}} := (\rho(ab)\lambda_n^{1/2}\varphi_n)_{n \in \mathbb{N}}, \quad a, b \in A,$$

defines a $*$-representation ρ_Φ of A on the subspace $\{(\rho(a)\lambda_n^{1/2}\varphi_n)_{n \in \mathbb{N}} : a \in A\}$ of $\mathcal{H}_\infty$ with algebraically cyclic vector $\Phi := (\lambda_n^{1/2}\varphi_n)_{n \in \mathbb{N}}$. Using (3.7) we derive

$$f_t(a) = \sum_{n=1}^\infty \langle \rho(a)\varphi_n, \lambda_n \varphi_n \rangle = \sum_{n=1}^\infty \langle \rho(a)\lambda_n^{1/2}\varphi_n, \lambda_n^{1/2}\varphi_n \rangle = \langle \rho_\Phi(a)\Phi, \Phi \rangle, \quad a \in A.$$

Therefore, by the uniqueness assertion in Theorem 4.38, the *GNS representation π_{f_t} is unitarily equivalent to ρ_Φ*.

Now we assume in addition that $t^{1/2}\mathcal{H}(\rho) \subseteq \mathcal{D}(\rho)$. We use the $*$-representation ρ_{HS} from Example 4.10 to obtain a realization of π_{f_t} on Hilbert–Schmidt operators. Since t is trace class, $t^{1/2}$ is Hilbert–Schmidt. First we prove that $t^{1/2}$ belongs to the domain $\mathcal{D}(\rho_{HS})$. For this we have to show that, for any $a \in A$, the operator $s := \rho(a)t^{1/2}$ is in $\mathbf{B}_2(\mathcal{H}(\rho))$. Clearly, $s^* \supseteq t^{1/2}\rho(a^+)$, so that $s^*s \supseteq t^{1/2}\rho(a^+)\rho(a)t^{1/2} = t^{1/2}\rho(a^+a)t^{1/2}$. But $t^{1/2}\rho(a^+a)t^{1/2}$ is defined on the whole Hilbert space $\mathcal{H}(\rho)$ by the assumption $t^{1/2}\mathcal{H}(\rho) \subseteq \mathcal{D}(\rho)$. Thus, $s^*s = t^{1/2}\rho(a^+a)t^{1/2}$ and this operator is trace class by Proposition 3.24(iii). Hence s is Hilbert–Schmidt, so that $t^{1/2} \in \mathcal{D}(\rho_{HS})$. Further, since $f_t(a) = \operatorname{Tr} t^{1/2}at^{1/2}$ by (3.10), we obtain

$$\langle \rho_{HS}(a)t^{1/2}, t^{1/2} \rangle_{HS} = \operatorname{Tr} t^{1/2}\rho(a)t^{1/2} = f_t(a) \quad \text{for} \quad a \in A. \tag{4.28}$$

This shows that the positive functional f_t is realized as a vector functional of the $*$-representation ρ_{HS}. Hence, as discussed above, the *GNS representation* π_{f_t} *is unitarily equivalent to the restriction of* ρ_{HS} *to the domain* $\rho_{HS}(A)t^{1/2}$.   ○

Example 4.49 (*Vector functionals of the Schrödinger representation*)
Let π be the Schrödinger representation of the Weyl algebra W on $L^2(\mathbb{R})$; see Example 4.32. Suppose $\varphi \in C_0^\infty(\mathbb{R})$ is a fixed function satisfying the following condition:

$$\varphi(x) \neq 0 \ \text{for} \ x \in (0, 1), \quad \varphi(x) = 0 \ \text{for} \ x \in (-\infty, 0] \cup [1, +\infty). \qquad (4.29)$$

Define $f(\cdot) := \langle \pi(\cdot)\varphi, \varphi \rangle$. Let π_0 denote the restriction of π to the dense domain

$$\mathcal{D}(\pi_0) = \big\{ \psi \in C^\infty([0, 1]) : \psi^{(k)}(0) = \psi^{(k)}(1) = 0 \ \text{for} \ k \in \mathbb{N}_0 \big\}$$

of the Hilbert space $L^2(0, 1)$. The following result says π_0 is unitarily equivalent to the GNS representation $\overline{\pi}_f$.

Statement: *There exists a unitary operator* U *of* $\mathcal{H}(\pi_f)$ *on* $L^2(0, 1)$, *given by* $U(\pi_f(a)\varphi) = \pi_0(a)\varphi$ *for* $a \in$ W, *such that* $\pi_0 = U\overline{\pi}_f U^{-1}$.

Proof Let ρ_φ denote the restriction of π to $\mathcal{D}(\rho_\varphi) = \pi(\mathsf{W})\varphi$. As noted above, the unitary operator U defined by $U(\pi_f(a)\varphi_f) = \rho_\varphi(a)\varphi \equiv \pi_0(a)\varphi$, $a \in$ W, provides the unitary equivalence $\overline{\rho}_\varphi = U\overline{\pi}_f U^{-1}$. Clearly, $\overline{\rho}_\varphi \subseteq \pi_0$, since π_0 is closed. To prove the statement it therefore suffices to show that $\pi(\mathsf{W})\varphi$ is dense in $\mathcal{D}(\pi_0)$ in the graph topology of $\pi_0(\mathsf{W})$. For this we will use Lemma 4.51 below.

Each element $a \in$ W is a finite sum of terms $f(q)p^n$, with $n \in \mathbb{N}_0$ and $f \in \mathbb{C}[q]$. Since the operators $\pi_0(f(q))$ are bounded, the graph topology $\mathsf{t}_{\pi_0(\mathsf{W})}$ is generated by the seminorms $\|\pi_0(ip)^n \cdot \|$, $n \in \mathbb{N}_0$, on $\mathcal{D}(\pi_0)$. Let $\psi \in \mathcal{D}(\pi_0)$.

First assume that ψ vanishes in neighborhoods of the points 0 and 1. Then, by Lemma 4.51, for any $m \in \mathbb{N}$ there is sequence $(f_n)_{n \in \mathbb{N}}$ of polynomials such that

$$\lim_n \pi_0((ip)^k)(\pi_0(f_n(q))\varphi - \psi) = \lim_n ((f_n\varphi)^{(k)} - \psi^{(k)}) = 0, \quad k = 0, \ldots, m,$$

in $L^2(0, 1)$. This shows that ψ is in the closure of $\pi_0(\mathsf{W})\varphi$ in the graph topology.

In the general case we set $\psi_\varepsilon(x) := \psi((1 + 2\varepsilon)x - \varepsilon)$ for $\varepsilon > 0$, where $\psi(x) := 0$ on $\mathbb{R}/[0, 1]$. Then ψ_ε vanishes in neighborhoods of 0 and 1, so it is in the closure of $\pi_0(\mathsf{W})\varphi$ by the preceding. By the dominated Lebesgue convergence theorem,

$$\lim_{\varepsilon \to +0} \pi_0((ip)^k)(\psi_\varepsilon - \psi) = \lim_{\varepsilon \to +0} (\psi_\varepsilon^{(k)} - \psi^{(k)}) = 0 \ \text{for} \ k \in \mathbb{N}_0$$

in $L^2(0, 1)$. Therefore, since ψ_ε is in the closure of $\pi_0(\mathsf{W})\varphi$, so is ψ.   □○

Lemma 4.50 *Let* $k \in \mathbb{N}$ *and* $g \in C^{(k)}([0, 1])$. *There exists a sequence* $(f_n)_{n \in \mathbb{N}}$ *of polynomials such that* $f_n^{(j)} \Longrightarrow g^{(j)}$ *uniformly on* $[0, 1]$ *for* $j = 0, \ldots, k$.

Proof By the Weierstrass theorem there is a sequence $(h_n)_{n\in\mathbb{N}}$ of polynomials such that $h_n \Longrightarrow g^{(k)}$ uniformly on $[0, 1]$. Set $h_{n,k} := h_n$. Then

$$h_{n,k-1}(x) := g^{(k)}(0) + \int_0^x h_{n,k}(y)dy \Longrightarrow g^{(k-1)}(x) = g^{(k)}(0) + \int_0^x g^{(k)}(y)dy.$$

Clearly, $(h_{n,k-1})_{n\in\mathbb{N}}$ is sequence of polynomials such that $h'_{n,k-1}(x) = h_{n,k}(x)$. Proceeding by induction we obtain sequences $(h_{n,k-j})_{n\in\mathbb{N}}$, $j = 0, \ldots, k$, of polynomials such that $h_{n,k-j} \Longrightarrow g^{(k-j)}$ uniformly on $[0, 1]$ and $h'_{n,k-j}(x) = h_{n,k+1-j}(x)$. Then the sequence $(f_n := h_{n,0})_{n\in\mathbb{N}}$ has the desired properties. $\qquad\square$

Lemma 4.51 *Suppose that $\varphi \in C_0^\infty(\mathbb{R})$ satisfies condition (4.29). Let $m \in \mathbb{N}$ and $\psi \in C_0^{(m)}([0, 1])$. Then there exists a sequence $(f_n)_{n\in\mathbb{N}}$ of polynomials such that $\lim_{n\to\infty} (f_n\varphi)^{(k)} = \psi^{(k)}$ in $L^2(0, 1)$ for $k = 0, \ldots, m$.*

Proof By the assumption, ψ vanishes in neighborhoods of 0 and 1. Hence $\psi\varphi^{-1}$ is in $C^{(m)}([0, 1])$. By Lemma 4.50, there exists a sequence $(f_n)_{n\in\mathbb{N}}$ of polynomials such that $f_n^{(j)} \Longrightarrow (\psi\varphi^{-1})^{(j)}$ for $j = 0, \ldots, m$ uniformly on $[0, 1]$. Then

$$(f_n\varphi)^{(k)} = \sum_{j=0}^k \binom{k}{j} f_n^{(j)} \varphi^{(k-j)} \Longrightarrow \sum_{j=0}^k \binom{k}{j} (\psi\varphi^{-1})^{(j)} \varphi^{(k-j)} = \psi^{(k)}$$

uniformly on $[0, 1]$ and hence in the Hilbert space $L^2(0, 1)$. $\qquad\square$

4.6 Positive Semi-definite Functions on Groups

In this section, we use the GNS construction to relate positive semi-definite functions on groups to unitary representations. First suppose G is an arbitrary group.

Definition 4.52 A *unitary representation* of a group G on a Hilbert space $\mathcal{H}$ is an algebra homomorphism U of G into the group of unitary operators on $\mathcal{H}(U) := \mathcal{H}$.

Since a unitary representation U is a homomorphism into the unitaries, we have

$$U(e) = I \quad \text{and} \quad U(g^{-1}) = U(g)^{-1} = U(g)^* \quad \text{for} \quad g \in G.$$

As noted in Example 2.40, there is a one-to-one correspondence between positive linear functionals on the group $*$-algebra $\mathbb{C}[G]$ and positive semi-definite functions on the group G. Recall that positive semi-definite functions on groups were introduced in Definition 2.41.

Proposition 4.53 *A function f on a group G is positive semi-definite if and only if there exist a unitary representation U of G and a vector $\xi \in \mathcal{H}(U)$ such that*

$$f(g) = \langle U(g)\xi, \xi \rangle, \quad g \in G. \tag{4.30}$$

Proof First suppose that f is a positive semi-definite function on G. Let F be the positive functional on $\mathbb{C}[G]$ given by $F(\sum_g \alpha_g g) = \sum_g \alpha_g f(g)$ and let π_F be its GNS representation. For $g \in G$ and $\varphi \in \mathcal{D}(\pi_F)$, we obtain

$$\begin{aligned}
\|\pi_F(g)\varphi\|^2 &= \langle \pi_F(g)\varphi, \pi_F(g)\varphi \rangle = \langle \pi_F(g^+)\pi_F(g)\varphi, \varphi \rangle \\
&= \langle \pi_F(g^{-1}g)\varphi, \varphi \rangle = \langle \pi_F(e)\varphi, \varphi \rangle = \|\varphi\|^2.
\end{aligned}$$

Therefore, since $\pi_F(g)\mathcal{D}(\pi_F) \supseteq \pi_F(g)\pi_F(g^{-1})\mathcal{D}(\pi_F) = \pi_F(e)\mathcal{D}(\pi_F) = \mathcal{D}(\pi_F)$ is dense in $\mathcal{H}(\pi_F)$, $\pi_F(g)$ extends by continuity to a unitary operator $U(g)$ on $\mathcal{H}(\pi_F)$. Since π_F is a ∗-representation of $\mathbb{C}[G]$, $g \mapsto U(g)$ is a unitary representation of G. Then, $f(g) = F(g) = \langle \pi_F(g)\varphi_F, \varphi_F \rangle = \langle U(g)\varphi_F, \varphi_F \rangle$ for $g \in G$, which proves (4.30) with $\xi = \varphi_F$.

Now suppose f is of the form (4.30). Define $\pi(\sum_g \alpha_g g) = \sum_g \alpha_g U(g)$. Since U is a unitary representation of G, π is a ∗-representation of the ∗-algebra $\mathbb{C}[G]$. Hence $F(a) = \langle \pi(a)\xi, \xi \rangle$, $a \in \mathbb{C}[G]$, is a positive linear functional on $\mathbb{C}[G]$. Then, $F(g) = \langle \pi(g)\xi, \xi \rangle = \langle U(g)\xi, \xi \rangle = f(g)$, $g \in G$, so the restriction of F to G is f. Hence f is a positive semi-definite function. $\qquad\square$

Now we turn to topological groups.

A *topological group* is a group G equipped with a topology such that the mappings $(g, h) \mapsto g \cdot h$ of $G \times G$ into G and $g \mapsto g^{-1}$ of G into G are continuous.

Definition 4.54 A *unitary representation* of a topological group G is a unitary representation U of the group G (according to Definition 4.52) such that for each vector $\varphi \in \mathcal{H}(U)$ the map $G \ni g \mapsto U(g)\varphi \in \mathcal{H}(U)$ is continuous.

The continuity condition in Definition 4.54 is the continuity in the strong operator topology. A *scalar* characterization of the continuity is contained in the next lemma.

Lemma 4.55 *Suppose G is a topological group and U is a homomorphism of G into the unitary group of a Hilbert space $\mathcal{H}$. Let $\mathcal{D}$ be a dense subset of $\mathcal{H}$. Then the following statements are equivalent:*

(i) *U is a unitary representation of the topological group G.*
(ii) *For all $\varphi, \psi \in \mathcal{H}$, the function $g \mapsto \langle U(g)\varphi, \psi \rangle$ is continuous on G.*
(iii) *For all $\varphi \in \mathcal{H}$, the function $g \mapsto \langle U(g)\varphi, \varphi \rangle$ is continuous at the unit e of G.*
(iv) *For all $\varphi \in \mathcal{D}$, the function $g \mapsto \langle U(g)\varphi, \varphi \rangle$ is continuous on G.*

Proof The implications (i)→(ii)→(iii)→(iv) are obvious, so it suffices to prove (iii)→(i) and (iv)→(iii).

(iii)→(i): Let $g, h \in G$. Using that U is a homomorphism and the operators $U(g), U(h)$ are unitary, we derive

$$\|U(g)\varphi - U(h)\varphi\|^2$$
$$= \langle U(g)\varphi, U(g)\varphi \rangle + \langle U(h)\varphi, U(h)\varphi \rangle - \langle U(g)\varphi, U(h)\varphi \rangle - \langle U(h)\varphi, U(g)\varphi \rangle$$
$$= 2\langle \varphi, \varphi \rangle - \langle U(h^{-1}g)\varphi, \varphi \rangle - \langle \varphi, U(h^{-1}h)\varphi \rangle = 2\|\varphi\|^2 - 2\,\mathrm{Re}\,\langle U(h^{-1}g)\varphi, \varphi \rangle.$$

If $g \to h$ in G, then $h^{-1}g \to e$ in G, hence $\langle U(h^{-1}g)\varphi, \varphi \rangle \to \langle U(e)\varphi, \varphi \rangle = \|\varphi\|^2$ by (iii), and therefore $U(g)\varphi \to U(h)\varphi$ in $\mathcal{H}$ by the preceding equality.

(iv)→(iii): Let $\varphi \in \mathcal{H}(U)$ and $\varepsilon > 0$. Since $\mathcal{D}$ is dense, we can find a vector $\varphi' \in \mathcal{D}$ such that $\|\varphi - \varphi'\| \leq \varepsilon$. Then

$$|\langle U(g)\varphi, \varphi \rangle - \langle U(g)\varphi', \varphi' \rangle| = |\langle U(g)\varphi, \varphi - \varphi' \rangle + \langle U(g)(\varphi - \varphi'), \varphi' \rangle|$$
$$\leq (\|\varphi\| + \|\varphi'\|)\,\|\varphi - \varphi'\| \leq (2\,\|\varphi\| + \varepsilon)\,\varepsilon.$$

Therefore, by (iv), the function $\langle U(g)\varphi, \varphi \rangle$ is uniform limit of continuous functions $\langle U(g)\varphi', \varphi' \rangle$. Hence $\langle U(g)\varphi, \varphi \rangle$ is continuous on G. □

The following is the counterpart of Proposition 4.53 for topological groups.

Theorem 4.56 *Suppose G is a topological group. A function f on G is positive semi-definite and continuous if and only if there exist a unitary representation U of the topological group G and a vector $\xi \in \mathcal{H}(U)$ such that (4.30) holds. The vector ξ can be chosen as a cyclic vector for the representation U.*

Proof The if part is clear by Proposition 4.53 and Definition 4.54.

Conversely, suppose that f is positive semi-definite and continuous. By Proposition 4.53, there are a unitary representation V of the group G and a vector $\xi \in \mathcal{H}(V)$ such that $f(g) = \langle V(g)\xi, \xi \rangle$, $g \in G$. Let $\mathcal{H}$ be the closure of the linear subspace $\mathcal{D} := \mathrm{Lin}\,\{V(g)\xi : g \in G\}$ in $\mathcal{H}(V)$. Then V leaves $\mathcal{D}$ invariant and so $\mathcal{H}$. Hence $U := V \upharpoonright \mathcal{H}$ is a unitary representation of the group G with cyclic vector ξ.

Let $\varphi \in \mathcal{D}$. Then φ is of the form $\varphi = \sum_j V(h_j)\xi$, where $h_j \in G$, and

$$\langle U(g)\varphi, \varphi \rangle = \sum_{j,k} \langle U(g)V(h_j)\xi, V(h_k)\xi \rangle$$
$$= \sum_{j,k} \langle V(h_k^{-1}gh_j)\xi, \xi \rangle = \sum_{j,k} f(h_k^{-1}gh_j)$$

is continuous on G, because f is continuous. Therefore, since $\mathcal{D}$ is dense in $\mathcal{H}$, it follows from Lemma 4.55,(iv)→(i), that U is a unitary representation of the topological group G. By construction, $f(g) = \langle V(g)\xi, \xi \rangle = \langle U(g)\xi, \xi \rangle$ for $g \in G$. □

Example 4.57 (*Positive semi-definite functions on $\mathbb{R}^d$*)
Suppose f is a continuous positive semi-definite function on the topological group $G = \mathbb{R}^d$. Then, by Bochner's theorem (see, e.g., [RS75]), there exists a unique finite Radon measure μ on $\mathbb{R}^d$ such that

$$f(x) = \int_{\mathbb{R}^d} e^{\mathrm{i}(x,y)} d\mu(y), \quad x \in \mathbb{R}^d, \tag{4.31}$$

where (x, y) denotes the inner product of $x, y \in \mathbb{R}^d$.

Let U be the unitary representation of $\mathbb{R}^d$ on $\mathcal{H}(U) = L^2(\mathbb{R}^d; \mu)$ defined by $(U(x)f)(y) = e^{\mathrm{i}(x,y)} f(y)$, $x, y \in \mathbb{R}^d$, and $\xi(y) \equiv 1$. Then (4.31) reads as $f(x) = \langle U(x)\xi, \xi \rangle$, which is Eq. (4.30) in this case.   ○

4.7 Pathologies with Unbounded Representations

It is well known that unbounded operators can lead to many technical difficulties and subtleties, but unbounded *-representations are even worse. In this short section we mention and briefly discuss some of these problems.

(1) *Orthogonal complements of subrepresentations*

If $\mathcal{K}$ is an invariant closed linear subspace for a bounded *-representation π on a Hilbert space $\mathcal{H} = \mathcal{D}(\pi)$, then $\mathcal{K}^\perp$ is also invariant and π is the direct sum of the subrepresentations $\pi \lceil \mathcal{K}$ and $\pi \lceil \mathcal{K}^\perp$.

In contrast, for unbounded *-representations these facts are no longer true, see Example 4.33. It may even happen that $\mathcal{E}$ is an invariant linear subspace of $\mathcal{D}(\pi)$ such that $\mathcal{E}^\perp \neq \{0\}$, but $\mathcal{E}^\perp \cap \mathcal{D}(\pi) = \{0\}$; see [Sch90, Example 8.3.8].

(2) *Decomposition into a direct sum of cyclic subrepresentations*

Each bounded *-representation π on a Hilbert space $\mathcal{H} = \mathcal{D}(\pi)$ is a direct sum of cyclic *-representations. The standard proof of this fact goes as follows: Let π_0 on $\mathcal{H}_0$ be a maximal direct sum of cyclic subrepresentations of π. If $\mathcal{H}_0 \neq \mathcal{H}$, then $(\mathcal{H}_0)^\perp$ is invariant and there is a nonzero vector $\varphi \in \mathcal{H}_0^\perp$. Then $\mathcal{K}_\varphi := \overline{\pi(\mathsf{A})\varphi}$ is an invariant subspace and $\pi \oplus (\pi \lceil \mathcal{K}_\varphi)$ is a larger sum of cyclic subrepresentations, which contradicts the maximality of the chosen family. Thus, $\mathcal{H}_0 = \mathcal{H}$.

This procedure fails for unbounded representations, since it can happen that $(\mathcal{H}_0)^\perp \cap \mathcal{D}(\pi) = \{0\}$, as noted in (1). In fact, there exists a closed *-representation of the *-algebra $\mathsf{A} = \mathbb{C}[x_1, x_2]$ that is not a direct sum of cyclic representations [Sch90, Corollary 11.6.8]. In general it is a difficult task to prove that all representations of some class of representations are direct sums of cyclic representations of this class.

(3) *Irreducible representations of commutative algebras*

If a bounded *-representation π of *-commutative algebra on a Hilbert space $\mathcal{H} = \mathcal{D}(\pi)$ is irreducible, then dim $\mathcal{H} = 1$. This is no longer true for unbounded representations. In Example 7.6 we construct an irreducible self-adjoint representation of $\mathsf{A} = \mathbb{C}[x_1, x_2]$ that acts on an infinite-dimensional Hilbert space.

(4) *Self-adjoint representations are not necessarily well behaved*

As noted in (3), there exists an irreducible self-adjoint representation of the commutative *-algebra $\mathsf{A} = \mathbb{C}[x_1, x_2]$ on an infinite-dimensional Hilbert space.

However, for an irreducible integrable $*$-representation (Definition 7.7) of a commutative $*$-algebra, the Hilbert space is always one-dimensional (Corollary 7.15).

(5) *Pure states and irreducibility of GNS representations*

A state on a C^*-algebra is pure if and only if its GNS representation is irreducible. For states on general unital $*$-algebras this is not true in general.

A state f on a unital $*$-algebra A is pure if and only the weak commutant $\pi_f(A)'_w = \overline{\pi}_f(A)'_w$ is trivial (Corollary 5.4), while the GNS representation π_f (resp. $\overline{\pi}_f$) is irreducible if and only if the strong commutant $\pi_f(A)'_s$ (resp. $\overline{\pi}_f(A)'_s$) does not contain a nontrivial projection (Proposition 4.26). A nonpure state for which the GNS representations π_f and $\overline{\pi}_f$ are irreducible will be constructed in Example 5.6. In this example, $\overline{\pi}_f$ is an irreducible closed $*$-representation of the polynomial algebra $\mathbb{C}[x]$ in one variable acting on an infinite-dimensional Hilbert space!

(6) *Positivity for representations of commutative algebras*

Let $\mathcal{X}$ be a compact topological Hausdorff space. Suppose π is a bounded $*$-representation of the C^*-algebra $A = C(\mathcal{X})$ on a Hilbert space $\mathcal{H}$. Then, for each function $a \in C(\mathcal{X})$ such that $a(x) > 0$ on $\mathcal{X}$, we have $\pi(a) \geq 0$ on $\mathcal{H}$. Further, if $(a_{ij}(x))_{i,j=1}^n$ is a matrix of entries $a_{ij} \in C(\mathcal{X})$ which is positive semi-definite for each $x \in \mathcal{X}$, then the matrix $(\pi(a_{ij}))_{i,j=1}^n$ of Hilbert space operators is positive semi-definite on the Hilbert space $\oplus_{i=1}^n \mathcal{H}$.

Again, all this is no longer true for unbounded representations of $A = \mathbb{C}[x_1, x_2]$. It will be shown in Example 7.5 that there exist a positive functional f on A and a polynomial $p \in A$ such that $p(x) \geq 0$ on $\mathbb{R}^2$, but $f(p) = \langle \pi_f(p)\varphi_f, \varphi_f \rangle < 0$.

Even more, there exists a $*$-representation π of A which is 1-positive, but not 2-positive [Sch90, Theorem 11.6.7]. That π is 1-positive means that $\pi(p) \geq 0$ for each nonnegative polynomial p on $\mathbb{R}^2$. Being not 2-positive means that there is a matrix $(p_{ij}(x))_{i,j=1}^2$ with $p_{ij} \in A$ which is positive semi-definite for each $x \in \mathbb{R}^2$, but the operator matrix $(\pi(p_{ij}))_{i,j=1}^2$ is not positive semi-definite on $\mathcal{D}(\pi) \oplus \mathcal{D}(\pi)$. An example of such a matrix is given by formula (14.19) in Chap. 14.

4.8 Exercises

1. Suppose A is a commutative real unital $*$-algebra with the identity map as involution. Let $A_{\mathbb{C}}$ be its complexification. Show that for each positive functional $f : A \mapsto \mathbb{R}$ there exist a nondegenerate $*$-representation π of $A_{\mathbb{C}}$ and a vector $\varphi \in \mathcal{D}(\pi)$ such that $f(a) = \langle \pi(a)\varphi, \varphi \rangle$ for $a \in A$.

2. Let A be a $*$-algebra, π an algebra homomorphism of A into $L(\mathcal{D})$ for some complex inner product space $\mathcal{D}$ and A_0 a subset of A which generates A as a $*$-algebra. Show that if (4.1) holds for $a \in A_0$, it does for all $a \in A$ and π is a $*$-representation.

3. In the notation of Remark 4.13, show that $\pi_{\mathbb{C}}$ is a $*$-representation of the complex $*$-algebra $A_{\mathbb{C}}$ on the complex inner product space $\mathcal{D}_{\mathbb{C}}$.

4. Let π be a ∗-representation of $\mathsf{A} = \mathbb{C}[x]$ on $L^2(\mathbb{R})$ such that $\mathcal{D}(\pi) \subseteq C^\infty(\mathbb{R})$ and $\pi(x) = -i\frac{d}{dx}$. Describe the ∗-representation ρ from Theorem 4.18 and decide whether or not it is self-adjoint:

   a. $\mathsf{B} = \{x\}$, $\mathcal{D}(\pi) = C_0^\infty(\mathbb{R})$.
   b. $\mathsf{B} = \{x\}$, $\mathcal{D}(\pi) = \{\varphi \in C_0^\infty(\mathbb{R}) : \varphi^{(n)}(0) = 0, \ n \in \mathbb{N}_0\}$.
   c. $\mathsf{B} = \{x\}$, $\mathcal{D}(\pi) = \{\varphi \in C_0^\infty(\mathbb{R}) : \varphi^{(n)}(0) = 0, \ n \in \mathbb{N}\}$.
   d. $\mathsf{B} = \{x, x^2\}$, $\mathcal{D}(\pi) = \{\varphi \in C_0^\infty(\mathbb{R}) : \varphi^{(n)}(0) = 0, \ n \in \mathbb{N}\}$.

5. (*q-Oscillator algebra and Fock representation, see also* Sect. 11.7)
   Suppose that $q > 0, q \neq 1$. Let $\mathsf{A} = \mathbb{C}\langle x, x^+ | xx^+ - qx^+x = 1\rangle$.

   a. Show that there exists a ∗-representation π of A on $l^2(\mathbb{N}_0)$ with domain $\mathcal{D}(\pi) = \mathrm{Lin}\,\{e_k : k \in \mathbb{N}_0\}$ such that $\pi(x)e_k = \sqrt{\frac{1-q^k}{1-q}}\,e_{k-1}, k \in \mathbb{N}_0$, where $\{e_k\}$ is the standard orthonormal basis of $l^2(\mathbb{N}_0)$ and $e_{-1} := 0$.
   b. Show that the corresponding ∗-representation ρ for $\mathsf{B} = \{x, x^+\}$ from Theorem 4.18 is self-adjoint.

6. Let π be a ∗-representation of a ∗-algebra A acting by bounded operators on a Hilbert space $\mathcal{D}(\pi) = \mathcal{H}(\pi)$. Show that the following are equivalent:

   (i) $\mathcal{H}(\pi)$ and $\{0\}$ are the only closed linear subspaces of $\mathcal{H}(\pi)$ which are invariant under π.
   (i) If $\pi = \pi_1 \oplus \pi_2$ is a direct sum of ∗-representations π_1 and π_2 of A, then $\mathcal{H}(\pi_1)$ or $\mathcal{H}(\pi_2)$ is $\{0\}$.
   (iii) $\pi(\mathsf{A})' = \mathbb{C} \cdot I$.

7. Let A be the (nonunital!) ∗-algebra $C_c(\mathbb{R})$ with pointwise multiplication and complex conjugation as involution. Show that the positive functional f on A defined by $f(\varphi) = \int_\mathbb{R} \varphi(x)dx$, $\varphi \in \mathsf{A}$, is not extendable. Describe the ∗-representation π_f from Proposition 4.36.
   Hint: Look for $\varphi \in \mathsf{A}$ such that $\varphi(x) = 1$ if $|x| \leq n$, $\varphi(x) = 0$ if $|x| \geq n + 1$.

8. Suppose $\varphi \in C_c(\mathbb{R}^d)$. Define a positive linear functional f on the polynomial algebra $\mathsf{A} = \mathbb{C}_d[\underline{x}]$ by $f(p) = \int p(x)|\varphi(x)|^2dx$, $p \in \mathsf{A}$.
   Describe the GNS representation $\overline{\pi}_f$. What is the domain of $\overline{\pi}_f$?

9. Let π_S be the Schrödinger representation of the Weyl algebra. Suppose that $\varphi_1, \varphi_2 \in C_0^\infty(\mathbb{R})$, $\varphi_1(x) \neq 0$ if and only if $x \in (0, 1)$, and $\varphi_2(x) \neq 0$ if and only if $x \in (2, 3)$. Describe the GNS representation $\overline{\pi}_{f_t}$ for $f_t(\cdot) = \mathrm{Tr}\,\pi_S(\cdot)t$:

   a. $t = \varphi_1 \otimes \varphi_1 + \varphi_2 \otimes \varphi_2$.
   b. $t = (\varphi_1 + \varphi_2) \otimes (\varphi_1 + \varphi_2)$.
   c. $t = \varphi \otimes \varphi$, where $\varphi(x) := e^{-\alpha x^2}, \alpha > 0$.

10. (*Unitary implementation of ∗-automorphisms for invariant positive functionals*)
    Let $g \mapsto \alpha_g$ be a homomorphism of a group G in the group of ∗-automorphisms of a unital ∗-algebra A. Suppose f is a positive functional on A such that

$f(\alpha_g(a)) = f(a)$ for $a \in A$, $g \in G$. Show that there exists a unitary representation U of G on $\mathcal{H}(\pi_f)$ such that $U(g)\mathcal{D}(\pi_f) = \mathcal{D}(\pi_f)$, $U(g)\varphi_f = \varphi_f$,

$$\pi_f(\alpha_g(a)) = U(g)\pi_f(a)U(g)^{-1} \quad \text{for} \quad a \in A, \ g \in G.$$

Hint: Define $U(g)(\pi_f(a)\varphi_f) = \pi_f(\alpha_g(a))\varphi_f$.

11. Let f be a positive functional on a $*$-algebra A and $a \in A$. Show that $p_r(x) := f(a^+x^+xa)^{1/2}$ and $p_l(x) := f(x^+a^+ax)^{1/2}$ define seminorms p_r and p_l on A.

12. Show that the polynomial algebra $\mathbb{C}[x_1, \ldots, x_d]$ is $*$-semisimple.

13. Let f be a positive semi-definite function on a group G and let $g, h \in G$. Use Eq. (4.30) to prove the following:

   a. $|f(g)| \leq f(e)$ and $f(g^{-1}) = \overline{f(g)}$.
   b. $|f(g) - f(h)|^2 \leq 2f(e)(f(e) - \operatorname{Re} f(g^{-1}h))$.

In Exercises 14–16, A_1 and A_2 are unital $*$-algebras and $A = A_1 \otimes A_2$ is the tensor product $*$-algebra with involution determined by $(a_1 \otimes a_2)^+ = a_1^+ \otimes a_2^+$.

14. Suppose that ρ_1 and ρ_2 are nondegenerate $*$-representations of A_1 and A_2, respectively, acting on the same domain $\mathcal{D}(\rho_1) = \mathcal{D}(\rho_2)$ such that $\rho_1(a_1)\rho_2(a_2) = \rho_2(a_2)\rho_1(a_1)$ for $a_1 \in A_1, a_2 \in A_2$. Show that there exists a unique $*$-representation ρ of A on $\mathcal{D}(\rho) := \mathcal{D}(\rho_1)$ such that $\rho(a_1 \otimes a_2) = \rho_1(a_1)\rho_2(a_2)$ for $a_1 \in A_1, a_2 \in A_2$. Show that each nondegenerate $*$-representation of A arises in this manner.

15. Show that for nondegenerate $*$-representations π_1 of A_1 and π_2 of A_2 there is a unique $*$-representation $\pi_1 \otimes \pi_2$ of A on $\mathcal{D}(\pi_1 \otimes \pi_2) := \mathcal{D}(\pi_1) \otimes \mathcal{D}(\pi_2)$ such that $(\pi_1 \otimes \pi_2)(a_1 \otimes a_2) = \pi_1(a_1) \otimes \pi_2(a_2)$, $a_1 \in A_1, a_2 \in A_2$.

16. Let f_1 and f_2 be positive linear functionals on A_1 and A_2, respectively. Show that there is a well-defined positive linear functional $f_1 \otimes f_2$ on A such that $(f_1 \otimes f_2)(a_1 \otimes a_2) = f_1(a_1)f_2(a_2)$ for $a_1 \in A_1, a_2 \in A_2$.
   Hint: Apply Exercise 15 to the GNS representations π_{f_1} and π_{f_2}.

4.9 Notes

The GNS construction for bounded representations and normed algebras was discovered by I.M. Gelfand and M.A. Naimark in their seminal paper [GN43]. It was made explicit by Segal [Se47b] who invented also the term "state." This construction was used for the field algebra by Borchers [B62] and Uhlmann [U62]. For unbounded representations of general $*$-algebras it was elaborated by Powers in [Pw71]. The representability of positive functionals as vector functionals of *bounded* $*$-representations was studied in [Sb84, Sb86]. Representations of symmetric $*$-algebras were investigated in [In77, Bh84]. Theorem 4.18 was obtained in [Sch90].

A natural question is when an abstract unital $*$-algebra admits a faithful $*$-representation. A sufficient condition was given in Corollary 4.45. This problem was studied for general representations in [Po08] and for bounded representations in [T99, La97, GM90].

Positive semi-definite functions play an important role in representation theory of locally compact groups; see, e.g., [Di77b, Part II] and [F89].

This chapter and the next follow partly the monograph [Sch90, Chap. 8], but a number of new results have been also added.

Chapter 5
Positive Linear Functionals

Positive functionals and states are fundamental objects in the theory of $*$-algebras. The GNS construction provides a powerful method for their study which is the aim of this chapter. This can lead to difficult problems already for very simple algebras. For instance, the one-dimensional classical moment problem is, in fact, the study of positive functionals on the polynomial algebra $\mathbb{R}[x]$ (or $\mathbb{C}[x]$) in one variable!

In Sect. 5.1, we develop an ordering of positive functionals which is related to operators in the weak commutant⊦or of the GNS representation (Theorem 5.3). Section 5.2 deals with orthogonal positive functionals (Theorem 5.13). Sections 5.3 and 5.4 provide a detailed study of the transition probability of two positive functionals. We characterize it in terms of the symmetrized commutant (Theorem 5.18) and obtain explicit formulas for trace functionals or functionals which are defined as integrals (Theorems 5.23 and 5.25). Section 5.5 contains a Radon–Nikodym theorem for positive functionals. In Sect. 5.6, we use the Choquet theory to represent positive functionals as integrals over pure states (Theorems 5.35 and 5.36). Section 5.7 is about quadratic modules defined by families of $*$-representations.

Throughout this chapter, A is a **complex unital** $*$-algebra. Recall that $\mathcal{P}(A)^*$ denotes the set of positive linear functionals and $\mathcal{S}(A)$ is the set of states on A.

5.1 Ordering of Positive Functionals

For positive linear functionals f and g on A, we define

$$f \leq g \quad \text{if} \quad f(a^+a) \leq g(a^+a) \ \text{ for all } \ a \in A.$$

Let $[0, f]$ denote the set of all functionals $g \in \mathcal{P}(A)^*$ such that $g \leq f$.

© The Editor(s) (if applicable) and The Author(s), under exclusive license
to Springer Nature Switzerland AG 2020

K. Schmüdgen, *An Invitation to Unbounded Representations of $*$-Algebras on Hilbert Space*, Graduate Texts in Mathematics 285,
https://doi.org/10.1007/978-3-030-46366-3_5

Definition 5.1 A positive functional f on the unital $*$-algebra A is said to be *pure* if $[0, f]$ consists only of multiples λf of f, where $\lambda \in [0, 1]$.

It is easy to check that a *state* on A is pure according to Definition 5.1 if and only if it is an extreme point of the convex set $\mathcal{S}(A)$, so Definitions 2.60 and 5.1 coincide for states.

Remark 5.2 For a *nonunital* $*$-algebra, a positive functional f is called *pure* if each *extendable* positive functional g satisfying $g \leq f$ is a multiple λf, with $\lambda \in [0, 1]$. But in this chapter we study only positive functionals on unital $*$-algebras.   ○

By $[0, I]$ we denote the set $\{T = T^* \in \mathbf{B}(\mathcal{H}) : 0 \leq T \leq I\}$, equipped with the usual order relation of self-adjoint operators.

Theorem 5.3 *Let $f \in \mathcal{P}(A)^*$. If $T = T^* \in \pi_f(A)'_w$ and $T \geq 0$, then the equation*

$$f_T(a) := \langle T\pi_f(a)\varphi_f, \varphi_f \rangle, \quad a \in A,$$

defines a positive linear functional f_T on A.
   The map $T \mapsto f_T$ is an order isomorphism of $\pi_f(A)'_w \cap [0, I]$ onto $[0, f]$, that is, this map is bijective and we have $T_1 \leq T_2$ if and only if $f_{T_1} \leq f_{T_2}$ for any $T_1, T_2 \in \pi_f(A)'_w \cap [0, I]$.

Proof Let $a \in A$. Using that T in the weak commutant $\pi_f(A)'_w$ we derive

$$f_T(a^+a) = \langle T\pi_f(a^+)\pi_f(a)\varphi_f, \varphi_f \rangle = \langle T\pi_f(a)\varphi_f, \pi_f(a)\varphi_f \rangle. \tag{5.1}$$

Since $T \geq 0$, this gives $f_T(a^+a) \geq 0$, so f_T is a positive linear functional on A.
   Now suppose $T \in \pi_f(A)'_w \cap [0, I]$. Since $T \leq I$, (5.1) and (4.25) imply that

$$f_T(a^+a) \leq \langle \pi_f(a)\varphi_f, \pi_f(a)\varphi_f \rangle = f(a^+a), \quad a \in A,$$

that is, $f_T \leq f$. Since $\mathcal{D}(\pi_f) = \pi_f(A)\varphi_f$ is dense in $\mathcal{H}(\pi_f)$, it follows from (5.1) that $f_{T_1} \leq f_{T_2}$ is equivalent to $T_1 \leq T_2$ for $T_1, T_2 \in \pi_f(A)'_w \cap [0, I]$ and that the map $T \mapsto f_T$ is injective. It remains to prove that this map is surjective.
   For suppose $g \in \mathcal{P}(A)^*$ and $g \leq f$. We show that there is a well-defined bounded sesquilinear form $\langle \cdot, \cdot \rangle_g$ on the complex inner product space $\mathcal{D}(\pi_f)$ such that

$$\langle \pi_f(x)\varphi_f, \pi_f(y)\varphi_f \rangle_g = g(y^+x), \quad x, y \in A. \tag{5.2}$$

First we prove that $\langle \cdot, \cdot \rangle_g$ is well defined by (5.2). Suppose $\pi_f(x)\varphi_f = \pi_f(x')\varphi_f$, where $x, x' \in A$. Then, by (4.25),

$$f((x - x')^+(x - x')) = \|\pi_f(x - x')\varphi_f\|^2 = 0.$$

Hence $g((x - x')^+(x - x')) = 0$, since $g \le f$, and therefore

$$|g(y^+(x - x'))|^2 \le g(y^+y)g((x - x')^+(x - x')) = 0$$

by the Cauchy–Schwarz inequality (2.34). Thus, $g(y^+x) = g(y^+x')$. The same reasoning works for the variable y as well. Hence $\langle \cdot, \cdot \rangle_g$ is well defined by (5.2).

The functional g is positive and hence hermitian, so is the sesquilinear form $\langle \cdot, \cdot \rangle_g$. It is bounded on the complex inner product space $\mathcal{D}(\pi_f)$, since for $x \in \mathsf{A}$,

$$\langle \pi_f(x)\varphi_f, \pi_f(x)\varphi_f \rangle_g = g(x^+x) \le f(x^+x) = \langle \pi_f(x)\varphi_f, \pi_f(x)\varphi_f \rangle. \qquad (5.3)$$

Hence $\langle \cdot, \cdot \rangle_g$ extends to a bounded positive sesquilinear form on the Hilbert space $\mathcal{H}(\pi_f)$. Then there is a positive self-adjoint operator $T \in \mathbf{B}(\mathcal{H}(\pi_f))$ such that

$$g(y^+x) = \langle T\pi_f(x)\varphi_f, \pi_f(y)\varphi_f \rangle, \quad x, y \in \mathsf{A}. \qquad (5.4)$$

Combining (5.3) and (5.4) yields $T \le I$. For $a, x, y \in \mathsf{A}$, we derive

$$\begin{aligned}
\langle T\pi_f(a)\pi_f(x)\varphi_f, \pi_f(y)\varphi_f \rangle &= \langle T\pi_f(ax)\varphi_f, \pi_f(y)\varphi_f \rangle \\
&= g(y^+ax) = g((a^+y)^+x) = \langle T\pi_f(x)\varphi_f, \pi_f(a^+y)\varphi_f \rangle \\
&= \langle T\pi_f(x)\varphi_f, \pi_f(a^+)\pi_f(y)\varphi_f \rangle.
\end{aligned}$$

Therefore, $T \in \pi_f(\mathsf{A})'_w$. Thus we have shown that $T \in \pi_f(\mathsf{A})'_w \cap [0, I]$ and $g = f_T$ by (5.4). This proves the surjectivity of the map $T \mapsto f_T$. $\qquad \square$

Corollary 5.4 *A positive linear functional f on A is pure if and only if the weak commutant $\pi_f(\mathsf{A})'_w$ is trivial, that is, $\pi_f(\mathsf{A})'_w = \mathbb{C} \cdot I$.*

Proof Theorem 5.3 implies that f is pure if and only if $\pi_f(\mathsf{A})'_w \cap [0, I]$ is equal to $\{\lambda \cdot I : 0 \le \lambda \le 1\}$. Since $\pi_f(\mathsf{A})'_w$ is a $*$-invariant vector space, the latter is obviously equivalent to $\pi_f(\mathsf{A})'_w = \mathbb{C} \cdot I$. $\qquad \square$

Corollary 5.5 *If a positive linear functional f is pure, the GNS representations π_f and $\overline{\pi}_f$ are irreducible.*

Proof Since f is pure, $\pi_f(\mathsf{A})'_w = \mathbb{C} \cdot I$ by Corollary 5.4. Proposition 3.15 implies that $\pi_f(\mathsf{A})'_w = \overline{\pi}_f(\mathsf{A})'_w$. Since the weak commutants $\pi_f(\mathsf{A})'_w$ and $\overline{\pi}_f(\mathsf{A})'_w$ are trivial, so are the corresponding strong commutants. Hence π_f and $\overline{\pi}_f$ are irreducible by Proposition 4.26. $\qquad \square$

In contrast to C^*-algebras, the converse direction in Corollary 5.5 is not true, as shown by the next example. In this example we use the some advanced results from the theory of indeterminate Hamburger moment problems (see [Sch17, Chap. 7]).

Example 5.6 (*A nonpure state with irreducible GNS representation*)
Suppose μ is a von Neumann solution, or equivalently an N-extremal solution, of an indeterminate Hamburger moment problem such that $\mu(\mathbb{R}) = 1$. Then $\mathbb{C}[x]$ is dense in the Hilbert space $\mathcal{H} := L^2(\mathbb{R}; \mu)$. Hence there exists a $*$-representation of the $*$-algebra $\mathsf{A} := \mathbb{C}[x]$ with domain $\mathcal{D}(\pi) := \mathbb{C}[x]$ on $\mathcal{H}$ defined by $(\pi(p)q)(x) = p(x)q(x)$, $p, q \in \mathbb{C}[x]$. Since $\pi(\mathbb{C}[x])1 = \mathcal{D}(\pi) = \mathbb{C}[x]$ and

$$f_{\pi,1}(p) = \langle \pi(p)1, 1 \rangle = \int_{\mathbb{R}} p(x)d\mu(x), \quad p \in \mathbb{C}[x], \tag{5.5}$$

it follows from Theorem 4.38 that π is (unitarily equivalent to) the GNS representation of the state $f_{\pi,1}$.

Statement 1: *The state* $f \equiv f_{\pi,1}$ *on the $*$-algebra* $\mathbb{C}[x]$ *is not pure.*

Proof For any Borel set M, $f_M(p) := \int_M p(x)d\mu(x)$, $p \in \mathbb{C}[x]$, defines a positive functional such that $f_M \leq f$. If $\mu(M) \neq 0$ and M is bounded, $\pi_{f_M}(x)$ is a nonzero bounded operator. Since $\pi_f(x)$ is unbounded, f_M is not a multiple of f. $\square$

For the next result we recall some facts about the indeterminate Hamburger moment problem [Sch17, Theorem 7.7]. The symmetric operator $T := \pi(x)$ has deficiency indices $(1, 1)$ and its self-adjoint extensions A_t of T on $\mathcal{H}$ are parametrized by $\mathbb{R} \cup \{\infty\}$. Each operator A_t has a discrete spectrum consisting of eigenvalues of multiplicity one. For $t \in \mathbb{R}$ these eigenvalues are zeros of the function $B(z) + tD(z)$, where A, B, C, D are the Nevanlinna entire functions of the indeterminate moment problem. The operators A_{t_1} and A_{t_2} for different reals t_1, t_2 have no common eigenvalue (otherwise there is a λ such that $B(\lambda) = D(\lambda) = 0$, which contradicts the relation $A(z)B(z) - C(z)D(z) = 1$.) It is not difficult to verify that all bounded functions of any self-adjoint extension A_t belong to the *weak commutant* $\pi(\mathbb{C}[x])'_w$.

Statement 2: π_f *and* $\overline{\pi}_f$ *are irreducible.*

Proof By Corollary 4.27, it suffices to prove that $\overline{\pi}_f$ is irreducible. Assume that $\overline{\pi}_f = \pi_1 \oplus \pi_2$. Then $T = \overline{\pi_f(x)} = \overline{\pi}_f(x) = T_1 \oplus T_2$, where $T_j = \overline{\pi_j(x)}$. Let P_j be the projection on $\mathcal{H}(\pi_j)$ and set $\varphi_j = P_j(1)$. Since

$$\langle \pi_j(p)\varphi_j, \varphi_j \rangle = \int p(x)|\varphi_j(x)|^2 d\mu(x), \quad p \in \mathbb{C}[x],$$

the operator T_j comes from a moment problem with measure $d\mu_j = |\varphi_j|^2 d\mu$, so T_j has equal deficiency indices. Since T has deficiency indices $(1, 1)$, one of the operators T_j, say T_1, is self-adjoint. Clearly, $\pi_1(x) \subseteq \overline{\pi}_f(x) \subseteq A_t$ gives $T_1 \subseteq A_t$. This implies that $\sigma(T_1) \subseteq \sigma(A_t)$ for *all* $t \in \mathbb{R}$. Since A_{t_1} and A_{t_2} for $t_1 \neq t_2$ have no common eigenvalue, $\mathcal{H}(T_1) = \{0\}$. Thus, $\pi \cong \overline{\pi}_f$ is irreducible. $\square\bigcirc$

5.2 Orthogonal Positive Functionals

First we introduce another important technical notion in representation theory.

Definition 5.7 For representations π_1 and π_2 of A, the *intertwining space* $I(\pi_1, \pi_2)$ is the vector space of bounded linear operators T of $\mathcal{H}(\pi_1)$ in $\mathcal{H}(\pi_2)$ satisfying

$$T\varphi \in \mathcal{D}(\pi_2) \quad \text{and} \quad T\pi_1(a)\varphi = \pi_2(a)T\varphi \quad \text{for} \ a \in A, \ \varphi \in \mathcal{D}(\pi_1). \tag{5.6}$$

It is instructive to consider two special cases: If π is a $*$-representation, then $I(\pi, \pi)$ is just the strong commutant $\pi(A)'_s$ of the O^*-algebra $\pi(A)$ and $I(\pi, \pi^*)$ coincides with the weak commutant $\pi(A)'_w$.

Lemma 5.8 *If π_1 and π_2 are representations of A such that $\mathcal{H}(\pi_1^*) = \mathcal{H}(\pi_1)$ and $\mathcal{H}(\pi_2^*) = \mathcal{H}(\pi_2)$, then we have*

$$I(\pi_1, \pi_2)^* \subseteq I(\pi_2^*, \pi_1^*) \quad \text{and} \quad I(\pi_1, \pi_2) \subseteq I(\pi_1^{**}, \pi_2^{**}). \tag{5.7}$$

Proof Suppose $T \in I(\pi_1, \pi_2)$. Let $\psi \in \mathcal{D}(\pi_2^*)$. Then, for $a \in A$ and $\varphi \in \mathcal{D}(\pi_1)$,

$$\langle T^*\psi, \pi_1(a)\varphi \rangle - \langle \psi, T\pi_1(a)\varphi \rangle = \langle \psi, \pi_2(a)T\varphi \rangle = \langle T^*\pi_2^*(a^+)\psi, \varphi \rangle.$$

From this equation it follows that

$$T^*\psi \in \cap_{a \in A}\mathcal{D}(\pi_1(a)^*) = \mathcal{D}(\pi_1^*), \quad \pi_1(a)^*T^*\psi = \pi_1^*(a^+)T^*\psi = T^*\pi_2^*(a^+)\psi,$$

so that $T^* \in I(\pi_2^*, \pi_1^*)$. This proves the first inclusion of (5.7).

Since $\mathcal{H}(\pi_j^*) = \mathcal{H}(\pi_j)$, we have $\pi_j \subseteq \pi_j^{**}$ by Lemma 4.3(ii) and therefore $\mathcal{H}(\pi_j^{**}) = \mathcal{H}(\pi_j^*)$ for $j = 1, 2$. Hence, applying the preceding with T replaced by T^*, we get $T = T^{**} \in I(\pi_1^{**}, \pi_2^{**})$. This is the second inclusion of (5.7). $\qquad \square$

Corollary 5.9 *For any $*$-representation π of A, we have $\pi(A)'_s \subseteq \pi^{**}(A)'_s$ and $\pi(A)'_w \subseteq \pi^{**}(A)'_w$.*

Proof Using the second inclusion of (5.7) we derive

$$\pi(A)'_s = I(\pi, \pi) \subseteq I(\pi^{**}, \pi^{**}) = \pi^{**}(A)'_s,$$
$$\pi(A)'_w = I(\pi, \pi^*) \subseteq I(\pi^{**}, \pi^{***}) = \pi^{**}(A)'_w. \qquad \square$$

Corollary 5.10 *If π is a $*$-representation of A and essentially self-adjoint, then $\pi(A)'_w = \pi^*(A)'_w = \pi^*(A)'_s$.*

Proof Since π is essentially self-adjoint, π^* is self-adjoint, so $\pi^*(A)'_w = \pi^*(A)'_s$ and $\pi^* = \pi^{**}$. From $\pi \subseteq \pi^*$ we obtain $\pi^*(A)'_w = I(\pi^*, \pi^*) \subseteq I(\pi, \pi^*) = \pi(A)'_w$. Further, $\pi(A)'_w = I(\pi, \pi^*) \subseteq I(\pi^{**}, \pi^{***}) = I(\pi^*, \pi^*) = \pi^*(A)'_s$ by the second inclusion of (5.7). Therefore, $\pi(A)'_w = \pi^*(A)'_s$. $\qquad \square$

Now let $f, g \in \mathcal{P}(A)^*$. We say that g is *dominated* by f and write $g \preceq f$ if there exists a number $\lambda > 0$ such that $g \leq \lambda f$. In this case, we have

$$\|\pi_g(a)\varphi_g\|^2 = g(a^+a) \leq \lambda f(a^+a) = \lambda \|\pi_f(a)\varphi_f\|^2, \quad a \in A.$$

Therefore, the equation

$$K_{f,g}(\pi_f(a)\varphi_f) = \pi_g(a)\varphi_g, \quad a \in A, \tag{5.8}$$

defines unambiguously a bounded linear operator of $\mathcal{D}(\pi_f)$ onto $\mathcal{D}(\pi_g)$. Its continuous extension to a bounded operator of the Hilbert space $\mathcal{H}(\pi_f)$ in $\mathcal{H}(\pi_g)$ is also denoted by $K_{f,g}$. Some simple properties are collected in the next proposition.

Proposition 5.11 *Suppose $f, g \in \mathcal{P}(A)^*$ and $g \preceq f$. Then:*

(i) $K_{f,g} \in I(\pi_f, \pi_g)$ *and* $K_{f,g}\varphi_f = \varphi_g$.
(ii) $T := (K_{f,g})^* K_{f,g} \in \pi_f(A)'_w$ *and* $g = f_T$.
(iii) *If $g \leq f$, then* $I = (K_{f,g})^* K_{f,g} + (K_{f,f-g})^* K_{f,f-g}$.

Proof (i): Setting $a = 1$ in (5.8) yields $K_{f,g}\varphi_f = \varphi_g$. For $a, b \in A$,

$$K_{f,g}\pi_f(a)(\pi_f(b)\varphi_f) = K_{f,g}\pi_f(ab)\varphi_f = \pi_g(ab)\varphi_g = \pi_g(a)K_{f,g}(\pi_f(b)\varphi_f),$$

which proves that $K_{f,g} \in I(\pi_f, \pi_g)$.

(ii): From $K_{f,g} \in I(\pi_f, \pi_g)$ and Lemma 5.8, $(K_{f,g})^* \in I((\pi_g)^*, (\pi_f)^*)$. Since $\pi_g \subseteq (\pi_g)^*$, this implies that $(K_{f,g})^* \in I(\pi_g, (\pi_f)^*)$. Therefore,

$$(K_{f,g})^* K_{f,g} \in I(\pi_g, (\pi_f)^*) \cdot I(\pi_f, \pi_g) \subseteq I(\pi_f, (\pi_f)^*) = \pi_f(A)'_w.$$

Further, for $a \in A$ we derive

$$\begin{aligned} f_T(a) &= \langle (K_{f,g})^* K_{f,g} \pi_f(a)\varphi_f, \varphi_f \rangle \\ &= \langle K_{f,g}\pi_f(a)\varphi_f, K_{f,g}\varphi_f \rangle = \langle \pi_g(a)\varphi_g, \varphi_g \rangle = g(a). \end{aligned}$$

(iii): Set $T := (K_{f,g})^* K_{f,g}$ and $S := (K_{f,f-g})^* K_{f,f-g}$. Then $g = f_T$ and $f - g = f_S$ by (ii). Therefore, $f = f_T + f_S$ and hence $I = T + S$. $\square$

Definition 5.12 We say two positive functionals f, g on A are *orthogonal* and write $f \perp g$ if $h \preceq f$ and $h \preceq g$ for some positive functional h on A implies that $h = 0$.

This notion is justified by Eq. (5.9) in the following theorem. Recall that $|T|$ denotes the operator $(T^*T)^{1/2}$ for any densely defined closed operator T.

Theorem 5.13 *Suppose f and g are positive functionals on A. If $|K_{f+g,f}|$ is an orthogonal projection, then $f \perp g$. Conversely, if $f \perp g$ and $\overline{\pi}_{f+g}$ is self-adjoint, then $\overline{\pi}_f$ and $\overline{\pi}_g$ are self-adjoint, $P := |K_{f+g,f}|$ is an orthogonal projection, and*

$$\overline{\pi}_{f+g} \cong \overline{\pi}_f \oplus \overline{\pi}_g \quad on \quad \mathcal{H}(\pi_{f+g}) = P\mathcal{H}(\pi_{f+g}) \oplus (I - P)\mathcal{H}(\pi_{f+g}). \tag{5.9}$$

Proof Throughout this proof we set $F := f + g$. Clearly, $F \in \mathcal{P}(A)^*$, $f \leq F$, and $g \leq F$.

Suppose $|K_{F,f}|$ is a projection. Then $|K_{F,f}| = |K_{F,f}|^2 = (K_{F,f})^* K_{F,f} =: T$. By Proposition 5.11(iii), $S := (K_{F,g})^* K_{F,g} = I - T$, so S is also a projection. Further, $f = F_T$ and $g = F_S$ by Proposition 5.11(iii). Now let $h \in \mathcal{P}(A)^*$ and suppose that $h \preceq f$ and $h \preceq g$. Upon scaling h we can assume that $h \leq f$ and $h \leq g$. Then $h \leq F$. By Theorem 5.3, there exists an operator $R \in \pi_F(A)'_w$, $0 \leq R \leq I$, such that $h = F_R$. Since $h = F_R \leq f = F_T$ and $h = F_R \leq g = F_S$, we obtain $0 \leq R \leq T$ and $0 \leq R \leq S = I - T$ again by Theorem 5.3. Since T is a projection, this implies $R = 0$ and so $h = 0$. This proves that $f \perp g$.

Conversely, assume that $\overline{\pi}_F$ is self-adjoint and $f \perp g$. Then $\pi_F(A)'_w = \overline{\pi}_F(A)'_w = \overline{\pi}_F(A)'_s$ is a von Neumann algebra. Hence, since $T := (K_{F,f})^* K_{F,f} \in \pi_F(A)'_w$ by Proposition 5.11(ii), $Z := T(I - T)$ is also in $\pi_F(A)'_w$. Proposition 5.11(iii) implies that $0 \leq T \leq I$, so $0 \leq Z \leq I$. By Theorem 5.3, F_Z is a positive linear functional on A. From $Z \leq T$ and $Z \leq I - T$ it follows that $F_Z \leq F_T = f$ and $F_Z \leq F_{I-T} = g$. Hence, by $f \perp g$, we obtain $F_Z = 0$ and so $Z = 0$. Since $Z = T(I - T) = 0$ has the spectrum $\{0\}$, T is an orthogonal projection.

Set $\mathcal{K} := T\mathcal{H}(\pi_F)$. Since $T = P_\mathcal{K} \in \overline{\pi}_F(A)'_s$ is a projection, it follows from Proposition 4.25 that $\overline{\pi}_F = \pi_\mathcal{K} \oplus \pi_{\mathcal{K}^\perp}$. Then $T\varphi_F \in \mathcal{K} \cap \mathcal{D}(\overline{\pi}_F)$. Further, we have $f = F_T$ by Proposition 5.11(ii). For $a \in A$, we derive

$$\langle \overline{\pi}_F(a)T\varphi_F, T\varphi_F \rangle = \langle T\overline{\pi}_F(a)\varphi_F, T\varphi_F \rangle = \langle T\pi_F(a)\varphi_F, \varphi_F \rangle = F_T(a) = f(a).$$

Since φ_F is cyclic for $\overline{\pi}_F$, the vector space $T\overline{\pi}_F(A)\varphi_F = \pi_\mathcal{K}(A)T\varphi_F$ is dense in $\mathcal{D}(\pi_\mathcal{K})$ with respect to the graph topology of $\pi_\mathcal{K}$. This means that $T\varphi_F$ is cyclic for $\pi_\mathcal{K}$. Therefore, by the uniqueness assertion for the GNS representation, $\pi_\mathcal{K}$ is (unitarily equivalent to) the GNS representation $\overline{\pi}_f$. Similarly, $\pi_{\mathcal{K}^\perp}$ has the cyclic vector $(I - T)\varphi_F$ and is (unitarily equivalent to) the GNS representation $\overline{\pi}_g$. Now the relation $\overline{\pi}_F = \pi_\mathcal{K} \oplus \pi_{\mathcal{K}^\perp}$ yields $\overline{\pi}_F \cong \overline{\pi}_f \oplus \overline{\pi}_g$, which proves (5.9). Since $\overline{\pi}_F$ is self-adjoint, it is clear that $\pi_\mathcal{K} \cong \overline{\pi}_f$ and $\pi_{\mathcal{K}^\perp} \cong \overline{\pi}_g$ are also self-adjoint. $\square$

5.3 The Transition Probability of Positive Functionals

Let $\mathrm{Rep}\, A$ denote the family of all nondegenerate $*$-representations of A. Given $\pi \in \mathrm{Rep}\, A$ and $f \in \mathcal{P}(A)^*$, let $\mathcal{S}(\pi, f)$ be the set of representing vectors for f in $\mathcal{D}(\pi)$, that is, $\mathcal{S}(\pi, f)$ is the set of vectors $\varphi \in \mathcal{D}(\pi)$ such that $f(a) = \langle \pi(a)\varphi, \varphi \rangle$, $a \in A$. Note that $\mathcal{S}(\pi, f)$ may be empty, but by the GNS construction for each $f \in \mathcal{P}(A)^*$ there exists a $\pi \in \mathrm{Rep}\, A$ for which $\mathcal{S}(\pi, f)$ is not empty. Clearly, if f is a state, all vectors $\varphi \in \mathcal{S}(\pi, f)$ are unit vectors.

The following definition introduces two closely related notions associated with a pair of positive linear functionals.

Definition 5.14 Let $f, g \in \mathcal{P}(\mathsf{A})^*$. The *transition probability* $P_{\mathsf{A}}(f, g)$ is

$$P_{\mathsf{A}}(f, g) := \sup_{\pi \in \mathrm{Rep}\,\mathsf{A}} \quad \sup_{\varphi \in \mathcal{S}(\pi, f), \psi \in \mathcal{S}(\pi, g)} |\langle \varphi, \psi \rangle|^2 \qquad (5.10)$$

and the *Bures distance* $d_{\mathsf{A}}(f, g)$ is

$$d_{\mathsf{A}}(f, g) := \inf_{\pi \in \mathrm{Rep}\,\mathsf{A}} \quad \inf_{\varphi \in \mathcal{S}(\pi, f), \psi \in \mathcal{S}(\pi, g)} \| \varphi - \psi \|. \qquad (5.11)$$

The quantity $P_{\mathsf{A}}(f, g)$ has the following interpretation in quantum physics: For vector states with representing vectors φ, ψ, the number $|\langle \varphi, \psi \rangle|^2$ is considered as their transition probability.[1] If f and g are states on an (abstract) observable algebra A, then $P_{\mathsf{A}}(f, g)$ is the supremum of transition probabilities for all realizations of f and g as vector states in the same nondegenerate $*$-representation of A.

In the literature on quantum information theory the number $\sqrt{P_{\mathsf{A}}(f, g)}$ is called the *fidelity* of f and g.

Proposition 5.15 *The Bures distance and the transition probability are related by*

$$d_{\mathsf{A}}(f, g)^2 = f(1) + g(1) - 2\sqrt{P_{\mathsf{A}}(f, g)}, \quad f, g \in \mathcal{P}(\mathsf{A})^*. \qquad (5.12)$$

Proof Clearly, $\varphi \in \mathcal{S}(\pi, f)$ if and only if $z\varphi \in \mathcal{S}(\pi, f)$ for $z \in \mathbb{T}$. Hence we can restrict ourselves in (5.10) to vectors φ, ψ such that $\langle \varphi, \psi \rangle \geq 0$. Then, since

$$\| \varphi - \psi \|^2 = \| \varphi \|^2 + \| \psi \|^2 - 2 \, \mathrm{Re}\, \langle \varphi, \psi \rangle = f(1) + g(1) - 2|\langle \varphi, \psi \rangle|,$$

the infimum of $\| \varphi - \psi \|$ corresponds to the supremum of $|\langle \varphi, \psi \rangle|$, which in turn implies (5.12). $\qquad \square$

Let us begin with a simple special case.

Lemma 5.16 *Let f and g be states on A and suppose there is a projection $p \in \mathsf{A}$ (that is, $p = p^* = p^2$) such that*

$$f(p) = 1 \quad \text{and} \quad pap = f(a)p \quad \text{for} \quad a \in \mathsf{A}. \qquad (5.13)$$

Then $P_{\mathsf{A}}(f, g) = g(p)$ and for arbitrary $\varphi \in \mathcal{S}(\pi, f)$ and $\psi \in \mathcal{S}(\pi, g)$ we have

$$P_{\mathsf{A}}(f, g) = |\langle \widetilde{\varphi}, \psi \rangle|^2,$$

where we set $\widetilde{\varphi} := \varphi$ if $P_{\mathsf{A}}(f, g) = 0$ and $\widetilde{\varphi} := \| \pi(p)\psi \|^{-1} \pi(p)\psi \in \mathcal{S}(\pi, f)$ if $P_{\mathsf{A}}(f, g) \neq 0$.

---

[1] The expression $|\langle \varphi, \psi \rangle|^2$ for the transition probability was proposed by Max Born (1925/1926).

Proof Let $\varphi \in \mathcal{S}(\pi, f)$ and $\psi \in \mathcal{S}(\pi, g)$. Using that p is a projection we derive

$$\|\pi(p)\varphi - \varphi\|^2 = \langle \pi(p)\varphi, \pi(p)\varphi \rangle + \langle \varphi, \varphi \rangle - \langle \pi(p)\varphi, \varphi \rangle - \langle \varphi, \pi(p)\varphi \rangle$$
$$= \langle \varphi, \varphi \rangle - \langle \pi(p)\varphi, \varphi \rangle = 1 - f(p) = 0$$

by (5.13). Hence $\pi(p)\varphi = \varphi$ and

$$|\langle \varphi, \psi \rangle|^2 = |\langle \pi(p)\varphi, \psi \rangle|^2 = |\langle \varphi, \pi(p)\psi \rangle|^2 \leq \|\pi(p)\psi\|^2 = g(p). \tag{5.14}$$

By taking the supremum over φ, ψ this yields $P_A(f, g) \leq g(p)$.

Therefore, if $g(p) = 0$, then $|\langle \widetilde{\varphi}, \psi \rangle|^2 = |\langle \varphi, \psi \rangle|^2 = 0 = P_A(f, g)$ by (5.14).

Now suppose $g(p) \neq 0$. Since $\|\pi(p)\psi\|^2 = \langle \pi(p)\psi, \psi \rangle = g(p) \neq 0$, the vector $\widetilde{\varphi} = \|\pi(p)\psi\|^{-1}\pi(p)\psi$ has norm 1 and satisfies $\pi(p)\widetilde{\varphi} = \widetilde{\varphi}$. Hence, by (5.13),

$$f(a) = f(a)\langle \widetilde{\varphi}, \widetilde{\varphi} \rangle = \langle \pi(f(a)p)\widetilde{\varphi}, \widetilde{\varphi} \rangle = \langle \pi(pap)\widetilde{\varphi}, \widetilde{\varphi} \rangle = \langle \pi(a)\widetilde{\varphi}, \widetilde{\varphi} \rangle, \quad a \in A,$$

so $\widetilde{\varphi} \in \mathcal{S}(\pi, f)$. Now we replace φ by $\widetilde{\varphi} = g(p)^{-1/2}\pi(p)\psi$ in (5.14) and obtain

$$|\langle \widetilde{\varphi}, \psi \rangle|^2 = |\langle g(p)^{-1/2}\pi(p)\psi, \psi \rangle|^2 = g(p)^{-1}g(p)^2 = g(p),$$

so that $g(p) \leq P_A(f, g)$. Thus, $P_A(f, g) = g(p) = |\langle \widetilde{\varphi}, \psi \rangle|^2$. $\qquad\square$

Example 5.17 (*Vector states on O^*-algebras*)
Suppose $\mathcal{A}$ is a unital O^*-algebra on $\mathcal{D}$. For a unit vector $\xi \in \mathcal{D}$, let f_ξ denote the vector state defined by $f_\xi(a) = \langle a\xi, \xi \rangle$, $a \in \mathcal{A}$. Suppose the rank one projection $p_\xi := \xi \otimes \xi$ is in $\mathcal{A}$. Then f_ξ satisfies condition (5.13). Hence, by Lemma 5.16, $P_\mathcal{A}(f_\xi, g) = g(p_\xi)$ for each state g on $\mathcal{A}$. In particular, setting $g = f_\eta$, this gives

$$P_\mathcal{A}(f_\xi, f_\eta) = f_\eta(p_\xi) = |\langle \xi, \eta \rangle|^2 \quad \text{for} \quad \xi, \eta \in \mathcal{D}, \ \|\xi\| = \|\eta\| = 1. \tag{5.15}$$

That is, for vector states f_ξ and f_η we obtain the expected expression $|\langle \xi, \eta \rangle|^2$. $\quad\bigcirc$

Our main result in this section is the following theorem. It expresses the transition probability in terms of the symmetrized commutant $\pi(A)'_{\text{sym}}$ (see Definition 3.12). Recall that $\pi(A)'_{\text{sym}}$ is always a von Neumann algebra (Proposition 3.17(iv)).

Theorem 5.18 *Let $f, g \in \mathcal{P}(A)^*$. Suppose the GNS representations π_f and π_g are essentially self-adjoint. Let π be a nondegenerate biclosed $*$-representation of A such that $\mathcal{S}(\pi, f)$ and $\mathcal{S}(\pi, g)$ are not empty. Fix $\varphi \in \mathcal{S}(\pi, f)$, $\psi \in \mathcal{S}(\pi, g)$. Then*

$$P_A(f, g) = \sup_{T \in \pi(A)'_{\text{sym}}, \|T\| \leq 1} |\langle T\varphi, \psi \rangle|^2 \tag{5.16}$$

and the supremum at the right-hand side of (5.16) is a maximum.

Proof First, let $\pi_0 \in \text{Rep A}$ and suppose that there exist vectors $\varphi_0 \in \mathcal{S}(\pi_0, f)$ and $\psi_0 \in \mathcal{S}(\pi_0, g)$. Set $F(a) := \langle \pi_0(a)\varphi_0, \psi_0 \rangle$, $a \in \text{A}$. Then, for $a, b \in \text{A}$,

$$|F(b^+a)|^2 = |\langle \pi_0(a)\varphi_0, \pi_0(b)\psi_0 \rangle|^2 \leq \|\pi_0(a)\varphi_0\|^2 \|\pi_0(b)\psi_0\|^2$$
$$= f(a^+a)g(b^+b) = \|\pi_f(a)\varphi_f\|^2 \|\pi_g(b)\varphi_g\|^2. \tag{5.17}$$

From this inequality it follows that the map $(\pi_f(a)\varphi_f, \pi_g(b)\varphi_g) \mapsto F(b^+a)$ gives a densely defined bounded sesquilinear form on $\mathcal{H}(\pi_f) \times \mathcal{H}(\pi_g)$. Hence there exists a bounded linear operator T_0 of $\mathcal{H}(\pi_f)$ into $\mathcal{H}(\pi_g)$ such that

$$F(b^+a) = \langle T_0\pi_f(a)\varphi_f, \pi_g(b)\varphi_g \rangle \quad \text{for} \quad a, b \in \text{A}. \tag{5.18}$$

By (5.17), we have $\|T_0\| \leq 1$. Further, for $a, b, c \in \text{A}$, we derive

$$\langle T_0\pi_f(a)\varphi_f, \pi_g(c)\pi_g(b)\varphi_g \rangle = F((cb)^+a) = F(b^+(c^+a))$$
$$= \langle T_0\pi_f(c^+)\pi_f(a)\varphi_f, \pi_g(b)\varphi_g \rangle.$$

Therefore, $T_0\pi_f(a)\varphi_f \in \mathcal{D}(\pi_g(c)^*)$ and $\pi_g(c)^*T_0\pi_f(a)\varphi_f = T_0\pi_f(c^+)\pi_f(a)\varphi_f$. Because $c \in \text{A}$ is arbitrary, we conclude that $T_0\pi_f(a)\varphi_f \in \mathcal{D}((\pi_g)^*)$ and hence $(\pi_g)^*(c^+)T_0(\pi_f(a)\varphi_f) = T_0\pi_f(c^+)(\pi_f(a)\varphi_f)$ for $a, c \in \text{A}$. Since $a, c \in \text{A}$ are arbitrary, the latter implies that $T_0 \in I(\pi_f, (\pi_g)^*)$.

Next we consider the $*$-representation π appearing in Theorem 5.18 and set $\rho_f := \pi \lceil \pi(\text{A})\varphi$ and $\rho_g := \pi \lceil \pi(\text{A})\psi$. Since $\varphi \in \mathcal{S}(\pi, f)$ and $\psi \in \mathcal{S}(\pi, g)$ by assumption, Theorem 4.41 implies that ρ_f and ρ_g are unitarily equivalent to the GNS representations π_f and π_g, respectively. For notational simplicity we identify ρ_f with π_f and ρ_g with π_g. Since $\rho_f \subseteq \pi$ and π is biclosed, Lemma 4.4(iii) implies $(\rho_f)^{**} \subseteq \pi^{**} = \pi$. By assumption, $(\pi_f)^* = (\rho_f)^*$ is self-adjoint, so $(\rho_f)^* = (\rho_f)^{**}$ is a self-adjoint subrepresentation of π. Hence, by Corollary 4.31, there is a subrepresentation ρ_1 of π such that $\pi = (\rho_f)^* \oplus \rho_1$. Similarly, $(\rho_g)^* = (\rho_g)^{**}$ and $\pi = (\rho_g)^* \oplus \rho_2$.

Now, as above, let $\varphi_0 \in \mathcal{S}(\pi_0, f)$ and $\psi_0 \in \mathcal{S}(\pi_0, g)$ be arbitrary. As shown in the paragraph before last, the operator T_0 is in $I(\pi_f, (\pi_g)^*) = I(\rho_f, (\rho_g)^*)$. Hence it follows from Lemma 5.8 that

$$T_0^* \in I((\rho_g)^{**}, (\rho_f)^*) = I((\rho_g)^*, (\rho_f)^*), \tag{5.19}$$
$$T_0 \in I((\rho_f)^{**}, (\rho_g)^{***}) = I((\rho_f)^*, (\rho_g)^*). \tag{5.20}$$

Next we define an operator T of $\mathcal{H}(\pi)$ into $\mathcal{H}(\pi)$ by

$$T : \mathcal{H}((\rho_f)^*) \oplus \mathcal{H}(\rho_1) \mapsto \mathcal{H}((\rho_g)^*) \oplus \mathcal{H}(\rho_2), \quad T(\xi_f, \xi_1) = (T_0\xi_f, 0),$$

where $\xi_f \in \mathcal{H}((\rho_f)^*)$ and $\xi_1 \in \mathcal{H}(\rho_1)$. Then, $T^*(\eta_g, \eta_2) = (T_0^*\eta_g, 0)$. Recall that $\pi = (\rho_f)^* \oplus \rho_1 = (\rho_g)^* \oplus \rho_2$. From (5.19) and (5.20) we derive that $T, T^* \in \pi(\text{A})'_s$. Hence $T \in \pi(\text{A})'_{\text{sym}}$ by Proposition 3.17(i). Clearly, $\|T\| = \|T_0\| \leq 1$. By (5.18),

$$\langle \varphi_0, \psi_0 \rangle = \langle \pi_0(1)\varphi_0, \psi_0 \rangle = F(1) = \langle T_0\varphi_f, \varphi_g \rangle = \langle T\varphi_f, \varphi_g \rangle = \langle T\varphi, \psi \rangle.$$

Taking the supremum of $|\langle \varphi_0, \psi_0 \rangle|^2$ over φ_0, ψ_0, it follows that $P_A(f, g)$ is less than or equal to the supremum of $|\langle T\varphi, \psi \rangle|^2$ over $T \in \pi(A)'_{sym}$, $\|T\| \le 1$.

To prove the converse inequality, we fix an operator $T \in \pi(A)'_{sym}$, $\|T\| \le 1$. By the Russo–Dye theorem [Dv96, Theorem 1.8.4], the unit ball of the unital C^*-algebra $\pi(A)'_{sym}$ is the closed convex hull of unitaries of $\pi(A)'_{sym}$. Thus, given $\varepsilon > 0$, there exist unitaries $U_k \in \pi(A)'_{sym}$ and numbers $\lambda_k \in [0, 1]$, $k = 1, \ldots, n$, such that

$$\sum_{k=1}^{n} \lambda_k = 1 \quad \text{and} \quad \left\| T - \sum_{k=1}^{n} \lambda_k U_k \right\| \le \varepsilon. \tag{5.21}$$

Since π is biclosed, hence closed, we have $\pi(A)'_{sym} \subseteq \pi(A)'_s$ by Proposition 3.17(ii). Since $U_k \in \pi(A)'_s$ is unitary and $\varphi \in \mathcal{S}(\pi, f)$, it follows that $U_k\varphi \in \mathcal{S}(\pi, f)$. Hence $|\langle U_k\varphi, \psi \rangle|^2 \le P_A(f, g)$. Therefore, by (5.21),

$$|\langle T\varphi, \psi \rangle| \le \varepsilon \|\varphi\| \|\psi\| + \sum_{k=1}^{n} \lambda_k |\langle U_k\varphi, \psi \rangle| \le \varepsilon \|\varphi\| \|\psi\| + P_A(f, g)^{1/2}.$$

Letting $\varepsilon \to +0$, we get $|\langle T\varphi, \psi \rangle| \le P_A(f, g)^{1/2}$, which gives the converse inequality. Now the proof of the equality (5.16) is complete.

The unit ball of the von Neumann algebra $\pi(A)'_{sym}$ is compact in the weak operator topology [KR83]. Therefore, the supremum of the continuous function $T \mapsto |\langle T\varphi, \psi \rangle|$ on the unit ball is a maximum. Hence the supremum in (5.16) is a maximum. □

Theorem 5.18 states that if the two GNS representations π_f and π_g are essentially self-adjoint, then the transition probability $P_A(f, g)$ is given by (5.16) in *each* biclosed $*$-representation π for which the sets $\mathcal{S}(\pi, f)$ and $\mathcal{S}(\pi, g)$ are not empty and for *arbitrary* vectors $\varphi \in \mathcal{S}(\pi, f)$ and $\psi \in \mathcal{S}(\pi, g)$! For instance, we may set $\pi := (\pi_f)^* \oplus (\pi_g)^*$, $\varphi := \varphi_f$, $\psi := \varphi_g$.

Corollary 5.19 *Let us retain the assumptions and the notation of Theorem 5.18. Let F_φ and F_ψ denote the vector functionals on the von Neumann algebra $M := (\pi(A)'_{sym})'$ given by $F_\varphi(x) = \langle x\varphi, \varphi \rangle$ and $F_\psi(x) = \langle x\psi, \psi \rangle$, $x \in M$. Then*

$$P_A(f, g) = P_M(F_\varphi, F_\psi). \tag{5.22}$$

Proof Since $\pi(A)'_{sym}$ is a von Neumann algebra, we have $T \in \pi(A)'_{sym}$ if and only if $T \in (\pi(A)'_{sym})'' = M'$. Therefore, applying formula (5.16) twice, it follows that the supremum of $|\langle T\varphi, \psi \rangle|^2$ over all operators $T \in \pi(A)'_{sym}$, $\|T\| \le 1$, is equal to $P_A(f, g)$ and also to $P_M(F_\varphi, F_\psi)$. This yields (5.22). □

Formula (5.22) reduces the computation of the transition probability $P_A(f, g)$ to the von Neumann algebra $M = (\pi(A)'_{\text{sym}})'$. We will use this fact in the proofs of Theorems 5.23 and 5.25 in the next section.

Corollary 5.20 *Suppose* A *is a unital* C^*-*algebra and* b, c *are elements of* A *such that* $c^+ b \in \sum A^2$. *Let* $h \in \mathcal{P}(A)^*$. *Define positive linear functionals* f, g *on* A *by* $f(a) = h(b^+ a b)$ *and* $g(a) = h(c^+ a c)$ *for* $a \in A$. *Then*

$$P_A(f, g) = h(c^+ b)^2.$$

Proof Since A is a C^*-algebra, the GNS representation $\overline{\pi}_h$ acts by *bounded* operators on $\mathcal{H}(\overline{\pi}_h)$, so it is obviously self-adjoint and biclosed. By the definitions of f and g, we have $\overline{\pi}_h(b)\varphi_h \in \mathcal{S}(\overline{\pi}_h, f)$ and $\overline{\pi}_h(c)\varphi_h \in \mathcal{S}(\overline{\pi}_h, g)$. The positive element $c^+ b$ of the C^*-algebra A has a positive square root $(c^+ b)^{1/2} \in A$. Then, for $T \in \overline{\pi}_h(A)'_{\text{sym}} \equiv \overline{\pi}_h(A)'$, we compute

$$|\langle T\overline{\pi}_h(c)\varphi_h, \overline{\pi}_h(b)\varphi_h \rangle| = |\langle T\varphi_h, \pi_h(c^+ b)\varphi_h \rangle|$$
$$= |\langle T\overline{\pi}_h((c^+ b)^{1/2})\varphi_h, \overline{\pi}_h((c^+ b)^{1/2})\varphi_h \rangle|. \tag{5.23}$$

Clearly, the supremum in (5.23) over the unit ball of $\overline{\pi}_h(A)'$ is attained at $T = I$ and it is $\langle \overline{\pi}_h((c^+ b)^{1/2})\varphi_h, \overline{\pi}_h((c^+ b)^{1/2})\varphi_h \rangle = \langle \overline{\pi}_h(c^+ b)\varphi_h, \varphi_h \rangle = h(c^+ b)$. Hence $P_A(f, g) = h(c^+ b)^2$ by Theorem 5.18 and (5.16), applied with $\pi = \overline{\pi}_h$. $\qquad \square$

5.4 Examples of Transition Probabilities

Our first application deals with trace functionals (see Sect. 3.3). We begin with some preparations and use the notation and facts established in Example 4.10. For a Hilbert space $\mathcal{H}$ and $T \in \mathbf{B}(\mathcal{H})$, we define an operator T_1 on $\mathbf{B}_2(\mathcal{H})$ by $T_1 x = xT$ for $x \in \mathbf{B}_2(\mathcal{H})$. Clearly, T_1 is a bounded operator on the Hilbert space $\mathbf{B}_2(\mathcal{H})$.

Lemma 5.21 *Suppose* π *is an irreducible self-adjoint* $*$-*representation of* A. *Then the* $*$-*representation* π_{HS} *from Example 4.10 is self-adjoint and we have*

$$\pi_{\text{HS}}(A)'_{\text{sym}} = \{T_1 : T \in \mathbf{B}(\mathcal{H}(\pi))\}. \tag{5.24}$$

Proof As shown in Example 4.10, π_{HS} is self-adjoint, so all three bounded commutants coincide by Proposition 3.17(iii). It only remains to prove that the equality (5.24) holds. Throughout this proof, we abbreviate $\mathcal{H} = \mathcal{H}(\pi)$.

Let $T \in \mathbf{B}(\mathcal{H})$ and $x \in \mathcal{D}(\pi_{\text{HS}})$. Then, for $a \in A$, $\pi(a)x$ and xT are also in $\mathcal{D}(\pi_{\text{HS}})$. Using the associativity of operator multiplication we get

$$T_1 \pi_{\text{HS}}(a)x = T_1(\pi(a) \cdot x) = (\pi(a) \cdot x)T = \pi(a) \cdot (xT) = \pi_{\text{HS}}(a)T_1 x,$$

which proves that $T_1 \in \pi_{\text{HS}}(A)'_s$.

Conversely, let $S \in \pi_{\mathrm{HS}}(\mathsf{A})'_s$. Fix $\varphi, \psi \in \mathcal{H}$. We define a sesquilinear form $\mathfrak{s}_{\varphi,\psi}$ on $\mathcal{H} \times \mathcal{H}$ by

$$\mathfrak{s}_{\varphi,\psi}(\eta, \xi) = \langle S(\varphi \otimes \eta), \psi \otimes \xi \rangle_{\mathrm{HS}}, \quad \eta, \xi \in \mathcal{H}. \tag{5.25}$$

Then, since

$$|\mathfrak{s}_{\varphi,\psi}(\eta, \xi)| = |\langle S(\varphi \otimes \eta), \psi \otimes \xi \rangle_{\mathrm{HS}}| = |\mathrm{Tr}\,[(\psi \otimes \xi)^* S(\varphi \otimes \eta)]\,|$$
$$\leq \|\psi \otimes \xi\| \, \|S(\varphi \otimes \eta)\| \leq \|S\| \, \|\varphi\| \, \|\eta\| \, \|\psi\| \, \|\xi\|, \tag{5.26}$$

the form $\mathfrak{s}_{\varphi,\psi}$ is bounded. Hence it is given by a bounded operator $S_{\varphi,\psi}$ on $\mathcal{H}$, so

$$\langle S(\varphi \otimes \eta), \psi \otimes \xi \rangle_{\mathrm{HS}} = \mathfrak{s}_{\varphi,\psi}(\eta, \xi) = \langle S_{\varphi,\psi}\eta, \xi \rangle, \quad \eta, \xi \in \mathcal{H}. \tag{5.27}$$

Let $\eta, \xi \in \mathcal{D}(\pi)$. Then $\varphi \otimes \eta, \psi \otimes \xi \in \mathcal{D}(\pi_{\mathrm{HS}})$. Using that $S \in \pi_{\mathrm{HS}}(\mathsf{A})'_s$ we derive

$$\langle S_{\varphi,\psi}\pi(a)\eta, \xi \rangle = \mathfrak{s}_{\varphi,\psi}(\pi(a)\eta, \xi) = \langle S(\varphi \otimes \pi(a)\eta), \psi \otimes \xi \rangle$$
$$= \langle S\,\pi_{\mathrm{HS}}(a)(\varphi \otimes \eta), \psi \otimes \xi \rangle_{\mathrm{HS}} = \langle \pi_{\mathrm{HS}}(a)S(\varphi \otimes \eta), \psi \otimes \xi \rangle_{\mathrm{HS}}$$
$$= \langle S(\varphi \otimes \eta), \pi_{\mathrm{HS}}(a^+)(\psi \otimes \xi) \rangle_{\mathrm{HS}} = \langle S(\varphi \otimes \eta), \psi \otimes \pi(a^+)\xi \rangle_{\mathrm{HS}}$$
$$= \mathfrak{s}_{\varphi,\psi}(\eta, \pi(a^+)\xi) = \langle S_{\varphi,\psi}\eta, \pi(a^+)\xi \rangle, \quad a \in \mathsf{A}.$$

This shows that the operator $S_{\varphi,\psi}$ belongs to the weak commutant $\pi(\mathsf{A})'_w$. Since π is self-adjoint, $\pi(\mathsf{A})'_w = \pi(\mathsf{A})'_s$ is a von Neumann algebra (Proposition 3.15). It is equal to $\mathbb{C} \cdot I$, because π is irreducible. Thus, $S_{\varphi,\psi} = c_{\varphi,\psi} \cdot I$ for some $c_{\varphi,\psi} \in \mathbb{C}$.

From (5.25) we derive that $S_{\varphi,\psi}$, hence $c_{\varphi,\psi}$, is linear in ψ and conjugate linear in φ. Hence $(\psi, \varphi) \mapsto c_{\varphi,\psi}$ is a sesquilinear form on $\mathcal{H} \times \mathcal{H}$. From (5.26) and (5.27) we deduce that $|c_{\varphi,\psi}| = \|S_{\varphi,\psi}\| \leq \|\varphi\| \, \|\psi\|$. Hence there exists an operator $T \in \mathbf{B}(\mathcal{H})$ such that $c_{\varphi,\psi} = \langle T\psi, \varphi \rangle$ for $\varphi, \psi \in \mathcal{H}$. For $\varphi, \psi, \eta, \xi \in \mathcal{H}$, we compute $(\psi \otimes \xi)^*(\varphi \otimes \eta)T = \langle \eta, \xi \rangle \, T^*\varphi \otimes \psi$ and

$$\langle S(\varphi \otimes \eta), \psi \otimes \xi \rangle_{\mathrm{HS}} = \mathfrak{s}_{\varphi,\psi}(\eta, \xi) = \langle c_{\varphi,\psi}\eta, \xi \rangle = \langle \psi, T^*\varphi \rangle \langle \eta, \xi \rangle$$
$$= \langle \eta, \xi \rangle \, \mathrm{Tr}\,(T^*\varphi \otimes \psi) = \mathrm{Tr}\,[(\psi \otimes \xi)^*(\varphi \otimes \eta)T] = \langle (\varphi \otimes \eta)T, \psi \otimes \xi \rangle_{\mathrm{HS}}.$$

Therefore, since the span of rank one operators $\psi \otimes \xi$ is dense in the Hilbert space $\mathbf{B}_2(\mathcal{H})$, we conclude that $S(\varphi \otimes \eta) = (\varphi \otimes \eta)T$ for $\varphi, \eta \in \mathcal{H}$ and hence $Sx = xT$ for all $x \in \mathbf{B}_2(\mathcal{H})$. This proves that $S = T_1$. $\qquad\square$

The next lemma determines the transition probability for normal states on $\mathbf{B}(\mathcal{H})$.

Lemma 5.22 *Let $\mathcal{H}$ be a Hilbert space and τ the $*$-representation of $\mathsf{B} := \mathbf{B}(\mathcal{H})$ on the Hilbert space $\mathbf{B}_2(\mathcal{H})$ defined by $\tau(x)y = x \cdot y, x \in \mathsf{B}, y \in \mathbf{B}_2(\mathcal{H})$, see Example 4.10. Suppose $s, t \in \mathbf{B}_1(\mathcal{H})_+$. Let $g_s(\cdot) = \mathrm{Tr}\,\tau(\cdot)s$ and $g_t(\cdot) = \mathrm{Tr}\,\tau(\cdot)t$ be the corresponding positive linear functionals on B. Then*

$$P_{\mathsf{B}}(g_t, g_s) = \left(\mathrm{Tr}\,|t^{1/2}s^{1/2}|\right)^2. \tag{5.28}$$

Proof [2] In this proof, we essentially use some properties of the polar decomposition of operators (see Proposition A.3). For $x \in \mathbf{B}(\mathcal{H})$ let $e(x)$ denote the projection on $\overline{\text{ran } x^*} = \overline{\text{ran } |x|}$. Note that if u is a partial isometry, then $x = u|x|$ is the polar decomposition of x if and only if $u^*u = e(x)$.

Since $\overline{\text{ran } (s^{1/2}t^{1/2})^*} = \overline{\text{ran } t^{1/2}s^{1/2}} \subseteq \overline{\text{ran } t}$, we have $e(t) \geq e(|s^{1/2}t^{1/2}|)$. Similarly, $e(s) \geq e(|t^{1/2}s^{1/2}|)$.

First we characterize the set $\mathcal{S}(\tau, g_t)$ and prove that

$$\mathcal{S}(\tau, g_t) = \{a \in \mathbf{B}_2(\mathcal{H}) : a = t^{1/2}v^*, \ v^*v = e(t), \ v \in \mathbf{B}(\mathcal{H})\}. \tag{5.29}$$

Indeed, let $a \in \mathbf{B}_2(\mathcal{H})$. Since $g_t(x) = \text{Tr } xt = \text{Tr } t^{1/2}xt^{1/2}$ for $x \in \mathbf{B}$, the equality $\langle \tau(x)a, a\rangle_{\text{HS}} = g_t(x) \equiv \langle xt^{1/2}, t^{1/2}\rangle_{\text{HS}}$ is equivalent to $\text{Tr } a^*xa = \text{Tr } t^{1/2}xt^{1/2}$ and so to $\text{Tr } xaa^* = \text{Tr } xt$. This holds for all $x \in \mathbf{B}$ if and only if $t = aa^*$, that is, $t^{1/2} = |a^*|$. Equivalently, the polar decomposition of a^* is the form $a^* = vt^{1/2}$ for a partial isometry v such that $v^*v = e(a^*) = e(t)$. Then $a = t^{1/2}v^*$. Since $v \in \mathbf{B}(\mathcal{H})$ is a partial isometry if and only if v^*v is a projection, this gives (5.29).

Let u denote the phase operator of $t^{1/2}s^{1/2}$. Then u^* is the phase operator of $(t^{1/2}s^{1/2})^* = s^{1/2}t^{1/2}$ and hence

$$u^*u = e(|t^{1/2}s^{1/2}|), \quad uu^* = e(|s^{1/2}t^{1/2}|). \tag{5.30}$$

Now suppose that $a = t^{1/2}v^* \in \mathcal{S}(\tau, g_t)$ and $b = s^{1/2}w^* \in \mathcal{S}(\tau, g_s)$, with $v, w \in \mathbf{B}(\mathcal{H})$. Since $t^{1/2}s^{1/2} = u|t^{1/2}s^{1/2}|$ and hence $s^{1/2}t^{1/2} = |t^{1/2}s^{1/2}|u^*$, using the commutation properties of the trace we derive

$$\langle a, b\rangle_{\text{HS}} = \text{Tr } b^*a = \text{Tr } ws^{1/2}t^{1/2}v^* = \text{Tr } u^*v^*w|t^{1/2}s^{1/2}|. \tag{5.31}$$

By (5.29), v and w can be taken as partial isometries. Then $\|u^*v^*w\| \leq 1$, so (5.31) implies that $|\langle a, b\rangle_{\text{HS}}|^2 \leq (\text{Tr } |t^{1/2}s^{1/2}|)^2$. Taking the supremum over a, b we obtain $P_{\mathbf{B}}(g_t, g_s) \leq (\text{Tr } |t^{1/2}s^{1/2}|)^2$.

Therefore, to complete the proof of formula (5.28) it suffices to construct operators $a = t^{1/2}v^* \in \mathcal{S}(\tau, g_t)$ and $b = s^{1/2}w^* \in \mathcal{S}(\tau, g_s)$ such that

$$u^*v^*w = e(|t^{1/2}s^{1/2}|). \tag{5.32}$$

Indeed, then it follows from (5.32) and (5.31) that

$$|\langle a, b\rangle_{\text{HS}}| = \left|\text{Tr } u^*v^*w|t^{1/2}s^{1/2}|\right| = \left|\text{Tr } e(|t^{1/2}s^{1/2}|)|t^{1/2}s^{1/2}|\right| = \text{Tr } |t^{1/2}s^{1/2}|,$$

which implies $P_{\mathbf{B}}(g_t, g_s) \geq (\text{Tr } |t^{1/2}s^{1/2}|)^2$, so (5.28) holds.

First assume that $\mathcal{H}$ has finite dimension. Then the initial space $u^*u\mathcal{H}$ and the final space $uu^*\mathcal{H}$ of the partial isometry u have the same codimensions. Hence u can be extended to a unitary operator u_0 which maps $(I - u^*u)\mathcal{H}$ on $(I - uu^*)\mathcal{H}$. Therefore,

---

[2]This elegant proof I owe to my colleague P.M. Alberti.

$u^*u_0 = u^*u$. Set $w := u_0 e(s)$. Then $w^*w = e(s)u_0^* u_0 e(s) = e(s)$ and hence $b := s^{1/2}w^* \in \mathcal{S}(\tau, g_s)$ by (5.29). Clearly, $a := t^{1/2}v \in \mathcal{S}(\tau, g_t)$ with $v = I$. Using (5.30) and the relation $e(s) \geq e(|t^{1/2}s^{1/2}|)$ noted above, we derive

$$u^*vw = u^*u_0 e(s) = u^*u e(s) = e(|t^{1/2}s^{1/2}|)e(s) = e(|t^{1/2}s^{1/2}|),$$

which proves (5.32).

Now we treat the case when $\mathcal{H}$ has infinite dimension. Since $e(t) \geq e(|s^{1/2}t^{1/2}|)$, $e(t) - e(|s^{1/2}t^{1/2}|)$ is a projection. Then there exist partial isometries v_1, w such that $v_1^*v_1 = e(t) - e(|s^{1/2}t^{1/2}|)$ and $w^*w = e(s)$. These requirements fix only the initial spaces of v_1 and w. Since dim $\mathcal{H} = \infty$, we can choose w, v_1 such that their final spaces are orthogonal, so that

$$w^*v_1 = v_1^*w = 0. \tag{5.33}$$

Since $u^*u = e(|t^{1/2}s^{1/2}|)$ by (5.30) and $e(|t^{1/2}s^{1/2}|) \leq e(s)$, we obtain

$$ue(s)u^* = ue(|t^{1/2}s^{1/2}|)u^* = uu^*uu^* = uu^* = e(|s^{1/2}t^{1/2}|) \tag{5.34}$$

again by (5.30). We set $v := v_1 + wu^*$. Then, using (5.33) and (5.34) we compute

$$v^*v = (v_1^* + uw^*)(v_1 + wu^*) = v_1^*v_1 + uw^*wu^*$$
$$= e(t) - e(|s^{1/2}t^{1/2}|) + ue(s)u^* = e(t) - e(|s^{1/2}t^{1/2}|) + e(|s^{1/2}t^{1/2}|) = e(t).$$

Therefore, $a := t^{1/2}v^* \in \mathcal{S}(\tau, g_t)$ and $b := s^{1/2}w^* \in \mathcal{S}(\tau, g_s)$ by (5.29).

Finally, from (5.33) and (5.30) we obtain

$$u^*v^*w = u^*(v_1^* + uw^*)w = u^*uw^*w = e(|t^{1/2}s^{1/2}|)e(s) = e(|t^{1/2}s^{1/2}|),$$

which gives (5.32) in the case dim $\mathcal{H} = \infty$. This completes the proof. $\qquad\square$

Our first main result in this section is the following theorem.

Theorem 5.23 *Let ρ be a nondegenerate self-adjoint irreducible $*$-representation of* A. *Suppose that $s, t \in \mathbf{B}_1(\rho(A))_+$, $s^{1/2}\mathcal{H}(\rho) \subseteq \mathcal{D}(\rho)$, and $t^{1/2}\mathcal{H}(\rho) \subseteq \mathcal{D}(\rho)$. We define positive functionals f_s, f_t on* A *by*

$$f_s(a) = \text{Tr } \rho(a)s, \quad f_t(a) = \text{Tr } \rho(a)t \quad \text{for} \quad a \in \mathsf{A}.$$

Suppose the GNS representations π_{f_s} and π_{f_t} are essentially self-adjoint. Then

$$P_\mathsf{A}(f_t, f_s) = \left(\text{Tr } |t^{1/2}s^{1/2}|\right)^2 = \left(\text{Tr } (s^{1/2}ts^{1/2})^{1/2}\right)^2. \tag{5.35}$$

In particular, if $\eta, \xi \in \mathcal{D}(\rho)$ are unit vectors and $s := \eta \otimes \eta$, $t := \xi \otimes \xi$, then

$$P_\mathsf{A}(f_t, f_s) = |\langle \xi, \eta \rangle|^2. \qquad (5.36)$$

Proof Set $\mathcal{H} := \mathcal{H}(\rho)$. By Lemma 5.24, ρ_{HS} is a self-adjoint $*$-representation of A and we have $\rho_{\mathrm{HS}}(\mathsf{A})'_{\mathrm{sym}} = \{T_1 : T \in \mathbf{B}(\mathcal{H})\}$. Let π denote the identity representation of $\mathsf{B} := \mathbf{B}(\mathcal{H})$ on $\mathcal{H}$, that is, $\pi(b)\varphi = b\varphi$, $b \in \mathsf{B}$, $\varphi \in \mathcal{H}$. Obviously, the representation π is irreducible and self-adjoint. Applying Lemma 5.24 to π and B we obtain $\pi_{\mathrm{HS}}(\mathsf{B})' = \{T_1 : T \in \mathbf{B}(\mathcal{H})\}$. Hence $(\rho_{\mathrm{HS}}(\mathsf{A})'_{\mathrm{sym}})' = \pi_{\mathrm{HS}}(\mathsf{B})'' =: \mathsf{M}$.

It is easily checked that $\pi_{\mathrm{HS}}(\mathsf{B})$ is closed in $\mathbf{B}_2(\mathcal{H})$ in the strong operator topology. Therefore, by the double commutant theorem [KR83, Theorem 5.3.1], $\pi_{\mathrm{HS}}(\mathsf{B})'' = \pi_{\mathrm{HS}}(\mathsf{B})$, so that $\mathsf{M} = (\rho_{\mathrm{HS}}(\mathsf{A})'_{\mathrm{sym}})' = \pi_{\mathrm{HS}}(\mathsf{B})$. As shown in Example 4.48, the functionals f_s, f_t on A are realized as vector functionals in ρ_{HS} with vectors $s^{1/2}, t^{1/2} \in \mathbf{B}_2(\mathcal{H})$. Likewise, $F_{s^{1/2}}(\cdot) := \mathrm{Tr} \,\cdot\, s$ and $F_{t^{1/2}}(\cdot) := \mathrm{Tr} \,\cdot\, t$ are vector functionals on M with vectors $s^{1/2}$ and $t^{1/2}$, respectively. Thus, Corollary 5.19 applies to ρ_{HS}, $\varphi = t^{1/2}$, $\psi = s^{1/2}$ and yields $P_\mathsf{A}(f_t, f_s) = P_\mathsf{M}(F_{t^{1/2}}, F_{s^{1/2}})$.

Clearly, $\tau := \pi_{\mathrm{HS}}$ is the $*$-representation of $\mathsf{B} := \mathbf{B}(\mathcal{H})$ from Lemma 5.22. It is a $*$-isomorphism of B and M and we have $g_s(\cdot) = F_{s^{1/2}}(\tau(\cdot)) = \mathrm{Tr}\,\tau(\cdot)s$ and $g_t(\cdot) = F_{t^{1/2}}(\tau(\cdot)) = \mathrm{Tr}\,\tau(\cdot)t$. Obviously, the transition probability is preserved under $*$-isomorphism. Hence $P_\mathsf{M}(F_{t^{1/2}}, F_{s^{1/2}}) = P_\mathsf{B}(g_t, g_s)$. By Lemma 5.22, $P_\mathsf{B}(g_t, g_s) = (\mathrm{Tr}\,|t^{1/2}s^{1/2}|)^2$. Combining the preceding equalities gives the first equality of (5.35). The second equality follows at once from the first by using that $|t^{1/2}s^{1/2}|^2 = (t^{1/2}s^{1/2})^* t^{1/2}s^{1/2} = s^{1/2}ts^{1/2}$.

Formula (5.36) is a special case of (5.35). First we compute $sts = |\langle \xi, \eta \rangle|^2 s$. Since η, ξ are unit vectors, s, t are projections and $\mathrm{Tr}\, s = 1$. Hence $s^{1/2}ts^{1/2} = sts$ and $|t^{1/2}s^{1/2}| = |\langle \xi, \eta \rangle|s$, so that $(\mathrm{Tr}\,|t^{1/2}s^{1/2}|)^2 = |\langle \xi, \eta \rangle|^2 (\mathrm{Tr}\, s)^2 = |\langle \xi, \eta \rangle|^2$. By (5.35), this gives (5.36). □

Example 5.24 (*Schrödinger representation of the Weyl algebra*)
Let W be the Weyl algebra and ρ the Schrödinger representation of W on $\mathcal{H} := L^2(\mathbb{R})$, see Example 4.32. As shown there, ρ is irreducible and self-adjoint.

Suppose $s, t \in \mathbf{B}_1(\rho(\mathsf{W}))_+$, $s^{1/2}\mathcal{H} \subseteq \mathcal{D}(\rho)$, $t^{1/2}\mathcal{H} \subseteq \mathcal{D}(\rho)$. Further, suppose that the GNS representations π_{f_s} and π_{f_t} are essentially self-adjoint.

Under these assumptions, it follows from Theorem 5.23 that the transition probability $P_\mathsf{W}(f_t, f_s)$ is given by (5.35). In particular, if $\eta, \xi \in \mathcal{D}(\rho)$ are unit vectors and $s = \eta \otimes \eta$, $t = \xi \otimes \xi$, then we have $f_s(\cdot) = \langle \rho(\cdot)\eta, \eta \rangle$, $f_t(\cdot) = \langle \rho(\cdot)\xi, \xi \rangle$ and formula (5.36) yields $P_\mathsf{W}(f_t, f_s) = |\langle \xi, \eta \rangle|^2$.

In general, the latter formula is no longer true if the assumption that π_{f_s} and π_{f_t} are essentially self-adjoint is omitted. To give a counter-example, we choose functions $\varphi_1 \in C_0^\infty(0, 1)$, $\varphi_2 \in C_0^\infty(2, 3)$ such that $\|\varphi_1\|^2 = \|\varphi_2\|^2 = \frac{1}{2}$. Then the unit vectors $\eta := \varphi_1 + \varphi_2$, $\xi := \varphi_1 - \varphi_2$ define the same state $f_\eta = f_\xi$, so that $P_\mathsf{W}(f_\xi, f_\eta) = 1$. But we have $\langle \xi, \eta \rangle = 0$.   ○

Now we turn to the second main application.

Theorem 5.25 *Let $\mathcal{X}$ be a locally compact topological Hausdorff space and A a $*$-subalgebra of $C(\mathcal{X})$ which contains the constant functions and separates the* points of $\mathcal{X}$. Suppose μ is a Radon measure on $\mathcal{X}$ such that $\mathsf{A} \subseteq L^1(\mathcal{X}; \mu)$ and let $\eta, \xi \in L^\infty(\mathcal{X}; \mu)$ be nonnegative functions. Define positive functionals f_η, f_ξ by

$$f_\eta(a) = \int_{\mathcal{X}} a(x)\eta(x)\, d\mu(x), \quad f_\xi(a) = \int_{\mathcal{X}} a(x)\xi(x)\, d\mu(x) \quad \text{for} \quad a \in \mathsf{A}.$$

If the GNS representations π_{f_η} and π_{f_ξ} are essentially self-adjoint, then

$$P_{\mathsf{A}}(f_\eta, f_\xi) = \left(\int_{\mathcal{X}} \sqrt{\eta(x)\xi(x)}\, d\mu(x) \right)^2. \tag{5.37}$$

Proof We define a $*$-representation π of A on $L^2(\mathcal{X}; \mu)$ by $\pi(a)\varphi = a \cdot \varphi$ for $a \in \mathsf{A}$ and φ in $\mathcal{D}(\pi) := \{\varphi \in L^2(\mathcal{X}; \mu) : a \cdot \varphi \in L^2(\mathcal{X}; \mu),\ a \in \mathsf{A}\}$. It is easily verified that π is self-adjoint. Hence π is biclosed.

First we show that $\pi(\mathsf{A})'_{\text{sym}} = L^\infty(\mathcal{X}; \mu)$, where the functions of $L^\infty(\mathcal{X}; \mu)$ act as multiplication operators. Let $\mathfrak{A}$ denote the $*$-subalgebra of $L^\infty(\mathcal{X}; \mu)$ generated by the functions $(a \pm i)^{-1}$, where $a \in \mathsf{A}_{\text{her}}$. Obviously, $L^\infty(\mathcal{X}; \mu) \subseteq \pi(\mathsf{A})'_{\text{sym}}$. Conversely, take $T \in \pi(\mathsf{A})'_{\text{sym}}$. Let $a \in \mathsf{A}_{\text{her}}$. Since multiplication by the function $(a \pm i)^{-1}$ leaves $\mathcal{D}(\pi)$ invariant, $(\pi(a) \pm iI)\mathcal{D}(\pi)$ is dense. Hence $\overline{\pi(a)}$ is self-adjoint (Proposition A.1) and it is the multiplication operator by the function a. By definition T commutes with $\overline{\pi(a)}$, hence with $(\overline{\pi(a)} \pm iI)^{-1} = (a \pm i)^{-1}$, and so with the whole algebra $\mathfrak{A}$. Since A separates the points of $\mathcal{X}$, so does the $*$-algebra $\mathfrak{A}$. From the Stone–Weierstrass theorem [Cw90, V.8.2], applied to the one point compactification of $\mathcal{X}$, it follows that $\mathfrak{A}$ is norm dense in $C_0(\mathcal{X})$. Hence T commutes with $C_0(\mathcal{X})$ and so with its closure $L^\infty(\mathcal{X}; \mu)$ in the weak operator topology. Therefore, $T \in L^\infty(\mathcal{X}; \mu)'$. Since $L^\infty(\mathcal{X}; \mu)' = L^\infty(\mathcal{X}; \mu)$, we have proved that $\pi(\mathsf{A})'_{\text{sym}} = L^\infty(\mathcal{X}; \mu)$.

Thus, $\mathsf{M} := (\pi(\mathsf{A})'_{\text{sym}})' = L^\infty(\mathcal{X}; \mu)$. Clearly, $\sqrt{\eta}$ and $\sqrt{\xi}$ are in $\mathcal{D}(\pi)$. By definition, $f_\eta(a) = \langle \pi(a)\sqrt{\eta}, \sqrt{\eta} \rangle$ and $f_\xi(a) = \langle \pi(a)\sqrt{\xi}, \sqrt{\xi} \rangle$. Then, by Corollary 5.19, $P_{\mathsf{A}}(f, g) = P_{\mathsf{M}}(F_{\sqrt{\eta}}, F_{\sqrt{\xi}})$. To compute the latter we apply Corollary 5.20 to the positive functional h on the C^*-algebra M given by $h(a) := \int a\, d\mu$, $a \in \mathsf{M}$, and elements $b := \sqrt{\eta}$, $c := \sqrt{\xi} \in \mathsf{M}$. Obviously, $c^+b \in \sum \mathsf{M}^2$. Hence $P_{\mathsf{M}}(F_{\sqrt{\eta}}, F_{\sqrt{\xi}}) = h(c^+b)^2$ by Corollary 5.20. Since $h(c^+b) = \int \sqrt{\eta\xi}\, d\mu$, this gives (5.37). $\square$

In the following two examples we specialize Theorem 5.25 to the case $\mathcal{X} = \mathbb{R}$, $\mathsf{A} = \mathbb{C}[x]$ and reconsider the one-dimensional Hamburger moment problem.

Example 5.26 (*Determinate Hamburger moment problem*)
Suppose that μ is a Radon measure on $\mathbb{R}$ such that $\mathsf{A} := \mathbb{C}[x] \subseteq L^1(\mathbb{R}; \mu)$ and $\eta, \xi \in L^\infty(\mathbb{R}; \mu)$ are nonnegative on $\mathbb{R}$. We define two Radon measures μ_η, μ_ξ on $\mathbb{R}$ by $d\mu_\eta = \eta d\mu$, $d\mu_\xi = \xi d\mu$ and two positive linear functionals f_η, f_ξ on A by $f_\eta(a) = \int a\, d\mu_\eta = \int a\eta\, d\mu$, $f_\xi(a) = \int a\, d\mu_\xi = \int a\xi\, d\mu$. Clearly, μ_η and μ_ξ have finite moments. Then, if both measures μ_η and μ_ξ are determinate, the GNS representations π_{f_η} and π_{f_ξ} are essentially self-adjoint (as shown in Example 4.47) and hence formula (5.37) holds by Theorem 5.25. $\bigcirc$

Example 5.27 (*Counter-example: Indeterminate Hamburger moment problem*)
Let ν be a solution of an indeterminate Hamburger moment problem and $\nu(\mathbb{R}) = 1$.

Let V_ν denote the set of all Radon measures μ on $\mathbb{R}$ which have the same moments as ν, that is, $\int x^n d\mu(x) = \int x^n d\nu(x)$ for all $n \in \mathbb{N}_0$. Since ν is indeterminate and V_ν is convex and vaguely compact [Sch17, Theorem 1.19], there exists a $\mu \in V_\nu$ that is not an extreme point of V_ν. Then there are measures $\mu_1, \mu_2 \in V_\nu$, $\mu_j \neq \mu$ for $j = 1, 2$, such that $\mu = \frac{1}{2}(\mu_1 + \mu_2)$. Since $\mu_j(M) \leq 2\mu(M)$ for all Borel sets M and $\mu_1 + \mu_2 = 2\mu$, there exists functions $\eta, \xi \in L^\infty(\mathbb{R}; \nu)$ satisfying

$$\eta(x) \geq 0, \quad \xi(x) \geq 0, \quad \eta(x) + \xi(x) = 2, \quad d\mu_1 = \eta d\mu, \quad d\mu_2 = \xi d\mu. \qquad (5.38)$$

We define $f(p) = \int p(x) d\mu(x)$ for $p \in \mathsf{A} := \mathbb{C}[x]$. Since $\mu_1, \mu_2, \mu \in V_\nu$, the corresponding positive linear functionals f_η, f_ξ on A coincide with f. Therefore, since $f(1) = \mu(\mathbb{R}) = \nu(\mathbb{R}) = 1$, we have $P_\mathsf{A}(f_\eta, f_\xi) = P_\mathsf{A}(f, f) = 1$.

Set $J := (\int \sqrt{\eta\xi}\, d\mu)^2$. Now (5.38) yields $\eta(x)\xi(x) = \eta(x)(2 - \eta(x)) \leq 1$, so $J \leq 1$, since $\mu(\mathbb{R}) = 1$. If J would be equal to 1, then $\eta(x)(2 - \eta(x)) = 1$, so $(\eta(x) - 1)^2 = 0$ and hence $\eta(x) = 1$ μ-a.e. on $\mathbb{R}$. Then $\mu_1 = \mu_2 = \mu$, which contradicts the choice of μ_1, μ_2. Therefore, $J \neq 1 = P_\mathsf{A}(f_\eta, f_\xi)$, that is, formula (5.37) is not valid in this case. $\qquad\qquad\bigcirc$

5.5 A Radon–Nikodym Theorem for Positive Functionals

Let f and g be positive linear functionals on A.

Definition 5.28 A sequence $(a_n)_{n\in\mathbb{N}}$ of elements $a_n \in \mathsf{A}$ is an (f, g)-*sequence* if

$$\lim_{n\to\infty} f(a_n^+ a_n) = 0 \quad \text{and} \quad \lim_{n,m\to\infty} g((a_n - a_m)^+(a_n - a_m)) = 0.$$

Definition 5.29 (i) g is called f-*absolutely continuous* if $f(a^+a) = 0$ for $a \in \mathsf{A}$ implies that $g(a^+a) = 0$.

(ii) g is called *strongly* f-*absolutely continuous* if $\lim_{n\to\infty} g(a_n^+ a_n) = 0$ for each (f, g)-sequence $(a_n)_{n\in\mathbb{N}}$.

Next we give reformulations of these notions in terms of the GNS representations π_f and π_g. It is obvious that the strong f-absolute continuity implies the f-absolute continuity (take constant sequences).

Recall that $f(a^+a) = \|\pi_f(a)\varphi_f\|^2$ and $g(a^+a) = \|\pi_g(a)\varphi_g\|^2$ for $a \in \mathsf{A}$. Therefore, g is f-*absolutely continuous* if and only if there exists a *well-defined* (!) linear map $K_{f,g} : \mathcal{D}(\pi_f) \mapsto \mathcal{D}(\pi_g)$ such that

$$K_{f,g}(\pi_f(a)\varphi_f) = \pi_g(a)\varphi_g, \quad a \in \mathsf{A}. \qquad (5.39)$$

(This operator appeared already in Eq. (5.8) in another context.) Further, for any sequence (a_n) of elements of A, we have

$$g((a_n - a_m)^+(a_n - a_m)) = \|\pi_g(a_n)\varphi_g - \pi_g(a_m)\varphi_g\|^2,$$
$$f(a_n^+ a_n) = \|\pi_f(a_n)\varphi_f\|^2, \quad g(a_n^+ a_n) = \|\pi_g(a_n)\varphi_g\|^2.$$

From these relations it follows that g is *strongly f-absolutely continuous* if and only if the operator $K_{f,g}$ of the Hilbert space $\mathcal{H}(\pi_f)$ into $\mathcal{H}(\pi_g)$ is *closable*.

Proposition 5.30 *Let $f, g \in \mathcal{P}(A)^*$. Then g is strongly f-absolutely continuous if and only if there exists a positive self-adjoint operator H on the Hilbert space $\mathcal{H}(\pi_f)$ such that $\mathcal{D}(\pi_f) = \pi_f(A)\varphi_f \subseteq \mathcal{D}(H)$ and*

$$g(b^+ a) = \langle H\pi_f(a)\varphi_f, H\pi_f(b)\varphi_f \rangle \quad \text{for} \quad a, b \in A. \tag{5.40}$$

The operator H can be choosen such that $\mathcal{D}(\pi_f)$ is a core for H. In this case, the positive self-adjoint operator H is uniquely determined by (5.40) and called the Radon–Nikodym derivative *of g with respect to f.*

Proof First suppose that g is strongly f-absolutely continuous. Then, as noted above, the linear operator $K_{f,g}$ defined by (5.39) is closable. Its domain is $\mathcal{D}(\pi_f)$, so it is densely defined. Let UH be the polar decomposition (see Proposition A.3) of the closure of the operator $K_{f,g}$. Since UH is the closure of $K_{f,g}$, $\mathcal{D}(\pi_f)$ is a core for H. Using (5.39) and the fact that U is isometric on the range of H we derive

$$g(b^+ a) = \langle \pi_g(a)\varphi_g, \pi_g(b)\varphi_g \rangle = \langle K_{f,g}\pi_f(a)\varphi_f, K_{f,g}\pi_f(b)\varphi_f \rangle$$
$$= \langle UH\pi_f(a)\varphi_f, UH\pi_f(b)\varphi_f \rangle = \langle H\pi_f(a)\varphi_f, H\pi_f(b)\varphi_f \rangle,$$

which is Eq. (5.40).

Next we prove the converse direction. Let $(a_n)_{n\in\mathbb{N}}$ be an (f, g)-sequence. Then we have $\lim_n \pi_f(a_n)\varphi_f = 0$ and using (5.40) and Definition 5.28 we obtain

$$\|H\pi_f(a_n)\varphi_f - H\pi_f(a_m)\varphi_f\|^2 = \langle H\pi_f(a_n - a_m)\varphi_f, H\pi_f(a_n - a_m)\varphi_f \rangle$$
$$= g((a_n - a_m)^+(a_n - a_m)) \to 0 \quad \text{as} \quad n, m \to \infty.$$

Hence $(H\pi_f(a_n)\varphi_f)_{n\in\mathbb{N}}$ is a Cauchy sequence in $\mathcal{H}(\pi_f)$, so it converges. Since the operator H is closed and $\lim_n \pi_f(a_n)\varphi_f = 0$, we get $\lim_n H\pi_f(a_n)\varphi_f = 0$. Applying once more (5.40) the latter yields

$$g(a_n^+ a_n) = \langle H\pi_f(a_n)\varphi_f, H\pi_f(a_n)\varphi_f \rangle \to 0.$$

By Definition 5.29 this proves that g is strongly f-absolutely continuous.

Now let G be another positive self-adjoint operator satisfying (5.40), with H replaced by G, such that $\mathcal{D} := \mathcal{D}(\pi_f)$ is a core for G. Then (5.40) implies that

$$\langle H\varphi, H\psi \rangle = \langle G\varphi, G\psi \rangle \tag{5.41}$$

for $\varphi, \psi \in \mathcal{D}$. Let $\varphi \in \mathcal{D}(G)$. Since $\mathcal{D}$ is a core for G, there exists a sequence (φ_n) of vectors $\varphi_n \in \mathcal{D}$ such that $\lim_n \varphi_n = \varphi$ and $\lim_n G\varphi_n = G\varphi$. By (5.41) we have $\|H\varphi_n - H\varphi_m\| = \|G\varphi_n - G\varphi_m\|$, so the sequence $(H\varphi_n)$ converges as well. Since H is closed, it follows that $\varphi \in \mathcal{D}(H)$ and $\lim_n H\varphi_n = H\varphi$. Thus, $\mathcal{D}(G) \subseteq \mathcal{D}(H)$ and it follows that (5.41) holds for all $\varphi, \psi \in \mathcal{D}(G)$.

Let $\psi \in \mathcal{D}(G^2)$. Then $\psi \in \mathcal{D}(G)$. Hence $\langle H\varphi, H\psi \rangle = \langle G\varphi, G\psi \rangle = \langle \varphi, G^2\psi \rangle$ for $\varphi \in \mathcal{D}$. Therefore, $H\psi \in \mathcal{D}((H\restriction\mathcal{D})^*) = \mathcal{D}(H)$ and $H(H\psi) = G^2\psi$, because $\mathcal{D}$ is a core for the self-adjoint operator H. Thus $G^2 \subseteq H^2$, so that $G^2 = H^2$, because H^2 and G^2 are self-adjoint. The positive self-adjoint operators G and H are square roots of $G^2 = H^2$. Hence $G = H$ by the uniqueness of the positive square root [Sch12, Proposition 5.13]. $\qquad\qquad\square$

Let $a, b, c \in \mathsf{A}$. Applying Eq. (5.40) twice, to both sides of the equality $g(c^+(ab)) = g((a^+c)^+b)$, we obtain

$$\langle H\pi_f(a)\pi_f(b)\varphi_f, H\pi_f(c)\varphi_f \rangle = \langle H\pi_f(b)\varphi_f, H\pi_f(a^+)\pi_f(c)\varphi_f \rangle. \tag{5.42}$$

Thus, (5.40) implies that the map π_f is also a $*$-representation of A on the vector space $\mathcal{D}(\pi_f)$ equipped with the new inner product $\langle \cdot, \cdot \rangle_H := \langle H\cdot, H\cdot \rangle + \langle \cdot, \cdot \rangle$.

Example 5.31 Let A be the C^*-algebra $C([0, 1])$ with the supremum norm and let f and g be the states on A defined by $f(a) = \int_0^1 a(x)dx$ and $g(a) = a(0)$, $a \in \mathsf{A}$.

Obviously, g is f-absolutely continuous. But g is not strongly f-absolutely continuous. Indeed, set $a_n(x) := 1 - nx$ for $x \in [0, 1/n]$, $a_n(x) := 0$ for $x \in (1/n, 1]$. Then $a_n \in \mathsf{A}$ for $n \in \mathbb{N}$, $\lim_n f(a_n^+a_n) = 0$, and $g((a_n - a_m)^+(a_n - a_m)) = 0$, so $(a_n)_{n\in\mathbb{N}}$ is an (f, g)-sequence. Since $g(a_n^+a_n) = 1$, g is not strongly f-absolutely continuous. $\qquad\qquad\bigcirc$

Example 5.32 Let A be the C^*-algebra of bounded measurable functions on a measurable space $(X, \mathfrak{A})$ with the supremum norm. Let μ and ν be probability measures on $(X, \mathfrak{A})$. We define states by $f(a) = \int a(x)d\mu$ and $g(a) = \int a(x)d\nu$, $a \in \mathsf{A}$. The GNS representation π_f acts on $L^2(X; \mu)$ by $\pi_f(a)b = a \cdot b$, $a, b \in \mathsf{A}$, with $\varphi_f = 1$.

One easily verifies that g is f-absolutely continuous if and only if the measure ν is absolutely continuous with respect to μ. It can be shown (see Exercise 7.) that this holds if and only if g is strongly f-absolutely continuous.

Suppose g is strongly f-absolutely continuous. Let $\frac{d\nu}{d\mu} \in L^1(X; \mu)$ be the corresponding Radon–Nikodym derivative [Ru74, Theorem 6.9] and let H denote the multiplication operator on the Hilbert space $L^2(X; \mu)$ by the function $(\frac{d\nu}{d\mu})^{1/2}$. Then $\mathcal{D}(\pi_f) = \pi_f(\mathsf{A})\varphi_f$ is a core for H and we have

$$g(b^+a) = \int_X b^+a \, dv = \int_X b^+a \, \frac{dv}{d\mu} \, d\mu = \langle H\pi_f(a)\varphi_f, H\pi_f(b)\varphi_f \rangle, \quad a, b \in \mathbf{A}.$$

This is Eq. (5.40) and the positive self-adjoint operator H on $L^2(X; \mu)$ is the Radon–Nikodym derivative of the state g with respect to the state f.   ○

5.6   Extremal Decomposition of Positive Functionals

In this section, we apply the Choquet theory to obtain extremal decompositions of positive linear functionals. To be precise, we use the *Bishop-de Leeuw theorem* stated as Lemma 5.33 below.

We begin with some preliminaries (see e.g. Appendix C). Let E be a real vector space. A point x of a convex set K of E is said to be an *extreme point* of K if each equation $x = \lambda x_1 + (1 - \lambda)x_2$ with $x_1, x_2 \in K$ and $\lambda \in (0, 1)$ implies $x_1 = x_2$. The set of all extreme points of K is denoted by ex K. By a *cone* in E we mean a subset C of E such that $C + C \subseteq C$ and $\lambda C \subseteq C$ for $\lambda > 0$.

Suppose now that E is a real locally convex Hausdorff space. The dual space of E is denoted by E'. A G_δ-subset of E is a countable intersection of open subsets of E. A subset of E is called a *Baire set* if it is contained in the σ-algebra generated by the compact G_δ-subsets of E. Each Baire set of E is obviously a Borel set.

Lemma 5.33 (*Bishop-de Leeuw theorem*) *Suppose K is a compact convex subset of a real locally convex space E. Let $x \in K$. Then there exists a positive measure v_x on the σ-algebra generated by* ex K *and the Baire subsets of K with $\mu(K) = 1$ such that $v_x(\text{ex } K) = 1$ and*

$$\omega(x) = \int_K \omega(y) \, dv_x(y) \quad \text{for all} \quad \omega \in E'. \tag{5.43}$$

Proof [BdL59] or [Ph01, p. 22].   □

If the set K in Lemma 5.33 is metrizable, then ex K is a G_δ-set [Ph01, p. 5] and there exists a Radon measure v_x on K supported on ex K such that (5.43) holds. This is Choquet's theorem [Ph01, p. 13]. However, if K is not metrizable, then ex K need not be a Borel set [Ph01, p. 5]; one way to circumvent this difficulty to change the σ-algebra as done in Lemma 5.33.

A *cap* of a cone C is a nonempty compact convex subset K of C such that the complement set $C \backslash K$ is also convex. The technical facts for applying the Bishop-de Leeuw theorem are contained in the following lemma.

Lemma 5.34 *Let C be a cone in a real locally convex space E and suppose that $h : C \mapsto [0, +\infty]$ is a positively homogeneous additive map (that is, we have $h(\lambda x) = \lambda h(x)$ and $h(x + y) = h(x) + h(y)$ for $x, y \in C$, $\lambda \geq 0$) such that the set $K_h := \{x \in C : h(x) \leq 1\}$ is compact in E. Then:*

(i) K_h is a cap of C and $0 \in K_h$.
(ii) Suppose that ω is a C-positive linear functional on E such that $\omega(x) \neq 0$ for $x \in K_h$, $x \neq 0$. If $y \neq 0$ is an extreme point of K_h, then $\omega(y)^{-1}y$ is an extreme point of the convex set $B_\omega := \{x \in C : \omega(x) = 1\}$.

Proof (i): We show that $C \backslash K_h$ is convex. Indeed, if $x, y \in C \backslash K_h$, then $h(x) > 1$ and $h(y) > 1$, so $h(\lambda x + (1 - \lambda)y) = \lambda h(x) + (1 - \lambda)h(y) > 1$ for $\lambda \in [0, 1]$ and hence $(\lambda x + (1 - \lambda)y) \in C \backslash K_h$. Thus, $C \backslash K_h$ is convex. Similarly, K_h is convex. Since $h(0) = h(\lambda \cdot 0) = \lambda h(0)$, we have $h(0) = 0$, so $0 \in K_h$.

(ii): First we note that $h(x) \neq 0$ for all nonzero $x \in K_h$. (Otherwise K_h contains the half-line $\{\lambda x : \lambda \geq 0\}$, which contradicts the compactness of K_h.)
Let y be a nonzero extreme point of K_h. We verify that $h(y) = 1$. As just shown, $h(y) \neq 0$, so $h(y)^{-1}y \in K_h$. We have $y = h(y)(h(y)^{-1}y) + (1 - h(y))0$ with $0 < h(y) \leq 1$ (by $y \in K_h$), where $0 \in K_h$ (by (i)) and $h(y)^{-1}y \in K_h$. Since $y \in \mathrm{ex}\, K_h$ and $y \neq 0$, the latter implies $h(y) = 1$.
Assume that $\omega(y)^{-1}y = \lambda y_1 + (1 - \lambda)y_2$ with $y_1, y_2 \in B_\omega$ and $\lambda \in (0, 1)$. Since h is positively homogeneous and additive, we obtain

$$1 = h(y) = \lambda \omega(y)h(y_1) + (1 - \lambda)\omega(y)h(y_2). \tag{5.44}$$

Let $j \in \{1, 2\}$. Since $\lambda \in (0, 1), \omega(y) > 0$ (by the C-positivity of ω) and $y_j \in C$, (5.44) yields $h(y_j) \in [0, +\infty)$. By $y_j \in B_\omega$, $y_j \neq 0$ and hence $h(y_j) \neq 0$ as noted in the first paragraph of this proof. Thus, $h(y_j) \in (0, +\infty)$. Therefore, it follows from (5.44) that $\lambda \omega(y)h(y_1) \in (0, 1)$ and $1 - \lambda \omega(y)h(y_1) = (1 - \lambda)\omega(y)h(y_2)$. Put $z_j := h(y_j)^{-1}y_j$. Then $z_1, z_2 \in K_h$ and $y = \lambda \omega(y)h(y_1)z_1 + (1 - \lambda)\omega(y)h(y_2)z_2$ is a convex combination of z_1, z_2. Since y is an extreme point of K_h, $z_1 = z_2 = y$. Hence $\omega(y) = \omega(z_j) = h(y_j)^{-1}\omega(y_j) = h(y_j)^{-1}$, because $y_j \in B_\omega$ for $j = 1, 2$. Since $y_j = h(y_j)z_j$ and $z_1 = z_2$, the latter yields $y_1 = y_2$. This shows that $\omega(y)^{-1}y$ is an extreme point of B_ω. $\square$

Now suppose A is a $*$-algebra. We denote by $\mathsf{A}^*_{\mathrm{her}}[\sigma]$ the real vector space $\mathsf{A}^*_{\mathrm{her}}$ of hermitian linear functionals on A, equipped with the locally convex topology $\sigma = \sigma(\mathsf{A}^*_{\mathrm{her}}, \mathsf{A}_{\mathrm{her}})$ given by the seminorms $f \mapsto |f(a)|$, $a \in \mathsf{A}_{\mathrm{her}}$, on $\mathsf{A}^*_{\mathrm{her}}$. By a slight abuse of notation we do not distinguish in what follows between hermitian functionals on A and their restrictions to $\mathsf{A}_{\mathrm{her}}$. Recall that $\mathcal{S}(\mathsf{A})$ denotes the set of states of A. The subset $\mathcal{S}(\mathsf{A})$ of $\mathsf{A}^*_{\mathrm{her}}$ will be endowed with the induced topology.
Let Q be a cone in $\mathsf{A}_{\mathrm{her}}$. The dual cone $Q^\wedge$ (see (C.3)) is the set of $f \in \mathsf{A}^*_{\mathrm{her}}$ satisfying $f(c) \geq 0, c \in Q$. We shall say that a subset Q_0 of Q is Q-dominating for A if, given $a \in \mathsf{A}_{\mathrm{her}}$, there exist $c_0 \in Q_0$ and $\lambda > 0$ such that $\lambda c_0 - a \in Q$.

Theorem 5.35 Suppose A is a complex unital $*$-algebra. Let Q be a cone in $\mathsf{A}_{\mathrm{her}}$ such that $1 \in Q$ and $f(1) = 0$ for $f \in Q^\wedge$ implies $f = 0$. Suppose that there exists a countable subset Q_0 of Q which is Q-dominating for A.

Then, for each linear functional $f_0 \in Q^\wedge$, there exist a topological subspace Ω of $\mathcal{S}(A)$, a subset Ω_0 of Ω and a positive measure μ on the σ-algebra generated by Ω_0 and the Baire subsets of Ω such that $\Omega_0 \subseteq ex\,(Q^\wedge \cap \mathcal{S}(A))$ and

$$f_0(a) = \int_{\Omega_0} f(a)\,d\mu(f) \quad for \ \ a \in A. \tag{5.45}$$

Proof In this proof we will use some notions and results from the theory of locally convex spaces, see e.g. Appendix C and [Sh71].

Our first aim is to apply Lemma 5.34 to $C = Q^\wedge$ and $E = A^*_{her}[\sigma]$. We write $Q_0 = \{c_n : n \in \mathbb{N}\}$ and choose numbers $\gamma_n > 0, n \in \mathbb{N}$, such that $\sum_n \gamma_n f_0(c_n) \leq 1$. Clearly, $h(f) = \sum_n \gamma_n f(c_n)$, $f \in Q^\wedge$, defines a positively homogeneous additive map $h : Q^\wedge \mapsto [0, +\infty]$. Set $K_h := \{f \in Q^\wedge : h(f) \leq 1\}$.

Let $a \in A_{her}$. Since Q_0 is Q-dominating, there are numbers $n, k \in \mathbb{N}, \alpha > 0$, and $\beta > 0$ such that $\alpha c_n - a \in Q$ and $\beta c_k + a \in Q$. Then, for $f \in K_h$,

$$|f(a)| = \max\{f(a), f(-a)\} \leq \max\{\alpha f(c_n), \beta f(c_k)\} \leq \max\{\alpha\gamma_n^{-1}, \beta\gamma_k^{-1}\}.$$

Hence K_h is bounded in the locally convex space $A^*_{her}[\sigma]$. We consider the dual pairing of the real vector spaces A^*_{her} and A_{her} and equip A_{her} with the finest locally convex topology τ_{st}. Since K_h is bounded in $A^*_{her}[\sigma]$, its polar is a neighborhood of zero in $A_{her}[\tau_{st}]$. Hence, by the Alaoglu–Bourbaki theorem [Sh71, Chap. III, 4.3], the bipolar $(K_h)^{\circ\circ}$ of K_h is σ-compact in $A^*_{her} = A_{her}[\tau_{st}]'$. From its definition it follows easily that K_h is σ-closed. Therefore, since $K_h \subseteq (K_h)^{\circ\circ}$, K_h is also σ-compact and hence a cap in $A^*_{her}[\sigma]$ by Lemma 5.34(i). By construction, $f_0 \in K_h$.

Now we apply the Bishop-de Leeuw theorem (Lemma 5.33) to the compact convex set K_h in $E = A^*_{her}[\sigma]$ and $x = f_0$. Let $\nu := \nu_x$ be the corresponding measure. Then $\nu(K_h) = \nu(ex\,K_h) = 1$ and for $a \in E'$ we have

$$f_0(a) = \int_{K_h} f(a)\,d\nu(f). \tag{5.46}$$

By the definition of the weak topology σ, we have $A_{her} = E'$. Hence (5.46) holds for $a \in A_{her}$. We extend each $\mathbb{R}$-linear functional $f \in A^*_{her}$ to a $\mathbb{C}$-linear functional, denoted also f, on A. Then, by linearity, (5.46) holds for all $a \in A$. Now we define a mapping T of $K_h \backslash \{0\}$ on a subspace Ω of $\mathcal{S}(A)$ by $\tilde{f} \equiv T(f) := f(1)^{-1}f$ and a measure μ on Ω by $\mu(\cdot) = \nu(T^{-1}(\cdot))$. Set $\Omega_0 := \{\tilde{f} : f \in ex\,K_h, f \neq 0\}$. Since ν is supported on $ex\,K_h$, we obtain for $a \in A$,

$$f_0(a) = \int_{ex\,K_h \backslash \{0\}} f(a)\,d\nu(f) = \int_{\Omega_0} \tilde{f}(a)f(1)\,d\nu(f) = \int_{\Omega_0} \tilde{f}(a)\,d\mu(\tilde{f}).$$

This proves (5.45).

To show that $\Omega_0 \subseteq ex\,(Q^\wedge \cap \mathcal{S}(A))$ we define a linear functional ω on E by $\omega(f) = f(1)$, $f \in E$. By the assumptions on Q, ω fulfills the assumptions of Lemma 5.34(ii). Therefore, if f is a nonzero extreme point of K_h, then by Lemma 5.34(ii), $\tilde{f} = \omega(f)^{-1} f$ is an extreme point of $Q^\wedge \cap \mathcal{S}(\mathsf{A})$. $\qquad \square$

It should be noted that $\mathrm{ex}\,(Q^\wedge \cap \mathcal{S}(\mathsf{A}))$ in Theorem 5.35 denotes the extreme points of the convex set of Q-*positive* states. Elements of $\mathrm{ex}\,(Q^\wedge \cap \mathcal{S}(\mathsf{A}))$ are not necessarily extreme points of the set of *all* states and Ω_0 is not a Borel set in general.

Our main application of Theorem 5.35 is the following result.

Theorem 5.36 *Suppose* A *is a complex unital* $*$*-algebra such that* $\sum \mathsf{A}^2$ *contains a countable subset* Q_0 *which is* $\sum \mathsf{A}^2$*-dominating for* A *(that is, for each* $a \in \mathsf{A}_{\mathrm{her}}$ *there exist an element* $c \in Q_0$ *and a number* $\lambda > 0$ *such that* $\lambda c - a \in \sum \mathsf{A}^2$*).*

Then, for each positive linear functional f_0 *on* A*, Eq.(5.45) represents* f_0 *as an integral over pure states of* A*.*

Proof Clearly, $Q := \sum \mathsf{A}^2$ satisfies all assumptions of Theorem 5.35. Then $Q^\wedge$ are the positive functionals and $\mathrm{ex}\,(Q^\wedge \cap \mathcal{S}(\mathsf{A}))$ are the extreme points of the set of all states on A. These are pure states, so (5.45) is an integral over pure states. $\qquad \square$

Corollary 5.37 *Suppose* A *is a countably generated complex unital* $*$*-algebra. Then each positive functional on* A *is an integral over pure states of* A*.*

Proof Let B be the real and imaginary parts of all finite products of a countable set of algebra generators of A and let Q_0 be the finite sums of elements $(b \pm 1)^2$, where $b \in \mathsf{B}$. Then Q_0 is a countable subset of $\sum \mathsf{A}^2$. Clearly, $\mathsf{A}_{\mathrm{her}}$ is the real span of B and so of Q_0, since $4b = (b+1)^2 - (b-1)^2$ for $b \in \mathsf{B}$. Hence each $a \in \mathsf{A}_{\mathrm{her}}$ is of the form $a = \sum_j \lambda_j c_j$, with $\lambda_j \in \mathbb{R}$, $c_j \in Q_0$. Then $\lambda := 1 + \max_j |\lambda_j| > 0$, $c := \sum_j c_j \in Q_0$ and $\lambda c - a = \sum_j (\lambda - \lambda_j) c_j \in \sum \mathsf{A}^2$. This shows that Q_0 is $\sum \mathsf{A}^2$-dominating. Thus the assertion follows from Theorem 5.36. $\qquad \square$

Corollary 5.38 *Let* $\mathcal{A}$ *be a unital* O^**-algebra on* $\mathcal{D}$ *with metrizable graph topology and* $\mathcal{A}_+ := \{a \in \mathcal{A}_{\mathrm{her}} : \langle a\varphi, \varphi \rangle \geq 0, \varphi \in \mathcal{D}\}$*. Then each* $\mathcal{A}_+$*-positive linear functional on* $\mathcal{A}$ *is an integral over extreme points of the set of* $\mathcal{A}_+$*-positive states.*

Proof The assertion follows from Theorem 5.35, applied to the cone $Q = \mathcal{A}_+$, once we have shown the existence of a countable $\mathcal{A}_+$-dominating subset.

The graph topology $t_\mathcal{A} >_\mathcal{A}$ is metrizable, so it is generated by countably many seminorms $\{\| \cdot \|_{a_n} : n \in \mathbb{N}\}$. By Lemma 3.5 we can assume that $\| \cdot \| \leq \| \cdot \|_{a_n} \leq \| \cdot \|_{a_{n+1}}$ for $n \in \mathbb{N}$. Let $a \in \mathcal{A}_{\mathrm{her}}$. Then there are numbers $n \in \mathbb{N}$ and $\lambda > 0$ such that $\|a\varphi\| \leq \lambda \|a_n \varphi\|$ for $\varphi \in \mathcal{D}$ and therefore

$$\langle a\varphi, \varphi \rangle \leq \|a\varphi\|\,\|\varphi\| \leq \lambda \|a_n\varphi\|\|\varphi\| \leq \lambda \|a_n\varphi\|^2 = \langle \lambda (a_n)^+ a_n \varphi, \varphi \rangle,$$

so $\lambda (a_n)^+ a_n - a \in \mathcal{A}_+$. Hence $Q_0 := \{(a_n)^+ a_n : n \in \mathbb{N}\}$ is $\mathcal{A}_+$-dominating. $\qquad \square$

5.7 Quadratic Modules and *-Representations

Recall that quadratic modules are introduced in Definition 2.25. An important class of quadratic modules is obtained by means of *-representations.

Definition 5.39 For a family $\mathcal{R}$ of *-representations of A, we define

$$A(\mathcal{R})_+ := \big\{ a \in A_{her} : \langle \pi(a)\varphi, \varphi \rangle \geq 0 \text{ for all } \varphi \in \mathcal{D}(\pi), \ \pi \in \mathcal{R} \big\}.$$

Then $A(\mathcal{R})_+$ is a *pre-quadratic module* of A. Indeed, condition (2.22) in Definition 2.25 is obvious and (2.23) follows from

$$\langle \pi(x^+ a x)\varphi, \varphi \rangle = \langle \pi(a)\pi(x)\varphi, \pi(x)\varphi \rangle \geq 0 \tag{5.47}$$

for $\pi \in \mathcal{R}, \varphi \in \mathcal{D}(\pi), x \in A, a \in A(\mathcal{R})_+$. Let $F(\mathcal{R})$ denote the set of vector functionals

$$f_{\pi,\varphi}(\cdot) := \langle \pi(\cdot)\varphi, \varphi \rangle, \quad \text{where } \varphi \in \mathcal{D}(\pi), \pi \in \mathcal{R}.$$

From the equality in (5.47) we conclude that

$$f \in F(\mathcal{R}) \ \text{ implies } \ f_x(\cdot) \equiv f(x^+ \cdot x) \in F(\mathcal{R}) \ \text{ for } \ x \in A. \tag{5.48}$$

Hence, by Definition 5.39, $a \in A_{her}$ is in $A(\mathcal{R})_+$ if and only if $f(a) \geq 0$ for all $f \in F(\mathcal{R})$. This fact and Proposition 5.43 below hold also for nonunital *-algebras. Clearly, if A is unital, then $1 \in A(\mathcal{R})_+$ and $A(\mathcal{R})_+$ is a *quadratic module* of A.

Now we suppose that A is unital and F is a set of positive functionals on A satisfying condition (5.48), that is, $f \in F$ implies $f_x \in F$ for all $x \in A$. Let $\mathcal{R}$ denote the family of GNS representations $\pi_f, f \in F$. Since A is unital, we have $f = f_{\pi_f, \varphi_f}$ by (4.26), so that $F \subseteq F(\mathcal{R})$. From the equality $f(x^+ \cdot x) = \langle \pi_f(\cdot)\pi_f(x)\varphi_f, \pi(x)\varphi_f \rangle$ by (4.25) and assumption (5.48) it follows that $F(\mathcal{R}) \subseteq F$. Thus, $F = F(\mathcal{R})$. Therefore, the quadratic module $A(\mathcal{R})_+$ is given by

$$A(\mathcal{R})_+ = \big\{ a \in A_{her} : f(a) \geq 0 \ \text{ for } \ f \in F \big\}.$$

Example 5.40 Suppose $A = \mathbb{C}[x_1, \ldots, x_d]$. Let K be a closed subset of $\mathbb{R}^d$ and let $\mathcal{R}_K$ be the set of point evaluations at points of K. Then $A(\mathcal{R}_K)_+$ is precisely the set of polynomials of $A_{her} = \mathbb{R}[x_1, \ldots, x_d]$ that are nonnegative on K. ○

Example 5.41 Suppose $\mathcal{A}$ is an O^*-algebra. If the family $\mathcal{R}$ consists only of the identity map, then $\mathcal{A}(\mathcal{R})_+$ is the pre-quadratic module defined by (3.1). ○

A natural question is to ask when $A(\mathcal{T})_+ \subseteq A(\mathcal{R})_+$ for two families $\mathcal{R}$ and $\mathcal{T}$. To answer this question by Proposition 5.43 we need the following definition.

Definition 5.42 Let $\mathcal{R}$ and $\mathcal{T}$ be families of $*$-representations of A. We say that $\mathcal{R}$ is *weakly contained* in $\mathcal{T}$ if each vector functional $f_{\pi,\varphi}$, with $\pi \in \mathcal{R}$, $\varphi \in \mathcal{D}(\pi)$, can be weakly approximated by a finite sum of vector functionals $f_{\rho,\psi}$, $\rho \in \mathcal{T}$, $\psi \in \mathcal{D}(\rho)$, that is, given $\varepsilon > 0$ and $a_1, \ldots, a_n \in$ A, there exist elements $\rho_1, \ldots, \rho_k \in \mathcal{T}$ and vectors $\psi_1 \in \mathcal{D}(\rho_1), \ldots, \psi_k \in \mathcal{D}(\rho_k)$ such that

$$\left| \langle \pi(a_j)\varphi, \varphi \rangle - \sum_{i=1}^{k} \langle \rho_i(a_j)\psi_i, \psi_i \rangle \right| < \varepsilon \quad \text{for} \quad j = 1, \ldots, n. \tag{5.49}$$

Proposition 5.43 *Suppose $\mathcal{R}$ and $\mathcal{T}$ are families of $*$-representations of a $*$-algebra* A. *Then $\mathcal{R}$ is weakly contained in $\mathcal{T}$ if and only if* $\mathsf{A}(\mathcal{T})_+ \subseteq \mathsf{A}(\mathcal{R})_+$.

The main ingredient of the proof of this result is the bipolar theorem on locally convex spaces. Before we prove Proposition 5.43 we note a preliminary lemma.

Let E be a real vector space and E^* the vector space of all linear functionals on E. Then (E, E^*) forms a dual pairing with pairing $(u, f) \mapsto f(u)$ and the weak topology σ on E^* is defined by the family of seminorms $f \mapsto |f(u)|, u \in E$. Recall that the dual cone $C^\wedge$ is defined by (C.3).

Lemma 5.44 *Suppose C is a cone in E^* and $f \in E^*$. If $f \in (C^\wedge)^\wedge$, then f is in the closure of C with respect to the weak topology σ of E^*.*

Proof First we recall the notion of a polar M° (see [Sh71, IV.1.3]) of a subset M of E^* with respect to the dual pairing (E, E^*). This is the set

$$M^\circ := \{ x \in E : f(x) \leq 1 \text{ for } f \in M \}.$$

In a similar manner the polar of a subset of E is defined. Since $\lambda \cdot C \subseteq C$ for *all* $\lambda > 0$, it is clear that $c \in C^\circ$ if and only if $f(c) \leq 0$ for all $f \in C$. This implies that $C^\wedge = (-C)^\circ$ and hence $(C^\wedge)^\wedge = (-(-C)^\circ))^\circ = (C^\circ)^\circ$.

Thus, the assumption means that f is in the bipolar of C. Therefore, by the bipolar theorem [Sh71, IV.1.5], f is in the weak closure of C. $\quad\square$

Proof of Proposition 5.43 First suppose that $\mathcal{R}$ is weakly contained in $\mathcal{T}$. Let $a \in \mathsf{A}(\mathcal{T})_+$. We apply (5.49) with $a = a_1$. Since $f_{\rho_i,\psi_i}(a) \geq 0$ (by $a \in \mathsf{A}(\mathcal{T})_+$) and $f_{\pi,\varphi}(a)$ is real, (5.49) implies that $f_{\pi,\varphi}(a) \geq -\varepsilon$. Since $\varepsilon > 0$ is arbitrary, $f_{\pi,\varphi}(a) \geq 0$. Therefore, $a \in \mathsf{A}(\mathcal{R})_+$. This proves that $\mathsf{A}(\mathcal{T})_+ \subseteq \mathsf{A}(\mathcal{R})_+$.

Now we suppose that $\mathsf{A}(\mathcal{T})_+ \subseteq \mathsf{A}(\mathcal{R})_+$. Since $f_{\pi,\varphi}$ and f_{ρ_i,ψ_i} are real-valued on A_{her}, we can assume that $a_1, \ldots, a_n \in \mathsf{A}_{\text{her}}$ in Definition 5.42. We will apply Lemma 5.44 to $E := \mathsf{A}_{\text{her}}$ and the cone C of finite sums of functionals $f_{\rho,\psi}$ on E, where $\rho \in \mathcal{T}$ and $\psi \in \mathcal{D}(\rho)$. Then $C^\wedge = \mathsf{A}(\mathcal{T})_+$ by the corresponding definitions. Let $\pi \in \mathcal{R}$ and $\varphi \in \mathcal{D}(\pi)$. Since $\mathsf{A}(\mathcal{T})_+ \subseteq \mathsf{A}(\mathcal{R})_+$ by assumption, we have $f_{\pi,\varphi} \in (C^\wedge)^\wedge$. Therefore, $f_{\pi,\varphi}$ is in the weak closure of C by Lemma 5.44. According to Definition 5.42, this means that $\mathcal{R}$ is weakly contained in $\mathcal{T}$. $\quad\square$

5.8 Exercises

1. Let $A = \mathbb{C}[x]$ and define $f(p) = \int_0^1 p(x)dx$, $p \in A$.

   a. Describe the GNS representation $\overline{\pi}_f$ and the weak commutant $\overline{\pi}_f(A)'_w$.
   b. Describe the states g on A such that $g \leq f$.
   c. Use b. to characterize those states $g_1, g_2 \in [0, f]$ for which $g_1 \perp g_2$.

2. Let π be the Schrödinger representation of the Weyl algebra W (Example 4.32) and $\varphi \in C_0^\infty(\mathbb{R})$. Suppose supp $\varphi \subseteq [0, 2]$ and $\varphi(t) \neq 0$ on $(0, 1) \cup (1, 2)$. Define $f(x) = \langle \pi(x)\varphi, \varphi \rangle$, $x \in W$. Let T be the multiplication operator by the characteristic function of $[0, 1]$.

   a. Describe the GNS representation $\overline{\pi}_f$.
   b. Is T in the weak commutant $\pi_f(A)'_w$?
   c. Is f is pure?

3. Let $f, g \in \mathcal{P}(A)^*$. Show that f is dominated by g if and only if there is an operator $T \in I(\pi_g, \pi_f)$ such that $T\varphi_g = \varphi_f$.

4. Let $f, g \in \mathcal{P}(A)^*$ and suppose $\overline{\pi}_f(A)'_w = \overline{\pi}_f(A)'_s$. Let Q be a quadratic module of A. Prove that if $g \leq f$ and f is Q-positive, then g is also Q-positive.
   Hint: Use Theorem 5.3. The main argument will be also used in the proof of Lemma 10.19 below.

5. Let A be a unital $*$-subalgebra of a unital $*$-algebra B and $f, g \in \mathcal{P}(B)^*$. Show that $P_B(f, g) \leq P_A(f \restriction A, g \restriction A)$.

6. Let f_1, f_2, f, g be positive functionals on A. Prove the following:

   a. $0 \leq P_A(f, g) = P_A(g, f) \leq f(1)g(1)$.
   b. $P_A(f, g) = 1$ if and only if $f = g$.
   c. $\lambda P_A(f_1, g) + (1 - \lambda) P_A(f_2, g) \leq P_A(\lambda f_1 + (1 - \lambda) f_2, g)$ for $\lambda \in [0, 1]$.

7. In Example 5.32, show that g is f-absolutely continuous if and only if g is strongly f-absolutely continuous.

8. Let $A = \mathbb{C}_d[\underline{x}]$. For closed subsets K, K' of $\mathbb{R}^d$, let $\mathcal{R}_K$ and $\mathcal{R}_{K'}$ be the point evaluations at points of K and K', respectively (Example 5.40). What does Proposition 5.43 say in the case $\mathcal{R} := \mathcal{R}_K$, $\mathcal{T} := \mathcal{R}_{K'}$? Prove Proposition 5.43 directly in this case.

5.9 Notes

The ordering of positive functionals is due to Powers [Pw71]. Examples 5.6 is new. Orthogonal states on $*$-algebras were first studied in [Sch90, Sect. 8.6]. The transition probability for general $*$-algebras was introduced by Uhlmann [U76]; it also called fidelity in the literature, see e.g. [Jo94]. Basic results on transition probabilities for C^*-algebras and von Neumann algebras were obtained by Alberti [Al83, Al03]. Lemma 5.22 is a classical result on von Neumann algebras, see [Ak72, p. 341] or [Al03, Corollary 1, p. 93]. The results of Sects. 5.3 and 5.4 were obtained by the author [Sch15]. The Radon–Nikodym theorem in Sect. 5.5 is due to Gudder [Gu79], see also [In83].

The extremal decomposition of ∗-representations and states was first studied by Borchers and Yngvason [BY75a, BY75b] using different methods. The approach via Choquet theory was proposed in [He85] and studied in [Ri84]. In Sect. 5.6 we followed [Sch90, Sect. 12.4], which contains a number of further results.

Chapter 6
Representations of Tensor Algebras

Free $*$-algebras $\mathbb{C}\langle x_i; i \in I \,|\, x_i^+ = x_i \rangle$, or equivalently tensor algebras $V_\otimes$, form the simplest class of $*$-algebras. This chapter gives a brief introduction into positive functionals and $*$-representations of these $*$-algebras. In Sect. 6.1, we define $*$-vector spaces and their tensor algebras and state elementary algebraic properties. Sections 6.2 and 6.3 deal with positive linear functionals on the $*$-algebra $V_\otimes$. We approximate positive functionals by vector functionals of finite-dimensional representations and develop some operations to construct new positive functionals from old ones. An important class of $*$-representations is constructed in Sect. 6.4, see Theorem 6.12. In particular, we show that each tensor algebra admits a faithful $*$-representation. Section 6.5 is about the tensor algebra of a topological $*$-vector space (V, τ). A topology $\tau_\otimes$ on $V_\otimes$ is defined such that its completion $\underline{V}_\otimes[\tau_\otimes]$ is a topological $*$-algebra and representations of $\underline{V}_\otimes$ are considered (Theorem 6.19).

Throughout this chapter, V denotes a $*$-**vector space** with involution $v \mapsto v^+$.

6.1 Tensor Algebras

The first main notion in this chapter is the following.

Definition 6.1 A $*$-*vector space* is a complex vector space V equipped with a mapping $v \mapsto v^+$ of V, called *involution*, such that

$$(\alpha u + \beta v)^+ = \overline{\alpha}\, u^+ + \overline{\beta}\, v^+ \quad \text{and} \quad (v^+)^+ = v \quad \text{for} \ \alpha, \beta \in \mathbb{C}, \ u, v \in V.$$

Set $V_0 := \mathbb{C}$ and $V_1 := V$. For $n \in \mathbb{N}, n \geq 2$, let V_n denote the n-fold complex tensor product $V \otimes \cdots \otimes V$ of vector spaces. The elements of V_n are finite sums

© The Editor(s) (if applicable) and The Author(s), under exclusive license
to Springer Nature Switzerland AG 2020
K. Schmüdgen, *An Invitation to Unbounded Representations of *-Algebras on Hilbert Space*, Graduate Texts in Mathematics 285,
https://doi.org/10.1007/978-3-030-46366-3_6

$v = \sum_i v_1^i \otimes \cdots \otimes v_n^i$ with $v_1^i, \ldots, v_n^i \in V$. Clearly, V_n is also a $*$-vector space with involution

$$\left(\sum_i v_1^i \otimes \cdots \otimes v_n^i \right)^+ := \sum_i (v_n^i)^+ \otimes \cdots \otimes (v_1^i)^+.$$

Throughout this chapter we adopt the following notational conventions:
Upper numbers such as i are always indices! They never refer to powers of elements.
Sums such as $\sum_i$ are always over finitely many indices i.

Now we consider the direct sum $V_\otimes$ of vector spaces V_n, $n \in \mathbb{N}_0$, that is,

$$V_\otimes := \sum_{n=0}^{\infty} \oplus V_n.$$

The elements of $V_\otimes$ are written as sequences (v_n), where $v_n \in V_n$ and only finitely many terms are nonzero. It is easily verified that $V_\otimes$ becomes a unital complex $*$-algebra with multiplication and involution

$$(u_n)(v_n) := \left(\sum_{j+k=n} u_j \otimes v_k \right), \quad (v_n)^+ := (v_n^+), \tag{6.1}$$

where $\lambda \otimes v_k = v_k \otimes \lambda = \lambda v_k$ and $\lambda^+ = \overline{\lambda}$ for $\lambda \in \mathbb{C}$ and $v_k \in V_k$. Note that the sequence $(1, 0, 0, \ldots)$ is the unit element of the $*$-algebra $V_\otimes$.

Definition 6.2 The $*$-algebra $V_\otimes$ is the *tensor algebra* of the $*$-vector space V.

Example 6.3 ($V = \mathbb{C}$) Then the map $(v_n) \mapsto \sum_n v_n x^n$ is a $*$-isomorphism of the tensor algebra $V_\otimes$ on the polynomial $*$-algebra $\mathbb{C}[x]$ in a single hermitian variable.   ◯

Example 6.4 ($V = \mathcal{S}(\mathbb{R}^d)$) We identify $\sum_i v_1^i \otimes \cdots \otimes v_n^i \in V_n$ with the function $\sum_i v_1^i(x_1) \cdots v_n^i(x_n)$ on $\mathbb{R}^{dn}$. Then V_n becomes a dense linear subspace of $\mathcal{S}(\mathbb{R}^{dn})$.   ◯

The tensor algebra $V_\otimes$ is the free unital $*$-algebra over the $*$-vector space V. This means that $V_\otimes$ has the following universal property.

Lemma 6.5 *Each $*$-preserving linear map π of V into a unital $*$-algebra A has a unique extension to a $*$-homomorphism, denoted π, of $V_\otimes$ into A such that $\pi(1) = 1$.*

Proof We extend π to V_n by defining $\pi(\sum_i v_1^i \otimes \cdots \otimes v_n^i) = \sum_i \pi(v_1^i) \cdots \pi(v_n^i)$ and $\pi(1) = 1$, and then by linearity to $V_\otimes$. From the definitions of the multiplication and involution of $V_\otimes$ it follows that this gives a $*$-homomorphism in A. The uniqueness assertion is obvious.   □

Recall that $\mathbb{C}\langle x_i; i \in I | (x_i)^+ = x_i \rangle$ denotes the free unital $*$-algebra with hermitian generators x_i, $i \in I$. This notion is equivalent to the notion of a tensor algebra. Indeed, let V be the complex span of x_i, $i \in I$, in $\mathbb{C}\langle x_i; i \in I | (x_i)^+ = x_i \rangle$, equipped with the involution defined by $(x_i)^+ = x_i$. Then V is a $*$-vector space and $\mathbb{C}\langle x_i; i \in I | (x_i)^+ = x_i \rangle$ is $*$-isomorphic to $V_\otimes$. Conversely, if V is a $*$-vector space, we choose a vector space basis $\{v_i : i \in I\}$ of V consisting of hermitian elements; then $V_\otimes$ is $*$-isomorphic to $\mathbb{C}\langle v_i; i \in I | (v_i)^+ = v_i \rangle$. In particular, the free $*$-algebra $\mathbb{C}\langle x_1, \ldots, x_n | (x_i)^+ = x_i \rangle$ is $*$-isomorphic to $V_\otimes$ with $V = \mathbb{C}^n$.

Next we develop some constructions with representations of tensor algebras.

Let $(\mathcal{D}_1, \langle \cdot, \cdot \rangle_1)$ and $(\mathcal{D}_2, \langle \cdot, \cdot \rangle_2)$ be complex inner product spaces. We denote by $\mathcal{D}_1 \otimes \mathcal{D}_2$ the tensor product vector space equipped with the inner product $\langle \cdot, \cdot \rangle$ which is given by $\langle \varphi_1 \otimes \varphi_2, \psi_1 \otimes \psi_2 \rangle = \langle \varphi_1, \psi_1 \rangle_1 \langle \varphi_2, \psi_2 \rangle_2$, where $\varphi_1, \psi_1 \in \mathcal{D}_1, \varphi_2, \psi_2 \in \mathcal{D}_2$. Thus, $(\mathcal{D}_1 \otimes \mathcal{D}_2, \langle \cdot, \cdot \rangle)$ is a complex inner product space.

(1) Let $\{v_i : i \in I\}$ be a vector space basis of hermitian elements for V. Then, for *every* (!) family $\{T_i : i \in I\}$ of symmetric operators of $\mathcal{L}^+(\mathcal{D})$, there is a unique $*$-representation π of $V_\otimes$ on $\mathcal{D}$ such that $\pi(v_i) = T_i, i \in I$, and $\pi(1) = I$. (Indeed, we define a $*$-preserving linear map π of V in $\mathcal{L}^+(\mathcal{D})$ by $\pi(v_i) = T_i$ and apply Lemma 6.5.) In particular, for *each* n-tuple of bounded self-adjoint operators $T_1, \ldots, T_n$ on a Hilbert space there is a unique $*$-representation π of the $*$-algebra $\mathbb{C}\langle x_1, \ldots, x_n | (x_i)^+ = x_i \rangle$ such that $\pi(x_i) = T_i, i = 1, \ldots, n$, and $\pi(1) = I$.

(2) Suppose that π_1 and π_2 are $*$-representations of $V_\otimes$. Then it is clear that $\pi(v) := \pi_1(v) \otimes I_{\mathcal{D}(\pi_2)} + I_{\mathcal{D}(\pi_1)} \otimes \pi_2(v), v \in V$, is a $*$-preserving linear map of $V_\otimes$ into $\mathcal{L}^+(\mathcal{D}(\pi_1) \otimes \mathcal{D}(\pi_2))$. Hence, by Lemma 6.5, this map extends uniquely to a $*$-representation π of $V_\otimes$ on $\mathcal{D}(\pi) := \mathcal{D}(\pi_1) \otimes \mathcal{D}(\pi_2)$ such that $\pi(1) = I$.

(3) Let π_1 be a $*$-representation of $V_\otimes$ and let T be a symmetric operator of $\mathcal{L}^+(\mathcal{D}_0)$. By a straightforward verification it follows that there is a unique $*$-representation π of $V_\otimes$ on $\mathcal{D}(\pi) := \mathcal{D}(\pi_1) \otimes \mathcal{D}_0$ given by $\pi(v_n) = \pi_1(v_n) \otimes T^n$ for $v_n \in V_n$ and $n \in \mathbb{N}_0$, where we set $T^0 := I$.

6.2 Positive Functionals on Tensor Algebras

For a $*$-vector space V, the space V^* of all $\mathbb{C}$-linear functionals on V is also a $*$-vector space with involution given by $F^+(v) := \overline{F(v^+)}, v \in V$.

First we describe the *characters* of the algebra $V_\otimes$. Let $F \in V^*$. Then there is a linear functional $F^{[n]}$ on $V_n, n \in \mathbb{N}$, defined by

$$F^{[n]}\left(\sum_i v_1^i \otimes \cdots \otimes v_n^i\right) = \sum_i F(v_1^i) \cdots F(v_n^i), \quad v_1^i, \ldots, v_n^i \in V,$$

and there is linear functional χ_F on $V_\otimes$ given by $\chi_F(v_n) = F^{[n]}(v_n), v_n \in V_n, n \in \mathbb{N}$, and $\chi_F(1) = 1$. It is easily verified that χ_F is a character of the algebra $V_\otimes$. If F is hermitian, so is the character χ_F. Conversely, if χ is a character of $V_\otimes$, then χ is of the form $\chi = \chi_F$ with $F := \chi \upharpoonright V$.

Next we prove some results concerning the approximation of positive functionals by vector functionals in finite-dimensional representations.

Proposition 6.6 *Suppose π is a nondegenerate $*$-representation of $V_\otimes$ such that* $\dim \pi(V) < \infty$. *For any $n \in \mathbb{N}$ and $\xi \in \mathcal{D}(\pi)$, there exists a nondegenerate $*$-representation $\rho_{\xi,n}$ of $V_\otimes$ on a finite-dimensional subspace $\mathcal{H}_{\xi,n}$ of $\mathcal{D}(\pi)$ such that*

$$\pi(u)\xi = \rho_{\xi,n}(u)\xi \quad \text{and} \quad \langle \pi(u)\xi, \xi \rangle = \langle \rho_{\xi,n}(u)\xi, \xi \rangle \quad \text{for } u \in \sum_{j=0}^{n} \oplus V_j. \quad (6.2)$$

Proof Fix $n \in \mathbb{N}$. Set $k := \dim \pi(V)$. We choose $v_1, \dots, v_k \in V$ such that $\pi(V) =$ Lin $\{\pi(v_j) : j = 1, \dots, k\}$. Further, we abbreviate $\mathsf{A}_n := \sum_{j=0}^{n} \oplus V_j$ and

$$\mathsf{B}_n := \text{Lin } \{1, v_{i_1} \cdots v_{i_r} : i_1, \dots, i_r \in \{1, \dots, k\}, r \leq n\}.$$

First we note that $\pi(\mathsf{A}_n) = \pi(\mathsf{B}_n)$. Indeed, for $u \in V$, $\pi(u)$ is in the linear span of operators $\pi(v_j)$. Hence, for elements $u_1, \dots, u_r \in V, r \leq n$, the operator $\pi(u_1) \cdots \pi(u_r) = \pi(u_1 \cdots u_r)$ is in the span of products $\pi(v_{i_1}) \cdots \pi(v_{i_r}) = \pi(v_{i_1} \cdots v_{i_r})$. Therefore, since each $u \in \sum_{j=1}^{n} \oplus V_n$ is a sum of products $u_1 \cdots u_r$, it follows that $\pi(\mathsf{A}_n) \subseteq \pi(\mathsf{B}_n)$. The opposite inclusion is obvious, since $\mathsf{B}_n \subseteq \mathsf{A}_n$.

Let $\xi \in \mathcal{D}(\pi)$. Since $\dim \mathsf{B}_n < \infty$ and $\pi(\mathsf{A}_n) = \pi(\mathsf{B}_n)$, the linear subspace $\mathcal{H}_{\xi,n} := \pi(\mathsf{A}_n)\xi$ of $\mathcal{D}(\pi)$ is finite-dimensional and hence closed in $\mathcal{H}(\pi)$. Let P be the projection of $\mathcal{H}(\pi)$ on $\mathcal{H}_{\xi,n}$. Clearly, $v \mapsto P\pi(v) \upharpoonright P\mathcal{H}(\pi)$ is a $*$-preserving linear map of V into $\mathbf{B}(\mathcal{H}_{\xi,n})$. Let $\rho \equiv \rho_{\xi,n}$ denote its unique extension to a nondegenerate $*$-representation of $V_\otimes$ on $\mathcal{H}_{\xi,n}$ according to Lemma 6.5. Since π is nondegenerate by assumption, $\pi(1) = I$, so $\xi \in \mathcal{H}_{\xi,n}$ and $\rho(1)\xi = \pi(1)\xi = \xi$. Let $u_1, \dots, u_r \in V, r \leq n$. Using that $\xi \in \mathcal{H}_{\xi,n}$ and $\pi(u_1) \cdots \pi(u_j)\xi \in \pi(\mathsf{A}_n)\xi = P\mathcal{H}$ for $1 \leq j \leq r$, we derive

$$\rho(u_1 \cdots u_r)\xi = \rho(u_1) \cdots \rho(u_r)\xi = P\pi(u_1)P \cdots P\pi(u_r)P\xi$$
$$= P\pi(u_1)P \cdots P\pi(u_{r-1})\pi(u_r)\xi = \pi(u_1) \cdots \pi(u_r)\xi = \pi(u_1 \cdots u_r)\xi.$$

It follows that $\rho(u)\xi = \pi(u)\xi$ for $u \in \mathsf{A}_n$, which implies (6.2). $\qquad \square$

Note that the assumption $\dim \pi(V) < \infty$ is trivially satisfied if $\dim V < \infty$.

Proposition 6.7 *Let π be a nondegenerate $*$-representation of $V_\otimes$ and $\xi \in \mathcal{D}(\pi)$. For any finite-dimensional subspace $\mathcal{U}$ of $V_\otimes$, there exists a nondegenerate $*$-representation $\rho_\mathcal{U}$ of $V_\otimes$ on a finite-dimensional linear subspace of $\mathcal{D}(\pi)$ such that*

$$\langle \pi(u)\xi, \xi \rangle = \langle \rho_\mathcal{U}(u)\xi, \xi \rangle \quad \text{for } u \in \mathcal{U}. \quad (6.3)$$

Proof Since $\dim \mathcal{U} < \infty$, there exist an $n \in \mathbb{N}$ and a finite-dimensional $*$-invariant subspace U of V such that $\mathcal{U} \subseteq \sum_{j=0}^{n} \oplus U_n \subseteq U_\otimes$. We choose a $*$-invariant subspace W of V such that V is the direct sum of U and W and define a $*$-preserving linear map ρ of V into $\mathcal{D}(\pi)$ by $\rho(u + w) = \pi(u)$, $u \in U, w \in W$. This map extends to a nondegenerate $*$-representation, denoted again ρ, of $V_\otimes$ on $\mathcal{D}(\pi)$ by

Lemma 6.5. By the definition of this extension we have $\rho(u) = \pi(u)$ for $u \in U_\otimes \subseteq V_\otimes$ and $\dim \rho(V) < \infty$. Then Proposition 6.6 applies to ρ, so there exists a subrepresentation $\rho_{\xi,n}$ of ρ on a finite-dimensional subspace of $\mathcal{D}(\rho) \subseteq \mathcal{D}(\pi)$ such that $\langle \rho(v)\xi, \xi \rangle = \langle \rho_{\xi,n}(v)\xi, \xi \rangle$ for $v \in \sum_{j=0}^n \oplus V_n$. Set $\rho_{\mathcal{U}} := \rho_{\xi,n}$. Since $\rho(u) = \pi(u)$ for $u \in U_\otimes$ and $\mathcal{U} \subseteq \sum_{j=0}^n \oplus U_j \subseteq \sum_{j=0}^n \oplus V_j$, the latter yields (6.3). $\qquad\square$

Corollary 6.8 *The set of all vector functionals of finite-dimensional nondegenerate $*$-representations of $V_\otimes$ is dense, in the weak topology, in the cone of all positive functionals on $V_\otimes$.*

Proof By the GNS construction (Theorem 4.38), each positive functional on the unital (!) $*$-algebra $V_\otimes$ is a vector functional of some algebraically cyclic, hence nondegenerate, $*$-representation. Then the assertion follows from (6.3). $\qquad\square$

6.3 Operations with Positive Functionals

First note that there is a one-to-one correspondence between linear functionals F on the vector space $V_\otimes$ and sequences $(F_n)_{n \in \mathbb{N}_0}$ of functionals $F_n \in (V_n)^*$ given by

$$F(u) = \sum_{n=0}^\infty F_n(u_n) \quad \text{for} \quad u = (u_n) \in V_\otimes. \tag{6.4}$$

We shall write $F = (F_n)_{n \in \mathbb{N}_0}$ if Eq. (6.4) holds.

1. Tensor product of positive linear functionals

Let $k, l \in \mathbb{N}$. By an *ordered decomposition* of $\{1, 2, \dots, k + l\}$ we mean a decomposition of this set into two disjoint subsets $i = \{i_1, \dots, i_k\}$, $j = \{j_1, \dots, j_l\}$ such that $i_1 < i_2 < \cdots < i_k$ and $j_1 < j_2 < \cdots < j_l$. The set of such ordered decompositions (i, j) is denoted by $P(k, l)$. For linear functionals $F_k \in (V_k)^*$ and $G_l \in (V_l)^*$, we define a linear functional $F_k \otimes_s G_l \in (V_{k+l})^*$ by

$$(F_k \otimes_s G_l)(v_1 \otimes \cdots \otimes v_{k+l}) = \sum_{(i,j) \in P(k,l)} F_k(v_{i_1} \otimes \cdots \otimes v_{i_k}) G_l(v_{j_1} \otimes \cdots \otimes v_{j_l}).$$

For $k, l \in \mathbb{N}_0$ and $F_0, G_0 \in (V_0)^*$, $F_k \in (V_k)^*$, $G_l \in (V_l)^*$, let $F_0 \otimes G_l \in (V_l)^*$ and $F_k \otimes G_0 \in (V_k)^*$ be the linear functionals defined by

$$(F_0 \otimes_s G_l)(\lambda \otimes v_l) = F_0(\lambda) G_l(v_l), \quad (F_k \otimes_s G_0)(v_k \otimes \lambda) = F_k(v_k) G_0(\lambda).$$

Now suppose $F = (F_n)_{n \in \mathbb{N}_0}$ and $G = (G_n)_{n \in \mathbb{N}_0}$ are linear functionals on $V_\otimes$. We define a linear functional $F \otimes_s G = ((F \otimes_s G)_n)_{n \in \mathbb{N}_0}$ on $V_\otimes$, called the *tensor product* of F and G, by

$$(F \otimes_s G)_n := \sum_{k=0}^{n} F_k \otimes_s G_{n-k}. \qquad (6.5)$$

Proposition 6.9 *Suppose F and G are positive linear functionals on $V_\otimes$. Let $\rho_{F,G}$ denote the nondegenerate $*$-representation of $V_\otimes$ on $\mathcal{D}(\rho_{F,G}) := \mathcal{D}(\pi_F) \otimes \mathcal{D}(\pi_G)$ which is associated (by Lemma 6.5) with the $*$-preserving linear map*

$$\rho(v) := \pi_F(v) \otimes I_{\mathcal{D}(\pi_G)} + I_{\mathcal{D}(\pi_F)} \otimes \pi_G(v), \quad v \in V,$$

of V into $\mathcal{L}^+(\mathcal{D}(\pi_F) \otimes \mathcal{D}(\pi_G))$. Then the linear functional $F \otimes_s G$ on $V_\otimes$ is the vector functional of the $*$-representation $\rho_{F,G}$ for the vector $\varphi_F \otimes \varphi_G$, that is,

$$(F \otimes_s G)(u) = \langle \rho_{F,G}(u)(\varphi_F \otimes \varphi_G), \varphi_F \otimes \varphi_G \rangle, \quad u \in V_\otimes. \qquad (6.6)$$

(Recall that φ_F and φ_G denote the algebraically cyclic vectors for the GNS representations π_F and π_G.)

Proof We abbreviate $I_1 = I_{\mathcal{D}(\pi_F)}$, $I_2 = I_{\mathcal{D}(\pi_G)}$. Let $v_1, \ldots, v_n \in V$, $n \geq 2$, and set $u = v_1 \cdots v_n$. Inserting the definition of $\rho(u)$ from Lemma 6.5 we compute

$$\langle \rho_{F,G}(u)(\varphi_F \otimes \varphi_G), \varphi_F \otimes \varphi_G \rangle$$
$$= \langle (\pi_F(v_1) \otimes I_2 + I_1 \otimes \pi_G(v_1))(\pi_F(v_2) \otimes I_2 + I_1 \otimes \pi_G(v_2))$$
$$\cdots (\pi_F(v_n) \otimes I_2 + I_1 \otimes \pi_G(v_n))(\varphi_F \otimes \varphi_G), \varphi_F \otimes \varphi_G \rangle$$
$$= \sum_{k=1}^{n-1} \sum_{(i,j) \in P_{k,n-k}} \langle (\pi_F(v_{i_1} \cdots v_{i_k}) \otimes \pi_G(v_{j_1} \cdots v_{j_{n-k}})(\varphi_F \otimes \varphi_G), \varphi_F \otimes \varphi_G \rangle$$
$$+ \langle (\pi_F(v_1 \cdots v_n) \otimes I_2)(\varphi_F \otimes \varphi_G), \varphi_F \otimes \varphi_G \rangle$$
$$+ \langle (I_1 \otimes \pi_G(v_1 \cdots v_n))(\varphi_F \otimes \varphi_G), \varphi_F \otimes \varphi_G \rangle$$
$$= \sum_{k=1}^{n-1} \sum_{(i,j) \in P_{k,n-k}} \langle \pi_F(v_{i_1} \cdots v_{i_k})\varphi_F, \varphi_F \rangle \langle \pi_G(v_{j_1} \cdots v_{j_{n-k}})\varphi_G, \varphi_G \rangle$$
$$+ \langle \pi_F(v_1 \cdots v_n)\varphi_F, \varphi_F \rangle \langle \varphi_G, \varphi_G \rangle + \langle \varphi_F, \varphi_F \rangle \langle \pi_G(v_1 \cdots v_n)\varphi_G, \varphi_G \rangle$$
$$= \sum_{k=1}^{n-1} \sum_{(i,j) \in P_{k,n-k}} F_k(v_{i_1} \cdots v_{i_k}) G_{n-k}(v_{j_1} \cdots v_{j_{n-k}})$$
$$+ F_n(v_1 \cdots v_n)G_0(1) + F_0(1)G_n(v_1 \cdots v_n)$$
$$= \sum_{k=0}^{n} (F_k \otimes_s G_{n-k})(v_1 \cdots v_n) = (F \otimes_s G)_n(v_1 \cdots v_n) = (F \otimes_s G)(u).$$

This proves (6.6) for $u = v_1 \cdots v_n$, $n \geq 2$. One easily verifies (6.6) for $u \in V_0$ and $u \in V_1$. Since both sides of (6.6) are linear in u, it holds for all $u \in V_\otimes$. $\qquad \square$

2. Multiplication of a positive functional by a positive semi-definite sequence

A real sequence $c = (c_n)_{n \in \mathbb{N}_0}$ is called *positive semi-definite* if

$$\sum_{j,k=0}^{n} c_{j+k} \zeta_j \zeta_k \geq 0 \quad \text{for all } (\zeta_0, \zeta_1, \ldots, \zeta_n)^T \in \mathbb{R}^{n+1}, n \in \mathbb{N}_0.$$

Proposition 6.10 *Suppose F is a positive linear functional on $V_\otimes$ and $c = (c_n)_{n \in \mathbb{N}_0}$ is a positive semi-definite sequence. Then $c \cdot F := (c_n F_n)_{n \in \mathbb{N}_0}$ is also a positive linear functional on $V_\otimes$.*

Further, there exist a vector ψ of a complex inner product space $\mathcal{D}$ and a symmetric operator $T \in \mathcal{L}^+(\mathcal{D})$ such that $c \cdot F$ is the vector functional of the vector $\varphi_F \otimes \psi$ for the nondegenerate $*$-representation $\rho_{c,F}$ of $V_\otimes$ on $\mathcal{D}(\rho_{c,F}) := \mathcal{D}(\pi_F) \otimes \mathcal{D}$ defined by $\rho_{c,F}(u_n) = \pi_F(u_n) \otimes T^n$, $u_n \in V_n, n \in \mathbb{N}_0$, that is,

$$(c \cdot F)(u) = \langle \rho_{c,F}(u)(\varphi_F \otimes \psi), \varphi_F \otimes \psi \rangle \quad \text{for } u \in V_\otimes. \tag{6.7}$$

Proof Since c is positive semi-definite, it is a Hamburger moment sequence (see e.g. [Sch17, Theorem 3.8]). This means that there exists a Radon measure μ on $\mathbb{R}$ such that $c_n = \int x^n d\mu(x)$ for $n \in \mathbb{N}_0$. Let T be the multiplication operator by the variable x on $\mathcal{D} = \mathbb{C}[x]$ in the Hilbert space $L^2(\mathbb{R}; \mu)$ and $\psi = 1 \in \mathcal{D}$. Then $c_n = \langle T^n \psi, \psi \rangle$ for $n \in \mathbb{N}_0$ and $T \in \mathcal{L}^+(\mathcal{D})$ is symmetric. It is easily verified that $\rho_{c,F}$, as defined above, is a nondegenerate $*$-representation of $V_\otimes$.

For $n \in \mathbb{N}$ and $u_n \in V_n$ we compute

$$\langle \rho_{c,F}(u_n)(\varphi_F \otimes \psi), \varphi_F \otimes \psi \rangle = \langle \pi_F(u_n)\varphi_F \otimes T^n \psi, \varphi_F \otimes \psi \rangle$$
$$= \langle \pi_F(u_n)\varphi_F, \varphi_F \rangle \langle T^n \psi, \psi \rangle = F(u_n)c_n = (c_n F_n)(u_n) = (c \cdot F)(u_n).$$

Let $u_0 \in V_0 = \mathbb{C}$. Using that $\|\psi\|^2 = \langle T^0 \psi, \psi \rangle = c_0$ we obtain

$$\langle \rho_{c,F}(u_0)(\varphi_F \otimes \psi), \varphi_F \otimes \psi \rangle = u_0 \|\varphi_F\|^2 \|\psi\|^2 = u_0 F(1)c_0 = (c \cdot F)(u_0).$$

Now Eq. (6.7) follows by linearity and we have shown that $c \cdot F$ is a vector functional of the $*$-representation $\rho_{c,F}$. Hence $c \cdot F$ is a positive functional. $\qquad\square$

6.4 Representations of Free Field Type

In this section, V is a $*$-vector space and $\langle \cdot, \cdot \rangle_1$ is an inner product on V satisfying

$$\langle u^+, v^+ \rangle_1 = \overline{\langle u, v \rangle_1} \quad \text{for } u, v \in V. \tag{6.8}$$

Let $\langle \cdot, \cdot \rangle_n$ denote the corresponding inner product on $V_n, n \in \mathbb{N}$, defined by

$$\left\langle \sum_i u_1^i \otimes \cdots \otimes u_n^i, \sum_j v_1^j \otimes \cdots \otimes v_n^j \right\rangle_n := \sum_{i,j} \langle u_1^i, v_1^j \rangle_1 \cdots \langle u_n^i, v_n^j \rangle_1.$$

Let $\mathcal{H}_n$ be the Hilbert space obtained by completing the complex inner product space $(V_n, \langle \cdot, \cdot \rangle_n)$ and let $\mathcal{D}$ denote the algebraic direct sum of Hilbert spaces $\mathcal{H}_n, n \in \mathbb{N}_0$, where $\mathcal{H}_0 := \mathbb{C}$. Then the elements of $\mathcal{D}$ are finite sequences $(u_n), (v_n)$ with $u_n, v_n \in \mathcal{H}_n$ and $\mathcal{D}$ is a complex inner product space with inner product

$$\langle (u_n), (v_n) \rangle := \sum_n \langle u_n, v_n \rangle_n.$$

Clearly, $V_\otimes$ is a dense linear subspace of the complex inner product space $(\mathcal{D}, \langle \cdot, \cdot \rangle)$.

Fix $u \in V$ and define $A^+(u)(v_n) = (u \otimes v_n)$ for $(v_n) \in \mathcal{D}$. Then, since

$$\|A^+(u)(v_n)\|^2 = \sum_n \|u \otimes v_n\|_{n+1}^2 = \sum_n \|u\|_1^2 \|v_n\|_n^2 = \|u\|_1^2 \|(v_n)\|^2,$$

$A^+(u)$ is a bounded linear operator on $(\mathcal{D}, \langle \cdot, \cdot \rangle)$ with norm $\|A^+(u)\| = \|u\|_1$.

Now, for $v_0 \in V_0$ and $v_n = \sum_i v_1^i \otimes \cdots \otimes v_n^i \in V_n$, we define $A^-(u)v_0 = 0$,

$$A^-(u)v_n = \sum_i \langle u, (v_1^i)^+ \rangle_1 v_2^i \otimes \cdots \otimes v_n^i, \quad n \in \mathbb{N}. \tag{6.9}$$

Clearly, this definition is independent of the particular representation of v_n. Note that $A^-(u)$ is a linear operator of V_n into V_{n-1} for $n \in \mathbb{N}$ and it is also linear in u.

Let $v_n \in V_n$ and $w_{n+1} \in V_{n+1}$. We write w_{n+1} as $w_{n+1} = \sum_i y_1^i \otimes z_n^i \in V_{n+1}$ with $y_1^i \in V, z_n^i \in V_n$. Inserting the corresponding definitions and (6.8) we derive

$$\langle A^+(u^+)v_n, w_{n+1} \rangle_{n+1} = \langle u^+ \otimes v_n, w_{n+1} \rangle_{n+1}$$

$$= \sum_i \langle u^+, y_1^i \rangle_1 \langle v_n, z_n^i \rangle_n = \sum_i \overline{\langle u, (y_1^i)^+ \rangle_1} \langle v_n, z_n^i \rangle_n$$

$$= \left\langle v_n, \sum_i \langle u, (y_1^i)^+ \rangle_1 z_n^i \right\rangle_n = \langle v_n, A^-(u)w_{n+1} \rangle_n. \tag{6.10}$$

Now we define $A^-(u)(v_n) = (A^-(u)v_n)$ for $(v_n) \in V_\otimes$. Equation (6.10) implies that $\langle A^+(u^+)v, w \rangle = \langle v, A^-(u)w \rangle$ for $v, w \in V_\otimes$. Therefore, the operator $A^-(u)$ on $V_\otimes$ is the restriction of the adjoint of the bounded operator $A^+(u^+)$ on $\mathcal{D}$. Hence $A^-(u)$ is bounded on $V_\otimes$ and its continuous extension $A^-(u)$ is a bounded operator on $\mathcal{D}$ such that $\|A^-(u)\| = \|A^+(u^+)\| \le \|u^+\|_1 = \|u\|_1$. Since $A^-(u)V_n \subseteq V_{n-1}$, this extension maps $\mathcal{H}_n$ into $\mathcal{H}_{n-1}$ and so $\mathcal{D}$ into $\mathcal{D}$. From its definition it follows that $A^+(u^+)$ leaves $\mathcal{D}$ invariant. Therefore, $A^+(u^+)$ and $A^-(u)$ belong to $\mathcal{L}^+(\mathcal{D})$ and are adjoints of each other in this $*$-algebra.

We summarize the preceding in the next lemma.

Lemma 6.11 *For any $u \in V$, the operators $A^+(u)$ and $A^-(u)$ defined above are bounded linear operators of $\mathcal{L}^+(\mathcal{D})$ and we have $A^+(u^+)^+ = A^-(u)$,*

$$\|A^+(u)\| \le \|u\|_1, \quad \|A^-(u)\| \le \|u\|_1. \tag{6.11}$$

Now we are ready for the main result of this section.

Theorem 6.12 *Let $(c_n)_{n \in \mathbb{N}_0}$ be a real sequence. There is a unique nondegenerate ∗-representation π of the tensor algebra $V_\otimes$ on $\mathcal{D}(\pi) := \mathcal{D}$ such that*

$$\pi(u)v_n = c_{n+1} A^+(u)v_n + c_n A^-(u^+)v_n, \quad u \in V, \; v_n \in \mathcal{H}_n, \; n \in \mathbb{N}_0. \quad (6.12)$$

If $c_n \neq 0$ for $n \in \mathbb{N}$, then π is faithful and $\pi(V_\otimes)1 = V_\otimes$ is dense in the Hilbert space $\mathcal{H}(\pi)$.

Proof Let $u \in V$. We extend $\pi(u)$ by linearity to $\mathcal{D} = \mathrm{Lin}\,\{\mathcal{H}_n : n \in \mathbb{N}_0\}$. From Lemma 6.11 it follows that π is a ∗-preserving linear map of V in the ∗-algebra $\mathcal{L}^+(\mathcal{D})$. By Lemma 6.5, it extends to a nondegenerate ∗-representation π of $V_\otimes$ on $\mathcal{D}$. The uniqueness assertion is clear. This completes the proof of the first assertion.

Now suppose that $c_n \neq 0$ for $n \in \mathbb{N}$. Let $v = (v_n) \in V_\otimes$. From (6.12) it follows that $(\pi(v_k)1)_k = c_1 \cdots c_k v_k$ and $(\pi(v_k)1)_n = 0$ for $n > k$.

Therefore, if $v \neq 0$ and k is the largest index n such that $v_n \neq 0$, the k-th component of $\pi(v)1$ is nonzero, so $\pi(v) \neq 0$. This shows that π is faithful.

Further, it follows by induction that each subspace V_k is contained in $\pi(V_\otimes)1$. Hence $V_\otimes \subseteq \pi(V_\otimes)1$, so that $\pi(V_\otimes)1 = V_\otimes$. Obviously, $V_\otimes$ is dense in $\mathcal{H}(\pi)$. □

The construction of the ∗-representations π in Theorem 6.12 resembles the definition of the free field in quantum field theory (see e.g. [SW00]). Therefore, we call them *representations of free field type*. Note that the representation π depends not only on the sequence (c_n), but also on the inner product $\langle \cdot, \cdot \rangle_1$ satisfying (6.8).

There are plenty of such inner products on each ∗-vector space V. For instance, take a vector space basis $\{v_i : i \in I\}$ of hermitian elements of V and define

$$\left\langle \sum_i \alpha_i v_i, \sum_j \beta_j v_j \right\rangle_1 := \sum_i \alpha_i \overline{\beta_i}.$$

Then $\langle \cdot, \cdot \rangle_1$ is an inner product on V and condition (6.8) holds.

Corollary 6.13 *Each ∗-algebra $V_\otimes$, so each ∗-algebra $\mathbb{C}\langle x_i; i \in I | (x_i)^+ = x_i \rangle$, admits a faithful nondegenerate ∗-representation and is ∗-isomorphic to a unital O^*-algebra.*

Proof Choose an inner product $\langle \cdot, \cdot \rangle_1$ as above and set $c_n = 1$ in Theorem 6.12. □

6.5 Topological Tensor Algebras

In the preceding sections only the algebraic structure of tensor algebras $V_\otimes$ was used. In this section we assume that V is a **topological ∗-vector space** according to the following definition.

Definition 6.14 A *topological $*$-vector space* (V, τ) is a $*$-vector space V together with a locally convex topology τ on V such that the involution is τ-continuous.

Let Γ be a family of seminorms which defines the locally convex topology τ, see Appendix C. Without loss of generality we can assume that the seminorms of Γ are $*$-invariant and Γ is directed (that is, given p_1, $p_2 \in \Gamma$, there exists a seminorm $p \in \Gamma$ such that $p_1(v) \leq p(v)$ and $p_2(v) \leq p(v)$ for $v \in V$).

There is no unique way to define a "reasonable" topology on the tensor product $E \otimes F$ of two locally convex spaces E, F; see [Tr67, Sect. 43] or [Sh71, Chap. IV, 9.]. We use the *projective topology*. It is the strongest locally convex topology on $E \otimes F$ for which the map $(e, f) \mapsto e \otimes f$ of $E \times F$ into $E \otimes F$ is continuous.

The projective topology on $V_n = V \otimes \cdots \otimes V$ will be denoted by τ_n. It is defined by the directed family of $*$-invariant seminorms $\{p^n : p \in \Gamma\}$, where for $u_n \in V_n$,

$$p^n(u_n) := \inf \left\{ \sum_{i=1}^{r} p(v_1^i) \cdots p(v_n^i) : u_n = \sum_{i=1}^{r} v_1^i \otimes \cdots \otimes v_n^i, \; v_j^i \in V \right\}. \quad (6.13)$$

These seminorms have the cross-properties

$$p^{k+m}(u_k \otimes u_m) \leq p^k(u_k) p^m(u_m) \quad \text{for} \quad u_k \in V_k, u_m \in V_m, \quad (6.14)$$

$$p^n(v_1 \otimes \cdots \otimes v_n) = p(v_1) \cdots p(v_n) \quad \text{for} \quad v_1, \ldots, v_n \in V. \quad (6.15)$$

Let $\underline{V}_n[\tau_n]$ denote the completion of the locally convex space $V_n[\tau_n]$ and let

$$\underline{V}_\otimes[\tau_\otimes] := \sum_{n=0}^{\infty} \oplus \, \underline{V}_n[\tau_n]$$

be the direct sum of locally convex spaces $\underline{V}_n[\tau_n]$, where $\underline{V}_0 := \mathbb{C}$, equipped with the *direct sum topology* $\tau_\otimes$. This is the locally convex topology on $\underline{V}_\otimes$ given by the family of seminorms $\| \cdot \|_{(\gamma_n),(p_n)}$:

$$\|v\|_{(\gamma_n),(p_n)} := \sum_{n=0}^{\infty} \gamma_n (p_n)^n (v_n), \quad v = (v_0, v_1, \ldots, v_k, 0, \ldots) \in \underline{V}_\otimes.$$

Here (γ_n) and (p_n) are arbitrary sequences of numbers $\gamma_n \geq 0$ and of seminorms $p_n \in \Gamma$, respectively. We abbreviate $V^n := \sum_{j=0}^{n} \oplus V_j$ and $\underline{V}^n := \sum_{j=0}^{n} \oplus \underline{V}_j$.

Lemma 6.15 (i) *The involution of $V_\otimes$ is continuous in the topology $\tau_\otimes$.*
  (ii) *The product of $V_\otimes$ is separately continuous, that is, the maps $v \mapsto v\,w$ and $v \mapsto w\,v$ of $V_\otimes$ are continuous for any $w \in V_\otimes$.*
  (iii) *For any $k, m \in \mathbb{N}$, the product of $V_\otimes$ is a continuous map of $V^k \times V^m$ into V^{k+m} with respect to the induced topologies of $\tau_\otimes$.*

Proof (i) is clear, since all seminorms p^n are $*$-invariant for $p \in \Gamma$ and $n \in \mathbb{N}$.
In the proofs of (ii) and (iii) we assume without loss of generality that $p_n \le p_{n+1}$, $n \in \mathbb{N}_0$. Let $v = (v_0, \ldots, v_k, 0, \ldots) \in V_\otimes$ and $w = (w_0, \ldots, w_m, 0, \ldots) \in V_\otimes$.

(ii): We fix w and prove the continuity of the map $v \mapsto v\, w$. Let $\| \cdot \|_{(\gamma_n),(p_n)}$ be a seminorm for the topology $\tau_\otimes$. Set $\delta_i := \sum_{j=0}^m \gamma_{i+j}(p_{i+j})^j(w_j)$ and $q_i = p_{i+m}$ for $i \in \mathbb{N}_0$. Using the definition (6.1) of the multiplication, the relations (6.14), and the inequalities $p_{i+j} \le p_{i+m} = q_i$ for $j \le m$ we derive

$$
\| v\, w \|_{(\gamma_n),(p_n)} = \sum_{n=0}^{k+m} \gamma_n(p_n)^n \Big(\sum_{i+j=n} v_i \otimes w_j \Big)
$$

$$
\le \sum_{n=0}^{k+m} \sum_{i+j=n} \gamma_n(p_n)^i(v_i)(p_n)^j(w_j) = \sum_{i=0}^{k} \sum_{j=0}^{m} \gamma_{i+j}(p_{i+j})^i(v_i)(p_{i+j})^j(w_j)
$$

$$
\le \sum_{i=0}^{k} \delta_i(q_i)^i(v_i) = \| v \|_{(\delta_n),(q_n)}.
$$

The proof for the map $v \mapsto w\, v$ is the same. It follows also from (i) by the chain of continuous maps $v \mapsto v^+ \mapsto w^+ v^+ \mapsto (w^+ v^+)^+ = v\, w$.

(iii): Now both numbers k and m are fixed. We define constant (!) sequences (q_n) and (δ_n) by $q_n := p_{k+m}$ and $\delta_n := (\max(\gamma_0, \ldots, \gamma_{k+m}))^{1/2}$ for $n \in \mathbb{N}_0$. Then, similarly as in the proof of (ii),

$$
\| v\, w \|_{(\gamma_n),(p_n)} \le \sum_{i=0}^{k} \sum_{j=0}^{m} \gamma_{i+j}(p_{i+j})^i(v_i)(p_{i+j})^j(w_j)
$$

$$
\le \sum_{i=0}^{k} \sum_{j=0}^{m} \delta_i \delta_j(q_i)^i(v_i)(q_j)^j(w_j) = \| v \|_{(\delta_n),(q_n)} \| w \|_{(\delta_n),(q_n)}. \qquad \square
$$

Recall that, according to Definition 2.67, a *topological $*$-algebra* is a $*$-algebra
A, equipped with a locally convex topology, such that the involution and the maps
$a \mapsto a\, b$ and $a \mapsto b\, a$ of A into A are continuous for each $b \in$ B.

By Lemma 6.15, (i) and (ii), $V_\otimes[\tau_\otimes]$ is a topological $*$-algebra. It is not difficult
to verify that $\underline{V}_\otimes[\tau_\otimes]$ is the completion of the locally convex space $V_\otimes[\tau_\otimes]$. As
noted in Exercise 2.22, the completion of a topological $*$-algebra is not necessarily
an algebra. That is, the separate continuity of the multiplication is not sufficient for
extending the multiplication to the completion. In order to remedy this failure we use
Lemma 6.15(iii).

Clearly, the involution extends by continuity to $\underline{V}_\otimes$. By Lemma 6.15(iii), the
multiplication is a continuous map of $V^k \times V^m$ into V^{k+m}, so it can be extended to a
continuous map of $\underline{V}^k \times \underline{V}^m$ into $\underline{V}^{k+m}$. From the definition of the topology it is clear
that this extension is consistent with restrictions to $\underline{V}^r \times \underline{V}^s$, $r \le k, s \le m$. Therefore

it follows easily that this extension defines a multiplication on $\underline{V}_\otimes = \sum_{n=0}^{\infty} \oplus \underline{V}_n$ such that $\underline{V}_\otimes[\tau_\otimes]$ is a topological $*$-algebra. Thus we have

Proposition 6.16 $V_\otimes[\tau_\otimes]$ and $\underline{V}_\otimes[\tau_\otimes]$ are topological $*$-algebras.

Definition 6.17 The topological $*$-algebra $\underline{V}_\otimes[\tau_\otimes]$ is called the *topological tensor algebra* of the topological $*$-vector space (V, τ).

One of the most interesting and important examples of topological tensor algebras is the field algebra $\underline{S}$, which appears in algebraic quantum field theory.

Example 6.18 (*Field algebra and tensor algebra over the Schwartz space* $S(\mathbb{R}^d)$) The locally convex space $S(\mathbb{R}^k)$ was defined in Example 3.11. First we introduce the field algebra $\underline{S}$ and then we relate it to the tensor algebra $\underline{V}_\otimes$ with $V = S(\mathbb{R}^d)$.

Fix $d \in \mathbb{N}$. Set $S_0 := \mathbb{C}$ and $S_n := S(\mathbb{R}^{dn})$ for $n \in \mathbb{N}$. Then the direct sum

$$\underline{S} = \sum_{n=0}^{\infty} \oplus\, S_n, \tag{6.16}$$

equipped with the direct sum topology of locally convex spaces S_n, is a complete locally convex space. The elements of $\underline{S}$ are finite sequences

$$f = (f_0, f_1, \ldots, f_k, 0, \ldots), \quad g = (g_0, g_1, \ldots, g_m, 0, \ldots), \quad f_j, g_j \in S_j.$$

By lengthy but straightforward computations one verifies that the vector space $\underline{S}$ becomes a complex unital $*$-algebra, with multiplication $f \cdot g$ and involution f^+ defined by

$$(f \cdot g)_n(x_1, \ldots, x_n) := \sum_{k=0}^{n} f_k(x_1, \ldots, x_k) g_{n-k}(x_{k+1}, \ldots, x_n), \quad x_j \in \mathbb{R}^d,$$

$$(f^+)_n(x_1, \ldots, x_n), := \overline{f_n(x_n, \ldots, x_1)}, \quad x_j \in \mathbb{R}^d,$$

(with obvious interpretations for $n = 0$, $k = 0$, $n = k$), and a topological $*$-algebra under the direct sum topology. The unit element of $\underline{S}$ is $1 = (1, 0, 0, \ldots)$. This topological $*$-algebra $\underline{S}$ is called the *field algebra*.

Suppose now that V is the topological $*$-vector space $S(\mathbb{R}^d)$ with involution

$$f^+(x_1, \ldots, x_d) := \overline{f(x_1, \ldots, x_d)}.$$

Similarly as in Example 6.4, we identify $\sum_i v_1^i \otimes \cdots \otimes v_n^i \in V_n$ with the function

$$\sum_i v_1^i(x_1, \ldots, x_d) \cdots v_n^i(x_{d(n-1)+1}, \ldots, x_{dn}) \in S(\mathbb{R}^{dn}), \quad x_j \in \mathbb{R}.$$

Then the vector space V_n becomes a subspace of $S(\mathbb{R}^{dn})$ and $V_\otimes$ a $*$-subalgebra of the field algebra $\underline{S}$. It is known [Tr67, Theorem 51.6] that V_n is dense in $S(\mathbb{R}^{dn})$ and that the "natural" topology of the Schwartz space $\mathcal{S}(\mathbb{R}^{dn})$, as described in Example 3.11, coincides with the topology τ_n on V_n. Taking these two facts for granted we conclude that *the topological tensor algebra* $\underline{V}_\otimes[\tau_\otimes]$ *with* $V = \mathcal{S}(\mathbb{R}^d)$ *coincides with the field algebra* $\underline{S}$.

Each Schwartz space $\mathcal{S}(\mathbb{R}^k)$, $k \in \mathbb{N}$, is a nuclear Frechet space [Tr67, p. 530]. This implies that the standard locally convex topologies on any tensor product with $\mathcal{S}(\mathbb{R}^k)$ coincide [Tr67, Theorem 50.1]. The locally convex space $\underline{V}_\otimes[\tau_\otimes] \cong \underline{S}$ with $V = \mathcal{S}(\mathbb{R}^d)$ is also a nuclear space. ○

Finally, we turn to $*$-representations of the $*$-algebra $\underline{V}_\otimes$. By Lemma 6.5, each $*$-preserving linear map of V into $\mathcal{L}^+(\mathcal{D})$ has an extension to a $*$-representation of $V_\otimes$. Any counterpart of this simple fact for $\underline{V}_\otimes$ would require additional considerations concerning topologies on $\mathcal{L}^+(\mathcal{D})$ that are beyond the scope of this book. Here we will treat only representations of free field type, see also Exercise 9c.

Theorem 6.19 *Let* (V, τ) *be a topological $*$-vector space. Suppose* $\langle \cdot, \cdot \rangle_1$ *is an inner product on* V *such that (6.8) holds and its norm is continuous on* $V[\tau]$. *Then, for any real sequence* $(c_n)_{n \in \mathbb{N}_0}$, *the $*$-representation* π *from Theorem 6.12 extends by continuity to a $*$-representation, denoted again by* π, *of* $\underline{V}_\otimes$.

If the sequence (c_n) *is bounded and* $c_n \neq 0$ *for* $n \in \mathbb{N}$, *then* π *is a bounded faithful $*$-representation of* $V_\otimes$ *with cyclic vector* 1 *and the operator norm* $\|\pi(\cdot)\|$ *is continuous on* $\underline{V}_\otimes[\tau_\otimes]$.

Since $\| \cdot \|_1$ is continuous on $V[\tau]$ and τ is defined by the directed family of seminorms Γ, there are $M > 0$ and $p_0 \in \Gamma$ such that $\|v\|_1 \leq M p_0(v)$, $v \in V$. Then $\|v\|_1 \leq p(v)$, $v \in V$, for the τ-continuous seminorm $p := M p_0$. Further, in the notation of Sect. 6.4, we set $\mathcal{H}^m := \sum_{j=0}^m \oplus \mathcal{H}_j$ and $b_n := \max\{|c_0|, \ldots, |c_n|\}$.

The main technical step of the proof is contained in the next lemma.

Lemma 6.20 $\|\pi(u_n)\varphi\| \leq (b_{m+n})^n \|\varphi\| p^n(u_n)$ *for* $\varphi \in \mathcal{H}^m$ *and* $u_n \in V_n$.

Proof Let $n = 1$ and $v \in V$. By (6.11), we have $\|A^+(v)\| \leq \|v\|_1 \leq p(v)$. Using this inequality it follows from (6.12) that

$$\|\pi(v)\varphi\| \leq b_{m+1}\|\varphi\| p(v), \quad \varphi \in \mathcal{H}^m. \tag{6.17}$$

Now suppose $u_n \in V_n, n \geq 2$. Let $u_n = \sum_i v_1^i \otimes \cdots \otimes v_n^i$ with $v_1^i, \ldots, v_n^i \in V$. Using (6.17) and the inequality $b_j \leq b_k$ for $j \leq k$ we derive for $\varphi \in \mathcal{H}^m$,

$$\|\pi(u_n)\varphi\| = \left\| \pi\left(\sum_i v_1^i \otimes \cdots \otimes v_n^i \right)\varphi \right\| \leq \sum_i \|\pi(v_1^i) \cdots \pi(v_n^i)\varphi\|$$

$$\leq \sum_i b_{m+n}\, p(v_1^i)\, \|\pi(v_2^i) \cdots \pi(v_n^i)\varphi\|$$

$$\leq \cdots \leq \sum_i b_{m+n}b_{m+n-1} \cdots b_{m+1}\, p(v_1^i)p(v_2^i) \cdots p(v_n^i)\, \|\varphi\|$$

$$\leq (b_{m+n})^n \|\varphi\| \sum_i p(v_1^i) \cdots p(v_n^i).$$

Taking the infimum over all representations of u_n and inserting the definition (6.13) of $p^n(u_n)$ the latter yields $\|\pi(u_n)\varphi\| \leq (b_{m+n})^n \|\varphi\| \, p^n(u_n)$.   □

Proof of Theorem 6.19 By Lemma 6.20, the map $u_n \mapsto \pi(u_n)$ of $V_n[\tau_n]$ into $(\mathbf{B}(\mathcal{H}^n, \mathcal{H}(\pi)), \|\cdot\|)$ is continuous, so it has a unique extension to a continuous map of $\underline{V}_n$ into $(\mathbf{B}(\mathcal{H}^n, \mathcal{H}(\pi)), \|\cdot\|)$. We extend this map by linearity first to $\underline{V}_\otimes$ and then to $\mathcal{D} = \mathrm{Lin}\,\{\mathcal{H}^m : m \in \mathbb{N}\}$. This extension to $\underline{V}_\otimes$ is also denoted by π.

We show that π is a *-representation of $\underline{V}_\otimes$ on $\mathcal{D}$. We fix $k, n, m \in \mathbb{N}_0$ and let $u_k \in V_k$, $w_n \in V_n$. Since π is a *-representation of $V_\otimes$ (by Theorem 6.12), we have

$$\pi(u_k w_n)\varphi = \pi(u_k)\pi(w_n)\varphi \quad \text{for} \quad \varphi \in \mathcal{H}^m. \tag{6.18}$$

Lemma 6.20 shows that each $\pi(y_i)$, $y_i \in V_i$, is a bounded operator of the Hilbert space $\mathcal{H}^j$ into $\mathcal{H}^{i+j}$ and the operator norm is continuous on $V_i[\tau_i]$. Therefore, passing to the limit and using Lemma 6.15(iii), Eq. (6.18) remains valid for $u_k \in \underline{V}_k$, $w_n \in \underline{V}_n$ and the corresponding operator $\pi(u_k w_n) = \pi(u_k)\pi(w_n)$ maps $\mathcal{H}^m$ into $\mathcal{H}^{k+n+m}$. Hence, by linearity, π is an algebra homomorphism of $\underline{V}_\otimes$ into $\mathcal{L}^+(\mathcal{D})$. Since π is *-preserving for $V_\otimes$ by Theorem 6.12, it is so for $\underline{V}_\otimes$. Thus, π is a *-representation of $\underline{V}_\otimes$.

Suppose now that (c_n) is bounded, say $|c_n| \leq c$, and $c_n \neq 0$ for $n \in \mathbb{N}_0$. Then $b_n \leq c$ and from Lemma 6.20 we obtain for $u = (u_n) \in V_\otimes$ and $\varphi \in \mathcal{H}^m$,

$$\|\pi(u)\varphi\| \leq \sum_n \|\pi(u_n)\varphi\| \leq \left(\sum_n c^n p^n(v_n) \right) \|\varphi\| = \|u\|_{(c^n)(p)} \|\varphi\|. \tag{6.19}$$

Since $\mathcal{D} = \cup_m \mathcal{H}^m$, (6.19) holds for all $\varphi \in \mathcal{D}$ and shows that the operator $\pi(u)$ is bounded and satisfies $\|\pi(u)\| \leq \|u\|_{(c^n)(p)}$. By continuity, the latter is valid for all $u \in \underline{V}_\otimes$ and the operator norm $\|\pi(\cdot)\|$ is continuous on $\underline{V}_\otimes[\tau_\otimes]$.

Since $\pi(V_\otimes)1 = V_\otimes$ is dense, 1 is a cyclic vector. The proof of the assertion about the faithfulness of π is verbatim the same as for $V_\otimes$, see Theorem 6.12.   □

6.6 Exercises

1. Let V and W be *-vector spaces. Show that any *-preserving linear bijection of V on W extends uniquely to a *-isomorphism of the *-algebras $V_\otimes$ and $W_\otimes$.
2. Write the free *-algebra $\mathbb{C}\langle x_1, \ldots, x_d, y_1, \ldots, y_d | (x_i)^+ = y_i, i = 1, \ldots, d\rangle$ as a tensor algebra $V_\otimes$ of some *-vector space V.
3. Show that the *-algebras $\mathbb{C}\langle x_1, \ldots, x_d, y_1, \ldots, y_d | (x_i)^+ = y_i, i = 1, \ldots, d\rangle$ and $\mathbb{C}\langle y_1, \ldots, y_{2d} | (y_i)^+ = y_i, i = 1, \ldots, 2d\rangle$ are *-isomorphic.
4. Show that each tensor algebra $V_\otimes$ or $\underline{V}_\otimes$ has no divisor of zero.
5. When is a tensor algebra $V_\otimes$ commutative?
6. Show that the center of a tensor algebra $V_\otimes$ is $\mathbb{C} \cdot 1$ if and only if $\dim V \geq 2$.

7. Let $V_\otimes$ be a tensor algebra and $f = (f_n) \in \sum(V_\otimes)^2$, $f \neq 0$. We denote by l_f the smallest n for which $f_n \neq 0$ and by m_f the largest n such that $f_n \neq 0$. Show that both numbers l_f and m_f are even.

8. Show each $*$-representation π in Theorem 6.12 is self-adjoint.

9. Let (V, τ) be a topological $*$-vector space.

   a. Show the multiplication of $\underline{V}_\otimes[\tau_\otimes]$ is jointly continuous (that is, the product is a continuous map of $\underline{V}_\otimes[\tau_\otimes] \times \underline{V}_\otimes[\tau_\otimes]$ into $\underline{V}_\otimes[\tau_\otimes]$) if and only if the locally convex topology τ on V can be given by a single norm.

   b. Show that if $\dim V < \infty$, then the topology $\tau_\otimes$ is the finest locally convex topology on the vector space $V_\otimes$.

   c. Let $\mathcal{H}$ be a Hilbert space and let π be a continuous $*$-preserving linear map of $V[\tau]$ into $(\mathbf{B}(\mathcal{H}), \| \cdot \|)$. Prove that π extends to a nondegenerate bounded $*$-representation of $\underline{V}_\otimes$ on $\mathcal{H}$.

   Hint for c.: Mimic the proof of Lemma 6.20.

10. Show that the multiplication of the field algebra $\underline{S}$ is not jointly continuous.

6.7 Notes

The field algebra $\underline{S}$ was first treated in algebraic quantum field theory by Borchers [B62] and Uhlmann [U62]. Since then the field algebra and general topological tensor algebras were studied in many papers of mathematical physics, see e.g. [B72, Yn73, Lr74, Sch84, Ac82, DH89, Ho90] and the references therein.

Chapter 7
Integrable Representations
of Commutative ∗-Algebras

In Sect. 4.7, we listed a number of technical pecularities of unbounded representation theory. As a consequence, additional regularity conditions such as the self-
adjointness were invented to circumvent these difficulties. However, in contrast to
single operators, the self-adjointness of a representation is not enough to rule out
all pathologies. As shown in Sect. 7.2, for the (commutative!) ∗-algebra $\mathbb{C}[x_1, x_2]$,
there exist a self-adjoint irreducible representation acting on an infinite-dimensional
Hilbert space and a state which is not an integral of characters. In this chapter, we
develop a class of "well-behaved" representations, called *integrable* representations,
of commutative ∗-algebras which excludes such pathological phenomena.

In Sect. 7.3, we define integrable representations of commutative ∗-algebras and
prove our main results concerning these representations (Theorems 7.11, 7.14, and
7.20). In Sect. 7.4, we represent integrable representations of finitely generated commutative ∗-algebras by spectral measures (Theorem 7.23) and apply this to moment
functionals. Section 7.1 contains some technical operator-theoretic facts.

Throughout this chapter, A denotes a **commutative unital complex** ∗-algebra, all
∗-representations of A are **nondegenerate**, and $\mathcal{D}$ is a complex inner product space.

7.1 Some Auxiliary Operator-Theoretic Results

The following technical lemmas will be used in Sects. 7.3 and 9.4.

Lemma 7.1 *Let* $x \in \mathcal{L}^+(\mathcal{D})$. *If* xx^+ *is essentially self-adjoint, then* $\overline{x^+} = x^*$.

Proof Since $x^+ \subseteq x^*$, we have $\overline{x^+} \subseteq x^*$. To prove that both closed operators are
equal it suffices to show that $(0, 0)$ is the only element of the graph of x^* that is
orthogonal to the graph of x^+. Suppose $(\zeta, x^*\zeta)$ is orthogonal to the graph of x^+ in
$\mathcal{H} \oplus \mathcal{H}$. This means that $\langle \zeta, \eta \rangle + \langle x^*\zeta, x^+\eta \rangle = 0$ for all $\eta \in \mathcal{D}$. Since $x \in \mathcal{L}^+(\mathcal{D})$

© The Editor(s) (if applicable) and The Author(s), under exclusive license
to Springer Nature Switzerland AG 2020
K. Schmüdgen, *An Invitation to Unbounded Representations of *-Algebras
on Hilbert Space*, Graduate Texts in Mathematics 285,
https://doi.org/10.1007/978-3-030-46366-3_7

and $x^+\eta \in \mathcal{D}$, we have $\langle x^*\zeta, x^+\eta \rangle = \langle \zeta, xx^+\eta \rangle$, so that $\langle \zeta, (I + xx^+)\eta \rangle = 0$ for $\eta \in \mathcal{D}$. Since $xx^+ \geq 0$ on $\mathcal{D}$ and xx^+ is essentially self-adjoint, it follows from Proposition A.1 that $(I + xx^+)\mathcal{D}$ is dense in $\mathcal{H}$. Hence $\zeta = 0$. $\qquad\square$

Recall that a densely defined linear operator x on a Hilbert space $\mathcal{H}$ is called *formally normal* if $\mathcal{D}(x) \subseteq \mathcal{D}(x^*)$ and $\|x\varphi\| = \|x^*\varphi\|$ for all $\varphi \in \mathcal{D}(x)$.

A formally normal operator x is *normal* if $\mathcal{D}(x) = \mathcal{D}(x^*)$.

Lemma 7.2 *Let x_1 and x_2 be commuting symmetric operators of $\mathcal{L}^+(\mathcal{D})$ and set $x := x_1 + ix_2$. Then:*

(i)   *The operator $\overline{x}$ is formally normal.*
(ii) *The operator $\overline{x}$ is normal if and only if $\overline{x^+} = x^*$. In this case, $\overline{x_1}$ and $\overline{x_2}$ are strongly commuting self-adjoint operators.*
(iii) *If the operator xx^+ is essentially self-adjoint, then $\overline{x}$ is normal and $\overline{xx^+} = \overline{x}\,\overline{x^+} = (\overline{x})^* \overline{x}$.*

Proof (i): Using that x_1 and x_2 are commuting symmetric operators we compute

$$\langle (x_1 \pm ix_2)\varphi, (x_1 \pm ix_2)\varphi \rangle = \|x_1\varphi\|^2 + \|x_2\varphi\|^2 \pm i\langle x_2\varphi, x_1\varphi \rangle \mp i\langle x_1\varphi, x_2\varphi \rangle$$
$$= \|x_1\varphi\|^2 + \|x_2\varphi\|^2 \pm i\langle (x_1x_2 - x_2x_1)\varphi, \varphi \rangle = \|x_1\varphi\|^2 + \|x_2\varphi\|^2 \qquad (7.1)$$

for $\varphi \in \mathcal{D}$. Since $x^+ = x_1 - ix_2 \subseteq x^*$, (7.1) yields $\|x\varphi\| = \|x^+\varphi\| = \|x^*\varphi\|$. From this it follows that $\mathcal{D}(\overline{x}) = \mathcal{D}(\overline{x^+}) \subseteq \mathcal{D}(x^*)$ and $\|\overline{x}\,\psi\| = \|x^*\psi\|$ for all $\psi \in \mathcal{D}(\overline{x})$. Therefore, since $(\overline{x})^* = x^*$, the operator $\overline{x}$ is formally normal.

(ii): First suppose that $\overline{x}$ is normal. Then $\mathcal{D}(\overline{x}) = \mathcal{D}(x^*)$. Since $\overline{x^+} \subseteq x^*$ and $\mathcal{D}(\overline{x}) = \mathcal{D}(\overline{x^+})$ by the proof of (i), we have $\mathcal{D}(\overline{x^+}) = \mathcal{D}(x^*)$ and hence $\overline{x^+} = x^*$.

Further, since $\overline{x}$ is normal, it is known that

$$a_1 := (\overline{x} + x^*)/2 \quad \text{and} \quad a_2 := (\overline{x} - x^*)/2i$$

are strongly commuting self-adjoint operators (see [Sch12, Proposition 5.30]). From $x = x_1 + ix_2$ and $x^+ = x_1 - ix_2 \subseteq x^*$ we obtain $x_j \subseteq a_j$ and hence $\overline{x_j} \subseteq a_j$ for $j = 1, 2$. The inequality $\|x_j\varphi\| \leq \|x\varphi\|$ by (7.1) implies $\mathcal{D}(\overline{x}) \subseteq \mathcal{D}(\overline{x_j})$. By definition, $\mathcal{D}(\overline{x}) = \mathcal{D}((\overline{x})^*)$ is a core for a_j, that is, $\overline{a_j \upharpoonright \mathcal{D}(\overline{x})} = a_j$. Therefore, since $\mathcal{D}(\overline{x}) \subseteq \mathcal{D}(\overline{x_j})$ and $\overline{x_j} \subseteq a_j$, it follows that $\overline{x_j} = \overline{a_j \upharpoonright \mathcal{D}(\overline{x_j})} = a_j$. Thus, $\overline{x_1} = a_1$ and $\overline{x_2} = a_2$ are strongly commuting self-adjoint operators. Conversely, assume that $\overline{x^+} = x^*$. As noted in the proof of (i), $\mathcal{D}(\overline{x}) = \mathcal{D}(\overline{x^+})$. Hence, since $\overline{x^+} = x^*$, we have $\mathcal{D}(\overline{x}) = \mathcal{D}(x^*) = \mathcal{D}((\overline{x})^*)$. Therefore, since $\overline{x}$ is formally normal by (i), the latter implies that $\overline{x}$ is normal.

(iii): Suppose xx^+ is essentially self-adjoint. Then, by Lemma 7.1 and (i), $\overline{x}$ is normal. Hence $\overline{x}\,(\overline{x})^* = (\overline{x})^* \overline{x}$ [Sch12, Proposition 3.25]. The closed symmetric operator $\overline{x}\,(\overline{x})^*$ is an extension of xx^+ and hence of $\overline{xx^+}$. Therefore, since $\overline{xx^+}$ is self-adjoint by assumption, $\overline{xx^+} = \overline{x}\,(\overline{x})^*$. These relations give the assertion. $\qquad\square$

Lemma 7.3 *Let x and y be commuting symmetric operators of $\mathcal{L}^+(\mathcal{D})$ such that $\langle x\varphi, \varphi \rangle \geq 0$ for $\varphi \in \mathcal{D}$. Suppose there exists a constant $c > 0$ such that*

$$\|\varphi\| + \|y\varphi\| \leq c\|x\varphi\| \quad \text{for} \quad \varphi \in \mathcal{D}. \tag{7.2}$$

Then, if $\overline{x}$ is self-adjoint, so is $\overline{y}$.

Proof By Proposition A.1, it suffices to show that $\ker(y^* - \lambda\,\mathrm{i}I) = \{0\}$ for nonzero $\lambda \in \mathbb{R}$. Let $\xi \in \ker(y^* - \lambda\,\mathrm{i}I)$.

Clearly, (7.2) implies that $\|\overline{x}\psi\| \geq c^{-1}\|\psi\|$, $\psi \in \mathcal{D}(\overline{x})$. Hence, since $x \geq 0$ and $\overline{x}$ is self-adjoint by assumption, we have $\overline{x} \geq c^{-1}I$. Therefore, $x\mathcal{D}$ is dense in $\mathcal{H}$ by Proposition A.1(iv), applied to the positive operator $T = x - c^{-1}I$ and $z = c^{-1}$. In particular, there exists a sequence (η_n) of vectors $\eta_n \in \mathcal{D}$ such that $\lim_n x\eta_n = \xi$. From (7.2) it follows that the sequences (η_n) and $(y\eta_n)$ converge, say $\eta = \lim_n \eta_n$. Then $\overline{x}\,\eta = \lim_n x\eta_n = \xi$ and $\overline{y}\,\eta = \lim_n y\eta_n$. Since $(\overline{y})^*\xi = y^*\xi = \lambda\,\mathrm{i}\xi$, we get

$$\lim_n \langle (y + \lambda\,\mathrm{i})\eta_n, x\eta_n \rangle = \langle \overline{y}\,\eta + \lambda\,\mathrm{i}\,\eta, \xi \rangle = \langle \eta, (\overline{y})^*\xi - \lambda\,\mathrm{i}\xi \rangle = 0. \tag{7.3}$$

Using the assumption that x and y commute we obtain

$$0 = \langle (xy - yx)\eta_n, \eta_n \rangle = \langle (y + \lambda\,\mathrm{i})\eta_n, x\eta_n \rangle - \langle x\eta_n, (y + \lambda\,\mathrm{i})\eta_n \rangle - 2\lambda\,\mathrm{i}\langle x\eta_n, \eta_n \rangle.$$

The first two terms converge to zero (by (7.3)) and the last to $-2\lambda\,\mathrm{i}\langle \overline{x}\,\eta, \eta \rangle$. Thus, $\langle \overline{x}\,\eta, \eta \rangle = 0$, because $\lambda \neq 0$. Since the self-adjoint operator $\overline{x}$ satisfies $\overline{x} \geq c^{-1}I$, we obtain $\eta = 0$ and $\xi = \overline{x}\,\eta = 0$. This proves that $\overline{y}$ is self-adjoint. $\qquad\square$

Lemma 7.4 *Suppose that A is a self-adjoint operator on a Hilbert space $\mathcal{H}$ and $\mathcal{D} \subseteq \mathcal{D}(A)$ is a dense linear subspace of $\mathcal{H}$. If $\mathcal{D}$ is invariant under the unitary group $U(t) = e^{\mathrm{i}tA}$, $t \in \mathbb{R}$, then $A{\upharpoonright}\mathcal{D}$ is essentially self-adjoint, that is, $\overline{A{\upharpoonright}\mathcal{D}} = A$.*

Proof [Sch12, Proposition 6.3]. $\qquad\square$

7.2 "Bad" Representations of the Polynomial Algebra $\mathbb{C}[x_1, x_2]$

In this section, A denotes the polynomial $*$-algebra $\mathbb{C}[x_1, x_2]$. Our aim is to develop two examples, which show pathological behavior of representations of A.

Example 7.5 *("Bad" representations: failure of positivity)*
Let $p_0 \in \mathbb{C}[x_1, x_2]$ be a polynomial which is nonnegative on $\mathbb{R}^2$, but not in $\sum \mathsf{A}^2$. (For instance, we may take the Motzkin polynomial from Example 2.30.)

Statement 1: *There exists a state f on A such that $f(p_0) < 0$.*

Proof The cone $\sum A^2$ is closed in the finest locally convex topology on **A**. This follows from Theorem 10.36 below. (We prove this result only in Sect. 10.7 after Chap. 9 in order to cover enveloping algebras as well.) By the separation of convex sets (Proposition C.2), applied to the closed (!) cone $\sum A^2$ and the singleton $\{p_0\}$, there exists an $\mathbb{R}$-linear functional $g : A_{her} \mapsto \mathbb{R}$ such that $g(p_0) < 0$ and $g(p) \geq 0$ for all $p \in \sum A^2$. We extend g to a $\mathbb{C}$-linear functional, denoted also by g, on **A** by setting $g(p_1 + ip_2) := g(p_1) + ig(p_2)$, $p_1, p_2 \in A_{her}$. Then $g(1) \neq 0$, since $g(1) = 0$ implies $g = 0$ by the Cauchy–Schwarz inequality, which contradicts $g(p_0) < 0$. Setting $f := g(1)^{-1}g$, f is a state on **A** and $f(p_0) < 0$. □

Statement 2:   *The state f from Statement 1 is not an integral over hermitian characters of the *-algebra* **A**.

Proof Each hermitian character χ of **A** is a point evaluation at some point $t \in \mathbb{R}^2$, that is, $\chi(p) = p(t)$, $p \in A$. If f were an integral over hermitian characters, then we would have $f(p_0) \geq 0$, since $p_0(t) \geq 0$. This contradicts the choice of f. □

Statement 3:   *There exists a pure state on **A** which is not a hermitian character.*

Proof By Corollary 5.37, f is an integral over pure states. From Statement 2 it follows that not all of them can be hermitian characters of **A**. □

Let π_f be the GNS representation of f. Since $\langle \pi_f(p_0)\varphi_f, \varphi_f \rangle = f(p_0) < 0$, the operator $\pi_f(p_0)$ is not nonnegative, though the polynomial p_0 is nonnegative on $\mathbb{R}^2$. That is, π_f is a "bad" representation in the sense that it does not respect the natural positivity of polynomials on $\mathbb{R}^2$. One may express the assertion of Statement 2 by saying that f is a "bad" state on **A**. ○

Example 7.6 (*"Bad" representations: failure of strong commutativity*)
Let $\mathcal{D}(\pi)$ be the set of functions $\varphi \in C_0^\infty(\mathcal{X})$, where $\mathcal{X} := \mathbb{R}^2 \setminus \{(0, x_2) : x_2 \geq 0\}$, such that $\frac{\partial^k}{\partial x_1^k} \frac{\partial^l \varphi}{\partial x_2^l} \in \mathcal{H} := L^2(\mathbb{R}^2)$ for $k, l \in \mathbb{N}_0$ and the limits $\frac{\partial^k \varphi}{\partial x_1^k}(\pm 0, x_2)$ exist and satisfy

$$\frac{\partial^k \varphi}{\partial x_1^k}(+0, x_2) = -\frac{\partial^k \varphi}{\partial x_1^k}(-0, x_2) \quad \text{for} \quad x_2 > 0, \ k \in \mathbb{N}_0.$$

We define a *-representation π of $A = \mathbb{C}[x_1, x_2]$ on the domain $\mathcal{D}(\pi)$ by $\pi(1) = I$,

$$\pi(x_1) = -i\frac{\partial}{\partial x_1}, \quad \pi(x_2) = -i\frac{\partial}{\partial x_2}.$$

Since $\pi(x_1)$ and $\pi(x_2)$ are commuting symmetric operators leaving the domain $\mathcal{D}(\pi)$ invariant, this gives indeed a *-representation of $\mathbb{C}[x_1, x_2]$.

Define one-parameter unitary groups on $\mathcal{H}$ by $(U_2(t)\varphi)(x_1, x_2) = \varphi(x_1, x_2+t)$,

$$(U_1(t)\varphi)(x_1, x_2) =$$
$$\left\{ \begin{array}{l} -\varphi(x_1 + t, x_2) \ \text{if} \ \ x_2 > 0, x_1 < 0, x_1 + t > 0 \ \text{or} \ x_1 > 0, x_1 + t < 0 \\ \varphi(x_1 + t, x_2) \quad \text{otherwise} \end{array} \right\}.$$

That is, U_1 is the left translation parallel to the x_1-axis such that the function is multiplied by -1 when the positive x_2-axis is crossed. By Stone's theorem (Proposition A.5), there are self-adjoint operators A_1 and A_2 on $\mathcal{H}$ such that $U_1(t) = e^{itA_1}$ and $U_2(t) = e^{itA_2}$, $t \in \mathbb{R}$.

Statement 1: π^* is a self-adjoint $*$-representation of $\mathbb{C}[x_1, x_2]$. Moreover, we have $A_j = \overline{\pi(x_j)} = \pi^*(x_j)$ for $j = 1, 2$.

Proof From the definitions of the unitary groups it follows that $A_j \varphi = -i \frac{\partial \varphi}{\partial x_j} = \pi(x_j)\varphi$ for $j = 1, 2$ and $\varphi \in \mathcal{D}(\pi)$.

Let $j \in \{1, 2\}$. We denote by $\mathcal{D}_j$ the set of functions of $\mathcal{D}(\pi)$ which vanish in some neighborhood of the x_j-axis. Then $\mathcal{D}_j$ is a dense linear subspace of $\mathcal{H}$ which is invariant under the unitary group $U_j(t)$. Therefore, by Lemma 7.4, $\mathcal{D}_j$, hence $\mathcal{D}(\pi)$, is a core for A_j. Since $A_j \lceil \mathcal{D}(\pi) = \pi(x_j)$, this implies $A_j = \overline{\pi(x_j)}$. Hence π^* is self-adjoint by Theorem 4.18(iii) and $\overline{\pi^*(x_j)} = \overline{\pi(x_j)}$ by Theorem 4.18(ii). (If we use [Sch12, Exercise 6.3] or Proposition 9.32 below instead of Lemma 7.4 we even obtain $A_j^n = \overline{\pi(x_j^n)}$, $j = 1, 2$, for *all* $n \in \mathbb{N}$.) $\qquad\square$

Statement 2: $(I - U_1(-t)U_2(-s)U_1(t)U_2(s))\varphi = 2\chi_{ts} \cdot \varphi$ for $\varphi \in \mathcal{H}, t > 0, s > 0$, where χ_{ts} is the characteristic function of the rectangle $[0, t] \times (0, s]$.

Proof The formula follows by a simple computation; we omit the details. $\qquad\sqcup$

Statement 3: π^* is irreducible.

Proof Let $T \in \pi^*(A)'_s$. Then $T \in \{\pi^*(x_j)\}'_s$ for $j = 1, 2$. From Lemma 3.16(i) it follows that T commutes with the self-adjoint operator $\overline{\pi^*(x_j)} = A_j$ (by Statement 1) and hence with the unitary group U_j. Then, by Statement 2, T commutes with all multiplication operators by characteristic functions χ_{ts} for $t > 0$, $s > 0$ and similarly for the other cases of real values of s, t. This implies that T commutes with the von Neumann algebra $L^\infty(\mathbb{R}^2)$ on $L^2(\mathbb{R}^2)$. Since $L^\infty(\mathbb{R}^2) = L^\infty(\mathbb{R}^2)'$, T is a multiplication operator by some function $\psi \in L^\infty(\mathbb{R}^2)$. Using once again that T commutes with U_1 and U_2 we conclude that ψ is constant a.e. on $\mathbb{R}^2$. Thus $T = \lambda \cdot I$ for some $\lambda \in \mathbb{C}$, so π^* is irreducible by Proposition 4.26. $\qquad\square$

By Statements 1 and 2, π^* is a self-adjoint $*$-representation of $\mathbb{C}[x_1, x_2]$ for which the operators $\overline{\pi^*(x_1)}$ and $\overline{\pi^*(x_2)}$ are self-adjoint, but the corresponding unitary groups U_1 and U_2 do not commute. Hence the *self-adjoint operators* $\overline{\pi^*(x_1)}$ and $\overline{\pi^*(x_2)}$ *do not strongly commute*. Further, π^* is an *irreducible* self-adjoint $*$-representation of the commutative $*$-algebra $\mathbb{C}[x_1, x_2]$ acting on an infinite-dimensional Hilbert space. In order to exclude such pathological behavior, a stronger notion than self-adjointness is needed. $\qquad\bigcirc$

In the preceding example, we constructed a self-adjoint $*$-representation π^* of $\mathbb{C}[x_1, x_2]$ such that the bicommutant $\pi^*(\mathbb{C}[x_1, x_2])'' := (\pi^*(\mathbb{C}[x_1, x_2])'_s)'$ is the von Neumann algebra $\mathbf{B}(L^2(\mathbb{R}^2))$. Here $\mathbf{B}(L^2(\mathbb{R}^2))$ can be replaced by an arbitrary properly infinite von Neumann algebra on a separable Hilbert space. More precisely, the following result was proved in [Sch86] (see also [Sch90, Theorem 9.4.1]):

Suppose $\mathcal{N}$ is a properly infinite von Neumann algebra on a separable Hilbert space $\mathcal{H}$. There exists a self-adjoint $*$-representation ρ of $\mathbb{C}[x_1, x_2]$ on $\mathcal{H}$ such that $\rho(\mathbb{C}[x_1, x_2])'' = \mathcal{N}$ and the operators $\overline{\rho(x_j^n)}$, $j = 1, 2, n \in \mathbb{N}$, are self-adjoint.

7.3 Integrable Representations of Commutative *-Algebras

The following definition introduces the two fundamental notions of this chapter.

Definition 7.7 Let π be a $*$-representation of the commutative unital complex $*$-algebra A. We shall say that

- *integrable* if π is closed and

$$\overline{\pi(a^+)} = \pi(a)^* \quad \text{for all} \quad a \in \mathsf{A}, \tag{7.4}$$

- *subintegrable* if there exists an integrable representation ρ of A such that $\pi \subseteq \rho$.

An explanation of the name "integrable" in case of the polynomial algebra $\mathbb{C}[x_1, \ldots, x_d]$ will be given in Corollary 9.28.

To find reasonable sufficient conditions for subintegrability is an important and difficult problem. For instance, the GNS representation of a positive functional on the polynomial algebra $\mathbb{C}_d[\underline{x}]$ is subintegrable if and only if the corresponding moment problem is solvable, as shown by Corollary 7.27 below.

Remark 7.8 1. Since $\pi(a^+) \subseteq \pi(a)^*$ and hence $\overline{\pi(a^+)} \subseteq \pi(a)^*$ holds for any $*$-representation π, condition (7.4) in Definition 7.7 can be replaced by

$$\pi(a)^* \subseteq \overline{\pi(a^+)} \quad \text{for} \quad a \in \mathsf{A}. \tag{7.5}$$

2. Clearly, a $*$-representation of A is integrable if and only if the O^*-algebra $\pi(\mathsf{A})$ is closed and $\overline{x^+} = x^*$ for all operators $x \in \pi(\mathsf{A})$. Hence the integrability of a $*$-representation depends only on the O^*-algebra $\pi(\mathsf{A})$. That is, if π_1 and π_2 are $*$-representations of commutative unital $*$-algebras A_1 and A_2 such that $\pi_1(\mathsf{A}_1) = \pi_2(\mathsf{A}_2)$, then π_1 is integrable if and only if π_2 is integrable.   ○

Proposition 7.9 *Each integrable $*$-representation of A is self-adjoint.*

Proof Using condition (7.4) and the assumption that π is closed we obtain

$$\mathcal{D}(\pi) = \mathcal{D}(\overline{\pi}) = \cap_{a \in \mathsf{A}} \mathcal{D}(\overline{\pi(a)}) = \cap_{a \in \mathsf{A}} \mathcal{D}(\pi(a^+)^*) = \mathcal{D}(\pi^*). \qquad \square$$

Example 7.10 *(Algebras of measurable functions: Example* 4.21 *continued)*
Let π be the $*$-representation of the $*$-algebra A from Example 4.21.

Statement: π *is integrable.*

Proof Suppose $f \in \mathsf{A}$. As noted in Example 4.21, we have $\pi(f)^*\psi = \overline{f} \cdot \psi$ for $\psi \in \mathcal{D}(\pi(f)^*) = \{\psi \in L^2(\mathbb{R}^d; \mu) : \overline{f} \cdot \psi \in L^2(\mathbb{R}^d; \mu)\}$.

Fix $\psi \in \mathcal{D}(\pi(f)^*)$. Let $(\chi_{K_n})_{n \in \mathbb{N}}$ be a sequence of characteristic functions of compact subsets K_n of $\mathbb{R}^d$ such that $\cup_n K_n = \mathbb{R}^d$. Set $\psi_n := \chi_{K_n}\psi$. Clearly, we have $\psi_n \in \mathcal{D}(\pi)$. Since $|\overline{f}\,\psi_n - \overline{f}\,\psi| \leq 2|\overline{f}\,\psi|$ on $\mathbb{R}^d$ and $\overline{f}\,\psi \in L^2(\mathbb{R}^d; \mu)$, we conclude from Lebesgue's dominated convergence theorem that $\psi_n \to \psi$ and $\pi(f^+)\psi_n = \overline{f} \cdot \psi_n \to \overline{f} \cdot \psi$ in $L^2(\mathbb{R}^d; \mu)$. Hence we have $\psi \in \mathcal{D}(\overline{\pi(f^+)})$ and $\overline{\pi(f^+)}\,\psi = \overline{f} \cdot \psi = \pi(f)^*\psi$, so that $\pi(f)^* \subseteq \overline{\pi(f^+)}$. Since π is self-adjoint (Example 4.21), π is integrable by Remark 7.8.1. □○

Theorem 7.11 *For each *-representation π of A the following are equivalent:*

(i) *The closure $\overline{\pi}$ of π is integrable.*
(ii) *The operator $\overline{\pi(a)}$ is normal for any $a \in \mathsf{A}$.*
(iii) *The operator $\overline{\pi(a)}$ is self-adjoint for any $a \in \mathsf{A}_{her}$.*
(iv) *$\overline{\pi(a_1)}$ and $\overline{\pi(a_2)}$ are strongly commuting self-adjoint operators for any $a_1, a_2 \in \mathsf{A}_{her}$.*

Proof Since $\pi(a) = \overline{\overline{\pi}(a)}$ for $a \in \mathsf{A}$, we can assume without loss of generality that π is closed. Let $a \in \mathsf{A}$. We write $a = a_1 + ia_2$, $a_1, a_2 \in \mathsf{A}_{her}$, and apply Lemma 7.2 with $x_1 := \pi(a_1)$, $x_2 := \pi(a_2)$. Then Lemma 7.2(ii) yields (i)↔(ii)→(iv), while Lemma 7.2(iii) gives (iii)→(ii). The implication (iv)→(iii) is trivial. ⊔

Corollary 7.12 *Every closed *-representation of a hermitian *-algebra (see Definition 2.69) is integrable.*

Proof By Corollary 4.12, condition (iii) of Theorem 7.11 is satisfied. □

Recall that for the self-adjoint *-representation π^* in Example 7.6 the self-adjoint operators $\overline{\pi^*(x)}$ and $\overline{\pi^*(y)}$ do not commute strongly. Hence π^* is not integrable by Theorem 7.11(iv), so the converse of Proposition 7.9 is not true.

Next we characterize integrable representations in terms of affiliated operators (Appendix B) with abelian von Neumann algebras. First we prove a simple lemma.

Lemma 7.13 *Let B be a subset of A_{her} such that $\mathsf{B} \cup \{1\}$ generates A as a *-algebra and let π be a *-representation of A. Suppose $\overline{\pi(b_1)}$ and $\overline{\pi(b_2)}$ are strongly commuting self-adjoint operators for all $b_1, b_2 \in \mathsf{B}$. Then $\pi(\mathsf{A})'_w$ is a von Neumann algebra with abelian commutant. Moreover, $\pi(\mathsf{A})'_w = \cap_{b \in \mathsf{B}} \{\overline{\pi(b)}\}'_s$.*

Proof Let $E(\lambda; b)$, $\lambda \in \mathbb{R}$, denote the spectral projections of the self-adjoint operator $\overline{\pi(b)}$. Since $\mathsf{B} \subseteq \mathsf{A}_{her}$, A is also generated as an algebra by $\mathsf{B} \cup \{1\}$. Hence $\pi(\mathsf{A})'_w = \cap_{b \in \mathsf{B}} \{\overline{\pi(b)}\}'_w$ by Corollary 4.17. Since $\overline{\pi(b)}$ is self-adjoint, we have $\{\overline{\pi(b)}\}'_w = \{\overline{\pi(b)}\}'_s = \{E(\lambda; b) : \lambda \in \mathbb{R}\}'$ by Lemma 3.16(iv). Therefore, we obtain $\pi(\mathsf{A})'_w = \{E(\lambda; b) : \lambda \in \mathbb{R}, b \in \mathsf{B}\}'$. The latter set is a von Neumann algebra. Its commutant is the von Neumann algebra generated by $E(\lambda; b)$, $\lambda \in \mathbb{R}$, $b \in \mathsf{B}$. Since the spectral projections $E(\lambda_1; b_1)$ and $E(\lambda_2; b_2)$ commute by the strong commutativity of $\overline{\pi(b_1)}$ and $\overline{\pi(b_2)}$, this von Neumann algebra is abelian. □

Theorem 7.14 *For any $*$-representation π of A the following are equivalent:*

(i) *The closure $\overline{\pi}$ of π is integrable.*
(ii) *$\overline{\pi}$ is self-adjoint and the von Neumann algebra $(\pi(A)'_w)'$ is abelian.*
(iii) *The von Neumann algebra $(\pi(A)'_{sym})'$ is abelian.*
(iv) *There exists an abelian von Neumann algebra $\mathcal{N}$ such that $\overline{\pi(a)}$ is affiliated with $\mathcal{N}$ for all $a \in$ A.*
(v) *There exists an abelian von Neumann algebra $\mathcal{N}$ such that $\overline{\pi(a)}$ is affiliated with $\mathcal{N}$ for all $a \in$ A$_{her}$.*

In particular, we can take $\mathcal{N} := (\pi(A)'_w)'$ in (iv) and (v).

Proof All conditions are preserved if π is replaced by its closure, so we can assume without loss of generality that π is closed.

(i)→(ii): From Theorem 7.11,(i)→(iv), it follows that π satisfies the assumptions of Lemma 7.13 with B $=$ A$_{her}$. Therefore, $(\pi(A)'_w)'$ is an abelian von Neumann algebra. By Proposition 7.9, π is self-adjoint.

(ii)→(iii): Since π is self-adjoint, $\pi(A)'_{sym} = \pi(A)'_w$ by Proposition 3.17(iii).

(iii)→(iv): By Proposition 3.17(ii), we can take $\mathcal{N} := (\pi(A)'_{sym})'$.

(iv)→(v) is trivial.

(v)→(i): Any closed symmetric operator affiliated with an abelian von Neumann algebra is self-adjoint [KR83, Theorem 5.6.15, (vii)]. Hence, for $a \in$ A$_{her}$, the operator $\overline{\pi(a)}$ is self-adjoint and the assertion follows from Theorem 7.11,(iii)→(i). $\square$

Corollary 7.15 *Suppose π is an integrable $*$-representation of A. If π is irreducible, then the Hilbert space $\mathcal{H}(\pi)$ is one-dimensional.*

Proof Since π is irreducible and self-adjoint, $\pi(A)' = \mathbb{C} \cdot I$ by Proposition 4.26 and hence $\pi(A)'' = \mathbf{B}(\mathcal{H}(\pi))$. On the other hand, because π is integrable, $\pi(A)''$ is abelian by Theorem 7.14(ii). This implies that $\mathcal{H}(\pi)$ has dimension one. $\square$

The preceding results are mainly of theoretical importance. To verify that a $*$-representation is integrable it is useful to have criteria involving algebra generators.

Theorem 7.16 *Let B be a subset of A$_{her}$ such that the $*$-algebra A is generated by the set B $\cup \{1\}$ and let π be a $*$-representation of A. Suppose $\overline{\pi(b_1)}$ and $\overline{\pi(b_1)}$ are strongly commuting self-adjoint operators for all $b_1, b_2 \in$ B. Then π^* is an integrable $*$-representation of A. Moreover, $\pi^*(A)'_w = \cap_{b \in B} \left\{ \overline{\pi(b)} \right\}'$.*

Proof Since B $\subseteq$ A$_{her}$, the set B $\cup \{1\}$ generates A as an algebra. Thus, by Theorem 4.18, π^* is self-adjoint. The relations $\pi(b) \subseteq \pi^*(b) \subseteq \pi(b)^* = \overline{\pi(b)}$ imply that $\overline{\pi^*(b)} = \overline{\pi(b)}$ for $b \in$ B. Hence Lemma 7.13 applies to π^*. It gives the description of the weak commutant $\pi^*(A)'_w$ and implies that $\pi^*(A)'_w$ is a von Neumann algebra with abelian commutant. Therefore, π^* is integrable by Theorem 7.14(ii). $\square$

Corollary 7.17 *Let* B *and* A *be as in Theorem 7.16. Suppose in addition that* π *is a self-adjoint *-representation of* A. *Then* π *is integrable if and only if* $\overline{\pi(b_1)}$ *and* $\overline{\pi(b_1)}$ *are strongly commuting self-adjoint operators for all* $b_1, b_2 \in$ B.

Proof Since $\pi = \pi^*$ by assumption, Theorem 7.16 yields the if part, while the only if part follows from Theorem 7.11(iv). □

Proposition 7.18 *Any self-adjoint subrepresentation of an integrable representation is integrable.*

Proof Let π be a self-adjoint subrepresentation of the integrable representation ρ. By Corollary 4.31, ρ decomposes as a direct sum $\rho = \pi \oplus \pi_0$. Clearly, $\overline{\rho(a^+)} = \rho(a)^*$ implies that $\overline{\pi(a^+)} = \pi(a)^*$. Since π is self-adjoint and hence closed, π is integrable by Definition 7.7. □

The following example shows that the main sufficient condition for integrability in Theorem 7.16 is not necessary.

Example 7.19 Let π_1 be a *-representation of $\mathbb{C}[x]$ such that $\overline{\pi_1(x)}$ is self-adjoint, but $\pi_1(x^2)$ and $\pi_1(x^2 + x)$ are not self-adjoint; the representation π_1 in Example 4.20 has this property. We define a *-representation π of the *-algebra $\mathbb{C}[x_1, x_2]$ on $\mathcal{D}(\pi) = \mathcal{D}(\pi_1)$ by $\pi(x_1) - \pi_1(x^2)$, $\pi(x_2) = \pi_1(x^2 + x)$. Since $\overline{\pi_1(x)}$ is self-adjoint, π_1^* is an integrable representation of $\mathbb{C}[x]$ by Theorem 7.16. Therefore, since $\pi(\mathbb{C}[x_1, x_2]) = \pi_1(\mathbb{C}[x])$, we have $\pi^*(\mathbb{C}[x_1, x_2]) = \pi_1^*(\mathbb{C}[x])$, so π^* is an integrable *-representation of $\mathbb{C}[x_1, x_2]$ by Remark 7.8.1.

But $\overline{\pi(x_1)} = \pi_1(x^2)$ and $\pi(x_2) = \pi_1(x^2 + x)$ are not self-adjoint. Thus, for B $= \{x_1, x_2\}$ the assumption in Theorem 7.16 does not hold. ○

The next theorem shows that GNS representations of A_+-positive functionals are subintegrable. From (2.56) we recall that

$$A_+ := \{a \in A_{her} : \chi(a) \geq 0 \text{ for } \chi \in \hat{A}\}.$$

Clearly, A_+ is a quadratic module of A.

Theorem 7.20 *Suppose* f *is an* A_+-*positive linear functional on* A. *Then the GNS representation* π_f *has an extension to an integrable* A_+-*positive *-representation of* A *on a possibly larger Hilbert space.*

If f *is a state and an extreme point of the* A_+-*positive states, then* f *is a hermitian character and hence a pure state of* A.

Proof We fix a set $\{y_j : j \in J\}$ of hermitian generators of the *-algebra A. Let P be the free commutative unital *-algebra generated by hermitian indeterminants $x_j, j \in J$. Further, let R denote the free commutative unital *-algebra generated by the indeterminants x_j and the inverses $(p \pm i)^{-1}$ for $p = p^+ \in$ P. The elements of P are complex polynomials in x_j. There is a surjective unital *-homomorphism $\theta : P \mapsto$ A given by $\theta(x_j) = y_j$, $j \in J$, and P is a *-subalgebra of R. Clearly, any character χ of P is determined by its values $\lambda_j = \chi(x_j) \in \mathbb{R}$, $j \in J$, and for arbitrary real numbers λ_j, $j \in J$, there is a character χ of P such that $\chi(x_j) = \lambda_j$, $j \in J$. Each character of P extends uniquely to a character of R. Let Q_R denote the quadratic module of elements $b \in R_{her}$ such that $\chi(b) \geq 0$ for all characters $\chi \in \widehat{R}$ for which there is a character $\chi' \in \widehat{A}$ satisfying $\chi(x_j) = \chi'(y_j)$, $j \in J$. Set $Q_P = Q_R \cap P_{her}$.

We define a linear functional f_P on P by $f_P(\cdot) = f(\theta(\cdot))$. Since f is A_+-positive, it follows from the definition of the quadratic module Q_P of P that f_P is Q_P-positive. It is easy to verify that P is cofinal in R with respect to the cone $\sum R^2$ and hence also with respect to the larger cone Q_R. Therefore, by Proposition C.4, f_P extends to a Q_R-positive linear functional f_R on the larger *-algebra R. The GNS representation π_{f_R} of this positive functional f_R on R will be crucial in what follows.

Let $p \in P$ be such that $\theta(p) = 0$. Using that A is commutative, we derive for $q \in R$,

$$\|\pi_{f_R}(p)\pi_{f_R}(q)\varphi_{f_R}\|^4 = f_R((pq)^+pq)^2 = f_R(p^+pq^+q)^2$$
$$\leq f_R((p^+p)^2)f_R((q^+q)^2) = f_P((p^+p)^2)f_R((q^+q)^2)$$
$$= f(\theta((p^+p)^2))f_R((q^+q)^2) = f(\theta(p^+)^2\theta(p)^2)f_R((q^+q)^2) = 0.$$

Therefore, $\rho(\theta(p)) := \pi_{f_R}(p)$, $p \in P$, gives a well-defined (!) *-representation ρ of A on the domain $\mathcal{D}(\pi_{f_R})$. Let $p = p^+ \in P$. Since $(p \pm i)^{-1} \in R$, it follows from Lemma 4.11 that $\overline{\pi_{f_R}(p)}$ is self-adjoint. Thus, since $A_{her} = \theta(P_{her})$, $\overline{\rho(a)}$ is self-adjoint for all $a \in A_{her}$. Hence $\pi := \overline{\rho}$ is integrable by Theorem 7.11.

Let $a \in A$. We choose an element $p \in P$ such that $a = \theta(p)$. Then

$$f(a) = f_P(p) = \langle \pi_{f_R}(p)\varphi_{f_R}, \varphi_{f_R} \rangle = \langle \rho(a)\varphi_{f_R}, \varphi_{f_R} \rangle = \langle \pi(a)\varphi_{f_R}, \varphi_{f_R} \rangle.$$

Hence the GNS representation π_f of A is unitarily equivalent to the subrepresentation $\pi \restriction \pi(A)\varphi_{f_R}$. For notational simplicity, we identify π_f with this subrepresentation and φ_{f_R} with φ_f. Then π is an integrable extension of π_f.

Next we prove that π is A_+-positive. Let $a \in A_+$. Then, by the definition of Q_P, $a = \theta(p)$ for some $p \in Q_P$. For $q \in P$ we derive

$$\langle \rho(a)\pi_{f_R}(q)\varphi_{f_R}, \pi_{f_R}(q)\varphi_{f_R} \rangle = \langle \pi_{f_R}(p)\pi_{f_R}(q)\varphi_{f_R}, \pi_{f_R}(q)\varphi_{f_R} \rangle = f_R(q^+pq) \geq 0,$$

since $q^+pq \in Q_R$ and f_R is Q_R-positive. Thus, $\rho(a) \geq 0$. Hence $\pi(a) = \overline{\rho}(a) \geq 0$.

This completes the proof of the first assertion. The second assertion is only a restatement of Corollary 2.64. In the following we give another proof of this result. Assume now that f is a state and an extreme point of the set of A_+-positive states.

Since π is self-adjoint, $\pi(A)' \equiv \pi(A)'_s = \pi(A)'_w = \pi(A)'_{sym}$ is a von Neumann algebra by Proposition 3.17. Let e be a projection of $\pi(A)'$ and set $T := Pe \restriction \mathcal{H}(\pi_f)$, where P is the projection of $\mathcal{H}(\pi)$ on $\mathcal{H}(\pi_f)$. For $a \in A$ and $\psi \in \mathcal{D}(\pi_f) \subseteq \mathcal{D}(\pi)$,

$$T\pi_f(a)\psi = Pe\pi(a)\psi = P\pi(a)e\psi = (\pi_f)^*(a)Pe\psi = (\pi_f)^*(a)T\psi.$$

Here, since $\pi_f \subseteq \pi \subseteq \pi^*$, the third equality follows from Lemma 4.4(i). This proves $T \in \pi_f(\mathsf{A})'_w$. Clearly, $0 \leq T \leq I$ on $\mathcal{H}(\pi_f)$. For $a \in \mathsf{A}$, we derive

$$
\begin{aligned}
f_T(a) &= \langle T\pi_f(a)\varphi_f, \varphi_f \rangle = \langle Pe\pi_f(a)\varphi_f, \varphi_f \rangle = \langle e\pi_f(a)\varphi_f, P\varphi_f \rangle \\
&= \langle e^2\pi(a)\varphi_{f_\mathsf{R}}, \varphi_{f_\mathsf{R}} \rangle = \langle \pi(a)e\varphi_{f_\mathsf{R}}, e\varphi_{f_\mathsf{R}} \rangle.
\end{aligned}
\tag{7.6}
$$

Since $e\varphi_{f_\mathsf{R}} \in \mathcal{D}(\pi)$ and π is A_+-positive as shown above, (7.6) implies that f_T is A_+-positive and $f_T(1) = \|e\varphi_{f_\mathsf{R}}\|^2$. Replacing e by $I - e$ it follows that f_{I-T} is also A_+-positive and $f_{I-T}(1) = \|(I-e)\varphi_{f_\mathsf{R}}\|^2 = 1 - \|e\varphi_{f_\mathsf{R}}\|^2$. If $e\varphi_{f_\mathsf{R}} = 0$, then $f_T = 0$ by (7.6) and hence $T = 0$. Similarly, $(I - e)\varphi_{f_\mathsf{R}} = 0$ implies $I - T = 0$. Now suppose both vectors $e\varphi_{f_\mathsf{R}}$ and $(I - e)\varphi_{f_\mathsf{R}}$ are nonzero and set $T_1 := \|e\varphi_{f_\mathsf{R}}\|^{-2}T$ and $T_2 := \|(I - e)\varphi_{f_\mathsf{R}}\|^{-2}(I - T)$. Then

$$
\|e\varphi_{f_\mathsf{R}}\|^2 f_{T_1} + \|(I - e)\varphi_{f_\mathsf{R}}\|^2 f_{T_2} = f_T + f_{I-T} = f
$$

represents f as a convex combination of two A_+-positive states f_{T_1} and f_{T_2}. Since f is an extreme point of the A_+-positive states, there exists a $\lambda \in [0, 1]$ such that $f_T = \lambda f$ and hence $T = \lambda \cdot I$. Thus, in any case, $T = \lambda \cdot I$ with $\lambda \in [0, 1]$.

Next, since π is integrable, $\mathcal{N} := (\pi(\mathsf{A})'_{\mathrm{sym}})'$ is an abelian von Neumann algebra by Theorem 7.14, so $\mathcal{N} \subseteq \mathcal{N}' = \pi(\mathsf{A})'_{\mathrm{sym}}$. Thus, if $e \in \mathcal{N}$ is a projection, then we have $e \in \pi(\mathsf{A})'_{\mathrm{sym}}$ and hence $T = Pe{\upharpoonright}\mathcal{H}(\pi_f) = \lambda \cdot I$, as shown in the preceding paragraph. Therefore, since $\mathcal{N}$ is generated by its projections, $P\mathcal{N}{\upharpoonright}\mathcal{H}(\pi_f) = \mathbb{C} \cdot I$.

Fix $a = a^+ \in \mathsf{A}$. Recall from Proposition 3.17(iv) that the operator $\overline{\pi(a)}$ is affiliated with the von Neumann algebra $\mathcal{N}$. Let e_n denote the spectral projection of the self-adjoint operator $\overline{\pi(a)}$ for the interval $[-n, n]$. Then $\overline{\pi(a)}e_n$ is a bounded operator. It belongs to $\mathcal{N}$, because $\overline{\pi(a)}$ is affiliated with $\mathcal{N}$, so there is a real number λ_n such that $P(\overline{\pi(a)}e_n)\varphi = \lambda_n\varphi$, $\varphi \in \mathcal{H}(\pi_f)$. For any $\varphi \in \mathcal{D}(\pi_f)$, we have $\varphi \in \mathcal{D}(\pi) \cap \mathcal{H}(\pi_f)$ and $\lim_n(\overline{\pi(a)}e_n)\varphi = \overline{\pi(a)}\,\varphi = \pi(a)\varphi$. Therefore, it follows that $P\pi(a)\varphi = \lambda\varphi$, $\varphi \in \mathcal{D}(\pi_f)$, for some $\lambda \in \mathbb{R}$.

The result of the preceding paragraph implies that $P\pi(\mathsf{A}){\upharpoonright}\mathcal{D}(\pi_f) = \mathbb{C} \cdot I_{\mathcal{D}(\pi_f)}$. On the other hand, $P\pi^*(a) \subseteq (\pi_f)^*(a)P$ for $a \in \mathsf{A}$ by Lemma 4.4(i). Hence, $P\pi(a)\varphi = (\pi_f)^*(a)\varphi = \pi_f(a)\varphi$ for $\varphi \in \mathcal{D}(\pi_f)$. Combining both facts, we conclude that $\pi_f(\mathsf{A}) = \mathbb{C} \cdot I_{\mathcal{D}(\pi_f)}$. Thus, for each $a \in \mathsf{A}$ there is a unique $\lambda(a) \in \mathbb{C}$ such that $\pi_f(a) = \lambda(a) \cdot I_{\mathcal{D}(\pi_f)}$. Then $\lambda(a) = \langle \pi_f(a)\varphi_f, \varphi_f \rangle = f(a)$, because f is a state. Since π_f is an algebra homomorphism, so is $\lambda(\cdot) = f(\cdot)$. That is, the state f is a hermitian character and hence a pure state by Corollary 2.61. $\square$

Corollary 7.21 *Suppose the quadratic module* A_+ *has a countable* A_+*-dominating subset* Q_0 *(that is, given* $a \in \mathsf{A}_{\mathrm{her}}$*, there exist an element* $c \in Q_0$ *and a number* $\lambda > 0$ *such that* $\lambda c - a \in \mathsf{A}_+$*). Then each* A_+*-positive functional* f *on* A *is an integral over hermitian characters.*

Proof Theorem 5.35 applies to $Q := \mathsf{A}_+$ and implies that f is an integral over $Q^\wedge$-extremal states. These are extreme points of the A_+-positive states and hence hermitian characters by the second assertion of Theorem 7.20. $\square$

Corollary 7.22 *Each A_+-positive linear functional on a countably generated *- algebra A is an integral over hermitian characters.*

Proof Since $\sum A^2$ admits a countable dominating subset, as shown in the proof of Corollary 5.37, so does A_+. Hence the assertion follows from Corollary 7.21. □

Note that if the set $\hat{A}$ is empty (see, for instance, Example 2.68), we obviously have $A_+ = A_{her}$ and there is no nonzero A_+-positive functional on A.

At the end of the next section, we will use the spectral theorem to study decompositions of positive functionals as integrals over hermitian characters.

7.4   Spectral Measures of Integrable Representations

In this section, A is a **finitely generated** commutative complex unital *-algebra.

The main result of this section associates a spectral measure with each integrable representation of A. The proof is based on the multi-dimensional spectral theorem (Proposition A.4).

From Sect. 2.7 we recall some simple facts on the set $\hat{A}$ of hermitian characters of A. Let us fix a set $\{a_1, \ldots, a_d\}$ of hermitian generators of A. Then each $\chi \in \hat{A}$ is uniquely determined by the point $\lambda_\chi := (\chi(a_1), \ldots, \chi(a_d)) \in \mathbb{R}^d$. We identify χ with λ_χ, so $\hat{A}$ becomes a real algebraic subset of $\mathbb{R}^d$, see formula (2.58). We equipp $\hat{A}$ with the induced topology from $\mathbb{R}^d$. It is not difficult to show that this is the weakest topology for which all functions $f_a(\chi) := \chi(a), a \in A$, on $\hat{A}$ are continuous. In particular, $\hat{A}$ is closed in $\mathbb{R}^d$ and hence locally compact.

Theorem 7.23 *Suppose π is an integrable representation of the finitely generated commutative unital *-algebra A.*

(i) *There exists a unique spectral measure E_π on $\hat{A}$, called the* spectral measure *associated with π, such that for all $a \in A$ and $\varphi \in \mathcal{D}(\pi)$ we have:*

$$\overline{\pi(a)} = \int_{\hat{A}} \chi(a)\, dE_\pi(\chi), \tag{7.7}$$

$$\langle \pi(a)\varphi, \varphi \rangle = \int_{\hat{A}} \chi(a)\, d\langle E_\pi(\chi)\varphi, \varphi \rangle. \tag{7.8}$$

(ii) *The spectral projections $E_\pi(\cdot)$ leave the domain $\mathcal{D}(\pi)$ invariant. Further, if $\{a_1, \ldots, a_d\}$ are hermitian generators of the *-algebra A, then*

$$\mathcal{D}(\pi) = \bigcap_{j=1}^{d} \mathcal{D}^\infty(\overline{\pi(a_j)}). \tag{7.9}$$

(iii) *Let Q be a quadratic module of* A. *If* $\pi(c) \geq 0$ *for* $c \in Q$, *then* E_π *is supported on the closed subset* $\hat{\mathsf{A}}(Q)_+$ *of* $\hat{\mathsf{A}} \subseteq \mathbb{R}^d$, *where*

$$\hat{\mathsf{A}}(Q)_+ := \left\{ \chi \in \hat{\mathsf{A}} : \chi(c) \geq 0 \text{ for } c \in Q \right\}. \tag{7.10}$$

Proof (i): Since π is integrable, $A_j := \overline{\pi(a_j)}$, $j = 1, \ldots, d$, are pairwise strongly commuting self-adjoint operators by Theorem 7.11. Hence, by Proposition A.4, there exists a spectral measure E_π on the Borel σ-algebra of $\mathbb{R}^d$ such that

$$A_j = \overline{\pi(a_j)} = \int_{\mathbb{R}^d} \lambda_j \, dE_\pi(\lambda_1, \ldots, \lambda_d), \quad j = 1, \ldots, d. \tag{7.11}$$

Let $a \in \mathsf{A}$. There exists a polynomial $p \in \mathbb{C}_d[\underline{x}]$ such that $a = p(a_1, \ldots, a_d)$. Let $\mathbb{I}(p) := \int_{\mathbb{R}^d} p(\lambda) \, dE_\pi(\lambda)$ denote the corresponding spectral integral (Appendix A). It follows from (7.11) that

$$\pi(a) = \pi(p(a_1, \ldots, a_d)) = p(\pi(a_1), \ldots, \pi(a_d)) \subseteq \mathbb{I}(p).$$

Hence $\overline{\pi(a)} \subseteq \mathbb{I}(p)$. The operator $\overline{\pi(a)}$ is normal by Theorem 7.11. The spectral integral $\mathbb{I}(p)$ is also normal [Sch12, Theorem 4.16(iv)]. Since normal operators are maximal normal, both operators coincide:

$$\overline{\pi(a)} = \overline{\pi(p(a_1, \ldots, a_d))} = \mathbb{I}(p) = \int_{\mathbb{R}^d} p(\lambda) \, dE_\pi(\lambda). \tag{7.12}$$

Now we prove that the spectral measure E_π is supported on $\hat{\mathsf{A}} \subseteq \mathbb{R}^d$. Let $\lambda_0 \in \operatorname{supp} E_\pi$ and assume to the contrary that $\lambda_0 \notin \hat{\mathsf{A}}$. The finitely generated commutative complex unital $*$-algebra A is $*$-isomorphic to a quotient algebra $\mathbb{C}_d[\underline{x}]/\mathcal{J}$ and $\hat{\mathsf{A}}$ is the zero set of the ideal $\mathcal{J}$, see (2.58). Since $\lambda_0 \notin \hat{\mathsf{A}}$, there exists a polynomial $p \in \mathbb{C}_d[x]$ such that $p(\lambda_0) \neq 0$ and $p(a_1, \ldots, a_d) = 0$. We choose an open ball U around λ_0 such that $p(\lambda) \neq 0$ on U. From $\lambda_0 \in \operatorname{supp} E_\pi$ it follows that $E_\pi(U) \neq 0$. Hence there exists a vector $\varphi \in E_\pi(U)\mathcal{H}(\pi)$ such that $\int_U p(\lambda) \, dE_\pi(\lambda)\varphi \neq 0$. From (7.12) we obtain

$$0 = \overline{\pi(p(a_1, \ldots, a_d))}\, \varphi = \int_U p(\lambda) \, dE_\pi(\lambda)\varphi \neq 0,$$

a contradiction. Thus, $\operatorname{supp} E_\pi \subseteq \hat{\mathsf{A}}$. We insert this fact into (7.12) and obtain (7.7). Obviously, (7.7) implies (7.8). The uniqueness of the spectral measure E_π follows at once from the corresponding assertion in Proposition A.4.

(ii): Since π is integrable, it is closed, so $\mathcal{D}(\pi) = \cap_{a \in \mathsf{A}} \mathcal{D}(\overline{\pi(a)})$. Therefore, by (7.12), $\mathcal{D}(\pi)$ is the intersection of domains $\mathcal{D}(\mathbb{I}(p))$, $p \in \mathbb{C}_d[\underline{x}]$. The spectral measure E_π leaves each domain $\mathcal{D}(\mathbb{I}(p))$ invariant and hence $\mathcal{D}(\pi)$. By definition,

$$\mathcal{D}(\mathbb{I}(p)) = \left\{ \varphi \in \mathcal{H}(\pi) : \int_{\mathbb{R}^d} |p(\lambda)|^2 d\langle E_\pi(\lambda)\varphi, \varphi \rangle < \infty \right\}. \tag{7.13}$$

The operator $\left(\overline{\pi(a_j)} \right)^n = \int \lambda_j^n \, dE_\pi(\lambda)$ has the domain

$$\mathcal{D}\left(\left(\overline{\pi(a_j)} \right)^n \right) = \left\{ \varphi \in \mathcal{H}(\pi) : \int_{\mathbb{R}^d} \lambda_j^2 \, d\langle E_\pi(\lambda)\varphi, \varphi \rangle < \infty \right\}. \tag{7.14}$$

From (7.13) and (7.14) it follows that the intersection of $\mathcal{D}(\mathbb{I}(p))$, $p \in \mathbb{C}_d[\underline{x}]$, is equal to the intersection of $\mathcal{D}\left(\left(\overline{\pi(a_j)} \right)^n \right)$, $j = 1, \ldots, d$, $n \in \mathbb{N}$. This implies (7.9).

(iii): Clearly, $\hat{\mathsf{A}}(Q)_+$ is closed in $\hat{\mathsf{A}}$. We prove that E_π is supported on $\hat{\mathsf{A}}(Q)_+$. Suppose $\lambda_0 \in \hat{\mathsf{A}}$ and $\lambda_0 \notin \hat{\mathsf{A}}(Q)_+$. Hence there exists an element $c \in Q$ such that $\lambda_0(c) < 0$. Then there is a polynomial $p \in \mathbb{C}_d[\underline{x}]$ such that $c = p(a_1, \ldots, a_d)$ and $\lambda_0(c) = p(\lambda_0(a_1), \ldots, \lambda_0(a_d)) = p(\lambda_0) < 0$. We can find an $\varepsilon > 0$ and a ball U around λ_0 such that $p(\lambda) \leq -\varepsilon$ on U. Since $c \in Q$, we have $\pi(c) \geq 0$ by assumption and hence $\overline{\pi(c)} \geq 0$. For $\varphi \in E_\pi(U)\mathcal{H}$, using (7.8) we derive

$$0 \leq \langle \overline{\pi(c)} \varphi, \varphi \rangle = \int_U p(\lambda) \, d\langle E_\pi(\lambda)\varphi, \varphi \rangle \leq -\varepsilon \, \|E_\pi(U)\varphi\|^2.$$

This implies $E_\pi(U)\varphi = 0$. Thus $E_\pi(U) = 0$. Therefore, $\lambda_0 \notin \text{supp } E_\pi$. Since $\text{supp } E_\pi \subseteq \hat{\mathsf{A}}$ as proved in (i), we have shown that $\text{supp } E_\pi \subseteq \hat{\mathsf{A}}(Q)_+$. $\qquad \square$

Corollary 7.24 *If the finitely generated commutative complex unital *-algebra* A *admits a faithful integrable *-representation, then* $\hat{\mathsf{A}}$ *separates the point of* A.

Proof Let π be a faithful integrable *-representation of A and $a \in \mathsf{A}$, $a \neq 0$. Since π is faithful, $\langle \pi(a)\varphi, \varphi \rangle \neq 0$ for some vector $\varphi \in \mathcal{D}(\pi)$. Then, by (7.8), there exists a $\chi \in \hat{\mathsf{A}}$ such that $\chi(a) \neq 0$. $\qquad \square$

Formula (7.8) represents vector states of integrable *-representations as integrals over *hermitian characters*. In particular, if A has a nontrivial integrable *-representation, it follows that A admits a hermitian character. The following example shows that this is not true in general if the *-algebra is not finitely generated.

Example 7.25 Let $\mathsf{B} = L^\omega(0, 1)$ be the Arens algebra (see Example 2.68) and let π be its representation by multiplication operators on $L^2(0, 1)$, that is, $\pi(f)\varphi = f \cdot \varphi$ for $f \in \mathsf{B}$ with domain

$$\mathcal{D}(\pi) = \left\{ \varphi \in L^2(0, 1) : f \cdot \varphi \in L^2(0, 1) \text{ for } f \in \mathsf{B} \right\}.$$

If we set the functions of $L^\omega(0, 1)$ zero on $\mathbb{R}\backslash[0, 1]$, this is a special case of Example 4.21. Hence π is integrable, as shown in Example 7.10. Clearly, we have $(f + \lambda)^{-1} \in \mathsf{B}$ for $f \in \mathsf{B}_{\mathrm{her}}$ and $\lambda \in \mathbb{C}\backslash\mathbb{R}$. Thus, B is even a hermitian $*$-algebra. Since B has no character, as proved in Example 2.68, the assertion of Theorem 7.23(i) does not hold for π and B. No vector state of π can be written as an integral over hermitian characters.   ○

At the end of this section, we briefly mention the relation to the moment problem.

Definition 7.26 A linear functional f on A is called a *moment functional* if there exists a Radon measure μ on $\hat{\mathsf{A}}$ such that

$$f(a) = \int_{\hat{\mathsf{A}}} \chi(a) \, d\mu(\chi) \quad \text{for} \quad a \in \mathsf{A}. \tag{7.15}$$

Corollary 7.27 *Let f be a positive linear functional on a finitely generated commutative complex unital $*$-algebra A. Then f is a moment functional if and only if the GNS representation π_f is subintegrable.*

Proof Suppose π_f is subintegrable. Let ρ be an integrable extension of π_f. Then (7.8), applied to the vector functional $f(\cdot) = \langle \pi_f(\cdot)\varphi_f, \varphi_f \rangle = \langle \rho(\cdot)\varphi_f, \varphi_f \rangle$ of ρ, yields (7.15), with $\mu(\cdot) = \langle E_\pi(\cdot)\varphi_f, \varphi_f \rangle$.

Conversely, let f be a moment functional. Then the representation from Example 7.10 is integrable and an extension of π_f. Thus, π_f is subintegrable.   □

Example 7.28 *(Classical moment problem on $\mathbb{R}^d$)*
Let $\mathsf{A} := \mathbb{C}_d[\underline{x}]$. Then the hermitean characters of A are given by point evaluations on $\mathbb{R}^d$, that is, $\hat{\mathsf{A}} \cong \mathbb{R}^d$. Hence Definition 7.26 gives the classical notion of a moment functional (see e.g. [Sch17]).

Let Q be a quadratic module of A generated by polynomials $p_1, \ldots, p_k \in \mathbb{R}_d[\underline{x}]$. Then a character $\chi \in \hat{\mathsf{A}}$ is in $\hat{\mathsf{A}}(Q)_+$ if and only if $\chi(p_j) \geq 0$ for $j = 1, \ldots, k$. Therefore, under the identification $\hat{\mathsf{A}} \cong \mathbb{R}^d$, the set $\hat{\mathsf{A}}(Q)_+$ becomes the semi-algebraic set $K(p_1, \ldots, p_k) := \{x \in \mathbb{R}^d : p_j(x) \geq 0, j = 1, \ldots, k\}$.   ○

The proof of Theorem 7.23 was based on Proposition A.4. If we use instead the spectral theorem for countably many self-adjoint operators [S91, Theorem 1], Theorem 7.23 and its subsequent applications remain valid for countably generated commutative unital $*$-algebras (see [SS13, Theorem 7]).

7.5 Exercises

1. Let π^* be the self-adjoint $*$-representation from Example 7.6. Is π^* integrable? Is π^* subintegrable?

2. Define a *-representation π of $\mathbb{C}[x_1, x_2]$ by $\pi(x_1) = -i\frac{\partial}{\partial x_1}, \pi(x_2) = -i\frac{\partial}{\partial x_2}$ acting on the following domains.

    a. Let $\mathcal{D}(\pi) = \{\varphi \in C^\infty(\mathbb{R}^2) : \frac{\partial^{n_1}}{\partial x_1^{n_1}} \frac{\partial^{n_2}}{\partial x_2^{n_2}} \varphi \in L^2(\mathbb{R}^2)$ for $n_1, n_2 \in \mathbb{N}_0\}$ and $\mathcal{H}(\pi) = L^2(\mathbb{R}^2)$. Show that π is integrable.

    b. Let $\mathcal{D}(\pi) = C_0^\infty(\mathbb{R}^2)$ and $\mathcal{H}(\pi) = L^2(\mathbb{R}^2)$. Prove that π is not integrable, but π^* is integrable.

    c. Suppose $\mathcal{O}$ is a domain in $\mathbb{R}^2$. Let $\mathcal{D}(\pi) = C_0^\infty(\mathcal{O})$ and $\mathcal{H}(\pi) = L^2(\mathcal{O})$. Show that π is subintegrable.

    d. Set $\mathcal{R} = [0, 1] \times \mathbb{R}$. Let $\mathcal{H}(\pi) = L^2(\mathcal{R})$ and

$$\mathcal{D}(\pi) = \left\{\varphi \in C^\infty(\mathbb{R}^2) : \frac{\partial^{n_1}}{\partial x_1^{n_1}} \frac{\partial^{n_2}}{\partial x_2^{n_2}} \varphi \in L^2(\mathcal{R}) \text{ for } n_1, n_2 \in \mathbb{N}_0; \right.$$

$$\left. \frac{\partial^n \varphi}{\partial x_1^n}(0, x_2) = \frac{\partial^n \varphi}{\partial x_1^n}(1, x_2) \text{ for } x_2 \in \mathbb{R}, n \in \mathbb{N}_0 \right\}. \quad (7.16)$$

    Show that π in integrable.

    e. Discuss other boundary conditions in (7.16) for which π is not integrable.

3. Show that each *-representation of the *-algebra $\mathbb{C}[x]$ is subintegrable.

4. Show that the polynomial *-algebra $\mathbb{C}[x_1, \ldots, x_d]$ for $d \geq 2$ admits a *-representation which is not subintegrable.

5. Let ρ be a *-representation of $\mathbb{C}[x]$. Define a *-representation π of $\mathsf{A} = \mathbb{C}[x_1, x_2]$ by $\pi(x_1) = \rho(x), \pi(x_2) = \rho(x^2)$ on $\mathcal{D}(\pi) = \mathcal{D}(\rho)$.

    a. Formulate conditions on ρ for π being integrable.

    b. Formulate conditions on ρ for π^* being integrable.

    c. Is π subintegrable?

6. Let A be the group algebra of the d-torus $\mathbb{T}^d$. Describe $\hat{\mathsf{A}}$ and show that each positive linear functional on A is a moment functional.

7. Let $\mathsf{A} = \mathbb{C}[x_1, \ldots, x_d]$ and $p \in \mathbb{R}[x_1, \ldots, x_d]$. Show that $\pi(p) \geq 0$ for all integrable *-representations of A if and only if $p(x) \geq 0$ for all $x \in \mathbb{R}^d$.
Hint: Using Theorem 7.23 or Proposition A.4.

8. Formulate and prove the counterpart of Exercise 7 for an arbitrary finitely generated commutative complex unital *-algebra A.

7.6 Notes

Integrable representations of commutative *-algebras were invented and studied by Powers [Pw71] under the name "standard representations"; additional results can be found in [Sch90, Chap. 9]. An important theorem of Powers [Pw71] (see [Sch90, Theorem 9.2.3]) states that any integrable representation with metrizable graph topology is a direct sum of *cyclic* integrable representations.

    The main part of Example 7.6 is taken from [Pw71]. The corresponding operator-theoretic phenomenon was discovered by Nelson [N59]. Theorem 7.23 was proved in [SS13].

Chapter 8
The Weyl Algebra and the Canonical Commutation Relation

In this chapter, we investigate well-behaved $*$-representations of the Weyl algebra

$$\mathsf{W} = \mathbb{C}\langle p, q \mid p=p^+, q=q^+, pq - qp = -\mathrm{i}\rangle$$

and the canonical commutation relation

$$PQ - QP = -\mathrm{i}I,$$

where P and Q are self-adjoint operators. We also study the isomorphic $*$-algebra $\mathbb{C}\langle a, a^+ \mid aa^+ - a^+a = 1\rangle$ and the commutation relation $AA^* = A^*A + I$.

Section 8.1 deals with algebraic properties of the Weyl algebra. In Sect. 8.2, we describe solutions of the operator relation $AA^* = A^*A + I$ (Theorem 8.4). In Sects. 8.3 and 8.4, we develop the Bargmann–Fock representation and the Schrödinger representation and show that the Segal–Bargmann transform provides a unitary equivalence of both representations.

Under additional regularity conditions, the Schrödinger representation, or the Bargmann–Fock representation, is up to unitary equivalence the only irreducible representation of the Weyl algebra. Our main aim in this chapter is to prove such uniqueness theorems. In Sects. 8.3 and 8.4, we derive two results (Theorems 8.8 and 8.9) for the Bargmann–Fock representation and the Rellich–Dixmier theorem (Theorem 8.17) for the Schrödinger representation. Sections 8.5 and 8.6 are devoted to two other uniqueness results for Schrödinger pairs, the famous Stone–von Neumann uniqueness theorem (Theorem 8.18) and Kato's theorem (Theorem 8.22).

In the final sections we touch two related topics. In Sect. 8.7, we treat the Heisenberg uncertainty relation. Section 8.8 is about the Groenewold–van Hove "no-go" theorem and van Hove's prequantization map (Theorems 8.28 and 8.30).

© The Editor(s) (if applicable) and The Author(s), under exclusive license
to Springer Nature Switzerland AG 2020
K. Schmüdgen, *An Invitation to Unbounded Representations of *-Algebras on Hilbert Space*, Graduate Texts in Mathematics 285,
https://doi.org/10.1007/978-3-030-46366-3_8

8.1   The Weyl Algebra

Let x denote the multipication operator by the variable x and y the derivation $\frac{d}{dx}$ acting on the polynomial algebra $\mathbb{C}[x]$. Then

$$yx(f) = \frac{d}{dx}(xf) = f + x\frac{d}{dx}f = f + xy(f) \quad \text{for} \quad f \in \mathbb{C}[x];$$

that is, the operators x and y satisfy the commutation rule

$$yx - xy = 1. \tag{8.1}$$

Obviously, the complex algebra generated by x and y is the algebra of differential operators $\sum_{j=0}^{k} p_j(x)(\frac{d}{dx})^j$, $k \in \mathbb{N}$, with polynomial coefficients $p_j \in \mathbb{C}[x]$.

Definition 8.1 The *Weyl algebra*, or the *Heisenberg–Weyl algebra*, is the unital complex algebra W with generators x, y and defining relation (8.1).

From (8.1) it follows easily that each element $f \in \mathsf{W}$ is of the form

$$f = \sum_{j,k=0}^{n} \alpha_{jk} x^j y^k, \tag{8.2}$$

where $n \in \mathbb{N}$ and $\alpha_{jk} \in \mathbb{C}$. Further, by (8.1), for each polynomial p we have

$$yp(x) = p(x)y + p'(x), \quad xp(y) = p(y)x - p'(y). \tag{8.3}$$

Proposition 8.2 (i) *The sets $\{x^j y^k : j, k \in \mathbb{N}_0\}$ and $\{y^j x^k : j, k \in \mathbb{N}_0\}$ are vector space bases of W.*

(ii) *The algebra W is simple; that is, W has no nontrivial two-sided ideal.*

(iii) *The algebra W has trivial center $\mathbb{C} \cdot 1$.*

(iv) *If π is a homomorphism of W into another algebra B, then either $\pi(\mathsf{W}) = \{0\}$ or $\ker \pi = \{0\}$.*

Proof Let f be of the form (8.2). From (8.3) we obtain

$$yf - fy = \sum_{j=1,k=0}^{n} \alpha_{jk} j x^{j-1} y^k \quad \text{and} \quad xf - fx = - \sum_{j=0,k=1}^{n} \alpha_{jk} k x^j y^{k-1}. \tag{8.4}$$

(i): We carry out the proof for the first set. It suffices to show the linear independence of elements $x^j y^k$. Assume to the contrary that there are $\alpha_{jk} \in \mathbb{C}$, not all zero, such that the element f in (8.2) is zero in W. Choose r, s such that $\alpha_{rs} \neq 0$ and $\alpha_{jk} = 0$ if $j > r$ or $k > s$. Since $yf - fy = xf - xf = 0$ in W by $f = 0$, it follows by a repeated application of (8.4) that $\alpha_{rs} r! s! \cdot 1 = 0$ in W, a contradiction.

(ii): Let $\mathcal{J} \neq \{0\}$ be a two-sided ideal of W. Then there exists an $f \in \mathcal{J}$, $f \neq 0$. We write f as in (8.2) and choose r, s as in the proof of of (i). Then $yf - fy \in \mathcal{J}$ and $xf - xf \in \mathcal{J}$. As in the proof of (i), a repeated application of (8.4) implies that $\alpha_{rs} r! s! \cdot 1 \in \mathcal{J}$. Therefore, $1 \in \mathcal{J}$, since $\alpha_{rs} \neq 0$. Hence $\mathcal{J} = W$.

(iii): Let f be in the center of W. Then $yf - fy = xf - xf = 0$. Therefore, since the set $\{x^j y^k : j, k \in \mathbb{N}_0\}$ is linearly independent by (i), it follows from (8.4) that $\alpha_{jk} = 0$ for all indices j, k such that $j + k > 0$. Thus, $f = \alpha_{00} \cdot 1$.

(iv): Since $\ker \pi$ is a two-sided ideal of W, (iv) follows at once from (ii). $\square$

The multiplication rule for the vector space basis $\{x^j y^k\}$ of W is given by a classical formula of Littlewood [Lt33], which can be proved by induction:

$$x^j y^k x^n y^m = \sum_{l=0}^{\min(k,n)} \frac{1}{l!} \binom{k}{l} \binom{n}{l} x^{j+n-l} y^{m+n-l}, \quad j, k, n, m \in \mathbb{N}_0. \quad (8.5)$$

For instance, $y^2 x^3 = x^3 y^2 + 6x^2 y + 6x$.

Next we consider automorphisms of the algebra W. For $n \in \mathbb{N}_0$ and $\lambda \in \mathbb{C}$, there exist automorphisms $\Phi_{n,\lambda}$, $\Phi'_{n,\lambda}$ of W such that

$$\Phi_{n,\lambda}(x) = x, \quad \Phi_{n,\lambda}(y) = y + \lambda x^n, \quad \Phi'_{n,\lambda}(y) = y, \quad \Phi'_{n,\lambda}(x) = x + \lambda y^n. \quad (8.6)$$

(Indeed, the images of x, y satisfy again the relation (8.1), so $\Phi_{n,\lambda}$, $\Phi'_{n,\lambda}$ define algebra homomorphisms of W into W. Since x, y are in the ranges of these mappings, the homomorphisms are surjective. By Proposition 8.2(iv), they are injective.)

By a theorem of Dixmier [Di68, Theorem 8.10], the set of automorphisms $\Phi_{n,\lambda}$, $\Phi'_{n,\lambda}$ generates the group of *all* algebra automorphisms of W. Further, the algebra W has also linear automorphisms

$$\Phi(x) = ax + by, \quad \Phi(y) = cx + dy, \quad \text{where } a, b, c, d \in \mathbb{C}, ad - bc = 1. \quad (8.7)$$

(Indeed, the assumption $ad - bc = 1$ implies that the images x', y' of x, y satisfy (8.1) and $x = dx' - by'$, $y = -cx' + ay'$. Hence Φ defines an automorphism of W.) From Dixmier's theorem it follows easily that the automorphism group of W is generated by the automorphisms (8.7) and $\Phi_{n,\lambda}$, where $n \in \mathbb{N}_0$, $\lambda \in \mathbb{C}$.

Remark 8.3 1. There is a similar result for the polynomial algebra $\mathbb{C}[x_1, x_2]$. A classical theorem of Jung [J42] states that the automorphism group of the algebra $\mathbb{C}[x_1, x_2]$ is generated by linear automorphisms

$$\Phi(x_1) = ax_1 + bx_2, \quad \Phi(x_2) = cx_1 + dx_2, \quad \text{where } a, b, c, d \in \mathbb{C}, ad - bc \neq 0,$$

and triangular automorphisms $\Phi(x_1) = x_1 + f(x_2)$, $\Phi(x_2) = x_2$, where $f \in \mathbb{C}[x_2]$.

2. We briefly mention two famous conjectures. The first is *Dixmier's conjecture*; see [Di68, Probleme 11.1]. He conjectured that each endomorphism of the Weyl algebra W is an automorphism of W, or equivalently, if x', y' are elements of W such that $x'y' - x'y' = 1$, then x', y' generate the algebra W.

The second is the *Jacobi conjecture*. It was first stated by Keller [Ke39] and says the following: If $f = (f_1, f_2) : \mathbb{C}^2 \mapsto \mathbb{C}^2$, where $f_1, f_2 \in \mathbb{C}[x_1, x_2]$, is a polynomial map such that the Jacobian $\det(\frac{\partial f_i}{\partial x_j})$ is equal 1 on $\mathbb{C}^2$, then f has a polynomial inverse on $\mathbb{C}^2$, or equivalently, $\mathbb{C}[f_1, f_2] = \mathbb{C}[x_1, x_2]$.

Both conjectures are equivalent as shown by Tsuchimoto [Ts05]. They are still open, and they have natural d-dimensional versions; see, e.g., [VE00].   ○

The algebra W carries two algebra involutions, one defined by $x^+ := y$, $y^+ := x$ and the other by $x^+ := x$, $y^+ := -y$. Renaming $a := y$, $a^+ := x$ in the first case and $q := x$, $p := -iy$ in the second case, we obtain the following two $*$-algebras:

$$W_{ca} = \mathbb{C}\langle a, a^+ \mid aa^+ - a^+a = 1 \rangle, \tag{8.8}$$

$$W_{pq} = \mathbb{C}\langle p, q \mid p = p^+, q = q^+, pq - qp = -i \rangle. \tag{8.9}$$

The $*$-algebras W_{ca} and W_{pq} are $*$-isomorphic. A $*$-isomorphism θ of W_{ca} on W_{pq} is given by $\theta(a) = \frac{1}{\sqrt{2}}(q + ip)$ and $\theta(a^+) = \frac{1}{\sqrt{2}}(q - ip)$. (To see that θ is a $*$-isomorphism it suffices to check that the relations for a, a^+ and p, q are equivalent.) The generators a, a^+ reflect the annihilation and creation operators and p, q correspond to the momentum and position operators in quantum mechanics.

Let Aut (W_{pq}) denote the group of $*$-automorphisms of the $*$-algebra W_{pq}. It is convenient to replace $\Phi_{n,\lambda}$, $\Phi'_{n,\lambda}$ by the automorphism $\Psi_{n,\lambda}$, $\Psi'_{n,\lambda}$ of W defined by

$$\Psi_{n,\lambda}(q) = q, \quad \Psi_{n,\lambda}(p) = p + \lambda q^n, \quad \Psi'_{n,\lambda}(p) = p, \quad \Psi'_{n,\lambda}(q) = q + \lambda p^n. \tag{8.10}$$

Then $\Psi_{n,\lambda}$, $\Psi'_{n,\lambda}$ are $*$-automorphisms of W_{pq} if and only if λ is real. Each algebra automorphism Ψ of W is a composition of $\Psi_{n,\lambda}$, $\Psi'_{n,\lambda}$. By induction on the number of factors it follows that Ψ preserves the involution if and only if all factors do. Thus, Aut (W_{pq}) is generated by the $*$-automorphisms $\Psi_{n,\lambda}$, $\Psi'_{n,\lambda}$ with $n \in \mathbb{N}_0$, $\lambda \in \mathbb{R}$.

An interesting $*$-automorphism Θ of $W_{p,q}$ is given by $\Theta(p) = q$, $\Theta(q) = -p$. Its importance stems from the fact that it exchanges the two generators p and q.

In what follows we identify the $*$-algebras W_{ca} and W_{pq} by identifying $f \in W_{ca}$ with $\theta(f) \in W_{pq}$. Then, by a slight abuse of notation, we denote the corresponding $*$-algebra by W and call it the *Weyl algebra*. In the $*$-algebra W we have

$$a = \frac{1}{\sqrt{2}}(q + ip), \quad a^+ = \frac{1}{\sqrt{2}}(q - ip), \tag{8.11}$$

$$q = \frac{1}{\sqrt{2}}(a^+ + a), \quad p = \frac{i}{\sqrt{2}}(a^+ - a).$$

An important element of the Weyl algebra W is $N := a^+a$. From $aa^+ - a^+a = 1$ we easily derive the identities

$$af(N) = f(N+1)a, \quad a^+ f(N) = f(N-1)a^+, \tag{8.12}$$

$$(a^+)^{n+1} a^{n+1} = N(N-1) \cdots (N-n), \tag{8.13}$$

$$a^n (a^+)^n = (N+1) \cdots (N+n) \tag{8.14}$$

for each polynomial $f \in \mathbb{C}[x]$ and $n \in \mathbb{N}$. Clearly, (8.9) and (8.11) imply that

$$N = \frac{1}{2}(q - ip)(q + ip) = \frac{1}{2}(p^2 + q^2 - 1). \tag{8.15}$$

8.2 The Operator Equation $AA^* = A^*A + I$

First we develop a model for operators satisfying the operator relation

$$AA^* = A^*A + I. \tag{8.16}$$

We say that a densely defined closed operator A on a Hilbert space satisfies the relation (8.16) if $\mathcal{D}(A^*A) = \mathcal{D}(AA^*)$ and $AA^*\varphi = AA^*\varphi + \varphi$ for $\varphi \in \mathcal{D}(A^*A)$.

Let $\mathcal{G}$ be an auxiliary Hilbert space with inner product $(\cdot, \cdot)$ and norm $|\cdot|$. We consider the Hilbert space $l^2(\mathbb{N}_0; \mathcal{G})$ of sequences $(\varphi_n)_{n \in \mathbb{N}_0}$ of elements $\varphi_n \in \mathcal{G}$ such that $\sum_{n=0}^{\infty} |\varphi_n|^2 < \infty$ with inner product given by

$$\langle (\varphi_n), (\psi_n) \rangle = \sum_{n=0}^{\infty} (\varphi_n, \psi_n) \quad \text{for} \ (\varphi_n), (\psi_n) \in l^2(\mathbb{N}_0; \mathcal{G}).$$

Now we define two operators $A_\mathcal{G}$ and $A_\mathcal{G}^+$ on $l^2(\mathbb{N}_0; \mathcal{G})$ by

$$A_\mathcal{G}(\varphi_0, \varphi_1, \varphi_2, \dots) = (\varphi_1, \sqrt{2}\,\varphi_2, \sqrt{3}\,\varphi_3, \dots), \tag{8.17}$$

$$A_\mathcal{G}^+(\varphi_0, \varphi_1, \varphi_2, \dots) = (0, \varphi_0, \sqrt{2}\,\varphi_1, \sqrt{3}\,\varphi_2, \dots) \tag{8.18}$$

with domains

$$\mathcal{D}(A_\mathcal{G}) = \mathcal{D}(A_\mathcal{G}^+) = \{(\varphi_n) \in l^2(\mathbb{N}_0; \mathcal{G}) : (\sqrt{n}\,\varphi_n) \in l^2(\mathbb{N}_0; \mathcal{G})\}.$$

It is straightforward to verify that the operators $A_\mathcal{G}$ and $A_\mathcal{G}^+$ are adjoints of each other, hence closed, and that $A_\mathcal{G}^+ A_\mathcal{G}$ and $A_\mathcal{G} A_\mathcal{G}^+$ act on the same domain

$$\mathcal{D}(A_\mathcal{G}^+ A_\mathcal{G}) = \mathcal{D}(A_\mathcal{G} A_\mathcal{G}^+) = \{(\varphi_n) \in l^2(\mathbb{N}_0; \mathcal{G}) : (n\varphi_n) \in l^2(\mathbb{N}_0; \mathcal{G})\}$$

as diagonal operators $A_\mathcal{G}^+ A_\mathcal{G}(\varphi_n) = (n\varphi_n)$ and $A_\mathcal{G} A_\mathcal{G}^+(\varphi_n) = ((n+1)\varphi_n)$. Thus, $A_\mathcal{G} A_\mathcal{G}^+(\varphi_n) = A_\mathcal{G}^+ A_\mathcal{G}(\varphi_n) + (\varphi_n)$ for $(\varphi_n) \in \mathcal{D}(A_\mathcal{G}^+ A_\mathcal{G}) = \mathcal{D}(A_\mathcal{G} A_\mathcal{G}^+)$. That is, the densely defined closed operator $A_\mathcal{G}$ satisfies the relation (8.16).

The following result is essentially *Tillmann's theorem*. It says that, up to unitary equivalence, the operator model $A_{\mathcal{G}}$ exhausts all solutions of equation (8.16).

Theorem 8.4 *Suppose A is a densely defined closed operator on a Hilbert space $\mathcal{H}$ which satisfies the relation (8.16). Then there exist a Hilbert space $\mathcal{G}$ and a unitary operator U of $\mathcal{H}$ on $l^2(\mathbb{N}_0; \mathcal{G})$ such that $U A U^{-1} = A_{\mathcal{G}}$ and $U A^* U^{-1} = A_{\mathcal{G}}^+$. Further, A is irreducible if and only if $\dim \mathcal{G} = 1$, or equivalently, $\dim \ker A = 1$.*

Proof In this proof we use some facts from spectral theory of self-adjoint operators; see, e.g., [Sch12]. Let $C := A^*A = \int \lambda \, dE(\lambda)$ be the spectral resolution of the self-adjoint operator A^*A.

Our first aim is to show that $\gamma \in \sigma(C)$ and $\gamma > 0$ imply that $\gamma - 1 \in \sigma(C)$. Set $J_n = (\gamma - \frac{1}{n}, \gamma + \frac{1}{n})$ for $n \in \mathbb{N}$. Since $\gamma \in \sigma(C)$, there exists a sequence (f_n) of unit vectors $f_n = E(J_n) f_n$ such that $g_n := (C - \gamma) f_n \to 0$ in $\mathcal{H}$. Then

$$\|A f_n\|^2 = \langle C f_n, f_n \rangle = \langle (C - \gamma) f_n, f_n \rangle + \gamma \to \gamma.$$

Clearly, $g_n = (C - \gamma) E(J_n) f_n = E(J_n)(C - \gamma) f_n = E(J_n) g_n \in \mathcal{D}(C)$ and

$$\|A g_n\|^2 = \langle C g_n, g_n \rangle = \int_{J_n} \lambda \, d\langle E(\lambda) g_n, g_n \rangle \to 0.$$

Using relation (8.16) we derive

$$(C - (\gamma - 1)) A f_n = (A^*A + 1 - \gamma) A f_n$$
$$= (A A^* - \gamma) A f_n = A(C - \gamma) f_n = A g_n \to 0.$$

Therefore, since $\|A f_n\|^2 \to \gamma > 0$, we conclude that $\gamma - 1 \in \sigma(C)$.

Since $C \geq 0$, we have $\sigma(C) \subseteq [0, \infty)$. Hence, the preceding result implies that $\sigma(C) \subseteq \mathbb{N}_0$ and $0 \in \sigma(C)$. Since all points of $\sigma(C)$ are isolated, they are eigenvalues. Set $\mathcal{G}_n := \ker(C - nI)$ for $n \in \mathbb{N}_0$ and $\mathcal{G} := \mathcal{G}_0$.

Let $f \in \mathcal{G}_n$. Then $f \in \mathcal{D}(C^2) \subseteq \mathcal{D}(CA) \cap \mathcal{D}(CA^*)$ by (8.16), so that $Af \in \mathcal{D}(C)$ and $A^* f \in \mathcal{D}(C)$. Therefore, using Eq. (8.16) we compute

$$CA^* f = A^* A A^* f = A^*(A^*A + 1) f = A^*(C + 1) f = (n + 1) A^* f, \qquad (8.19)$$
$$CAf = A^* A A f = (A A^* - 1) A f = A C f - A f = (n - 1) A f, \quad n \geq 1. \qquad (8.20)$$

Clearly, (8.19) implies $A^* \mathcal{G}_n \subseteq \mathcal{G}_{n+1}$. If $g \in \mathcal{G}_{n+1}$, then $h := (n + 1)^{-1} A g$ belongs to $\mathcal{G}_n$ by (8.20) and $A^* h = (n + 1)^{-1} C g = g$, so $g = A^* h \in A^* \mathcal{G}_n$. Thus, $A^* \mathcal{G}_n = \mathcal{G}_{n+1}$. Hence $(A^*)^n \mathcal{G} = \mathcal{G}_n$. Similarly, $A \mathcal{G}_{n+1} = \mathcal{G}_n$ and $A \mathcal{G}_0 = \{0\}$.

For $f \in \mathcal{G}_n$, we obtain

$$\|(n + 1)^{-1/2} A^* f\|^2 = (n + 1)^{-1} \langle A^* f, A^* f \rangle = (n + 1)^{-1} \langle (C + 1) f, f \rangle = \|f\|^2.$$

Therefore, the map $f \mapsto (n + 1)^{-1/2} A^* f$ is an isometric bijection of $\mathcal{G}_n$ on $\mathcal{G}_{n+1}$. By a repeated application of this fact it follows that $f \mapsto (n!)^{-1/2} (A^*)^n f$ is an isometric map of $\mathcal{G} = \mathcal{G}_0$ on $\mathcal{G}_n = (A^*)^n \mathcal{G}$. Since $\sigma(C) = \mathbb{N}_0$, $\mathcal{H}$ is the orthogonal sum of subspaces $E(\{n\})\mathcal{H} = \mathcal{G}_n$, $n \in \mathbb{N}_0$. Hence each $f \in \mathcal{H}$ is of the form

$$f = \sum_{n=0}^{\infty} (n!)^{-1/2}(A^*)^n f_n, \quad \text{where} \quad f_n \in \mathcal{G}, \tag{8.21}$$

and $Uf = (f_n)_{n \in \mathbb{N}_0}$ defines a unitary map of the Hilbert spaces $\mathcal{H}$ and $l^2(\mathbb{N}_0; \mathcal{G})$.

From operator theory it is known that $\mathcal{D}(A) = \mathcal{D}((A^*A)^{1/2}) = \mathcal{D}(C^{1/2})$. Hence $\mathcal{D}(A)$ is the set of vectors $f \in \mathcal{H}$ such that the sequence $(\|\sqrt{n}\, E(\{n\})f\|)$ is in $l^2(\mathbb{N}_0)$, or equivalently, $(\sqrt{n}\, f_n) \in l^2(\mathbb{N}_0; \mathcal{G})$, where $Uf = (f_n)$. The latter implies that $U\mathcal{D}(A) = \mathcal{D}(A_{\mathcal{G}})$. Let $f \in \mathcal{D}(A)$ be of the form (8.21). Then we compute

$$Af = \sum_{n=0}^{\infty}(n!)^{-1/2}A(A^*)^n f_n = \sum_{n=1}^{\infty}(n!)^{-1/2}(C+1)(A^*)^{n-1} f_n$$

$$= \sum_{n=1}^{\infty}(n!)^{-1/2}n(A^*)^{n-1} f_n = \sum_{n=1}^{\infty}((n-1)!)^{-1/2}\sqrt{n}\,(A^*)^{n-1} f_n$$

$$= \sum_{n=0}^{\infty}(n!)^{-1/2}(A^*)^n \sqrt{n+1}\, f_{n+1},$$

so that $UAf = (\sqrt{n+1}\, f_{n+1}) = A_{\mathcal{G}}Uf$. Since $U\mathcal{D}(A) = \mathcal{D}(A_{\mathcal{G}})$, this yields $UAU^{-1} = A_{\mathcal{G}}$. Applying the adjoint to both sides, we get $UA^*U^{-1} = A_{\mathcal{G}}^+$.

The last assertion is derived from the equality $\dim \mathcal{G} = \dim \ker A_{\mathcal{G}} = \dim \ker A$. Indeed, if $\mathcal{G} = \mathcal{G}_1 \oplus \mathcal{G}_2$, then $A_{\mathcal{G}} = A_{\mathcal{G}_1} \oplus A_{\mathcal{G}_2}$. Hence $A_{\mathcal{G}}$, so A, is not irreducible if $\dim \mathcal{G} > 1$. Conversely, suppose $\dim \mathcal{G} = 1$ and assume to the contrary that $A_{\mathcal{G}}$ is a nontrivial direct sum $A_1 \oplus A_2$. Then each A_j satisfies (8.16). Hence, by the preceding, A_j is unitarily equivalent to some operator $A_{\mathcal{G}_j}$. Since $\dim \mathcal{G}_j = \dim \ker A_{\mathcal{G}_j}$, we obtain $\dim \mathcal{G} = \dim \mathcal{G}_1 + \dim \mathcal{G}_2 \geq 2$, a contradiction. $\qquad \square$

Clearly, if $\mathcal{G}$ is an orthogonal sum of one-dimensional Hilbert spaces $\mathcal{G}_i$, $i \in J$, then $A_{\mathcal{G}} = \oplus_{i \in J} A_{\mathcal{G}_i}$ on $l^2(\mathbb{N}_0; \mathcal{G}) = \oplus_{i \in J} l^2(\mathbb{N}_0; \mathcal{G}_i)$ and the cardinality of J is $\dim \ker A_{\mathcal{G}}$. Combined with Theorem 8.4 this fact has the following corollary.

Corollary 8.5 *Suppose A is a densely defined closed operator such that the operator relation (8.16) holds. Then A is an orthogonal direct sum of $\dim \ker A$ irreducible densely defined closed operators satisfying (8.16).*

Let us specialize to the case $\mathcal{G} = \mathbb{C}$. Then the operators $A_{\mathbb{C}}$ and $A_{\mathbb{C}}^+$ act on the standard orthonormal basis $\{e_n : n \in \mathbb{N}_0\}$ of the Hilbert space $l^2(\mathbb{N}_0)$ by

$$A_{\mathbb{C}}e_n = \sqrt{n}\, e_{n-1} \quad \text{and} \quad A_{\mathbb{C}}^+ e_n = \sqrt{n+1}\, e_{n+1}, \quad n \in \mathbb{N}_0, \quad e_{-1} := 0. \tag{8.22}$$

By Theorem 8.4, the operator $A_{\mathbb{C}}$ is, up to unitary equivalence, the only densely defined closed *irreducible* operator satisfying the operator relation (8.16).

In quantum mechanics, the operators $A_\mathbb{C}$ and $A_\mathbb{C}^+$ are called *annihilation operator* and *creation operator*, respectively, $A_\mathbb{C}^+ A_\mathbb{C}$ is the *number operator* and the vector e_0 which is annihilated by the operator $A_\mathbb{C}$ is the *vacuum vector*. Annihilation and creation operators are basic constructions in quantum physics and also in representation theory of Lie algebras (see, for instance, Sect. 9.7).

Next we develop the *Bargmann–Fock representation* of the operator relation (8.16) acting on the Fock space $\mathcal{F}(\mathbb{C})$. Let μ denote the planar Lebesgue measure on $\mathbb{C}$. The *Fock space*

$$\mathcal{F}(\mathbb{C}) := \left\{ f(z) : \text{ holomorphic on } \mathbb{C} \text{ and } \int_\mathbb{C} |f(z)|^2 e^{-|z|^2} \, d\mu(z) < \infty \right\}$$

is a Hilbert space with inner product given by

$$\langle f, g \rangle = \frac{1}{\pi} \int_\mathbb{C} f(z)\overline{g(z)}\, e^{-|z|^2} \, d\mu(z), \quad f, g \in \mathcal{F}(\mathbb{C}). \tag{8.23}$$

Proposition 8.6 $\left\{ \eta_k(z) := \frac{1}{\sqrt{k!}} z^k : k \in \mathbb{N}_0 \right\}$ *is an orthonormal basis of* $\mathcal{F}(\mathbb{C})$.

Proof For $k, n \in \mathbb{N}_0$ we compute

$$\langle z^k, z^n \rangle = \frac{1}{\pi} \int_\mathbb{C} z^k \bar{z}^n e^{-|z|^2} \, d\mu(z) = \frac{1}{\pi} \int_0^\infty \int_0^{2\pi} e^{i\theta(k-n)} r^{k+n} e^{-r^2} r \, d\theta dr$$

$$= \delta_{k,n} 2 \int_0^\infty r^{2k} e^{-r^2} \, r dr = \delta_{k,n} 2 \int_0^\infty s^k e^{-s} \frac{1}{2} ds = \delta_{k,n} k! \, .$$

Hence $\{\eta_k(z) : k \in \mathbb{N}_0\}$ is orthonormal.

To prove that this set is complete, take $f \in \mathcal{F}(\mathbb{C})$ and let $f(z) = \sum_{n=0}^\infty a_n z^n$ be its Taylor expansion at zero. This series converges uniformly on each ball and by the preceding computation we have $\int_{|z| \leq R} \eta_n \overline{\eta_k} \, d\mu = 0$ for $k \neq n$ and $R > 0$. Therefore,

$$\frac{1}{\pi} \int_{|z| \leq R} f(z)\, \overline{\eta_k(z)} \, d\mu(z) = \frac{1}{\pi} \int_{|z| \leq R} \left(\sum_{n=0}^\infty a_n z^n \right) \overline{\eta_k(z)} \, d\mu(z)$$

$$= \sum_{n=0}^\infty \frac{a_n}{\pi} \sqrt{n!} \int_{|z| \leq R} \eta_n(z)\overline{\eta_k(z)} \, d\mu(z) = \frac{a_k}{\pi} \sqrt{k!} \int_{|z| \leq R} |\eta_k(z)|^2 d\mu(z).$$

Letting $R \to \infty$, we get $\langle f, \eta_k \rangle = \pi^{-1} a_k \sqrt{k!}$. Thus, $f \perp \eta_k$ for all k implies $a_k = 0$ for all k and $f = 0$. Hence $\{\eta_k\}$ is complete and an orthonormal basis of the Hilbert space $\mathcal{F}(\mathbb{C})$. Moreover, we have proved the expansion $f = \sum_{k=0}^\infty \pi^{-1} a_k \sqrt{k!} \, \eta_k$ of $f \in \mathcal{F}(\mathbb{C})$ with respect to the orthonormal basis $\{\eta_k\}$. $\qquad \square$

The next proposition says that the function $K(z, w) := e^{z\overline{w}}$ is a reproducing kernel of the Fock space $\mathcal{F}(\mathbb{C})$. The functions

$$\varphi_w(z) := K(z, w) = e^{z\overline{w}}, \quad w \in \mathbb{C}, \tag{8.24}$$

of the Fock space $\mathcal{F}(\mathbb{C})$ are called *coherent states*; see, e.g., [Ha13, p. 299].

Proposition 8.7 *For $f \in \mathcal{F}(\mathbb{C})$ and $w \in \mathbb{C}$ we have $\langle f, \varphi_w \rangle = f(w)$. In particular,*

$$\langle \varphi_z, \varphi_w \rangle = K(w, z) = e^{w\overline{z}} \quad \text{for} \quad z, w \in \mathbb{C} \tag{8.25}$$

and the set of functions φ_w, $w \in \mathbb{C}$, is total in $\mathcal{F}(\mathbb{C})$.

Proof By Proposition 8.6, $\{\eta_k : k \in \mathbb{N}_0\}$ is an orthonormal basis of the Hilbert space $\mathcal{F}(\mathbb{C})$. The corresponding expansion of $\varphi_w \in \mathcal{F}(\mathbb{C})$ is

$$\varphi_w(z) = \sum_{k=0}^{\infty} \frac{1}{k!}(z\overline{w})^k = \sum_{k=0}^{\infty} \overline{\eta_k(w)}\, \eta_k(z).$$

Let $f(z) = \sum_{k=0}^{\infty} c_k \eta_k(z)$ be the expansion of $f \in \mathcal{F}(\mathbb{C})$. The Parseval identity for the expansions of f and φ_w gives

$$\langle f, \varphi_w \rangle = \sum_{k=0}^{\infty} c_k \overline{\overline{\eta_k(w)}} = \sum_{k=0}^{\infty} c_k \eta_k(w) = f(w).$$

In particular, setting $f = \varphi_z$, we obtain $\langle \varphi_z, \varphi_w \rangle = \varphi_z(w) = K(w, z)$.

Now let $f \in \mathcal{F}(\mathbb{C})$ and suppose $f \perp \varphi_w$ for all $w \in \mathbb{C}$. Then $0 = \langle f, \varphi_w \rangle = f(w)$ for $w \in \mathbb{C}$, so $f = 0$. Hence the span of functions φ_w, $w \in \mathbb{C}$, is dense in $\mathcal{F}(\mathbb{C})$. □

Next we consider the operator $\frac{\partial}{\partial z}$ and the multiplication operator z on the Fock space $\mathcal{F}(\mathbb{C})$. They act on the orthonormal basis $\{\eta_n = (\sqrt{n!})^{-1}z^n\}$ of $\mathcal{F}(\mathbb{C})$ by

$$\frac{\partial}{\partial z}\eta_n(z) = \sqrt{n}\,\eta_{n-1}(z) \quad \text{and} \quad z\eta_n(z) = \sqrt{n+1}\,\eta_{n+1}(z), \quad n \in \mathbb{N}_0, \tag{8.26}$$

where we set $\eta_{-1} := 0$. Comparing (8.22) and (8.26) it follows that the operator pair $\{\frac{\partial}{\partial z}, z\}$ on $\mathcal{F}(\mathbb{C})$ is unitarily equivalent to the operator pair $\{A_{\mathbb{C}}, A_{\mathbb{C}}^+\}$ on $l^2(\mathbb{N}_0)$. Therefore, we have the (at first glance surprising) results that the operators $\frac{\partial}{\partial z}$ and z on the Fock space $\mathcal{F}(\mathbb{C})$ are adjoints of each other and that *the operator $A = \frac{\partial}{\partial z}$ on $\mathcal{F}(\mathbb{C})$ is, up to unitary equivalence, the unique closed irreducible operator satisfying the operator relation* (8.16)!

8.3   The Bargmann–Fock Representation of the Weyl Algebra

In this section we use form W_{ca}, defined by (8.8), of the Weyl algebra W.

Let us return to the operators $A_{\mathcal{G}}$ and $A_{\mathcal{G}}^+$ given by (8.17) and (8.18) and define

$$\mathcal{D}(\pi_{\mathcal{G}}) := \left\{ (\varphi_n) \in l^2(\mathbb{N}_0; \mathcal{G}) : (n^k \varphi_n) \in l^2(\mathbb{N}_0; \mathcal{G}) \text{ for } k \in \mathbb{N} \right\}. \tag{8.27}$$

Obviously, $A_{\mathcal{G}}$ and $A_{\mathcal{G}}^+$ leave $\mathcal{D}(\pi_{\mathcal{G}})$ invariant. Therefore, since $A_{\mathcal{G}}$ satisfies (8.16), there is a $*$-representation $\pi_{\mathcal{G}}$ of the $*$-algebra W on $\mathcal{D}(\pi_{\mathcal{G}})$ defined by $\pi_{\mathcal{G}}(1) = I$,

$$\pi_{\mathcal{G}}(a) = A_{\mathcal{G}} {\restriction} \mathcal{D}(\pi_{\mathcal{G}}), \quad \pi_{\mathcal{G}}(a^+) = A_{\mathcal{G}}^+ {\restriction} \mathcal{D}(\pi_{\mathcal{G}}).$$

Clearly,

$$\pi_{\mathcal{G}}(a^+ a)(\varphi_n) = \pi_{\mathcal{G}}(N)(\varphi_n) = (n \varphi_n) \quad \text{for } (\varphi_n) \in \mathcal{D}(\pi_{\mathcal{G}}). \tag{8.28}$$

Hence the operator $\pi_{\mathcal{G}}(N^k)$ for $k \in \mathbb{N}$ is self-adjoint and the diagonal operator with diagonal sequence $(n^k)_{n \in \mathbb{N}_0}$. This implies that $\mathcal{D}(\pi_{\mathcal{G}}) = \mathcal{D}((\pi_{\mathcal{G}})^*)$, so the $*$-representation $\pi_{\mathcal{G}}$ is *self-adjoint*. The closures of $\pi_{\mathcal{G}}(a)$ and $\pi_{\mathcal{G}}(a^+)$ are $A_{\mathcal{G}}$ and $(A_{\mathcal{G}})^* = A_{\mathcal{G}}^+$, respectively. Therefore, since $\pi_{\mathcal{G}}(1) = I$, it is easily verified that for the $*$-representation $\pi := \pi_{\mathcal{G}}$ of W we have

$$\overline{\pi(a)} \, \pi(a)^* = \pi(a)^* \, \overline{\pi(a)} + I. \tag{8.29}$$

Theorem 8.8 *Suppose π is a $*$-representation of the $*$-algebra W such that $\pi(1) = I$ and the operator $\pi(a^+ a)$ is essentially self-adjoint. Then (8.29) holds and there exists a Hilbert space $\mathcal{G}$ such that π is unitarily equivalent to a subrepresentation of $\pi_{\mathcal{G}}$ acting on a dense domain of $\mathcal{H}(\pi) \cong l^2(\mathbb{N}_0; \mathcal{G})$.*

Proof From $\pi(a^+ a) = \pi(a^+) \pi(a) \subseteq \pi(a)^* \overline{\pi(a)}$ we get $\overline{\pi(a^+ a)} \subseteq \pi(a)^* \overline{\pi(a)}$. By assumption, the operator $\pi(a^+ a)$ is self-adjoint. Since $\pi(a)^* \overline{\pi(a)}$ is also self-adjoint, it follows that $\overline{\pi(a^+ a)} = \pi(a)^* \overline{\pi(a)}$. Hence

$$\overline{\pi(a a^+)} = \overline{\pi(a^+ a + 1)} = \overline{\pi(a^+ a) + I} = \overline{\pi(a^+ a)} + I = \pi(a)^* \overline{\pi(a)} + I$$

is self-adjoint as well. On the other hand, $\overline{\pi(a a^+)}$ is a restriction of the self-adjoint operator $\overline{\pi(a)} \, \pi(a)^*$. Therefore, $\overline{\pi(a a^+)} = \overline{\pi(a)} \, \pi(a)^*$. By comparing both expressions for $\overline{\pi(a a^+)}$ we conclude that (8.29) holds.

Equation (8.29) means that the densely defined closed operator $A := \overline{\pi(a)}$ satisfies the relation (8.16). Therefore, by Theorem 8.4, A is, up to unitary equivalence, of the form $A_{\mathcal{G}}$ on $\mathcal{H}(\pi) \cong l^2(\mathbb{N}_0; \mathcal{G})$. This implies that π is a subrepresentation of the corresponding $*$-representation $\pi_{\mathcal{G}}$. $\qquad\square$

The next theorem contains several characterizations of representations $\pi_{\mathcal{G}}$.

Theorem 8.9 *Suppose that π is a self-adjoint representation of the $*$-algebra* W *such that $\pi(1) = I$. The following are equivalent:*

(i) *There exists a Hilbert space $\mathcal{G}$ such that π is unitarily equivalent to $\pi_{\mathcal{G}}$.*
(ii) *The operator relation (8.29) holds.*
(iii) *The symmetric operator $\pi(N) = \pi(a^+a)$ is essentially self-adjoint.*

Suppose that these conditions are satisfied. Then π is unitarily equivalent to a direct sum of $*$-representations $\pi_{\mathbb{C}}$. In particular, π is irreducible if and only if it is unitarily equivalent to $\pi_{\mathbb{C}}$.

Proof Since $\pi_{\mathcal{G}}(N)$ is essentially self-adjoint, (i)$\to$(iii). By the proof of Theorem 8.8, we have (iii)$\to$(ii) and (ii) implies that π is a subrepresentation of $\pi_{\mathcal{G}}$ on $\mathcal{H}(\pi) \cong l^2(\mathbb{N}_0; \mathcal{G})$. Since π is self-adjoint, it has no proper extension to a $*$-representation on the same Hilbert space (Corollary 4.31). Hence $\pi = \pi_{\mathcal{G}}$, so we have (ii)$\to$(i). This proves that conditions (i)–(iii) are equivalent. Corollary 8.5 and Theorem 8.4 imply that $\pi_{\mathcal{G}}$ is unitarily equivalent to a direct sum of representations $\pi_{\mathbb{C}}$ and $\pi_{\mathbb{C}}$ is irreducible. This gives the last assertion. $\qquad\square$

Thus, $\pi_{\mathbb{C}}$ is the unique (up to unitary equivalence) *irreducible* self-adjoint $*$-representation π of W for which the operator $\overline{\pi(N)}$ is self-adjoint.

Let $e := (1, 0, 0, \ldots) \in l^2(\mathbb{N}_0)$. It is easily seen that $\pi_{\mathbb{C}}(W)e$ is the space of finite vectors $\varphi = (\varphi_0, \ldots, \varphi_n, 0, \ldots)$ and a core for $\pi_{\mathbb{C}}(N^k)$, $k \in \mathbb{N}_0$, and that the norms $\|\pi_{\mathbb{C}}(N^k) \cdot \|, k \subset \mathbb{N}_0$, generate the graph topology of $\pi_{\mathbb{C}}$. Therefore, $\pi_{\mathbb{C}}(W)e$ is dense in $\mathcal{D}(\pi_{\mathcal{G}})$ in the graph topology $\mathsf{t}_{\pi_{\mathbb{C}}} >_{\pi_{\mathbb{C}}}$. Hence e is a *cyclic vector* for $\pi_{\mathbb{C}}$. It is called the *vacuum vector*.

As noted above, the pair $\{A_{\mathbb{C}}, A_{\mathbb{C}}^+\}$ on $l^2(\mathbb{N}_0)$ is unitarily equivalent to the pair $\{\frac{\partial}{\partial z}, z\}$ on $\mathcal{F}(\mathbb{C})$ and the equivalence is given by the unitary V of $l^2(\mathbb{N}_0)$ on $\mathcal{F}(\mathbb{C})$ defined by $V e_n = \eta_n, n \in \mathbb{N}_0$. Hence $\pi_{\mathcal{F}}(\cdot) := V \pi_{\mathbb{C}}(\cdot) V^{-1}$ defines a unitarily equivalent $*$-representation $\pi_{\mathcal{F}}$ on the Fock space $\mathcal{F}(\mathbb{C})$ such that $\pi(1) = I$,

$$\pi_{\mathcal{F}}(a)\xi = \frac{\partial \xi}{\partial z} \quad \text{and} \quad \pi_{\mathcal{F}}(a^+)\xi = z\xi \quad \text{for} \quad \xi \in \mathcal{D}(\pi_{\mathcal{F}}) := V\mathcal{D}(\pi_{\mathbb{C}}).$$

Definition 8.10 The $*$-representation $\pi_{\mathbb{C}}$ on $l^2(\mathbb{N}_0)$, and its unitarily equivalent $*$-representation $\pi_{\mathcal{F}}$ on $\mathcal{F}(\mathbb{C})$, are called the *Bargmann–Fock representation* of the Weyl algebra W. The state $\langle \pi_{\mathbb{C}}(\cdot)e, e \rangle$ on W is the *Fock state* or the *vacuum state*.

8.4   The Schrödinger Representation of the Weyl Algebra

In this section, we mainly use the version W_{pq} (see (8.9)) of the Weyl algebra W.

Clearly, the self-adjoint operators $P := -i\frac{d}{dx}$ and $Q := x$, the multiplication operator by the variable x, on the Hilbert space $L^2(\mathbb{R})$ satisfy the relation

$$PQ\varphi - QP\varphi = -i\varphi \tag{8.30}$$

for $\varphi \in \mathcal{D}(PQ) \cap \mathcal{D}(QP)$. In quantum physics, Q is the *position operator* and P is the *moment operator*. This operator pair $\{P, Q\}$ is called the *Schrödinger pair*. Note that the Schrödinger pair is irreducible, as proved in Example 4.32.

The operators P, Q give rise to the *Schrödinger representation* of W. This is the $*$-representation π_S of W on the Hilbert space $L^2(\mathbb{R})$ defined by $\pi_S(1) = I$,

$$\pi_S(p) = P{\upharpoonright}\mathcal{D}(\pi_S), \quad \pi_S(q) = Q{\upharpoonright}\mathcal{D}(\pi_S), \quad \text{with} \ \mathcal{D}(\pi_S) := \mathcal{S}(\mathbb{R}). \tag{8.31}$$

The Schrödinger representation π_S is faithful (by Proposition 8.2(iv)), and it has the important property that $*$-automorphisms of the Weyl algebra W are implemented by unitary operators on $L^2(\mathbb{R})$.

Example 8.11 Recall that there is a $*$-isomorphism Θ of W such that $\Theta(p) = q$ and $\Theta(q) = -p$. The Fourier transform $\mathcal{F}$ on $\mathbb{R}$ satisfies $\mathcal{F}P\mathcal{F}^{-1} = Q$, $\mathcal{F}Q\mathcal{F}^{-1} = -P$, and $\mathcal{F}\mathcal{S}(\mathbb{R}) = \mathcal{S}(\mathbb{R})$. Hence it follows that

$$\pi_S(\Theta(f)) = \mathcal{F}\pi_S(f)\mathcal{F}^{-1} \ \text{for} \ f \in W. \hspace{2cm} \bigcirc$$

Proposition 8.12 *For each $*$-automorphism Ψ of the $*$-algebra W there exists a unitary operator U on $L^2(\mathbb{R})$ such that $U\mathcal{S}(\mathbb{R}) = \mathcal{S}(\mathbb{R})$ and*

$$\pi_S(\Psi(f)) = U\pi_S(f)U^{-1} \ \text{for} \ f \in W. \tag{8.32}$$

Proof As noted in Sect. 8.1, Ψ is a composition of $*$-automorphisms $\Psi_{n,\lambda}$, $\Psi'_{n,\lambda}$, where $n \in \mathbb{N}_0$, $\lambda \in \mathbb{R}$. Hence it suffices to prove the assertion for $\Psi_{n,\lambda}$ and $\Psi'_{n,\lambda}$.

We carry out the proof for $\Psi_{n,\lambda}$. Recall that $\Psi_{n,\lambda}(q) = q$, $\Psi_{n,\lambda}(p) = p + \lambda q^n$. Let U be the multiplication operator by the function $\exp(-i\lambda(n+1)^{-1}x^{n+1})$. Clearly, U is a unitary operator on $L^2(\mathbb{R})$ such that U and U^{-1} leave $\mathcal{S}(\mathbb{R})$ invariant. Hence $U\mathcal{S}(\mathbb{R}) = \mathcal{S}(\mathbb{R})$. Let $\varphi \in \mathcal{S}(\mathbb{R})$. From

$$P\exp(i\lambda(n+1)^{-1}x^{n+1})\varphi$$
$$= -i \cdot i\exp(i\lambda(n+1)^{-1}x^{n+1})\lambda x^n \varphi + \exp(i\lambda(n+1)^{-1}x^{n+1})P\varphi$$

we obtain $PU^{-1}\varphi = U^{-1}(\lambda Q^n + P)\varphi$, so that $UPU^{-1}\varphi = \lambda Q^n\varphi + P\varphi$. This gives $U\pi_S(p)U^{-1}\varphi = \pi_S(p + \lambda q^n)\varphi = \pi_S(\Psi_{n,\lambda}(p))\varphi$. Obviously, we have $U\pi_S(q)U^{-1}\varphi = \pi_S(q)\varphi = \pi_S(\Psi_{n,\lambda}(q))\varphi$. Thus (8.32) holds for $f = p, q$. Since both sides of (8.32) are algebra homomorphism, (8.32) is valid for all $f \in W$.

Using the identity $\Psi'_{n,\lambda} = \Theta \circ \Psi_{n,\lambda(-1)^n} \circ \Theta^{-1}$ the proof for $\Psi'_{n,\lambda}$ follows from the preceding combined with Example 8.11. □

Next we show that the $*$-representations π_C and π_S are unitarily equivalent. For this we need the Hermite functions. First we recall the *Hermite polynomials*

$$h_n(x) := (-1)^n e^{x^2} \frac{d^n}{dx^n} e^{-x^2}, n \in \mathbb{N}_0, \quad \text{where } h_0(x) := 1.$$

It is easily verified that h_n is a polynomial of degree n,

$$h_n(x) = 2xh_{n-1}(x) - h'_{n-1}(x) \text{ and } h'_n(x) = 2nh_{n-1}(x), \quad n \in \mathbb{N}. \tag{8.33}$$

Then the *Hermite functions* H_n are given by

$$\widetilde{H}_n(x) := h_n(x)e^{-x^2/2}, \quad H_n(x) := (\sqrt[4]{\pi} \sqrt{2^n n!})^{-1} \widetilde{H}_n(x), \, n \in \mathbb{N}_0.$$

It is well known that the functions $H_n, n \in \mathbb{N}_0$, form an orthonormal basis of the Hilbert space $L^2(\mathbb{R})$. The constant $c_n := (\sqrt[4]{\pi} \sqrt{2^n n!})^{-1}$ ensures that $\| H_n \| = 1$. In fact, the functions H_n are obtained by applying the Gram–Schmidt orthogonalization procedure to the functions $x^n e^{-x^2/2}$ in $L^2(\mathbb{R})$. Then, for $n \in \mathbb{N}_0$,

$$\frac{1}{\sqrt{2}}\left(x + \frac{d}{dx}\right) H_n(x) = \sqrt{n}\, H_{n-1}(x), \quad H_{-1} := 0, \tag{8.34}$$

$$\frac{1}{\sqrt{2}}\left(x - \frac{d}{dx}\right) H_n(x) = \sqrt{n+1}\, H_{n+1}(x), \tag{8.35}$$

$$\left(x^2 - \frac{d^2}{dx^2}\right) H_n(x) = (2n+1) H_n(x). \tag{8.36}$$

Indeed, using (8.33) and the definition of $\widetilde{H}_n$ we compute $(x + \frac{d}{dx})\widetilde{H}_n = 2n\widetilde{H}_{n-1}$. Multiplying by the constant $c_n = (\sqrt{2n})^{-1}c_{n-1}$ this yields (8.34). Using the relation $x - \frac{d}{dx} \subseteq (x + \frac{d}{dx})^*$ formula (8.34) implies (8.35). Equation (8.36) follows by combining (8.34), (8.35) and the identity $x^2 - \frac{d^2}{dx^2} = (x - \frac{d}{dx})(x + \frac{d}{dx}) + 1$.

Proposition 8.13 *The unitary operator U of $L^2(\mathbb{R})$ on $l^2(\mathbb{N}_0)$ given by $U H_n = e_n$, $n \in \mathbb{N}_0$, provides a unitary equivalence of the $*$-representations π_S and π_C, that is,*

$$U\pi_S(f)U^{-1} = \pi_C(f) \quad \text{for} \quad f \in W. \tag{8.37}$$

Each representation π_G is unitarily equivalent to a direct sum of Schrödinger representations π_S.

Proof Equations (8.34) and (8.35) give

$$U\frac{1}{\sqrt{2}}(Q+iP)U^{-1}e_n = A_{\mathbb{C}}e_n \quad \text{and} \quad U\frac{1}{\sqrt{2}}(Q-iP)U^{-1}e_n = A_{\mathbb{C}}^{+}e_n. \quad (8.38)$$

Clearly, $H_n(x) \in \mathcal{S}(\mathbb{R})$ for $n \in \mathbb{N}_0$. It is well known [RS72, V. 3] that the Schwartz space $\mathcal{S}(\mathbb{R})$ is precisely the set of functions $\varphi \in L^2(\mathbb{R})$ whose Fourier coefficients with respect to the orthonormal basis $\{H_n\}$ are in $\mathcal{D}(\pi_{\mathbb{C}})$. Hence $UD(\pi_S) = \mathcal{D}(\pi_{\mathbb{C}})$ and it follows from (8.11) and (8.38) that $U\pi_S(a)U^{-1}\eta = \pi_{\mathbb{C}}(a)\eta$ and $U\pi_S(a^+)U^{-1}\eta = \pi_{\mathbb{C}}(a^+)\eta$ for $\eta \in \mathcal{D}(\pi_{\mathbb{C}})$. Since $U\pi_S(\cdot)U^{-1}$ and $\pi_{\mathbb{C}}(\cdot)$ are algebra homomorphisms, this implies (8.37).

The last assertion follows from (8.37) combined with the fact that $\pi_{\mathcal{G}}$ is unitarily equivalent to a direct sum of representations $\pi_{\mathbb{C}}$ by Theorem 8.9. $\square$

Let $V : l_2(\mathbb{N}_0) \mapsto \mathcal{F}(\mathbb{C})$ and $U : L^2(\mathbb{R}) \mapsto l_2(\mathbb{N}_0)$ be the unitaries given by $Ve_n = \eta_n$ and $UH_n = e_n$, $n \in \mathbb{N}_0$. Then, as noted above, $V\pi_{\mathbb{C}}V^{-1} = \pi_{\mathcal{F}}$ and $U\pi_S U^{-1} = \pi_{\mathbb{C}}$. Hence the unitary $VU : L^2(\mathbb{R}) \mapsto \mathcal{F}(\mathbb{C})$ satisfies $VUH_n = \eta_n$, $n \in \mathbb{N}_0$, and it provides the unitary equivalence of the $*$-representations π_S and $\pi_{\mathcal{F}}$ of W, that is, $VU\pi_S(VU)^{-1} = \pi_{\mathcal{F}}$.

We will show that this unitary VU is the *Segal–Bargmann transform* $\mathcal{B}$:

$$(\mathcal{B}f)(z) := \frac{1}{\sqrt[4]{\pi}}\int_{\mathbb{R}}\exp\left(-\frac{x^2+z^2}{2}+\sqrt{2}zx\right)f(x)\,dx, \quad f \in L^2(\mathbb{R}). \quad (8.39)$$

That is, by this definition, $(\mathcal{B}f)(z) = \langle f(x), \psi_z(x)\rangle_{L^2(\mathbb{R})}$, where we abbreviate

$$\psi_z(x) := \pi^{-1/4}e^{\left(-\frac{x^2+\bar{z}^2}{2}+\sqrt{2}\bar{z}x\right)}, \quad z \in \mathbb{C}. \quad (8.40)$$

Proposition 8.14 *The set of functions ψ_z, $z \in \mathbb{C}$, is total in $L^2(\mathbb{R})$, and we have*

$$\langle \psi_z, \psi_w\rangle_{L^2(\mathbb{R})} = e^{w\bar{z}} \quad \text{for } z, w \in \mathbb{C}. \quad (8.41)$$

Proof Let $f \in L^2(\mathbb{R})$. Suppose $f \perp \psi_z$ for all $z \in \mathbb{C}$. Setting $z = -it$, $t \in \mathbb{R}$, it follows that the Fourier transform of the function $e^{-x^2/2}f$ is zero. Hence $e^{-x^2/2}f = 0$ and so $f = 0$. This proves that the functions ψ_z, $z \in \mathbb{C}$, are total in $L^2(\mathbb{R})$.

From complex analysis we recall that $\int_{\mathbb{R}}e^{-(x-v)^2}dx = \int_{\mathbb{R}}e^{-x^2}dx = \sqrt{\pi}$ for each number $v \in \mathbb{C}$. Using this fact we compute for $z, w \in \mathbb{C}$,

$$\langle \psi_z, \psi_w\rangle_{L^2(\mathbb{R})} = \pi^{-1/2}\int_{\mathbb{R}}e^{-\frac{x^2+\bar{z}^2}{2}+\sqrt{2}\bar{z}x}\,e^{-\frac{x^2+w^2}{2}+\sqrt{2}wx}\,dx$$

$$= \pi^{-1/2}e^{w\bar{z}}\int_{\mathbb{R}}e^{-\left(x+\frac{w+\bar{z}}{\sqrt{2}}\right)^2}dx = e^{w\bar{z}}. \qquad \square$$

Combining (8.25) and (8.41) yields $\langle \varphi_z, \varphi_w\rangle_{\mathcal{F}(\mathbb{C})} = \langle \psi_z, \psi_w\rangle_{L^2(\mathbb{R})}$, $z, w \in \mathbb{C}$. Therefore, since the sets $\{\varphi_z : z \in \mathbb{C}\}$ and $\{\psi_z : z \in \mathbb{C}\}$ are total in $\mathcal{F}(\mathbb{C})$ and $L^2(\mathbb{R})$

by Propositions 8.7 and 8.14, there exists a unique unitary operator W of $L^2(\mathbb{R})$ on $\mathcal{F}(\mathbb{C})$ such that $W\psi_z = \varphi_z$, $z \in \mathbb{C}$. For $f \in L^2(\mathbb{R})$, by Proposition 8.7,

$$(Wf)(z) = \langle Wf, \varphi_z \rangle_{\mathcal{F}(\mathbb{C})} = \langle f, W^{-1}\varphi_z \rangle_{L^2(\mathbb{R})} = \langle f, \psi_z \rangle_{L^2(\mathbb{R})} = (\mathcal{B}f)(z), z \in \mathbb{C}.$$

That is, this unitary operator W is just the Segal–Bargmann transform $\mathcal{B}$.

Theorem 8.15 *The Segal–Bargmann transform $\mathcal{B}$ is a unitary operator of $L^2(\mathbb{R})$ on the Fock space $\mathcal{F}(\mathbb{C})$ such that $\mathcal{B}\psi_z = \varphi_z$ for $z \in \mathbb{C}$ and $\mathcal{B}H_n = \eta_n$ for $n \in \mathbb{N}_0$.*

Proof By the preceding, it only remains to show that $\mathcal{B}H_n = \eta_n$.

Clearly, $\frac{d^n}{dt^n}e^{-(x-t)^2}\big|_{t=0} = (-1)^n \frac{d^n}{dx^n}e^{-x^2}$. Therefore, using the Taylor expansion of the function $t \mapsto e^{-(x-t)^2}$ at $t = 0$ we derive

$$e^{-(x^2/2+2tx+t^2)} = e^{x^2/2}e^{-(x-t)^2} = \sum_{n=0}^{\infty} \frac{t^n}{n!}(-1)^n e^{x^2/2}\frac{d^n}{dx^n}e^{-x^2} = \sum_{n=0}^{\infty} \frac{t^n}{n!}\tilde{H}_n(x).$$

Setting $t = \frac{z}{\sqrt{2}}$ and using the equality $H_n(x) = (\sqrt[4]{\pi}\sqrt{2^n n!})^{-1}\tilde{H}_n(x)$ we obtain

$$\pi^{-1/4}e^{\left(-\frac{x^2+z^2}{2}+\sqrt{2}zx\right)} = \pi^{-1/4}\sum_{n=0}^{\infty}\frac{z^n}{\sqrt{2^n\,n!}}\tilde{H}_n(x) = \sum_{n=0}^{\infty}\frac{z^n}{\sqrt{n!}}H_n(x). \qquad (8.42)$$

Fix $z \in \mathbb{C}$. The function of x on the left-hand side is $\psi_{\bar{z}}(x)$, see (8.40), and Eq. (8.42) shows that it has the Fourier coefficients $\eta_n(z) = \frac{z^n}{\sqrt{n!}}$, $n \in \mathbb{N}_0$, in the expansion with respect to the orthonormal basis $\{H_n\}$ of $L^2(\mathbb{R})$. Therefore, since $H_n(x)$ is real on $\mathbb{R}$, we have $\eta_n(z) = \langle \psi_{\bar{z}}, H_n \rangle = \langle H_n, \psi_z \rangle = (\mathcal{B}H_n)(z)$. $\qquad \square$

Thus, $\mathcal{B}$ is the unitary VU of $L^2(\mathbb{R})$ on $\mathcal{F}(\mathbb{C})$ which gives the unitary equivalence of the *-representations π_S and $\pi_{\mathcal{F}}$ and of the corresponding operators. That is,

$$\mathcal{B}^{-1}\pi_{\mathcal{F}}\mathcal{B} = \pi_S, \quad \mathcal{B}^{-1}z\mathcal{B} = \frac{1}{\sqrt{2}}\left(x - \frac{d}{dx}\right), \quad \mathcal{B}^{-1}\frac{d}{dz}\mathcal{B} = \frac{1}{\sqrt{2}}\left(x + \frac{d}{dx}\right).$$

Since $\mathcal{B}$ is unitary, its inverse is equal to its adjoint. Therefore, since $\mathcal{B}$ is an integral operator with kernel $\psi_{\bar{z}}(x)$, see (8.39) and (8.40), $\mathcal{B}^{-1}$ is an integral operator with adjoint kernel $\psi_z(x)$. Thus, $(\mathcal{B}^{-1}g)(x) = \langle g(z), \psi_{\bar{z}}(x) \rangle_{\mathcal{F}(\mathbb{C})}$, so that

$$(\mathcal{B}^{-1}g)(x) := \pi^{-1}\pi^{-1/4}\int_{\mathbb{C}} e^{\left(-\frac{x^2+z^2}{2}+\sqrt{2}\bar{z}x\right)}e^{-|z|^2}g(z)\,d\mu(z), \quad g \in \mathcal{F}(\mathbb{C}).$$

For our next result we need the following preliminary lemma.

Lemma 8.16 *Let P and Q be symmetric operators on a Hilbert space $\mathcal{H}$. Suppose there exists a linear subspace $\mathcal{D} \subseteq \mathcal{D}(P^2) \cap \mathcal{D}(Q^2) \cap \mathcal{D}(PQ) \cap \mathcal{D}(QP)$ such that (8.30) holds on $\mathcal{D}$, that is, $PQ\varphi - QP\varphi = -i\varphi$ for $\varphi \in \mathcal{D}$. Then, if the operator $(P^2 + Q^2)\restriction\mathcal{D}$ is essentially self-adjoint, so are the operators $P\restriction\mathcal{D}$ and $Q\restriction\mathcal{D}$.*

Proof Fix $c \in \mathbb{R}$, $|c| > 1$, and set $T := P^2 + Q^2 + I$. Let $\xi \in \mathcal{H}$ be such that $\xi \perp (P + c\,\mathrm{i})\mathcal{D}$. Our aim is to prove that $\xi = 0$.

Since $(P^2 + Q^2){\upharpoonright}\mathcal{D}$ is positive and essentially self-adjoint, $T\mathcal{D}$ is dense by Proposition A.1(iv). Thus there exists a sequence (φ_n) of vectors $\varphi_n \in \mathcal{D}$ such that $\lim_n T\varphi_n = \xi$. Clearly, $\|T\varphi\|^2 \geq \|P\varphi\|^2 + \|\varphi\|^2$ for $\varphi \in \mathcal{D}$. Hence both sequences (φ_n) and $(P\varphi_n)$ are Cauchy sequences. Since the operators P and T are closable, there is a vector $\psi \in \mathcal{D}(\overline{P}) \cap \mathcal{D}(\overline{T})$ such that $\lim_n \varphi_n = \psi$, $\lim_n P\varphi_n = \overline{P}\,\psi$, and $\lim_n T\varphi_n = \xi = \overline{T}\psi$. Then, $\lim_n(P + c\,\mathrm{i})\varphi_n = (\overline{P} + c\,\mathrm{i})\psi$, hence $\xi \perp (\overline{P} + c\,\mathrm{i})\psi$.

Let $\varphi \in \mathcal{D}$. Using the symmetry of P, Q and the relation (8.30) we compute

$$\langle P\varphi, T\varphi \rangle - \langle T\varphi, P\varphi \rangle = \langle P\varphi, (P^2 + Q^2 + I)\varphi \rangle - \langle (P^2 + Q^2 + I)\varphi, P\varphi \rangle$$
$$= \langle P\varphi, P^2\varphi \rangle + \langle P\varphi, Q^2\varphi \rangle + \langle P\varphi, \varphi \rangle - \langle P^2\varphi, P\varphi \rangle - \langle Q^2\varphi, P\varphi \rangle - \langle \varphi, P\varphi \rangle$$
$$= \langle QP\varphi, Q\varphi \rangle - \langle Q\varphi, QP\varphi \rangle = \langle (PQ + \mathrm{i})\varphi, Q\varphi \rangle - \langle Q\varphi, QP\varphi \rangle$$
$$= \mathrm{i}\langle \varphi, Q\varphi \rangle + \langle Q\varphi, PQ\varphi \rangle - \langle Q\varphi, QP\varphi \rangle = 2\,\mathrm{i}\langle \varphi, Q\varphi \rangle.$$

Hence, since $|\langle \varphi, Q\varphi \rangle| \leq \|\varphi\|^2 + \|Q\varphi\|^2 \leq \langle \varphi, T\varphi \rangle$ for $\varphi \in \mathcal{D}$, it follows that

$$|\langle (P + c\,\mathrm{i})\varphi_n, T\varphi_n \rangle - \langle T\varphi_n, (P + c\,\mathrm{i})\varphi_n \rangle - 2c\,\mathrm{i}\langle \varphi_n, T\varphi_n \rangle|$$
$$= |\langle P\varphi_n, T\varphi_n \rangle - \langle T\varphi_n, P\varphi_n \rangle| = |2\,\mathrm{i}\langle \varphi_n, Q\varphi_n \rangle| \leq 2\,\langle \varphi_n, T\varphi_n \rangle.$$

Passing to the limit $n \to \infty$ and using the preceding facts, we get

$$|\langle (\overline{P} + c\,\mathrm{i})\psi, \xi \rangle - \langle \xi, (\overline{P} + c\,\mathrm{i})\psi \rangle - 2c\,\mathrm{i}\langle \psi, \overline{T}\psi \rangle| = |-2c\,\mathrm{i}\langle \psi, \overline{T}\psi \rangle| \leq 2\,\langle \psi, \overline{T}\psi \rangle.$$

Therefore, $\langle \psi, \overline{T}\psi \rangle = 0$ by $|c| > 1$. Since $\langle \varphi, T\varphi \rangle \geq \|\varphi\|^2$ for $\varphi \in \mathcal{D}$, we have $0 = \langle \psi, \overline{T}\psi \rangle \geq \|\psi\|^2$. Thus, $\psi = 0$ and $\xi = \overline{T}\psi = 0$. This shows that $(P + c\,\mathrm{i})\mathcal{D}$ is dense for any $c \in \mathbb{R}$, $|c| > 1$. By Proposition A.1, $P{\upharpoonright}\mathcal{D}$ is essentially self-adjoint. (A closely related argument has been also used in the proof of Lemma 7.3.)

The proof for $Q{\upharpoonright}\mathcal{D}$ is similar. $\square$

The following is essentially the classical *Rellich–Dixmier theorem*.

Theorem 8.17 *Suppose π is a $*$-representation of the Weyl algebra* W *such that $\pi(1) = I$ and the operator $\pi(p^2 + q^2)$ is essentially self-adjoint. Then the operators $\overline{\pi(p)}$ and $\overline{\pi(q)}$ are self-adjoint and the pair $\{\overline{\pi(p)}, \overline{\pi(q)}\}$ is unitarily equivalent to a direct sum of Schrödinger pairs.*

Further, π is a subrepresentation of a representation on $\mathcal{H}(\pi)$ which is unitarily equivalent to a direct sum of Schrödinger representations. If in addition π is self-adjoint, then π is unitarily equivalent to a direct sum of Schrödinger representations.

Proof From Lemma 8.16, applied with $P := \pi(p)$, $Q := \pi(q)$, $\mathcal{D} := \mathcal{D}(\pi)$, it follows that the operators $\overline{\pi(p)}$ and $\overline{\pi(q)}$ are self-adjoint.

Since $\pi(p^2 + q^2)$ is essentially self-adjoint, so is $\pi(a^+ a) = \frac{1}{2}\pi(p^2 + q^2) - \frac{1}{2}I$ by (8.15). Therefore, by Theorem 8.8 and Proposition 8.13, there is a $*$-representation ρ acting on $\mathcal{H}(\pi) = \mathcal{H}(\rho)$ such that $\pi \subseteq \rho$ and ρ is unitarily equivalent to a direct sum of Schrödinger representations. Then $\pi(p) \subseteq \rho(p)$. Hence, since $\overline{\pi(p)} \subseteq \overline{\rho(p)}$ and $\overline{\pi(p)}$ is self-adjoint, $\overline{\pi(p)} = \overline{\rho(p)}$. Similarly, $\overline{\pi(q)} = \overline{\rho(q)}$. Because $\{ \rho(p), \rho(q) \}$ is unitarily equivalent to a direct sum of Schrödinger pairs, so is $\{ \pi(p), \pi(q) \}$.

We have $\pi \subseteq \rho$ on $\mathcal{H}(\pi) = \mathcal{H}(\rho)$. Therefore, if π is self-adjoint, it follows from Corollary 4.31 that $\pi = \rho$. $\qquad \square$

8.5 The Stone–von Neumann Theorem

The main results in this section and the next (Theorems 8.18 and 8.22) are uniqueness theorems for the canonical commutation relation. They contain criteria on a pair $\{P, Q\}$ of self-adjoint operators on a Hilbert space $\mathcal{H}$ for being unitarily equivalent to a direct sum of Schrödinger pairs $\{P_0, Q_0\}$. The latter means that there exists a unitary operator U of $\mathcal{H}$ on a direct sum $\oplus_i \mathcal{H}_i$ of Hilbert spaces such that $P = U^{-1}(\oplus_i P_i)U$, $Q = U^{-1}(\oplus_i Q_i)U$, where $P_i = -i\frac{d}{dx}$, $Q_i = x$ on $\mathcal{H}_i = L^2(\mathbb{R})$ for each i. In this case, there is a (self-adjoint) $*$-representation π of W defined by $\pi = U^{-1}(\oplus_i \pi_i)U$ on $\mathcal{D}(\pi) := U^{-1}(\oplus_i \mathcal{D}(\pi_i))$, where each π_i is the Schrödinger representation π_S and $\pi(p) = P \upharpoonright \mathcal{D}(\pi)$, $\pi(q) = Q \upharpoonright \mathcal{D}(\pi)$. That is, π is unitarily equivalent to a direct sum of Schrödinger representations.

First we consider the one-parameter unitary groups for the *Schrödinger pair* $\{P_0, Q_0\}$. Clearly, the unitary groups $V_0(s) := e^{is P_0}$ and $W_0(t) := e^{it Q_0}$ of the self-adjoint operators $P_0 := -i\frac{d}{dx}$ and $Q_0 := x$ act on $\varphi \in L^2(\mathbb{R})$ by

$$(V_0(s)\varphi)(x) = \varphi(x + s), \quad (W_0(t)\varphi)(x) = e^{itx}\varphi(x), \quad s, t \in \mathbb{R}. \tag{8.43}$$

It is easily verified that these groups $V = V_0$, $W = W_0$ satisfy the *Weyl relation*

$$V(s)W(t) = e^{ist}W(t)V(s), \quad t, s \in \mathbb{R}. \tag{8.44}$$

The following *Stone–von Neumann theorem* is the main result of this section.

Theorem 8.18 *Suppose P and Q are self-adjoint operators on a Hilbert space $\mathcal{H}$ such that the unitary groups $V(s) := e^{is P}$, $s \in \mathbb{R}$, and $W(t) := e^{it Q}$, $t \in \mathbb{R}$, satisfy the Weyl relation (8.44). Then the pair $\{P, Q\}$ is unitarily equivalent to an orthogonal direct sum of Schrödinger pairs. Further, the pair $\{P, Q\}$ is irreducible if and only if it is unitarily equivalent to the Schrödinger pair $\{P_0, Q_0\}$.*

There is an elegant proof of this result based on Mackey's imprimitivity theorem (see [Ty86, p. 146–147]), but we prefer to present von Neumann's original proof.

First let us explain the main ideas behind the following proof. The Heisenberg group H is the set $\mathbb{R}^3$ with the group law

$$(s, t, z)(s', t', z') := (s + s', t + t', z + z' + (st' - ts')/2). \tag{8.45}$$

Then, H is a Lie group with unit $(0, 0, 0)$ and $(-s, -t, -z)$ is the inverse of (s, t, z).

Now suppose that V and W are unitary groups as in Theorem 8.18. A simple computation shows that $\rho(s, t, z) := e^{iz} e^{-ist/2} V(s) W(t)$, $s, t, z \in \mathbb{R}$, defines a unitary representation of the Lie group H on $\mathcal{H}$. (We will not use this fact; it is only stated to mention the connection to the Heisenberg group and to explain where the factor $e^{-its/2}$ in formula (8.46) comes from.) Set

$$S(s, t) := \rho(s, t, 0) = e^{-its/2} V(s) W(t), \quad s, t \in \mathbb{R}. \tag{8.46}$$

From the Weyl relation (8.44) we obtain

$$S(s, t) S(s', t') = e^{i(st' - s't)/2} S(s + s', t + t'). \tag{8.47}$$

For $f \in L^1(\mathbb{R})$ the Bochner integral

$$T_f \varphi = \int_{\mathbb{R}^2} f(s, t) S(s, t) \varphi \, ds dt, \quad \varphi \in \mathcal{H}, \tag{8.48}$$

defines a bounded operator T_f on $\mathcal{H}$. We call f the *symbol* of the operator T_f. From (8.47) we derive by straightforward computations:

$$S(a, b) T_f = T_g \text{ with } g(s, t) := f(s - a, t - b) e^{i(at - bs)/2},$$

$$T_f T_g = T_h \text{ with } h(s, t) := \int_{\mathbb{R}^2} e^{i(st' - ts')/2} f(s - s', t - t') g(s', t') \, ds' dt'.$$

In what follows let T denote the operator T_{f_0} with $f_0(s, t) := (2\pi)^{-1} e^{-(s^2 + t^2)/4}$. Fix $a, b \in \mathbb{R}^2$. By combining the two preceding formulas it follows that the symbol $f(s, t)$ of the operator $T S(a, b) T \equiv T_{f_0}(S(a, b) T_{f_0})$ is

$$(2\pi)^{-2} \int_{\mathbb{R}^2} e^{i(st' - ts')/2} e^{-((s - s')^2 + (t - t')^2)/4} e^{-((s' - a)^2 + (t' - b)^2)/4} e^{i(at' - bs')/2} ds' dt'.$$

Recall the formula for the Fourier transform of Gaussians [RW91, p. 309]:

$$(2\pi)^{-1/2} \int_{\mathbb{R}} e^{-ixy} e^{-c(y - d)^2} dy = (2c)^{-1/2} e^{-ixd} e^{-\frac{x^2}{4c}}, \quad c > 0. \tag{8.49}$$

Using this formula a simple computation gives

$$f(s, t) = (2\pi)^{-1} e^{-(a^2 + b^2 + s^2 + t^2)/4} = e^{-(a^2 + b^2)/4} f_0(s, t).$$

Therefore,

$$T S(a, b) T = e^{-(a^2 + b^2)/4} T \quad \text{for } a, b \in \mathbb{R}. \tag{8.50}$$

From (8.46) it follows that $(T_f)^* = T_g$ with $g(s, t) := f(-s, -t)$. Hence $T = T_{f_0}$ is self-adjoint. Setting $a = b = 0$ in (8.50) yields $T^2 = T$. Thus, T is an orthogonal projection! Let $\mathcal{K}$ denote the range of this projection. To understand the role of this space it is instructive to determine T and $\mathcal{K}$ for the Schrödinger pair.

Example 8.19 Let V, W be the unitary groups V_0, W_0 defined by (8.43). Then

$$S_0(s, t) := e^{-its/2} V_0(s) W_0(t) = e^{-its/2} e^{is P_0} e^{it Q_0} = e^{i(s P_0 + t Q_0)}, \quad s, t \in \mathbb{R}.$$

Note that in this case the operator $T_{\hat{f}}$ in (8.48), with f replaced by its Fourier transform $\hat{f}$, is just the *Weyl quantization* of the function f; see, e.g., [F89, p. 79]. Now we determine T and $\mathcal{K}$. Inserting (8.43) in the definition of T_{f_0} we obtain

$$(T\varphi)(x) = (2\pi)^{-1} \int_{\mathbb{R}^2} e^{-(s^2 + t^2)/2} e^{-ist/2} e^{isx} e^{ist} \varphi(x + t)\, ds dt, \quad \varphi \in L^2(\mathbb{R}).$$

Next we compute the Fourier transform with respect to s by using (8.49) and substitute $y = x + t$. Then

$$(T\varphi)(x) = \pi^{-1/2} e^{-x^2/2} \int_{\mathbb{R}} e^{-y^2/2} \varphi(y)\, dy.$$

Thus, T is the rank one projection $H_0 \otimes H_0$ and $\mathcal{K} = \mathbb{C} \cdot H_0(x)$ for the Hermite function $H_0(x) = \pi^{-1/4} e^{-x^2/2}$. Note that H_0 is a unit vector in $L^2(\mathbb{R})$.   ○

In the proof of Theorem 8.4 the cyclic subspace $\ker \Lambda^* \Lambda = \ker \Lambda$ was important. For the Schrödinger representation this space is spanned by $H_0(x) = \pi^{-1/4} e^{-x^2/2}$, see (8.34). Since $\mathcal{K} = \mathbb{C} \cdot H_0$ for the Schrödinger pair, as shown by the preceding example, it is natural that the space $\mathcal{K}$ plays a crucial role in the following proof.

Proof of Theorem 8.18 Let $\varphi, \psi \in \mathcal{K}$ and $s, t, x, y \in \mathbb{R}$. Since $S(s, t)^* = S(-s, -t)$, it follows from (8.47) and (8.50) that

$$\langle S(a, b)\varphi, S(s, t)\psi \rangle = \langle S(-s, -t) S(a, b) T\varphi, T\psi \rangle$$
$$= e^{i(ta - sb)/2} \langle T S(a - s, b - t) T\varphi, \psi \rangle$$
$$= e^{-(a-s)^2/4 - (b-t)^2/4 + i(ta - sb)/2} \langle \varphi, \psi \rangle. \quad (8.51)$$

Now we choose an orthonormal basis $\{\varphi_i : i \in J\}$ of the Hilbert space $\mathcal{K}$. For $i \in J$, let $\mathcal{H}_i$ denote the closed linear span of $S(s, t)\varphi_i$, where $s, t \in \mathbb{R}$. If $i \neq j$, then $\mathcal{H}_i \perp \mathcal{H}_j$ by (8.51). Since $\mathcal{H}_i$ is invariant under all operators $S(s, t)$ and $S(s, t)^* = S(-s, -t)$, $\mathcal{H}_i$ is reducing for all operators $S(x, y)$. Therefore, the subspace $\oplus_{i \in J} \mathcal{H}_i$ and hence its orthogonal complement $\mathcal{G}$ in $\mathcal{H}$ are reducing for the unitary operators $S(s, t)$. By construction, $\mathcal{K} \cap \mathcal{G} = \{0\}$, that is, $T \restriction \mathcal{G} = 0$.

We show that $T \restriction \mathcal{G} = 0$ implies $\mathcal{G} = \{0\}$. Let $\eta \in \mathcal{G}$. Then $S(a, b)\eta \in \mathcal{G}$, hence $T S(a, b)\eta = 0$, and using (8.47) we compute

$$
0 = \langle S(-a, -b)T S(a, b)\eta, \eta \rangle = \int_{\mathbb{R}^2} f_0(s, t) \langle S(-a, -b)S(s, t)S(a, b)\eta, \eta \rangle \, ds dt
$$

$$
= \int_{\mathbb{R}^2} e^{i(sb-ta)} f_0(s, t) \langle S(s, t)\eta, \eta \rangle \, ds dt
$$

for all $a, b \in \mathbb{R}$. This means that the Fourier transform of $f_0(s, t) \langle S(s, t)\eta, \eta \rangle$ is zero. Hence the continuous function $\langle S(s, t)\eta, \eta \rangle$ is zero a.e.. Since $S(0, 0) = I$, we conclude that $\eta = 0$. Thus $\mathcal{G} = \{0\}$ and therefore $\mathcal{H} = \oplus_{i \in J} \mathcal{H}_i$.

Since both pairs $\{V, W\}$ and $\{V_0, W_0\}$ of unitary groups satisfy the Weyl relation, the preceding applies to $S(s, t)$ and also to its counterpart $S_0(s, t)$ for $\{V_0, W_0\}$. As noted in Example 8.19, the space $\mathcal{K}$ for $S_0(s, t)$ is spanned by the unit vector $\psi(x) = \pi^{-1/4}e^{-x^2/2}$. Fix $i \in J$. For $a, b, s, t \in \mathbb{R}$, we have by (8.51),

$$
\langle S(a, b)\varphi_i, S(s, t)\varphi_i \rangle = \langle S_0(a, b)\psi, S_0(s, t)\psi \rangle = e^{-(a-s)^2/4-(b-t)^2/4+i(ta-sb)/2}.
$$

Hence there exists a (well-defined!) isometric linear map U_i of the vector space $\mathrm{Lin}\, \{S(s, t)\psi : s, t \in \mathbb{R}\}$ on $\mathrm{Lin}\, \{S_0(s, t)\varphi_i : s, t \in \mathbb{R}\}$ given by

$$
U_i S(s, t)\psi = S_0(s, t)\varphi_i, \quad s, t \in \mathbb{R}.
$$

Since these spans are dense in $\mathcal{H}_i$ and $L^2(\mathbb{R})$, U_i extends to a unitary operator of $\mathcal{H}_i$ on $L^2(\mathbb{R})$. Let $x, y \in \mathbb{R}$. Then, for $s, t \in \mathbb{R}$,

$$
U_i S(a, b)U_i^{-1} S_0(s, t)\varphi_i = U_i S(a, b)S(s, t)\psi = e^{i(at-sb)/2}U_i S(a + s, b + t)\psi
$$

$$
= e^{i(at-sb)/2}S_0(a + s, b + t)\varphi_i = S_0(a, b)S_0(s, t)\varphi_i.
$$

Hence $U_i S(a, b)U_i^{-1}\eta = S_0(a, b)\eta$ for $\eta \in L^2(\mathbb{R})$. This shows that $S \restriction \mathcal{H}_i$ on $\mathcal{H}_i$ and S_0 on $L^2(\mathbb{R})$ are unitarily equivalent. Set $V_i := V \restriction \mathcal{H}_i$ and $W_i := W \restriction \mathcal{H}_i$. Since $V(s) = S(s, 0)$, $W(t) = S(0, t)$ and $V_0(s) = S_0(s, 0)$, $W_0(t) = S_0(0, t)$, the pairs of unitary groups $\{V_i, W_i\}$ on $\mathcal{H}_i$ and $\{V_0, W_0\}$ on $L^2(\mathbb{R})$ are unitarily equivalent. From $V = \oplus_i V_i$ it follows that $P = \oplus_i P_i$, where P_i is the self-adjoint operator on $\mathcal{H}_i$ such that $V_i(s) = e^{isP_i}$, $s \in \mathbb{R}$. Similarly, $Q = \oplus_i Q_i$ and $W_i(t) = e^{itQ_i}$, $t \in \mathbb{R}$. Since U_i gives a unitary equivalence of $\{V_i, W_i\}$ and $\{V_0, W_0\}$, it does for the pairs $\{P_i, Q_i\}$ and $\{P_0, Q_0\}$ as well. Hence the unitary operator $\oplus_i U_i$ provides a unitary equivalence of $\{P, Q\}$ and a direct sum of Schrödinger pairs $\{P_0, Q_0\}$.

The last assertion follows at once from the irreducibility of the Schrödinger pair. We develop another proof of the fact that each pair $\{P_i, Q_i\}$ is irreducible. Assume to the contrary that $\{P_i, Q_i\}$ is a direct sum of two pairs acting on nontrivial Hilbert spaces $\mathcal{G}_1, \mathcal{G}_2$. Then $\{V_i, W_i\}$ is a direct sum of two pairs of unitary groups satisfying the Weyl relation. As shown above, $\mathcal{G}_i \neq \{0\}$ implies $\mathrm{ran}\, T \cap \mathcal{G}_i \neq \{0\}$. Since $\dim(\mathrm{ran}\, T \cap \mathcal{H}_i) = 1$ by construction, this is impossible. $\qquad\square$

8.6 A Resolvent Approach to Schrödinger Pairs

In this section, we give another characterization of direct sums of Schrödinger pairs.

For a closed operator T, $R_\lambda(T) = (T - \lambda)^{-1}$ denotes its resolvent at $\lambda \in \rho(T)$.

Let $\{P, Q\}$ be a pair of self-adjoint operators on a Hilbert space $\mathcal{H}$ and let $\mathcal{D}$ be a linear subspace of $\mathcal{D}(PQ) \cap \mathcal{D}(QP)$ such that

$$PQ\varphi - QP\varphi = -\mathrm{i}\varphi \quad \text{for} \quad \varphi \in \mathcal{D}. \tag{8.52}$$

Lemma 8.20 *Suppose that $\alpha, \beta \in \mathbb{C}\backslash\mathbb{R}$. Then, for vectors $\psi \in (Q - \beta)(P - \alpha)\mathcal{D}$ and $\eta \in (P - \alpha)(Q - \beta)\mathcal{D}$, we have*

$$R_\alpha(P)R_\beta(Q)\psi - R_\beta(Q)R_\alpha(P)\psi = -\mathrm{i}\, R_\beta(Q)R_\alpha(P)^2 R_\beta(Q)\psi, \tag{8.53}$$

$$R_\alpha(P)R_\beta(Q)\eta - R_\beta(Q)R_\alpha(P)\eta = -\mathrm{i}\, R_\alpha(P)R_\beta(Q)^2 R_\alpha(P)\eta. \tag{8.54}$$

Proof Let $\psi = (Q - \beta)(P - \alpha)\varphi$ with $\varphi \in \mathcal{D}$. Then $\psi = (P - \alpha)(Q - \beta)\varphi + \mathrm{i}\varphi$ by (8.52). Hence we obtain

$$R_\alpha(P)R_\beta(Q)\psi = \varphi = R_\beta(Q)R_\alpha(P)(\psi - \mathrm{i}\varphi)$$
$$= R_\beta(Q)R_\alpha(P)\psi - \mathrm{i}R_\beta(Q)R_\alpha(P)^2 R_\beta(Q)\psi,$$

which gives (8.53). The proof of (8.54) is similar. $\square$

In particular, since the resolvents are bounded operators, if $(Q - \beta)(P - \alpha)\mathcal{D}$ in Lemma 8.20 is dense in $\mathcal{H}$, then (8.53) holds for all $\psi \subset \mathcal{H}$. Likewise, if the space $(P - \alpha)(Q - \beta)\mathcal{D}$ is dense, then (8.54) is valid for all $\eta \in \mathcal{H}$.

Example 8.21 *(Schrödinger pair $\{P = -\mathrm{i}\frac{d}{dx}, Q = x\}$ on $L^2(\mathbb{R})$)*
Then the commutation relation (8.52) is satisfied for the *invariant* dense domain $\mathcal{D} := \mathcal{S}(\mathbb{R})$. Since $P \upharpoonright C_0^\infty(\mathbb{R})$ is essentially self-adjoint, by Proposition A.1 the space $(P - \alpha)C_0^\infty(\mathbb{R})$ is dense in $L^2(\mathbb{R})$ and so is the larger space $(P - \alpha)(Q - \beta)\mathcal{D}$. Using this fact and applying the Fourier transform it follows that $(Q - \beta)(P - \alpha)\mathcal{D}$ is also dense in $L^2(\mathbb{R})$. Hence, by Lemma 8.20, both equations (8.53) and (8.54) are valid for all $\alpha, \beta \in \mathbb{C}\backslash\mathbb{R}$ and $\psi, \eta \in L^2(\mathbb{R})$.

Note that Eq. (8.52) and the corresponding density properties hold also for the larger domain $\mathcal{D} := \mathcal{D}(PQ) \cap \mathcal{D}(QP)$. $\bigcirc$

Suppose the pair $\{P, Q\}$ is unitarily equivalent of a direct sum of Schrödinger pairs. Then it follows from Example 8.21 that for $\mathcal{D} := \mathcal{D}(PQ) \cap \mathcal{D}(QP)$ the spaces $(P - \alpha)(Q - \beta)\mathcal{D}$ and $(Q - \beta)(P - \alpha)\mathcal{D}$ are dense. Hence equations (8.53) and (8.54) hold on the whole Hilbert space. In particular, for $\alpha, \beta \in \mathbb{C}\backslash\mathbb{R}$,

$$R_\alpha(P)R_\beta(Q) - R_\beta(Q)R_\alpha(P) = -\mathrm{i}\, R_\alpha(P)R_\beta(Q)^2 R_\alpha(P). \tag{8.55}$$

The following result is *Kato's theorem*. It says that each of the density conditions (8.56) implies that a pair of self-adjoint operators satisfying the canonical commutation relation (8.52) is a direct sum of Schrödinger pairs.

Theorem 8.22 *Let P and Q be self-adjoint operators on a Hilbert space $\mathcal{H}$ and $\alpha_0, \beta_0 \in \mathbb{C} \backslash \mathbb{R}$. Suppose there exists a linear subspace $\mathcal{D}$ of $\mathcal{D}(PQ) \cap \mathcal{D}(QP)$ such that (8.52) holds and*

$$(P - \alpha_0)(Q - \beta_0)\mathcal{D}, \quad \text{or} \quad (Q - \beta_0)(P - \alpha_0)\mathcal{D}, \quad \text{is dense in } \mathcal{H}. \qquad (8.56)$$

Then the pair $\{P, Q\}$ on $\mathcal{H}$ is unitarily equivalent to an orthogonal direct sum of Schrödinger pairs.

In the proof we use the following lemma from operator semigroup theory.

Lemma 8.23 *If T is a self-adjoint operator on a Hilbert space $\mathcal{H}$, then*

$$\lim_{n \to \infty} (I - n^{-1}(\mathrm{i}s)T)^{-n}\varphi = e^{\mathrm{i}sT}\varphi \quad \text{for } \varphi \in \mathcal{H}, s \in \mathbb{R}. \qquad (8.57)$$

Proof [Ka67, pp. 481–482]. $\qquad\qquad\qquad\qquad\qquad\qquad\qquad\qquad\qquad\qquad\qquad\qquad\qquad$ $\square$

Proof of Theorem 8.22 We assume that $(Q - \beta_0)(P - \alpha_0)\mathcal{D}$ is dense in $\mathcal{H}$. The proof in the case, when $(P - \alpha_0)(Q - \beta_0)\mathcal{D}$ is dense, is verbatim the same.

Our first aim is to prove that, for $\alpha = \alpha_0$, $\beta = \beta_0$ and $n \in \mathbb{N}$, we have

$$R_\alpha(P)^n R_\beta(Q) - R_\beta(Q) R_\alpha(P)^n = -\mathrm{i}\, n R_\beta(Q) R_\alpha(P)^{n+1} R_\beta(Q). \qquad (8.58)$$

We proceed by induction on n. Since $(Q - \beta_0)(P - \alpha_0)\mathcal{D}$ is dense, Lemma 8.20 implies that (8.53) holds for all $\psi \in \mathcal{H}$. This is relation (8.58) for $n = 1$.

Assume now that (8.58) is satisfied for n. To shorten the notation we write $R(P) = R_{\alpha_0}(P)$ and $R(Q) = R_{\beta_0}(Q)$ throughout the following computation. Using relation (8.58) for n and also for $n = 1$ we derive

$$
\begin{aligned}
R(P)^{n+1} &R(Q) - R(Q)R(P)^{n+1} \\
&= R(P)^n[R(P)R(Q) - R(Q)R(P)] + [R(P)^n R(Q) - R(Q)R(P)^n]R(P) \\
&= R(P)^n[-\mathrm{i}R(Q)R(P)^2 R(Q)] + [-\mathrm{i}\,n R(Q)R(P)^{n+1} R(Q)]R(P) \\
&= -\mathrm{i}\,[R(P)^n R(Q)]R(P)^2 R(Q) - \mathrm{i}\,n R(Q)R(P)^{n+1}[R(Q)R(P)] \\
&= -\mathrm{i}\,[R(Q)R(P)^n - \mathrm{i}\,n R(Q)R(P)^{n+1} R(Q)]R(P)^2 R(Q) \\
&\quad - \mathrm{i}\,n\, R(Q)R(P)^{n+1}[R(P)R(Q) + \mathrm{i}R(Q)R(P)^2 R(Q)] \\
&= -\mathrm{i}\,(n+1)\, R(Q)R(P)^{n+2} R(Q),
\end{aligned}
$$

which is relation (8.58) for $n + 1$. This completes the induction proof of (8.58).

Now we fix $\beta = \beta_0$, and let α vary. Consider the two Neumann series for the resolvents

$$R_\alpha(P) = \sum_{n=1}^{\infty} (\alpha - \alpha_0)^{n-1} R_{\alpha_0}(P)^n, \quad R_\alpha(P)^2 = \sum_{n=1}^{\infty} n(\alpha - \alpha_0)^{n-1} R_{\alpha_0}(P)^{n+1}.$$

They converge strongly if $|\alpha - \alpha_0| < \varepsilon$ for small $\varepsilon > 0$. Let $\mathbb{H}_{\alpha_0}$ and $\mathbb{H}_{\beta_0}$ denote the open half planes of $\mathbb{C} \backslash \mathbb{R}$ which contain α_0 and β_0, respectively. We show that (8.58) holds for $n \in \mathbb{N}$ and $\alpha \in \mathbb{H}_{\alpha_0}$. First let $n = 1$. We insert the Neumann series into both sides of (8.58) for $n = 1$ (which is (8.55)) and use that (8.58) is valid for $n \in \mathbb{N}$, $\alpha = \alpha_0$, as shown in the preceding paragraph. Then we conclude that (8.58) holds for $n = 1$, $|\alpha - \alpha_0| < \varepsilon$. Both sides of (8.58) for $n = 1$ are strongly analytic on $\mathbb{H}_{\alpha_0}$. Hence (8.58) holds for $n = 1$ and $\alpha \in \mathbb{H}_{\alpha_0}$. Differentiating this equation $n - 1$ times it follows that relation (8.58) is valid for all $n \in \mathbb{N}$ and $\alpha \in \mathbb{H}_{\alpha_0}$.

Fix $s \in \mathbb{R}$ such that $\alpha := n(is)^{-1} \in \mathbb{H}_{\alpha_0}$. We multiply (8.58) by $(-\alpha)^n$ and get

$$(I - n^{-1}(is)P)^{-n} R_\beta(Q) \tag{8.59}$$

$$= R_\beta(Q)(I - n^{-1}(is)P)^{-n} - s R_\beta(Q)(I - n^{-1}(is)P)^{-n-1} R_\beta(Q). \tag{8.60}$$

Note that $\lim_n (I - n^{-1}(is)P)^{-1} \varphi = \varphi$, $\lim_n (I - n^{-1}(is)P)^{-n} \varphi = e^{isP} \varphi$ by Lemma 8.23, and $I - s R_\beta(Q) = (Q - s - \beta) R_\beta(Q)$. Therefore, passing to limit $n \to \infty$ in (8.59)–(8.60) yields

$$e^{isP} R_\beta(Q) = R_\beta(Q)e^{isP} - s R_\beta(Q)e^{isP} R_\beta(Q) = R_\beta(Q)e^{isP}(Q-s-\beta) R_\beta(Q).$$

Now we multiply this equation by $R_{s+\beta}(Q)$ from the right. Then we obtain $e^{isP} R_{s+\beta}(Q) R_\beta(Q) = R_\beta(Q)e^{isP} R_\beta(Q)$. Since $R_\beta(Q)\mathcal{H} = \mathcal{D}(Q)$ is dense,

$$e^{isP} R_{s+\beta}(Q) = R_\beta(Q)e^{isP}. \tag{8.61}$$

Repeated multiplication of (8.61) by $R_{s+\beta}(Q)$ from the right gives

$$e^{isP} R_{s+\beta}(Q)^n = R_\beta(Q)^n e^{isP} \quad \text{for } n \in \mathbb{N}. \tag{8.62}$$

In the preceding $\beta = \beta_0$ was fixed. To extend (8.62) to all $\beta \in \mathbb{H}_{\beta_0}$ we argue in a similar manner as above, with P replaced by Q. Using the Neumann series for $(Q - \beta)^{-1}$ and $(Q - s - \beta)^{-1}$ around β_0 we conclude that (8.61) is valid for small $|\beta - \beta_0|$ and then for all $\beta \in \mathbb{H}_{\beta_0}$ by analytic continuation. By differentiating (8.61) with respect to β it follows that (8.62) is fulfilled for $\beta \in \mathbb{H}_{\beta_0}$.

Let $t \in \mathbb{R}$ be such that $\beta := n(it)^{-1} \in \mathbb{H}_{\beta_0}$. Similarly as above, we multiply (8.62) by $(-\beta)^n$, pass to the limit $n \to \infty$, and obtain $e^{isP} e^{it(Q-s)} = e^{itQ} e^{isP}$. Hence $e^{isP} e^{itQ} = e^{ist} e^{itQ} e^{isP}$ for s and t from half-lines. This extends easily to arbitrary reals s, t. Hence the unitary groups $V(s) := e^{isP}$, $W(t) := e^{itQ}$ satisfy the Weyl relation

(8.44). Therefore, by the Stone–von Neumann Theorem 8.18, the pair $\{P, Q\}$ is unitarily equivalent to a direct sum of Schrödinger pairs. $\qquad\square$

A large class of $*$-representations π of W for which the operators $P := \overline{\pi(p)}$ and $Q := \overline{\pi(q)}$ are self-adjoint, but the Weyl relation fails, is constructed in [Sch83c]. Let π be such a representation. Then condition (8.56) cannot hold; otherwise $\{P, Q\}$ is unitarily equivalent to a direct sum of Schrödinger pairs by Theorem 8.22 and hence the Weyl relation is fulfilled, a contradiction. But, since P and Q are self-adjoint, $(P - \alpha)\mathcal{D}(\pi)$ and $(Q - \beta)\mathcal{D}(\pi)$ are dense in $\mathcal{H}(\pi)$ for any $\alpha, \beta \in \mathbb{C}\backslash\mathbb{R}$ by Proposition A.1. This shows that the assumption (8.56) of Theorem 8.22 cannot be weakened by requiring instead the density of $(P - \alpha)\mathcal{D}$ and $(Q - \beta)\mathcal{D}$.

8.7 The Uncertainty Principle

For a linear operator T on a Hilbert space and $\varphi \in \mathcal{D}(T)$ we define

$$\langle T\rangle_\varphi := \langle T\varphi, \varphi\rangle.$$

Clearly, if the operator T is symmetric, then the number $\langle T\rangle_\varphi$ is real.

Definition 8.24 The *standard deviation* of a symmetric operator T in a unit vector $\varphi \in \mathcal{D}(T^2)$ is defined by

$$\Delta_\varphi(T) := \left(\langle T^2\rangle_\varphi - \langle T\rangle_\varphi^2\right)^{1/2} \equiv \left(\langle T^2\varphi, \varphi\rangle - \langle T\varphi, \varphi\rangle^2\right)^{1/2}. \tag{8.63}$$

Its square $\Delta_\varphi(T)^2 = \langle T^2\rangle_\varphi - \langle T\rangle_\varphi^2$ is called the *variance* of T in φ.

Suppose that T is a symmetric operator and φ is a unit vector of $\mathcal{D}(T^2)$. Then $0 \leq \langle T\varphi, \varphi\rangle^2 \leq \|T\varphi\|^2 = \langle T^2\varphi, \varphi\rangle$ and we easily derive the formulas

$$\begin{aligned}
\Delta_\varphi(T)^2 &= \|T\varphi\|^2 - \langle T\varphi, \varphi\rangle^2 = \langle (T - \langle T\rangle_\varphi I)^2\rangle_\varphi \\
&= \|(T - \langle T\rangle_\varphi I)\varphi\|^2.
\end{aligned} \tag{8.64}$$

From (8.64) we conclude that $\Delta_\varphi(T) = 0$ if and only if φ is an eigenvector of T; in this case, the corresponding eigenvalue is $\langle T\rangle_\varphi$.

In quantum mechanics, the number $\Delta_\varphi(T)$ describes the *uncertainty* of an "observable" T in the "state" φ.

The following inequalities (8.65) and (8.70) are called *uncertainty relations*.

Proposition 8.25 *Suppose a and b are symmetric operators on a Hilbert space and φ is a unit vector in $\mathcal{D}\{a, b\} := \mathcal{D}(a^2) \cap \mathcal{D}(b^2) \cap \mathcal{D}(ab) \cap \mathcal{D}(ba)$. Then*

$$\Delta_\varphi(a)\,\Delta_\varphi(b) \geq \frac{1}{2}\,|\langle [a, b]\rangle_\varphi|, \tag{8.65}$$

where $[a, b] := ab - ba$. Further, there is equality in (8.65) if and only if there are real numbers γ, δ, not both zero, such that

$$\gamma(a - \langle a \rangle_\varphi)\varphi + i\delta(b - \langle b \rangle_\varphi)\varphi = 0. \tag{8.66}$$

Proof Let $\alpha, \beta \in \mathbb{R}$. Using that a and b are symmetric operators we derive

$$|\langle [a, b]\varphi, \varphi \rangle| = |\langle (ab - ba)\varphi, \varphi \rangle| = |\langle b\varphi, a\varphi \rangle - \langle a\varphi, b\varphi \rangle|$$
$$= |\langle (b - \beta I)\varphi, (a - \alpha I)\varphi \rangle - \langle (a - \alpha I)\varphi, (b - \beta I)\varphi \rangle|$$
$$= |2i \operatorname{Im} \langle (b - \beta I)\varphi, (a - \alpha I)\varphi \rangle| \leq 2 |\langle (b - \beta I)\varphi, (a - \alpha I)\varphi \rangle| \tag{8.67}$$
$$\leq 2 \|(b - \beta I)\varphi\| \, \|(a - \alpha I)\varphi\|. \tag{8.68}$$

Now set $\alpha = \langle a \rangle_\varphi$ and $\beta = \langle b \rangle_\varphi$. Then, by (8.64), we have $\|(a - \alpha I)\varphi\| = \Delta_\varphi(a)$ and $\|(b - \beta I)\varphi\| = \Delta_\varphi(b)$. Therefore, the preceding inequality gives (8.65).

Now let us look when the inequality (8.65) is an equality. By the preceding proof this holds if and only if the two inequalities in (8.67) and (8.68) are equalities. It is well known that there is equality in the Cauchy–Schwarz inequality (8.68) if and only if the vectors $(b - \beta I)\varphi$ and $(a - \alpha I)\varphi$ are linearly dependent, that is,

$$\gamma(a - \alpha\varphi)\varphi + \delta'(b - \beta)\varphi = 0 \tag{8.69}$$

for some numbers $\gamma, \delta' \in \mathbb{C}$, not both zero. Upon multiplying by a constant we can assume that $\gamma \in \mathbb{R}$. It remains to consider (8.67). First note that if $(a - \alpha)\varphi = 0$ or $(b - \beta)\varphi = 0$, we obviously have equality in (8.67) and condition (8.66) holds. If both vectors are nonzero, we insert (8.69) into (8.67) and check that there is equality if and only $\delta' = i\delta$ with δ real. $\square$

An obvious consequence of the inequality (8.65) is the following.

Corollary 8.26 *Let P and Q be symmetric operators on a Hilbert space such that $(PQ - QP)\varphi = -i\hbar\varphi$ for $\varphi \in \mathcal{D}\{P, Q\}$. Then, for each unit vector $\varphi \in \mathcal{D}\{P, Q\}$,*

$$\Delta_\varphi(P)\,\Delta_\varphi(Q) \geq \frac{\hbar}{2}. \tag{8.70}$$

In particular, (8.70) holds for the operators $P = -i\hbar\frac{d}{dx}$ and $Q = x$ on $L^2(\mathbb{R})$.

The inequality (8.70) is the famous *Heisenberg uncertainty relation*. It implies that the uncertainties of the position operator and the moment operator in the same state cannot be arbitrarily small. This means that the position and moment operators cannot be measured with arbitrary accuracy simultaneously in the same state. Note that Planck's constant has the value $\hbar = 1.05457 \cdot 10^{-34}$ Js.

A unit vector φ for which there is equality in the uncertainty relation (8.70) is called a *minimum uncertainty state*. Now we determine these states in the case where $P = -i\hbar\frac{d}{dx}$ and $Q = x$ on $L^2(\mathbb{R})$.

By Proposition 8.25, we have to solve equation (8.66) with $a = Q, b = P$. Since Q and P have no eigenvectors, we have $\gamma \neq 0$, so we can assume without loss of generality that $\gamma = 1$, $\delta \neq 0$. Then (8.66) reads as

$$(Q - \langle Q \rangle_\varphi)\varphi + i\delta(P - \langle P \rangle_\varphi)\varphi = 0. \tag{8.71}$$

Let u, v, δ, where $\delta \neq 0$, be fixed real numbers and consider the first-order ordinary differential equation $(Q - u)\varphi + i\delta(P - v)\varphi = 0$. It has the solution

$$\varphi_{\delta,u,v}(x) = c_{\delta,u,v} \exp\left(-\frac{(x-u)^2}{2\delta\hbar} + ivx\right). \tag{8.72}$$

Here $c_{\delta,u,v}$ is a constant which will be chosen such that $\varphi_{\delta,u,v}$ has norm one. Clearly, $\varphi_{\delta,u,v} \in L^2(\mathbb{R})$ if and only if $\delta > 0$. Suppose now that $\delta > 0$. By some computation it follows that for the function given by (8.72) we have

$$u = \langle Q \rangle_{\varphi_{\delta,u,v}}, \quad v = \langle P \rangle_{\varphi_{\delta,u,v}}, \quad \Delta_{\varphi_{\delta,u,v}}(Q) = \delta\Delta_{\varphi_{\delta,u,v}}(P). \tag{8.73}$$

Hence $\varphi_{\delta,u,v}$ satisfies equation (8.71). Therefore, *for arbitrary numbers u, v, $\delta \in \mathbb{R}$, $\delta > 0$, the function $\varphi_{\delta,u,v}$ is a minimum uncertainty state.*

All three numbers u, v, δ can be recovered from the state $\varphi_{\delta,u,v}$ by (8.73). For δ this follows from the fact that $\Delta_{\varphi_{\delta,u,v}}(P) \neq 0$, because P has no eigenvalue.

Now we set $\hbar = 1$, $\delta = 1$. As above, let u, $v \in \mathbb{R}$. Then, by the preceding,

$$\varphi_{u,v}(x) := \pi^{-1/4} \exp((x - u)^2/2 + ivx - iuv/2) \tag{8.74}$$

is a minimum uncertainty state for $P = -i\frac{d}{dx}$, $Q = x$ on $L^2(\mathbb{R})$ and we have $u = \langle Q \rangle_{\varphi_{u,v}}$, $v = \langle P \rangle_{\varphi_{u,v}}$. The constant $-iuv/2$ in the exponential of (8.74) gives only a factor of modulus one, so it does not change the corresponding physical state. It was only added in order to formulate Exercise 13 below.

8.8 The Groenewold–van Hove Theorem

The *Poisson bracket* of functions $f, g \in C^\infty(\mathbb{R}^2)$ is the function defined by

$$\{f, g\} := \frac{\partial f}{\partial q}\frac{\partial g}{\partial p} - \frac{\partial f}{\partial p}\frac{\partial g}{\partial q}. \tag{8.75}$$

Here, as usual in classical mechanics, we denote the variables of the phase space by p, q. It is well known and easily verified that $C^\infty(\mathbb{R}^2)$, equipped with the Poison bracket, becomes a real Lie algebra and

$$[X_f, X_g] = X_{\{f,g\}}. \tag{8.76}$$

Real functions on the phase space of classical mechanics correspond to symmetric (or self-adjoint) operators on a Hilbert space in quantum mechanics. The counterpart of the Poisson bracket of functions is the commutator $[a, b] := ab - ba$ of observables, multiplied by $-i$. Roughly speaking, a "quantization" is a mapping $f \mapsto A(f)$ of real functions to symmetric operators satisfying $A(1) = I$ and

$$A(\{f, g\}) = -i\,[A(f), A(g)]. \tag{8.77}$$

(Strictly speaking, there should be $\frac{i}{\hbar}$ instead of i in (8.77), but we shall set $\hbar = 1$.)

Let $\mathcal{P}$ denote the polynomial algebra $\mathbb{R}[p, q]$, equipped with the Poisson bracket (8.75), and $\mathcal{P}_n$ the polynomials of degree at most n. A natural attempt for such a mapping is to set $A(1) = I$, $A(q) = Q$, $A(p) = P$ for $P = -i\frac{d}{dx}$, $Q = x$ on $L^2(\mathbb{R})$. (Then (8.77) holds for $f = q, g = p$, since $\{q, p\} = 1$ and $[Q, P] = iI$.) It was discovered by Groenewold [Gr46] that this does *not* give a consistent quantization of the algebra $\mathcal{P}$. In fact, as shown by Theorem 8.28 below, (8.77) is already violated for polynomials of $\mathcal{P}_3$.

For simplicity and by a slight abuse of notation, we shall write $P := -i\frac{d}{dx}\lceil \mathcal{S}(\mathbb{R})$ and $Q := x\lceil \mathcal{S}(\mathbb{R})$ in what follows.

Lemma 8.27 *Suppose $T \in \mathcal{L}^+(\mathcal{S}(\mathbb{R}))$. If T commutes with P and Q, then $T = \lambda \cdot I$ for some $\lambda \in \mathbb{C}$.*

Proof The Hermite functions H_n, $n \in \mathbb{N}_0$, belong to $\mathcal{S}(\mathbb{R})$ and form an orthonormal basis of $L^2(\mathbb{R})$. From (8.36) we recall that $(x^2 - \frac{d^2}{dx^2})H_n = (2n + 1)H_n$. Therefore, since T commutes with $\frac{d}{dx}$ and x, we have $T H_n = \lambda_n H_n$ for some $\lambda_n \in \mathbb{C}$. By (8.35), $(x - \frac{d}{dx})H_n = \sqrt{n + 1}\,H_{n+1}$. Applying T to both sides, we conclude that $\lambda_n = \lambda_{n+1}$ for all $n \in \mathbb{N}_0$. Thus $T = \lambda \cdot I$ on $\text{Lin}\{H_n : n \in \mathbb{N}_0\}$ and hence on $\mathcal{S}(\mathbb{R})$, since $T \in \mathcal{L}^+(\mathcal{S}(\mathbb{R}))$ is closable. $\square$

The following result is the *Groenewald–van Hove "no-go" theorem.*

Theorem 8.28 *There is no linear map $f \mapsto A(f)$ of $\mathcal{P}_4$ into $\mathcal{L}^+(\mathcal{S}(\mathbb{R}))$ such that (8.77) holds for all $f, g \in \mathcal{P}_3$ and*

$$A(1) = I, \quad A(q) = Q, \quad A(p) = P \quad \text{on } \mathcal{S}(\mathbb{R}). \tag{8.78}$$

Proof Assume to the contrary that such a map exists.

Suppose $A(q^{n-1}) = Q^{n-1}$ and $A(p^{n-1}) = P^{n-1}$ for $n = 2$ or $n = 3$. Then, using (8.78) and (8.77) we derive

$$[Q, A(q^n)] = [A(q), A(q^n)] = iA(\{q, q^n\}) = 0 = [Q, Q^n],$$

$$[P, A(q^n)] = [A(p), A(q^n)] = iA(\{p, q^n\}) = iA(-nq^{n-1}) = -inQ^{n-1} = [P, Q^n].$$

Hence $T_n := A(q^n) - Q^n \in \mathcal{L}^+(\mathcal{S}(\mathbb{R}))$ commutes with Q and P. By Lemma 8.27, $T_n = \lambda_n \cdot I$, so $A(q^n) = Q^n + \lambda_n \cdot I$, with $\lambda_n \in \mathbb{C}$. Similarly, $A(p^n) = P^n + \mu_n \cdot I$ for some $\mu_n \in \mathbb{C}$. By assumption (8.78), the preceding holds for $n = 2$. Hence

$$A(pq) = \frac{1}{4} A(\{q^2, p^2\}) = -\frac{i}{4}[A(q^2), A(p^2)] = -\frac{i}{4}[Q^2 + \lambda_2 \cdot I, P^2 + \mu_2 \cdot I]$$

$$= -\frac{i}{4}[Q^2, P^2] = \frac{1}{2}(PQ + QP).$$

Using this equality and again assumption (8.77) we conclude that

$$A(q^n) = -\frac{1}{n} A(\{pq, q^n\}) = \frac{i}{n} [A(pq), A(q^n)] = \frac{i}{2n}[PQ + QP, Q^n + \lambda_n \cdot I]$$

$$= \frac{i}{2n}[PQ + QP, Q^n] = Q^n.$$

Since $A(q) = Q$ by (8.78), we can set $n = 2$ in the preceding and obtain $A(q^2) = Q^2$. Similarly, $A(p^2) = P^2$. Therefore, the preceding applies also with $n = 3$ and yields $A(q^3) = Q^3$ and $A(p^3) = P^3$. Further,

$$A(qp^2) = -\frac{1}{6} A(\{p^3, q^2\}) = \frac{i}{6} [A(p^3), A(q^2)] = \frac{i}{6} [P^3, Q^2] = \frac{1}{2}(P^2Q + QP^2),$$

$$A(q^2p) = -\frac{1}{6} A(\{p^2, q^3\}) = \frac{i}{6} [A(p^2), A(q^3)] = \frac{i}{6} [P^2, Q^3] = \frac{1}{2}(Q^2P + PQ^2).$$

The polynomial p^2q^2 has two representations $\frac{1}{9}\{q^3, p^3\} = \frac{1}{3}\{q^2p, qp^2\}$ as a Poisson bracket. The comparison of (8.77) for these representations will lead to a contradiction. Using the formulas obtained in the preceding paragraphs we derive

$$A(p^2q^2) = \frac{1}{9} A(\{q^3, p^3\}) = -\frac{i}{9}[A(q^3), A(p^3)] = -\frac{i}{9}[Q^3, P^3],$$

$$A(p^2q^2) = \frac{1}{3} A(\{q^2p, qp^2\}) = -\frac{i}{3} [A(q^2p), A(qp^2)]$$

$$= -\frac{i}{12} [Q^2P + PQ^2, P^2Q + QP^2],$$

so that

$$4[Q^3, P^3] = 3[Q^2P + PQ^2, P^2Q + QP^2]. \tag{8.79}$$

But (8.79) does not hold! For instance, take $\varphi \in \mathcal{S}(\mathbb{R})$ such that $\varphi(x) = 1$ on $(0, 1)$. Then $4[Q^3, P^3]\varphi = -4P^3Q^3\varphi = -4P^3x^3\varphi = -24(-i)^3$ on $(0, 1)$. A similar slightly longer computation gives $3[Q^2P + PQ^2, P^2Q + QP^2,]\varphi = -12(-i)^3$ on $(0, 1)$. This is the desired contradiction.   □

We state a byproduct of the above proof separately as a corollary.

Corollary 8.29 *Let $f \mapsto A(f)$ be a linear map of $\mathcal{P}_3$ into $\mathcal{L}^+(\mathcal{S}(\mathbb{R}))$. If (8.78) and (8.77) hold for $f, g \in \mathcal{P}_2$, then we have*

$$A(q^2) = Q^2, \quad A(p^2) = P^2, \quad A(pq) = (PQ + QP)/2. \tag{8.80}$$

Theorem 8.28 says that there is no quantization map $f \mapsto A(f)$ of $\mathcal{P}$ into $\mathcal{L}^+(\mathcal{D})$ satisfying (8.77) and $A(1) = I$ such that $\{\overline{A(p)}, \overline{A(q)}\}$ is the Schrödinger pair. However, if infinite sums of Schrödinger pairs are allowed, Eq. (8.81) gives such a map. It was discovered by van Hove, and it is called the *prequantization map*.

For $f \in C^{\infty}(\mathbb{R}^2)$ we define operators acting on the Hilbert space $L^2(\mathbb{R}^2)$ (!) by

$$X_f := \frac{\partial f}{\partial q}\frac{\partial}{\partial p} - \frac{\partial f}{\partial p}\frac{\partial}{\partial q}, \quad \theta_f := f - p\frac{\partial f}{\partial p},$$

$$A(f) := iX_f + \theta(f) = i\{f, \cdot\} + f - p\frac{\partial f}{\partial p}. \tag{8.81}$$

Note that $\theta(f)$ acts as multiplication operator by the function $f - p\frac{\partial f}{\partial p}$ and we have $X_f g = \{f, g\}$ by the definition of the Poisson bracket.

Theorem 8.30 *Then* $f \mapsto A'(f) := A(f)\!\restriction\!\mathcal{S}(\mathbb{R}^2)$ *is a linear map of* $\mathcal{P}$ *into the symmetric operators of* $\mathcal{L}^+(\mathcal{S}(\mathbb{R}^2))$ *such that* $A'(1) = I$ *and the Eq. (8.77) is satisfied.*

Proof Since $i\frac{\partial}{\partial p}$ and $i\frac{\partial}{\partial q}$ are symmetric operators on $L^2(\mathbb{R}^2)$ and $f \in \mathcal{P}$ is real-valued on $\mathbb{R}^2$, $A'(f)$ is symmetric. Clearly, $A(f)$ leaves $\mathcal{S}(\mathbb{R}^2)$ invariant. Hence $A'(f) \in \mathcal{L}^+(\mathcal{S}(\mathbb{R}^2))$. It is obvious that $A'(1) = I$. We prove (8.77). The following three equations are obtained by straightforward computations based on the definition of the Poisson bracket. We have

$$\{\theta(f), g\} = \{f, g\} - \left\{p\frac{\partial f}{\partial p}, g\right\} = \{f, g\} + \frac{\partial g}{\partial q}\frac{\partial f}{\partial p} + p\left\{g, \frac{\partial f}{\partial p}\right\},$$

$$\{f, \theta(g)\} = \{f, g\} - \left\{f, p\frac{\partial g}{\partial p}\right\} = \{f, g\} - \frac{\partial f}{\partial q}\frac{\partial g}{\partial p} - p\left\{f, \frac{\partial g}{\partial p}\right\},$$

and hence

$$\{f, \theta(g)\} + \{\theta(f), g\} = \{f, g\} - p\frac{\partial}{\partial p}\{f, g\} = \theta(\{f, g\}). \tag{8.82}$$

Using that $X_{f_1}f_2 = \{f_1, f_2\}$ and the Eqs. (8.76) and (8.82) we derive

$$[A'(f), A'(g)] = -[X_f, X_g] + i[X_f, \theta(g)] + i[\theta(f), X_g] + [\theta(f), \theta(g)]$$
$$= -X_{\{f,g\}} + i\{f, \theta(g)\} + i\{\theta(f), g\} + 0$$
$$= -X_{\{f,g\}} + i\theta(\{f, g\}) = iA'(\{f, g\}). \qquad \square$$

In the special cases $f(p, q) = p, q$ the operators $A(f)$ from (8.81) are

$$A(p) = -i\frac{\partial}{\partial q}, \quad A(q) = q + i\frac{\partial}{\partial p}. \tag{8.83}$$

Let us return to representation theory, which is the main subject of this book. The operators (8.83) define a representation of the Weyl algebra. In order to discuss it we return to our earlier notation and denote points of $\mathbb{R}^2$ by (x, y) and the generators of the 2-dimensional Weyl algebra $W(2)$ by p_1, q_1, p_2, q_2. We define operators

$$P_1 = -i\frac{\partial}{\partial x}, \quad Q_1 = x + i\frac{\partial}{\partial y}, \quad P_2 = -i\frac{\partial}{\partial y}, \quad Q_2 = i\frac{\partial}{\partial x} + y$$

and one-parameter unitary groups

$$(V_1(t)\varphi)(x, y) = \varphi(x + t, y), \quad (W_1(t)\varphi)(x, y) = e^{itx}\varphi(x, y - t),$$
$$(V_2(t)\varphi)(x, y) = \varphi(x, y + t), \quad (W_2(t)\varphi)(x, y) = e^{ity}\varphi(x - t, y)$$

on the Hilbert space $L^2(\mathbb{R}^2)$. By differentiation of the groups at $t = 0$ it follows that

$$V_1(t) = e^{itP_1}, \quad W_1(t) = e^{itQ_1}, \quad V_2(t) = e^{itP_2}, \quad W_2(t) = e^{itQ_2}, \quad t \in \mathbb{R}.$$

The operators P_1, Q_1, P_2, Q_2 are self-adjoint on the corresponding domains in $L^2(\mathbb{R}^2)$. By a simple computation we verify that $\{V_1(s), W_1(t)\}$ and $\{V_2(s), W_2(t)\}$ are commuting pairs of unitary groups both satisfying the Weyl relation (8.44). In terms of the generators this means that for $\varphi \in \mathcal{S}(\mathbb{R}^2)$ we have

$$P_1 Q_1 \varphi - Q_1 P_1 \varphi = P_2 Q_2 \varphi - Q_2 P_2 \varphi = -i\varphi,$$
$$P_1 P_2 \varphi = P_2 P_1 \varphi, \quad P_1 Q_2 \varphi = Q_2 P_1 \varphi, \quad Q_1 P_2 \varphi = Q_1 P_2 \varphi, \quad Q_1 Q_2 \varphi = Q_2 Q_1 \varphi.$$

Clearly, the operators P_1, Q_1, P_2, Q_2 leave the Schwartz space $\mathcal{S}(\mathbb{R}^2)$ invariant. Hence there is a $*$-representation π of the Weyl algebra $W(2)$ defined by

$$\pi(p_1)\varphi = P_1\varphi, \quad \pi(q_1)\varphi = Q_1\varphi, \quad \pi(p_2)\varphi = P_2\varphi, \quad \pi(q_2)\varphi = Q_2\varphi$$

for $\varphi \in \mathcal{D}(\pi) = \mathcal{S}(\mathbb{R}^2)$. Recall that P_1, Q_1 are the operators $A(p), A(q)$ from (8.83). Since the unitary groups of P_1, Q_1 satisfy the Weyl relation, the pair $\{A(p) = P_1, A(q) = Q_1\}$ is a direct sum of Schrödinger pairs by the Stone–von Neumann Theorem 8.18.

Now we consider the unitary operator $U := \mathcal{F}e^{-ixy}\mathcal{F}^{-1}$ of $L^2(\mathbb{R}^2)$. Here e^{-ixy} denotes the corresponding multiplication operator and $\mathcal{F}$ is the Fourier transform:

$$(\mathcal{F}\varphi)(x, y) = (2\pi)^{-1}\int_{\mathbb{R}^2} e^{-i(xt+ys)}\varphi(t, s)\, dt ds, \quad \varphi \in L^2(\mathbb{R}^2).$$

Then *the unitary U transforms the operators P_1, Q_1, P_2, Q_2 in its canonical form*:

$$U^{-1}P_1 U = -i\frac{\partial}{\partial x}, \quad U^{-1}Q_1 U = x, \quad U^{-1}P_2 U = -i\frac{\partial}{\partial y}, \quad U^{-1}Q_2 U = y. \quad (8.84)$$

Indeed, as a sample, we verify the second equality of (8.84) and compute

$$U^{-1}Q_1U = \mathcal{F}e^{ixy}\mathcal{F}^{-1}\left(x + i\frac{\partial}{\partial y}\right)\mathcal{F}e^{-ixy}\mathcal{F}^{-1}$$

$$= \mathcal{F}e^{ixy}\left(-i\frac{\partial}{\partial x} + y\right)e^{-ixy}\mathcal{F}^{-1} = \mathcal{F}\left(-i\frac{\partial}{\partial x}\right)\mathcal{F}^{-1} = x.$$

The proofs of the other three equalities are similar; we omit the details.

8.9 Exercises

1. Show that the Weyl algebra W has no zero divisor.
2. Let $a, b, c, d \in \mathbb{R}$, $ad - bc = 1$. Represent the $*$-automorphism Ψ of W, defined by $\Psi(p) = ap + bq$, $\Psi(q) = cp + dq$, by a unitary operator U on $L^2(\mathbb{R})$, as in formula (8.32).
3. (*Wintner's theorem* [Wi47]) Let x and y be elements of a unital Banach algebra such that $xy - yx = \alpha \cdot 1$ for some $\alpha \in \mathbb{C}$. Prove that $\alpha = 0$.
   Hint: Verify that $x^n y - yx^n = \alpha n x^{n-1}$ and estimate the norms on both sides.
4. Prove that there is no representation π of W on a Hilbert space $\mathcal{H} \neq \{0\}$ such that $\pi(1) = I$ and both operators $\pi(p)$ and $\pi(q)$ are bounded.
5. Show that the Fock state $f(\cdot) := \langle \pi_C(\cdot)e, e\rangle$ (see Definition 8.10) is a pure state on W and $\pi_f(W)'_w = \bar{\pi}_f(W)'_w = \pi_C(W)'_s = \mathbb{C} \cdot I$.
6. Show that the creation operator A_C^+ is subnormal; that is, there exist a Hilbert space $\mathcal{K}$ and a normal operator T on $\mathcal{K}$ such that $l^2(\mathbb{N}_0) \subseteq \mathcal{K}$ and $A_C^+ \subseteq T$.
7. Let A be a densely defined closed operator such that $AA^* = A^*A + I$. Show that $(A - \lambda I)(A - \lambda I)^* = (A - \lambda I)^*(A - \lambda I) + I$ for any $\lambda \in \mathbb{C}$.
8. Let $\lambda \in \mathbb{C}$. Determine an orthonormal basis $\{f_n : n \in \mathbb{N}_0\}$ of $l^2(\mathbb{N}_0)$ such that

$$(A_C - \lambda I)f_n = \sqrt{n}\, f_{n-1}, \quad (A_C^+ - \bar{\lambda}I)f_n = \sqrt{n+1}\, f_{n+1}, \quad n \in \mathbb{N}_0, \quad f_{-1} := 0.$$

   Hint: Set $f_0 := e^{-|\lambda|^2/2}\sum_{n=0}^{\infty}\frac{\lambda^n}{\sqrt{n!}}e_n$ and $f_n := \frac{1}{\sqrt{n!}}(A_C^+ - \bar{\lambda}I)^n f_0, n \in \mathbb{N}$.

9. Let $\mathcal{D}(\pi) = \{\varphi \in C^{\infty}([0, 1]) : \varphi^{(k)}(0) = \varphi^{(k)}(1) = 0, k \in \mathbb{N}_0\}$ in $L^2(0, 1)$.

   a. Show that there is a $*$-representation π of W on $\mathcal{D}(\pi)$ such that $\pi(1) = I$, $\pi(p) = -i\frac{d}{dx}, \pi(q) = x$. (Note that the operator $\pi(q)$ is bounded!)
   b. Is there an extension of π to a larger Hilbert space such that the Weyl relation holds?

10. Suppose P and Q are self-adjoint operators on a Hilbert space. Let $V(s) = e^{isP}$ and $W(t) = e^{itQ}$ denote their one-parameter unitary groups, and let E_P and E_Q be their spectral measures. Show that the following are equivalent:

   (i) V and W satisfy the Weyl relation (8.44).

(ii) $V(s)QV(-s) = s \cdot I + Q$ for $t, s \in \mathbb{R}$.

(iii) $W(-t)PW(t) = t \cdot I + P$ for $t, s \in \mathbb{R}$.

(iv) $V(s)E_Q(\lambda)V(-s) = E_Q(\lambda - s)$ for $s, \lambda \in \mathbb{R}$.

(v) $W(-t)E_P(\lambda)W(t) = E_P(\lambda - t)$ for $t, \lambda \in \mathbb{R}$.

11. Find a *bounded* self-adjoint operator Q and a densely defined symmetric operator P such that $W(-t)PW(t)\varphi = (t \cdot I + P)\varphi$ for $\varphi \in \mathcal{D}(P), t \in \mathbb{R}$, where $W(t) := e^{itQ}$. Is it possible that P is self-adjoint?

12. Suppose two unitary groups $V(s) = e^{isP}$ and $W(t) = e^{itQ}$ on a Hilbert space satisfy the Weyl relation (8.44). Show without using the Stone-von Neumann theorem that $PQ\varphi - QP\varphi = -i\varphi$ for $\varphi \in \mathcal{D}(PQ) \cap \mathcal{D}(QP)$.

13. Relate the minimum uncertainty states (8.74) to the coherent states (8.24).

14. Show that the pair $\{A(p), A(q)\}$ of self-adjoint operators in (8.83) is unitarily equivalent to a direct sum of countably many Schrödinger pairs.

15. Show that there exists a linear map of $\mathcal{P}_3$ into $\mathcal{L}^+(\mathcal{S}(\mathbb{R}))$ satisfying (8.78) and (8.77) for $f, g \in \mathcal{P}_2$ (!). Is this map unique?

8.10   Notes

As noted in the introduction, the one-dimensional canonical commutation relation was first formulated by M. Born. It was first published by Born and Jordan [BJ26] and Dirac [D25].

Pioneering work on the Weyl algebra was done by Littlewood [Lt33] (who called it Dirac's quantum algebra) and Dixmier [Di68]. The name "Weyl algebra" (after Hermann Weyl) was appearently first used in [Di68]. Now there is an extensive literature about algebraic properties of the d-dimensional Weyl algebra. This algebra appears in the theory of enveloping algebras [Di77a], noncommutative rings [Lm99] and D-modules [Co95].

Theorem 8.4 was proved by Tillmann [Ti63]. Theorem 8.17 is due to Rellich [Re46] (under an additional assumption) and Dixmier [Di58]. The Stone–von Neumann Theorem 8.18 was announced by Stone [St30] and proved by von Neumann [vN31]. Theorem 8.22 is due to Kato [Ka63]. Further results are in [FGN60, FG63, Pu67, Fu82, Sch83b, In98]. Interesting discussions around the Stone–von Neumann theorem and the canonical commutation relations can be found in [S01, Rg03]. Exercise 8 is taken from [SS02]; this paper contains also a characterization of the creation operator among weighted shift operators.

The Fock space goes back to the Russian physicist Fock [F32]. It can be constructed over a general Hilbert space; our Fock space $\mathcal{F}(\mathbb{C})$ is the one-dimensional version realized as holomorphic functions. The Segal–Bargmann transform was discovered independently by Segal [Se62] in 1960 and Bargmann [Ba61] in 1961. A very readable book on these and related matters is [F89].

Theorem 8.28 was discovered by Groenewold [Gr46]. A gap in the proof was filled by van Hove [vH51]; see [Ch81, GGT96, Gt99] for modern and rigorous treatments. Geometric quantization and prequantization are out of the scope of this book; see, e.g., [GS84, Ha13, Wa07].

Most constructions and results of this chapter carry over to the d-dimensional Weyl algebra $W(d)$ and to the canonical commutation relations for finitely many degrees of freedom. In particular, the Stone–von Neumann uniqueness theorem remains valid; see, e.g., [F89, Theorem 6.49]. However, as first noted by Friedrichs [Fr53] and van Hove [vH52], this uniqueness theorem fails for infinitely many degrees of freedom; see, e.g., [BR97]. Haag [Hg55] showed that *physical* requirements imply that there are other representations than the Fock representation.

Chapter 9
Integrable Representations of Enveloping Algebras

Throughout this chapter, G is a finite-dimensional real Lie group, $\mathfrak{g}$ is its (real) Lie algebra, and $\mathcal{E}(\mathfrak{g})$ is the complexified universal enveloping algebra of $\mathfrak{g}$. For each unitary representation U of the Lie group G on a Hilbert space, there is an associated infinitesimal $*$-representation dU of the $*$-algebra $\mathcal{E}(\mathfrak{g})$ on the domain $\mathcal{D}^{\infty}(U)$ of C^{∞}-vectors. This chapter is about this important class of $*$-representations.

In Sect. 9.2, we define and develop the infinitesimal representation dU. The graph topology t_{dU} is studied in Sect. 9.3. We prove (Theorem 9.12) that each U-invariant dense subspace of $\mathcal{D}^{\infty}(U)$ is a core for all operators $dU(x)$, $x \in \mathcal{E}(\mathfrak{g})$. In Sect. 9.4, we use elliptic regularity theory as a powerful tool. We obtain results on descriptions of the domain $\mathcal{D}^{\infty}(U)$ (Theorems 9.22 and 9.30) and on the essential self-adjointness of (certain) symmetric operators (Theorems 9.22 and 9.23). In Sect. 9.5, we illustrate these general results for two examples. Section 9.6 gives an introduction into the theory of analytic vectors.

An important but difficult problem is the integrability problem. This is the question of when a representation of $\mathcal{E}(\mathfrak{g})$ is of the form dU for some unitary representation U. In Sect. 9.6.3, we state two fundamental integrability theorems, one due to Nelson and the other due to Flato, Simon, Snellman, and Sternheimer. The second result is based on analytic vectors, and it provides a useful criterion for proving the integrability of $*$-representations of enveloping algebras.

In Sect. 9.7, we treat K-finite vectors for a quasisimple unitary representation of $SL(2, \mathbb{R})$ and show that they are analytic vectors. As an application, we derive the oscillator representation.

In this chapter, we assume that the reader is familiar with the basics of Lie theory (see, e.g., [Va64, HN12, Kn96]) and invariant integration (see [F95]). In Sect. 9.1, we collect the notation and a number of facts that are used later in the exposition.

Throughout this chapter, U denotes a **unitary representation** of the Lie group G (according to Definition 4.54) on a Hilbert space $\mathcal{H}(U)$.

© The Editor(s) (if applicable) and The Author(s), under exclusive license
to Springer Nature Switzerland AG 2020
K. Schmüdgen, *An Invitation to Unbounded Representations of *-Algebras
on Hilbert Space*, Graduate Texts in Mathematics 285,
https://doi.org/10.1007/978-3-030-46366-3_9

9.1   Preliminaries on Lie Groups and Enveloping Algebras

Let μ_l denote a *left Haar measure* of G. Then, for any integrable function f on G,

$$\int_G f(hg)\,d\mu_l(g) = \int_G f(g)\,d\mu_l(g), \quad h \in G.$$

Let $\mathfrak{g}$ be the *Lie algebra* of G with Lie bracket $[\cdot,\cdot]$ and $x \mapsto \exp x$ the *exponential map* of $\mathfrak{g}$ into G. Each $x \in \mathfrak{g}$ acts as a *right-invariant vector field* $\tilde{x}$ on G by

$$(\tilde{x} f)(g) = \frac{d}{dt} f(\exp(-tx)g)\,|_{t=0}, \quad f \in C^\infty(G). \tag{9.1}$$

The map $x \mapsto \tilde{x}$ is a Lie algebra homomorphism; that is, it is $\mathbb{R}$-linear and satisfies

$$\widetilde{[x,y]} = \tilde{x}\tilde{y} - \tilde{y}\tilde{x}, \quad x,y \in \mathfrak{g}. \tag{9.2}$$

The group G acts on the Lie algebra $\mathfrak{g}$ by the *adjoint representation* Ad. For $g \in G$, $\mathrm{Ad}(g)$ is the differential of the inner automorphism $h \mapsto ghg^{-1}$. Then

$$\exp \mathrm{Ad}(g)x = g(\exp x)g^{-1} \quad \text{for } g \in G,\ x \in \mathfrak{g}. \tag{9.3}$$

The *convolution* of two functions $f_1, f_2 \in L^1(G;\mu_l)$ is

$$(f_1 * f_2)(g) := \int_G f_1(h) f_2(h^{-1}g)\,d\mu_l(h) = \int_G f_1(gh) f_2(h^{-1})\,d\mu_l(h). \tag{9.4}$$

There exists a unique continuous homomorphism Δ_G of G in the multiplicative group $(0,\infty)$, the *modular function* of G, such that $\mu_l(Mg) = \Delta_G(g)\mu_l(M)$ for $g \in G$ and Borel sets M. Then, for integrable functions $\varphi, f_1, f_2,$ and $g \in G$,

$$\int_G \varphi(hg)\,d\mu_l(h) = \Delta_G(g)^{-1} \int_G \varphi(h)\,d\mu_l(h), \tag{9.5}$$

$$\int_G \varphi(h^{-1})\,d\mu_l(h) = \int_G \varphi(h)\Delta_G(h)^{-1}\,d\mu_l(h), \tag{9.6}$$

$$(f_1 * f_2)(g) = \int_G f_1(gh^{-1}) f_2(h)\Delta_G(h)^{-1}\,d\mu_l(h).$$

Further, $d\mu_r(h) := \Delta_G(h)^{-1}d\mu_l(h)$ defines a right Haar measure μ_r on G.

The Banach space $L^1(G;\mu_l)$ is a Banach $*$-algebra with the convolution product (9.4) and the involution $f \mapsto f^+$, where

$$f^+(g) := \Delta_G(g)^{-1}\,\overline{f(g^{-1})}, \quad g \in G. \tag{9.7}$$

For the Lie group G, we have $\Delta_G(g) = |\det \operatorname{Ad}(g^{-1})|$ and hence $\Delta_G \in C^\infty(G)$. Therefore, by (9.7), the map $f \mapsto f^+$ leaves $C_0^\infty(G)$ invariant and it follows that $C_0^\infty(G)$ is a $*$-subalgebra of the $*$- algebra $L^1(G; \mu_l)$.

In exponential coordinates, the Haar measure has a C^∞-density with respect to the Lebesgue measure dx on $\mathfrak{g} \cong \mathbb{R}^d$ (see, e.g., [BGV04, Proposition 5.1]). Set $H(x) :=$ $|\det \zeta(\operatorname{ad} x)|$ for $x \in \mathfrak{g}$, where ζ is the holomorphic function $\zeta(z) = z^{-1}(1 - e^{-z})$ and $\operatorname{ad}(x)y = [x, y], x, y \in \mathfrak{g}$. Then $H(0) = 1$, and there is a neighborhood $\mathcal{O}$ of 0 in $\mathfrak{g}$ such that

$$H \in C^\infty(\mathcal{O}) \quad \text{and} \quad H(x) > 0, \quad d\mu_l(\exp x) = H(x)dx \quad \text{for } x \in \mathcal{O}. \tag{9.8}$$

Let $\mathcal{E}(\mathfrak{g})$ denote the complex universal enveloping algebra, briefly the *enveloping algebra*, of $\mathfrak{g}$. By definition, $\mathcal{E}(\mathfrak{g})$ is the quotient algebra of the tensor algebra over the complexification $\mathfrak{g}_{\mathbb{C}}$ of the real Lie algebra $\mathfrak{g}$ by the two-sided ideal generated by the elements $x \otimes y - y \otimes x - [x, y]$, where $x, y \in \mathfrak{g}$. We consider $\mathfrak{g}$ as a subspace of $\mathcal{E}(\mathfrak{g})$ by identifying $\mathfrak{g}$ with its image under the quotient map. The algebra $\mathcal{E}(\mathfrak{g})$ is a unital $*$-algebra with involution determined by $x^+ := -x$ for $x \in \mathfrak{g}$.

Let A be a unital algebra. A *homomorphism* of the Lie algebra $\mathfrak{g}$ into A is an $\mathbb{R}$-linear map $\theta : \mathfrak{g} \mapsto A$ such that $\theta([x, y]) = \theta(x)\theta(y) - \theta(y)\theta(x)$ for $x, y \in \mathfrak{g}$. By the universal property of the enveloping algebra, each Lie algebra homomorphism of $\mathfrak{g}$ into A has a unique extension to a unit preserving algebra homomorphism of $\mathcal{E}(\mathfrak{g})$ into A. In particular, by (9.2), the map $x \mapsto \tilde{x}$ defined by (9.1) extends to an algebra homomorphism of $\mathcal{E}(\mathfrak{g})$. Further, if A is a $*$-algebra and $\theta(x)^+ = \theta(x^+)$ for $x \in \mathfrak{g}$, then the extension of θ is a $*$-homomorphism.

Throughout this chapter, $\{x_1, \ldots, x_d\}$ denotes a fixed basis of the vector space $\mathfrak{g}$ and we set

$$x^{\mathfrak{n}} := x_1^{n_1} \cdots x_d^{n_d} \quad \text{for} \quad \mathfrak{n} = (n_1, \ldots, n_d) \in \mathbb{N}_0^d, \quad \text{where } x_j^0 := 1.$$

Then the *Poincare–Birkhoff–Witt theorem* says that the set $\{x^{\mathfrak{n}} : \mathfrak{n} \in \mathbb{N}_0^d\}$ is a basis of the complex vector space $\mathcal{E}(\mathfrak{g})$.

9.2 Infinitesimal Representations of Unitary Representations

Since the Lie group G is a topological group, Definition 4.54 applies to G. According to Definition 4.54, a *unitary representation* of the Lie group G on a Hilbert space $\mathcal{H}(U)$ is a homomorphism U of G into the group of unitary operators on $\mathcal{H}(U)$ such that for each vector $\varphi \in \mathcal{H}(U)$ the map $G \ni g \mapsto U(g)\varphi \in \mathcal{H}(U)$ is continuous. Equivalent forms of the continuity condition have been given in Lemma 4.55.

Definition 9.1 A vector $\varphi \in \mathcal{H}(U)$ is called a C^∞ -*vector* for U if the mapping

$$G \ni g \mapsto U(g)\varphi \in \mathcal{H}(U)$$

is of the class C^∞. The vector space of C^∞-vectors for U is denoted by $\mathcal{D}^\infty(U)$.

Let $h \in G$ and $\varphi \in \mathcal{D}^\infty(U)$. Being the composition of the two C^∞-maps $g \mapsto gh$ and $g \mapsto U(g)\varphi$, the map $g \mapsto U(g)U(h)\varphi = U(gh)\varphi$ is also C^∞, so it follows that $U(h)\varphi \in \mathcal{D}^\infty(U)$. This shows that $\mathcal{D}^\infty(U)$ is invariant under the representation U.

Now we turn to $*$-representations of the Lie algebra $\mathfrak{g}$.

Definition 9.2 Let $(\mathcal{D}, \langle \cdot, \cdot \rangle)$ be a complex inner product space. A $*$-*representation* of the Lie algebra $\mathfrak{g}$ on $\mathcal{D}$ is a linear mapping $\pi : \mathfrak{g} \mapsto L(\mathcal{D})$ such that for $x, y \in \mathfrak{g}$ and $\varphi, \psi \in \mathcal{D}$:

$$\pi([x, y]) = \pi(x)\pi(y) - \pi(y)\pi(x), \tag{9.9}$$

$$\langle \pi(x)\varphi, \psi \rangle = - \langle \varphi, \pi(x)\psi \rangle. \tag{9.10}$$

Recall that $L(\mathcal{D})$ denotes the algebra of linear operators mapping $\mathcal{D}$ into itself.

From the universal property of the enveloping algebra $\mathcal{E}(\mathfrak{g})$ and condition (9.9) it follows that π extends uniquely to an algebra homomorphism, denoted also π, of $\mathcal{E}(\mathfrak{g})$ into $L(\mathcal{D})$ such that $\pi(1) = I$. Since $x^+ = -x$ for $x \in \mathfrak{g}$, condition (9.10) means $\langle \pi(x)\varphi, \psi \rangle = \langle \varphi, \pi(x^+)\psi \rangle$. Since π is an algebra homomorphism, it follows easily (see Exercise 4.2) that the latter equation holds for all $x \in \mathcal{E}(\mathfrak{g})$. Hence π is a $*$-representation of the $*$-algebra $\mathcal{E}(\mathfrak{g})$ with domain $\mathcal{D}$. Summarizing, the extension of *each Lie algebra $*$-representation is a nondegenerate $*$-representation of the enveloping algebra $\mathcal{E}(\mathfrak{g})$.*

Let $x \in \mathfrak{g}$. Because $t \mapsto \exp tx$ is a continuous homomorphism of $\mathbb{R}$ into G, $t \mapsto U(\exp tx)$ is a strongly continuous one-parameter unitary group on $\mathcal{H}(U)$. Let $\partial U(x)$ denote its infinitesimal generator. Then $U(\exp tx) = \exp(t\partial U(x))$ for $t \in \mathbb{R}$. By Stone's theorem (Proposition A.5), $i\partial U(x)$ is a self-adjoint operator on $\mathcal{H}(U)$, so $\partial U(x)$ is skew-adjoint, and we have

$$\mathcal{D}(\partial U(x)) = \left\{ \varphi \in \mathcal{H}(U) : \partial(Ux)\varphi := \lim_{t \to 0} t^{-1}[U(\exp tx) - I]\varphi \text{ exists} \right\}.$$

From the latter we conclude that $\mathcal{D}^\infty(U) \subseteq \mathcal{D}(\partial U(x))$. For $x \in \mathfrak{g}$ we define

$$dU(x) = \partial U(x) \!\upharpoonright\! \mathcal{D}^\infty(U).$$

Then, by the definition of $\partial U(x)\varphi$ given above,

$$dU(x)\varphi = \frac{d}{dt}\big(U(\exp tx)\varphi\big)\,|_{t=0} \quad \text{for } \varphi \in \mathcal{D}^\infty(U),\ x \in \mathfrak{g}. \tag{9.11}$$

Proposition 9.3 *The map* $x \mapsto dU(x)$ *is a ∗-representation of the Lie algebra* $\mathfrak{g}$. *Its extension to the enveloping algebra* $\mathcal{E}(\mathfrak{g})$ *is a nondegenerate ∗-representation* dU *of the ∗-algebra* $\mathcal{E}(\mathfrak{g})$ *on the subspace* $\mathcal{D}(dU) := \mathcal{D}^\infty(U)$ *of* $\mathcal{H}(U)$. *Further,*

$$dU(\mathrm{Ad}(g)x) = U(g)dU(x)U(g^{-1}) \quad \text{for} \quad g \in G, \ x \in \mathfrak{g}. \tag{9.12}$$

Proof Throughout this proof, let $\varphi \in \mathcal{D}^\infty(U)$ and $x, y, z \in \mathfrak{g}$. From (9.11),

$$\lim_{t \to 0} t^{-1}[U(g \exp tx) - U(g)]\varphi = U(g) \lim_{t \to 0} t^{-1}[U(\exp tx) - I]\varphi = U(g)dU(x)\varphi.$$

Hence, since $g \mapsto U(g)\varphi$ is a C^∞-map of G into $\mathcal{H}(U)$, so is $g \mapsto U(g)dU(x)\varphi$. Therefore, $dU(x)\varphi \in \mathcal{D}^\infty(U)$, that is, $\mathcal{D}^\infty(U)$ is invariant under $dU(x)$.

As noted above, $\partial U(x)$, hence $dU(x)$, is skew-symmetric on $\mathcal{D}^\infty(U)$. Thus, condition (9.10) in Definition 9.2 is satisfied.

Next we show that condition (9.9) holds. Let $\psi \in \mathcal{H}(U)$. Since the inversion $g \mapsto g^{-1}$ is a C^∞-map of G, the function

$$f_{\varphi,\psi}(g) := \langle U(g^{-1})\varphi, \psi \rangle = \langle \varphi, U(g)\psi \rangle$$

belongs to $C^\infty(G)$. Then, using (9.1) and (9.11) we derive

$$\tilde{z}f_{\varphi,\psi}(g) = \frac{d}{dt}\big(\langle \varphi, U(\exp(-tz)g)\psi \rangle\big)\,|_{t=0} = \frac{d}{dt}\big(\langle U(\exp(tz))\varphi, U(g)\psi \rangle\big)\,|_{t=0}$$
$$= \langle dU(z)\varphi, U(g)\psi \rangle \equiv f_{dU(z)\varphi,\psi}(g). \tag{9.13}$$

Since $dU(z)\varphi \in \mathcal{D}^\infty(U)$, the function $f_{dU(z)\varphi,\psi}$ is of the same form as $f_{\varphi,\psi}$, so we can apply $\tilde{x}$ or $\tilde{y}$ to this function. Then, a repeated use of formula (9.13) yields

$$(\tilde{x}\tilde{y} - \tilde{y}\tilde{x})f_{\varphi,\psi}(g) = \langle (dU(x)dU(y) - dU(y)dU(x))\varphi, U(g)\psi \rangle, \tag{9.14}$$
$$\widetilde{[x,y]}f_{\varphi,\psi}(g) = \langle dU([x,y])\varphi, U(g)\psi \rangle. \tag{9.15}$$

By (9.2), the left-hand sides of (9.14) and (9.15) coincide, so do the right-hand sides. Setting $g = e$ and using that $\psi \in \mathcal{H}(U)$ was arbitrary, we obtain

$$dU([x,y]) = dU(x)dU(y) - dU(y)dU(x),$$

which proves (9.9). A similar, even simpler reasoning shows the linearity of the map $x \mapsto dU(x)$. Thus, dU is a ∗-representation of Lie algebra $\mathfrak{g}$. As discussed after Definition 9.2, dU extends uniquely to a nondegenerate ∗-representation of $\mathcal{E}(\mathfrak{g})$.

Finally, we verify formula (9.12). From (9.3) it follows that

$$U(\exp(t\,\mathrm{Ad}(g)x))\varphi = U(\exp(\mathrm{Ad}(g)tx))\varphi = U(g)U(\exp tx)U(g^{-1})\varphi. \tag{9.16}$$

Since $\mathcal{D}^{\infty}(U)$ is invariant under U, we have $U(g^{-1})\varphi \in \mathcal{D}^{\infty}(U)$. Hence we can differentiate both sides of equation (9.16) at $t = 0$ by using (9.11) and obtain $dU(\mathrm{Ad}(g)x)\varphi = U(g)dU(x)U(g^{-1})\varphi$. This proves (9.12).   □

Definition 9.4 The $*$-representation dU of the $*$-algebra $\mathcal{E}(\mathfrak{g})$ from Proposition 9.3 is called the *infinitesimal representation*, or the *derived representation*, or the *differential*, of the unitary representation U of the Lie group G.

In Proposition 9.3 it was only shown that dU is a $*$-representation of $\mathcal{E}(\mathfrak{g})$ on the complex inner product space $(\mathcal{D}^{\infty}(U), \langle \cdot, \cdot \rangle)$. From Proposition 9.6, (iv) and (v), below it follows that $\mathcal{D}^{\infty}(U)$ is dense in the Hilbert space $\mathcal{H}(U)$. Thus, dU is indeed a $*$-representation on the Hilbert space $\mathcal{H}(dU) = \mathcal{H}(U)$ in the sense of Sect. 4.1.

The representations of the form dU are the most important $*$-representations of enveloping algebras; they are called *integrable*; see also Definition 9.47 below.

A nice class of C^{∞}-vectors are the vectors in the range of the following "smoothing operator" U_f. Let $f \in C_0^{\infty}(G)$ and $\varphi \in \mathcal{H}(U)$. Since $g \mapsto f(g)U(g)\varphi$ is a continuous mapping of G into $\mathcal{H}(U)$, the $\mathcal{H}(U)$-valued Bochner integral

$$U_f\varphi := \int_G f(g)U(g)\varphi \, d\mu_l(g)$$

exists. Clearly, U_f is a bounded operator on $\mathcal{H}(U)$ and $\|U_f\| \leq \int_G |f(g)|d\mu_l$.

Definition 9.5 The linear span of vectors $U_f\varphi$, where $f \in C_0^{\infty}(G)$ and $\varphi \in \mathcal{H}(U)$, is called the *Gårding domain* of U and denoted by $\mathcal{D}_G(U)$.

Basic properties of these vectors $U_f\varphi$ are collected in the next proposition.

Proposition 9.6 *Suppose* $g \in G$, $x \in \mathcal{E}(\mathfrak{g})$, *and* $f, f_1, f_2 \in C_0^{\infty}(G)$. *Then:*

(i) $U(g)U_f = U_{f(g^{-1}\cdot)}$.
(ii) $U(g)\mathcal{D}_G(U) \subseteq \mathcal{D}_G(U)$.
(iii) $dU(x)U_f = U_{\tilde{x}f}$.
(iv) $dU(x)\mathcal{D}_G(U) \subseteq \mathcal{D}_G(U)$ *and* $\mathcal{D}_G(U) \subseteq \mathcal{D}^{\infty}(U)$.
(v) $\mathcal{D}_G(U)$ *is a dense linear subspace of* $\mathcal{H}(U)$.
(vi) $U_{f_1}U_{f_2} = U_{f_1 * f_2}$ *and* $U_{f^+} = (U_f)^*$.

Proof Throughout this proof, let $\varphi \in \mathcal{H}(U)$.

(i): Using the left invariance of the measure μ_l we obtain

$$U(g)U_f\varphi = \int_G f(h)U(gh)\varphi \, d\mu_l(h) = \int_G f(g^{-1}h)U(h)\varphi \, d\mu_l(h) = U_{f(g^{-1}\cdot)}\varphi.$$

(ii): Follows at once from (i), since $f(g^{-1}\cdot)$ is also in $C_0^{\infty}(G)$.
(iii) and (iv): First suppose $x \in \mathfrak{g}$. Let $t \in \mathbb{R}$. Then, by the preceding formula,

$$t^{-1}[U(\exp tx) - I]U_f\varphi = \int_G t^{-1}[f(\exp(-tx)g) - f(g)]U(g)\varphi\,d\mu_l(g).$$

$$(9.17)$$

Since $f \in C_0^\infty(G)$, we have $\lim_{t\to 0} t^{-1}[f(\exp(-tx)g) - f(g)] = (\tilde{x}f)(g)$ uniformly on G. Hence we can interchange integration and $\lim_{t\to 0}$ in Eq. (9.17) and obtain $U_f \in \mathcal{D}(\partial U(x))$ and $\partial U(x)U_f\varphi = U_{\tilde{x}f}\varphi$. Set $f_1(\cdot) := (\tilde{x}f)(g^{-1}\cdot)$. Then, $f_1 \in C_0^\infty(G)$ and

$$\frac{d}{dt}(U(g\exp tx)\,U_f\varphi)\,|_{t=0} = U(g)\partial U(x)U_f\varphi = U(g)U_{(\tilde{x}f)}\varphi = U_{f_1}\varphi.$$

This shows that all first-order partial derivatives of the map $g \mapsto U(g)U_f\varphi$ exist and are again of the form $U_{f_1}\varphi$ with $f_1 \in C_0^\infty(G)$. By a repeating application of this fact it follows that partial derivatives of arbitrary order exist. Hence the map $g \mapsto U(g)U_f\varphi$ is C^∞, so $U_f \in \mathcal{D}^\infty(U)$ and $\mathcal{D}_G(U) \subseteq \mathcal{D}^\infty(U)$. Since $U_f \in \mathcal{D}^\infty(U)$, we have $dU(x)U_f\varphi = \partial U(x)U_f\varphi = U_{\tilde{x}f}\varphi \in \mathcal{D}_G(U)$ for $x \in \mathfrak{g}$. Clearly, the relation $dU(x)U_f = U_{\tilde{x}f} \in \mathcal{D}_G(U)$ for $x \in \mathfrak{g}$ extends to elements x of the enveloping algebra $\mathcal{E}(\mathfrak{g})$.

(v): If $f \in C_0^\infty(G)$ is nonnegative and $\int f(g)d\mu_l = 1$, then

$$\|U_f\varphi - \varphi\| = \left\| \int f(g)(U(g) - U(e))\varphi\,d\mu_l(g) \right\| \le \sup_{g\in\mathrm{supp}\,f} \|(U(g) - U(e))\varphi\|.$$

Hence, by the continuity of U, we have $U_f\varphi \to \varphi$ in $\mathcal{H}(U)$ as supp f shrinks to $\{e\}$. This shows that each vector $\varphi \in \mathcal{H}(U)$ is in the closure of $\mathcal{D}_G(U)$.

(vi): Let $\varphi, \psi \in \mathcal{H}(U)$. Using (9.4), the left invariance of the Haar measure μ_l, and Fubini's theorem we compute

$$U_{f_1}U_{f_2}\varphi = \int_G f_1(g)U(g)\left(\int_G f_2(h)U(h)\varphi\,d\mu_l(h)\right)d\mu_l(g)$$

$$= \int_G f_1(g)U(g)\left(\int_G f_2(g^{-1}h)U(g^{-1}h)\varphi\,d\mu_l(h)\right)d\mu_l(g)$$

$$= \int_G \left(\int_G f_1(g)f_2(g^{-1}h)\,d\mu_l(g)\right)U(h)\varphi\,d\mu_l(h) = U_{f_1*f_2}\varphi.$$

Using that $U(g)^* = U(g^{-1})$ and formulas (9.7) and (9.6) we derive

$$\langle U_{f^+}\varphi, \psi \rangle = \int_G \overline{f(g^{-1})}\,\Delta_G(g)^{-1}\langle U(g)\varphi, \psi \rangle\,d\mu_l(g)$$

$$= \int_G \langle \varphi, f(g^{-1})\Delta_G(g)^{-1}U(g^{-1})\psi \rangle\,d\mu_l(g)$$

$$= \int_G \langle \varphi, f(g)U(g)\psi \rangle\,d\mu_l(g) = \langle \varphi, U_f\psi \rangle = \langle (U_f)^*\varphi, \psi \rangle,$$

which yields $U_{f^+} = (U_f)^*$. □

An immediate consequence of Proposition 9.6, (v) and (vi), is

Corollary 9.7 *The map $f \mapsto U_f$ is a nondegenerate $*$-representation of the $*$-algebra $C_0^\infty(G)$ on the Hilbert space $\mathcal{H}(U)$ such that $\|U_f\| \leq \int_G |f(g)| \, d\mu_l(g)$.*

The Gårding domain $\mathcal{D}_G(U)$ of a unitary representation U is better manageable than the space of C^∞-vectors, because it is closer related to the Lie group G itself. For instance, by Proposition 9.6, (i) and (iii), the actions of $U(g)$ and $dU(x)$ on a vector $U_f\varphi$ correspond to the left regular representation $U_l(g)$ (see (9.18) below) and the differential operator $\tilde{x}$ on the function f, respectively.

Remark 9.8 By a deep theorem of Dixmier and Malliavin [DM78], for each unitary representation U of a connected Lie group G the *Gårding domain $\mathcal{D}_G(U)$ is equal to the space $\mathcal{D}^\infty(U)$ of C^∞-vectors.* We will not use this result in this book.

More precisely, by [DM78, Theorem 3.1], each $f \in C_0^\infty(G)$ is a finite sum of functions of the form $f_1 * f_2$, where $f_1, f_2 \in C_0^\infty(G)$. (The case $G = \mathbb{R}^d$ is also treated in [RST78].) As a consequence, it is shown [DM78, Theorem 3.3] that each $\varphi \in \mathcal{D}^\infty(U)$ is a finite sum of vectors $U_f\psi$, where $f \in C_0^\infty(G)$, $\psi \in \mathcal{D}^\infty(U)$.   ○

We illustrate the preceding with an important example.

Example 9.9 (*Left and right regular representations*)
The *left regular representation* of the Lie group G is defined by

$$(U_l(g)\varphi)(h) = \varphi(g^{-1}h) \quad \text{for } g \in G, \ \varphi \in \mathcal{H}(U_l) := L^2(G; \mu_l). \tag{9.18}$$

Statement 1: *U_l is a unitary representation of the Lie group G.*

Proof The left invariance of the measure μ_l implies that each operator $U_l(g)$ is unitary. The map $g \mapsto U_l(g)$ is a group homomorphism, since

$$U_l(g_1)(U_l(g_2)\varphi)(h) = (U_l(g_2)\varphi)(g_1^{-1}h) = \varphi(g_2^{-1}g_1^{-1}h) = (U_l(g_1g_2)\varphi)(h).$$

The continuity is proved by the following standard argument. Using Lebesgue's dominated convergence theorem we first show that $\lim_{g \to e} U_l(g)\psi = \psi$ for ψ in $C_c(G)$. For vectors $\varphi \in L^2(G; \mu_l)$ we use the famous "3ε–trick." Given $\varepsilon > 0$ we can choose $\psi \in C_c(G)$ such that $\|\varphi - \psi\| \leq \varepsilon$ (since $C_c(G)$ is dense in $L^2(G; \mu_l)$) and a neighborhood $\mathcal{V}$ of e such that $\|U_l(g)\psi - \psi\| \leq \varepsilon$ on $\mathcal{V}$. Then, for $g \in \mathcal{V}$,

$$\|U_l(g)\varphi - \varphi\| = \|U_l(g)(\varphi - \psi) + (U_l(g)\psi - \psi) + (\psi - \varphi)\|$$
$$\leq 2\|\varphi - \psi\| + \|U_l(g)\psi - \psi\| \leq 3\varepsilon. \qquad \square$$

Statement 2: *$C_0^\infty(G) \subseteq \mathcal{D}^\infty(U_l)$ and $dU_l(x)f = \tilde{x}f$ for $f \in C_0^\infty(G)$, $x \in \mathcal{E}(\mathfrak{g})$. In particular, the $*$-representation dU_l of $\mathcal{E}(\mathfrak{g})$ is faithful.*

Proof First let $x \in \mathfrak{g}$. Since $\tilde{x} f \in C_0^\infty(G)$, $M := \sup\{|(\tilde{x} f)(h)| : h \in G\} < \infty$. Using the mean value theorem we conclude that

$$|f(\exp(-tx)g) - f(g)| \leq |t| \sup\left\{\left|\left(\frac{d}{ds} f\right)(\exp(-sx)g)\right| : |s| \leq |t|\right\}$$

$$\leq |t| \sup\{|(\tilde{x} f)(\exp(-sx)g)| : s \in \mathbb{R}\} \leq |t| M$$

and therefore

$$|t^{-1}[f(\exp(-tx)g) - f(g)] - (\tilde{x} f)(g)|^2$$

$$\leq 2|t^{-1}[f(\exp(-tx)g) - f(g)]|^2 + 2|(\tilde{x} f)(g)|^2 \leq 4M^2.$$

Hence the assumptions of Lebesgue's dominated convergence theorem are satisfied, so we can interchange integral and limit in the norm of $L^2(G; \mu_l)$ and obtain

$$\lim_{t \to 0} \|t^{-1}[f(\exp(-tx)\cdot) - f] - \tilde{x} f\| = \lim_{t \to 0} \|t^{-1}[U_l(\exp tx) - I)f] - \tilde{x} f\| = 0$$

in $L^2(G; \mu_l)$. Thus, $f \in \mathcal{D}(\partial U_l(x))$ and $\partial U_l(x) f = \tilde{x} f$. Since $\tilde{x} f$ is also in $C_0^\infty(G)$, arguing as in the proof of Proposition 9.6(iv), we derive $f \in \mathcal{D}^\infty(U_l)$. Then $dU_l(x) = \tilde{x} f$. By repeated application this extends to elements of $\mathcal{E}(\mathfrak{g})$.

A basic fact from Lie theory [Va64, Theorem 3.4.1] says that the map $x \mapsto \tilde{x}$ of $\mathcal{E}(\mathfrak{g})$ is injective on $C_0^\infty(G)$. Hence, if $dU_l(x) = 0$, then $dU_l(x)f = \tilde{x} f = 0$ for all $f \in C_0^\infty(G)$, so $\tilde{x} = 0$ and therefore $x = 0$. This proves that dU_l is faithful. $\square$

Let $f \in C_0^\infty(G)$ and $\varphi \in L^1(G; \mu_l) \cap L^2(G; \mu_l)$. Then we compute

$$(U_l)_f \varphi = \int_G f(h) U_l(h) \varphi \, d\mu_l(h) = \int_G f(h) \varphi(h^{-1} \cdot) \, d\mu_l(h) = f * \varphi. \quad (9.19)$$

That is, for the left regular representation U_l the element $(U_l)_f \varphi$ of the Gårding domain $\mathcal{D}_G(U_l)$ is just the convolution of the functions f and φ on G.

It can be shown [Pu72, Example 5.2] that the space of C^∞-vectors for U_l is

$$\mathcal{D}^\infty(U_l) = \{f \in C^\infty(G) : \tilde{x} f \in L^2(G; \mu_l) \text{ for all } x \in \mathcal{E}(\mathfrak{g})\}.$$

The *right regular representation* is the unitary representation U_r of the Lie group G acting on the Hilbert space $\mathcal{H}(U_r) := L^2(G; \mu_l)$ by

$$(U_r(g)\varphi)(h) = \Delta_G(g)^{1/2} \varphi(hg) \quad \text{for } g \in G, \; \varphi \in L^2(G; \mu_l). \quad (9.20)$$

The representations U_l and U_r are unitarily equivalent, where the equivalence is given by the unitary V defined by $(V\varphi)(h) := \Delta_G(h)^{-1/2} \varphi(h^{-1})$. That the operators $U_r(g)$ and V are unitary on $L^2(G; \mu_l)$ follows from (9.5) and (9.6). $\bigcirc$

9.3   The Graph Topology of the Infinitesimal Representation

The following proposition contains a description of the space $\mathcal{D}^\infty(U)$ and the graph topology t_{dU} (see Definition 4.8) in terms of the infinitesimal generators $\partial U(x_j)$. The proof uses only "elementary" Lie group techniques. A much stronger result based on elliptic regularity theory will be given in Theorem 9.30 below.

Proposition 9.10 *The locally convex space $\mathcal{D}^\infty(U)[t_{dU}]$ is a Frechet space,*

$$\mathcal{D}^\infty(U) = \bigcap_{k=1}^{\infty} \bigcap_{j_1,\ldots,j_k \in \{1,\ldots,d\}} \mathcal{D}(\partial U(x_{j_1}) \cdots \partial U(x_{j_k})), \tag{9.21}$$

and the graph topology t_{dU} on $\mathcal{D}^\infty(U)$ is given by the seminorms

$$\varphi \mapsto \|dU(x_{j_1} \cdots x_{j_k})\varphi\| = \|\partial U(x_{j_1}) \cdots \partial U(x_{j_k})\varphi\|, \tag{9.22}$$

where $j_1,\ldots,j_k \in \{1,\ldots,d\}$, $k \in \mathbb{N}$, together with the Hilbert space norm $\|\cdot\|$.

Proof We denote the right-hand side of (9.21) by $\mathcal{D}$. Let $\varphi \in \mathcal{D}^\infty(U)$. Since $\mathcal{D}^\infty(U)$ is invariant under dU by the proof of Proposition 9.3, φ is in the domain of $dU(x_{j_1}) \cdots dU(x_{j_k})$, hence of $\partial U(x_{j_1}) \cdots \partial U(x_{j_k})$. Thus $\varphi \in \mathcal{D}$.

We prove the converse inclusion $\mathcal{D} \subseteq \mathcal{D}^\infty(U)$. Let $\mathcal{D}_n$ denote the intersection of domains in (9.21) with $k \leq n$. For $t = (t_1,\ldots,t_d) \in \mathbb{R}^d$ and $j = 1,\ldots,d+1$, set

$$g_1(t) = e, \quad g_j(t) := \exp t_1 x_1 \cdots \exp t_{j-1} x_{j-1}, \ j \geq 2, \quad \text{and} \quad g(t) := g_{d+1}(t).$$

Let $\varphi \in \mathcal{D}_1$. We prove that the map $\mathbb{R}^d \ni t \mapsto F_\varphi(t) := U(g(t))\varphi \in \mathcal{H}(U)$ is differentiable at $t = 0$. Since U is a group homomorphism, we have

$$U(g(t)) - I = \sum_{j=1}^{d} U(g_j(t))[U(\exp t_j x_j) - I].$$

This equation implies the identity

$$F_\varphi(t) - F_\varphi(0) - \sum_{j=1}^{d} t_j \partial U(x_j)\varphi =$$

$$\sum_{j=1}^{d} \Big(U(g_j(t))[(U(\exp t_j x_j) - I)\varphi - t_j \partial U(x_j)\varphi] + [U(g_j(t)) - I]t_j \partial U(x_j)\varphi \Big).$$

Since $U(\cdot)$ is unitary and $|t_j| \leq \|t\|$, we therefore obtain

$$\|t\|^{-1}\left\|F_\varphi(t) - F_\varphi(0) - \sum_{j=1}^d t_j \partial U(x_j)\varphi\right\|$$

$$\leq \sum_{j=1}^d \Big(\|t_j^{-1}(U(\exp t_j x_j) - I)\varphi - \partial U(x_j)\varphi\| + \|(U(g_j(t)) - I)\partial U(x_j)\varphi\| \Big).$$

Since $\varphi \in \cap_j \mathcal{D}(\partial U(x_j))$ and U is continuous, the right side tends to 0 as $\|t\| \to 0$. Hence the derivative $F_\varphi'(0)$ exists and is the linear map of $\mathbb{R}^d$ into $\mathcal{H}(U)$ given by $F_\varphi'(0)(t) = \sum_{j=1}^d t_j \partial U(x_j)\varphi$. In particular, since $U(h)F_\varphi(t) = U(hg(t))\varphi$ and U is continuous, it follows that

$$\frac{\partial}{\partial t_j}(U(hg(t))\varphi)\,|_{t=0} = U(h)\partial U(x_j)\varphi, \quad j = 1, \ldots, d, \qquad (9.23)$$

for $h \in G$ and these partial derivatives are continuous. Therefore, since $t \mapsto hg(t)$ is a C^∞-isomorphism of a neighborhood of 0 in $\mathbb{R}^d$ on a neighborhood of h in G, we have shown that the map $g \mapsto U(g)\varphi$ of G into $\mathcal{H}(U)$ is C^1.

Suppose that $\varphi \in \mathcal{D}_2$. Let $i \in \{1, \ldots, d\}$. Then $\partial U(x_i)\varphi \in \mathcal{D}_1$ and hence

$$\frac{\partial}{\partial t_j}(U(hg(t))\partial U(x_i)\varphi)\,|_{t=0} = U(h)\partial U(x_j)\partial U(x_i)\varphi$$

by (9.23). Thus, $U(hg(t))$ has continuous second-order partial derivatives. Hence the map $g \mapsto U(g)\varphi$ is C^2. Now assume $\varphi \in \mathcal{D}_n$. Let $j_1, \ldots, j_n \in \{1, \ldots, d\}$. Then, by the definition of $\mathcal{D}_n$ we have $\varphi \in \mathcal{D}(\partial U(x_{j_1}) \cdots \partial U(x_{j_n}))$. By a repeated application of (9.23) it follows that the partial derivative $\frac{\partial}{\partial t_{j_1}} \cdots \frac{\partial}{\partial t_{j_n}}$ of $U(hg(t))\varphi$ at $t = 0$ is $U(h)\partial U(x_{j_1}) \cdots \partial U(x_{j_n})\varphi$. This implies that the map $g \mapsto U(g)\varphi$ is C^n. Therefore, $\mathcal{D} = \cap_n \mathcal{D}_n \subseteq \mathcal{D}^\infty(U)$, so that $\mathcal{D} = \mathcal{D}^\infty(U)$.

We turn to the graph topology $\mathfrak{t}_{dU}$. Since $dU(x_j) \subseteq \partial U(x_j)$ and $\mathcal{E}(\mathfrak{g})$ is the span of elements $1, x_{j_1} \cdots x_{j_k}$, the Hilbert space norm and the seminorms (9.22) generate the topology $\mathfrak{t}_{dU}$. This is a countable family of seminorms. Hence $\mathfrak{t}_{dU}$ is metrizable.

Let $(\varphi_n)_{n\in\mathbb{N}}$ be a Cauchy sequence in $\mathcal{D}^\infty(U)[\mathfrak{t}_{dU}]$. Then $(\varphi_n)_{n\in\mathbb{N}}$ and each sequence $(dU(x_{j_1}) \cdots x_{j_k})\varphi_n)_{n\in\mathbb{N}} = (\partial U(x_{j_1}) \cdots \partial U(x_{j_k})\varphi_n)_{n\in\mathbb{N}}$ are Cauchy sequences in $\mathcal{H}(U)$. Let φ denote the limit of the sequence $(\varphi_n)_{n\in\mathbb{N}}$ in $\mathcal{H}(U)$. Since each operator $\partial U(x_j)$ is closed, it follows that $\varphi \in \mathcal{D}(\partial U(x_{j_1}) \cdots \partial U(x_{j_k}))$ and $\lim_n \partial U(x_{j_1}) \cdots \partial U(x_{j_k})\varphi_n = \partial U(x_{j_1}) \cdots \partial U(x_{j_k})\varphi$. Hence $\varphi \in \mathcal{D}^\infty(U)$ by (9.21), and we have $\lim_n \varphi_n = \varphi$ in the graph topology $\mathfrak{t}_{dU}$. This proves that $\mathcal{D}^\infty(U)[\mathfrak{t}_{dU}]$ is complete and hence a Frechet space. $\qquad\square$

Proposition 9.11 *For each vector $\varphi \in \mathcal{D}^\infty(U)$, the map $g \mapsto U(g)\varphi$ of G into the locally convex space $\mathcal{D}^\infty(U)[\mathfrak{t}_{dU}]$ is a C^∞-mapping.*

Proof Fix $\varphi \in \mathcal{D}^\infty(U)$. Let $h \in G$ and put $\psi := U(h)\varphi \in \mathcal{D}^\infty(U)$. Note that $t = (t_1, \ldots, t_d) \mapsto g(t) = \exp t_1 x_1 \cdots \exp t_d x_d$ is a C^∞-map of $\mathbb{R}^d$ into G. The composition $(s, t) \mapsto U(g(s)g(t))\psi$ of the C^∞-mappings $(s, t) \mapsto g(s)g(t)$ of $\mathbb{R}^{2d}$ into G and $g \mapsto U(g)\psi$ of G into $\mathcal{H}(U)$ is a C^∞-mapping of $\mathbb{R}^{2d}$ into $\mathcal{H}(U)$.

For $\mathfrak{n} = (n_1, \ldots, n_d) \in \mathbb{N}_0^d$ we abbreviate $D_s^\mathfrak{n} := (\frac{\partial}{\partial s_1})^{n_1} \cdots (\frac{\partial}{\partial s_d})^{n_d}$. By (9.11), for $\mathfrak{n} \in \mathbb{N}_0^d$ and $t \in \mathbb{R}^d$ we obtain

$$dU(x^\mathfrak{n})U(g(t))\psi = D_s^\mathfrak{n}\big(U(g(s))U(g(t))\psi\big)\,|_{s=0} = D_s^\mathfrak{n}\big(U(g(s)g(t))\psi\big)\,|_{s=0}.$$

Hence the map $t \mapsto dU(x^\mathfrak{n})U(g(t))\psi$ of $\mathbb{R}^d$ into $\mathcal{H}(U)$ is C^∞. Since the $x^\mathfrak{n}$ span $\mathcal{E}(\mathfrak{g})$, the same is true for $t \mapsto dU(x)U(g(t))\psi$ with $x \in \mathcal{E}(\mathfrak{g})$. Since the operators $dU(x)$ are closable and $\mathcal{D}^\infty(U)[\mathfrak{t}_{dU}]$ is complete, each derivative $D_t^\mathfrak{n} U(g(t))\psi$ in the Hilbert space norm is equal to the derivative $D_t^\mathfrak{n}(U(g(t))\psi$ in the graph topology $\mathfrak{t}_{dU}$. Hence the map $t \mapsto U(g(t))\psi = U(g(t)h)\varphi$ of $\mathbb{R}^d$ into $\mathcal{D}^\infty(U)[\mathfrak{t}_{dU}]$ is C^∞ and so is the map $g \mapsto U(g)\varphi$ in a neighborhood of $h \in G$. $\qquad\square$

The following important and very useful result is *Poulsen's theorem*.

Theorem 9.12 *Let $\mathcal{D}$ be a linear subspace of $\mathcal{D}^\infty(U)$ which is dense in $\mathcal{H}(U)$. Suppose $\mathcal{D}$ is invariant under $U(g)$ for all elements g of the connected component G_0 of the unit element of G. Then $\mathcal{D}$ is dense in $\mathcal{D}^\infty(U)$ with respect to the graph topology $\mathfrak{t}_{dU}$ and $\mathcal{D}$ is a core for each operator $dU(x)$, $x \in \mathcal{E}(\mathfrak{g})$, that is,*

$$\overline{dU(x)\restriction\mathcal{D}} = \overline{dU(x)} \quad \text{for } x \in \mathcal{E}(\mathfrak{g}). \tag{9.24}$$

Proof Let $\overline{\mathcal{D}}$ denote the closure of $\mathcal{D}$ in the space $\mathcal{D}^\infty(U)[\mathfrak{t}_{dU}]$. Fix $\varphi \in \mathcal{D}$ and $f \in C_0^\infty(G_0)$. Let $x \in \mathcal{E}(\mathfrak{g})$. By Proposition 9.11, the map $g \mapsto U(g)\varphi$ is C^∞ in the graph topology $\mathfrak{t}_{dU}$. Hence the map $g \mapsto dU(x)U(g)\varphi$ of G into $\mathcal{H}(U)$ is continuous, so the Bochner integral $\int f(g)dU(x)U(g)\varphi \, d\mu_l$ exists and we have

$$dU(x)U_f\varphi = \int_{G_0} f(g)dU(x)U(g)\varphi \, d\mu_l(g). \tag{9.25}$$

Thus, the Bochner integral $U_f\varphi = \int f(g)U(g)\varphi \, d\mu_l$ exists in the graph topology. Hence $U_f\varphi$ is the $\mathfrak{t}_{dU}$-limit of Riemann sums for this integral. Since $\mathcal{D}$ is invariant under $U(g)$ for $g \in G_0$, these sums are in $\mathcal{D}$ and therefore $U_f\varphi \in \overline{\mathcal{D}}$.

Now let $\varphi \in \mathcal{D}^\infty(U)$. We prove that $U_f\varphi \in \overline{\mathcal{D}}$ for $f \in C_0^\infty(G_0)$. By assumption, $\mathcal{D}$ is dense in $\mathcal{H}(U)$. Hence there is a sequence $(\varphi_n)_{n \in \mathbb{N}}$ of vectors $\varphi_n \in \mathcal{D}$ such that $\lim_n \varphi_n = \varphi$ in $\mathcal{H}(U)$. Let $x \in \mathcal{E}(\mathfrak{g})$. Using Proposition 9.6(iii) and the fact that the smoothing operator $U_{\tilde{x}f}$ is bounded we obtain

$$\lim_{n \to \infty} dU(x)U_f\varphi_n = \lim_{n \to \infty} U_{\tilde{x}f}\varphi_n = U_{\tilde{x}f}\varphi = dU(x)U_f\varphi.$$

Hence $\lim_n U_f\varphi_n = U_f\varphi$ in $\mathcal{D}^\infty(U)[\mathfrak{t}_{dU}]$. In the preceding paragraph we have shown that $U_f\varphi_n \in \overline{\mathcal{D}}$. Thus $U_f\varphi \in \overline{\mathcal{D}}$.

Next we mimic the proof of Proposition 9.6(v) and choose a sequence $(f_n)_{n \in \mathbb{N}}$ of nonnegative functions $f_n \in C_0^\infty(G_0)$ such that $\int f_n(g) d\mu_l = 1$ for $n \in \mathbb{N}$ and supp f_n shrinks to $\{e\}$ as $n \to \infty$. Using (9.25) we derive

$$\|dU(x)(U_{f_n}\varphi - \varphi)\|$$
$$= \left\| dU(x) \int_{G_0} f_n(g) U(g) \varphi \, d\mu_l(g) - dU(x) \int_{G_0} f_n(g) \varphi \, d\mu_l(g) \right\|$$
$$= \left\| \int_{G_0} f_n(g) dU(x)(U(g) - I)\varphi \, d\mu_l(g) \right\|$$
$$\leq \sup \left\{ \|dU(x)(U(g) - I)\varphi\| : g \in \text{supp } f_n \right\}. \tag{9.26}$$

Since, by Proposition 9.11, the map $g \mapsto U(g)\varphi$ of G into $\mathcal{D}^\infty(U)[t_{dU}]$ is continuous, the expression in (9.26) tends to zero as $n \to \infty$. Therefore, $\lim_n U_{f_n}\varphi = \varphi$ in the locally convex space $\mathcal{D}^\infty(U)[t_{dU}]$. Since $U_{f_n}\varphi \in \overline{\mathcal{D}}$ as proved in the preceding paragraph, $\varphi \in \overline{\mathcal{D}}$. Thus, $\overline{\mathcal{D}} = \mathcal{D}^\infty(U)$, that is, $\mathcal{D}$ is dense in $\mathcal{D}^\infty(U)[t_{dU}]$. By the definition of the graph topology the latter means that $\mathcal{D}$ is a core for each operator $dU(x), x \in \mathcal{E}(\mathfrak{g})$. $\qquad\square$

By Proposition 9.6, (ii) and (v), the Gårding domain satisfies the assumptions of Theorem 9.12. Thus, Theorem 9.12 has the following corollary.

Corollary 9.13 *The Gårding domain $\mathcal{D}_G(U)$ is dense in the locally convex space $\mathcal{D}^\infty(U)[t_{dU}]$, and it is a core for each operator $dU(x), x \in \mathcal{E}(\mathfrak{g})$.*

According to Remark 9.8, if G is connected, we even have $\mathcal{D}_G(U) = \mathcal{D}^\infty(U)$ by the Dixmier–Malliavin theorem [DM78]. But Corollary 9.13 suffices for most applications.

9.4 Elliptic Elements

If $a = a^+ \in \mathcal{E}(\mathfrak{g})$, then $dU(a)$ is a symmetric operator. It is natural to ask whether this operator is essentially self-adjoint. For this and other problems elliptic elements play an important role. Let us begin with some technical preparations.

9.4.1 Preliminaries on Elliptic Operators

The first lemma says that weak C^∞-mappings into Hilbert spaces are (strong) C^∞-mappings.

Lemma 9.14 *Suppose $\varphi : \mathcal{O} \mapsto \mathcal{H}$ is a mapping from an open set $\mathcal{O}$ of $\mathbb{R}^d$ into a Hilbert space $\mathcal{H}$. Set $f_\eta(t) = \langle \eta, \varphi(t) \rangle$ for $\eta \in \mathcal{H}$ and $t \in \mathcal{O}$.*

(i) *If $f_\eta \in C^2(\mathcal{O})$ for all $\eta \in \mathcal{H}$, then φ is a C^1-mapping from $\mathcal{O}$ into $\mathcal{H}$.*
(ii) *If $f_\eta \in C^\infty(\mathcal{O})$ for all $\eta \in \mathcal{H}$, then φ is a C^∞-mapping from $\mathcal{O}$ into $\mathcal{H}$.*

Proof (i): Let D_k denote the directional derivative in the direction of the basis vector e_k of $\mathbb{R}^d$. Let $t \in \mathcal{O}$ and $\eta \in \mathcal{H}$. Then

$$\lim_{\lambda \to 0} \langle \eta, \lambda^{-1}(\varphi(t + \lambda e_k) - \varphi(t)) \rangle = D_k f_\eta(t).$$

Hence, by the Banach–Steinhaus theorem, $\eta \mapsto D_k f_\eta(t)$ is a continuous linear functional on $\mathcal{H}$, so by the Riesz theorem there exists a vector $\xi_k(t) \in \mathcal{H}$ such that

$$D_k f_\eta(t) = \langle \eta, \xi_k(t) \rangle, \quad \eta \in \mathcal{H}. \tag{9.27}$$

We fix $t \in \mathcal{O}$ and show that the map $s \mapsto \varphi(s)$ of $\mathcal{O}$ into $\mathcal{H}$ is continuous at t. We choose a compact convex neighborhood U of t in $\mathcal{O}$. Since $f_\eta \in C^2(\mathcal{O})$ by assumption, $D_k f_\eta(\cdot)$ is continuous, so $s \mapsto \xi_k(s)$ is weakly continuous. Therefore $\xi_k(U)$ is weakly compact, hence norm bounded in $\mathcal{H}$, so there is a $c > 0$ such that

$$\|\xi_k(s)\| \le c \quad \text{for } s \in U, \, k = 1, \dots, d. \tag{9.28}$$

Since U is convex, there exists an $\varepsilon > 0$ such that $t + a \in U$ for all $a = \sum_{i=1}^d \lambda_i e_i$ with $|\lambda_i| \le \varepsilon$. We fix such a vector a and set $a_j := \sum_{i=1}^{j-1} \lambda_i e_i$ for $j = 2, \dots, d$ and $a_1 := 0$. Using the mean value theorem, (9.27), and (9.28) we derive for $\eta \in \mathcal{H}$,

$$|\langle \eta, \varphi(t + a) - \varphi(t) \rangle| = |f_\eta(t + a) - f_\eta(t)|$$

$$= \left| \sum_{i=1}^d f_\eta(t + a_i + \lambda_i e_i) - f_\eta(t + a_i) \right|$$

$$\le \sum_{i=1}^d |\lambda_i| \sup \left\{ |D_i f_\eta(t + a_i + \gamma_i e_i)| : |\gamma_i| \le |\lambda_i| \right\}$$

$$= \sum_{i=1}^d |\lambda_i| \sup \left\{ |\langle \eta, \xi_i(t + a_i + \gamma_i e_i) \rangle| : |\gamma_i| \le |\lambda_i| \right\} \le \sum_{i=1}^d |\lambda_i| \|\eta\| c.$$

Therefore, $\|\varphi(t + a) - \varphi(t)\| \le c \sum_{i=1}^d |\lambda_i|$. This implies the continuity of φ at t.

Since $f_\eta \in C^2(\mathcal{O})$, the same reasoning applies to the map $t \mapsto \xi_k(t)$ of $\mathcal{O}$ into $\mathcal{H}$ and shows that this map is continuous. Hence, to complete the proof of (i) it suffices to show that $\xi_k(t) = D_k \varphi(t)$. By (9.27) we derive for small $\lambda \ne 0$,

$$|\langle \eta, \lambda^{-1}(\varphi(t + \lambda e_k) - \varphi(t)) - \xi_k(t) \rangle| = |\lambda^{-1}[f_\eta(t + \lambda e_k) - f_\eta(t)] - D_k f_\eta(t)|$$

$$= \left| \lambda^{-1} \int_0^\lambda (D_k f_\eta(t + \gamma e_k) - D_k f_\eta(t)) \, d\gamma \right|$$

$$\leq \|\eta\| \sup \left\{ \|\xi_k(t + \gamma e_k) - \xi_k(t)\| : |\gamma| \leq |\lambda| \right\},$$

so that

$$\|\lambda^{-1}(\varphi(t + \lambda e_k) - \varphi(t)) - \xi_k(t)\| \leq \sup \left\{ \|\xi_k(t + \gamma e_k) - \xi_k(t)\| : |\gamma| \leq |\lambda| \right\}.$$

Hence, by the continuity of $\xi_k(\cdot)$, $\lim_{\lambda \to 0} \lambda^{-1}(\varphi(t + \lambda e_k) - \varphi(t)) = \xi_k(t)$ in the norm of $\mathcal{H}$, that is, $\xi_k(t) = D_k \varphi(t)$.

(ii) follows from the following result. Using (i) one proves by induction on n:

If $f_\eta \in C^{n+1}(\mathcal{O})$ for all $\eta \in \mathcal{H}$, then φ is a C^n-mapping from $\mathcal{O}$ into $\mathcal{H}$. $\square$

The next proposition gives a *scalar characterization* of C^∞-vectors. The obvious implication (i)→(iii) was already used in the proof of Proposition 9.3.

Proposition 9.15 *Let $\varphi \in \mathcal{H}(U)$ and let $\mathcal{U}$ be an open subset of G. The following statements are equivalent:*

(i) $\varphi \in \mathcal{D}^\infty(U)$.

(ii) *For each $\psi \in \mathcal{H}(U)$, the function $\langle \psi, U(g)\varphi \rangle$ is in $C^\infty(\mathcal{U})$.*

(iii) *For each $\psi \in \mathcal{H}(U)$, the function $\langle U(g)\psi, \varphi \rangle$ is in $C^\infty(\mathcal{U})$.*

Proof (i)→(ii) is trivial. Clearly, $\langle \psi, U(hg)\varphi \rangle = \langle U(h^{-1})\psi, U(g)\varphi \rangle$ for $g, h \in G$. Hence (ii) implies $\langle \psi, U(g)\varphi \rangle \in C^\infty(G)$. Since $\langle \psi, U(g^{-1})\varphi \rangle = \langle U(g)\psi, \varphi \rangle$ and the map $g \mapsto g^{-1}$ of G is C^∞, (ii) for $\mathcal{U}^{-1}$ is equivalent to (iii) for $\mathcal{U}$. From these two facts it follows that it suffices to show that $\langle \psi, U(g)\varphi \rangle \in C^\infty(G)$ for all $\psi \in \mathcal{H}(U)$ implies (i).

We fix $g_0 \in G$ and choose a diffeomorphism $t \mapsto g(t)$ of an open subset $\mathcal{O} \subseteq \mathbb{R}^d$ on a neighborhood $\mathcal{U}_0$ of g_0. Set $\varphi(t) := U(g(t))\varphi$. Since $\langle \psi, U(g)\varphi \rangle \in C^\infty(\mathcal{U}_0)$ by the assumption, the map $\varphi(t) : \mathcal{O} \mapsto \mathcal{H}(U)$ is C^∞ by Lemma 9.14(ii). Hence $\mathcal{U}_0 \ni g \mapsto U(g)\varphi \in \mathcal{H}(U)$ is C^∞, which proves that $\varphi \in \mathcal{D}^\infty(U)$. $\square$

Now we consider elliptic operators on a bounded domain $\mathcal{O}$ of $\mathbb{R}^d$. Recall the notation $D_t^\alpha := (\frac{\partial}{\partial t_1})^{\alpha_1} \cdots (\frac{\partial}{\partial t_d})^{\alpha_d}$ for $\alpha \in \mathbb{N}_0^d$, $t \in \mathbb{R}^d$. Let $m \in \mathbb{N}$.

Let $b_\alpha(t) \in C^\infty(\mathcal{O})$ be given for $\alpha \in \mathbb{N}_0^d$, $|\alpha| \leq m$. The differential operator

$$T := \sum_{|\alpha| \leq m} b_\alpha(t) D_t^\alpha$$

is called *uniformly elliptic* of order m on $\mathcal{O}$ if there exists a constant $c > 0$ such that

$$\left| \sum_{|\alpha| = m} b_\alpha(t) \xi^\alpha \right| \geq c |\xi|^m \quad \text{for} \quad \xi \in \mathbb{R}^d, \ t \in \mathcal{O}. \tag{9.29}$$

The following result is essentially based on the *elliptic regularity theorem* combined with the embedding of Sobolev spaces.

Lemma 9.16 *Suppose* $T = \sum_{|\alpha| \leq m} b_\alpha(t) D_t^\alpha$ *is a uniformly elliptic differential operator of order m on the bounded domain $\mathcal{O}$ of $\mathbb{R}^d$. Let $u, v \in C(\mathcal{O})$. If*

$$\int_{\mathcal{O}} (Tf)(t)u(t)\, dt = \int_{\mathcal{O}} f(t)v(t)\, dt \quad \text{for all } f \in C_0^\infty(\mathcal{O}), \tag{9.30}$$

then $u \in C^k(\mathcal{O})$ for $k \in \mathbb{N}_0$, $k < m - d/2$. In particular, $u \in C^{m-d}(\mathcal{O})$ if $m \geq d$.

Proof The formal adjoint $T^\dagger := \sum_\alpha (-1)^{|\alpha|} D_t^\alpha b_\alpha$ has the homogeneous part $(-1)^m \sum_{|\alpha|=m} b_\alpha D_t^\alpha$ of highest degree, so $T^\dagger$ is also uniformly elliptic on $\mathcal{O}$. Now (9.30) implies that $\int f(T^\dagger u)\, dt = \int f\, v\, dt$ for $f \in C_0^\infty(\mathcal{O})$. This means the equation $T^\dagger u = v$ holds in the sense of distributions. Since $v \in C(\mathcal{O})$, it follows from the elliptic regularity theorem (see [BJS66, p. 190, Theorem 1] or [Ag15, Theorem 6.2.3]), applied to $T^\dagger$, that u belongs to the local Sobolev space $H_{\text{loc}}^m(\mathcal{O})$. Therefore, by the Sobolev embedding theorem [Ag15], we get $u \in C^k(\mathcal{O})$ for $k < m - d/2$. In particular, if $m - d \geq 0$, then $k := m - d < m - d/2$ and hence $u \in C^{m-d}(\mathcal{O})$. $\square$

Next we turn to elliptic elements of $\mathcal{E}(\mathfrak{g})$. Let $a \in \mathcal{E}(\mathfrak{g}), a \neq 0$. Recall that, by the Poincare–Birkhoff–Witt theorem, $\{x^n : n \in \mathbb{N}_0^d\}$ is a vector space basis of $\mathcal{E}(\mathfrak{g})$. Hence a can be written as

$$a = \sum_{n \in \mathbb{N}_0^d, |n| \leq m} \gamma_n x^n, \quad m \in \mathbb{N}_0, \tag{9.31}$$

where $\gamma_n \in \mathbb{C}$ are uniquely determined by a. We say that a has *degree m* and write $m = \deg(a)$ if there is an $n \in \mathbb{N}_0^d$ such that $|n| = m$ and $\gamma_n \neq 0$. To any element a of degree m we associate a homogeneous polynomial $p_{a,m} \in \mathbb{C}[\xi_1, \ldots, \xi_d]$ by

$$p_{a,m}(\xi) := \sum_{|n|=m} \gamma_n \xi^n.$$

Definition 9.17 The element $a \in \mathcal{E}(\mathfrak{g})$ in (9.31) is called *elliptic* if a has degree $m \in \mathbb{N}$ and the polynomial $p_{a,m}(\xi)$ is nonzero for all nonzero $\xi \in \mathbb{R}^d$.

A typical example of an elliptic element is $a = x_1^{2k} + \cdots + x_d^{2k}, k \in \mathbb{N}$. Note that according to Definition 9.17 elliptic elements have always positive degrees.

Lemma 9.18 *If $a, b \in \mathcal{E}(\mathfrak{g})$ are elliptic, so are ab and a^+.*

Proof Let $k = \deg(a)$ and $n = \deg(b)$. Since $x^n x^m - x^{n+m}$ has a lower degree than $x^n x^m$, we have $kn = \deg(ab)$ and $p_{kn,ab}(\xi) = p_{k,a}(\xi)p_{n,b}(\xi)$, so ab is elliptic. Since $x_j^+ = -x_j$, $(-1)^{|n|} x^n - x^n$ has a lower degree than x^n. This implies that $k = \deg(a^+)$ and $p_{k,a^+}(\xi) = (-1)^k \overline{p_{k,a}(\xi)}$ on $\mathbb{R}^d$, so a^+ is also elliptic. $\square$

An element $a = \sum_n \gamma_n x^n \in \mathcal{E}(\mathfrak{g})$ of degree $m \in \mathbb{N}$ is elliptic if and only if there exists a constant $c > 0$ such that

$$|p_{a,m}(\xi)| \geq c|\xi|^m \quad \text{for} \quad \xi \in \mathbb{R}^d. \tag{9.32}$$

The if part is trivial. Conversely, let c be the minimum of the continuous function $p_{a,m}(\xi)$ on the unit sphere. Since a is elliptic, $c > 0$ and (9.32) follows by scaling.

The following lemma applies Lemma 9.16 to elliptic elements of $\mathcal{E}(\mathfrak{g})$.

Lemma 9.19 *Suppose $a \in \mathcal{E}(\mathfrak{g})$ is elliptic and $\deg(a) = m \geq d$. Let $\varphi \in \mathcal{H}(U)$. Suppose for each $\psi \in \mathcal{H}(U)$ there exist vectors $\zeta, \eta \in \mathcal{H}(U)$ such that*

$$\int_G (\tilde{a} f)(g) \, \langle U(g)\psi, \varphi \rangle \, d\mu_l(g) = \int_G f(g) \, \langle U(g)\zeta, \eta \rangle \, d\mu_l(g) \tag{9.33}$$

for $f \in C_0^\infty(G)$. Then, for all $\psi \in \mathcal{H}(U)$, the function $\langle U(g)\psi, \varphi \rangle$ is in $C^{m-d}(G)$.

Proof Set $F(g) := \langle U(g)\psi, \varphi \rangle$ and $F_0(g) := \langle U(g)\zeta, \eta \rangle$. We rewrite (9.33) in terms of exponential coordinates. The C^∞-map $t \mapsto g(t) := \exp(t_1 x_1 + \cdots + t_d x_d)$ is a diffeomorphism of a neighborhood of 0 in $\mathbb{R}^d$ on a neighborhood of e in G. The following statements hold for all t in some appropriate neighborhood of 0 in $\mathbb{R}^d$ and for all $f \in C_0^\infty(G)$ that are supported on some appropriate neighborhood of e in G.

We express the Haar measure μ_l and the operators $\widetilde{x^\alpha}$ in exponential coordinates $g(t)$. From (9.8) we recall that in some neighborhood of 0 in $\mathfrak{g}$ there exists a C^∞-function $H(t) > 0$ such that $d\mu_l(g(t)) = H(t)dt$. For $\alpha \in \mathbb{N}_0^d$, there exists a C^∞-function b_α such that $b_\alpha(0) = (-1)^{|\alpha|}$ and $(\widetilde{x^\alpha} f)(g)_{|g=g(t)} = b_\alpha(t)(D_t^\alpha(f \circ g))(t)$. Let $a = \sum_{|\alpha| \leq m} \gamma_\alpha x^\alpha$. We define $T := \sum_{|\alpha| \leq m} \gamma_\alpha b_\alpha(t) D_t^\alpha$. Then we have $(\tilde{a} f)(g)_{|g=g(t)} = (T(f \circ g))(t)$. Since a is elliptic, there is a constant $c' > 0$ such that $|\sum_{|\alpha|-m} \gamma_\alpha \xi^\alpha| \geq c'|\xi|^m, \xi \in \mathbb{R}^d$. Since the functions $b_\alpha(t)$ are continuous and $b_\alpha(0) = (-1)^m$ for $|\alpha| = m$, it follows that (9.29) holds, with a positive constant $c < c'$, for t in some neighborhood of 0. Now we choose open neighborhoods $\mathcal{O}$ of 0 and $\mathcal{U}$ of e such that $\mathcal{U} = g(\mathcal{O})$ and all preceding facts hold for $t \in \mathcal{O}$ and $f \in C_0^\infty(\mathcal{U})$. Then the differential operator T is uniformly elliptic on $\mathcal{O}$ and for $f \in C_0^\infty(\mathcal{U})$ Eq. (9.33) reads as

$$\int_{\mathcal{O}} (T(f \circ g))(t) \, F(g(t)) H(t) \, dt = \int_{\mathcal{O}} (f \circ g)(t)) \, F_0(g(t)) H(t) dt. \tag{9.34}$$

All functions of $C_0^\infty(\mathcal{O})$ are of the form $f \circ g$ with $f \in C_0^\infty(\mathcal{U})$. Since the representation U is continuous, so are the functions $F(g), F_0(g)$ on G and hence $F(g(t)), F_0(g(t))$ on $\mathcal{O}$. Thus, the assumptions of Lemma 9.16 are satisfied, with $u(t) = F(g(t))H(t), \ v(t) = F_0(g(t))H(t)$. Then, $F(g(t))H(t) \in C^{m-d}(\mathcal{O})$ by Lemma 9.16. Since $H(t) > 0$ on $\mathcal{O}$ and $H \in C^\infty(\mathcal{O})$, $F(g(t)) \in C^{m-d}(\mathcal{O})$ and so $F(g) = \langle U(g)\psi, \varphi \rangle \in C^{m-d}(\mathcal{U})$. Replacing ψ by $U(h)\psi$, we conclude that $F(gh) \in C^{m-d}(\mathcal{U})$ for all $h \in G$. Therefore, $F(g) \in C^{m-d}(G)$. $\square$

9.4.2 Main Results on Elliptic Elements

The next lemma contains the crucial technical steps for our first main theorem.

Lemma 9.20 (i) *Suppose $a \in \mathcal{E}(\mathfrak{g})$ is an elliptic element and $m := \deg(a) \geq d$. If $\varphi \in \mathcal{D}(dU(a)^*)$ and $\psi \in \mathcal{H}(U)$, then the function $\langle U(g)\psi, \varphi \rangle$ is in $C^{m-d}(G)$.*
(ii) *Suppose $a_n \in \mathcal{E}(\mathfrak{g})$, $n \in \mathbb{N}$, are elliptic and $\sup_{n \in \mathbb{N}} \deg(a_n) = +\infty$. Then*

$$\bigcap_{n=1}^{\infty} \mathcal{D}(dU(a_n)^*) \subseteq \mathcal{D}^{\infty}(U). \tag{9.35}$$

Proof (i): Let $f \in C_0^{\infty}(G)$. Using Proposition 9.6(iii) we derive

$$\int_G (\tilde{a}f)(g)\langle U(g)\psi, \varphi \rangle \, d\mu_l(g) = \left\langle \int_G (\tilde{a}f)(g)U(g)\psi \, d\mu_l(g), \varphi \right\rangle$$

$$= \langle U_{\tilde{a}f}\psi, \varphi \rangle = \langle dU(a)U_f\psi, \varphi \rangle = \langle U_f\psi, dU(a)^*\varphi \rangle$$

$$= \int_G f(g)\langle U(g)\psi, dU(a)^*\varphi \rangle \, d\mu_l(g).$$

for all $\psi \in \mathcal{H}(U)$. Therefore, $\langle U(g)\psi, \varphi \rangle \in C^{m-d}(G)$ by Lemma 9.19.
(ii): Suppose $\varphi \in \cap_{n=1}^{\infty} \mathcal{D}(dU(a_n)^*)$. Let $\psi \in \mathcal{H}(U)$. Then, by (i), we have $\langle U(g)\psi, \varphi \rangle \in C^{m_n - d}(G)$ if $m_n := \deg(a_n) \geq d$. Since $\sup_n m_n = \infty$, it follows that $\langle U(g)\psi, \varphi \rangle \in C^{\infty}(G)$ for all $\psi \in \mathcal{H}(U)$. Therefore, $\varphi \in \mathcal{D}^{\infty}(U)$ by Proposition 9.15, (iii)$\rightarrow$(i). $\qquad\square$

Corollary 9.21 *The $*$-representation dU of $\mathcal{E}(\mathfrak{g})$ is self-adjoint.*

Proof Take an elliptic element $a \in \mathcal{E}(\mathfrak{g})$. Then, by Lemma 9.20(ii), with $a_n := a^n$,

$$\mathcal{D}((dU)^*) \subseteq \cap_n \mathcal{D}(dU(a^n)^*) \subseteq \mathcal{D}^{\infty}(U) = \mathcal{D}(dU).$$

Since dU is a $*$-representation, $dU \subseteq (dU)^*$. Thus $dU = (dU)^*$. $\qquad\square$

The following theorem collects our main results concerning elliptic elements. Recall that for a Hilbert space operator T we abbreviate $\mathcal{D}^{\infty}(T) := \cap_{n=1}^{\infty} \mathcal{D}(T^n)$.

Theorem 9.22 *Suppose $a \in \mathcal{E}(\mathfrak{g})$ is an elliptic element. Then*

$$\overline{dU(a^+)} = dU(a)^*. \tag{9.36}$$

In particular, if $a = a^+$, then $\overline{dU(a)}$ is self-adjoint. Further,

$$\mathcal{D}^{\infty}(U) = \mathcal{D}^{\infty}(\overline{dU(a)}), \tag{9.37}$$

the locally convex space $\mathcal{D}^{\infty}(U)[\mathsf{t}_{dU}]$ is a Frechet space, and the graph topology t_{dU} is given by the family of seminorms $\varphi \mapsto \|dU(a^n)\varphi\|, n \in \mathbb{N}_0$.

Proof First assume $a = a^+$. Let $\xi \in \ker(dU(a)^* - \lambda I)$ for some $\lambda \in \mathbb{C}\backslash\mathbb{R}$. Then $\xi \in \mathcal{D}((dU(a)^*)^n) \subseteq \mathcal{D}((dU(a)^n)^*) = \mathcal{D}(dU(a^n)^*)$ for $n \in \mathbb{N}$. Hence $\xi \in \mathcal{D}^\infty(U)$ by Lemma 9.20(ii), applied with $a_n = a^n$. Therefore, since $a = a^+$, it follows that $dU(a)\xi = dU(a)^*\xi = \lambda\xi$. Since $\lambda \notin \mathbb{R}$ and $dU(a)$ is symmetric, $\xi = 0$. Hence $dU(a)$ is essentially self-adjoint by Proposition A.1, so $\overline{dU(a)}$ is self-adjoint.

Now let a be an arbitrary elliptic element. Then aa^+ is hermitian and elliptic by Lemma 9.18, so $dU(aa^+) = dU(a)dU(a)^+$ is essentially self-adjoint, as shown in the preceding paragraph. Hence $\overline{dU(a^+)} = \overline{dU(a)}^+ = dU(a)^*$ by Lemma 7.1, applied with $x = dU(a)$. This proves (9.36).

Set $a_n := (a^+)^n$ for $n \in \mathbb{N}$. By Lemma 9.18, a_n is also elliptic. Then, since $\sup_n \deg a_n = +\infty$, Lemma 9.20(ii) yields

$$\bigcap_{n=1}^{\infty} \mathcal{D}(dU((a^+)^n)^*) \subseteq \mathcal{D}^\infty(U). \tag{9.38}$$

On the other hand, $dU(a) \subseteq dU(a^+)^*$, so that $\overline{dU(a)} \subseteq dU(a^+)^*$. Therefore, $(\overline{dU(a)})^n \subseteq (dU(a^+)^*)^n \subseteq (dU(a^+)^n)^* = dU((a^+)^n)^*$, and

$$\mathcal{D}^\infty(\overline{dU(a)}) \subseteq \bigcap_{n=1}^{\infty} \mathcal{D}(dU((a^+)^n)^*). \tag{9.39}$$

Combining (9.38) and (9.39) yields $\mathcal{D}^\infty(\overline{dU(a)}) \subseteq \mathcal{D}^\infty(U)$. Since the opposite inclusion holds trivially, we obtain (9.37).

Let $\mathfrak{t}$ denote the locally convex topology on $\mathcal{D}^\infty(U)$ generated by the seminorms $\varphi \mapsto \|dU(a^n)\varphi\|, n \in \mathbb{N}_0$. Since $\mathcal{D}^\infty(U) = \mathcal{D}^\infty(\overline{dU(a)})$ for the closed operator $\overline{dU(a)}$, $\mathcal{D}^\infty(U)[\mathfrak{t}]$ is a Frechet space. From the closed graph theorem it follows that $\mathfrak{t}_{dU} \subseteq \mathfrak{t}$; see also Lemma 3.6. Thus, $\mathfrak{t}_{dU} = \mathfrak{t}$. □

9.4.3 Applications of Elliptic Elements

Our main application to self-adjointness questions is the following theorem.

Theorem 9.23 *Suppose a is an elliptic element of $\mathcal{E}(\mathfrak{g})$ and $T \in \mathcal{L}^+(\mathcal{D}^\infty(U))$ such that T and T^+ commute with $dU(a)$ on $\mathcal{D}^\infty(U)$. Then TT^+ is essentially self-adjoint and $\overline{T^+} = T^*$. In particular, each symmetric operator on $\mathcal{D}^\infty(U)$, which leaves $\mathcal{D}^\infty(U)$ invariant and commutes with $dU(a)$, is essentially self-adjoint.*

Proof The closed graph theorem implies that the operator TT^+ maps the Frechet space $\mathcal{D}^\infty(U)[\mathfrak{t}_{dU}]$ continuously in the Hilbert space $\mathcal{H}(U)$. By Lemma 9.18, the element $b := a^+a + 1$ is also elliptic. Therefore, by Theorem 9.22, the graph topology $\mathfrak{t}_{dU}$ is generated by the family of seminorms $\|dU(b^n) \cdot \|, n \in \mathbb{N}_0$. It follows easily that $\|dU(b^n) \cdot \| \leq \|dU(b^{n+1}) \cdot \|$ for $n \in \mathbb{N}_0$. Hence there exist numbers $n \in \mathbb{N}$ and $c > 0$ such that $\|\varphi\| + \|TT^+\varphi\| \leq c\|dU(b^n)\varphi\|$ for $\varphi \in \mathcal{D}^\infty(U)$.

Since T and T^+ commute with $dU(a)$ in the $*$-algebra $\mathcal{L}^+(\mathcal{D}^\infty(U))$, both operators commute also with $dU(a^+) = dU(a)^+$. Hence TT^+ commutes with

$dU(b^n) = (dU(a)dU(a)^+ + I)^n$. Since b^n is elliptic and $(b^n)^+ = b^n$, $\overline{dU(b^n)}$ is self-adjoint by Theorem 9.22. Obviously, $dU(b^n) \geq 0$. Thus, the assumptions of Lemma 7.3, with $y = TT^+$ and $x = dU(b^n)$, are fulfilled. Therefore, TT^+ is essentially self-adjoint by Lemma 7.3 and hence $\overline{T^+} = T^*$ by Lemma 7.1. If T is symmetric, the latter means that T is essentially self-adjoint. $\square$

An equivalent form of the commutativity assumption in Theorem 9.23 is that T commutes with $dU(a)$ and $dU(a^+)$. This formulation is used in the next corollary.

Corollary 9.24 *Let $a, b \in \mathcal{E}(\mathfrak{g})$, and suppose a is elliptic. If $dU(b)dU(a) = dU(a)dU(b)$ and $dU(b)dU(a^+) = dU(a^+)dU(b)$, then $\overline{dU(b^+)} = dU(b)^*$.*

Proof Apply Theorem 9.23 with $T = dU(b)$. $\square$

We derive a number of interesting corollaries of Theorem 9.23.

Corollary 9.25 *If G is abelian, then $\overline{dU(x^+)} = dU(x)^*$ for all $x \in \mathcal{E}(\mathfrak{g})$.*

Proof Since G is abelian, $\mathcal{E}(\mathfrak{g})$ is commutative. Hence $dU(x)$, $x \in \mathcal{E}(\mathfrak{g})$, commutes with $dU(a)$ for each elliptic element $a \in \mathcal{E}(\mathfrak{g})$, so Corollary 9.24 applies. $\square$

Corollary 9.26 *Let $\mathcal{Z}$ be the center of $\mathcal{E}(\mathfrak{g})$. Then $\overline{dU(x^+)} = dU(x)^*$ for $x \in \mathcal{Z}$. If x_1 and x_2 are hermitian elements of $\mathcal{Z}$, then $\overline{dU(x_1)}$ and $\overline{dU(x_2)}$ are strongly commuting self-adjoint operators.*

Proof Since $x \in \mathcal{Z}$ commutes with all elliptic elements, Corollary 9.24 gives the first assertion. The second assertion follows by combining the first assertion with Lemma 7.2(ii), applied to $x := x_1 + ix_2 \in \mathcal{Z}$. $\square$

Corollary 9.27 *Suppose $x \in \mathfrak{g}$. Then, for each polynomial $p \in \mathbb{C}[x]$,*

$$\overline{dU(p(-ix)^+)} = dU(p(-ix))^*. \tag{9.40}$$

In particular, for $n \in \mathbb{N}$, $dU(-ix)^n = dU((-ix)^n)$ is essentially self-adjoint,

$$\overline{dU(x)^n} = \partial U(x)^n \quad \text{and} \quad U(\exp tx) = \exp t\,\overline{dU(x)}, \; t \in \mathbb{R}. \tag{9.41}$$

Proof There is a unitary representation U_1 of the Lie group $G_1 = \mathbb{R}$ defined by $U_1(t) = U(\exp tx)$, $t \in \mathbb{R}$. Obviously, we can assume that $x \neq 0$. Then we identify the Lie algebra $\mathfrak{g}_1$ of G_1 with $\mathbb{R} \cdot x$ and consider $\mathcal{E}(\mathfrak{g}_1)$ as a $*$-subalgebra of $\mathcal{E}(\mathfrak{g})$. Note that $y := -ix$ is hermitian in $\mathcal{E}(\mathfrak{g})$. Fix $p \in \mathbb{C}[x]$ and set $c := p(y) \in \mathcal{E}(\mathfrak{g}_1)$. By the corresponding definitions, $\mathcal{D}^\infty(U) \subseteq \mathcal{D}^\infty(U_1)$ and $\partial U_1(x) = \partial U(x)$. Hence $dU(c) \subseteq dU_1(c)$ and $dU(c^+) \subseteq dU_1(c^+)$. From Corollary 9.25, applied to U_1 and G_1, we obtain

$$\overline{dU_1(c^+)} = dU_1(c)^*. \tag{9.42}$$

Since $U_1(t)U_f\varphi = U(\exp tx)U_f\varphi = U_{f(\exp(-tx)\cdot)}\varphi$ by Proposition 9.6(i), U_1 leaves the Gårding domain $\mathcal{D}_G(U)$ invariant, so Theorem 9.12 applies to U_1 and $\mathcal{D} = \mathcal{D}_G(U)$. Thus, $\mathcal{D}_G(U)$, hence $\mathcal{D}^\infty(U)$, is a core for $dU_1(c^+)$ and $dU_1(c)$, so

$$\overline{dU(c^+)} \equiv \overline{dU_1(c^+)\restriction\mathcal{D}^\infty(U)} = \overline{dU_1(c^+)} \tag{9.43}$$

and similarly, $\overline{dU(c)} = \overline{dU_1(c)}$. Hence $dU(c)^* = dU_1(c)^*$. Combining the latter with (9.42) and (9.43) we obtain $\overline{dU(c^+)} = dU(c)^*$, which is (9.40).

Now set $p(x) = x^n$. Then $p(-ix) = (-ix)^n$ is hermitian, so Eq. (9.40) implies that $\overline{dU(-ix)^n}$ is self-adjoint.

Since $\partial U(x)$ is the infinitesimal generator of a unitary group, $-i\partial U(x)$ is self-adjoint by Stone's theorem, so is $(-i\partial U(x))^n \supseteq \overline{dU(-ix)^n}$. Because $\overline{dU(-ix)^n}$ is self-adjoint, this implies $(-i\partial U(x))^n = \overline{dU(-ix)^n}$. Hence $\partial U(x)^n = \overline{dU(x)^n}$. Setting $n = 1$ and using the relation $U(\exp tx) = \exp t\,\partial U(x)$ it follows that $U(\exp tx) = \exp t\,\overline{dU(x)}$. This proves (9.41). $\qquad\square$

The following corollary says that in the case $G = \mathbb{R}^d$ the integrable representations of the commutative $*$-algebra $\mathcal{E}(\mathfrak{g})$, as defined in Sect. 7.3, are precisely those representations that are integrable with respect to the Lie group $\mathbb{R}^d$. This justifies the name "integrable representation" in Definition 7.7.

Corollary 9.28 *Suppose $G = \mathbb{R}^d$. We identify $\mathcal{E}(\mathfrak{g})$ with the polynomial $*$-algebra $\mathbb{C}[x_1,\dots,x_d]$, with involution $x_j^+ = -x_j$, $j = 1,\dots,d$. Then a $*$-representation π of $\mathcal{E}(\mathfrak{g}) = \mathbb{C}[x_1,\dots,x_d]$ is integrable (according to Definition 7.7) if and only if there exists a unitary representation U of the Lie group $G = \mathbb{R}^d$ such that $\pi = dU$.*

Proof First suppose π is an integrable $*$-representation of $A = \mathbb{C}[x_1,\dots,x_d]$ in the sense of Definition 7.7. Then $\overline{\pi(-ix_1)},\dots,\overline{\pi(-ix_d)}$ are strongly commuting self-adjoint operators by Theorem 7.11. Hence the multi-dimensional spectral theorem (Proposition A.4) applies, and there exists a spectral measure E on $\mathbb{R}^d$ such that $\overline{\pi(-ix_j)} = \int_{\mathbb{R}^d} \lambda_j\, dE(\lambda)$ for $j = 1,\dots,d$. We define

$$U(x) = \int_{\mathbb{R}^d} e^{ix\lambda} E(\lambda), \quad x \in \mathbb{R}^d,$$

where $x\lambda$ denotes the inner product of $x, \lambda \in \mathbb{R}^d$. Using properties of spectral integrals (see [Sch12]) it is easily shown that U is a unitary representation of the Lie group $\mathbb{R}^d$ on $\mathcal{H}(\pi)$ and each one-parameter unitary group $U_j(t) = \int_{\mathbb{R}^d} e^{it\lambda_j} dE(\lambda)$, $t \in \mathbb{R}$, has the infinitesimal generator $\partial U(x_j) = \int_{\mathbb{R}^d} i\lambda_j\, dE(\lambda)$. Therefore, $\partial U(x_j) = i\,\overline{\pi(-ix_j)} = \overline{\pi(x_j)} \supseteq \pi(x_j)$ for $j = 1,\dots,d$. Hence $\mathcal{D}(\pi)$ is invariant under all operators $\partial U(x_j)$. This implies $\mathcal{D}(\pi) \subseteq \mathcal{D}^\infty(U)$ and $\pi \subseteq dU$. Then, since π is integrable and hence self-adjoint, $\pi = dU$ by Corollary 4.31.

Conversely, suppose U is a unitary representation of the Lie group $G = \mathbb{R}^d$. Then, by Corollary 9.25, $\overline{dU(x^+)} = dU(x)^*$ for $x \in \mathcal{E}(\mathfrak{g})$, so (7.4) is satisfied. Hence, since dU is self-adjoint by Corollary 9.21, $\pi := dU$ is integrable according to Definition 7.7. $\qquad\square$

Proposition 9.29 *Let G_0 denote the connected component of the unit element of G. Then $dU(\mathcal{E}(\mathfrak{g}))' = U(G_0)'$. In particular, $U \!\restriction\! G_0$ is irreducible if and only if dU is.*

Proof From operator theory it is well known that a bounded operator commutes with a one-parameter unitary group if and only if it does with its infinitesimal generator. Let $x \in \mathfrak{g}$ and $C \in \mathbf{B}(\mathcal{H}(U))$. Then, by (9.41), C commutes with $U(\exp tx)$ for all $t \in \mathbb{R}$ if and only if C commutes with $\overline{dU(x)} = \partial U(x)$.

Let $C \in dU(\mathcal{E}(\mathfrak{g}))'$. For $x \in \mathfrak{g}$, C commutes with $dU(x)$, hence with $\overline{dU(x)}$ by Lemma 3.16(i) and then with $U(\exp x)$ by the result stated in the preceding paragraph. Therefore, since G_0 consists of finite products of elements $\exp x$ with $x \in \mathfrak{g}$, C commutes with $U(G_0)$.

Now let $C \in U(G_0)'$. As noted above, C commutes with $\overline{dU(x)}$ for $x \in \mathfrak{g}$. Further, C leaves all domains $\mathcal{D}(\partial U(x_{j_1}) \cdots \partial U(x_{j_d}))$ invariant and so $\mathcal{D}^\infty(dU)$ by (9.21). Hence C commutes with $dU(x)$ for $x \in \mathfrak{g}$ and therefore with $dU(\mathcal{E}(\mathfrak{g}))$.

This proves that $dU(\mathcal{E}(\mathfrak{g}))' = U(G_0)'$. The assertion about the irreducibility follows at once from Proposition 4.26. $\qquad\square$

The next theorem, due to *R. Goodman*, is also based on elliptic regularity theory. It sharpens Proposition 9.10 by showing that mixed products $\partial U(x_{j_1}) \cdots \partial U(x_{j_k})$ as in (9.21) are not needed to describe the domain $\mathcal{D}^\infty(U)$ and the graph topology $\mathfrak{t}_{dU}$.

Theorem 9.30 *The space of C^∞-vectors is given by*

$$\mathcal{D}^\infty(U) = \bigcap_{k=1}^d \mathcal{D}^\infty(\partial U(x_k)) = \bigcap_{k=1}^d \bigcap_{n=1}^\infty \mathcal{D}\big(\overline{dU(x_k)^n}\big) \qquad (9.44)$$

and the graph topology $\mathfrak{t}_{dU}$ on $\mathcal{D}^\infty(U)$ is generated by the family of seminorms

$$\varphi \mapsto \|dU(x_k^n)\varphi\|, \quad k = 1, \dots, d,\ n \in \mathbb{N}_0. \qquad (9.45)$$

Proof Suppose $\varphi \in \bigcap_j \mathcal{D}^\infty(\partial U(x_j))$, $\psi \in \mathcal{H}(U)$, and $f \in C_0^\infty(G)$. Let $n \in \mathbb{N}$ and $j = 1, \dots, d$. Then $(x_j^{2n} f)(g) = (\frac{d}{dt})^{2n} f(\exp(-tx_j)g)\,|_{t=0}$ by (9.1) and

$$\left(\frac{d}{dt}\right)^{2n} U(\exp(-tx_j))\varphi\,|_{t=0} = \left(\frac{d}{dt}\right)^{2n} \exp(-t\partial U(x_j))\varphi\,|_{t=0} = \partial U(x_j)^{2n}\varphi$$

by (9.41). Using these facts we derive

$$\int_G \left(\widetilde{x_j^{2n}} f\right)(g) \, \langle U(g)\psi, \varphi \rangle \, d\mu_l(g)$$

$$= \int_G \left(\frac{d}{dt}\right)^{2n} f((exp(-tx_j))g) \, \langle U(g)\psi, \varphi \rangle \mid_{t=0} d\mu_l(g)$$

$$= \int_G \left(\frac{d}{dt}\right)^{2n} f(g) \, \langle U((exp \, tx_j)g)\psi, \varphi \rangle \mid_{t=0} d\mu_l(g)$$

$$= \int_G \left(\frac{d}{dt}\right)^{2n} f(g) \, \langle U(g)\psi, U(exp(-tx_j))\varphi \rangle \mid_{t=0} d\mu_l(g)$$

$$= \int_G f(g) \, \langle U(g)\psi, \partial U(x_j)^{2n}\varphi \rangle \, d\mu_l(g).$$

Set $a := x_1^{2n} + \cdots + x_d^{2n} \in \mathcal{E}(\mathfrak{g})$. From the preceding calculation we obtain

$$\int_G (\widetilde{a}f)(g) \, \langle U(g)\psi, \varphi \rangle \, d\mu_l(g) = \int_G f(g) \Big\langle U(g)\psi, \sum_{j=1}^d \partial U(x_j)^{2n}\varphi \Big\rangle d\mu_l(g)$$

for $f \in C_0^\infty(G)$. Suppose $2n \geq d$. Then Lemma 9.19 applies to the elliptic element a of degree $2n$ and yields $\langle U(g)\psi, \varphi \rangle \in C^{2n-d}(G)$. Hence, since $n \geq d/2$ is arbitrary, $\langle U(g)\psi, \varphi \rangle \in C^\infty(G)$ for all $\psi \in \mathcal{H}(U)$. Therefore, $\varphi \in \mathcal{D}^\infty(U)$ by Proposition 9.15, (iii)$\rightarrow$(i). This proves $\cap_j \mathcal{D}^\infty(\partial U(x_j)) \subseteq \mathcal{D}^\infty(U)$. The opposite inclusion is obvious, since $dU(x_j) = \partial U(x_j) \restriction \mathcal{D}^\infty(U)$. This gives the first equality in (9.44); the second restates only the first equality in (9.41).

The assertion about the graph topology is proved by verbatim the same reasoning as in the proof of Theorem 9.22; see also Lemma 3.6. $\qquad\square$

It is instructive to discuss the simplest case when G is the Lie group $\mathbb{R}$.

Example 9.31 Let $G = \mathbb{R}$. Then $\mathfrak{g} = \mathbb{R} \cdot x$, with $x^+ = -x$, and $\mathcal{E}(\mathfrak{g})$ is the polynomial algebra $\mathbb{C}[y]$, with $y^+ = y$ and $y := -ix$. Clearly, a unitary representation U of the Lie group $G = \mathbb{R}$ is just a one-parameter unitary group. By Stone's theorem (Proposition A.5), U is of the form $U(t) = \exp(itA)$, $t \in \mathbb{R}$, for some self-adjoint operator A and $\partial U(x) = iA$ is its infinitesimal generator. Then formula (9.21) yields

$$\mathcal{D}^\infty(U) = \bigcap_{n=1}^\infty \mathcal{D}(A^n) = \mathcal{D}^\infty(A). \qquad (9.46)$$

Using that $dU(y) = dU(-ix) = -i\partial U(x) \restriction \mathcal{D}^\infty(U) = A \restriction \mathcal{D}^\infty(A)$ we derive

$$dU(p(y)^+) = p^+(dU(y)) = p^+(A) \restriction \mathcal{D}^\infty(A) \subseteq p^+(A)$$
$$= p(A)^* \subseteq (p(A) \restriction \mathcal{D}^\infty(A))^* = p(dU(y))^* = dU(p(y))^*$$

for $p \in \mathbb{C}[y]$. Since $\overline{dU(p(y)^+)} = dU(p(y))^*$ by (9.40), the latter gives

$$\overline{p^+(A){\restriction}\mathcal{D}^\infty(A)} = p^+(A) = (p(A){\restriction}\mathcal{D}^\infty(A))^* \qquad (9.47)$$

for each polynomial p. Of course, the equality (9.47) can be also derived from the spectral theorem. We leave this an exercise for the reader.   ○

We closed this section with an operator-theoretic application of Theorem 9.12.

Proposition 9.32 *Suppose A is a self-adjoint operator on a Hilbert space $\mathcal{H}$. Let $U(t) := \exp(it A), t \in \mathbb{R}$, be the corresponding unitary group. Suppose that $\mathcal{D}$ is a linear subspace of $\mathcal{D}^\infty(A) := \cap_{n=1}^\infty \mathcal{D}(A^n)$ such that $\mathcal{D}$ is dense in $\mathcal{H}$ and invariant under $U(t)$ for all $t \in \mathbb{R}$. Then, for any polynomial $p \in \mathbb{C}[x]$, we have*

$$\overline{p^+(A){\restriction}\mathcal{D}} = p^+(A) = (p(A){\restriction}\mathcal{D})^*. \qquad (9.48)$$

If $p \in \mathbb{R}[x]$, then $p(A){\restriction}\mathcal{D}$ is essentially self-adjoint. In particular, $A^n{\restriction}\mathcal{D}$ is essentially self-adjoint for any $n \in \mathbb{N}$.

Proof We apply Theorem 9.12 to the unitary representation U of the Lie group $\mathbb{R}$ and use the notation and facts from Example 9.31. Recall that $p^+(A){\restriction}\mathcal{D}^\infty(A) = dU(p(y)^+)$ and $p(A){\restriction}\mathcal{D}^\infty(A) = dU(p(y))$. By Theorem 9.12 (see (9.24)), $\mathcal{D}$ is a core for $dU(p(y)^+)$ and $dU(p(y))$. Therefore, since $\mathcal{D}^\infty(U) = \mathcal{D}^\infty(A)$ by (9.46), we can replace $\mathcal{D}^\infty(A)$ in (9.47) by $\mathcal{D}$ and obtain (9.48). If $p \in \mathbb{R}[x]$, then $p^+ = p$ and (9.48) means that $p(A){\restriction}\mathcal{D}$ is essentially self-adjoint.   □

9.5   Two Examples

In this section we elaborate some of the preceding results for two Lie groups.

Example 9.33   (*Heisenberg group*)
Let G be the three-dimensional Heisenberg group; that is, G is the Lie group of matrices

$$g(a, b, c) := \begin{pmatrix} 1 & a & c \\ 0 & 1 & b \\ 0 & 0 & 1 \end{pmatrix}, \quad \text{where } a, b, c \in \mathbb{R}.$$

The group law is obtained by matrix multiplication, that is,

$$g(a_1, b_1, c_1)g(a_2, b_2, c_2) = g(a_1 + a_2, b_1 + b_2, c_1 + c_2 + a_1 b_2).$$

(Note that this version of the Heisenberg group is isomorphic to the one used in Sect. 8.5; see formula (8.45).)

The Lie algebra $\mathfrak{g}$ of G has a basis $\{x_1, x_2, x_3\}$ with commutation relations

$$[x_1, x_2] = x_3, \quad [x_1, x_3] = [x_2, x_3] = 0.$$

The corresponding one-parameter groups in G are

$$\exp t x_1 = g(t, 0, 0), \quad \exp t x_2 = g(0, t, 0), \quad \exp t x_3 = g(0, 0, t), \quad t \in \mathbb{R}.$$

The Lie group G has a family U_λ, $\lambda \in \mathbb{R}\backslash\{0\}$, of infinite-dimensional irreducible unitary representations [Ty86, Chap. 1, Theorem 2.5], which exhaust the infinite-dimensional irreducible unitary representations of G. They act by

$$(U_\lambda(g(a, b, c))\varphi)(t) := \exp(i\,\lambda(tb + c))\,\varphi(t + a), \quad \varphi \in \mathcal{H}(U_\lambda) = L^2(\mathbb{R}). \tag{9.49}$$

The unitary group $U_\lambda(\exp t x_1)$ acts by translation; hence, its infinitesimal generator is $\partial U_\lambda(x_1) = \frac{d}{dt}$, with domain of all absolutely continuous functions $\varphi \in L^2(\mathbb{R})$ such that $\varphi' \in L^2(\mathbb{R})$. Similarly, the generator of the unitary group $U_\lambda(\exp t x_2)$ is the multiplication operator $\partial U_\lambda(x_2) - i\lambda t$, with domain of $\varphi \in L^2(\mathbb{R})$ for which $t\varphi(t) \in L^2(\mathbb{R})$. Obviously, $\partial U_\lambda(x_3) = i\lambda \cdot I$. From this it follows that $dU_\lambda(\mathcal{E}(\mathfrak{g}))$ consists of all ordinary differential operators on $\mathbb{R}$ with polynomial coefficients.

From Proposition 9.10 we easily conclude that $\mathcal{D}^\infty(U_\lambda)$ is the Schwartz space

$$\mathcal{S}(\mathbb{R}) = \left\{\varphi \in C^\infty(\mathbb{R}) : \sup\{|t^k \varphi^{(n)}(t)| : t \in \mathbb{R}\} < \infty \quad \text{for } k, n \in \mathbb{N}_0\right\}.$$

Theorem 9.30 gives an apparently weaker, but equivalent characterization of the Schwartz space, which is of interest in itself: *A function $\varphi \in C^\infty(\mathbb{R})$ belongs to $\mathcal{S}(\mathbb{R})$ if and only if $t^k \varphi(t)$ and $\varphi^{(n)}(t)$ are in $L^2(\mathbb{R})$ for all $k, n \in \mathbb{N}_0$.*

By (9.49), $C_0^\infty(\mathbb{R})$ is invariant under U_λ. Thus, since $C_0^\infty(\mathbb{R}) \subseteq \mathcal{S}(\mathbb{R})$ is dense in $L^2(\mathbb{R})$, it follows from Theorem 9.12 that $C_0^\infty(\mathbb{R})$ is a core for all operators $dU_\lambda(x)$, $x \in \mathcal{E}(\mathfrak{g})$. This restates a known fact from analysis: For each differential operator $T = \sum_{k=0}^n p_k(t)(\frac{d}{dt})^k$, $p_k \in \mathbb{C}[t]$, and $\varphi \in \mathcal{S}(\mathbb{R})$ there is a sequence (φ_n) of functions $\varphi_n \in C_0^\infty(\mathbb{R})$ such that $\lim_n \varphi_n = \varphi$ and $\lim_n T\varphi_n = T\varphi$ in $L^2(\mathbb{R})$.

The element $a = -x_1^2 - x_2^2 - x_3^2$ of $\mathcal{E}(\mathfrak{g})$ is elliptic and hermitian. Hence, by Theorem 9.22, the operator $dU_\lambda(a) = -\frac{d^2}{dt^2} + \lambda^2 t^2 + \lambda^2$ is essentially self-adjoint on $\mathcal{S}(\mathbb{R})$ and we have $\mathcal{D}^\infty(U_\lambda) = \mathcal{D}^\infty(\overline{dU_\lambda(a)})$ by (9.37). In particular, this gives the well-known description (see, e.g., [RS72]) of the Schwartz space as $\mathcal{S}(\mathbb{R}) = \mathcal{D}^\infty(T)$ for the self-adjoint operator $T = -\frac{d^2}{dt^2} + t^2 + 1$.

The image $T := dU_1(i x_2 x_1 x_2) = -t^2 \frac{d}{dt} - it$ of the hermitian element $i x_2 x_1 x_2$ of $\mathcal{E}(\mathfrak{g})$ under the $*$-representation dU_1 is a symmetric operator which is not essentially self-adjoint. It can be shown that T has deficiency indices $(1, 1)$. (In fact, we have $\varphi_\pm \in \ker(T^* \pm i)$, where $\varphi_\pm$ are defined by $\varphi_+(t) = t^{-1}\exp(-t^{-1})$ for $t > 0$, $\varphi_+(t) = 0$ for $t \le 0$, $\varphi_-(t) = t^{-1}\exp t^{-1}$ for $t < 0$, $\varphi_-(t) = 0$ for $t \ge 0$.) ○

Example 9.34 (*Affine group of the real line*)
The affine group of the real line, or the $ax + b$-group, is the Lie group G of matrices

$$g(a, b) := \begin{pmatrix} a & b \\ 0 & 1 \end{pmatrix}, \quad \text{where } a, b \in \mathbb{R}, a > 0.$$

The group law is $g(a_1, b_1)g(a_2, b_2) = g(a_1 a_2, a_1 b_2 + b_1)$. The Lie algebra $\mathfrak{g}$ has a basis $\{x_1, x_2\}$ with commutation relation $[x_1, x_2] = x_2$, and the corresponding exponentials are $\exp t x_1 = g(e^t, 0)$ and $\exp t x_2 = g(1, t), t \in \mathbb{R}$.

The Lie group G has (up to unitary equivalence) precisely two infinite-dimensional irreducible unitary representations $U_\pm$ (see, e.g., [Ty86, p. 150]). They are defined by

$$(U_\pm(g(a, b))\varphi)(t) = \exp(\pm i\, e^t b)\, \varphi(t + \log a), \quad \varphi \in \mathcal{H}(U_\pm) := L^2(\mathbb{R}). \quad (9.50)$$

Then $\partial U_\pm(x_1) = \frac{d}{dt}$ and $\partial U_\pm(x_2) = \pm i e^t$. Now Theorem 9.30 yields

$$\mathcal{D}^\infty(U_\pm) = \{\varphi \in C^\infty(\mathbb{R}) : e^{tn}\varphi(t) \in L^2(\mathbb{R}), \varphi^{(n)}(t) \in L^2(\mathbb{R}) \text{ for } n \in \mathbb{N}_0\}.$$

From Theorem 9.22 it follows that the image $dU_\pm(a) = -\frac{d^2}{dt^2} + e^{2t}$ of the hermitian elliptic element $a := -x_1^2 - x_2^2 \in \mathcal{E}(\mathfrak{g})$ is essentially self-adjoint and $\mathcal{D}^\infty(U_\pm) = \mathcal{D}^\infty(\overline{dU_\pm(a)})$.

From (9.50) it is clear that $C_0^\infty(\mathbb{R}) \subseteq \mathcal{D}^\infty(U_\pm)$ is invariant under $U_\pm$. Hence $C_0^\infty(\mathbb{R})$ is a core for all operators $dU_\pm(x), x \in \mathcal{E}(\mathfrak{g})$, by Theorem 9.12.

On the other hand, the image $dU_\pm(x) = \pm 2i\, e^t \frac{d}{dt} \pm i\, e^t$ of the hermitian element $x := x_1 x_2 + x_2 x_1$ has no self-adjoint extension on the Hilbert space $L^2(\mathbb{R})$. It can be shown that the symmetric operators $dU_+(x)$ and $dU_-(x)$ have deficiency indices $(0, 1)$ and $(1, 0)$, respectively.

In the literature, the representations $U_\pm$ are often realized on the Hilbert spaces $\mathcal{H}(U_+) := L^2(0, +\infty)$ and $\mathcal{H}(U_-) := L^2(-\infty, 0)$ by the formulas

$$(U_\pm(g(a, b))\varphi)(t) = \exp(i\, tb)\, \sqrt{a}\, \varphi(at). \qquad \bigcirc$$

9.6 Analytic Vectors

Our aim is to study analytic vectors for representations. But let us begin with analytic vectors for one operator.

9.6.1 Analytic Vectors for Single Operators

Let T be a linear operator on a Hilbert space $\mathcal{H}$.

Definition 9.35 A vector $\varphi \in \mathcal{H}$ is called an *analytic vector* for T if $\varphi \in \mathcal{D}(T^n)$ for all $n \in \mathbb{N}$ and there exists a constant $C > 0$ such that

$$\|T^n \varphi\| \leq C^n n! \quad \text{for } n \in \mathbb{N}_0. \tag{9.51}$$

The set of analytic vectors for T is denoted by $\mathcal{D}^a(T)$.

It is clear that $\mathcal{D}^a(T)$ is a linear space. Obviously, eigenvectors are analytic vectors. Large classes of examples are obtained from the following proposition.

Proposition 9.36 *Let T be a self-adjoint operator, and let $T = \int_{\mathbb{R}} \lambda \, dE(\lambda)$ be its spectral resolution. For any $\alpha > 0$, the ranges of the operators $E([-\alpha, \alpha])$ and $e^{-\alpha T^2} := \int_{\mathbb{R}} e^{-\alpha \lambda^2} \, dE(\lambda)$ are contained in $\mathcal{D}^a(T)$. The set $\mathcal{D}^a(T)$ is dense in $\mathcal{H}$.*

Proof First note that $e^{-\alpha T^2}$ is a bounded operator defined on $\mathcal{H}$, because the function $e^{-\alpha \lambda^2}$ is bounded on $\mathbb{R}$. By Stirling's formula [RW91, p. 45] there exists a null sequence (ε_n) such that $n! = \sqrt{2\pi}\, n^{n+1/2}\, e^{-n}(1 + \varepsilon_n)$ for $n \in \mathbb{N}$. Hence there is a constant $c > 0$ such that $e^{-n} n^n \leq c\, n!$ for $n \subset \mathbb{N}$. It is not difficult to verify that the function $\lambda^{2n} e^{-2\alpha \lambda^2}$ has the maximum $(2\alpha)^{-n} e^{-n} n^n$ on $\mathbb{R}$. Let $\psi \in \mathcal{H}$. Then

$$\int_{\mathbb{R}} \lambda^{2n} e^{-2\alpha \lambda^2} d\langle E(\lambda)\psi, \psi \rangle \leq \int_{\mathbb{R}} (2\alpha)^{-n} e^{-n} n^n \, d\langle E(\lambda)\psi, \psi \rangle \tag{9.52}$$

$$\leq c(2\alpha)^{-n} n! \int_{\mathbb{R}} d\langle E(\lambda)\psi, \psi \rangle = c\|\psi\|^2 (2\alpha)^{-n} n! .$$

Hence, by the functional calculus, $\psi \in \mathcal{D}(T^n e^{-\alpha T^2})$, $\varphi := e^{-\alpha T^2}\psi \in \mathcal{D}(T^n)$, and the first integral in (9.52) is $\|T^n e^{-\alpha T^2}\psi\|^2 = \|T^n \varphi\|^2$. Thus, we have shown that $\|T^n \varphi\|^2 \leq c\|\psi\|^2 (2\alpha)^{-n} n!$ for $n \in \mathbb{N}$. This implies $\varphi = e^{-\alpha T^2}\psi \in \mathcal{D}^a(T)$.

Now let $\varphi \in E([-\alpha, \alpha])\mathcal{H}$. Using again the functional calculus we derive

$$\|T^n \varphi\|^2 = \int_{-\alpha}^{\alpha} \lambda^{2n} d\langle E(\lambda)\varphi, \varphi \rangle \leq \alpha^{2n} \|\varphi\|^2 \leq (\alpha(1 + \|\varphi\|))^{2n}, \quad n \in \mathbb{N},$$

so that $\varphi \in \mathcal{D}^a(T)$. Here we even have $\|T^n \varphi\| \leq M^n$ for some constant $M > 0$; such vectors are called *bounded vectors*.

Since $\psi = \lim_{\alpha \to \infty} E([-\alpha, \alpha])\psi$ for all $\psi \in \mathcal{H}$, $\mathcal{D}^a(T)$ is dense in $\mathcal{H}$. $\qquad \square$

Example 9.37 Let T be the multiplication operator by the variable x on $L^2(\mathbb{R})$. Then, for each polynomial $p \in \mathbb{C}[x]$ and $\beta > 0$, $\varphi(x) := p(x)e^{-\beta x^2}$ is in $\mathcal{D}^a(T)$. (Indeed, setting $\alpha := \beta/2$, we have $\psi := p(x)e^{-\alpha x^2} \in L^2(\mathbb{R})$ and $\varphi = e^{-\alpha x^2}\psi$, so Proposition 9.36 applies.) $\qquad \bigcirc$

The next proposition gives an explanation for the terminology "analytic vector."

Proposition 9.38 *Suppose T is a self-adjoint operator and $\varphi \in \mathcal{D}^a(T)$. Let C denote the constant from (9.51). Then, for $|z| < C^{-1}$, we have $\varphi \in \mathcal{D}(e^{zT})$ and*

$$e^{zT}\varphi = \lim_{n\to\infty} \sum_{k=0}^{n} \frac{z^k}{k!} T^k \varphi. \tag{9.53}$$

For $\psi \in \mathcal{H}$, the function $z \mapsto \langle e^{izT}\varphi, \psi \rangle$ is holomorphic on $\{z : |\operatorname{Im} z| < C^{-1}\}$.

Proof [Sch12, Proposition 7.8 and Corollary 7.10]. □

By Proposition 9.36, $\mathcal{D}^a(T)$ is dense if T is self-adjoint. The following *Nelson theorem* says that the denseness of $\mathcal{D}^a(T)$ implies the essential self-adjointness.

Theorem 9.39 *Suppose T is a symmetric operator on $\mathcal{H}$ such that the set $\mathcal{D}^a(T)$ of analytic vectors for T is dense in $\mathcal{H}$. Then T is essentially self-adjoint. In particular, if in addition T is closed, then T is self-adjoint.*

Proof [N59, Lemma 5.1], see, e.g., [Sch12, Theorem 7.16]. □

Example 9.40 (*Analytic vectors for the self-adjoint operator $T = -\mathrm{i}\frac{d}{dx}$ on $L^2(\mathbb{R})$*) Using the Paley–Wiener theorem [Kz68, Theorem 7.1] it can be shown [Sch12, Example 7.8] that $\varphi \in \mathcal{D}(T)$ is in $\mathcal{D}^a(T)$ if and only if φ is the restriction to $\mathbb{R}$ of a holomorphic function F on a strip $\{z \in \mathbb{C} : |\operatorname{Im} z| < c\}$ for some $c > 0$ satisfying

$$\sup_{|y|<c} \int_{\mathbb{R}} |F(x+\mathrm{i}y)|^2 \, dx < \infty.$$

In particular, $\mathcal{D}^a(T) \cap C_0^\infty(\mathbb{R}) = \{0\}$. Hence $\mathcal{D}^a(T_0) = \{0\}$ for $T_0 := T \!\restriction\! C_0^\infty(\mathbb{R})$.

But, since the translation group $e^{\mathrm{i}tT}$ leaves the dense linear subspace $C_0^\infty(\mathbb{R})$ of $L^2(\mathbb{R})$ invariant, T_0 is essentially self-adjoint according to Proposition 9.32. This shows that the converse of Nelson's Theorem 9.39 does not hold. ○

Example 9.41 Let $\{e_n : n \in \mathbb{N}_0\}$ be an orthonormal basis of a Hilbert space and $\mathcal{D} := \operatorname{Lin}\{e_n : n \in \mathbb{N}_0\}$. Let $(\alpha_n)_{n\in\mathbb{N}_0}, (\beta_n)_{n\in\mathbb{N}_0}, (\gamma_n)_{n\in\mathbb{N}_0}, (\delta_n)_{n\in\mathbb{N}_0}, (\eta_n)_{n\in\mathbb{N}_0}$ be complex sequences. Suppose there exist constants $a > 0$ and $b > 0$ such that

$$|\alpha_j|, \ |\beta_j|, \ |\gamma_j|, \ |\delta_j|, \ |\eta_j| \leq an + b \quad \text{for all} \quad j \leq n, n \in \mathbb{N}. \tag{9.54}$$

We define a linear operator T with domain $\mathcal{D}(T) := \mathcal{D}$ by

$$Te_k = \alpha_{k-2}e_{k-2} + \beta_{k-1}e_{k-1} + \gamma_k e_k + \delta_{k+1}e_{k+1} + \eta_{k+2}e_{k+2}, \tag{9.55}$$

where we set $e_j = 0$ if $j < 0$. We prove that $\mathcal{D}^a(T) = \mathcal{D}$.

Since $\mathcal{D}^a(T)$ is a vector space, it suffices to show that $e_k \in \mathcal{D}^a(T), k \in \mathbb{N}_0$. From (9.55) it follows that $T^n e_k$ is a sum of at most 5^n terms of the form $\zeta_m e_m$, where

$m \leq k + 2n$ and ζ_m is a product of n factors $\alpha_j, \beta_j, \gamma_j, \delta_j, \eta_j$ with $j \leq k + 2n$. Therefore, by (9.54), $\|T^n e_k\| \leq 5^n(a(2n + k) + b)^n$.

As noted in the proof of Proposition 9.36, we have $n^n \leq c_0 e^n n!$ for $n \in \mathbb{N}$ and some constant $c_0 > 0$. Set $c := \frac{ak+b}{2a}$. Then we derive

$$\|T^n e_k\| \leq 5^n(a(2n + k) + b)^n = (10a)^n(1 + cn^{-1})^n n^n \leq (10a)^n e^c c_0 e^n n!.$$

Hence $e_k \in \mathcal{D}^a(T)$, which completes the proof of the equality $\mathcal{D}^a(T) = \mathcal{D}$. ◯

Example 9.42 (*Application to the canonical commutation relation*)
Consider the subspace of elements of degree at most two of the Weyl algebra W:

$$\mathsf{B} := \mathrm{Lin}\,\{1, a, a^+, a^2, (a^+)^2, a^+a, aa^+\} = \mathrm{Lin}\,\{1, p, q, p^2, q^2, pq, qp\}.$$

Let $x \in \mathsf{B}$. From the definition of the Bargmann–Fock representation $\pi_{\mathbb{C}}$ and (8.22) it follows that $\pi_{\mathbb{C}}(x)$ is of the form (9.55) and that (9.54) is satisfied. Hence, by Example 9.41, all vectors of $\mathcal{D}_{\mathbb{C}} := \mathrm{Lin}\,\{e_n : n \in \mathbb{N}_0\}$ are analytic for $\pi_{\mathbb{C}}(x)$.

Recall from Proposition 8.13 that the unitary equivalence of the Bargmann–Fock representation $\pi_{\mathbb{C}}$ and the Schrödinger representation π_S is provided by a unitary that maps the Hermite function $H_n \in L^2(\mathbb{R})$ to $e_n \in l^2(\mathbb{N}_0)$. Therefore, for $x \in \mathsf{B}$, all vectors of $\mathcal{D}_S := \mathrm{Lin}\,\{H_n : n \in \mathbb{N}_0\}$ are analytic vectors for $\pi_S(x)$. ◯

9.6.2 Analytic Vectors for Unitary Representations

Recall that U is a unitary representation of the Lie group G on a Hilbert space $\mathcal{H}(U)$. The following definition uses the *real analytic* structure of the Lie group G.

Definition 9.43 A vector $\varphi \in \mathcal{H}(U)$ is called *analytic* for U if the map $g \mapsto U(g)\varphi$ of the real analytic manifold G into $\mathcal{H}(U)$ is real analytic.

The analytic vectors for U form a vector space which is denoted $\mathcal{D}^a(U)$. The following *infinitesimal characterization* can be also taken as a definition of the space $\mathcal{D}^a(U)$.

Proposition 9.44 *A vector $\varphi \in \mathcal{H}(U)$ is an analytic vector for U if and only if $\varphi \in \mathcal{D}^\infty(U)$ and there exists a constant $C > 0$ such that*

$$\|dU(x_{j_1} \cdots x_{j_n})\varphi\| \leq C^n n! \quad \text{for } j_1, \ldots, j_n \in \{1, \ldots, d\}, \ n \in \mathbb{N}.$$

Proof [N59, Lemma 7.1]. □

Set $\Delta := x_1^2 + \cdots + x_d^2$. Then $1 - \Delta$ is a hermitian elliptic element of $\mathcal{E}(\mathfrak{g})$. Therefore, the operator $A := \overline{dU(1 - \Delta)}$ is self-adjoint by Theorem 9.22. Note that $A \geq I$, since $(x_j)^+ = -x_j$, hence $(x_j)^+ x_j = -x_j^2$, and $dU(-\Delta) \geq 0$.

Fundamental results on the space $\mathcal{D}^a(U)$ are collected in the next theorem.

Theorem 9.45 (i) *The space $\mathcal{D}^a(U)$ of analytic vectors is dense in $\mathcal{H}(U)$.*
(ii) *Each analytic vector for A is an analytic vector for U: $\mathcal{D}^a(A) \subseteq \mathcal{D}^a(U)$.*
(iii) $\mathcal{D}^a(A^{1/2}) = \mathcal{D}^a(U)$.
(iv) *A vector $\varphi \in \mathcal{D}^\infty(U)$ is in $\mathcal{D}^a(U)$ if and only if there exists a constant $M > 0$ such that $\|dU(\Delta^n)\varphi\| \leq M^n(2n)!$ for $n \in \mathbb{N}_0$.*

Proof (ii) follows at once from (iii), and (iv) is easily derived from (iii).

(i): [CD58] or [N59, Theorem 7.3] or [Sch90, Corollary 10.4.12] or [Wr72, Theorem 4.4.5.7].
(ii): [N59, Theorem 7.3] or [Sch90, Theorem 10.4.4].
(iii): [G69b] or [Wr72, Theorem 4.4.6.1].
(iv): [Wr72, Corollary 4.4.6.4]. $\square$

The following interesting result was proved in [FS73]: *There exists a basis $\{x_1, \ldots, x_d\}$ of the Lie algebra $\mathfrak{g}$ such that $\mathcal{D}^a(U) = \cap_{j=1}^d \mathcal{D}^a\left(\overline{dU(x_j)}\right)$ for any unitary representation U of the Lie group G.*

Proposition 9.44 suggests the following definition of analytic vectors for general $*$-representations of $\mathcal{E}(\mathfrak{g})$.

Definition 9.46 For a $*$-representation π of $\mathcal{E}(\mathfrak{g})$, let $\mathcal{D}^a(\pi)$ denote the set of vectors $\varphi \in \mathcal{D}(\pi)$ for which there exists a constant $C > 0$ such that

$$\|\pi(x_{j_1} \cdots x_{j_n})\varphi\| \leq C^n n! \quad \text{for} \quad j_1, \ldots, j_n \in \{1, \ldots, d\}, \ n \in \mathbb{N}.$$

Obviously, $\mathcal{D}^a(\pi)$ is a vector space. It is invariant under π (Exercise 16) and does not depend on the particular basis of $\mathfrak{g}$. By Proposition 9.44, $\mathcal{D}^a(U) = \mathcal{D}^a(dU)$.

9.6.3 Exponentiation of Representations of Enveloping Algebras

In this section, we assume that the Lie group G is **connected** and **simply connected**.
Given a $*$-representation π of $\mathcal{E}(\mathfrak{g})$, it is natural to ask:

Does there exist a unitary representation U such that $\pi \subseteq dU$ or $\pi = dU$?

This question leads to the following definitions.

Definition 9.47 A $*$-representation π of $\mathcal{E}(\mathfrak{g})$ is called *exponentiable* if there exists a unitary representation U of the Lie group G such that $\overline{\pi(x_j)} = \overline{dU(x_j)}$ for $j = 1, \ldots, d$ and it is called *integrable* if $\pi = dU$.

Obviously, integrable representations are exponentiable. The converse is not true; see Exercise 11.

Proposition 9.48 *Suppose π is an exponentiable $*$-representation of $\mathcal{E}(\mathfrak{g})$. Then the corresponding unitary representation U from Definition 9.47 is uniquely determined by π, and we have $\pi \subseteq dU$ and $\pi^* = dU$. If π is self-adjoint, then $\pi = dU$. Moreover, all operators $\overline{\pi(-ix_j)}$, $j = 1, \ldots, d$, are self-adjoint.*

Proof Let V be another unitary representation of the Lie group G such that $\overline{\pi(x_j)} = \overline{dV(x_j)}$ for $j = 1, \dots, d$. Then $\overline{dU(x_j)} = \overline{dV(x_j)}$ and therefore

$$U(\exp t x_j) = \exp t\, \overline{dU(x_j)} = \exp t\, \overline{dV(x_j)} = V(\exp t x_j), \quad t \in \mathbb{R},$$

by (9.41). From this equation it follows that U and V coincide in a neighborhood of e and hence on the whole group G, since G is connected.

Corollaries 4.16, applied with $\mathsf{B} = \{1, x_1, \dots, x_d\}$, and 9.21 imply that $\pi^* = (dU)^* = dU$. Hence $\pi \subseteq dU$, and $\pi \subseteq \pi^*$. Therefore, if π is self-adjoint, we have $\pi = \pi^* = dU$. By Corollary 9.27, $\pi(-ix_j) = \overline{dU(-ix_j)}$ is self-adjoint. □

Now we state two fundamental exponentiation theorems without proofs. The first result is *Nelson's theorem.*

Theorem 9.49 *Suppose π is a $*$-representation of $\mathcal{E}(\mathfrak{g})$ such that the symmetric operator $\pi(\Delta)$ is essentially self-adjoint. Then π is exponentiable.*

Proof [N59, Theorem 9.5] or [Sch90, Theorem 10.5.6] or [Wr72, Theorem 4.4.6.6]. □

The second exponentiation theorem is due to M. Flato, J. Simon, H. Snellmann, and D. Sternheimer; in the literature it is often called the FS^3–*theorem.*

Theorem 9.50 *Let π be a $*$-representation of $\mathcal{E}(\mathfrak{g})$. If each vector $\varphi \in \mathcal{D}(\pi)$ is an analytic vector for the operator $\pi(x_j)$, $j = 1, \dots, d$, then π is exponentiable.*

Proof [FSSS72] or [Sch90, Theorem 10.5.4]. □

9.7 Analytic Vectors and Unitary Representations of $SL(2, \mathbb{R})$

Throughout this section, G denotes the Lie group $SL(2, \mathbb{R})$. The Lie algebra $\mathfrak{g} = sl(2, \mathbb{R})$ of G consists of the real 2×2 matrices of trace zero. Its basis

$$x_0 := \begin{pmatrix} 0 & 1 \\ -1 & 0 \end{pmatrix}, \quad x_1 := \begin{pmatrix} -1 & 0 \\ 0 & 1 \end{pmatrix}, \quad x_2 := \begin{pmatrix} 0 & 1 \\ 1 & 0 \end{pmatrix}$$

satisfies the commutation relations

$$[x_0, x_1] = 2x_2, \quad [x_0, x_2] = -2x_1, \quad [x_2, x_1] = 2x_0. \tag{9.56}$$

In the representation theory of G the compact abelian subgroup

$$K := SO(2) = \left\{ k(\theta) := \begin{pmatrix} \cos\theta & \sin\theta \\ -\sin\theta & \cos\theta \end{pmatrix} : \theta \in \mathbb{R} \right\}$$

plays a crucial role. Clearly, $\exp t x_0 = k(t)$, $t \in \mathbb{R}$. It is convenient to set

$$x_+ := x_1 - \mathrm{i} x_2, \quad x_- := x_1 + \mathrm{i} x_2. \tag{9.57}$$

Then $(x_\pm)^+ = -x_\mp$. Note that x_+ and x_- are not in $\mathfrak{g}$, but they are in the complexification $\mathfrak{g}_{\mathbb{C}}$ and in $\mathcal{E}(\mathfrak{g})$. It is known and easily verified that the *Casimir element*

$$q := x_0^2 - x_1^2 - x_2^2 = x_0^2 - 2\mathrm{i}\, x_0 - x_+ x_- = x_0^2 + 2\mathrm{i}\, x_0 - x_- x_+ \tag{9.58}$$

is a hermitian element belonging to the center of the algebra $\mathcal{E}(\mathfrak{g})$.

Definition 9.51 A unitary representation U of the Lie group $G = SL(2, \mathbb{R})$ is called *quasisimple* if $dU(q) = c \cdot I$ for some $c \in \mathbb{R}$.

   For instance, irreducible unitary representations are quasisimple.
   From now on we suppose that U is a **quasisimple unitary representation**.
   The characters of the compact abelian group K are $\chi_n(\exp \theta x_0) = e^{\mathrm{i}n\theta}$, $n \in \mathbb{Z}$. Hence the restriction of U to K is a direct sum of subrepresentations acting on

$$\mathcal{H}_n := \left\{ \varphi \in \mathcal{H}(U) : U(\exp t x_0)\varphi = e^{\mathrm{i}nt}\varphi, \ t \in \mathbb{R} \right\}. \tag{9.59}$$

Since $U(\exp t x_0) = \exp t\, \overline{dU(x_0)}$, we have $\overline{dU(x_0)}\, \varphi = \mathrm{i}n\varphi$ for $\varphi \in \mathcal{H}_n$. Therefore, $\mathcal{H}_n \perp \mathcal{H}_m$ if $n \neq m$. It is not difficult to verify that the orthogonal projection of $\mathcal{H}(U)$ on the closed subspace $\mathcal{H}_n$ is given by

$$E_n = (2\pi)^{-1} \int_0^{2\pi} e^{-\mathrm{i}nt} U(\exp t x_0)\, dt.$$

Let $\mathcal{H}_K$ denote the linear span of spaces $\mathcal{H}_n$, $n \in \mathbb{Z}$. It is easily shown that a vector $\varphi \in \mathcal{H}(U)$ belongs to $\mathcal{H}_K$ if and only if $\dim U(K)\varphi$ is finite. For this reason, vectors of $\mathcal{H}_K$ are called *K–finite*. These vectors play a fundamental role in the representation theory of semisimple Lie groups.

Proposition 9.52 $\mathcal{H}_K$ *is a subspace of* $\mathcal{D}^\infty(U)$ *and invariant under* $dU(\mathcal{E}(\mathfrak{g}))$. *Moreover,* $dU(x_\pm)\mathcal{H}_n \subseteq \mathcal{H}_{n\pm2}$ *for* $n \in \mathbb{Z}$.

Proof Let $f \in C_0^\infty(G)$. The function $f_0(g) := (2\pi)^{-1} \int_0^{2\pi} e^{-\mathrm{i}nt} f(\exp(-t x_0)g)dt$ is also in $C_0^\infty(G)$, and we have $E_n U_f \varphi = U_{f_0}\varphi$ for $\varphi \in \mathcal{H}(U)$ by Proposition 9.6(i). Hence E_n leaves the Gårding space $\mathcal{D}_G(U)$ invariant. Therefore, since $\mathcal{D}_G(U)$ is dense in $\mathcal{H}(U)$, $\mathcal{D}_n := E_n \mathcal{D}_G(U)$ is dense in $\mathcal{H}_n = E_n \mathcal{H}(U)$. Fix $n \in \mathbb{Z}$. Let $\varphi \in \mathcal{D}_n$. Since $x_0 x_\pm = x_\pm(x_0 \pm 2\mathrm{i})$ by (9.56) and (9.57),

$$dU(x_0)dU(x_\pm)\varphi = dU(x_\pm)(dU(x_0) \pm 2\mathrm{i})\varphi = (n \pm 2)\mathrm{i}\, dU(x_\pm)\varphi,$$

so $dU(x_\pm) : \mathcal{D}_n \mapsto \mathcal{D}_{n\pm2}$ and hence $dU(x_j) : \mathcal{D}_n \mapsto \mathcal{D}_{n-2} + \mathcal{D}_{n+2}$. By (9.58),

$$\|dU(x_1)\varphi\|^2 + \|dU(x_2)\varphi\|^2 = \langle dU(-x_1^2 - x_2^2)\varphi, \varphi \rangle$$
$$= \langle dU(q - x_0^2)\varphi, \varphi \rangle = (c + n^2)\|\varphi\|^2, \quad \varphi \in \mathcal{D}_n.$$

Hence $dU(x_j)$, $j = 1, 2$, is bounded on $\mathcal{D}_n$. Therefore, since $\mathcal{D}_n$ is dense in $\mathcal{H}_n$, the closure $\overline{dU(x_j)}$ is defined on $\mathcal{H}_n$ and maps $\mathcal{H}_n$ into $\mathcal{H}_{n-2} \oplus \mathcal{H}_{n+2}$. Hence $\overline{dU(x_j)}\mathcal{H}_K \subseteq \mathcal{H}_K$. Clearly, $\overline{dU(x_0)}\mathcal{H}_K \subseteq \mathcal{H}_K$. Since $\mathcal{D}^\infty(U)$ is the largest subspace invariant under $\partial U(x_j) = \overline{dU(x_j)}$, $j = 0, 1, 2$, we conclude that $\mathcal{H}_K \subseteq \mathcal{D}^\infty(U)$. The latter implies also that $\mathcal{D}_n = \mathcal{H}_n$.

Let $\psi \in \mathcal{H}_K$ and $x \in \mathfrak{g}$. Since $\psi \in \mathcal{D}^\infty(U)$ as just shown, it follows from (9.12) that $U(k)dU(x)\psi = dU(\mathrm{Ad}(k)x)U(k)\psi$ for $k \in K$, so $U(K)dU(x)\varphi$ is contained in the image of the finite-dimensional space $U(K)\psi$ under $dU(\mathfrak{g})$. Hence $U(K)dU(x)\varphi$ is finite-dimensional. Therefore, $dU(x)\psi$ is K–finite, that is, $dU(x)\psi \in \mathcal{H}_K$. Thus $\mathcal{H}_K$ is invariant under $dU(\mathfrak{g})$ and hence under $dU(\mathcal{E}(\mathfrak{g}))$. $\qquad\square$

Proposition 9.53 $\mathcal{H}_K \subseteq \mathcal{D}^a(U)$, that is, all K-finite vectors are analytic vectors.

Proof Since $\mathcal{D}^a(U)$ is a linear space, it suffices to prove this for $\varphi \in \mathcal{H}_k, k \in \mathbb{Z}$. We choose $C \geq (2k^2 + |c| + 2)^{1/2}$. By induction on $n \in \mathbb{N}$ we prove that

$$\|dU(x_{j_n} \cdots x_{j_1})\varphi\| \leq \|\varphi\|C^n n! \quad \text{for all} \quad j_n, \ldots, j_1 \in \{0, 1, 2\}. \tag{9.60}$$

Assume that the inequality (9.60) holds for $n \in \mathbb{N}_0$, with the obvious interpretation for $n = 0$. Let $j_n, \ldots, j_1 \in \{0, 1, 2\}$. First suppose that $j_{n+1} = 1, 2$. Since $\varphi \in \mathcal{H}_k$, we can write $\psi := dU(x_{j_n} \cdots x_{j_1})\varphi = \sum_{j \leq k \mid n} \varphi_j$ with $\varphi_j \in \mathcal{H}_j$. From (9.58) and $(x_\pm)^+ = -x_\mp$ we obtain

$$\|dU(x_\pm)\varphi_j\|^2 = \langle dU(-x_\mp x_\pm)\varphi_j, \varphi_j \rangle = \langle dU(-x_0^2 \mp 2\mathrm{i}x_0 + q)\varphi_j, \varphi_j \rangle$$
$$= (j^2 \pm 2j + c)\|\varphi_j\|^2 = ((j \pm 1)^2 + c - 1)\|\varphi_j\|^2. \tag{9.61}$$

Recall that $x_1 = \frac{1}{2}(x_+ + x_-)$, $x_2 = \frac{1}{2}(x_+ - x_-)$ and $\|\psi\| \leq \|\varphi\|C^n n!$ by the induction hypothesis. Further, we have $(C^2 - 2)(n + 1)^2 \geq C^2 - 2 \geq 2k^2 + |c|$ by the choice of C. Note that $c - 1 \leq |c|$ and $\varphi_j \perp \varphi_l$ if $j \neq l$. Using these facts and (9.61) we derive

$$\|dU(x_j)\psi\|^2 \leq \frac{1}{4} \sum_{j \leq k+n} \left(\|dU(x_+)\varphi_j\| + \|dU(x_-)\varphi_j\| \right)^2$$

$$\leq \sum_{j \leq k+n} [(j + 1)^2 + c - 1]\|\varphi_j\|^2 \leq [(k + n + 1)^2 + |c|)]\,\|\psi\|^2$$

$$\leq [2k^2 + 2(n + 1)^2 + |c|]\,\|\varphi\|^2 C^{2n}(n!)^2 \leq \|\varphi\|^2 C^{2n+2}((n + 1)!)^2,$$

which gives (9.60) for $j_{n+1} = 1, 2$. The case $j_{n+1} = 0$ is even simpler. This completes the induction proof of (9.60). Clearly, (9.60) implies $\varphi \in \mathcal{D}^a(U)$. $\qquad\square$

In the rest of this section we sketch two applications of the FS^3-Theorem 9.50. They illustrate the exceptional usefulness and power of this result.

The standard classification of irreducible unitary representations of the Lie group $SL(2, \mathbb{R})$ (see, for instance, [Lg85, Ty86], or [HT92]) is based on $*$-representations π of $\mathcal{E}(\mathfrak{g})$ for which the Casimir acts as a scalar multiple of the identity. Then one finds an orthonormal basis $\{v_n\}$ with $n \in \mathbb{Z}$ such that $\pi(x_0)v_n = i n v_n$,

$$\pi(x_1)v_n = a_n v_{n+2} - \overline{a_{n-2}}\, v_{n-2}, \quad \text{and} \quad \pi(x_2)v_n = i(a_n v_{n+2} + \overline{a_{n-2}}\, v_{n-2}),$$

where $|a_n| = ((n+1)^2 + \gamma)^{1/2}$ and γ depends only on the value of the Casimir; see, e.g., [Ty86, p. 185]. Thus, $\pi(x_j)$, $j = 0, 1, 2$, are operators of the form (9.55) and assumption (9.54) is satisfied. From Example 9.41 we know that all vectors in $\text{Lin}\{v_n\}$ are analytic vectors for the images of the basis. Therefore, the FS^3-Theorem 9.50 applies and π is exponentiable to a unitary representation of the universal covering group $\widetilde{G}$ of $SL(2, \mathbb{R})$. Looking at the kernel of the canonical projection $\widetilde{G} \mapsto SL(2, \mathbb{R})$ leads then to unitary representations of $SL(2, \mathbb{R})$; see [Ty86, pp. 181–188] or [Lg85] for details.

The second application concerns the oscillator representation of the universal covering group of $SL(2, \mathbb{R})$. Recall that W denotes the Weyl algebra.

Lemma 9.54 *There is a unital $*$-homomorphism $\vartheta : \mathcal{E}(sl(2, \mathbb{R})) \mapsto \mathsf{W}$ given by*

$$\vartheta(x_0) = \frac{i}{2}(q^2 + p^2), \quad \vartheta(x_2) = \frac{i}{2}(q^2 - p^2), \quad \vartheta(x_1) = -\frac{i}{2}(pq + qp). \quad (9.62)$$

Proof To prove that Eq. (9.62) defines an algebra homomorphism, we have to show that the elements $\vartheta(x_j)$ satisfy the relations (9.56), with $\vartheta(x_j)$ in place of x_j. These are straightforward computations; we carry out the proof of the first relation. Then, since $\vartheta(x_j^+) = -\vartheta(x_j)^+$ by (9.62), ϑ is a $*$-homomorphism.

Since $[p, q] = -i$ in the Weyl algebra W, we have $[p^2, pq] = [p^2, qp] = -2i\, p^2$ and $[q^2, pq] = [q^2, qp] = 2i\, q^2$. Therefore, the first relation of (9.56) follows by

$$4\,[\vartheta(x_0), \vartheta(x_1)] = [q^2 + p^2, pq + qp] = 4 i\, q^2 - 4 i\, p^2 = 8\, \vartheta(x_2).$$

The proof of the third relation of (9.56) is almost the same. In the proof of the second relation we use that $[q^2, p^2] = -[p^2, q^2] = 2 i\, (pq + qp)$ in the algebra W. $\square$

Recall that π_S denotes the Schrödinger representation (8.31) of the Weyl algebra W. Its composition with the $*$-homomorphism ϑ gives a nondegenerate $*$-representation ρ of $\mathcal{E}(sl(2, \mathbb{R}))$ on the domain $\mathcal{D}(\rho) = \mathcal{S}(\mathbb{R})$ of the Hilbert space $L^2(\mathbb{R})$. Inserting $\pi_S(q) = t$, $\pi_S(p) = -i\frac{d}{dt}$ into (9.62) yields

$$\rho(x_0) = \frac{i}{2}\left(t^2 - \frac{d^2}{dt^2}\right), \quad \rho(x_2) = \frac{i}{2}\left(t^2 + \frac{d^2}{dt^2}\right), \quad \rho(x_1) = -t\frac{d}{dt} - \frac{1}{2}.$$

Let $\mathcal{D}_S$ denote the span of Hermite functions H_n. As shown in Example 9.42, all vectors of $\mathcal{D}_S$ are analytic for the operators $\rho(x_j)$. Therefore, the assumptions of the FS^3-Theorem 9.50 are fulfilled, so the $*$-representation $\rho \lceil \mathcal{D}_S$ is exponentiable. Since $\mathcal{D}_S$ is a core for the operators $\rho(x_j)$, there exists a unitary representation U of the universal covering group $\widetilde{SL(2, \mathbb{R})}$ such that $\overline{\rho(x_j)} = \overline{dU(x_j)}$, $j = 0, 1, 2$.

The representation U appears in the literature under different names such as Shale–Weil representation, metaplectic representation, and *oscillator representation*. In n dimensions it is a representation of the universal covering group $\widetilde{Sp(n, \mathbb{R})}$ of the symplectic group. Note that $Sp(1, \mathbb{R}) = SL(2, \mathbb{R})$.

9.8 Exercises

1. What is the exponential map for the Lie group $G = \mathbb{R}^d$?
2. Describe the Haar measures of the torus $\mathbb{T}$ and the multiplicative group $\mathbb{R} \setminus \{0\}$.
3. (*Haar measure and modular function of the ax + b-group, see Example 9.34*)

   Let G be the Lie group of matrices $g(a, b) := \begin{pmatrix} a & b \\ 0 & 1 \end{pmatrix}$, where $a, b \in \mathbb{R}, a > 0$.

   a. Show that $d\mu_l(g(a, b)) = a^{-2} da\, db$ is a left Haar measure of G.
   b. Show that $d\mu_r(g(a, b)) = a^{-1} da\, db$ is a right Haar measure of G.
   c. Show that the modular function of G is given by $\Delta_G(g(a, b)) = a^{-1}$.

4. Show that the left and right regular representations, defined by (9.18) and (9.20), respectively, are unitarily equivalent.
5. Let $\mathcal{U}$ be an open subset of G and $\varphi \in \mathcal{H}(U)$. Show that if the function $\langle U(g)\varphi, \psi \rangle$ on G is in $C^\infty(\mathcal{U})$ for all $\psi \in \mathcal{H}(U)$, then $\varphi \in \mathcal{D}^\infty(U)$.
6. Suppose $a_n \in \mathcal{E}(\mathfrak{g})$, $n \in \mathbb{N}$, are elliptic elements and $\sup_n \deg(a_n) = +\infty$. Prove that $\mathcal{D}^\infty(U) = \cap_{n=1}^\infty \mathcal{D}(\overline{dU(a_n)})$.
7. Let π be a $*$-representation of $\mathbb{C}[x_1, x_2]$ such that the operator $\pi(x_1^2 + x_2^2)$ is essentially self-adjoint. Use Theorem 9.49 for $G = \mathbb{R}^2$ to prove that the operators $\overline{\pi(x_1)}$ and $\overline{\pi(x_2)}$ are strongly commuting self-adjoint operators.
8. Use the Stone–von Neumann Theorem 8.18 to prove that each irreducible unitary representation U of the Heisenberg group such that $dU(x_3) \neq 0$ is unitarily equivalent to a representation U_λ, $\lambda \in \mathbb{R}^\times$, given by (9.49) in Example 9.33.
   Hint: $U(t) := U(g(t, 0, 0))$, $V_\lambda(s) := U(g(0, \lambda^{-1}s, 0))$, $\lambda \in \mathbb{R} \setminus \{0\}$, are unitary groups satisfying the Weyl relation, where $dU(x_3) = i\lambda I$.
9. Let π be a $*$-representation of the Weyl algebra such that all vectors of $\mathcal{D}(\pi)$ are analytic vectors for $\pi(p)$ and $\pi(q)$. Show that $\overline{\pi(p)}$ and $\overline{\pi(q)}$ are self-adjoint operators and their unitary groups fulfill the Weyl relation.
   Hint: Use Theorem 9.39. Verify the Weyl relation (8.44) for small $|s|, |t|$.
10. Let π be a $*$-representation of the Weyl algebra W such that $\pi(p^2 + q^2)$ is essentially self-adjoint. Use Theorem 9.49 to prove that the operators $\overline{\pi(p)}$ and $\overline{\pi(q)}$ are self-adjoint and their unitary groups satisfy the Weyl relation.

11. Suppose $G = \mathbb{R}$ and let $\mathcal{E}(\mathfrak{g}) \cong \mathbb{C}[x]$, with involution $x^+ = -x$. Define a $*$-representation π of $\mathcal{E}(\mathfrak{g})$ on the Hilbert space $L^2(0, 1)$ by $\pi(\mathrm{i}x) = -\mathrm{i}\frac{d}{dt}$, with

$$\mathcal{D}(\pi) = \{\varphi \in C^\infty([0, 1]) : \varphi(0) = \varphi(1); \varphi^{(k)}(0) = \varphi^{(k)}(1) = 0, k \in \mathbb{N}\}.$$

    a. Show that π is exponentiable, but not integrable.

    b. Describe the corresponding unitary representation U of the Lie group $\mathbb{R}$.

12. Let A be a self-adjoint operator on $\mathcal{H}$ and define $U(t) := e^{\mathrm{i}tA}, t \in \mathbb{R}$. Let $\psi \in \mathcal{H}$ and $f \in L^1(\mathbb{R})$. Show that if the Fourier transform of f has compact support, then $U_f \psi := \int_{\mathbb{R}} f(t)U(t)\psi\,dt$ is a bounded vector for A.

13. Show that for the Lie group $G = \mathbb{R}/\mathbb{Z}$ the assertion of Theorem 9.49 is not true. What is the reason for this?

14. Suppose G is connected and simply connected. Set $\Delta := x_1^2 + \cdots + x_d^2$. Show that a $*$-representation π of $\mathcal{E}(\mathfrak{g})$ is integrable if and only if $\overline{\pi(\Delta)}$ is self-adjoint and $\mathcal{D}(\pi) = \mathcal{D}^\infty\big(\overline{\pi(\Delta)}\big)$.

15. Let T be a symmetric operator on a Hilbert space. Prove that $\mathcal{D}^a(T^2) \subseteq \mathcal{D}^a(T)$.

16. Let π be a $*$-representation of $\mathcal{E}(\mathfrak{g})$. Show that $\mathcal{D}^a(\pi)$ is invariant under π.

17. Let G be a connected and simply connected Lie group and f a positive linear functional on the $*$-algebra $\mathcal{E}(\mathfrak{g})$. Suppose $\varphi_f \in \mathcal{D}^a(\pi_f)$, see Definition 9.46. Prove the following:

    a. The GNS representation π_f is exponentiable, so that, by Proposition 9.48, $(\pi_f)^* = dU$ for some unique unitary representation U of the Lie group G.

    b. The positive functional f is pure if and only if U is irreducible.

Hint: For a., use Exercise 16 and Theorem 9.50. For b., verify that $U(G)' = \pi_f(\mathcal{E}(\mathfrak{g}))'_w$ by using Corollary 5.10 and Proposition 9.29.

18. Show that each irreducible unitary representation of $SL(2, \mathbb{R})$ is quasisimple.

9.9 Notes

The Gårding subspace was introduced in Gårding's short note [Ga47]. Proposition 9.11 and Theorem 9.12 are due to N.S. Poulsen [Pu72]. Lemma 9.14 is a well-known classical result; see, e.g., [Gk73, p. 134].

    The pioneering work on elliptic elements of $\mathcal{E}(\mathfrak{g})$ is due to Nelson and Stinespring [NS59, N59] and Langlands [Ls60a]. The three main theorems of Langlands' unpublished thesis [Ls60a] were announced in [Ls60b]. A first group of fundamental results concerns self-adjointness properties. The important equality (9.36) in Theorem 9.22 was proved in [NS59, Ls60a]. Corollary 9.24 was obtained in [NS59]. The general Theorem 9.23 is from Poulsen [Pu72]. Some results on the essential self-adjointness of $dU(x)$ for $\mathrm{i}x \in \mathfrak{g}$ and for central hermitian elements $x \in \mathcal{E}(\mathfrak{g})$ (contained in Corollaries 9.26 and 9.27) were proved much earlier by Segal [Se51, Se52]. The second major result is the description (9.37) of the C^∞-domain given in Theorem 9.22. It was proved independently by Nelson [N59] for the Laplacian and by Langlands [Ls60a] for general elliptic operators. Another basic result of Langlands [Ls60a] states that uniformly elliptic elements generate holomorphic semigroups. Elliptic operators on Lie groups are treated in [Rn91]; see also the references therein. Theorem 9.30 was proved by Goodman [G69a].

Analytic vectors were introduced by Harish-Chandra [Hs53] under the name "well-behaved vectors." Fundamental and deep results on analytic vectors and the exponentiation problem (such as Theorems 9.39, 9.45(ii), and 9.49) were obtained by E. Nelson in his seminal paper [N59]. The denseness of analytic vectors (Theorem 9.45(i)) was proved by Harish-Chandra [Hs53] in special cases and by Cartier and Dixmier [CD58] in full generality; it is also contained in [N59]. The assertions (iii), (iv) of Theorem 9.45 are due to Goodman [G69b]. Theorem 9.50 was proved in [FSSS72]. This result remains valid if the Lie algebra basis is replaced by a set of Lie algebra generators, as shown in [S72]. Other integrability results for Lie algebra representations are given in [Rk87, BGJR88]. Exercise 17 is contained in [Ds71].

The oscillator representation is due to Shale [Sh62] and Weil [Wi64]; see [Ho88, F89] for detailed presentations in n dimensions.

Infinitesimal representations are treated in [Wr72, BR77, Sch90, Nb00]. A number of results of this chapter remain valid for Banach space representations of Lie groups; see, e.g., [Wr72, CD58, Pu72].

Chapter 10
Archimedean Quadratic Modules and Positivstellensätze

Quadratic modules are basic objects of real algebraic geometry. If Q is an Archimedean quadratic module of the polynomial algebra $\mathbb{R}[x_1, \ldots, x_d]$, then the Archimedean Positivstellensatz [Ms08, Theorem 5.4.4] says that strictly positive polynomials on the corresponding semi-algebraic set belong to Q. This chapter deals with Archimedean quadratic modules and Positivstellensätze in *noncommutative* *-algebras.

In Sects. 10.1 and 10.2, we study the *-algebra $A_b(Q)$ of bounded elements with respect to a quadratic module Q. It carries a natural C^*-seminorm (Theorem 10.5) which can be also characterized by means of Q-positive *-representations.

Section 10.3 contains two abstract "Stellensätze" for Archimedean quadratic modules. As an application, we derive in Sect. 10.4 a Positivstellensatz for matrix algebras of polynomials (Theorem 10.25).

The main result of Sect. 10.6 is a Positivstellensatz for the Weyl algebra (Theorem 10.33). The proof of this result relies essentially on the description of * representations of a related *-algebra of fractions, which is studied in Sect. 10.5.

In Sect. 10.7, we prove the closedness of the cone $\sum A^2$ of finite sums of hermitian squares for certain countably generated *-algebras (Theorems 10.35 and 10.36). This result is used to characterize elements of $\sum A^2$ by the positivity in *-representations or states (Theorem 10.37).

Throughout this chapter, A is a **real or complex unital** *-algebra and Q is a **quadratic module** of A. We write λ instead of $\lambda \cdot 1$ for λ in $\mathbb{K} = \mathbb{R}$ or $\mathbb{K} = \mathbb{C}$.

10.1 Archimedean Quadratic Modules and Bounded Elements

According to Definition 2.25, a *quadratic module* of A is a subset Q of A_{her} such that $1 \in Q$ and $a + b \in Q$, $\lambda a \in Q$, $x^+ax \in Q$ for $a, b \in Q$, $\lambda \geq 0$, and $x \in A$.

© The Editor(s) (if applicable) and The Author(s), under exclusive license to Springer Nature Switzerland AG 2020
K. Schmüdgen, *An Invitation to Unbounded Representations of *-Algebras on Hilbert Space*, Graduate Texts in Mathematics 285,
https://doi.org/10.1007/978-3-030-46366-3_10

Recall that a *seminorm* on A is a mapping $p : A \rightarrow [0, +\infty)$ such that

$$p(\lambda a) = |\lambda| p(a) \quad \text{and} \quad p(a + b) \le p(a) + p(b) \quad \text{for} \quad a, b \in A, \ \lambda \in \mathbb{K}. \quad (10.1)$$

Definition 10.1 A C^*-*seminorm* on A is a seminorm p on A such that for $a, b \in A$,

$$p(a^+) = p(a) \quad \text{and} \quad p(ab) \le p(a)p(b), \quad (10.2)$$
$$p(a^+a) = p(a)^2. \quad (10.3)$$

The second condition in (10.2) is called *submultiplicativity* and Eq. (10.3) is the C^*-*condition*.

Remark 10.2 Z. Sebestyen ([Sb79]; see also [DB86, Theorem 38.1] or [Pl01, Theorem 9.5.14]) has proved the following interesting theorem: Each seminorm on a complex $*$-algebra which satisfies the C^*-condition (10.3) is automatically submultiplicative and a C^*-seminorm. We will not use this result in this book. ○

Definition 10.3 An element $a \in A$ is called *bounded with respect to the quadratic module* Q if there exists a number $\lambda > 0$ such that $\lambda^2 - a^+a \in Q$.
  The set of such elements is denoted by $A_b(Q)$. For $a \in A_b(Q)$ we define

$$\|a\|_Q := \inf \left\{ \lambda > 0 : \lambda^2 - a^+a \in Q \right\}. \quad (10.4)$$

A quadratic module Q is called *Archimedean* if $A_b(Q) = A$.

In general the infimum in (10.4) is not attained, as shown in Example 10.11 below. The following lemma contains some simple but useful facts.

Lemma 10.4 *Suppose* $a \in A$, $c \in A_{\text{her}}$ *and* $\lambda > 0$.

 (i)   *If* $\lambda > \|a\|_Q$, *then* $\lambda^2 - a^+a \in Q$.
 (ii) $\lambda^2 - c^2 \in Q$ *if and only if* $\lambda \pm c \in Q$.

Proof (i) By (10.4), there exists $\mu > 0$ such that $\|a\|_Q < \mu < \lambda$ and $\mu^2 - a^+a \in Q$. Then $\lambda^2 - a^+a = (\lambda^2 - \mu^2) + (\mu^2 - a^+a) \in Q$.
(ii) If $\lambda \pm c \in Q$, then

$$\lambda^2 - c^2 = \frac{1}{2\lambda} \left[(\lambda + c)^+ (\lambda - c)(\lambda + c) + (\lambda - c)^+ (\lambda + c)(\lambda - c) \right] \in Q.$$

Conversely, if $\lambda^2 - c^2 \in Q$, then

$$\lambda \pm c = \frac{1}{2\lambda} \left[(\lambda^2 - c^2) + (\lambda \pm c)^2 \right] \in Q. \qquad \square$$

The main result of this section is the following.

Theorem 10.5 *The set* $A_b(Q)$ *is a unital $*$-subalgebra of* A *and* $\| \cdot \|_Q$ *is a C^*-seminorm on* $A_b(Q)$. *Moreover,*

$$\|a\|_Q^2 \leq \|a^+ a + b^+ b\|_Q \quad \text{for} \quad a, b \in A. \tag{10.5}$$

Proof Let $a \in A_b(Q)$. Take a number λ such that $\lambda > \|a\|_Q$. Then $\lambda^2 - a^+ a \in Q$ by Lemma 10.4(i) and hence

$$(\lambda^2/2)^2 - (\lambda^2/2 - aa^+)^2 = a(\lambda^2 - a^+ a)a^+ \in Q.$$

Therefore, from Lemma 10.4(ii), applied with $c = \lambda^2/2 - aa^+$, it follows that

$$\lambda^2/2 + c = \lambda^2/2 + (\lambda^2/2 - aa^+) = \lambda^2 - aa^+ \in Q.$$

Thus, $a^+ \in A_b(Q)$, so $A_b(Q)$ is $*$-invariant, and $\|a^+\|_Q \leq \lambda$. Letting $\lambda \searrow \|a\|_Q$ yields $\|a^+\|_Q \leq \|a\|_Q$. Replacing a by a^+ we obtain $\|a\|_Q \leq \|a^+\|_Q$. Thus, $\|a\|_Q = \|a^+\|_Q$.

Next suppose $a, b \in A_b(Q)$ and $\|a\|_Q < \lambda$, $\|b\|_Q < \mu$. By Lemma 10.4(i), we have $\lambda^2 - a^+ a \in Q$ and $\mu^2 - b^+ b \in Q$. Therefore, we conclude that

$$(\lambda\mu)^2 - (ab)^+ ab = \lambda(\mu^2 - b^+ b)\lambda + b^+(\lambda^2 - a^+ a)b \in Q. \tag{10.6}$$

Hence $ab \in A_b(Q)$ and $\|ab\|_Q \leq \lambda\mu$. Taking the infimum over λ, μ, we obtain $\|ab\|_Q \leq \|a\|_Q \|b\|_Q$.

Now we consider $a + b$. Since $\|b^+ a\|_Q \leq \|b^+\|_Q \|a\|_Q = \|b\|_Q \|a\|_Q < \lambda\mu$ as shown in the preceding paragraphs, $(\lambda\mu)^2 - (b^+ a)^+ b^+ a = (\lambda\mu)^2 - a^+ bb^+ a \in Q$. Similarly, $(\lambda\mu)^2 - b^+ aa^+ b \in Q$. Putting these facts together, we obtain

$$4(\lambda\mu)^2 - (a^+ b + b^+ a)^2$$
$$= 2((\lambda\mu)^2 - a^+ bb^+ a) + 2((\lambda\mu)^2 - b^+ aa^+ b) + (a^+ b - b^+ a)^+ (a^+ b - b^+ a) \in Q.$$

From Lemma 10.4(ii), with $c = a^+ b + b^+ a$, we get $2\lambda\mu \pm (a^+ b + b^+ a) \in Q$. Then

$$(\lambda + \mu)^2 - (a + b)^+ (a + b)$$
$$= (\lambda^2 - a^+ a) + (\mu^2 - b^+ b) + (2\lambda\mu - (a^+ b + b^+ a)) \in Q.$$

so $a + b \in Q$ and $\|a + b\|_Q \leq \lambda + \mu$. This implies $\|a + b\|_Q \leq \|a\|_Q + \|b\|_Q$.

It is easily checked that $\alpha a \in A_b(Q)$ and $\|\alpha a\|_Q = |\alpha| \|a\|_Q$ for $\alpha \in \mathbb{K}$. Obviously, $1 \in A_b(Q)$. Summarizing the preceding, we have shown that $A_b(Q)$ is a unital $*$-subalgebra and $\| \cdot \|_Q$ is a seminorm on $A_b(Q)$ satisfying (10.2).

Next we prove (10.5). Take $\beta > \|a^+ a + b^+ b\|_Q$. Then $\beta^2 - (a^+ a + b^+ b)^2 \in Q$, so $\beta - (a^+ a + b^+ b) \in Q$ by Lemma 10.4(ii). Hence $\beta - a^+ a \in Q$ and $\|a\|_Q \leq \sqrt{\beta}$, so that $\|a\|_Q^2 \leq \beta$. Taking the infimum over β yields (10.5). $\qquad \blacksquare$

Clearly, $\|a^+a\|_Q \leq \|a^+\|_Q \|a\|_Q = \|a\|_Q^2 \leq \|a^+a\|_Q$, where the last inequality follows from (10.5), applied with $b = 0$. Hence we have equality throughout, which gives the C^*-condition (10.3). This completes the proof of Theorem 10.5. $\square$

Since the seminorm $\|\cdot\|_Q$ satisfies the conditions in (10.2) (by Theorem 10.5),

$$J_Q := \{a \in \mathsf{A}_b(Q) : \|a\|_Q = 0\} \tag{10.7}$$

is a $*$-ideal of $\mathsf{A}_b(Q)$. By the definition of $\|\cdot\|_Q$, an element $a \in \mathsf{A}$ belongs to J_Q if and only if $\varepsilon - a^+a \in Q$ for each $\varepsilon > 0$. The quotient $*$-algebra $\mathsf{A}_b(Q)/J_Q$ carries the C^*-norm $\|\cdot\|_Q$ given by $\|a + J_Q\|_Q := \|a\|_Q, a \in \mathsf{A}_b(Q)$.

Corollary 10.6 *The completion* A_Q *of* $(\mathsf{A}_b(Q)/J_Q, \|\cdot\|_Q)$ *is a real or complex C^*-algebra, respectively.*

Proof By definition, A_Q is a $*$-algebra, equipped with a C^*-norm $\|\cdot\|_Q$, and complete in this norm. Thus, for $\mathbb{K} = \mathbb{C}$, A_Q is a complex C^*-algebra (Definition B.1). In the case $\mathbb{K} = \mathbb{R}$, condition (10.5) implies that A_Q is a real C^*-algebra according to Definition B.3. $\square$

Note that complex and *real* C^*-algebras are discussed in Appendix B.

The following corollaries show that it suffices to verify the Archimedean condition for sets of generators.

Corollary 10.7 *Suppose* $\{a_i : i \in I\}$ *is a set of hermitian elements of* A *which generate the $*$-algebra* A. *Then* Q *is Archimedean if and only if for each* a_i *there exists a number* $\lambda_i > 0$ *such that* $\lambda_i \pm a_i \in Q$.

Proof By Lemma 10.4(ii), $\lambda_i \pm a_i \in Q$ is equivalent to $\lambda_i^2 - a_i^2 \in Q$ and so to $a_i \in \mathsf{A}_b(Q)$. Since $\mathsf{A}_b(Q)$ is a $*$-algebra (by Theorem 10.5) and $\{a_i : i \in I\}$ is a set of generators of A, the latter holds if and only if $\mathsf{A}_b(Q) = \mathsf{A}$. $\square$

Corollary 10.8 *Suppose that the $*$-algebra* A *is generated by finitely many elements* $a_1, \ldots, a_k$. *Then the following are equivalent:*

  (i) Q *is Archimedean.*
 (ii)   *There exists a number* $\lambda > 0$ *such that* $\lambda - \sum_{j=1}^{k}(a_j)^+a_j \in Q$.
(iii)   *For each* $j = 1, \ldots, k$, *there exists a* $\lambda_j > 0$ *such that* $\lambda_j - (a_j)^+a_j \in Q$.

Proof (i)$\rightarrow$(ii) is clear by definition. If $\lambda - \sum_{i=1}^{d}(a_i)^+a_i \in Q$, then

$$\lambda - (a_j)^+a_j = \lambda - \sum_i (a_i)^+a_i + \sum_{i \neq j}(a_i)^+a_i \in Q.$$

This proves (ii)$\rightarrow$(iii). If (iii) holds, then $a_j \in \mathsf{A}_b(Q)$ and hence $\mathsf{A}_b(Q) = \mathsf{A}$, since $\mathsf{A}_b(Q)$ is a $*$-algebra by Theorem 10.5. Thus, (iii)$\rightarrow$(i). $\square$

We illustrate these considerations with three examples of *commutative* $*$-algebras.

Example 10.9 Let A be the (real or complex) $*$-algebra generated by the functions

$$a_{ij} := x_i x_j (1 + x_1^2 + \cdots + x_d^2)^{-1}, \quad i, j = 0, \ldots, d, \quad \text{where } x_0 := 1,$$

on $\mathbb{R}^d$, with complex conjugation as involution. Obviously, $(a_{ij})^+ = a_{ij}$. Since $1 = \sum_{i,j=0}^d a_{ij}^2$ as easily verified, A is unital and $1 - \sum_{i,j=0}^d a_{ij}^2 = 0 \in \sum A^2$. Thus, by Corollary 10.8, the quadratic module $\sum A^2$ is Archimedean.

Let K be a nonempty closed subset of $\mathbb{R}^d$. We define a quadratic module of A by

$$Q := \left\{ a \in A_{\text{her}} : a(x) \geq 0, \ x \in K \right\}.$$

Then Q is also Archimedean, because $\sum A^2 \subset Q$. The corresponding seminorm is

$$\|a\|_Q = \sup \left\{ |a(x)| : x \in K \right\}, \quad a \in A. \qquad \bigcirc$$

Example 10.10 (*Archimedean quadratic modules of* $\mathbb{R}[x_1, \ldots, x_d]$)
Let $A = \mathbb{R}[x_1, \ldots, x_d]$ and fix polynomials $p_1, \ldots, p_k \in A$, $k \in \mathbb{N}$. Let

$$K := \left\{ x \in \mathbb{R}^d : p_1(x) \geq 0, \ldots, p_k(x) \geq 0 \right\}$$

denote the semi-algebraic set and Q the quadratic module generated by $p_1, \ldots, p_k$. By the definition of K, all polynomials of Q are nonnegative on K.

Suppose Q is Archimedean. Since $\lambda_j - x_j^2 \in Q$ by Corollary 10.8, $\lambda_j - x_j^2 \geq 0$ on K. Hence K is bounded. Since K is closed, K is compact.

Let $\| \cdot \|_K$ be the supremum seminorm on K. We prove that

$$\|f\|_Q = \|f\|_K \quad \text{for } f \in A. \qquad (10.8)$$

Fix $\lambda > \|f\|_K$. Then the polynomial $\lambda^2 - f(x)^2$ is strictly positive on K. Hence, by the Archimedean Positivstellensatz (see [Ms08, Theorem 5.4] or [Sch17, Theorem 12.36]), it belongs to Q. Thus, $\lambda > \|f\|_Q$. Taking the infimum over λ, we get $\|f\|_K \geq \|f\|_Q$. Now let $\lambda < \|f\|_K$. Then there exists an $x \in K$ such that $\lambda^2 - f(x)^2 < 0$. Hence $\lambda^2 - f^2 \notin Q$, because elements of Q are nonnegative on K. Thus, $\lambda \leq \|f\|_Q$. Letting $\lambda \nearrow \|f\|_K$ gives $\|f\|_K \leq \|f\|_Q$, which completes the proof of (10.8). $\bigcirc$

Example 10.11 Let $A = \mathbb{R}[x]$ and $Q := (1 - x^2)^3 \sum A^2 + \sum A^2$. This is a special case of Example 10.10: $d = 1$, $p_1(x) = (1 - x^2)^3$, $K = [-1, 1]$.

From the identity $4 - 3x^2 = 4(1 - x^2)^3 + x^2(2x^2 - 3)^2 \in Q$ it follows that $\sqrt{3}\, x \in A_b(Q)$, so Q is Archimedean by Corollary 10.8. Therefore, formula (10.8) holds, as shown in Example 10.10.

Let $f(x) = x$. Then $\|f\|_Q = \|f\|_K = 1$ by (10.8). We prove that $1 - f^+ f = 1 - x^2$ is not in Q; that is, the infimum in (10.4) is not attained.

Assume to the contrary $1 - x^2 \in Q$, say $1 - x^2 = (1 - x^2)^3 \sum p_i^2 + \sum q_j^2$, with $p_i, q_j \in A$. This implies $q_j(\pm 1) = 0$. Hence $q_j = (1 - x^2)r_j$ with $r_j \in \mathbb{R}[x]$. Then $1 - x^2 = (1 - x^2)^3 \sum p_i^2 + (1 - x^2)^2 \sum r_j^2$, so $1 = (1 - x^2)^2 \sum p_i^2 + (1 - x^2) \sum r_j^2$. Setting $x = 1$, we obtain a contradiction.   ○

In the remaining part of this chapter, A is a **complex unital** $*$-algebra. The next result characterizes bounded elements in terms of their real and imaginary parts.

Proposition 10.12 *Suppose* A *is a complex unital* $*$-*algebra. An element* $a \in A$ *belongs to* $A_b(Q)$ *if and only if there exists a* $\lambda > 0$ *such that*

$$(\lambda \pm \operatorname{Re} a) \in Q \quad \text{and} \quad (\lambda \pm \operatorname{Im} a) \in Q. \tag{10.9}$$

Proof First assume that (10.9) holds. Setting $a_1 = \operatorname{Re} a$ and $a_2 = \operatorname{Im} a$, we have $a = a_1 + ia_2$. Then $\lambda \pm a_j \in Q$ for $j = 1, 2$ and since Q is a quadratic module,

$$(a + 2\lambda)^+ (\lambda - a_1)(a + 2\lambda) + (a - 2\lambda)^+ (\lambda + a_1)(a - 2\lambda)$$
$$+ (a + 2\lambda i)^+ (\lambda - a_2)(a + 2\lambda i) + (a - 2\lambda i)^+ (\lambda + a_2)(a - 2\lambda i)$$

belongs to Q. A straightforward computation shows that this element is equal to $4\lambda(4\lambda^2 - a^+ a)$. Hence $4\lambda^2 - a^+ a \in Q$, so that $a \in A_b(Q)$.

Conversely, let $a \in A_b(Q)$. Then $\beta - a^+ a \in Q$ for some $\beta > 0$ and hence

$$\beta - a^+ a + (a \pm \alpha)^+ (a \pm \alpha) = \beta + \overline{\alpha}\alpha \pm (\overline{\alpha}a + \alpha a^+) \in Q \tag{10.10}$$

for all $\alpha \in \mathbb{C}$. Set $\lambda := \beta + \overline{\alpha}\alpha$. Since we have $(\overline{\alpha}a + \alpha a^+) = \operatorname{Re} a$ for $\alpha = \frac{1}{2}$ and $(\overline{\alpha}a + \alpha a^+) = \operatorname{Im} a$ for $\alpha = \frac{i}{2}$, (10.10) yields (10.9).   □

The hermitian part A_{her} is an ordered real vector space with ordering defined by

$$a \succeq b \quad \text{if and only if} \quad a - b \in Q.$$

Let $a \in A_{\mathrm{her}}$. By Definition 10.3 and Lemma 10.4(ii), we have $a \in A_b(Q)$ if and only if $\lambda \pm a \in Q$ for some $\lambda > 0$. This implies that the unit element 1 of A is an *order unit* for the cone $Q_b := Q \cap A_b(Q)_{\mathrm{her}}$ in the real vector space $A_b(Q)_{\mathrm{her}}$; see Appendix C.

Let $Q^\wedge$ denote the set of Q-positive functionals f on A, that is,

$$Q^\wedge := \left\{ f \in A^* : f(a) \geq 0 \text{ for } a \in Q \right\}.$$

Lemma 10.13 *Suppose that* Q *is an Archimedean quadratic module of the complex unital* $*$-*algebra* A. *If* $c \in A_{\mathrm{her}}$ *and* $c \notin Q$, *then there exists an extremal functional* $f \in Q^\wedge$ *such that* $f(1) = 1$ *and* $f(c) \leq 0$.

Proof Since Q is Archimedean, $A_b(Q) = A$ and $Q = Q_b$. As noted above, 1 is an order unit for $Q = Q_b$. Hence, by Proposition C.5, applied to $E = A_{\mathrm{her}}$, there exists an extremal Q-positive $\mathbb{R}$-linear functional f on A_{her} such that $f(1) = 1$ and $f(c) \leq 0$. Since $\mathsf{A} = \mathsf{A}_{her} + i\mathsf{A}_{her}$, we can extend f to a $\mathbb{C}$-linear functional on A, which has the desired properties. $\qquad\square$

10.2 Representations of ∗-Algebras with Archimedean Quadratic Modules

From now on, A is a **complex unital** ∗-algebra.

The following definition appeared already in Sect. 4.4.

Definition 10.14 A ∗-representation π of A is Q-*positive* if $\pi(a) \geq 0$ for $a \in Q$, that is, $\langle \pi(a)\varphi, \varphi \rangle \geq 0$ for all $a \in Q$ and $\varphi \in \mathcal{D}(\pi)$.

The family of Q-positive nondegenerate ∗-representations of A is denoted by $\mathrm{Rep}_Q\, \mathsf{A}$.

Clearly, each ∗-representation of A is $(\sum \mathsf{A}^2)$-positive. By Corollary 4.37, the GNS representation π_f of a Q-positive linear functional f on A is also Q-positive.

Proposition 10.15 *Suppose Q is an Archimedean quadratic module of A. Then, for any $\pi \in \mathrm{Rep}_Q\, \mathsf{A}$, all operators $\pi(a)$, $a \in \mathsf{A}$, are bounded and we have*

$$\|a\|_Q = \sup\,\{\|\pi(a)\| : \pi \in \mathrm{Rep}_Q\, \mathsf{A}\}, \quad a \in \mathsf{A}, \tag{10.11}$$

$$J_Q = \{a \in \mathsf{A} : \pi(x) = 0 \ \text{ for } \ \pi \in \mathrm{Rep}_Q\, \mathsf{A}\}. \tag{10.12}$$

Proof Fix $a \in \mathsf{A}$. Let $\pi \in \mathrm{Rep}_Q\, \mathsf{A}$. Take $\lambda > \|a\|_Q$. Then $\lambda^2 - a^+a \in Q$. Therefore, since π is Q-positive and $\pi(1) = I$, for $\varphi \in \mathcal{D}(\pi)$ we have

$$\langle \pi(\lambda^2 - a^+a)\varphi, \varphi \rangle = \lambda^2 \|\varphi\|^2 - \|\pi(a)\varphi\|^2 \geq 0,$$

so $\|\pi(a)\varphi\| \leq \lambda \|\varphi\|$. Thus $\pi(a)$ is bounded and $\|\pi(a)\| \leq \lambda$. Taking the infimum over λ and the supremum over π, we get $\sup\,\{\|\pi(a)\| : \pi \in \mathrm{Rep}_Q\, \mathsf{A}\} \leq \|a\|_Q$.

Now we prove the converse inequality. It is obviously true if $\|a\|_Q = 0$, so we can assume that $\|a\|_Q > 0$. We choose $\alpha > 0$ such that $\alpha < \|a\|_Q$. Then, $\alpha^2 - a^+a \notin Q$ by the definition of $\|a\|_Q$. From Lemma 10.13 it follows that there exists a Q-positive linear functional f on A such that $f(1) = 1$ and $f(\alpha^2 - a^+a) \leq 0$. Then

$$\alpha^2 \|\varphi_f\|^2 - \|\pi_f(a)\varphi_f\|^2 = \alpha^2 f(1) - f(a^+a) = f(\alpha^2 - a^+a) \leq 0,$$

which gives $\alpha \leq \|\pi_f(a)\|$. Since $\pi_f \in \mathrm{Rep}_Q\, \mathsf{A}$ by Corollary 4.37 and $\alpha < \|a\|_Q$ was arbitrary, we conclude that $\|a\|_Q \leq \sup\,\{\|\pi(a)\| : \pi \in \mathrm{Rep}_Q\, \mathsf{A}\}$.

Combining both paragraphs we obtain (10.11), and this implies (10.12). $\qquad\square$

Suppose Q is an Archimedean quadratic module of A. Then $\mathsf{A} = \mathsf{A}_b(Q)$.

Let $\mathrm{Rep}\,\mathsf{A}_Q$ denote the family of nondegenerate $*$-representations ρ of the C^*-algebra A_Q on a Hilbert space, that is, $\mathcal{D}(\rho) = \mathcal{H}(\rho)$.

Recall that for any $\pi \in \mathrm{Rep}_Q\,\mathsf{A}$ all operators $\pi(a)$ are bounded by Proposition 10.15. Hence, π is closed if and only if $\mathcal{D}(\pi) = \mathcal{H}(\pi)$.

Now let $\pi \in \mathrm{Rep}_Q\,\mathsf{A}$ be closed. From (10.12) and (10.11) it follows that π annihilates J_Q and π is $\|\cdot\|_Q$-continuous. Therefore, π passes to a continuous $*$-representation of the normed $*$-algebra $(\mathsf{A}/\mathsf{J}_Q, \|\cdot\|_Q)$ and, since $\mathcal{D}(\pi) = \mathcal{H}(\pi)$, this $*$-representation extends by continuity to a $*$-representation $\breve{\pi} \in \mathrm{Rep}\,\mathsf{A}_Q$ of the completion A_Q of $(\mathsf{A}/\mathsf{J}_Q, \|\cdot\|_Q)$ on $\mathcal{D}(\breve{\pi}) = \mathcal{H}(\pi)$.

Conversely, let $\rho \in \mathrm{Rep}\,\mathsf{A}_Q$. It is easily verified that $\pi(a) = \rho(a + \mathsf{J}_Q), a \in \mathsf{A}$, defines a closed $*$-representation $\pi \in \mathrm{Rep}_Q\,\mathsf{A}$ such that $\breve{\pi} = \rho$.

We summarize the outcome of the preceding discussion in the next proposition.

Proposition 10.16 *Suppose Q is an Archimedean quadratic module of A. Then the map $\pi \mapsto \breve{\pi}$ is a bijection of the closed representations of $\mathrm{Rep}_Q\,\mathsf{A}$ and $\mathrm{Rep}\,\mathsf{A}_Q$.*

We now turn to the simplest and most important quadratic module $Q = \sum \mathsf{A}^2$.

Definition 10.17 A unital $*$-algebra A is called *algebraically bounded*, or briefly *bounded*, if the quadratic module $\sum \mathsf{A}^2$ is Archimedean, or equivalently, for each $a \in \mathsf{A}$ there exists a number $\lambda > 0$ such that $\lambda - a^+ a \in \sum \mathsf{A}^2$.

The $*$-algebra in Example 10.9 is a bounded $*$-algebra.

Suppose A is a bounded $*$-algebra. Then, since $\sum \mathsf{A}^2$ is Archimedean and $*$-representations are always $(\sum \mathsf{A}^2)$-positive, all nondegenerate $*$-representations of A act by bounded operators (Proposition 10.15). Further, $\mathsf{J}_{\sum \mathsf{A}^2}$ is the intersection of kernels of nondegenerate $*$-representations (by (10.12)) and hence of all $*$-representations of A (by Lemma 4.9(ii)). Thus, by Definition 4.43, *the $*$-ideal $\mathsf{J}_{\sum \mathsf{A}^2}$ is equal to the $*$-radical $\mathrm{Rad}\,\mathsf{A}$ of the bounded $*$-algebra A.*

Proposition 10.18 *Suppose A is a bounded commutative complex unital $*$-algebra. Then each extremal state f on A is a hermitian character.*

Proof Since each state is hermitian, it remains to prove that f is a character.

Let $x \in \mathsf{A}$. First we note that is suffices to show that

$$f(x^+ a x) = f(a)f(x^+ x) \quad \text{for} \quad a \in \mathsf{A}. \tag{10.13}$$

Indeed, if (10.13) holds, then $f(ax^+ x) = f(a)f(x^+ x)$, because A is commutative. Since $\mathsf{A} = \mathrm{Lin}\sum \mathsf{A}^2$ by (2.32), this extends by linearity to $f(ab) = f(a)f(b)$ for all $a, b \in \mathsf{A}$, so f is a character.

Next, if $f(x^+ x) = 0$, then by the Cauchy–Schwarz inequality (2.34),

$$|f(x^+ a x)|^2 = |f((a^+ x)^+ x)|^2 \le f(x^+ x)f((a^+ x)^+ a^+ x) = 0.$$

Hence (10.13) holds, since both sides are zero. Thus we can assume that $f(x^+ x) \ne 0$.

That the ∗-algebra A is bounded means that the quadratic module $\sum A^2$ is Archimedean. Hence there exists a $\lambda > 0$ such that $\lambda - x^+x \in \sum A^2$. Therefore we have $\lambda - f(x^+x) \geq 0$. Upon replacing λ by $\lambda + 1$ and x by $x\lambda^{-1/2}$ we can assume without loss of generality that $1 - x^+x \in \sum A^2$ and $1 - f(x^+x) > 0$. Then $1 - x^+x = \sum_j (y_j)^+ y_j$ with $y_j \in$ A. Recall that $f(1) = 1$, because f is a state. For $a \in$ A we obtain

$$f((1 - x^+x)a^+a) = f\left(\sum_j (y_j)^+ y_j a^+ a\right) = \sum_j f((y_j a)^+ y_j a) \geq 0.$$

Hence there are states f_1 and f_2 on A defined by

$$f_1(a) := \frac{f(x^+ a x)}{f(x^+ x)} \quad \text{and} \quad f_2(a) := \frac{f((1 - x^+x)a)}{1 - f(x^+x)}, \quad a \in A,$$

and $f = f(x^+x)f_1 + (1 - f^+(xx))f_2$ is a convex combination of f_1 and f_2. By assumption, f is an extremal state. Therefore, $f = f_1$, which gives (10.13). □

10.3   Stellensätze for Archimedean Quadratic Modules

Throughout this section, Q is an **Archimedean** quadratic module of a complex unital ∗-algebra A.

A crucial technical fact is contained in the following lemma.

Lemma 10.19 *If a state f on A is an extremal element of $Q^\wedge$, then f is a pure state.*

Proof Suppose $f = \lambda_1 f_1 + \lambda_2 f_2$, where f_1, f_2 are states of A, $\lambda_1, \lambda_2 \in (0, 1)$, and $\lambda_1 + \lambda_2 = 1$. Fix $j \in \{1, 2\}$. Then $\lambda_j f_j \leq f$. Since $\lambda_j > 0$, it follows from Theorem 5.3 that there exists a positive operator $T_j \in \pi_f(A)'_w$ such that $f_j(\cdot) = \langle T_j \pi_f(\cdot)\varphi_f, \varphi_f \rangle$. Since $f \in Q^\wedge$, Corollary 4.37 implies that π_f, hence $\overline{\pi}_f$, is Q-positive. Hence all operators $\pi_f(a)$ are bounded by Proposition 10.15, so $\overline{\pi}_f(A)$ is a ∗-algebra of bounded operators on $\mathcal{H}(\pi_f)$. Therefore, $\pi_f(A)'_w = \overline{\pi}_f(A)'_w = \overline{\pi}_f(A)'$ is a von Neumann algebra. Hence $T_j^{1/2} \in \overline{\pi}_f(A)'$. Then, for $a \in$ A,

$$f_j(a) = \langle T_j \pi_f(a)\varphi_f, \varphi_f \rangle = \langle T_j^{1/2}\overline{\pi}_f(a)\varphi_f, T_j^{1/2}\varphi_f \rangle = \langle \overline{\pi}_f(a)T_j^{1/2}\varphi_f, T_j^{1/2}\varphi_f \rangle.$$

Therefore, since $\overline{\pi}_f$ is Q-positive, f_j is Q-positive, so that $f_j \in Q^\wedge$. Since f is an extremal element of $Q^\wedge$ and f, f_1, f_2 are states, we conclude that $f_j = f$. This proves that f is pure. □

The following results give algebraic certificates for the images in all Q-positive representations of a hermitian element to be positive, in Theorem 10.20, and to be not negative, in Proposition 10.21.

Theorem 10.20 *For any element* $a \in \mathsf{A}_{\mathrm{her}}$ *the following are equivalent:*

(i) $a + \varepsilon \in Q$ *for each* $\varepsilon > 0$.
(ii) $f(a) \geq 0$ *for each Q-positive linear functional f on* A.
(iii) $f(a) \geq 0$ *for each Q-positive pure state f of* A.
(iv) $\pi(a) \geq 0$ *for each $*$-representation $\pi \in \mathrm{Rep}_Q\mathsf{A}$.*
(v) $\pi(a) \geq 0$ *for each closed irreducible $*$-representation $\pi \in \mathrm{Rep}_Q\mathsf{A}$.*

Proof The implications (i)→(ii)→(iii) and (iv)→(v) are obvious.

(ii)→(iv): If $\pi \in \mathrm{Rep}_Q\mathsf{A}$ and $\varphi \in \mathcal{D}(\pi)$, then $\langle \pi(\cdot)\varphi, \varphi \rangle$ is a Q-positive functional on A, so that $\langle \pi(a)\varphi, \varphi \rangle \geq 0$ by (ii). Thus, $\pi(a) \geq 0$.

(iv)→(ii): If f is a Q-positive linear functional, then the GNS representations π_f is in $\mathrm{Rep}_Q\mathsf{A}$ by Corollary 4.37, so that $f(a) = \langle \pi_f(a)\varphi_f, \varphi_f \rangle \geq 0$ by (iv).

(v)→(iii): If f is a Q-positive pure state, then the GNS representation $\overline{\pi}_f$ is Q-positive and irreducible by Corollaries 4.37 and 5.5. Therefore, by (v), we have $f(a) = \langle \overline{\pi}_f(a)\varphi_f, \varphi_f \rangle \geq 0$.

(iii)→(i): Assume to the contrary that $a + \varepsilon \notin Q$ for some $\varepsilon > 0$. By Lemma 10.13, there exists an extremal element $f \in Q^\wedge$ such that $f(1) = 1$ and $f(a + \varepsilon) \leq 0$. Since $f(1) = 1$, f is a state. Lemma 10.19 implies that the extremal element f is a pure state. Then $f(a) \leq -f(\varepsilon \cdot 1) = -\varepsilon < 0$, which contradicts (iii).   □

Proposition 10.21 *For any $a \in \mathsf{A}_{\mathrm{her}}$ the following are equivalent:*

(i) *There exist elements* $b_1, \ldots, b_k \in \mathsf{A}$, $k \in \mathbb{N}$, *such that* $\sum_{j=1}^{k}(b_j)^+ab_j$ *belongs to* $1 + Q := \{1 + c : c \in Q\}$.
(ii) *For each $*$-representation* $\pi \in \mathrm{Rep}_Q\mathsf{A}$ *with domain* $\mathcal{D}(\pi) \neq \{0\}$ *there exists a vector* $\eta \in \mathcal{D}(\pi)$ *such that* $\langle \pi(a)\eta, \eta \rangle > 0$.

Proof (i)→(ii): Suppose $\sum_j (b_j)^+ab_j = 1 + c$ with $c \in Q$. Let $\pi \in \mathrm{Rep}_Q\mathsf{A}$ and suppose that $\mathcal{D}(\pi) \neq \{0\}$. Then $\pi(1) = I$ (because π is nondegenerate) and there exists a vector $\varphi \in \mathcal{D}(\pi)$, $\varphi \neq 0$. Hence

$$\sum_j \langle \pi(a)\pi(b_j)\varphi, \pi(b_j)\varphi \rangle = \sum_j \langle \pi((b_j)^+ab_j)\varphi, \varphi \rangle$$
$$= \langle \pi(1 + c)\varphi, \varphi \rangle \geq \langle \pi(1)\varphi, \varphi \rangle = \|\varphi\|^2 > 0.$$

Therefore, at least one summand $\langle \pi(a)\pi(b_j)\varphi, \pi(b_j)\varphi \rangle$ has to be positive.

(ii)→(i): Let C denote the finite sums of elements b^+ab, where $b \in \mathsf{A}$, and set $\tilde{Q} := \{\alpha + c : \alpha > 0, c \in Q\}$. Assume that (i) does not hold. Then, upon scaling the elements b_j, it follows that C and $\tilde{Q}$ are disjoint. Both sets C and $\tilde{Q}$ are convex. Since Q is Archimedean, 1 is an order unit, hence an algebraically interior point, of Q and so of $\tilde{Q}$. Thus, Proposition C.1 applies and there exists a linear functional $f : \mathsf{A}_{\mathrm{her}} \mapsto \mathbb{R}$, $f \neq 0$, such that $\sup\{f(c) : c \in C\} \leq \inf\{f(q) : q \in \tilde{Q}\}$. Since $\tilde{Q}$ is a cone, this infimum is zero. Therefore, f is Q-positive and $f(c) \leq 0$ for $c \in C$. We extend f to a $\mathbb{C}$-linear functional, denoted also f, on A. Since $\sum \mathsf{A}^2 \subseteq Q$, f is a positive functional on A. The GNS representation π_f is in $\mathrm{Rep}_Q \mathsf{A}$ by Corollary 4.37 and we have $\langle \pi_f(a)\pi_f(b)\varphi_f, \pi(b)\varphi_f \rangle = f(b^+ab) \leq 0$ for all $b \in \mathsf{A}$. Clearly, $\mathcal{D}(\pi_f) \neq \{0\}$, because $f \neq 0$ and hence $f(1) \neq 0$. Then, since $\mathcal{D}(\pi_f) = \pi_f(\mathsf{A})\varphi_f$, the condition in (ii) is not fulfilled for $\pi = \pi_f$. $\qquad\square$

We illustrate Theorem 10.20 by proving that every *strictly positive* trigonometric polynomial on the d-torus is a finite sum of hermitian squares. (This is also an immediate consequence of the Archimedean Positivstellensatz ([Ms08, Theorem 5.4.4] or [Sch17, Theorem 12.35]) from real algebraic geometry.)

The additive group $\mathbb{Z}^d$, $d \in \mathbb{N}$, is a $*$-semigroup with involution $\mathrm{n}^+ = -\mathrm{n}$ and the map $\mathrm{n} = (n_1, \ldots, n_d) \mapsto z^{\mathrm{n}} = z_1^{n_1} \cdots z_d^{n_d}$ is a $*$-isomorphism of the group $*$-algebra $\mathbb{C}[\mathbb{Z}^d]$ on the $*$-algebra of trigonometric polynomials in d variables. For notational simplicity we identify $\mathbb{C}[\mathbb{Z}^d]$ with the latter $*$-algebra, that is,

$$\mathsf{A} := \mathbb{C}[\mathbb{Z}^d] = \mathbb{C}[z_1, \bar{z}_1, \ldots, z_d, \bar{z}_d \,|\, z_1\bar{z}_1 = \bar{z}_1 z_1 = 1, \ldots, z_d\bar{z}_d = \bar{z}_d z_d = 1].$$

Since $1 - (z_j)^+ z_j = 0 \in \sum \mathsf{A}^2$, the $*$-algebra A is bounded by Corollary 10.8. The characters of A are precisely the point evaluations at points of the d-torus

$$\mathbb{T}^d = \big\{ z = (z_1, \ldots, z_d) \in \mathbb{C}^d : |z_1| = \cdots = |z_d| = 1 \big\}.$$

Proposition 10.22 *Suppose $p = p^+ \in \mathbb{C}[\mathbb{Z}^d]$. If $p(z) > 0$ for all $z \in \mathbb{T}^d$, then we have $p \in \sum \mathbb{C}[\mathbb{Z}^d]^2$.*

Proof Since $\mathbb{T}^d$ is compact, we have $\varepsilon := \inf\{ p(z) : z \in \mathbb{T}^d \} > 0$. We apply Theorem 10.20 with $Q = \sum \mathsf{A}^2$ and $a := p - \varepsilon$. Let f be an extremal state of A. By Proposition 10.18, f is a character. Hence f is a point evaluation at some $z \in \mathbb{T}^d$. Then $f(a) = p(z) - \varepsilon \geq 0$, so $p = a + \varepsilon \in Q$ by Theorem 10.20, (iii)$\rightarrow$(i). $\qquad\square$

10.4 Application to Matrix Algebras of Polynomials

We begin with a general result which is of interest in its own. It gives a one-to-one correspondence between $*$-representations of the matrix algebra $M_n(\mathsf{A})$ and A.

Proposition 10.23 *Let ρ be a nondegenerate $*$-representation of a complex unital $*$-algebra A and $n \in \mathbb{N}$. Set $\mathcal{D}(\rho_n) = \oplus_{k=1}^n \mathcal{D}(\rho)$. For $A = (a_{ij}) \in M_n(\mathsf{A})$ and $\varphi = (\varphi_1, \ldots, \varphi_n) \in \mathcal{D}(\rho_n)$, define*

$$(\rho_n(A)\varphi)_k = \sum_{j=1}^n \rho(a_{kj})\varphi_j, \quad k = 1, \ldots, n. \tag{10.14}$$

Then ρ_n is a nondegenerate $*$-representation of the $*$-algebra $M_n(\mathsf{A})$. Each nondegenerate $*$-representation π of $M_n(\mathsf{A})$ is, up to unitary equivalence, of this form.

Further, ρ_n is irreducible if and only if ρ is irreducible.

Proof The first assertion follows by straightforward computations; we omit the details. We carry out the proof of the second assertion.

Let π be a nondegenerate $*$-representation of $M_n(\mathsf{A})$. Since A is unital, $M_n(\mathbb{C})$ is a $*$-subalgebra of $M_n(\mathsf{A})$. It is well known that the restriction of π to $M_n(\mathbb{C})$ is a direct sum of identity representations. Hence $\mathcal{D}(\pi)$ is an n-fold direct sum of copies of a vector space $\mathcal{D}$ such that for $(\varphi_1, \ldots, \varphi_n) \in \mathcal{D}(\pi)$, $\varphi_j \in \mathcal{D}$, we have

$$(\pi(\Lambda)\varphi)_i = \sum_j \lambda_{ij}\varphi_j \quad \text{for } \Lambda = (\lambda_{ij}) \in M_n(\mathbb{C}), \ i = 1, \ldots, n. \qquad (10.15)$$

Let e_{ij} denote the matrix with 1 in the (i, j)-position and zeros elsewhere. Clearly, the operators $E_i = \overline{\pi(e_{ii})}$, $i = 1, \ldots, n$, are pairwise orthogonal projections such that $E_1 + \cdots + E_n = I$, since π is nondegenerate. Thus $\mathcal{D}(\pi) = \oplus_{i=1}^n E_i \mathcal{D}(\pi)$. From (10.15) we get $E_i \mathcal{D}(\pi) = \pi(e_{ii})\mathcal{D}(\pi) = \mathcal{D}$. The map $a \mapsto e_{11}ae_{11}$ is a $*$-homomorphism of A into $M_n(\mathsf{A})$. Hence $\rho(a) := \pi(e_{11}ae_{11}){\restriction}\mathcal{D}$ defines a nondegenerate $*$-representation ρ of A on $\mathcal{D}(\rho) := \mathcal{D} = E_1\mathcal{D}(\pi)$. Let $[a]_{ij} \in M_n(\mathsf{A})$ be the matrix with (i, j)-entry $a \in \mathsf{A}$ and zeros elsewhere. Suppose $\varphi = (\varphi_1, \ldots, \varphi_n) \in \mathcal{D}(\rho)$. Using the relations $e_{ij}e_{kl} = \delta_{jk}e_{il}$ and (10.15) we compute for $a \in \mathsf{A}$,

$$\pi([a]_{ij})\varphi = \pi(e_{i1}[a]_{11}e_{1j})\varphi = \pi(e_{i1})\pi(e_{11}ae_{11})\pi(e_{1j})\varphi$$
$$= \pi(e_{i1})\pi(e_{11}ae_{11})(\varphi_j, 0, \ldots, 0) = \pi(e_{i1})(\rho(a)\varphi_j, 0, \ldots, 0),$$

so that $(\pi([a]_{ij})\varphi)_k = \delta_{ik}\rho(a)\varphi_j$. Hence $\pi([a]_{ij}) = \rho_n([a]_{ij})$ by (10.14). Since $A = \sum_{i,j=1}^n [a_{ij}]_{ij}$ for $A = (a_{ij}) \in M_n(\mathsf{A})$, it follows that $\pi(A) = \rho_n(A)$. Thus, $\pi = \rho_n$.

We prove the last assertion. By straightforward computations it follows that the strong commutant $\rho_n(M_n(\mathsf{A}))'_s$ consists of all diagonal operators $(T, \ldots, T)$, with $T \in \rho(\mathsf{A})'_s$. Hence, by Proposition 4.26, ρ_n is irreducible if and only if so is ρ. $\quad\square$

Now we specialize to the case of the polynomial algebra $\mathsf{A} = \mathbb{C}_d[\underline{x}]$.

Let $t \in \mathbb{R}^d$. Clearly, there is a $*$-representation π_t of $M_n(\mathsf{A})$ on $\mathcal{D}(\pi_t) = \mathbb{C}^n$ given by $\pi_t(a) = a(t)$, $a \in M_n(\mathsf{A})$. The last assertion of Proposition 10.23 implies that π_t is irreducible. Obviously, π_t is closed.

For $\Lambda = \Lambda^+ \in M_n(\mathbb{C})$ we write $\Lambda \succeq 0$ if the matrix Λ is positive semi-definite.

Corollary 10.24 *Suppose Q is an Archimedean quadratic module of $M_n(\mathsf{A})$, where $\mathsf{A} = \mathbb{C}_d[\underline{x}]$. Then, up to unitary equivalence, the closed irreducible $*$-representations π in $\operatorname{Rep}_Q M_n(\mathsf{A})$ are precisely the representations π_t with $t \in K(Q)$, where*

$$K(Q) := \left\{ t \in \mathbb{R}^d : a(t) \succeq 0 \ \text{for all} \ a \in Q \right\}.$$

Proof Let $\pi \in \operatorname{Rep}_Q M_n(\mathsf{A})$ be irreducible. By Proposition 10.23, there exists an irreducible $*$-representation ρ of A such that $\pi = \rho_n$. Since the operators $\pi(a)$, $a \in M_n(\mathsf{A})$, are bounded by Proposition 10.15, so are the operators $\rho(b)$, $b \in \mathsf{A}$. Then ρ is a closed irreducible $*$-representation of the commutative $*$-algebra $\mathsf{A} = \mathbb{C}_d[\underline{x}]$

by bounded operators. Hence ρ acts on a one-dimensional space (this follows, for instance, from Corollary 7.15) and, up to unitary equivalence, ρ is the evaluation at some point $t \in \mathbb{R}^d$. Then $\pi = \rho_n$ is just the $*$-representation π_t defined above.

Let $t \in \mathbb{R}^d$. By definition, π_t is Q-positive if and only if $a(t) \succeq 0$ for all $a \in Q$, that is, $t \in K(Q)$. This gives the description stated in the corollary.   □

Combining Corollary 10.24 and Theorem 10.20, (i)↔(v), we obtain the following Positivstellensatz for matrix algebras of polynomials.

Theorem 10.25 *Suppose Q is an Archimedean quadratic module of the matrix $*$-algebra $M_n(\mathbb{C}_d[\underline{x}])$. For any $a = a^+ \in M_n(\mathbb{C}_d[\underline{x}])$ the following are equivalent:*

(i) $a + \varepsilon I \in Q$ for all $\varepsilon > 0$.
(ii) $a(t) \succeq 0$ for all $t \in K(Q)$.

10.5   A Bounded $*$-Algebra Related to the Weyl Algebra

In this section and the next, we are concerned with the Weyl algebra

$$\mathsf{W} := \mathbb{C}\langle a, a^+ \mid aa^+ - a^+a = 1 \rangle,$$

N denotes the element $N := a^+a \in \mathsf{W}$, and α is a positive number such that $\alpha \notin \mathbb{N}$.

Let us recall the Bargmann–Fock representation $\pi_{\mathbb{C}}$ of W from Sect. 8.3. It acts on the standard orthonormal basis $\{e_n : n \in \mathbb{N}_0\}$ of the Hilbert space $l^2(\mathbb{N}_0)$ by

$$\pi_{\mathbb{C}}(a)e_n = \sqrt{n}\, e_{n-1}, \quad \pi_{\mathbb{C}}(a^+)e_n = \sqrt{n+1}\, e_{n+1}, \quad \pi_{\mathbb{C}}(N)e_n = ne_n, \qquad (10.16)$$

where $e_{-1} := 0$. For notational simplicity we shall write x instead of $\overline{\pi_{\mathbb{C}}(x)}$ in what follows. Then N is a self-adjoint operator on $l^2(\mathbb{N}_0)$ with spectrum $\mathbb{N}_0$. Define

$$u := a(N+\alpha)^{-1}, \quad u^* := a^+(N+\alpha+1)^{-1},$$
$$v := a^2(N+\alpha)^{-1}, \quad v^* := (a^+)^2(N+\alpha+1)^{-1},$$
$$y_n := (N+\alpha+n)^{-1}, \quad n \in \mathbb{Z}, \quad \text{and} \quad y := y_0 = (N+\alpha)^{-1}.$$

Using (10.16) we easily verify that these are bounded operators defined on $l^2(\mathbb{N}_0)$ such that u^* and v^* are indeed the adjoints of u and v, respectively, y_n is self-adjoint and the following relations are satisfied:

$$uy_n = y_{n+1}u, \quad u^*y_n = y_{n-1}u^*, \quad vy = u^2(1-y), \quad v^*y_1 = (u^*)^2, \qquad (10.17)$$

$$y_k - y_n = (n-k)y_ky_n, \qquad (10.18)$$

$$u^*u = y - \alpha y^2, \quad uu^* = y_1 - \alpha y_1^2, \qquad (10.19)$$

$$v^*v = 1 - (2\alpha-1)y + \alpha(1+\alpha)y^2, \quad vv^* = 1 - (2\alpha-1)y_2 + \alpha(1+\alpha)y_2^2. \quad (10.20)$$

Let B be the unital $*$-subalgebra of $\mathbf{B}(l^2(\mathbb{N}_0))$ generated by u, v, and y_n, $n \in \mathbb{N}_0$. This auxiliary $*$-algebra and its $*$-representations (Proposition 10.30) are essentially used in the proof of the Positivstellensatz (Theorem 10.33) given in the next section. For $b, c \in B_{\mathrm{her}}$ we write $b \succeq c$ if $b - c \in \sum B^2$.

Lemma 10.26 *The $*$-algebra B is bounded.*

Proof Let B_b denote the bounded elements with respect to the quadratic module $\sum B^2$. By (10.19), $y = \alpha y^2 + u^* u$ and $\alpha^{-1} - y = \alpha(\alpha^{-1} - y)^2 + u^* u$, so that

$$\alpha^{-1} \succeq y \succeq 0. \tag{10.21}$$

Thus, $y \in B_b$. Further, $\alpha^{-1} \succeq y = \alpha y^2 + u^* u \succeq u^* u$ implies that $u \in B_b$.

Let $n \in \mathbb{N}$. We prove that $y_n \in B_b$. From $y = y_n + n y_n y$ (by (10.18)) and $y \succeq 0$ we obtain $y y_n = y_n^2 + n y_n y y_n \succeq 0$ and so $y = y_n + n y y_n \succeq y_n$. Further, from $y_n = y(1 - n y_n)$ and $y \succeq 0$ it follows that $y_n - n y_n^2 = (1 - n y_n) y (1 - n y_n) \succeq 0$. Hence $y_n \succeq n y_n^2 \succeq 0$. Combined with $y \succeq y_n$ and $y \in B_b$ we get $y_n \in B_b$.

Since the generators of B are in B_b and B_b is a $*$-algebra (by Theorem 10.5), we have $B = B_b$. $\square$

Since the $*$-algebra B is bounded, each nondegenerate $*$-representation π of B acts by bounded operators (Proposition 10.15). For notational simplicity we write U, U^*, V, V^*, Y, Y_n for the images of u, u^*, v, v^*, y, y_n under π. Clearly, these operators satisfy the corresponding relations of the generators stated above.

Lemma 10.27 *There is no nondegenerate $*$-representation of B on a Hilbert space $\mathcal{H} \neq \{0\}$ such that $\ker Y = \{0\}$ and $\ker(Y - \alpha^{-1} I) = \{0\}$.*

Proof Assume to the contrary that there is a such representation of B on a Hilbert space $\mathcal{H} \neq \{0\}$. Clearly, Eq. (10.21) implies that $\alpha^{-1} I \geq Y \geq 0$ on $\mathcal{H}$. Hence $\sigma(Y) \subseteq [0, \alpha^{-1}]$. Therefore, by the assumptions, there exists a $\lambda_0 \in (0, \alpha^{-1})$ in $\sigma(Y)$. From $Y - Y_1 = Y Y_1$ (by (10.18)) we obtain $Y_1 = Y(I + Y)^{-1}$. By the spectral mapping theorem, the map $\lambda_0 \mapsto \lambda_0(1 + \lambda_0)^{-1}$ is a bijection of $\sigma(Y)$ on $\sigma(Y_1)$ and $\lambda_0(1 + \lambda_0)^{-1} \in \sigma(Y_1)$. We have $0 < \sup \sigma(Y_1) < \sup \sigma(Y)$, so there exists an interval $J \subseteq (0, \alpha^{-1})$ such that $J \cap \sigma(Y_1) = \emptyset$ and $J \subseteq \sigma(Y)$. For the corresponding spectral projections we have $E_{Y_1}(J) = 0$ and $E_Y(J) \neq 0$, since $\mathcal{H} \neq \{0\}$. The relation $U Y = Y_1 U$ (by (10.17)) implies $U g(Y) = g(Y_1) U$ for $g \in L^\infty(\mathbb{R})$. In particular, $U E_Y(J) = E_{Y_1}(J) U = 0$. Hence $0 = U^* U E_Y(J) = Y(I - \alpha Y) E_Y(J)$ by (10.19). Since $E_Y(J) \neq 0$ and $0, \alpha^{-1} \notin J$, this is a contradiction. $\square$

In the following two lemmas we suppose that π is a nondegenerate $*$-representation of B on a Hilbert space $\mathcal{H}$.

Lemma 10.28 *$\mathcal{H}_0 := \ker Y$ is an invariant subspace for π. For the subrepresentation $\pi_0 := \pi \restriction \mathcal{H}_0$, we have $\pi_0(u) = \pi_0(u^*) = \pi_0(y_n) = 0$ and $\pi_0(v)$ is unitary.*

Proof From the relations $Y - Y_n = nY_nY = nYY_n$ we derive $\mathcal{H}_0 := \ker Y = \ker Y_n$. Combined with the relations $YU = Y_1U$ and $U^*Y_1 = YU^*$ it follows that $\mathcal{H}_0$ is invariant under U and U^*. From $U^*U = Y - \alpha Y^2$ (by (10.19)) it follows that $|U|^2 = U^*U = 0$ on $\mathcal{H}_0$. Therefore, $|U| = 0$ and hence $U = 0$ on $\mathcal{H}_0$ by the polar decomposition of U. Taking the adjoints of the last two equations of (10.17) we obtain $YV^* = (I - Y)(U^*)^2$ and $Y_1V = U^2$. Since U^* and U are zero on $\mathcal{H}_0$, these relations imply that $\mathcal{H}_0$ is invariant under V^* and V. Thus we have shown that the algebra generators of $\mathbf{B}$ leaves $\mathcal{H}_0$ invariant and so does the representation π.

Since $Y = Y_1 = 0$ on $\mathcal{H}_0$, (10.20) implies that $V \restriction \mathcal{H}_0$ is unitary on $\mathcal{H}_0$. □

Lemma 10.29 *Suppose $\varphi \in \ker(Y - \alpha^{-1}I)$ and $\|\varphi\| = 1$. Then, for $k, n \in \mathbb{N}_0$,*

$$Y_n\varphi = (\alpha + n)^{-1}\varphi, \tag{10.22}$$

$$U^n(U^*)^n\varphi = n!((\alpha + 1)_n)^{-2}, \tag{10.23}$$

$$\langle (U^*)^k\varphi, (U^*)^n\varphi \rangle = \delta_{kn} n!((\alpha + 1)_n)^{-2}. \tag{10.24}$$

Here $(\cdot)_n$ denotes the Pochhammer symbol: $(z)_n := z(z + 1) \cdots (z + n - 1)$ for $n \in \mathbb{N}$ and $(z)_0 := 1$.

Proof First we show that $U\varphi = 0$. Recall that, since $\alpha^{-1} \succeq y \succeq 0$ by (10.21), the operator Y satisfies $\alpha^{-1} \geq Y \geq 0$ and so $\sigma(Y) \subseteq [0, \alpha^{-1}]$. Next, by (10.18),

$$(I + Y)U\varphi = (I + Y)\alpha Y\varphi = \alpha(I + Y)Y_1U\varphi = \alpha YU\varphi,$$

hence $YU\varphi = (\alpha - 1)^{-1}U\varphi$. Since $(\alpha - 1)^{-1} \notin [0, \alpha^{-1}]$, this implies $U\varphi = 0$.

Equation (10.22) follows from $Y - Y_{n+1} = (n + 1)Y_{n+1}Y$ by (10.18) and $Y\varphi = \alpha^{-1}\varphi$ by the assumption $\varphi \in \ker(Y - \alpha^{-1}I)$.

Now we prove (10.23) by induction on n. The case $n = 0$ is clear. Suppose that (10.23) holds for $n \in \mathbb{N}_0$. Using (10.22) we derive

$$Y_{n+1}(I - \alpha Y_{n+1})\varphi = (\alpha + n + 1)^{-1}(1 - \alpha(\alpha + n + 1)^{-1})\varphi$$
$$= (\alpha + n + 1)^{-2}(n + 1)\varphi.$$

Therefore, by (10.19) and (10.18), we obtain

$$U^{n+1}(U^*)^{n+1}\varphi = U^n(UU^*)(U^*)^n\varphi = U^n(Y_1(I - \alpha Y_1))(U^*)^n\varphi$$
$$= U^n(U^*)^nY_{n+1}(I - \alpha Y_{n+1})\varphi = (\alpha + n + 1)^{-2}(n + 1)U^n(U^*)^n\varphi. \tag{10.25}$$

Since $(\alpha + n + 1)(\alpha + 1)_n = (\alpha + 1)_{n+1}$, by induction (10.25) yields (10.23) for $n + 1$.

Finally, we verify (10.24). In the case $n = k$, (10.24) follows at once from (10.23). To treat the case $n \neq k$, we can assume without loss of generality that $n > k$.

Let $k \in \mathbb{N}$. We use the relations (10.19), (10.18), and (10.17). By induction on k we prove that $u^k(u^*)^k = f_k(y_1, \ldots, y_k)$ for some polynomial f_k. The elements

y_j pairwise commute by (10.18). Since $U\varphi = 0$ as shown above, using (10.17) we derive

$$\langle (U^*)^k\varphi, (U^*)^n\varphi \rangle = \langle U^n(U^*)^k\varphi, \varphi \rangle = \langle U^{n-k}f_k(Y_1, \ldots, Y_k)\varphi, \varphi \rangle \quad (10.26)$$
$$= \langle U^{n-k-1}f_k(Y_2, \ldots, Y_{k+1})U\varphi, \varphi \rangle = 0. \quad (10.27)$$

For $k = 0$, we have $\langle \varphi, (U^*)^n\varphi \rangle = \langle U^{n-1}U\varphi, \varphi \rangle = 0$. $\square$

The main result of this section is the following proposition.

Proposition 10.30 *Each nondegenerate $*$-representation π of B on $\mathcal{H}$ is a direct sum of the subrepresentation π_0 from Lemma 10.28 and possibly of representations which are unitarily equivalent to the identity representation of B on $l^2(\mathbb{N}_0)$.*

Proof Since π_0 is a subrepresentation, the orthogonal complement $\mathcal{H}_1 = \mathcal{H} \ominus \mathcal{H}_0$ is invariant under π and $\pi_1 := \pi \restriction \mathcal{H}_1$ is a $*$-representation of B. (Note that we are dealing with bounded $*$-representations acting on the whole Hilbert space!) Consider the closed subspace $\mathcal{K} := \ker(Y - \alpha^{-1}I)$ of $\mathcal{H}_1$. If $\mathcal{K} = \{0\}$, then Lemma 10.27, applied to π_1, implies that $\mathcal{H}_1 = \{0\}$, so $\pi = \pi_0$ and the proof is finished.

Now assume that $\mathcal{K} \neq \{0\}$. Take a unit vector $\varphi \in \mathcal{K}$ and let $\mathcal{H}_\varphi$ denote the closed span of vectors $(U^*)^n\varphi$, $n \in \mathbb{N}_0$. Obviously, $\mathcal{H}_\varphi$ is invariant under U^*. From (10.22) and the commutation rules it follows that $\mathcal{H}_\varphi$ is invariant under U and Y_n, $n \in \mathbb{N}_0$. The last two equations of (10.17) imply that $VY\mathcal{H}_\varphi \subseteq \mathcal{H}_\varphi$ and $V^*Y_1\mathcal{H}_\varphi \subseteq \mathcal{H}_\varphi$. Since the kernel of $Y \restriction \mathcal{H}_1$ is trivial, $Y\mathcal{H}_\varphi$ and $Y_1\mathcal{H}_\varphi$ are dense in $\mathcal{H}_\varphi$. Therefore, $\mathcal{H}_\varphi$ is invariant under V and V^*. Thus $\mathcal{H}_\varphi$ is invariant under π_1.

From Lemma 10.29 it follows therefore that the vectors

$$\varphi_n := (\alpha + 1)_n (n!)^{-1/2}(U^*)^n\varphi, \quad n \in \mathbb{N}_0, \quad (10.28)$$

form an orthonormal basis of $\mathcal{H}_\varphi$, and by definition, we have

$$U^*\varphi_n = \sqrt{n+1}\,(\alpha + 1 + n)^{-1}\varphi_{n+1}, \quad n \in \mathbb{N}_0. \quad (10.29)$$

Using (10.29) and (10.23) we conclude that the operators U^* and Y_k act on the orthonormal basis $\{\varphi_n : n \in \mathbb{N}_0\}$ of $\mathcal{H}_\varphi$ by the same formulas as the identity representation of B does on the standard orthonormal basis of $l^2(\mathbb{N}_0)$. The same holds for U, since U is the adjoint of U^*, and for V and V^*, as it follows from the last formulas of (10.17). Thus, the cyclic subrepresentation of B on $\mathcal{H}_\varphi$ is unitarily equivalent to the identity representation of B.

Now we consider the subrepresentation on $\mathcal{H}_2 := \mathcal{H}_1 \ominus \mathcal{H}_\varphi$. If $\mathcal{H}_2 \neq \{0\}$, we repeat the same procedure with a unit vector of $\mathcal{K} \cap \mathcal{H}_2$. We continue in this manner. By Zorn's lemma, we obtain a direct sum $\oplus_i \pi_i$ on a subspace $\widetilde{\mathcal{H}}_1$ of subrepresentations π_i all of which are equivalent to the identity representation such that $\mathcal{K} \cap (\widetilde{\mathcal{H}}_1)^\perp = \{0\}$. Since $\ker Y \cap \mathcal{H}_1 = \{0\}$, it follows from Lemma 10.27, applied to $\pi \restriction (\widetilde{\mathcal{H}}_1)^\perp$, that $(\widetilde{\mathcal{H}}_1)^\perp = \{0\}$. Hence $\pi_1 = \oplus_i \pi_i$, which is the assertion. $\square$

10.6 A Positivstellensatz for the Weyl Algebra

From Proposition 8.2(i) we recall that the set $\{a^j (a^+)^k : j, k \in \mathbb{N}_0\}$ is a vector space basis of the Weyl algebra W. Hence each element $c \in \mathsf{W}$ is of the form

$$c = \sum_{j,k=0}^{n} \alpha_{jk} a^j (a^+)^k, \tag{10.30}$$

where $\alpha_{jk} \in \mathbb{C}$ are uniquely determined by c. For $c \neq 0$ we define the degree $m = \deg(c) := \max\{j + k : \alpha_{jk} \neq 0\}$ and the polynomial $c_m \in \mathbb{C}[z, \bar{z}]$ by

$$c_m(z, \bar{z}) = \sum_{j+k=m} \alpha_{jk} z^j \bar{z}^k.$$

The Bargmann–Fock representation $\pi_{\mathbb{C}}$ of W acts on the dense invariant domain

$$\mathcal{D}^\infty(N) = \cap_{k=1}^\infty \mathcal{D}(N^k) = \{(\varphi_n)_{n\in\mathbb{N}_0} \in l^2(\mathbb{N}_0) : (n^k \varphi_n) \in l^2(\mathbb{N}_0) \;\; \text{for } k \in \mathbb{N}\}$$

of $l^2(\mathbb{N}_0)$. The action of an element (10.30) is determined by the action of the generators a, a^+ given by (10.16). Since $(N + \alpha + n)^{-1}$ leaves $\mathcal{D}^\infty(N)$ invariant for $n \in \mathbb{N}$, so do all elements of the algebra B. Therefore, since the action of W is faithful, we can consider W and B as $*$-subalgebras of the larger $*$-algebra $\mathcal{L}^+(\mathcal{D}^\infty(N))$.

Let D denote the set of all finite products of elements $x_n := N + \alpha + n, n \in \mathbb{Z}$. The following two lemmas are used in the proof of our main theorem.

Lemma 10.31 *Let $c \in \mathsf{W}$ and $\deg(c) \leq 4k$, $k \in \mathbb{N}$. Then $b_c := (y_{2k})^k c (y_{2k})^k \in \mathsf{B}$. If π_0 is a $*$-representation as in Lemma 10.28 and $c = a^n (a^+)^m$ with $n, m \in \mathbb{N}_0$, then $\pi_0(b_c) = 0$ if $n + m < 4k$ and $\pi_0(b_c) = \pi_0(v^n (v^*)^m)$ if $n + m = 4k$.*

Proof It suffices to prove the assertions for elements $c = a^n (a^+)^m$. Let us write $n = 2r + i$, $m = 2s + j$ with $i, j \in \{0, 1\}$ and $r, s \in \mathbb{N}_0$, $2r + 2s + i + j \leq 4k$. Clearly, (10.16) implies $y_j a = a y_{j-1}$ and $y_j a^+ = a^+ y_{j+1}$. Hence

$$b_c = y_{2k}^k a^n (a^+)^m y_{2k}^k = a^2 y_{i_1} \cdots a^2 y_{i_r} x (a^+)^2 y_{i_{r+1}} \cdots (a^+)^2 y_{i_{r+s}}, \tag{10.31}$$

where $i_1, \ldots, i_{r+s} \in \mathbb{N}_0$ and x is an element of the form $a^i (a^+)^j y_{n_1}^{l_1} y_{n_2}^{l_2}$ with l_1, l_2, $n_1, n_2 \in \mathbb{N}_0$, $l_1 + l_2 + r + s = 2k$. Since $y_l = y_0(1 - l y_l) = (1 - l y_l) y_0$ and $v = a^2 y_0$, $v^* = a^+ y_2$, we derive $a^2 y_{i_l} = v(1 - i_l y_{i_l})$ and $(a^+)^2 y_{i_l} = (1 - i_l y_{i_l}) v^*$.

First let $n + m = 4k$. Then $x = 1$ or $x = aa^+ y_l$, so in any case $x = 1 - \alpha_l y_l$ for some $\alpha_l \in \mathbb{R}$. Since $\pi_0(y_l) = 0$ by Lemma 10.28, it follows from the preceding and (10.31) that $b_c \in \mathsf{B}$ and $\pi_0(b_c) = \pi_0(v^n (v^*)^m)$.

Now assume that $n + m < 4k$. Then we have $x = y_{n_1}^{l_1} y_{n_2}^{l_2}$, $x = a y_{n_1}^{l_1} y_{n_2}^{l_2}$, $x = a^+ y_{n_1}^{l_1} y_{n_2}^{l_2}$ with $l_1 > 0$ or $x = aa^+ y_{n_1}^{l_1} y_{n_2}^{l_2}$ with $l_1 + l_2 \geq 2$. In all these cases, x is a product of elements y_l, u, or u^*. Thus $x \in \mathsf{B}$, hence $b_c \in \mathsf{B}$ by the preceding, and $\pi_0(b_c) = 0$, since $\pi_0(u) = \pi_0(u^*) = \pi_0(y_l) = 0$. $\qquad \square$

Lemma 10.32 *For each $b \in \mathsf{B}$ there exists an element $d \in \mathsf{D}$ such that $bd \in \mathsf{W}$.*

Proof Suppose the assertion holds for $b_1, b_2 \in \mathsf{B}$, say $b_1 d_1, b_2 d_2 \in \mathsf{W}$. Then we have $d_1 d_2 \in \mathsf{D}$ and $(b_1 + b_2)d_1 d_2 = (b_1 d_1 d_2 + b_2 d_2 d_1) \in \mathsf{W}$, hence the assertion holds for $b_1 + b_2$. The same reasoning is valid for finite sums.

Since $u = ay, u^* = ya^+, v = a^2 y, v^* = y(a^+)^2$, each element of B is a finite sum of products of elements a, a^+, y_n with $n \in \mathbb{N}_0$. Note that $x_n y_n = y_n x_n = 1$. Using the relations $y_k a = a y_{k-1}$ and $y_k a^+ = a^+ y_{k+1}$ for $k \in \mathbb{Z}$, it follows that we can multiply such a product by products of elements $x_k \in \mathsf{D}$ with $k \in \mathbb{Z}$ (!) such that all factors y_n cancel and the corresponding element belongs to W. $\square$

Now we are prepared to state and prove the following strict Positivstellensatz for the Weyl algebra. It says that, under the positivity assumptions (i) and (ii), hermitian elements, multiplied by certain "denominators", are sums of hermitian squares.

Theorem 10.33 *Suppose c is a hermitian element of the Weyl algebra W of degree $2n, n \in \mathbb{N}$, satisfying the following assumptions:*

(i)   *There exists $\varepsilon > 0$ such that $\langle \pi_{\mathbb{C}}(c - \varepsilon)\varphi, \varphi \rangle \geq 0$ for $\varphi \in \mathcal{D}(\pi_{\mathbb{C}}) = \mathcal{D}^\infty(N)$.*
(ii)   *$c_{2n}(z, \overline{z}) > 0$ for all $z \in \mathbb{T}$.*

If n is even, then there exists an element $d \in \mathsf{D}$ such that $dcd \in \sum \mathsf{W}^2$. If n is odd, then there is an element $d \in \mathsf{D}$ such that $daca^+d \in \sum \mathsf{W}^2$.

Proof First we suppose that n is even, say $2n = 4k$.

By Lemma 10.32, $b_c = (y_{2k})^k c y_{2k}^k \in \mathsf{B}$. First we prove that $b_c \in \sum \mathsf{B}^2$.

Assume the contrary. The $*$-algebra B is bounded by Lemma 10.26, so the quadratic module $\sum \mathsf{B}^2$ is Archimedean. Therefore, since $b_c \notin \sum \mathsf{B}^2$, it follows from Lemma 10.13 that there exists a positive functional f on B such that $f(1) = 1$ and $f(b_c) \leq 0$. (We could even have f to be extremal, but this does not really simplify the following proof.) From Proposition 10.30 it follows that the GNS representation $\overline{\pi}_f$ is a direct sum of a representation π_0 as in Lemma 10.28 and possibly of representation $\pi_i, i \in J$, which are unitarily equivalent to the identity representation of B on $l^2(\mathbb{N}_0)$. Then, for any $b \in \mathsf{B}$, we have

$$f(b) = \langle \pi_f(b)\varphi_f, \varphi_f \rangle = \langle \pi_0(b)\psi, \psi \rangle + \sum_i \langle b\psi_i, \psi_i \rangle, \tag{10.32}$$

where $\psi \in \mathcal{D}(\pi_0)$ and $\psi_i \in l^2(\mathbb{N}), i \in J$. Using assumption (i) we derive

$$\langle b_c \psi_i, \psi_i \rangle = \langle \pi_{\mathbb{C}}(c) y_{2k}^k \psi_i, y_{2k}^k \psi_i \rangle \geq \varepsilon \langle y_{2k}^k \psi_i, y_{2k}^k \psi_i \rangle = \varepsilon \| y_{2k}^k \psi_i \|^2 \geq 0. \tag{10.33}$$

The operator $\pi_0(v)$ is unitary by Lemma 10.28. Let $\pi_0(v) = \int_{\mathbb{T}} z \, dE(z)$ be its spectral decomposition and let $c = \sum_{i,j} \alpha_{ij} a^i (a^+)^j$. Assumption (ii) implies that $\delta := \inf\{c_{4n}(z, \overline{z}) : z \in \mathbb{T}\} > 0$. Applying Lemma 10.31 we obtain

$$\langle \pi_0(b_c)\psi, \psi \rangle = \sum_{i+j=4n} \alpha_{ij} \langle \pi_0(v^i (v^*)^j)\psi, \psi \rangle$$

$$= \int_{\mathbb{T}} c_{4n}(z, \overline{z}) \, d\langle E(z)\psi, \psi \rangle \geq \delta \langle E(\mathbb{T})\psi, \psi \rangle \geq 0. \tag{10.34}$$

Now we apply (10.32) with $b = b_c$. Since $f(b_c) \leq 0$, it follows from (10.33) and (10.34) that all summands in (10.32) vanish. Therefore, since $\ker y_{2k} = \{0\}$ on $l^2(\mathbb{N}_0)$, (10.33) implies $\psi_i = 0$. Similarly, (10.34) leads to $\psi = 0$. Then, (10.32) for $b = 1$ yields $f(1) = 0$, a contradiction. Thus we have shown that $b_c \in \sum \mathsf{B}^2$.

Since $b_c = y_{2k}^k c y_{2k}^k \in \sum \mathsf{B}^2$, there exist elements $b_1, \ldots, b_m \in \mathsf{B}$ such that $y_{2k}^k c y_{2k}^k = \sum_{j=1}^m (b_j)^+ b_j$. Let $d \in \mathsf{D}$. We multiply this equation by $dx_{2k}^k = x_{2k}^k d$ from both sides. Since $y_{2k}^k x_{2k}^k = x_{2k}^k y_{2k}^k = 1$, we then obtain

$$dx_{2k}^k \, y_{2k}^k c y_{2k}^k \, x_{2k}^k d = dcd = \sum_{j=1}^m (b_j dx_{2k}^k)^+ b_j dx_{2k}^k. \tag{10.35}$$

From Lemma 10.32 we can find elements $d_j \in \mathsf{D}$ such that $b_j d_j \in \mathsf{W}$. We set $d := d_1 \cdots d_m \in \mathsf{D}$ and $a_j := b_j dx_{2k}^k = (b_j d_j) d_1 \cdots d_{j-1} d_{j+1} \cdots d_m x_{2k}^k$. Then $a_j \in \mathsf{W}$ and (10.35) yields $dcd = \sum_j (a_j)^+ a_j$. This proves the assertion for even n.

Now we suppose that n is odd, say $n = 2k - 1$. Put $\check{c} := aca^+$. Let $\varphi \in \mathcal{D}(\pi_{\mathbb{C}})$. Using assumption (i) we derive

$$\langle \pi_{\mathbb{C}}(\check{c})\varphi, \varphi \rangle = \langle \pi_{\mathbb{C}}(c)\pi_{\mathbb{C}}(a^+)\varphi, \pi_{\mathbb{C}}(a^+)\varphi \rangle \geq \varepsilon \langle \pi_{\mathbb{C}}(a^+)\varphi, \pi_{\mathbb{C}}(a^+)\varphi \rangle$$

$$= \varepsilon \langle \pi_{\mathbb{C}}(aa^+)\varphi, \varphi \rangle = \varepsilon \langle \pi_{\mathbb{C}}(N+1)\varphi, \varphi \rangle \geq \varepsilon \|\varphi\|^2.$$

Further, $\check{c}_{4k}(z, \overline{z}) = c_{2n}(z, \overline{z})$. Thus $\check{c}$ has degree $4k$ and satisfies both assumptions (i) and (ii). Hence the assertion follows from the preceding result, applied to $\check{c}$. $\square$

We illustrate the assertion of Theorem 10.33 with an example.

Example 10.34 $c_\varepsilon := (N-1)(N-2) + \varepsilon$, where $\varepsilon > 0$. Since the self-adjoint operator N on $l^2(\mathbb{N}_0)$ has the spectrum $\mathbb{N}_0$, it follows that $(N-1)(N-2) \geq 0$. Clearly, c_ε has degree 4 and $(c_\varepsilon)_4(z, \overline{z}) = (z\overline{z})^2$, so $(c_\varepsilon)_4 = 1$ on $\mathbb{T}$. Thus the assumptions (i) and (ii) of Theorem 10.33 are satisfied. Hence there exists an element $d \in \mathsf{D}$ such that $dc_\varepsilon d \in \sum \mathsf{W}^2$.

We determine such an element d explicitly. For $\alpha \in \mathbb{R}$ we have the identity

$$(N+\alpha)c_\varepsilon(N+\alpha) = \frac{1}{2}\alpha^2 (N-1)^2(N-2)^2 + \left(1 - \frac{1}{2}\alpha^2\right)N(N-1)(N-2)(N-3)$$

$$+ (2\alpha + 3)N(N-1)(N-2) + \varepsilon(N+\alpha)^2.$$

Recall from (8.13) that $(a^+)^k a^k = N(N-1)\cdots(N-(k-1))$ for $k \in \mathbb{N}$. Therefore, if $\alpha^2 \leq 2$, the right-hand side belongs to $\sum \mathsf{W}^2$ and we can take

$d = N + \alpha$. It can be shown that the element c_ε itself is not in $\sum W^2$ if $0 < \varepsilon < \frac{1}{4}$, so in this case the "denominator" factor $N + \alpha$ is needed in order to get a sum of hermitian squares.   ○

10.7   A Theorem About the Closedness of the Cone $\sum A^2$

The following theorem gives sufficient conditions on a $*$-algebra A which ensure that the cone $\sum A^2$ is closed in the finest locally convex topology τ_{st} (see Appendix C for the definition of the topology τ_{st}). It should be noted that we do not assume in Theorem 10.35 that the $*$-algebra is unital.

Theorem 10.35 *Let A be a complex $*$-algebra that has a faithful $*$-representation. Suppose A is the union of a sequence of finite-dimensional subspaces E_n, $n \in \mathbb{N}$, such that $E_n \subseteq E_{n+1}$ for $n \in \mathbb{N}$ and for any $n \in \mathbb{N}$ there exists a number $k_n \in \mathbb{N}$ such that the following condition is satisfied:*
(∗) *Each element $a \in E_n \cap \sum A^2$ is of the form $a = \sum_j (a_j)^+ a_j$, with $a_j \in E_{k_n}$.*
Then the cone $\sum A^2$ is closed in the finest locally convex topology τ_{st} of A.

Proof Since $\dim E_{k_n} < \infty$, there exists a number $s_n \in \mathbb{N}$ such that $u^+ v \in E_{s_n}$ for $u, v \in E_{k_n}$. We can assume without loss of generality that $n \leq k_n \leq s_n$ for $n \in \mathbb{N}$. Set $d_n := \dim E_{s_n}$. We divide the proof into four steps.

The main content of the first statement is also known as the *Carathéodory lemma*.

Statement 1: *Each $a \in E_n \cap \sum A^2$ is a sum of d_n squares $(a_j)^+ a_j$, with $a_j \in E_{k_n}$.*

Proof Let $a \in E_n \cap \sum A^2$. By assumption (∗) we can write $a = \sum_{j=1}^m (a_j)^+ a_j$, where $a_1, \ldots, , a_m \in E_{k_n}$ and $m \in \mathbb{N}$. If $m < d_n$, we add zeros.

Now suppose $m > d_n$. Then, since $(a_j)^+ a_j \in E_{s_n}$ and $d_n = \dim E_{s_n}$, the elements $(a_j)^+ a_j, j = 1, \ldots, m$, are linearly dependent, so there is a vector $(\lambda_1, \ldots, \lambda_m) \neq 0$ of $\mathbb{R}^m$ such that $\sum_{j=1}^m \lambda_j (a_j)^+ a_j = 0$. Without loss of generality we assume that $\lambda_m \geq |\lambda_j|$ for $j = 1, \ldots, m - 1$. Then $\lambda_m \neq 0$. Setting $\tilde{a}_j := (1 - \lambda_j \lambda_m^{-1})^{1/2} a_j$ for $j = 1, \ldots, m - 1$, a simple computation yields $a = \sum_{j=1}^{m-1} (\tilde{a}_j)^+ \tilde{a}_j$. Continuing this procedure we arrive at d_n squares.   □

Since A has a faithful $*$-representation, we can assume that A is a $*$-subalgebra of $\mathcal{L}^+(\mathcal{D})$ for some complex inner product space $(\mathcal{D}, \langle \cdot, \cdot \rangle)$.

Statement 2: *There exist sequences $(\varphi_k)_{k \in \mathbb{N}}$ of unit vectors $\varphi_k \in \mathcal{D}$ and $(r_k)_{k \in \mathbb{N}}$ of natural numbers such that*

$$\|a\|_n := \max \{|\langle a\, \varphi_l, \varphi_l \rangle| : l = 1, \ldots, r_n\}, \quad a \in E_n,$$

defines a norm on the finite-dimensional vector space E_n.

Proof Let S be the unit sphere of some norm on E_n. For any $a \in S$, we have $a \neq 0$, so there exists a unit vector $\psi_a \in \mathcal{D}$ such that $\langle a\psi_a, \psi_a \rangle \neq 0$. Clearly, the sets $S_a :=$ $\{x \in S : \langle x\psi_a, \psi_a \rangle \neq 0\}$, $a \in S$, form an open cover of the compact set S. Hence there exists a finite subcover, say $\{S_{a_1}, \cdots, S_{a_p}\}$. Then $\max\{ |\langle \cdot \psi_{a_l}, \psi_{a_l} \rangle| : l = 1, \ldots, p\}$ is a norm on E_n. We write the vectors ψ_{a_l} for the subspaces $E_1, E_2, \ldots$ as a sequence $(\varphi_k)_{k\in\mathbb{N}}$. This sequence has the desired properties. $\square$

The crucial step of the proof is the following result.

Statement 3: *The set $E_n \cap \sum A^2$ is closed in the normed space $(E_n, \|\cdot\|_n)$.*

Proof Since each norm on a finite-dimensional normed space is continuous, there is a constant $c_n > 0$ such that $\|a\|_{k_n} \leq c_n \|a\|_n$ for all $a \in E_n \subseteq E_{k_n}$.

Let $b \in E_n$ be in the closure of $E_n \cap \sum A^2$. Then there is a sequence $(b_m)_{m\in\mathbb{N}}$ of elements $b_m \in E_n \cap \sum A^2$ such that $b = \lim_m b_m$ in E_n. By Statement 2 there exist elements $a_{m1}, \ldots, a_{md_n} \in E_{k_n}$, $m \in \mathbb{N}$, such that

$$b_m = \sum_{j=1}^{d_n} (a_{mj})^+ a_{mj}. \qquad (10.36)$$

Then, for $l = 1, \ldots, r_{k_n}$, $j = 1, \ldots, d_n$, and $m \in \mathbb{N}$, we obtain

$$|\langle a_{mj}\varphi_l, \varphi_l \rangle|^2 \leq \|a_{mj}\varphi_l\|^2 = \langle (a_{mj})^+ a_{mj}\varphi_l, \varphi_l \rangle \leq \langle b_m\varphi_l, \varphi_l \rangle$$

and hence

$$\|a_{mj}\|_{k_n}^2 \leq \|b_m\|_{k_n} \leq c_n \sup_{i\in\mathbb{N}} \|b_i\|_n < \infty.$$

This shows that for each $j = 1, \ldots, d_n$ the sequence $(a_{mj})_{m\in\mathbb{N}}$ is bounded in the finite-dimensional normed space $(E_{k_n}, \|\cdot\|_{k_n})$. Therefore, by a diagonal procedure it follows that there exist a sequence $(m_k)_{k\in\mathbb{N}}$ of natural numbers $m_{k+1} > m_k$ and elements $a_j \in E_{k_n}$ such that $a_j = \lim_k a_{m_k,j}$ in E_{k_n} for $j = 1, \ldots, d_n$.

Recall that $u^+ v \in E_{s_n}$ for $u, v \in E_{k_n}$. Since bilinear mappings of finite-dimensional normed spaces are continuous and the norm $\|\cdot\|_{k_n}$ is $*$-invariant, there is a constant $\gamma_n > 0$ such that $\|u^+ v\|_{s_n} \leq \gamma_n \|u\|_{k_n} \|v\|_{k_n}$ for $u, v \in E_{k_n}$. Then, for $x, y \in E_{k_n}$, we derive

$$\|x^+ x - y^+ y\|_{s_n} = \|(x-y)^+ x + y^+(x-y)\|_{s_n} \leq \gamma_n \|x-y\|_{k_n}(\|x\|_{k_n} + \|y\|_{k_n}).$$

Therefore, $a_j = \lim_k a_{m_k,j}$ in E_{k_n} implies $(a_j)^+ a_j = \lim_k (a_{m_k,j})^+ a_{m_k,j}$ in E_{s_n} for $j = 1, \ldots, d_n$. Now we consider equation (10.36) for $m = m_k$. Then, passing to the limit $k \to \infty$ we obtain $b = \sum_{j=1}^{d_n} (a_j)^+ a_j \in E_n \cap \sum A^2$. $\square$

Statement 4: *$\sum A^2$ is closed in the locally convex space $A[\tau_{\text{st}}]$.*

Proof In this proof we use some notions and advanced results from the theory of locally convex spaces (all of them can be found in [Sh71]). The locally convex space

$E := A[\tau_{st}]$ is the strong inductive limit of the family of normed spaces E_k, $k \in \mathbb{N}$. Hence its strong dual E' is a reflexive Frechet space. If U is a neighborhood of zero in E', the polar U° of U in E is bounded. Hence U° is contained in some space E_n. Since $E_n \cap \sum A^2$ is closed in E_n by Statement 3 and U° is closed in E_n, so is $U^\circ \cap \sum A^2$. Therefore, the Krein–Shmulian theorem [Sh71, Theorem 6.4], applied to the Frechet space E', implies that $\sum A^2$ is closed in $(E')' = E = A[\tau_{st}]$.   □

A longer, but completely elementary proof (without using the Krein–Shmulian theorem) of the fact that Statement 3 implies Statement 4 is given in [Sch17, Theorem A.28] and [Ms08, Theorem 3.6.1].

Now the proof of Theorem 10.35 is complete.   □

Theorem 10.36 *Let* A *be the polynomial ∗-algebra* $\mathbb{C}[x_1, \ldots, x_d]$ *or the Weyl algebra* $W(d)$ *or the free polynomial ∗-algebra* $\mathbb{C}\langle x_1, \ldots, x_d | (x_j)^+ = x_j\rangle$ *or the enveloping algebra* $\mathcal{E}(\mathfrak{g})$ *of a finite-dimensional real Lie algebra* $\mathfrak{g}$, *where* $d \in \mathbb{N}$.
Then the cone $\sum A^2$ *is closed in the finest locally convex topology* τ_{st} *of* A .

Proof The assertion will be derived from Theorem 10.35. Thus, it suffices to verify that the assumptions of this theorem are satisfied for each of the four ∗-algebras.

First we note that each of these ∗-algebras has a faithful ∗-representation. For the polynomial algebra $\mathbb{C}[x_1, \ldots, x_d]$ and the Weyl algebra $W(d)$ this was stated in Examples 4.21 and 3.11, respectively. For the free polynomial algebra we refer to Corollary 6.13 and for the enveloping algebra to Statement 2 in Example 9.9.

Next we show that condition (∗) holds. We carry out this proof for $A = \mathcal{E}(\mathfrak{g})$. The proofs for the other three algebras follow by similar, even simpler reasonings.

We fix a basis $\{x_1, \ldots, x_d\}$ of the real vector space $\mathfrak{g}$. Put $x^{\mathfrak{k}} := x_1^{k_1} \cdots x_d^{k_d}$ and $|\mathfrak{k}| := k_1 + \cdots + k_d$ for $\mathfrak{k} = (k_1, \ldots, k_d) \in \mathbb{N}_0^d$. By the Poincare–Birkhoff–Witt theorem (Sect. 9.1), the set $\{x^{\mathfrak{k}} : \mathfrak{k} \in \mathbb{N}_0^d\}$ is a vector space basis of $\mathcal{E}(\mathfrak{g})$. Set $E_n := \text{Lin}\,\{x^{\mathfrak{k}} : |\mathfrak{k}| \le n\}$. Clearly, $E_n \subseteq E_{n+1}$ and $\mathcal{E}(\mathfrak{g}) = \cup_{n=1}^\infty E_n$. For $a \in \mathcal{E}(\mathfrak{g})$ let $d(a)$ denote the smallest number $k \in \mathbb{N}_0$ such that $a \in E_k$.

To prove condition (∗) we suppose $a \in E_{2n} \cap \sum A^2$. Then $a = \sum_j (a_j)^+ a_j$ with $a_j \in A$. We write a_j as a linear combination $a_j = \sum_{\mathfrak{k}} \alpha_{j,\mathfrak{k}} x^{\mathfrak{k}}$ of basis elements $x^{\mathfrak{k}}$. From the commutation rules of the Lie algebra $\mathfrak{g}$ and the definition of the involution $(x^+ = -x$ for $x \in \mathfrak{g})$ it follows that there are numbers $c_{\mathfrak{l}}^{n,\mathfrak{k}}$ (depending on the structure constants of the Lie algebra) such that

$$(x^n)^+ x^{\mathfrak{k}} = (-1)^{|n|} x^{n+\mathfrak{k}} + \sum_{|\mathfrak{l}| < |n| + |\mathfrak{k}|} c_{\mathfrak{l}}^{n,\mathfrak{k}} x^{\mathfrak{l}}.$$

Inserting the preceding formulas we obtain

$$a = \sum_j (a_j)^+ a_j = \sum_{j,\mathfrak{k},n} \overline{\alpha}_{j,n}\, \alpha_{j,\mathfrak{k}} (x^n)^+ x^{\mathfrak{k}}$$

$$= \sum_{j,\mathfrak{k},n} \overline{\alpha}_{j,n}\, \alpha_{j,\mathfrak{k}} \left((-1)^{|n|} x^{n+\mathfrak{k}} + \sum_{|\mathfrak{l}| < |n| + |\mathfrak{k}|} c_{\mathfrak{l}}^{n,\mathfrak{k}} x^{\mathfrak{l}} \right). \qquad (10.37)$$

We show that $a_j \in E_n$ for all j. Assume the contrary. Then $d(a_j) > n$ for some j. Let a_{j_0} be an element a_j with largest number $k := d(a_{j_0}) > n$. Further, let $\mathfrak{n}_0$ be the largest with respect to the lexicographic ordering among elements $\mathfrak{n} \in \mathbb{N}_0^d$ for which $|\mathfrak{n}| = k$ and $\alpha_{j_0,\mathfrak{n}} \neq 0$. The coefficient of $(-1)^k x^{2\mathfrak{n}_0} = (-1)^{|\mathfrak{n}_0|} x^{2\mathfrak{n}_0}$ in (10.37) is

$$\sum_{d(a_j)=k} |\alpha_{j,\mathfrak{n}_0}|^2 \geq |\alpha_{j_0,\mathfrak{n}_0}|^2 > 0.$$

Since $|2\mathfrak{n}_0| = 2k > 2n$, this gives $d(a) > 2n$, which contradicts $a \in E_{2n}$. $\qquad\square$

The following results show that the closedness of the cone $\sum A^2$ can be used to characterize its elements in terms of representations or states.

Theorem 10.37 *Suppose* A *is a countably generated complex unital $*$-algebra such that* $\sum A^2$ *is τ_{st}-closed in* A. *For any element* $a \in A_{her}$ *the following are equivalent:*

(i) $a \in \sum A^2$.

(ii) $\pi(a) \geq 0$ *for all nondegenerate $*$-representations* π *of* A.

(iii) $\pi(a) \geq 0$ *for all irreducible nondegenerate $*$-representations* π *of* A.

(iv) $f(a) \geq 0$ *for each state* f *of* A.

(v) $f(a) \geq 0$ *for each pure state* f *of* A.

(Recall that $\pi(a) \geq 0$ *means* $\langle \pi(a)\varphi, \varphi \rangle \geq 0$ *for all* $\varphi \in \mathcal{D}(\pi)$.)

Proof (i)→(ii): Let $a = \sum_j (a_j)^+ a_j$. Then, for $\varphi \in \mathcal{D}(\pi)$,

$$\langle \pi(a)\varphi, \varphi \rangle = \left\langle \pi\left(\sum_j (a_j)^+ a_j\right)\varphi, \varphi \right\rangle = \sum_j \langle \pi(a_j)\varphi, \pi(a_j)\varphi \rangle \geq 0.$$

(ii)→(iv): We apply (ii) to the GNS representation π_f of the state f and obtain $f(a) = \langle \pi_f(a)\varphi_f, \varphi_f \rangle \geq 0$.

(iii)→(v): If the state f is pure, the GNS representation π_f is irreducible by Corollary 5.5, so the same reasoning yields (iii)→(v).

(iv)→(i): Assume to the contrary that $a \notin \sum A^2$. By the assumption, $\sum A^2$ is a closed (!) convex cone of the real locally convex space $A_{her}[\tau_{st}]$. Hence, by the separation of convex sets (Proposition C.2), applied to $\sum A^2$ and $\{a\}$, there exists an $\mathbb{R}$-linear functional g on A_{her} such that $g(a) < \inf \{g(c) : c \in \sum A^2\}$. Since $\sum A^2$ is a cone, the infimum is zero. Therefore, $g(a) < 0$ and $g(c) \geq 0$ for $c \in \sum A^2$. We extend the $\mathbb{R}$-linear functional g on A_{her} to a $\mathbb{C}$-linear positive functional, denoted again g, on A. Since $g \neq 0$, we have $g(1) > 0$. Then $f := g(1)^{-1} g$ is a state such that $f(a) < 0$, which contradicts (iv).

(v)→(iv): Since A is countably generated, it follows from Corollary 5.36 that each state of A is an integral over *pure* states. This yields (v)→(iv).

(ii)→(iii) is trivial. By the preceding, the equivalence of (i)–(v) is proved. $\qquad\square$

For the free polynomial $*$-algebra one can do even better and restrict to finite-dimensional representations.

Corollary 10.38 *Let* $\mathsf{A} = \mathbb{C}\langle x_1, \ldots, x_d | (x_j)^+ = x_j \rangle$, $d \in \mathbb{N}$, *be the free polynomial* $*$-*algebra and* $a \in \mathsf{A}_{\mathrm{her}}$. *If* $\pi(a) \geq 0$ *for all finite-dimensional nondegenerate* $*$-*representations* π *of* A, *then* $a \in \sum \mathsf{A}^2$.

Proof Let π be a nondegenerate $*$-representation of A and $\varphi \in \mathcal{D}(\pi)$. By Theorem 10.37, (ii)$\rightarrow$(i), it suffices to show that there exists a *finite-dimensional* nondegenerate $*$-representation ρ such that $\varphi \in \mathcal{D}(\rho)$ and $\langle \pi(a)\varphi, \varphi \rangle = \langle \rho(a)\varphi, \varphi \rangle$. This follows at once from Proposition 6.7 and formula (6.3). □

Combining Theorems 10.36 and 10.37 gives the following corollary.

Corollary 10.39 *Suppose* A *is one of the four* $*$-*algebras from Theorem 10.36 and* $a \in \mathsf{A}_{\mathrm{her}}$. *Then*, $a \in \sum \mathsf{A}^2$ *if and only if* $\pi(a) \geq 0$ *for all nondegenerate* $*$-*representations* π *of* A, *or equivalently*, $f(a) \geq 0$ *for all states* f *of* A.

This result might be surprising, at least at first glance, because there is no difference therein between the commutative polynomial $*$-algebra $\mathbb{C}[x_1, \ldots, x_d]$ and the free polynomial $*$-algebra $\mathbb{C}\langle x_1, \ldots, x_d | (x_j)^+ = x_j \rangle$! The reason is that it is required in Corollary 10.39 that $\pi(a) \geq 0$ (or $f(a) \geq 0$) for *all* nondegenerate $*$-representations π (or *all* states f).

If we allow only *integrable* representations (Definition 7.7) of $\mathbb{C}[x_1, \ldots, x_d]$, the equivalence in Corollary 10.39 is no longer true. In fact, we have the following result (see Exercise 7.7): *Let* $p \in \mathbb{R}[x_1, \ldots, x_d]$. *Then*, $\pi(p) \geq 0$ *for all integrable representations* π *of* $\mathbb{C}[x_1, \ldots, x_d]$ *if and only if* $p(x)$ *is nonnegative on* $\mathbb{R}^d$.

10.8 Exercises

1. Decide whether or not the quadratic module $\sum \mathsf{A}^2$ is Archimedean:

   a. $\mathsf{A} = \mathbb{R}[\frac{1}{1+x^2}]$.

   b. $\mathsf{A} = \mathbb{R}[\frac{1}{1+x^2}, \frac{x}{1+x^2}]$.

2. Let A be the complex $*$-algebra of functions on $\mathbb{R}^d$ generated by the constants and the functions $(1 + x_j^2)^{-1}, x_j(1 + x_j^2)^{-1}$ for $j = 1, \ldots, d$.

   a. Show that A is bounded.

   b. Describe the norm $\| \cdot \|_Q$ for the Archimedean quadratic module $Q = \sum \mathsf{A}^2$.

3. Let $p(x) = (x - a)(b - x)$, where $a, b \in \mathbb{R}$, and let Q be the quadratic module of $\mathbb{C}[x]$ of generated by p.

   a. Is Q Archimedean?

   b. What changes if p is replaced by $(x - a)(x - b)$?

4. (*Strict Positivstellensatz for d-dimensional compact intervals*)
   Let $a_j, b_j \in \mathbb{R}, a_j < b_j$, for $j = 1, \ldots, d$ and set $J := [a_1, b_1] \times \cdots \times [a_d, b_d]$. Let Q be the quadratic module of $\mathsf{A} := \mathbb{C}_d[\underline{x}]$ generated by the polynomials $f_{2j-1} := b_j - x_j$, $f_{2j} := x_j - a_j$, for $j = 1, \ldots, d$.

   a. Show that Q is Archimedean.

b. Use Theorem 10.20 to prove that each polynomial p satisfying $p(x) > 0$ for $x \in J$ belongs to Q.
(b. is a special case of the Archimedean Positivstellensatz of real algebraic geometry; see, e.g., [Ms08, Theorem 5.4.4] or [Sch17, Theorem 12.37].)

5. Show that the Weyl algebra W has no Archimedean quadratic module.
6. Show that the bijection $\pi \mapsto \tilde{\pi}$ in Proposition 10.16 preserves irreducibility.
7. Generalize the assertion of Proposition 10.22 to the $*$-algebra $M_n(\mathbb{C}) \otimes \mathbb{C}[\mathbb{Z}^d]$.
   Hint: Describe the $*$-representations of $M_n(\mathbb{C}) \otimes \mathbb{C}[\mathbb{Z}^d]$.
8. Show that each unital C^*-algebra is a bounded $*$-algebra.
9. Let A be the quotient $*$-algebra of $\mathbb{C}[x]$ by the ideal generated by x^2. Show that the cone $\sum A^2$ is not closed in $A[\tau_{st}]$. Where does the proof of Theorem 10.35 fail in this case?
10. Show that the free $*$-algebra $\mathbb{C}\langle x_1, \ldots, x_d | (x_j)^+ = x_j \rangle$ satisfies condition $(*)$ in Theorem 10.35.
    Hint: Modify the reasoning for $\mathcal{E}(\mathfrak{g})$ given in the proof of Theorem 10.36.

10.9 Notes

The results on the bounded $*$-algebra $A_b(Q)$ are taken from [Sch05, C09, Po08]; we followed mainly [C09]. Bounded elements in topological $*$-algebras were investigated in [An67, AT02]. Theorem 10.20 was proved in [Sch09]. The "Nichtnegativstellensatz" Proposition 10.21 was discovered by Cimprič [C08].

Theorem 10.33 is a special case of the main result in [Sch05]. Positivstellensätze for the Weyl algebra with other sets of denominators can be found in [Sch10, Z13].

The three Theorems 10.35–10.37 are due to the author; Theorem 10.36 was proved in [Sch84], Theorem 10.35 in [Sch90], and Theorem 10.37 in [Sch09]. Corollary 10.38 is due to Helton [H02]. The problem of whether a quadratic module is closed is already difficult for the polynomial algebras $\mathbb{R}_d[\underline{x}]$ and $\mathbb{C}_d[\underline{x}]$; see [Sch17, Sect. 13.8] for a study of this question.

An introduction to the world of noncommutative real algebraic geometry of given in [Sch09]. There is an extensive literature on noncommutative Positivstellensätze for *free* algebras by W.J. Helton, S. McCullough and their coworkers. This topic is not treated in this book.

Chapter 11
The Operator Relation $XX^* = F(X^*X)$

If a $*$-algebra is defined in terms of generators and relations, it is natural to study $*$-representations in terms of the corresponding operator relations. In this chapter we restrict ourselves to a single important operator relation:

$$XX^* = F(X^*X). \tag{11.1}$$

Here F is a real-valued Borel (in most cases continuous) function on $\mathbb{R}$ and X is a densely defined closed operator on a Hilbert space. Then X^*X is a positive self-adjoint operator and $F(X^*X)$ is a well-defined operator by the functional calculus.

If the involved operators are unbounded, one has to specify the rigorous meaning of an operator relation. This is done for the relation $AB = BF(A)$ in Sect. 11.2 and for $XX^* = F(X^*X)$ in Sect. 11.3. Roughly speaking, the "good" representations of these relations are characterized in terms of the spectral projections of the possibly unbounded operators A and X^*X (Theorems 11.6 and 11.8), respectively.

Let $X = UC$ be the polar decomposition of X. Then (11.1) is equivalent to $UC^2U^* = F(C^2)$. Let E_{C^2} be the spectral measure of the self-adjoint operator C^2. For a "good" representation of (11.1) we require that $\ker U^* \subseteq \ker F(C^2)$ and

$$E_{C^2}(\cdot)U^* = U^*E_{C^2}(F^{-1}(\cdot)). \tag{11.2}$$

The corresponding finite-dimensional representations are treated in Sect. 11.4, while the infinite-dimensional representations are studied in Sect. 11.5.

Equation (11.2) and the representation theory of (11.1) are closely linked to properties of the *dynamical system* $\lambda \mapsto F(\lambda)$. Finite-dimensional irreducible representations correspond to cycles of F (Theorems 11.13 and 11.15). Infinite-dimensional irreducible representations with unitary, isometric, or co-isometric operators U are related to orbits or semiorbits of F (Theorems 11.18, 11.19, and 11.21).

© The Editor(s) (if applicable) and The Author(s), under exclusive license   251
to Springer Nature Switzerland AG 2020
K. Schmüdgen, *An Invitation to Unbounded Representations of *-Algebras on Hilbert Space*, Graduate Texts in Mathematics 285,
https://doi.org/10.1007/978-3-030-46366-3_11

The partial isometry U appearing in the polar decomposition of X is a power partial isometry. Section 11.1 gives a brief introduction into this class of operators.

In Sects. 11.6 and 11.7, we consider two applications which are of interest in themselves. These are the Hermitian quantum plane (with $F(\lambda) = q\lambda$) and the q-oscillator algebra (with $F(\lambda) = q\lambda + 1$), where $q > 0, q \neq 1$.

The real quantum plane $AB = qBA$, with A, B self-adjoint operators and $q \neq 1$ of modulus one, cannot be treated by this theory. In Sect. 11.8 we develop some "good" representations of this relation.

Throughout this chapter, F is **real Borel function** on $\mathbb{R}$ and X is a **densely defined closed operator** on a Hilbert space $\mathcal{H}$ with **polar decomposition** $X = UC$; see Proposition A.3. Further, $e_k = (\delta_{nk})$ is the kth basis vector of $l^2(\mathbb{N}_0)$ or $l^2(\mathbb{Z})$.

11.1   A Prelude: Power Partial Isometries

First we recall some standard facts on partial isometries

Suppose $\mathcal{H}$ is a Hilbert space and $\mathcal{G}$ and $\mathcal{K}$ are closed subspaces of $\mathcal{H}$. A linear operator $U : \mathcal{H} \mapsto \mathcal{H}$ is called a *partial isometry* with *initial space* $\mathcal{G}$ and *final space* $\mathcal{K}$ if U is an isometric mapping of $\mathcal{G}$ on $\mathcal{K}$ and $U\varphi = 0$ for $\varphi \in \mathcal{G}^\perp$. In this case, $U^*U = P_{\mathcal{G}}$ and $UU^* = P_{\mathcal{K}}$, where $P_{\mathcal{G}}$ and $P_{\mathcal{K}}$ are the projections on $\mathcal{G}$ and $\mathcal{K}$, and the adjoint U^* is also a partial isometry with initial space $\mathcal{K}$ and final space $\mathcal{G}$.

Let U be a bounded operator defined on $\mathcal{H}$. Then U is a partial isometry if and only if U^*U is a projection, or equivalently, UU^* is a projection, or equivalently, $U = UU^*U$, or equivalently, $U^* = U^*UU^*$.

Definition 11.1 A linear operator U on a Hilbert space is called a *power partial isometry* if each power U^k, $k \in \mathbb{N}$, is a partial isometry.

Clearly, the adjoint of a power partial isometry is again a power partial isometry.

Lemma 11.2 *If the product UV of partial isometries U and V is a partial isometry, then the initial projection U^*U of U and the final projection VV^* of V commute.*

Proof We abbreviate $a := (U^*UVV^* - VV^*U^*U)V$. By a straightforward calculation we verify that

$$a^*a = V^*(VV^*U^*U - U^*UVV^*)(U^*UVV^* - VV^*U^*U)V = 0.$$

Therefore, $a = 0$, so $0 = aV^* = U^*UVV^*VV^* - VV^*UU^*VV^*$. Hence, since VV^* is a projection, $U^*UVV^* = VV^*UU^*VV^*$. The operator on the right is self-adjoint, so is the operator on the left. This gives $U^*UVV^* = VV^*U^*U$.   □

The converse direction in Lemma 11.2 is also valid, but we shall not need this.

Suppose U is a power partial isometry. Let P_k and P_{-k} denote the projections on the initial and final spaces of the partial isometry U^k, that is,

$$P_0 := I, \quad P_k := (U^*)^k U^k, \quad P_{-k} := U^k (U^*)^k \quad \text{for } k \in \mathbb{N}. \tag{11.3}$$

Because U and U^* are partial isometries and hence contractions, we conclude that

$$P_k \geq P_{k+1} \quad \text{and} \quad P_{-k} \geq P_{-k-1} \quad \text{for } k \in \mathbb{N}. \tag{11.4}$$

Hence $P_j P_k = P_k P_j$ and $P_{-j} P_{-k} = P_{-k} P_{-j}$ for $j, k \in \mathbb{N}$. Since $U^j U^k = U^{j+k}$ is a partial isometry, Lemma 11.2 implies that $P_j P_{-k} = P_{-k} P_j$, $j, k \in \mathbb{N}$. Thus, all projections P_k, $k \in \mathbb{Z}$, pairwise commute.

Let $k \in \mathbb{N}$. Using the equation $U = UU^*U$ and the permutability relations $P_{-1} P_k = P_k P_{-1}$ and $P_1 P_{-(k-1)} = P_{-(k-1)} P_1$ we compute

$$\begin{aligned}
P_k U &= (U^*)^k U^k U = (U^*)^k U^k U U^* U = U U^* (U^*)^k U^k U \\
&= U (U^*)^{k+1} U^{k+1} = U P_{k+1}, \\
P_{-k} U &= U^k (U^*)^k U = U U^{k-1} (U^*)^{k-1} U^* U \\
&= U U^* U U^{k-1} (U^*)^{k-1} = U U^{k-1} (U^*)^{k-1} = U P_{-k+1}.
\end{aligned}$$

Clearly, $P_0 U = I \cdot U = U(U^*U) = U P_1$ and $P_{-1} U = (UU^*)U = U \cdot I = U P_0$. Summarizing the preceding, we have proved the following proposition.

Proposition 11.3 *Suppose that U is power partial isometry. Then the projections P_k, $k \in \mathbb{Z}$, defined by (11.3), pairwise commute and satisfy*

$$P_k U = U P_{k+1} \quad \text{for } k \in \mathbb{Z}. \tag{11.5}$$

Now we turn to the structure of power partial isometries.

Clearly, isometries and co-isometries, hence unitaries, are power partial isometries. Let $\mathcal{K}$ be a Hilbert space and let $\mathcal{K}_n$, $n \in \mathbb{N}$, denote the n-fold direct sum Hilbert space $\mathcal{K}_n = \mathcal{K} \oplus \cdots \oplus \mathcal{K}$. The operator U on $\mathcal{K}_n$, given by

$$U(\varphi_1, \ldots, \varphi_n) = (0, \varphi_1, \ldots, \varphi_{n-1}), \quad (\varphi_1, \ldots, \varphi_n) \in \mathcal{K}_n,$$

is called a *truncated shift* of index n. It is easily verified that U is also a power partial isometry and its adjoint is unitarily equivalent to a truncated shift of index n.

It can be shown that each power partial isometry is a direct sum of unitaries, isometries, co-isometries, and truncated shifts; see [HW63] for a proof. The next proposition clarifies the structure of irreducible power partial isometries.

Proposition 11.4 *Up to unitary equivalence, the irreducible power partial isometries U are precisely the following operators:*

  (i) *One-dimensional unitary on $\mathcal{H} = \mathbb{C}$: $U = \alpha$, $|\alpha| = 1$.*
 (ii) *Unilateral shift on $l^2(\mathbb{N}_0)$: $U(\varphi_0, \varphi_1, \dots) = (0, \varphi_0, \varphi_1, \dots)$.*
(iii) *Adjoint of the unilateral shift on $l^2(\mathbb{N}_0)$: $U(\varphi_0, \varphi_1, \dots) = (\varphi_1, \varphi_2, \dots)$.*
 (iv) *Truncated shift of index n on $\mathcal{H} = \mathbb{C}^n$: $U(\varphi_1, \dots, \varphi_n) = (0, \varphi_1, \dots, \varphi_{n-1})$.*

Proof Clearly, each irreducible unitary U is of the form given in (i). Now assume that U is not unitary. Then $\ker U \cup \ker U^* \neq \{0\}$. We carry out the case when $\ker U \neq \{0\}$; the case $\ker U^* \neq \{0\}$ is treated in a similar manner.

We take a unit vector $e \in \ker U$. Then there are two possible cases.

Case I $U^k(U^*)^k e = e$ for all $k \in \mathbb{N}$.
Then $P_{-k}e = e$ and $P_k e = 0$ for $k \in \mathbb{N}$, because $Ue = 0$. Hence $f := e$ is a joint unit eigenvector of the projections P_n, $n \in \mathbb{Z}$, such that $Uf = 0$.

Case II There is a number $n \in \mathbb{N}$ such that $U^k(U^*)^k e = e$ for $k = 0, \dots, n - 1$ and $U^n(U^*)^n e \neq e$.

Set $f' := e - U^n(U^*)^n e = (I - P_{-n})e$ and $f = f' \|f'\|^{-1}$. Since P_{-n} is a projection, $P_{-n}f = 0$ and $Uf = 0$ by (11.5). Because the projections P_j pairwise commute, f is again a joint unit eigenvector of P_j, $j \in \mathbb{Z}$.

That is, in both cases, $Uf = 0$ and f is an eigenvector of P_j for all $j \in \mathbb{Z}$. Since P_j is a projection, it follows that $P_j f = f$ or $P_j f = 0$.

Let $\mathcal{G}$ denote the closed span of vectors $(U^*)^k f$, $k \in \mathbb{N}_0$. Obviously, $\mathcal{G}$ is invariant under U^* and $Uf = 0 \in \mathcal{G}$. Since $P_j U^* = U^* P_{j-1}$ by (11.5), we derive $U(U^*)^k f = P_{-1}(U^*)^{k-1} f = (U^*)^{k-1} P_{-1-(k-1)} f \in \mathcal{G}$ for $k \in \mathbb{N}$. Thus, $\mathcal{G}$ is invariant under U, U^* and hence equal to $\mathcal{H}$, because U is irreducible.

Let $k \in \mathbb{N}_0$. Since $P_{-k}f = f$ or $P_{-k}f = 0$, we have $UP_{-k}f = 0$. For $j > k$,

$$\langle (U^*)^k f, (U^*)^j f \rangle = \langle U^{j-k} U^k (U^*)^k f, f \rangle = \langle U^{j-k} P_{-k} f, f \rangle = 0.$$

Further, $\|(U^*)^k f\|^2 = \langle P_{-k} f, f \rangle = \|P_{-k} f\|^2$ is 1 or 0. Thus, the vectors $(U^*)^k f$, $k \in \mathbb{N}_0$, are mutually orthogonal and either unit vectors or zero. From these facts it follows that, up to unitary equivalence, U is the adjoint of a unilateral shift in Case I and a truncated shift of index n in Case II. $\square$

11.2   The Operator Relation $AB = BF(A)$

If all involved operators are bounded and defined on the whole Hilbert space, one usually requires that an operator relation holds for all vectors of the space. However, if some operators are unbounded, one has to explain how the relation is meant.

Let us discuss this in the simplest case of the relation

$$AB = BA \tag{11.6}$$

for self-adjoint operators A, B. An obvious candidate for a meaning of (11.6) is to require that $AB\varphi = BA\varphi$ for $\varphi \in \mathcal{D}(AB) \cap \mathcal{D}(BA)$. But, by a result of von Neumann (see [Sch83a, Theorem 5.2]) there exist unbounded self-adjoint operators A and B such that $\mathcal{D}(A) \cap \mathcal{D}(B) = \{0\}$. Then, trivially, $AB\varphi = BA\varphi$ for all $\varphi \in \mathcal{D}(AB) \cap \mathcal{D}(BA) = \{0\}$, but it is absurd to say that (11.6) holds in this case.

The next natural meaning of (11.6) is to suppose that $AB\varphi = BA\varphi$ for vectors φ of a common core $\mathcal{D}$ for A and B. But even this does not exclude pathological behavior, as the following example shows. We only sketch the main lines of proof; more details concerning this example can be found in [Sch12, Example 5.5].

Example 11.5 Let S denote the unilateral shift on $l^2(\mathbb{N}_0)$ given by $Se_n = e_{n+1}$, where $\{e_n : n \in \mathbb{N}_0\}$ is the canonical basis of $l^2(\mathbb{N}_0)$. Then $X := S + S^*$ and $Y := -i(S - S^*)$ are bounded self-adjoint operators with trivial kernels. Therefore, $A := X^{-1}$ and $B := Y^{-1}$ are self-adjoint operators on $l^2(\mathbb{N}_0)$.

Let P be the rank projection $e_0 \otimes e_0$. Set $\mathcal{D} := XY(I - P)l^2(\mathbb{N}_0)$. One verifies that $XY(I - P) = YX(I - P)$. Therefore, $AB\varphi = BA\varphi$ for $\varphi \in \mathcal{D}$.

It is easily checked that $e_0 \notin \operatorname{ran} X$ and $e_0 \notin \operatorname{ran} Y$. From this it follows that $A\mathcal{D} = Y(I - P)l^2(\mathbb{N}_0)$ and $B\mathcal{D} = X(I - P)l^2(\mathbb{N}_0)$ are dense in $l^2(\mathbb{N}_0)$. Hence $\mathcal{D}$ is a core for A and B. Thus, A and B commute pointwise on the common core $\mathcal{D}$.

But the inverses $X = A^{-1}$ and $Y = B^{-1}$ do not commute, since otherwise S and S^* would commute, a contradiction.   ○

The "correct" meaning of (11.6) is to require that the self-adjoint operators *strongly commute*; that is, the spectral projections $E_A(M)$ and $E_B(N)$ commute for Borel sets M, N, or equivalently, their resolvents commute [Sch12, Section 5.6].

For a general operator relation it can be difficult to find an appropriate rigorous meaning and neither a unique nor a canonical way for this is known.

Now we will treat the operator relation

$$AB = BF(A). \tag{11.7}$$

To be precise, we consider pairs of a *self-adjoint* operator A and a *bounded* operator B on a Hilbert space $\mathcal{H}$ such that

$$BF(A) \subseteq AB. \tag{11.8}$$

This is the rigorous operator-theoretic meaning of relation (11.7) for unbounded A and bounded B. If the operators A and $F(A)$ are also bounded, then (11.8) is just the equation $AB = BF(A)$.

The operator relation (11.7), with bounded operator B, is crucial for the study of the relation $XX^* = F(X^*X)$ in Sect. 11.3; see Theorem 11.8 and (11.16).

Relation (11.8) says that for any vector $\varphi \in \mathcal{D}(F(A))$ we have $B\varphi \in \mathcal{D}(A)$ and $BF(A)\varphi = AB\varphi$. The operator $F(A)$ is defined by the functional calculus of self-adjoint operators [Sch12, Chap. 4]: If E_A is the spectral measure of A, then

$$\mathcal{D}(F(A)) := \left\{ \varphi \in \mathcal{H} : \int_{\mathbb{R}} |F(\lambda)|^2 \, d\langle E_A(\lambda)\varphi, \varphi\rangle < \infty \right\},$$

$$F(A)\varphi := \int_{\mathbb{R}} F(\lambda) \, dE_A(\lambda)\varphi, \quad \varphi \in \mathcal{D}(F(A)).$$

The following theorem collects a number of conditions that are equivalent to (11.8). Note that (iii) and (iv) deal only with *bounded* operators and the spectral measure $E_{F(A)}$ of the self-adjoint operator $F(A)$ is given by $E_A(F^{-1}(\cdot))$.

Theorem 11.6 *Suppose A is a self-adjoint operator and B is a bounded operator on a Hilbert space $\mathcal{H}$ with $\mathcal{D}(B) = \mathcal{H}$. Let $B = U|B|$ be the polar decomposition of B and P the projection on* $\ker B = \ker |B|$. *Then the following are equivalent:*

(i) $BF(A) \subseteq AB$.
(ii) $B^*A \subseteq F(A)B^*$.
(iii) $E_A(M)B = BE_A(F^{-1}(M))$ *for each Borel subset M of* $\mathbb{R}$.
(iv) $f(A)B = Bf(F(A))$ *for all real bounded Borel functions f on* $\mathbb{R}$.
(v) $|B|F(A) \subseteq F(A)|B|$ *and* $UF(A)(I - P) \subseteq AU(I - P)$.

If one of these statements is true, then

$$AB\varphi = BF(A)\varphi \quad \text{for} \quad \varphi \in \mathcal{D}(F(A)) = \mathcal{D}(AB) \cap \mathcal{D}(BF(A)). \tag{11.9}$$

Proof (i)↔(ii): First suppose (i) holds. Applying the adjoint to $BF(A) \subseteq AB$ yields $(AB)^* \subseteq (BF(A))^*$. Clearly, $B^*A \subseteq (AB)^*$. Since B is bounded (!), we have $(BF(A))^* = F(A)B^*$. Combining these relations we obtain (ii).

The converse implication (ii)→(i) follows by the same reasoning, now with $B, F(A), A$ replaced by $B^*, A, F(A)$, respectively.

(i)↔(iii): Suppose B is a bounded operator and A, C are self-adjoint operators on $\mathcal{H}$. Then, by [Sch12, Proposition 5.15], $BC \subseteq AB$ if and only if $BE_C(M) = E_A(M)B$ for all Borel sets M. We apply this result with $C = F(A)$. Since $F(A)$ has the spectral measure $E_A(F^{-1}(\cdot))$, this gives the equivalence of (i) and (iii).

(iii)→(iv): Since the function f is bounded on $\mathbb{R}$, $f(A)$ and $f(F(A))$ are bounded self-adjoint operators by the functional calculus. Let $S_{f(A)} := \sum_j f(\lambda_j)E_A(M_j)$ be an integral sum for the spectral integral $f(A) = \int f(\lambda) \, dE_A(\lambda)$, where $\lambda_j \in M_j$ and $M_i \cap M_j = \emptyset$ for $i \neq j$. Then $S_{f(F(A))} := \sum_j f(\lambda_j)E_{F(A)}(M_j)$ is an integral sum for the spectral integral $f(F(A)) = \int f(\lambda) \, dE_{F(A)}(\lambda)$. Since $E_A(F^{-1}(M_j)) = E_{F(A)}(M_j)$, it follows from (iii) that $S_{f(A)}B = BS_{f(F(A))}$. Passing to the limit we obtain $f(A)B = Bf(F(A))$.

(iv)→(iii): Let f denote the characteristic function of M. Then (iv) means that $E_A(M)B = BE_A(F^{-1}(M))$.

(i)→(v): Since $BF(A) \subseteq AB$ by (i) and $B^*A \subseteq B^*F(A)$ by (i)→(ii), we have $B^*BF(A) \subseteq B^*AB \subseteq F(A)B^*B$. Hence $(B^*B)^n F(A) \subseteq F(A)(B^*B)^n$ by induction, so $p(B^*B)F(A) \subseteq F(A)p(B^*B)$ for any polynomial p. Since the operator

B^*B is bounded and positive, we can find a sequence $(p_n)_{n\in\mathbb{N}}$ of polynomials such that $\lim_n p_n(B^*B) = \sqrt{B^*B} = |B|$ in the operator norm.

Let $\varphi \in \mathcal{D}(F(A))$. As noted above, $p_n(B^*B)F(A)\varphi = F(A)p_n(B^*B)\varphi$. Then, $\lim_n F(A)p_n(B^*B)\varphi = \lim_n p_n(B^*B)F(A)\varphi = |B|F(A)\varphi$ and also $\lim_n p_n(B^*B)\varphi = |B|\varphi$. Hence, since the operator $F(A)$ is closed, we obtain $|B|\varphi \in \mathcal{D}(F(A))$ and $|B|F(A)\varphi = F(A)|B|\varphi$, which is the first relation of (v).

Next we turn to the second relation of (v). By (i) and the polar decomposition, $BF(A) = U|B|F(A) \subseteq AB = AU|B|$. Let $\varphi \in \mathcal{D}(F(A))$. Then, as shown in the preceding paragraph, $|B|\varphi \in \mathcal{D}(F(A))$ and $|B|F(A)\varphi = F(A)|B|\varphi$. Combining these facts yields $UF(A)|B|\varphi = AU|B|\varphi$, that is,

$$UF(A)\eta = AU\eta \quad \text{for} \quad \eta \in |B|\mathcal{D}(F(A)). \tag{11.10}$$

Let $|B| = \int_0^\infty \lambda\, dE(\lambda)$ be the spectral resolution of the positive self-adjoint operator $|B|$. Since $|B|F(A) \subseteq F(A)|B|$ as shown above, all spectral projections and bounded functions of $|B|$ commute with $F(A)$ and leave $\mathcal{D}(F(A))$ invariant. We abbreviate $P_\varepsilon := E([\varepsilon, +\infty))$ for $\varepsilon > 0$. Clearly, $E(\{0\}) = P$ and $\lim_{\varepsilon\to+0} P_\varepsilon\xi = E((0,\infty))\xi = (I - P)\xi$ for $\xi \in \mathcal{H}$.

Fix $\varphi \in \mathcal{D}(F(A)(I - P))$. Then $\psi := (I - P)\varphi$ and $P_\varepsilon\psi$ are in $\mathcal{D}(F(A))$. Set $C_\varepsilon := \int_\varepsilon^\infty \lambda^{-1}E(\lambda)$. Since C_ε is a function of $|B|$, we have $C_\varepsilon P_\varepsilon\psi \in \mathcal{D}(F(A))$ and $P_\varepsilon\psi = |B|C_\varepsilon P_\varepsilon\psi \in |B|\mathcal{D}(F(A))$. Thus, (11.10) applies to $\eta = P_\varepsilon\psi$, so

$$UF(A)P_\varepsilon\psi = AUP_\varepsilon\psi \quad \text{for} \quad \varepsilon > 0. \tag{11.11}$$

We have $\lim_{\varepsilon\to+0} P_\varepsilon\psi = (I - P)\psi = (I - P)\varphi$ and

$$\lim_{\varepsilon\to+0} F(A)P_\varepsilon\psi = \lim_{\varepsilon\to+0} P_\varepsilon F(A)\psi = (I - P)F(A)\psi = F(A)(I - P)\varphi.$$

Hence, passing to the limit $\varepsilon \to +0$ in (11.11) by using that the operator A is closed, we get $UF(A)(I - P)\varphi = AU(I - P)\varphi$. This proves the second half of (v).

(v)$\to$(i): Since ran $|B| \perp \ker |B| = P\mathcal{H}$, we have $|B| = (I - P)|B|$, so by (v),

$$BF(A) = U|B|F(A) \subseteq UF(A)|B| = UF(A)(I - P)|B|$$
$$\subseteq AU(I - P)|B| = AU|B| = AB.$$

This completes the proof of the equivalence of (i)–(v).

Finally, we prove the last assertion. Let $\varphi \in \mathcal{D}(F(A))$. Then $F(A)\varphi \in \mathcal{D}(AB)$ by (i). This implies the equality of domains $\mathcal{D}(F(A)) = \mathcal{D}(AB) \cap \mathcal{D}(BF(A))$.

For $n \in \mathbb{N}$ we define a bounded function f_n by $f_n(\lambda) = \lambda$ if $|\lambda| \leq n$ and $f_n(\lambda) = 0$ otherwise. Let $\varphi \in \mathcal{D}(AB) \cap \mathcal{D}(BF(A))$. By (iv), we have

$$f_n(A)B\varphi = Bf_n(F(A))\varphi. \tag{11.12}$$

Since $\lim f_n(A)B\varphi = AB\varphi$ by $\varphi \in \mathcal{D}(AB)$ and $\lim_n f_n(F(A))\varphi = F(A)\varphi$ by $\varphi \in \mathcal{D}(F(A))$, we pass to the limit in (11.12) and obtain $AB\varphi = BF(A)\varphi$. $\square$

11.3 Strong Solutions of the Relation $XX^* = F(X^*X)$

In this section, we begin our study of the operator relation

$$XX^* = F(X^*X) \tag{11.13}$$

for a densely defined closed operator X on a Hilbert space $\mathcal{H}$.

Let $X = UC$ be the polar decomposition of X; see Proposition A.3. Then $X^*X = C^2$, $X^* = CU^*$, and $XX^* = UC^2U^*$, so (11.13) can be restated as

$$UC^2U^* = F(C^2). \tag{11.14}$$

Suppose for a moment that the operator X is *bounded* and the function F is continuous on $\mathbb{R}_+$. Then the operator $F(XX^*)$ is also bounded. We multiply (11.14) by U^* from the left. Since U^*U is the projection on $\overline{\operatorname{ran} C}$, we obtain the relation

$$C^2U^* = U^*F(C^2), \tag{11.15}$$

which holds on the whole Hilbert space $\mathcal{H}$. However, Eq. (11.15) alone does not imply (11.14). (If X is the unilateral shift and $F(t) = 1$, then $X = U$, $C = F(C^2) = I$, so (11.15) is trivially true, but (11.14) and (11.13) do not hold.)

Further, multiplying (11.15) by U from the left and comparing with (11.14) yield $(I - UU^*)F(C^2) = 0$, that is, $\operatorname{ran} F(C^2) \subseteq UU^*\mathcal{H}$, or equivalently, $\ker F(C^2) \supseteq \ker UU^* = \ker U^*$. Conversely, from $\ker U^* \subseteq \ker F(C^2)$ and (11.15) we easily derive Eq. (11.14) and therefore (11.13).

For general operators X we require the following rigorous meaning of (11.13).

Definition 11.7 We say that a densely defined closed operator X with polar decomposition $X = UC$ is a *strong solution* of the operator Eq. (11.13) if

$$U^*F(C^2) \subseteq C^2U^* \quad \text{and} \quad \ker U^* \subseteq \ker F(C^2). \tag{11.16}$$

If the operator X is bounded and the function F is continuous on $\mathbb{R}_+$, then, by the preceding discussion, X is a strong solution of (11.13) if and only if Eq. (11.13) holds for all vectors of the Hilbert space. Another justification for Definition 11.7 is the fact that it implies relation (11.17) in Theorem 11.8.

Note that, by the theory of spectral integrals [Sch12, Corollary 4.14], the domain $\mathcal{D}(F(C^2))$ is dense in $\mathcal{H}$ and a core for $C^2 = X^*X$ and $C = |X|$, hence for X.

The following theorem contains our main results about strong solutions of the operator relation (11.13).

Theorem 11.8 *Suppose X is a densely defined closed operator, and let $X = UC$ be the polar decomposition of X. Then the following are equivalent:*

   (i) $U^*F(C^2) \subseteq C^2U^*$.
  (ii) $UC^2 \subseteq F(C^2)U$.
 (iii) $E_{C^2}(M)U^* = U^*E_{C^2}(F^{-1}(M))$ *for any Borel subset M of $\mathbb{R}$.*
 (iv) $f(X^*X)U^* = U^*f(F(X^*X))$ *for all bounded Borel functions f on $\mathbb{R}$.*

If X is a strong solution of (11.13), then $\ker U^* = \ker F(C^2)$ and

$$XX^*\varphi = F(X^*X)\varphi \text{ for } \varphi \in \mathcal{D}(F(C^2)) = \mathcal{D}(XX^*) \cap \mathcal{D}(F(X^*X)). \quad (11.17)$$

Proof The equivalence of (i)–(iv) is a special case of Theorem 11.6, applied with $B = U^*$, $A = C^2$. It remains to prove the last assertions.

Since $XX^* = UC^2U^*$ and $F(X^*X) = F(C^2)$, the domain equality on (11.17) follows at once from the corresponding equality in (11.9) or directly from (i).

Let $\varphi \in \mathcal{D}(F(C^2))$. Then, by (i), $U^*\varphi \in \mathcal{D}(C^2)$ and $C^2U^*\varphi = U^*F(C^2)\varphi$. Hence $UC^2U^*\varphi = UU^*F(C^2)\varphi$. Since $\ker U^* \subseteq \ker F(C^2)$ by assumption (11.16), $\operatorname{ran} F(C^2) \subseteq U^*\mathcal{H} = UU^*\mathcal{H}$ and therefore $UU^*F(C^2)\varphi = F(C^2)\varphi$. Thus, $UC^2U^*\varphi = F(C^2)\varphi$, that is, $XX^*\varphi = F(X^*X)\varphi$, which proves (11.17).

Now let $\psi \in \ker F(C^2)$. Then $\psi \in E_{C^2}(F^{-1}(\{0\}))\mathcal{H}$. Hence $E_{C^2}(\{0\})U^*\psi = U^*\psi$ by (iii), so that $U^*\psi \in \ker C^2 = \ker C$. But $\operatorname{ran} U^* = (\ker C)^\perp$ by the polar decomposition; see (A.2). Thus, $U^*\psi = 0$. This proves that $\ker F(C^2) \subseteq \ker U^*$. Since $\ker U^* \subseteq \ker F(C^2)$ by Definition 11.16, we have $\ker U^* = \ker F(C^2)$. $\square$

Remark 11.9 The following weakening of Theorem 11.8(i) is often useful: Let $\mathcal{D}_0$ be a subset of $\mathcal{D}(F(C^2))$ such that $\mathcal{D} := \operatorname{Lin}\mathcal{D}_0$ is a core for $F(C^2)$. Suppose for $\varphi \in \mathcal{D}_0$ we have $U^*\varphi \in \mathcal{D}(C^2)$ and $U^*F(C^2)\varphi = C^2U^*\varphi$. Then condition (i) in Theorem 11.8 is fulfilled. (Indeed, the relation $U^*F(C^2)\varphi = C^2U^*\varphi$ extends by linearity to vectors φ of $\mathcal{D}$ and then of $\mathcal{D}(F(C^2))$, because the operator $F(C^2)$ is closed.) A similar weakening holds for condition (ii) in Theorem 11.8. $\bigcirc$

Proposition 11.10 *Suppose $X = UC$ is a strong solution of (11.13). Then the phase operator U is a power partial isometry and the projections $P_k, k \in \mathbb{Z}$, defined by (11.3), pairwise commute. For $k \in \mathbb{Z}$ and all Borel sets M of $\mathbb{R}$, we have*

$$P_kE_{C^2}(M) = E_{C^2}(M)P_k \text{ and } P_kU = UP_{k+1}. \quad (11.18)$$

Proof In this proof we abbreviate $E := E_{C^2}$ and we use essentially Theorem 11.8(iii) and properties of the polar decomposition (Proposition A.3).

Define $N := \mathbb{R}\backslash\{0\}$. From spectral theory it is known that $E(N)$ is the projection on $\overline{\operatorname{ran} C^2} = \overline{\operatorname{ran} C}$. Therefore, $E(N) = U^*U = P_1$.

Set $N_1 = F^{-1}(N)$. By the same reason, with C^2 replaced by $F(C^2)$, the projection on $\overline{\operatorname{ran} F(C^2)}$ is $E_{F(C^2)}(N) = E(F^{-1}(N)) = E(N_1)$. By Theorem 11.8, $\ker UU^* = \ker U^* = \ker F(C^2)$, so that $\overline{\operatorname{ran} UU^*} = \overline{\operatorname{ran} F(C^2)} = E(N_1)\mathcal{H}$. Since UU^* and $E(N_1)$ are projections with the same range, they coincide, so that

$$P_{-1} = UU^* = E(N_1). \tag{11.19}$$

We show that there exist Borel sets N_k, $k \in \mathbb{N}$, such that $P_{-k} = E(N_k)$. For $k = 1$ this is (11.19). If it holds for k, we set $N_{k+1} := N_1 \cap F^{-1}(N_k)$ and derive

$$P_{-(k+1)} = U P_{-k} U^* = UU^* E(F^{-1}(N_k)) = E(N_1) E(F^{-1}(N_k)) = E(N_{k+1}),$$

which completes the induction proof.

Thus, $P_{-k} = U^k (U^*)^k$ is a projection for each $k \in \mathbb{N}$. Therefore, $(U^*)^k$, hence U^k, is a partial isometry. This means that U is a power partial isometry. Then, by Proposition 11.3, the operators P_k, $k \in \mathbb{Z}$, are pairwise commuting projections and the second equality of (11.18) holds.

We prove the first equation of (11.18). Let $k \in \mathbb{N}$. Then $P_{-k} = E(N_k)$, so P_{-k} commutes with $E(M)$. To prove that P_k commutes with $E(M)$, we proceed by induction. For $k = 1$ this is true, since $P_1 = E(N)$. Suppose that it holds for k. From Theorem 11.8(iii), $E(M)U^* = U^* E(F^{-1}(M))$. Taking the adjoint we obtain $U E(M) = E(F^{-1}(M))U$. Therefore,

$$E(M)P_{k+1} = E(M)U^* P_k U = U^* E(F^{-1}(M)) P_k U = U^* P_k E(F^{-1}(M))U$$
$$= U^* P_k U E(M) = P_{k+1} E(M),$$

which is the assertion for $k + 1$. Thus, $P_n E(M) = E(M) P_n$ for all $n \in \mathbb{Z}$. $\qquad \square$

We shall say that an operator on a Hilbert space is *irreducible* if it cannot be written as a direct sum of two operators acting on nonzero Hilbert spaces. In our irreducibility proofs below we will use the following simple lemma; we leave its proof as an exercise to the reader. Note that the commutants in (ii) (by Lemma 3.16(ii)) and (iii) are von Neumann algebras.

Lemma 11.11 *Let $X = UC$ be the polar decomposition of the closed operator X on $\mathcal{H}$, and let E_C denote the spectral measure of C. The following are equivalent:*

(i) *X is irreducible.*
(ii) *$\{T \in \mathbf{B}(\mathcal{H}) : TX \subseteq XT, TX^* \subseteq X^*T\}' = \mathbb{C} \cdot I$.*
(iii) *$\{U, U^*, E_C(M) : M \text{ Borel set of } \mathbb{R}\}' = \mathbb{C} \cdot I$.*
(iv) *Any nonzero closed subspace of $\mathcal{H}$ which is invariant under U, U^*, E_C is $\mathcal{H}$.*

A spectral measure E is called *ergodic* with respect to F if for any Borel set $\Delta \subseteq \mathbb{R}$ such that $F^{-1}(\Delta) = \Delta$ we have $E(\Delta) = 0$ or $E(\mathbb{R} \backslash \Delta) = 0$.

Corollary 11.12 *If $X = UC$ is an irreducible strong solution of relation (11.13), then the spectral measure E_{C^2} is ergodic with respect to F.*

Proof Suppose Δ is a Borel set of $\mathbb{R}$ such that $F^{-1}(\Delta) = \Delta$. Then, by Theorem 11.8(iii), $E_{C^2}(\Delta)U^* = U^* E_{C^2}(\Delta)$. By taking the adjoint, $U E_{C^2}(\Delta) = E_{C^2}(\Delta)U$. Hence the decomposition $\mathcal{H} = E_{C^2}(\Delta)\mathcal{H} \oplus E_{C^2}(\mathbb{R}\backslash\Delta)\mathcal{H}$ reduces U and U^*. It obviously reduces the spectral measure of C^2 and hence of C. Therefore, it reduces the operator $X = UC$. Hence, since X is irreducible, $E_{C^2}(\Delta) = 0$ or $E_{C^2}(\mathbb{R}\backslash\Delta) = 0$, so E_{C^2} is ergodic.   □

We will see in the subsequent sections that it is useful to interpret the function $\lambda \mapsto F(\lambda)$ as a one-dimensional *dynamical system*. Thus, it is convenient to adopt some notions on dynamical systems (even if F is not necessarily continuous).

For $\lambda \in \mathbb{R}$ we define

$$F^{\circ k}(\lambda) := F(F^{\circ(k-1)}(\lambda)) \text{ for } k \in \mathbb{N}, \quad F^{\circ 0}(\lambda) := \lambda, \tag{11.20}$$

$$\mathbb{O}^+(\lambda) := \{F^{\circ k}(\lambda) : k \in \mathbb{N}_0\}. \tag{11.21}$$

The set $\mathbb{O}^+(\lambda)$ is called the *positive semiorbit* of λ.

A number $\lambda \in \mathbb{R}$ is said to be a *periodic point* of F if $F^{\circ n}(\lambda) = \lambda$ for some $n \in \mathbb{N}$. Then the smallest $n \in \mathbb{N}$ such that $F^{\circ n}(\lambda) = \lambda$ is called the *period* of λ and the set $\{\lambda, F(\lambda), \ldots, F^{\circ(n-1)}(\lambda)\}$ is called a *cycle* of period n.

For a Borel subset M of $\mathbb{R}$ we define inductively $F^{-1}(F^{\circ(-k)}(M))$ by

$$F^{\circ(-k-1)}(M) := F^{-1}(F^{\circ(-k)}(M)) \quad \text{for } k \in \mathbb{N}_0, \quad F^{\circ 0}(M) := M.$$

If $F : \mathbb{R} \mapsto \mathbb{R}$ is bijective, we set $F^{\circ(-k-1)}(\lambda) := F^{-1}(F^{\circ(-k)}(\lambda))$ for $\lambda \in \mathbb{R}$ and $k \in \mathbb{N}_0$, so that $F^{\circ n}(\lambda)$ is defined for all $n \in \mathbb{Z}$. Then the *orbit* of λ is the set

$$\mathbb{O}(\lambda) := \{F^{\circ k}(\lambda) : k \in \mathbb{Z}\}. \tag{11.22}$$

The iterates $F^{\circ k}$ play a crucial role in describing the spectrum of C^2. To explain this let X be a *strong solution* of (11.13). Let $e \in \mathcal{D}(C^2)$ and assume that $C^2 e = \lambda e$, where $\lambda \in \mathbb{R}$. Then $e \in \mathcal{D}(F(C^2))$ and $F(C^2)e = F(\lambda)e$. Hence, by Theorem 11.8(i), $U^* F(C^2)e = F(\lambda)U^* e = C^2 U^* e$. Proceeding by induction we obtain

$$C^2 (U^*)^k e = F^{\circ k}(\lambda)(U^*)^k e, \quad k \in \mathbb{N}. \tag{11.23}$$

Thus, if $(U^*)^k e \neq 0$, then $(U^*)^k e$ is an eigenvector of C^2 with eigenvalue $F^{\circ k}(\lambda)$. This simple reasoning indicates how representations of (11.13) could look like.

11.4   Finite-Dimensional Representations

Throughout this section, we assume that $F : \mathbb{R}_+ \mapsto \mathbb{R}_+$ is a **continuous** map.

As noted after Definition 11.7, a *bounded* operator X on $\mathcal{H}$ is a strong solution of (11.13) if and only if $XX^*\varphi = F(X^*X)\varphi$ for all $\varphi \in \mathcal{H}$. In particular, this holds for operators acting on finite-dimensional Hilbert spaces.

Theorem 11.13 *Suppose that $\lambda \geq 0$ is a periodic point of period n of F such that $F^{\circ j}(\lambda) > 0$ for $j = 1, \ldots, n-1$. Let $\theta \in [0, 2\pi)$. Then the operator*

$$X_{\lambda,\theta} := \begin{pmatrix} 0 & F(\lambda)^{1/2} & \cdots & & 0 \\ \vdots & \vdots & \ddots & & \vdots \\ 0 & 0 & \cdots & F^{\circ(n-1)}(\lambda)^{1/2} \\ e^{i\theta}\lambda & 0 & \cdots & & 0 \end{pmatrix} \tag{11.24}$$

on the Hilbert space $\mathbb{C}^n$ is irreducible and fulfills (11.13).

Each irreducible operator X on a finite-dimensional Hilbert space $\mathcal{H}$ satisfying (11.13) is unitarily equivalent to an operator $X_{\lambda,\theta}$ in (11.24).

Clearly, the operator $X_{\lambda,\theta}$ has the polar decomposition $X_{\lambda,\theta} = U_{\lambda,\theta} C_{\lambda,\theta}$ with

$$U_{\lambda,\theta} := \begin{pmatrix} 0 & 1 & \cdots & 0 \\ \vdots & \vdots & \ddots & \vdots \\ 0 & 0 & \cdots & 1 \\ c_{\lambda,\theta} & 0 & \cdots & 0 \end{pmatrix}, \quad C_{\lambda,\theta} := \begin{pmatrix} \lambda^{1/2} & 0 & \cdots & & 0 \\ 0 & F(\lambda)^{1/2} & \cdots & & 0 \\ \vdots & \vdots & \ddots & & \vdots \\ 0 & 0 & \cdots & F^{\circ(n-1)}(\lambda)^{1/2} \end{pmatrix},$$

where $c_{\lambda,\theta} := e^{i\theta}$ if $\lambda \neq 0$ and $c_{0,\theta} := 0$.

If $\lambda > 0$, then Eq. (11.24) gives a one-parameter family $X_{\lambda,\theta}$, $\theta \in [0, 2\pi)$, of inequivalent representations with unitary operator $U_{\lambda,\theta}$. If $\lambda = 0$, then (11.24) yields a single representation $X_{0,0}$ and $U_{0,0}$ is a truncated shift of index n. Note that if $F(0) = 0$, then $\lambda = 0$, $n = 1$ gives the trivial representation $X_{0,0} = 0$ on $\mathbb{C}$.

Proof of Theorem 11.13 A direct computation shows that $X_{\lambda,\theta}$ satisfies (11.13). We prove that $X_{\lambda,\theta}$ is irreducible. Let T be an operator commuting with $U_{\lambda,\theta}$ and the spectral projections of $C_{\lambda,\theta}$. Then T commutes with $C_{\lambda,\theta}$. Since λ has period n, the diagonal entries of $C_{\lambda,\theta}$ are pairwise different. Hence T is diagonal. Since T commutes with $U_{\lambda,\theta}$, $T = c \cdot I$ with $c \in \mathbb{C}$. By Lemma 11.11, $X_{\lambda,\theta}$ is irreducible.

Conversely, let $X = UC$ be an irreducible operator on $\mathcal{H}$ such that (11.13) holds. We denote the spectral measure of C^2 by E.

Case I U is unitary.
Consider an eigenvalue t with eigenvector e_t of the self-adjoint operator C^2. Then $C^2(U^*)^j e_t = F^{\circ j}(t)(U^*)^j e_t$ by (11.23). Since U is unitary, $(U^*)^j e_t \neq 0$, so $(U^*)^j e_t$ is an eigenvector with eigenvalue $F^{\circ j}(t)$ of C^2 for $j \in \mathbb{N}$. Since $\mathcal{H}$ has finite dimension, C^2 has only finitely many eigenvalues. Hence there exist an eigenvalue λ with unit eigenvector e of C^2 and a number $n \in \mathbb{N}$ such that $F^{\circ n}(\lambda) = \lambda$. We choose the smallest such n; then λ is a periodic point with period n. Since U is unitary, $\ker C = \{0\}$ and therefore $\lambda \neq 0$. The linear span $\mathcal{H}_0$ of eigenvectors $e_k := (U^*)^k e$, $k \in \mathbb{N}_0$, for C^2 is invariant under U^* and C^2, hence under C as well. Since U is unitary and $\mathcal{H}_0$ has finite dimension, we have $U^*\mathcal{H}_0 = \mathcal{H}_0$, so $U\mathcal{H}_0 = (U^*)^{-1}\mathcal{H}_0 = \mathcal{H}_0$. Since X is irreducible and $\mathcal{H}_0$ is invariant under U, U^*, C, it follows that $\mathcal{H}_0 = \mathcal{H}$ by Lemma 11.11.

The operator U^n commutes with C (by $F^{\circ n}(\lambda) = \lambda$) and obviously with U, U^*. Hence, by the irreducibility of X, we have $U^n = e^{i\theta}I$ for some $\theta \in [0, 2\pi)$. Then it follows that $Ue_k = e_{k-1}$ for $k = 1, \ldots, n - 1$, $Ue_0 = e^{i\theta}e_0$, $\{e_0, \ldots, e_{n-1}\}$ is an orthonormal basis of $\mathcal{H}$, and the operator X is given by the matrix (11.24).

Case II U is not unitary.
Then, since U is a partial isometry and $\mathcal{H}$ has finite dimension, $\ker U = \ker C \neq \{0\}$. Take a unit vector $e \in \ker C$, and set $e_k := (U^*)^k e$, $k \in \mathbb{N}_0$. By Proposition 11.10, U is power partial isometry. Therefore, if $n > k$, using (11.5) we derive

$$\langle e_n, e_k \rangle = \langle (U^*)^n e, (U^*)^k e \rangle = \langle e, U^{n-k} P_{-k} e \rangle = \langle e, P_{-k-(n-k)} U^{n-k} e \rangle = 0.$$

Thus, the vectors e_j, $j \in \mathbb{N}_0$, are mutually orthogonal, so there is a smallest $n \in \mathbb{N}$ such that $e_n = 0$. Then, $e_k = (U^*)^k e \neq 0$ for $k = 0, \ldots, n - 1$. Since $C^2 e = 0$, e_k is an eigenvector of C^2 with eigenvalue $F^{\circ k}(0)$ by (11.23). Hence $F^{\circ k}(0) \geq 0$ and $Ce_k = F^{\circ k}(0)^{1/2} e_k$. Let $\mathcal{H}_0 = \mathrm{Lin}\{e_k : k = 0, \ldots, n - 1\}$. Clearly, $\mathcal{H}_0$ is invariant under U^* and C, hence under E. Further, by (11.19), $Ue_{k+1} = UU^*(U^*)^k e = E(N_1)e_k$ for $k \in \mathbb{N}$. Since $Ue = 0$ and $E(N_1)$ leaves $\mathcal{H}_0$ invariant, so does U. Therefore, $\mathcal{H}_0 = \mathcal{H}$ by the irreducibility of X (Lemma 11.11). By construction, $\ker U^* = \mathbb{C} \cdot e_{n-1}$. Hence U^* is isometric on $(e_{n-1})^\perp$, so that $\|e_k\| = \|U^* e_{k-1}\| = \|e_{k-1}\| = \cdots = \|e\| = 1$ for $0 \leq k \leq n - 1$. Putting the preceding together we have shown that X is of the form (11.24) with $\lambda = 0$.

Finally, we prove that 0 is a periodic point with period n. By (11.24), XX^* and X^*X are diagonal matrices with diagonal entries $F(0), \ldots, F^{\circ n-1}(0), 0$ and $0, F(0), \ldots, F^{\circ n-1}(0)$, respectively. Hence $F(X^*X)$ has the diagonal entries $F(0), \ldots, F^{\circ n}(0)$. Since $XX^* = F(X^*X)$, $F^{\circ n}(0) = 0$. Further, $F^{\circ k}(0) > 0$ for $k = 0, \ldots, n - 1$. Indeed, $F^{\circ k}(0) = 0$ implies that the vectors $(x_0, \ldots, x_k, 0, \ldots, 0)$ form a nontrivial invariant subspace for X and X^*. This subspace is reducing. This is a contradiction, because X is irreducible. Thus, 0 is periodic with period n. □

Theorem 11.13 allows one to apply results on periods from the theory of dynamical systems to the existence of finite-dimensional irreducible representations of the operator relation (11.13). Before we do this we introduce the *Sharkovsky ordering* "$\prec$" of the positive integers:

$$1 \prec 2^2 \prec 2^3 \prec \cdots \prec 2^n \prec \cdots \prec 2^m(2k+1) \prec \cdots \prec 2^m 7 \prec 2^m 5 \prec 2^m 3 \prec \cdots$$
$$\prec 2(2k+1) \prec \cdots \prec 2 \cdot 7 \prec 2 \cdot 5 \prec 2 \cdot 3 \prec \cdots \prec (2k+1) \prec \cdots \prec 7 \prec 5 \prec 3.$$

Let us explain this ordering of the set $\mathbb{N}$ and denote by $\mathbb{O}$ the odd positive integers greater than 1. Then, from the right to the left, we first take the set $\mathbb{O}$ in increasing order, then $2 \cdot \mathbb{O}$ in increasing order, then $2^2 \cdot \mathbb{O}$ in increasing order, etc., and at the end the powers 2^n of 2 in decreasing order. Note that $n \prec m$ if and only if $2n \prec 2m$.

Recall that a number $n \in \mathbb{N}$ is called a period of F if F has a periodic point λ of period n, that is, $F^{\circ n}(\lambda) = \lambda$ and $F^{\circ k}(\lambda) \neq \lambda$ for $0 < k < n$.

Then following fundamental result about periods is *Sharkovsky's theorem*.

Proposition 11.14 *Suppose F is a continuous map of a closed interval J into itself. If $m \in \mathbb{N}$ is a period of F and $n \prec m$, then n is also a period of F.*

Further, for any $m \in \mathbb{N}$ there exists a continuous mapping $F : J \mapsto J$ such that m is a period of F, but no n such that $m \prec n$ is a period of F.

Proof [Sh64, Sh65], see [BH11]. The first assertion is usually called Sharkovsky theorem and proved in most books on dynamical systems; see, e.g., [BV13].   □

An immediate consequence of Proposition 11.14 and the Sharkovsky ordering is the following: If 3 is a period of F, then all numbers $n \in \mathbb{N}$ are periods of F.

Combining Theorem 11.13 and Proposition 11.14, applied to $J = \mathbb{R}_+$, gives the following theorem. Recall that we assumed that $F : \mathbb{R}_+ \mapsto \mathbb{R}_+$ is *continuous*.

Theorem 11.15 *Let $m \in \mathbb{N}$. If there exists an irreducible operator on $\mathbb{C}^m$ such that (11.13) holds, then for any $n \in \mathbb{N}$ such that $n \prec m$ there exists an irreducible operator on $\mathbb{C}^n$ satisfying (11.13).*

This theorem has the following really surprising corollary.

Corollary 11.16 *If there exists an irreducible operator on $\mathbb{C}^3$ satisfying (11.13), then for any $n \in \mathbb{N}$ there is an irreducible operator on $\mathbb{C}^n$ such that (11.13) holds.*

Proof Since $n \prec 3$ for all $n \neq 3$, the assertion follows from Theorem 11.15.   □

The following example illustrates that quadratic polynomials F can lead to relations with an interesting finite-dimensional representation theory.

Example 11.17 (*Quadratic polynomial $F(\lambda) = (\lambda - q)^2, q \in \mathbb{R}$*)
The function $G(\lambda) = \alpha\lambda(1 - \lambda)$, with $\alpha \in \mathbb{R}$, is a well-studied dynamical system. In order to obtain a function which leaves $\mathbb{R}_+$ invariant, we use $F(\lambda) = (\lambda - q)^2$. (By a linear change of variables F and G can be transformed into each other.) All results on the existence of periods used in this example can be found in texts on dynamical systems [SKSF, HK03]. We only restate them here and follow [OS99].

We are looking for the existence of an *irreducible* operator X on a finite-dimensional Hilbert space satisfying the relation

$$XX^* = (X^*X - qI)^2.$$

The existence of such an operator X depends on the value of the parameter $q \in \mathbb{R}$.

$q = -\frac{1}{4}$: In this case F has the fix point $\frac{1}{4}$ and no other cycles. This leads to the one-dimensional representations $X = \frac{1}{2}e^{i\theta}$, $\theta \in [0, 2\pi)$.

$-\frac{1}{4} < q < \frac{3}{4}$: Then F has two fix points $2q + 1 \pm \frac{1}{2}\sqrt{4q + 1}$ and no other cycles. Thus we get two one-parameter families of one-dimensional representations and there are no other finite-dimensional irreducible representations.

$\frac{3}{4} < q < q_1 \approx 1.4$: There are cycles of all periods 2^n for $n \in \mathbb{N}$ and hence irreducible representations on $\mathbb{C}^{2^n}$ for any $n \in \mathbb{N}$.

$q = q_1 \approx 1.4$: There exist cycles of period 2^n for all $n \in \mathbb{N}$ and no other cycles. Hence for each $n \in \mathbb{N}$ there is an irreducible representation on $\mathbb{C}^{2^n}$ and there is no other irreducible finite-dimensional representation.

$q > q_2 \approx 1.75$: There are cycles of period 3 and hence cycles of each order $n \in \mathbb{N}$. Thus, for any $n \in \mathbb{N}$ there exists an irreducible representation on $\mathbb{C}^n$.   ○

11.5   Infinite-Dimensional Representations

For the classification of irreducible strong solutions of (11.13) we use formula (11.23) as a model for building the operator X. The corresponding irreducible representation operators are weighted shifts. First we briefly discuss these operators.

Suppose $(\lambda_n)_{n \in \mathbb{N}_0}$ is a complex sequence. There is a closed linear operator X on $l^2(\mathbb{N}_0)$ defined by $X(\varphi_n) := (\lambda_{n+1}\varphi_{n+1})$ for (φ_n) in the domain

$$\mathcal{D}(X) := \big\{(\varphi_n)_{n \in \mathbb{N}_0} \in l^2(\mathbb{N}_0) : (\lambda_{n+1}\varphi_{n+1})_{n \in \mathbb{N}_0} \in l^2(\mathbb{N}_0)\big\}.$$

The adjoint operator X^* acts by $X^*(\varphi_n) = (\overline{\lambda}_n\varphi_{n-1})$ and has the domain

$$\mathcal{D}(X^*) = \big\{(\varphi_n)_{n \in \mathbb{N}_0} \in l^2(\mathbb{N}_0) : (\overline{\lambda}_n\varphi_{n-1})_{n \in \mathbb{N}_0} \in l^2(\mathbb{N}_0)\big\},$$

where $\varphi_{-1} := 0$. In particular, for the base vector $e_k := (\delta_{nk})_{n \in \mathbb{N}_0}$ this yields

$$Xe_k = \lambda_k e_{k-1}, \ k \in \mathbb{N}, \ Xe_0 = 0, \ \text{and} \ X^*e_k = \overline{\lambda}_{k+1}e_{k+1}, \ k \in \mathbb{N}_0. \qquad (11.25)$$

Conversely, since X and X^* are linear and closed, from (11.25) we derive

$$X(\varphi_n)_{n \in \mathbb{N}_0} = X\Big(\sum_n \varphi_n e_n\Big) = \sum_n \varphi_n X e_n = \sum_n \varphi_n \lambda_n e_{n-1}$$
$$= \sum_n \varphi_{n+1} \lambda_{n+1} e_n = (\lambda_{n+1}\varphi_{n+1})_{n \in \mathbb{N}_0}$$

and similarly for X^*. Hence $\mathrm{Lin}\,\{e_k : k \in \mathbb{N}_0\}$ is a core for X and X^* and both operators are uniquely determined by their actions (11.25) on the base vectors e_k. Therefore, since $\langle Xe_k, e_n \rangle = \lambda_k\langle e_{k-1}, e_n \rangle = \lambda_k\delta_{k-1,n} = \lambda_{n+1}\delta_{k,n+1} = \langle e_k, X^*e_n \rangle$ for $k, n \in \mathbb{N}_0$ by (11.25), we conclude that X^* is indeed the adjoint operator of X.

Other types of weighted shift operators are defined and treated similarly.

The type of the power partial isometry U (Proposition 11.4) plays a crucial role in the infinite-dimensional irreducible representations. We distinguish three cases.

Case I $\ker U \neq \{0\}$.
Suppose that $F^{\circ k}(0) \geq 0$ for all $k \in \mathbb{N}$. We define a closed linear operator X_0 and its adjoint X_0^* on the Hilbert space $l^2(\mathbb{N}_0)$ by their actions on the standard basis:

$$X_0 e_k = F^{\circ k}(0)^{1/2} e_{k-1}, \quad X_0^* e_k = F^{\circ(k+1)}(0)^{1/2} e_{k+1}, \quad k \in \mathbb{N}_0, \qquad (11.26)$$

where $e_{-1} := 0$. Let $X_0 = U_0 C_0$ be the polar decomposition of X_0.

Note that $X_0 e_0 = 0$. If $F^{\circ k}(0) > 0$ for all $k \in \mathbb{N}$, then X_0 is called the *Fock representation* of (11.13). This is justified by the following theorem.

Theorem 11.18 (Fock representation) *The closed operator X_0 defined by (11.26) is irreducible if and only if 0 is not a period of F, or equivalently, $F^{\circ k}(0) > 0$ for all $k \in \mathbb{N}$. In this case, $X_0 = U_0 C_0$ is a strong solution of (11.13) and $\ker U_0 \neq \{0\}$. Each* *irreducible strong solution $X = UC$ of (11.13) acting on an infinite-dimensional Hilbert space $\mathcal{H}$ such that* $\ker U \neq \{0\}$ *is unitarily equivalent to the Fock representation operator X_0.*

Proof If $F^{\circ n}(0) = 0$ for some $n \in \mathbb{N}$, then $\mathrm{Lin}\,\{e_j : j = 0, \dots, n\}$ is a reducing subspace of X_0, so X_0 is not irreducible.

Now suppose that $F^{\circ k}(0) > 0$ for $k \in \mathbb{N}$. Let C_0 be the diagonal operator with $C_0 e_k = F^{\circ k}(0)^{1/2}\, e_k$, $k \in \mathbb{N}_0$, and let U_0 denote the adjoint of the unilateral shift. Then $X_0 = U_0 C_0$ is the polar decomposition of X_0. Obviously, $\ker U_0 \neq \{0\}$. Since $C_0^2 e_k = F^{\circ k}(0)e_k$, we have $F(C_0^2)e_k = F^{\circ(k+1)}(0)e_k$ and therefore $U_0^* F(C_0^2)e_k = F^{\circ(k+1)}(0)e_{k+1} = C_0^2 U_0^* e_k$ for all $k \in \mathbb{N}_0$. Further, $\ker U_0^* = \{0\}$. Hence, since the span of vectors e_k, $k \in \mathbb{N}_0$, is a core for $F(C_0^2)$, it follows from Remark 11.9 that (11.16) holds. Thus, X_0 is a strong solution of (11.13).

We show that X_0 is irreducible. Let T be a bounded operator which commutes with X_0 and X_0^*. Since $X_0 X_0^* e_k = F^{\circ(k+1)}(0)e_k$ and $F^{\circ(k+1)}(0) > 0$ for $k \in \mathbb{N}_0$, $\mathrm{ran}\, X_0 X_0^* = \mathrm{ran}\,|X_0^*|$ is dense. Therefore, since T commutes with $X_0 X_0^*$, hence with $|X_0^*|$, and with $X^* = U_0^*|X_0^*|$, it follows that T commutes with U_0^*. It is well known that U_0 and U_0^* are irreducible, so T is a scalar multiple of I. Then, by Lemma 11.11, X_0 is irreducible.

Next we prove the last assertion. Let $X = UC$ be as in the theorem. Set $E := E_{C^2}$. Since $\ker U \neq \{0\}$, we can find a unit vector $e \in \ker U$. Let $\mathcal{H}_0$ be the closed span of vectors $e_k := (U^*)^k e$, $k \in \mathbb{N}_0$. Since $\ker C = \ker U$, we have $e = E(\{0\})e$, so $E(M)e = e$ if $0 \in M$ and $E(M)e = 0$ if $0 \notin M$. Hence, by Theorem 11.8(iii), $E(M)(U^*)^k e = (U^*)^k E(F^{\circ(-k)}(M))e \in \mathcal{H}_0$ for any $k \in \mathbb{N}_0$. From (11.19) we recall that $UU^* = E(N_1)$. Therefore, $U e_n = UU^*(U^*)^{n-1}e = E(N_1)(U^*)^{n-1}e \in \mathcal{H}_0$ for $n \in \mathbb{N}$. Obviously, $Ue = 0 \in \mathcal{H}_0$ and $\mathcal{H}_0$ is invariant under U^*. Thus we have shown that $\mathcal{H}_0$ is invariant under U, U^*, and $E(\cdot)$. Hence, again by Lemma 11.11, we have $\mathcal{H}_0 = \mathcal{H}$, because X is irreducible.

Now we mimic the proof of Proposition 11.4. Assume for a moment that we would have Case II therein. Then the corresponding finite-dimensional subspace is also invariant under E (because the spectral projections $E(M)$ commute with P_k by (11.18)) and hence equal to $\mathcal{H}$ by the irreducibility of X. This is impossible, since $\mathcal{H}$ has infinite dimension. Thus Case I occurs, so U is (unitarily equivalent to) the adjoint of the unilateral shift and $e_k = (U^*)^k e$ has norm one.

From $Ue = 0$ we get $C^2 e = 0$. Hence $C^2 e_k = C^2 (U^*)^k e = F^{\circ k}(0)e_k$ by (11.23). Since the operator C^2 is positive and $e_k \neq 0$, we have $F^{\circ k}(0) \geq 0$ and $Xe_k = UCe_k = UF^{\circ k}(0)^{1/2}e_k = F^{\circ k}(0)^{1/2}e_{k-1}$ for $k \in \mathbb{N}_0$. This shows that X is unitarily equivalent to X_0. Since X is irreducible, $F^{\circ k}(0) > 0$ for all $k \in \mathbb{N}$, as shown above. Thus, X_0 is indeed the Fock representation operator. $\qquad \square$

Case II $\ker U^* \neq \{0\}$.

Suppose $(\lambda_k)_{k \in \mathbb{N}}$ is a sequence such that $\lambda_k = F(\lambda_{k+1}) \in \mathbb{R}_+$ for $k \in \mathbb{N}_0$ and $F(\lambda_0) = 0$. We define a closed linear operator X_1 on $l^2(\mathbb{N}_0)$ and its adjoint X_1^* by

$$X_1 e_k = \lambda_k^{1/2} e_{k+1} \quad \text{and} \quad X_1^* e_k = \lambda_{k-1}^{1/2} e_{k-1}, k \in \mathbb{N}_0, \tag{11.27}$$

where we set $e_{-1} := 0$. Let $X_1 = U_1 C_1$ be the polar decomposition of X_1. If $\lambda_k > 0$ for all $k \in \mathbb{N}_0$, then X_1 is called an *anti-Fock representation* of (11.13).

If F is a bijection of $\mathbb{R}$, then $\lambda_k = F^{\circ(-k-1)}(0)$ for $k \in \mathbb{N}_0$, so that

$$X_1 e_k = F^{\circ(-k-1)}(0)^{1/2} e_{k+1}, \quad X_1^* e_k = F^{\circ(-k)}(0)^{1/2} e_{k-1}, \, k \in \mathbb{N}_0. \qquad (11.28)$$

Theorem 11.19 (Anti-Fock representations) *The operator X_1 defined by (11.27) is irreducible if and only if $\lambda_k > 0$ for all $k \in \mathbb{N}_0$. In this case, $X_1 = U_1 C_1$ is an irreducible strong solution of the relation (11.13) and we have $\ker U_1^* \neq \{0\}$.*

Suppose F is a bijection of $\mathbb{R}$. If $X = UC$ is an irreducible strong solution of (11.13) acting on an infinite-dimensional Hilbert space such that $\ker U^* \neq \{0\}$, then X is unitarily equivalent to the operator X_1 defined by (11.28).

Proof Interchanging X, U, C^2 and $X^*, U^*, F(C^2)$ this proof is similar to the proof of Theorem 11.18. We do not carry out the details. $\qquad \square$

Case III $\ker U^* = \{0\}$ and $\ker U^* = \{0\}$.

Then the phase operator U is *unitary*. By the polar decomposition (Proposition A.3), Case III holds if and only if we have $\ker X = \{0\}$ and $\overline{\operatorname{ran} X} = \mathcal{H}$.

We begin with a simple preliminary lemma.

Lemma 11.20 *Suppose B is a self-adjoint operator on a Hilbert space $\mathcal{H}$ such that its spectrum consists of k distinct eigenvalues with eigenspaces $\mathcal{H}_1, \ldots, \mathcal{H}_k$. Let U be a unitary operator on $\mathcal{H}$ such that $U\mathcal{H}_j = \mathcal{H}_{j+1}$ for $j = 1, \ldots, k$, where $\mathcal{H}_{k+1} := \mathcal{H}_1$. If the operator UB is irreducible, then $\dim \mathcal{H}_j = 1$ for all j.*

Proof Clearly, $U_j := U \restriction \mathcal{H}_j \mapsto \mathcal{H}_{j+1}$ is a unitary operator of $\mathcal{H}_j$ on $\mathcal{H}_{j+1}$ and $U := U_1 \cdots U_k$ is a unitary operator of $\mathcal{H}_1$. Assume to the contrary that we have $\dim \mathcal{H}_j \geq 2$ for some j. Then $\dim \mathcal{H}_1 \geq 2$. Hence there is a projection $T_1 \neq 0, I$ of the Hilbert space $\mathcal{H}_1$ which commutes with U. We define $T_{j+1} = U_j T_j U_j^{-1}$ for $j = 1, \ldots, k-1$. Let T be the diagonal operator on $\mathcal{H} = \mathcal{H}_1 \oplus \cdots \oplus \mathcal{H}_k$ with diagonal entries $T_1, \ldots, T_k$. By a straightforward computation we verify that T commutes with U, U^*, B and hence with UB and $(UB)^*$. Therefore, since T is a nontrivial projection, UB is not irreducible, a contradiction. $\qquad \square$

Suppose that F is bijective on $\mathbb{R}$. Then $F^{\circ k}(\lambda)$ is well defined for all $k \in \mathbb{Z}$ and $\mathbb{O}(\lambda) := \{F^{\circ k}(\lambda) : k \in \mathbb{Z}\}$ is the orbit of $\lambda \in \mathbb{R}$.

Let $\lambda_0 \in \mathbb{R}_+$ and suppose that $F^{\circ k}(\lambda_0) \geq 0$ for all $k \in \mathbb{Z}$. Let X_2 denote the closed linear operator on $l_2(\mathbb{Z})$ and X_2^* its adjoint given by

$$X_2 e_k = F^{\circ k}(\lambda_0)^{1/2} e_{k-1} \quad \text{and} \quad X_2^* e_k = F^{\circ(k+1)}(\lambda_0)^{1/2} e_{k+1}, \quad k \in \mathbb{Z}. \qquad (11.29)$$

Let $X_2 = U_2 C_2$ be the polar decomposition of X_2. Then, for $k \in \mathbb{Z}$ we obtain

$$C_2^2 e_k = X_2^* X_2 e_k = F^{\circ k}(\lambda_0) e_k, \quad |X_2^*|^2 e_k = X_2 X_2^* e_k = F^{\circ(k+1)}(\lambda_0) e_k.$$

From the first equality it follows that the set $\sigma_p(C_2^2)$ of eigenvalues of the self-adjoint operator C_2^2 is just the orbit $\mathbb{O}(\lambda_0)$. By Proposition A.3, U_2 is unitary if and only if $\ker C_2 = \ker |X_2^*| = \{0\}$. Therefore, U_2 is unitary if and only if $F^{\circ k}(\lambda_0) > 0$ for all $k \in \mathbb{Z}$. Clearly, in this case, $U_2 e_k = e_{k-1}$ for $k \in \mathbb{Z}$.

Theorem 11.21 *Suppose F is a bijection of $\mathbb{R}$. Let $\lambda_0 \in \mathbb{R}_+$ be such that $F^{\circ k}(\lambda_0) > 0$ for $k \in \mathbb{Z}$. Then the operator $X_2 = U_2 C_2$ on $l^2(\mathbb{Z})$ given by (11.29) is a strong solution of (11.13) such that U_2 is unitary and $\sigma_p(C_2^2) = \mathbb{O}(\lambda_0)$. The operator X_2 is irreducible if and only if λ_0 is not a periodic point of F, that is,*

$$F^{\circ j}(\lambda_0) \neq F^{\circ k}(\lambda_0) \quad \text{for} \quad j \neq k, \ j, k \in \mathbb{Z}. \tag{11.30}$$

Conversely, if $X = UC$ is an irreducible strong solution of (11.13) on an infinite-dimensional Hilbert space such that U is unitary and C^2 has at least one eigenvalue, then X is unitarily equivalent to an operator X_2 defined by (11.29).

Proof Clearly, $U_2^* F(C_2^2) e_k = F^{\circ(k+1)}(\lambda_0) e_{k+1} = C_2^2 U_2^* e_k$ for $k \in \mathbb{Z}$ and $\ker U_2^* = \{0\}$. Therefore, by Remark 11.9, (11.16) holds, so X_0 is a strong solution of (11.13). Next we prove the assertion about the irreducibility.

First suppose (11.30) holds. Let P be the projection on a reducing subspace of X_2. Then P commutes with X_2 and X_2^*, hence with $C_2^2 = X_2^* X_2$. From this we obtain $C_2^2 P e_k = P C_2^2 e_k = F^{\circ k}(\lambda_0) P e_k$. Combined with (11.30) we conclude that for each $k \in \mathbb{Z}$ there exists a number $\alpha_k \in \mathbb{C}$ such that $P e_k = \alpha_k e_k$. Then

$$X_2(P e_k - \alpha_{k-1} e_k) = (P - \alpha_{k-1}) X_2 e_k = (P - \alpha_{k-1}) F^{\circ k}(\lambda_0) e_{k-1} = 0.$$

Therefore, since U_2 is unitary and hence $\ker X_2 = \{0\}$, we get $P e_k - \alpha_{k-1} e_k = 0$, so that $P e_k = \alpha_k e_k = \alpha_{k-1} e_k$ for $k \in \mathbb{Z}$. Hence α_k is constant, say $\alpha_k = \alpha$ for $k \in \mathbb{Z}$. Then $P = \alpha I$, so either $P = 0$ or $P = I$. Hence X_2 is irreducible.

To prove the reverse direction we assume the contrary. Then $F^{\circ k}(\lambda_0) = F^{\circ n}(\lambda_0)$ for some $k > n$. Upon replacing λ_0 by $F^{\circ n}(\lambda_0)$ we can assume that $n = 0$. There is a smallest $k \in \mathbb{N}$ such that $F^{\circ k}(\lambda_0) = \lambda_0$. Since X_2 is irreducible, Lemma 11.20, applied with $B := C_2^2$, implies that $l^2(\mathbb{Z})$ has finite dimension, a contradiction.

Finally, we prove the last assertion. Let $X = UC$ be as stated in the theorem. Then C^2 has an eigenvalue, so there exist a $\lambda_0 \in \mathbb{R}_+$ and a unit vector e such that $C_2^2 e = \lambda_0 e$. From (11.23) we obtain $C^2 U^{-k} e = F^{\circ k}(\lambda_0) U^{-k} e$ for $k \in \mathbb{N}$. Theorem 11.8(ii) gives $UC^2 e = \lambda U e = F(C^2) U e$. Then, by induction we derive $U^j C^2 e = \lambda U^j e = F^{\circ j}(C^2) U^j e$ for $j \in \mathbb{N}$. Hence $U^j e$ is an eigenvector of $F^{\circ j}(C^2)$ with eigenvalue λ, so it is an eigenvector of $F^{\circ(-j)}(F^{\circ j}(C^2)) = C^2$ with eigenvalue $F^{\circ(-j)}(\lambda_0)$ for $j \in \mathbb{N}$. Set $e_k := U^{-k} e$ for $k \in \mathbb{Z}$. Since U is unitary, $\|e_k\| = 1$. Then, $U e_k = e_{k-1}$, and by the preceding, $C^2 e_k = F^{\circ k}(\lambda_0) e_k$ for all $k \in \mathbb{Z}$. Hence $F^{\circ k}(\lambda_0) \geq 0$, $C e_k = F^{\circ k}(\lambda_0)^{1/2} e_k$, and $X e_k = U C e_k = F^{\circ k}(\lambda_0)^{1/2} e_{k-1}$. Clearly, the closed span $\mathcal{H}_0$ of vectors $e_k, k \in \mathbb{Z}$, is invariant under U, U^*, E_C. Therefore, since X is irreducible, Lemma 11.11 yields $\mathcal{H}_0 = \mathcal{H}$.

Next we note that (11.30) holds. Indeed, if (11.30) would fail, then, by arguing as in the paragraph before last, C^2 has a finite spectrum and Lemma 11.20 implies that $\mathcal{H}$ has finite dimension. This contradicts the assumption.

For $k \neq n$, $U^k e$ and $U^n e$ are eigenvectors of C^2 with different eigenvalues by condition (11.30), so that $U^k e \perp U^n e$. Therefore, $\{e_k : k \in \mathbb{Z}\}$ is an orthonormal basis of $\mathcal{H}$. Since U is unitary, $F^{\circ k}(\lambda_0) > 0$ for $k \in \mathbb{Z}$. Putting all things together, we have shown that X is unitarily equivalent to an operator X_2. $\qquad\square$

Now we suppose that $X = UC$ is an *irreducible* strong solution of (11.13) acting on an *infinite-dimensional* Hilbert space. We summarize Theorems 11.18, 11.19, 11.21 and the outcomes in all three cases.

Case I $\ker U \neq \{0\}$.
Then we have a unique irreducible representation, called the *Fock representation*: U is the adjoint of the unilateral shift on $l^2(\mathbb{N}_0)$, hence $\ker U^* = \{0\}$, and C acts as a diagonal operator by $Ce_k = \lambda_k^{1/2} e_k$, so $Xe_k = \lambda_k^{1/2} e_{k-1}$, for $k \in \mathbb{N}_0$. The sequence $(\lambda_k)_{k \in \mathbb{N}_0}$ is defined inductively by $\lambda_0 = 0$, $\lambda_{k+1} = F(\lambda_k)$ and $\lambda_{k+1} > 0$ for $k \in \mathbb{N}_0$. Thus, $\lambda_k = F^{\circ k}(0)$ for $k \in \mathbb{N}_0$. The set $\sigma_p(C^2)$ of eigenvalues of C^2 is the positive semiorbit $\mathbb{O}^+(0)$.

Case II $\ker U^* \neq \{0\}$.
Then we have a class of irreducible representations, called *anti-Fock representations*: U is the unilateral shift on $l^2(\mathbb{N}_0)$, hence $\ker U = \{0\}$, and the operators C and X act by $Ce_k = \lambda_k^{1/2} e_k$ and $Xe_k = \lambda_k^{1/2} e_{k+1}$, $k \in \mathbb{N}_0$. The sequence $(\lambda_k)_{k \in \mathbb{N}_0}$ satisfies $\lambda_k = F(\lambda_{k+1}) > 0$ for $k \in \mathbb{N}_0$ and $F(\lambda_0) = 0$.

Case III $\ker U = \ker U^* = \{0\}$, that is, U is unitary.
Suppose $F : \mathbb{R} \mapsto \mathbb{R}$ is a bijection. There are infinite-dimensional irreducible representations corresponding to numbers $\lambda_0 \in \mathbb{R}_+$ such that $\lambda_k := F^{\circ k}(\lambda_0) > 0$ for all $k \in \mathbb{Z}$ and condition (11.30) is satisfied: U is the adjoint of the bilateral shift on $l^2(\mathbb{Z})$ and $Ce_k = \lambda_k^{1/2} e_k$, $Xe_k = \lambda_k^{1/2} e_{k-1}$ for $k \in \mathbb{Z}$. The set $\sigma_p(C^2)$ of eigenvalues of C^2 is the orbit $\mathbb{O}(\lambda_0)$. Condition (11.30) means that the orbit has no periodic point. One easily verifies that if $\lambda_0, \lambda_0' \in \mathbb{R}_+$ belong to the same orbit, the corresponding representations are unitarily equivalent.

It should be emphasized that it may happen that there is no infinite-dimensional irreducible representation for some of the Cases I–III. For the relation (11.31) in Sect. 11.6 only Case III occurs, while for the relation (11.41) in Sect. 11.7 we have Cases I and III. An example for Case II is given in Exercise 11.

As we have seen, the nonunitary phase operators of irreducible strong solutions of (11.13) are precisely the three nonunitary irreducible power partial isometries from Proposition 11.4. This is not accidental as shown in [OS99, Theorem 19, p. 98]. The case of unitary phase operators can be very complicated. It is possible that the corresponding dynamical system has no measurable section and that nontype I representations occur; see, e.g., [VS94, Section 3.3].

11.6   The Hermitian Quantum Plane

In this section, q is a **positive** real number. Our aim is to study the operator relation

$$XX^* = qX^*X. \tag{11.31}$$

This is one of the simplest relations (11.13) with $F(\lambda) = q\lambda$, and the theory from Sect. 11.3 can be applied.

Definition 11.22 A densely defined closed operator X on a Hilbert space is called q-*normal* if $\mathcal{D}(XX^*) = \mathcal{D}(X^*X)$ and $XX^*\varphi = qX^*X\varphi$ for $\varphi \in \mathcal{D}(XX^*)$.

Clearly, a q-normal operator with $q = 1$ is normal.

The next proposition collects various characterizations of q-normal operators. It shows that q-normal operators are precisely the strong solutions of relation (11.31).

Proposition 11.23 *Let $X = UC$ be the polar decomposition of a densely defined closed operator X and E_C the spectral measure of C. The following are equivalent:*

(i) *X is q-normal.*
(ii) *$UC^2U^* = qC^2$.*
(iii) *$UCU^* = q^{1/2}C$.*
(iv) *X is a strong solution of the relation (11.31) according to Definition 11.7.*

If one of these statements holds, then the closed linear subspace

$$\mathcal{H}_0 := \ker U = \ker U^* = \ker C = \ker X = \ker X^* \tag{11.32}$$

is reducing for the operators U, U^*, C, X, X^* and U is unitary on $\mathcal{H} \ominus \mathcal{H}_0$. If U is unitary on $\mathcal{H}$, then each of these statements is also equivalent to

(v) *$UE_C(M)U^* = E_C(q^{-1/2}M)$ for each Borel set M of $\mathbb{R}$.*

Proof (i)$\leftrightarrow$(ii) follows from the relations $X^*X = C^2$ and $XX^* = UC^2U^*$.

By the formulas (A.2) and (A.3) for the polar decomposition (Proposition A.3), we have $\mathcal{H}_0 := \ker U = \ker C = \ker X$ and $\ker X^* = \ker U^*$. Therefore, to prove (11.32) it suffices to verify that $\mathcal{H}_0 = \ker U^*$. First we show that each of the conditions (ii)–(iv) implies $\mathcal{H}_0 = \ker U^*$.

First assume (ii). Then, for $f \in \mathcal{D}(C^2)$, we have $U^*f \in \mathcal{D}(C^2) \subseteq \mathcal{D}(C)$ and

$$q\|Cf\|^2 = q\langle C^2f, f\rangle = \langle UC^2U^*f, f\rangle = \langle CU^*f, CU^*f\rangle = \|CU^*f\|^2. \tag{11.33}$$

Hence $\ker U^* \subseteq \ker C = \mathcal{H}_0$. Conversely, if $f \in \ker C$, it follows from (11.33) that $U^*f \in \ker C = \ker U$, so $UU^*f = 0$ and hence $U^*f = 0$. Thus, $\ker U^* = \mathcal{H}_0$.

Since $\ker C = \ker C^2$, the same reasoning, with C^2 replaced by C, is valid if we assume (iii) instead of (ii). If (iv) holds, then $\ker U^* = \ker F(C^2) = \ker (qC)^2$ by

Theorem 11.8 and hence ker $U^* = \mathcal{H}_0$, because ker $C = \text{ker}(qC^2)$. Thus, in all three Cases (ii)–(iv), the proof of (11.32) is complete.

By (11.32), $\mathcal{H}_0$ is reducing for U, U^*, C, X and these operators are zero on $\mathcal{H}_0$. Hence the equivalence of (ii)–(iv) holds trivially on $\mathcal{H}_0$.

On the subspace $\mathcal{H} \ominus \mathcal{H}_0$, U and U^* have trivial kernels, so U is unitary. Thus it remains to prove the equivalence of (ii)–(v) in the case when U is unitary.

First we note that for unitary U it follows from Theorem 11.8 that (iv) is equivalent to $E_{C^2}(M)U^* = U^*E_{C^2}(q^{-1}M)$ and so to $UE_{C^2}(M)U^* = E_{C^2}(q^{-1}M)$. From operator theory we know that square roots and spectral projections are preserved under unitary equivalence. Hence $UC^2U^* = qC^2$ is equivalent to $UCU^* = q^{1/2}C$, to $UE_{C^2}U^* = E_{qC^2}$, and to $UE_CU^* = E_{q^{1/2}C}$. Since $E_{qC^2}(M) = E_{C^2}(q^{-1}M)$ and $E_{q^{1/2}C}(M) = E_C(q^{-1/2}M)$ for all Borel sets M, this gives the equivalence of conditions (ii)–(v) in the unitary case. $\square$

From now on in this section we assume that $q \neq 1$. Next we construct a basic example of a q-normal operator. Let us abbreviate

$$\Delta_q := [1, q^{1/2}) \quad \text{if} \quad q > 1 \quad \text{and} \quad \Delta_q := [q^{1/2}, 1) \quad \text{if} \quad q < 1. \tag{11.34}$$

Suppose that μ is a Radon measure on $\mathbb{R}_+ = [0, +\infty)$ such that

$$\mu(M) = \mu(q^{1/2}M) \quad \text{for all Borel subsets } M \subseteq \mathbb{R}_+. \tag{11.35}$$

Since $(0, +\infty) = \cup_{k\in\mathbb{Z}} q^{k/2}\Delta_q$, such a measure μ is uniquely defined by its restrictions to Δ_q and $\{0\}$. Moreover, given a Radon measure μ^0 on Δ_q and a number $c_0 \geq 0$, there is a unique Radon measure μ on $\mathbb{R}_+$ satisfying (11.35) such that $\mu(M) = \mu^0(M)$ for all Borel sets $M \subseteq \Delta_q$ and $\mu(\{0\}) = c_0$.

Now we define an operator X_μ on the Hilbert space $\mathcal{H} := L_2(\mathbb{R}_+; \mu)$ by

$$(X_\mu\varphi)(t) := q^{1/2}t\varphi(q^{1/2}t), \quad \mathcal{D}(X_\mu) = \{\varphi(t) \in \mathcal{H} : t\varphi(t) \in \mathcal{H}\}. \tag{11.36}$$

Set $\mathcal{H}_0 = \{f \in \mathcal{H} : f = 0, \mu\text{-a.e. on } (0, +\infty)\}$. Obviously, $\mu(\{0\}) = 0$ if and only if $\mathcal{H}_0 = \{0\}$.

Proposition 11.24 *Suppose $q \neq 1$. The operator X_μ defined by (14.43) is q-normal with $\ker X_\mu = \mathcal{H}_0$, and its adjoint operator acts by*

$$(X_\mu^*\varphi)(t) = t\varphi(q^{-1/2}t) \quad \text{for} \quad \varphi \in \mathcal{D}(X_\mu^*) = \mathcal{D}(X_\mu).$$

Proof We define operators U_μ, C_μ on $\mathcal{H}$ by

$$(C_\mu\psi)(t) := t\psi(t), \quad (U_\mu\varphi)(t) := \begin{Bmatrix} \varphi(q^{1/2}t) & \text{for } t \neq 0 \\ 0 & \text{for } t = 0 \end{Bmatrix}, \tag{11.37}$$

where $\varphi \in \mathcal{H}$, $\psi \in \mathcal{D}(C_\mu) := \mathcal{D}(X_\mu)$. Using (11.35) we verify that U_μ is a partial isometry on $\mathcal{H}$. Further, $X_\mu = U_\mu C_\mu$, $\ker C_\mu = \ker U_\mu = \mathcal{H}_0$, and C_μ is a positive self-adjoint operator. It follows that $X_\mu = U_\mu C_\mu$ is the polar decomposition of X_μ. Since U_μ is bounded, $X_\mu^* = C_\mu U_\mu^*$. This implies that X_μ^* has the form stated above. Further, it is easily checked that $U C_\mu^2 U^* = q C_\mu^2$; that is, X_μ is q-normal. $\square$

Let μ_i, $i \in I$, be a family of Radon measures on $\mathbb{R}_+$ satisfying (11.35). Since the operators X_{μ_i} are q-normal by Proposition 11.24, so is their orthogonal direct sum. It can be shown [CSS14, Theorem 1] that each q-normal operator is unitarily equivalent to a direct sum of such operators X_{μ_i}.

For irreducible q-normal operators there is the following description.

Proposition 11.25 *Suppose $q \neq 1$. A densely defined closed operator X on $\mathcal{H}$ is an irreducible q-normal operator if and only if either $X = 0$ on $\dim \mathcal{H} = 1$ or there exist a $\lambda \in \Delta_q$ and an orthonormal basis $\{e_k : k \in \mathbb{Z}\}$ of $\mathcal{H}$ such that*

$$X e_k = \lambda q^{k/2} e_{k-1}, \quad X^* e_k = \lambda q^{(k+1)/2} e_{k+1}, \quad k \in \mathbb{Z}. \tag{11.38}$$

If $\lambda, \lambda' \in \Delta_q$ and $\lambda \neq \lambda'$, the corresponding operators are not unitarily equivalent.

Proof Obviously, $X = 0$ is q-normal. One easily verifies that the operator X in (11.38) is q-normal and irreducible. Since $\sigma_p(C) = \{\lambda q^{k/2} : k \in \mathbb{Z}\}$, it follows that the operators in (11.38) for different $\lambda, \lambda' \in \Delta_q$ are not unitarily equivalent.

Conversely, suppose $X = UC$ is an irreducible q-normal operator on $\mathcal{H}$. Then the spectral measure E_{C^2} is ergodic with respect to F by Corollary 11.12 and hence supported on an orbit. The dynamical system $F(\lambda) = q\lambda$ on $\mathbb{R}_+$ has precisely the following orbits: $\mathbb{O}(0) = \{0\}$ and $\mathbb{O}(\lambda^2) = \{F^{\circ k}(\lambda) = q^k \lambda^2 : k \in \mathbb{Z}\}$ for $\lambda \in \Delta_q$. Since $\ker C$ is reducing by Proposition 11.23 and X is irreducible, $\mathbb{O}(0) = \{0\}$ yields $X = 0$ and $\dim \mathcal{H} = 1$.

Now suppose $\lambda \in \Delta_q$. Since E_{C^2} is supported on $\mathbb{O}(\lambda^2)$, $E_{C^2}(\{\lambda^2\}) \neq 0$, so λ^2 is an eigenvalue of C^2. Hence there is a unit vector $e_0 \in \mathcal{H}$ such that $C^2 e_0 = \lambda^2 e_0$, so $C e_0 = \lambda e_0$. Since $\ker C = \{0\}$, U is unitary by (11.32). Set $e_k := (U^*)^k e_0$ for $k \in \mathbb{Z}$. Since U is unitary, Proposition 11.23(v) implies that $C e_k = \lambda q^{k/2} e_k$. Hence $e_k \perp e_n$ if $k \neq n$. Moreover, $U e_k = e_{k-1}$. Thus, the closed span $\mathcal{H}_0$ of vectors $e_k, k \in \mathbb{Z}$, is invariant under U, U^*, and E_C. Therefore, $\mathcal{H}_0 = \mathcal{H}$ by Lemma 11.11, because X is irreducible. Then $\{e_k : k \in \mathbb{Z}\}$ is an orthonormal basis of $\mathcal{H}$, and we have $X e_k = U C e_k = \lambda q^{k/2} e_{k-1}$ for $k \in \mathbb{Z}$. Hence X is of the form (11.38) and so is X^*. (Note that the preceding mimics the proof of Theorem 11.21 for $F(\lambda) = q\lambda$ and (11.29) gives (11.38) in this special case.) $\square$

Proposition 11.26 *Let $X = UC$ be a q-normal operator on a Hilbert space $\mathcal{H}$. Then $\mathcal{D}(X^{*m} X^n) = \mathcal{D}(C^{m+n})$ for $m, n \in \mathbb{N}_0$. The domain*

$$\mathcal{D}_\infty(X) := \bigcap_{n, m \in \mathbb{N}_0} \mathcal{D}(X^{*m} X^n) = \bigcap_{k \in \mathbb{N}_0} \mathcal{D}(C^k) \tag{11.39}$$

is dense in $\mathcal{H}$ and an invariant core for all operators $X^{*m} X^n$, $m, n \in \mathbb{N}_0$.

Proof By Proposition 11.23 we can assume that U is unitary. Then, we conclude from Proposition 11.23(iii) that $UC = q^{1/2}CU$ and $U^*C = q^{-1/2}CU^*$, so

$$X^{*m}X^n = (CU^*)^m(UC)^n = q^{\frac{m(m+1)}{4} - \frac{n(n-1)}{4}} U^{*m}C^m U^n C^n$$

$$= q^{(m^2+m-n^2+n-2mn)/4} U^{n-m} C^{m+n}. \tag{11.40}$$

Since U^{n-m} is unitary, (11.40) implies $\mathcal{D}(X^{*m}X^n) = \mathcal{D}(C^{m+n})$ and so (11.39). From (11.31) it follows at once that $\mathcal{D}_\infty(X)$ is invariant under X and X^*.

Let $\mathcal{D}_0 := \cup_{n \in \mathbb{N}} E_C([0, n])\mathcal{H}$. From the spectral theory we easily derive that $\mathcal{D}_0 \subseteq \cap_{k \in \mathbb{N}_0} \mathcal{D}(C^k)$ and that $\mathcal{D}_0$ is a core for each power C^k. By (11.40), $\mathcal{D}_0$, hence $\mathcal{D}_\infty(X)$, is a core for all operators $X^{*m}X^n$ as well. $\qquad\square$

Let $\mathsf{A} = \mathbb{C}\langle x, x^+ | xx^+ = qx^+x \rangle$ be the unital $*$-algebra with a single generator x and defining relation $xx^+ = qx^+x$. In quantum group theory, A is considered as the coordinate algebra of the *Hermitian quantum plane*. From Proposition 11.26 it follows that each q-normal operator X gives rise to a unique $*$-representation π of A on the domain $\mathcal{D}(\pi) := \mathcal{D}_\infty(X)$ defined by $\pi(x) := X \lceil \mathcal{D}_\infty(X)$.

11.7 The q-Oscillator Algebra

In this section, q is a fixed **positive** number and we treat the operator relation

$$XX^* = qX^*X + I. \tag{11.41}$$

This is relation (11.13) for the function $F(\lambda) := q\lambda + 1$. Note that F is a bijection of $\mathbb{R}$. The special case $q = 1$ was studied in detail in Sect. 8.2.

In terms of the polar decomposition $X = UC$ relation (11.41) reads as

$$UC^2U^* = qC^2 + I. \tag{11.42}$$

The following theorem collects a number of slight variations and sharpenings of Theorem 11.8 in the special case $F(\lambda) = q\lambda + 1$.

Theorem 11.27 *Let $X = UC$ be the polar decomposition of a densely defined closed operator X on a Hilbert space. Then the following statements are equivalent:*

(i) *X is a strong solution of $XX^* = qX^*X + I$ according to Definition 11.7.*
(ii) *$\mathcal{D}(XX^*) = \mathcal{D}(X^*X)$ and $XX^*\varphi = qX^*X\varphi + \varphi$ for $\varphi \in \mathcal{D}(XX^*)$.*
(iii) *$\mathcal{D}(UC^2U^*) = \mathcal{D}(C^2)$ and $UC^2U^*\varphi = qC^2\varphi + \varphi$ for $\varphi \in \mathcal{D}(C^2)$.*
(iv) *$\ker U^* = \{0\}$ and $U^*(qC^2 + I) \subseteq C^2U^*$.*
(v) *$\ker U^* = \{0\}$ and $UC^2 \subseteq (qC^2 + I)U$.*
(vi) *$\ker U^* = \{0\}$ and $E_{C^2}(M)U^* = U^*E_{C^2}(q^{-1}(M-1))$ for all Borel sets M.*

If one of these conditions holds, we have $UU^* = I$ and $\ker X^* = \{0\}$. Moreover, if $q \neq 1$ and we set $B := (q-1)C^2 + I$, then (v) is equivalent to

(vii) $\ker U^* = \{0\}$ *and* $UB \subseteq qBU$.

Proof It is clear that $F(C^2) = \ker(qC^2 + I) = \{0\}$. Hence the condition $\ker U^* \subseteq \ker F(C^2)$ in Definition 11.7 is equivalent to $\ker U^* = \{0\}$ and Theorem 11.8 yields the equivalence of (i), (iv), (v), (vi). Further, since $X^*X = C^2$ and $XX^* = UC^2U^*$, (ii) and (iii) are equivalent.

(iii)$\to$(iv): Let $\varphi \in \ker U^*$. Then, obviously, $\varphi \in \mathcal{D}(UC^2U^*)$, so $\varphi \in \mathcal{D}(C^2)$ and $0 = UC^2U^*\varphi = qC^2\varphi + \varphi$ by (iii). Hence $0 = q\langle C\varphi, C\varphi \rangle + \|\varphi\|^2$ which implies $\varphi = 0$. Thus, $\ker U^* = \{0\}$.

Now let $\varphi \in \mathcal{D}(qC^2 + I)$. Then, $\varphi \in \mathcal{D}(C^2)$. Hence $qC^2\varphi + \varphi = UC^2U^*\varphi$ by (iii). Applying U^* yields $U^*(qC^2 + I)\varphi = U^*UC^2U^*\varphi$. Since U^*U is the projection on $\overline{\operatorname{ran} C}$ by (A.2), we have $U^*UC\psi = C\psi$ for $\psi \in \mathcal{D}(C)$. Therefore, $U^*(qC^2 + I)\varphi = C^2U^*\varphi$. This proves that $U^*(qC^2 + I) \subseteq C^2U^*$.

(iv)$\to$(iii): Suppose $\varphi \in \mathcal{D}(C^2) = \mathcal{D}(qC^2 + I)$. Then, it follows from (iv) that $U^*(qC^2 + I)\varphi = C^2U^*\varphi$ and hence $UU^*(qC^2 + I)\varphi = UC^2U^*\varphi$. Further, since $\ker U^* = \{0\}$ by (iv) and U is a partial isometry, $UU^* = I$. Therefore, $(qC^2 + I)\varphi = UC^2U^*\varphi$. By the preceding we have shown that $qC^2 + I \subseteq UC^2U^*$. Since the symmetric operator UC^2U^* is an extension of the self-adjoint operator $qC^2 + I$, we conclude that $qC^2 + I = UC^2U^*$, which is condition (iii).

This completes the proof of the equivalence of (i)–(vi). If $\ker U^* = \{0\}$, then the partial isometry U satisfies $UU^* = I$ and $\ker X^* = \ker U^* = \{0\}$.

A straightforward computation shows that $(q-1)UC^2 \subseteq (q-1)(qC^2 + I)U$ is equivalent to $UB \subseteq qBU$. Thus, if $q \neq 1$, then (v) and (vii) are equivalent. □

From now on assume that $q \neq 1$. Since $F(\lambda) = q\lambda + 1$, $F^{-1}(\lambda) = (\lambda - 1)q^{-1}$. A simple induction argument shows that

$$F^{\circ n}(\lambda) = \lambda q^n + (1 - q^n)(1 - q)^{-1}, \quad n \in \mathbb{Z}, \; \lambda \in \mathbb{R}. \tag{11.43}$$

Obviously, the dynamical system F on $\mathbb{R}$ has a fix point $\lambda_0 = (1-q)^{-1}$. For $\lambda \neq (1-q)^{-1}$ we easily derive from (11.43) that $F^{\circ k}(\lambda) \neq F^{\circ n}(\lambda)$ for $k \neq n$; that is, λ is not a periodic point of F.

The next theorem describes all irreducible strong solutions of the relation (11.41).

Theorem 11.28 *Suppose that $q > 0, q \neq 1$. An irreducible densely defined closed operator X is a strong solution of (11.41) if and only if, up to unitary equivalence, X is one of the following operators:*

(i) *Fock representation X_F for any $q > 0$:*
$$X_F e_k = \sqrt{\tfrac{1-q^k}{1-q}}\, e_{k-1}, \; k \in \mathbb{N}_0, \text{ on } \mathcal{H} = l^2(\mathbb{N}_0), \text{ where } e_{-1} := 0.$$

(ii) *Non-Fock representation X_γ for $q < 1$:*
$$X_\gamma e_k = \sqrt{\tfrac{1+q^k\gamma}{1-q}}\, e_{k-1}, \; k \in \mathbb{Z}, \text{ on } \mathcal{H} = l^2(\mathbb{Z}), \text{ where } \gamma \in [q, 1).$$

(iii) *Degenerate representation for $q < 1$:*
  $X_\theta = e^{i\theta}(1-q)^{-1/2}$ on $\mathcal{H} = \mathbb{C}$, where $\theta \in [0, 2\pi)$.

Proof By straightforward computations we verify that all operators X stated above satisfy condition (ii) of Theorem 11.27 and are irreducible.

Now let X be an irreducible operator satisfying the conditions in Theorem 11.27.

First suppose that the operator U is unitary. Then, since $UC^2U^* = qC^2 + 1$ by Theorem 11.27(iii) and the spectrum is invariant under unitary transformations, $\sigma(C^2)$ is invariant under F and hence under all mappings $F^{\circ n}, n \in \mathbb{Z}$.

If $\sigma(C^2)$ consists of one point λ_0, this is the fix point $\lambda_0 = (1-q)^{-1} \in \sigma(C^2)$. Hence $q < 1$ and $C = (1-q)^{-1/2}I$. Then U and C commute, so $\dim \mathcal{H} = 1$, and X is as in (iii). (This is the case of a cycle of period $n = 1$ in Theorem 11.13.)

Assume that $\sigma(C^2)$ is not a singleton. Then there is a $\lambda \neq (1-q)^{-1}$ in $\sigma(C^2)$. If we would have $q > 1$, then (11.43) yields $\lim_{n \to \infty} F^{\circ(-n)}(\lambda) = (1-q)^{-1} < 0$. Since $F^{\circ(-n)}(\lambda) \in \sigma(C^2)$, this is a contradiction. Therefore, $q < 1$.

By (11.43), $F^{\circ n}(\lambda) = q^n(\lambda - (1-q)^{-1}) + (1-q)^{-1}$. If $\lambda < (1-q)^{-1}$, then $\lim_{n \to \infty} F^{\circ(-n)}(\lambda) = -\infty$, which is a contradiction. Thus $\lambda > (1-q)^{-1}$ and therefore $\lim_{n \to \infty} F^{\circ(-n)}(\lambda) = +\infty$, so $\sigma(C^2)$ is unbounded. Since $\sigma(C^2)$ is also closed, there is a smallest $\lambda_0 \in \sigma(C^2)$ such that $\lambda_0 \geq (1+q)(1-q)^{-1}$. We verify that $\lambda_0 < 2(1-q)^{-1}$. Assume to the contrary that $\lambda_0 \geq 2(1-q)^{-1}$. Then we derive $(\lambda_0 q + 1)(1-q) \geq 1+q$, so $F(\lambda_0) \geq (1+q)(1-q)^{-1}$. But, by $\lambda_0 > (1-q)^{-1}$, we have $F(\lambda_0) < \lambda_0$ and $F(\lambda_0) \in \sigma(C^2)$. This contradicts the choice of λ_0 and proves our claim.

Consider the function $g(\lambda) = (1 + \lambda)(1-q)^{-1}$. As shown in the preceding paragraph, $(1+q)(1-q)^{-1} \leq \lambda_0 < 2(1-q)^{-1}$. Hence there exists a unique number $\gamma \in [q, 1)$ such that $\lambda_0 = g(\gamma)$. Further,

$$F(g(\lambda)) = (1+\lambda)(1-q)^{-1}q + 1 = (1 + \lambda q)(1-q)^{-1} = g(\lambda q).$$

Hence $F^{\circ n}(g(\lambda)) = g(q^n \lambda)$ for $n \in \mathbb{Z}$ and $\lambda \in \mathbb{R}$, so since $\lambda_0 = g(\gamma) \in \sigma(C^2)$,

$$\lambda_n := F^{\circ n}(\lambda_0) = F^{\circ n}(g(\gamma)) = g(q^n \gamma) = (1 + q^n \gamma)(1-q)^{-1} \in \sigma(C^2).$$

Therefore, $\lambda_n^{1/2} \in \sigma(C)$ for $n \in \mathbb{Z}$.

Because X is irreducible, the spectral measure E_{C^2} is ergodic (Corollary 11.12) and hence supported on the orbit $\mathbb{O}(\lambda_0) = \{\lambda_n : n \in \mathbb{Z}\}$. Since this orbit is discrete, its points are eigenvalues of C^2. We choose a unit vector e_0 such that $C^2 e_0 = \lambda_0 e_0$. Recall that $UC^2 \subseteq F(C^2)U$ and $U^* F(C^2) \subseteq C^2 U^*$ by Theorem 11.27. The first relation implies $U F^{-1}(C^2) \subseteq C^2 U$. From this and the second relation we derive by induction that $U^{*n} F^{\circ n}(C^2) \subseteq C^2 U^{*n}$ for all $n \in \mathbb{Z}$. Hence

$$U^{*n} F^{\circ n}(C^2)e_0 = U^{*n} F^{\circ n}(\lambda_0)e_0 = \lambda_n U^{*n} e_0 = C^2 U^{*n} e_0,$$

so $C^2 e_n = \lambda_n e_n$ for $e_n := U^{*n} e_0, n \in \mathbb{Z}$. Then $U e_n = e_{n-1}$, because U is unitary, and $C e_n = \lambda_n^{1/2} e_n$. From $\lambda_k \neq \lambda_n$ for $k \neq n$ we obtain $e_k \perp e_n$. The closed span $\mathcal{H}_0$

of vectors $e_n, n \in \mathbb{Z}$, is invariant under U, U^*, E_C. Hence, since X is irreducible, we have $\mathcal{H}_0 = \mathcal{H}$ again by Lemma 11.11. Thus $\{e_n : n \in \mathbb{Z}\}$ is an orthonormal basis of $\mathcal{H}$. Now $Xe_k = UCe_n = \lambda_n^{1/2}e_{n-1}$ for $n \in \mathbb{Z}$, so X is unitarily equivalent to the operator X_γ in (ii). (Except for the explicit formula for $F^{on}(\lambda_0)$ this is just a special case of Theorem 11.21.)

Now suppose U is not unitary. Then $\ker U \neq \{0\}$, since $\ker U^* = \{0\}$ by Theorem 11.27. Therefore, it follows from Theorem 11.18 that X is unitarily equivalent to the Fock representation (11.26). Since $F^{ok}(0) = (1 - q^k)(1 - q)^{-1}$ by (11.43), X is unitarily equivalent to the Fock operator X_F in (i). $\square$

Theorem 11.28 shows that the representation theory of the operator relation (11.41) depends essentially on the values of the parameter $q > 0$.

Case $q < 1$: The Fock operator X_F is bounded with norm $\|X_F\| = (1 - q)^{-1/2}$, and all operators $X_\gamma, \gamma \in [q, 1)$, are unbounded. It can be shown [BK91] that the symmetric operator $X_\gamma + (X_\gamma)^*$ is not self-adjoint and has deficiency indices $(1, 1)$.

Case $q > 1$: Then (11.41) has no bounded representation. The Fock representation X_F is unbounded, and it is the only irreducible strong solution of (11.41).

It is instructive to look at these representations from the perspective of the operator $B = (q - 1)C^2 + I$. By (11.41) we have $B = XX^* - X^*X$. The corresponding operators B_F, B_γ, B_θ for X_F, X_γ, X_θ, respectively, are self-adjoint and act by

$$B_F e_j = q^j e_j, \ j \in \mathbb{N}_0, \quad B_\gamma e_k = -q^k \gamma e_k, \ k \in \mathbb{Z}, \ \text{and} \ B_\theta = 0.$$

In particular, $B_F \geq 0$, $\ker B_F = \{0\}$, and $B_\gamma \leq 0$, $\ker B_\gamma = \{0\}$.

In the quantum group literature, the q-oscillator relation appears often in the form

$$YY^* - q^2Y^*Y = (1 - q^2)I. \tag{11.44}$$

For instance, the generator a of the coordinate algebra of the quantum group $SU_q(2)$ satisfies (11.44); see, e.g., [KS97, p. 102, formula (22)]. In fact, both operator relations (11.44) and (11.41) are equivalent; see Exercise 10.

11.8 The Real Quantum Plane

Throughout this section, q is a complex number of **modulus one** such that $q^2 \neq 1$. Our aim is to study the operator relation

$$AB = qBA \tag{11.45}$$

for *self-adjoint* operators A and B on a Hilbert space $\mathcal{H}$.

The relation (11.45) is of the form (11.7) with $F(\lambda) = q\lambda$, But, since q is not real, F does not map $\mathbb{R}$ into itself, so the method of Sect. 11.2 does not apply here. Nevertheless we will develop some "good" representations of relation (11.45).

Recall that $P = -i\frac{d}{dx}$ and $Q = x$ are self-adjoint operators on the Hilbert space $L^2(\mathbb{R})$. Therefore, by the functional calculus, $e^{\alpha Q}$ and $e^{\beta P}$ are positive self-adjoint operators for any $\alpha, \beta \in \mathbb{R}$. In our construction of representations of (11.45) the operator $e^{-\beta P}$ plays a crucial role. The following lemma is based on the classical Paley–Wiener theorem and describes the operator $e^{-\beta P}$ for $\beta > 0$.

Lemma 11.29 (i) *Suppose that $f(z)$ is a holomorphic function on the strip $\mathcal{I}_\beta := \{z \in \mathbb{C} : 0 < \mathrm{Im}\, z < \beta\}$ such that*

$$\sup_{0 < y < \beta} \int_{-\infty}^{+\infty} |f(x + iy)|^2 \, dx < \infty. \tag{11.46}$$

Set $f_y(x) := f(x + iy)$ for $0 < y < \beta$. Then the limits $f_0 := \lim_{y \downarrow 0} f_y$ and $f_\beta := \lim_{y \uparrow \beta} f_y$ exist in $L^2(\mathbb{R})$ and we have $f_0 \in \mathcal{D}(e^{-\beta P})$ and $e^{-\beta P} f_0 = f_\beta$.
(ii) *For each function $f_0 \in \mathcal{D}(e^{-\beta P})$ there exists a unique function f as in (i) such that $f_0 := \lim_{y \downarrow 0} f_y$ in $L^2(\mathbb{R})$ and $e^{-\beta P} f_0 = f_\beta$.*

Proof Recall that the Fourier transform $(\mathcal{F}\varphi)(x) = (2\pi)^{-1/2} \int_\mathbb{R} e^{-ixt} \varphi(t) dt$ is a unitary operator on $L^2(\mathbb{R})$ such that $\mathcal{F} Q \mathcal{F}^{-1} = -P$. Hence $\mathcal{F}\varphi(Q)\mathcal{F}^{-1} = \varphi(-P)$ for any Borel function φ on $\mathbb{R}$. For notational simplicity we set $\beta = 2$.

(i): By the assumptions, $f_1(x) := f(x + i)$ satisfies the assumptions of the Paley–Wiener theorem [Kz68, p. 174]. Hence $\mathcal{F}^{-1} f_1 \in \mathcal{D}(e^{|Q|}) = \mathcal{D}(e^Q) \cap \mathcal{D}(e^{-Q})$ by this theorem. From $\mathcal{F} e^{\pm Q} \mathcal{F}^{-1} = e^{\mp P}$ we obtain $f_1 \in \mathcal{D}(e^{-P}) \cap \mathcal{D}(e^P)$. Setting $g = e^{-P} f_1$ and $f = e^P f_1$, we have $e^{-2P} f = g$. Since $\mathcal{F}^{-1} f_1 \in \mathcal{D}(e^{|Q|})$,

$$(2\pi)^{-1/2} \int e^{-itz} (\mathcal{F}^{-1} f_1)(t) \, dt = (\mathcal{F}(e^{yQ}\mathcal{F}^{-1} f_1))(z), \quad z = x + iy,$$

and $f(z + i) = f_{y+1}(x)$ are holomorphic functions for $|\mathrm{Im}\, z| < 1$. For $y = 0$, both functions are equal to f_1. Hence they coincide on the strip $|\mathrm{Im}\, z| < 1$. Therefore, $f_{y+1} = \mathcal{F} e^{yQ} \mathcal{F}^{-1} f_1 = e^{-yP} f_1$ for $|y| < 1$, that is, $f_y = \mathcal{F} e^{(y-1)Q} \mathcal{F}^{-1} f_1$ for $y \in (0, 2)$. By definition, $f = e^P f_1 = \mathcal{F} e^{-Q} \mathcal{F}^{-1} f_1$. Then

$$\|f - f_y\| = \|\mathcal{F} e^{-Q} \mathcal{F}^{-1} f_1 - \mathcal{F} e^{(y-1)Q} \mathcal{F}^{-1} f_1\| = \|(e^{-Q} - e^{(y-1)Q}) \mathcal{F}^{-1} f_1\|.$$

Now we pass to the limit $y \downarrow 0$. Since $\mathcal{F}^{-1} f_1 \in \mathcal{D}(e^{|Q|})$, Lebesgue's dominated convergence theorem applies and yields $f - f_y \to 0$. Similarly, $g - f_y \to 0$ as $y \uparrow 2$. Thus, $f_0 = f$ and $f_2 = g = e^{-2P} f$, so that $e^{-2P} f_0 = f_2$.

(ii): Since $f_0 \in \mathcal{D}(e^{-2P})$, we have $f_1 := e^P f_0 \in \mathcal{D}(e^P) \cap \mathcal{D}(e^{-P})$ and therefore $\mathcal{F}^{-1} f_1 \in \mathcal{D}(e^{|Q|})$. Then the assertion follows from the converse direction of the Paley–Wiener theorem [Kz68]. $\qquad\square$

If f is as in Lemma 11.29(i), we shall write simply $f(x)$ for $f_0(x)$ and $f(x + i\beta)$ for $f_\beta(x)$. Then Lemma 11.29 says that the operator $e^{-\beta P}$ acts by

$$(e^{-\beta P} f)(x) = f(x + i\beta) \quad \text{for } f \in \mathcal{D}(e^{-\beta P}). \tag{11.47}$$

The case $\beta < 0$ is treated in a similar manner, and (11.47) remains valid in this case. For $\beta = 0$ it is trivial. Hence formula (11.47) holds for all real numbers β.

Corollary 11.30 *Suppose α, $\beta \in \mathbb{R}$. We define self-adjoint operators $A := e^{\alpha Q}$ and $B := e^{-\beta P}$ on the Hilbert space $L^2(\mathbb{R})$. For $f \in \mathcal{D}(AB) \cap \mathcal{D}(BA)$, we have*

$$ABf = e^{-i\alpha\beta} BAf. \tag{11.48}$$

Proof We apply formula (11.47) twice, first to $e^{\alpha x} f$ and then to f, and obtain

$$e^{-i\alpha\beta}(BAf)(x) = e^{-i\alpha\beta} B(e^{\alpha x} f) = e^{-i\alpha\beta} e^{\alpha(x+i\beta)} f(x+i\beta) = (ABf)(x). \qquad \square$$

Let $A = e^{\alpha Q}$ and $B = e^{-\beta P}$ be as in Corollary 11.30. Then the linear space

$$\mathcal{D}_0 = \text{Lin} \{e^{-\varepsilon x^2 + \gamma x} : \varepsilon > 0, \gamma \in \mathbb{C}\} \tag{11.49}$$

is contained in $\mathcal{D}(AB) \cap \mathcal{D}(BA)$. It is invariant under A, B and under the Fourier transform and its inverse. It can be also shown that $\mathcal{D}_0$ is a core for A and B.

Note that if $\alpha\beta = 2\pi k$ for some $k \in \mathbb{Z}$, then $ABf = BAf$ for $f \in \mathcal{D}_0$ by (11.48). That is, the self-adjoint operators A and B commute pointwise on the common invariant core $\mathcal{D}_0$ for A and B. But they do not commute strongly (because their functions $\log A = \alpha Q$ and $\log B = \beta P$ do not commute strongly)! In particular, for any $\alpha \in \mathbb{R}$, $\alpha \neq 0$, the operators $A := e^{\alpha Q}$ and $B := e^{\alpha^{-1} 2\pi P}$ form another couple of Nelson type, as constructed in Example 11.5.

Next we turn to representations of the relation (11.45). If $A = 0$ and B is arbitrary self-adjoint or if A is arbitrary self-adjoint and $B = 0$, then (11.45) holds trivially. We call such representations trivial and omit them in the following discussion.

Now we use Corollary 11.30 to construct representations of the operator relation (11.45) such that $\ker A = \ker B = \{0\}$. First we fix an argument of the number q, $q^2 \neq 1$, of modulus one and define real numbers θ_0, θ_1 by

$$q = e^{-i\theta_0}, \quad \text{with } 0 < |\theta_0| < \pi, \tag{11.50}$$

$$\theta_1 := \theta_0 - \pi \quad \text{if } \theta_0 > 0, \quad \theta_1 := \theta_0 + \pi \quad \text{if } \theta_0 < 0. \tag{11.51}$$

Then $-q = e^{-i\theta_1}$ and $0 < |\theta_1| < \pi$. Further, we choose $\alpha, \beta, \alpha_1, \beta_1 \in \mathbb{R}$ such that $\alpha\beta = \theta_0$ and $\alpha_1\beta_1 = \theta_1$. Let $\mathcal{K}$ be an auxiliary Hilbert space.

Let u and v be commuting self-adjoint unitaries on $\mathcal{K}$. We define self-adjoint operators A, B on the Hilbert space $\mathcal{K} \otimes L^2(\mathbb{R})$ by

$$A = u \otimes e^{\alpha Q}, \quad B = v \otimes e^{-\beta P}, \tag{11.52}$$

and on the Hilbert space $(\mathcal{K} \oplus \mathcal{K}) \otimes L^2(\mathbb{R})$ by the operator matrices

$$A = \begin{pmatrix} I \otimes e^{\alpha_1 Q} & 0 \\ 0 & -I \otimes e^{\alpha_1 Q} \end{pmatrix}, \quad B = \begin{pmatrix} 0 & I \otimes e^{-\beta_1 P} \\ I \otimes e^{-\beta_1 P} & 0 \end{pmatrix}. \tag{11.53}$$

Let $\mathsf{A} := \mathbb{C}\langle a, b | a = a^+, b = b^+, ab = qba \rangle$ be the unital $*$-algebra with hermitian generators a, b and defining relation $ab = qba$. In quantum group theory, this $*$-algebra is the coordinate algebra of the *real quantum plane*. The relation $ab = qba$ appears also in the definition of the quantum group $SL_q(\mathbb{R})$; see, e.g., [Sch94].

Now let $\{A, B\}$ be one of the pairs (11.52) or (11.53). From Corollary 11.30 we easily derive the relation $ABf = qBAf$ for $f \in \mathcal{D}(AB) \cap \mathcal{D}(BA)$. The corresponding dense domains $\mathcal{D}_1 = \mathcal{K} \otimes \mathcal{D}_0$ and $\mathcal{D}_1 = (\mathcal{K} \oplus \mathcal{K}) \otimes \mathcal{D}_0$, where $\mathcal{D}_0$ is given by (11.49), are invariant under A and B and cores for both operators. Then there exists a unique $*$-representation π of the $*$-algebra A on the domain $\mathcal{D}(\pi) := \mathcal{D}_1$ such that $\pi(1) = I$, $\pi(a) = A \upharpoonright \mathcal{D}_1$, $\pi(b) = B \upharpoonright \mathcal{D}_1$. Note that we also have $\ker A = \ker B = \{0\}$, $A\mathcal{D}_1 = \mathcal{D}_1$, and $B\mathcal{D}_1 = \mathcal{D}_1$.

Next we are looking for operator-theoretic characterizations of these pairs $\{A, B\}$. Let $A = U_A|A|$ and $B = U_B|B|$ be the polar decompositions of A and B. It is straightforward to verify that for the pair $\{A, B\}$ from (11.52) we have

$$|A|^{it}B \subseteq e^{\theta_0 t}B|A|^{it}, \ t \in \mathbb{R}, \quad \text{and} \quad U_A B \subseteq BU_A, \tag{11.54}$$

and that the pair $\{A, B\}$ from (11.53) satisfies the relations

$$|A|^{it}B \subseteq e^{\theta_1 t}B|A|^{it}, \ t \in \mathbb{R}, \quad \text{and} \quad U_A B \subseteq -BU_A. \tag{11.55}$$

Further, the relation $U_A B \subseteq \pm BU_A$ holds if and only if $U_A|B| \subseteq |B|U_A$ and $U_A U_B = \pm U_B U_A$.

The above formulas can be taken as the definition of well-behaved representations of the operator relation (11.45). That is, if A and B are self-adjoint operators on a Hilbert space such that $\ker A = \ker B = \{0\}$ and conditions (11.54) or (11.55) hold, we call the pair $\{A, B\}$ a *well-behaved representation* of relation (11.45). Thus, the pairs (11.52) and (11.53) are well-behaved representations of (11.45).

Clearly, the pairs (11.52) and (11.53) are *irreducible* if and only if the space $\mathcal{K}$ has dimension one, that is, $\mathcal{K} = \mathbb{C}$. In this case, $u = \varepsilon_1$, $v = \varepsilon_2$, where $\varepsilon_1, \varepsilon_2 \in \{1, -1\}$. If we assume that $\ker A = \ker B = \{0\}$, then, up to unitary equivalence, there are precisely five such well-behaved irreducible representations of relation (11.45). These are the four pairs $\{A = \varepsilon_1 e^{\alpha Q}, B = \varepsilon_2 e^{-\beta P}\}$, $\varepsilon_1, \varepsilon_2 \in \{1, -1\}$, on $L^2(\mathbb{R})$ and the pair (11.53) on $\mathbb{C}^2 \otimes L^2(\mathbb{R})$. A characterization of these representations in terms of resolvents is given in [OS14].

Note that the phase operators for the pair (11.53), $\mathcal{K} = \mathbb{C}$, are the Pauli matrices

$$U_A = \begin{pmatrix} 1 & 0 \\ 0 & -1 \end{pmatrix}, \quad U_B = \begin{pmatrix} 0 & 1 \\ 1 & 0 \end{pmatrix}.$$

Finally, it should be mentioned that the relation (11.45) leads to interesting and surprising operator-theoretic phenomena. For instance, if a pair $\{A, B\}$ of self-adjoint operators satisfies (11.45) on an invariant core and $\ker A = \ker B = \{0\}$, then there is no dense set of *common* analytic vectors for A and B.

11.9 Exercises

1. Show that two self-adjoint operators A and B on a Hilbert space commute strongly if and only if $E_A(M)B \subseteq BE_A(M)$ for all Borel sets M of $\mathbb{R}$.
   Hint: Verify $E_A(M)(B + iI)^{-1} = (B + iI)^{-1}E_A(M)$, and apply Theorem 11.6.
2. Prove Lemma 11.11.
3. Suppose F is a bounded Borel function on $\mathbb{R}$ and X is a *bounded* operator on a Hilbert space satisfying the relation $XX^* = F(X^*X)$. Show that any two of the operators $(X^*)^j X^j$ and $X^k (X^*)^k$ for $j, k \in \mathbb{N}$ commute. (An operator X obeying this property is called *centered*.)
4. Suppose F is continuous on $\mathbb{R}_+$ and $X = UC$ is a strong solution of (11.13) such that $\ker U^* = \{0\}$. Prove that F maps the spectrum of C^2 into itself.
5. Treat the operator relation $XX^* = X^*X + I$ with the methods of Sects. 11.3 and 11.5, and compare the outcome with the results of Sect. 8.2.
6. Let $q \in (0, 1)$. Let U be a unitary and A a self-adjoint operator on a Hilbert space $\mathcal{H}$ such that $A \geq 0$, $\ker A = \{0\}$. Suppose that $UA \subseteq qAU$. Show that, up to unitary equivalence, the pair $\{U, A\}$ is of the following form:
   There is a self-adjoint operator B on a Hilbert space $\mathcal{K}$ such that $\sigma(B) \subseteq [q, 1]$, $\ker(B - qI) = \{0\}$, $\mathcal{H} = \oplus_{n\in\mathbb{Z}}\mathcal{K}$, and $U(\varphi_n) = (\varphi_{n-1})$, $A(\eta_n) = (q^n B\eta_n)$ for $(\varphi_n) \in \mathcal{H}$ and $(\eta_n) \in \mathcal{D}(A)$.
7. Suppose $q > 0$. Show that a densely defined closed operator X is q-normal if and only if $\mathcal{D}(X) = \mathcal{D}(X^*)$ and $\|X^* f\| = q^{1/2}\|Xf\|$ for $f \in \mathcal{D}(X)$.
8. Let X_μ be the q-normal operator defined by (11.36), and let φ_0 denote the characteristic function of the set Δ_q. Show that if $\mu(\{0\}) = 0$, then the linear span of $\{(X_\mu^*)^m X_\mu^n \varphi_0 : m, n \in \mathbb{N}_0\}$ is dense in $\mathcal{H}_\mu$.
9. Write the q-normal operator X in (11.38) as an operator X_μ of the form (11.36).
10. Show that the relations (11.41) and (11.44) for $q > 0, q \neq 1$, are equivalent.
    Hint: Transform $Y = \alpha X + \beta I$ or $Y = \alpha X^* + \beta I$ and q if necessary.
11. Show the operator relation $XX^* = q^{-1}(X^*X - I)$, where $q > 0, q \neq 1$, has an anti-Fock representation as in Theorem 11.19.
12. Suppose $q > 0, q \neq 1$. Use the operator $B = (q - 1)C^2 + I$ and Theorem 11.27 to derive the classification of irreducible representations of Theorem 11.28.
13. Suppose $q \in (0, 1)$. Let S be the unilateral shift operator on $l^2(\mathbb{N}_0)$. Show that:

    a. $T := \sum_{n=1}^{\infty} q^{n-1} S^n S^{*n}$ is a bounded positive self-adjoint operator on $l^2(\mathbb{N}_0)$.
    b. $S^*TS = I + qT$ and $Te_0 = 0$, $Te_k = \frac{1-q^k}{1-q}e_k$ for $k \in \mathbb{N}$.
    c. $X := S^*T^{1/2}$ is the polar decomposition of X and $XX^* = qX^*X + I$.
    d. X is the Fock operator X_F from Theorem 11.28(i).

  e. The C^*-algebra generated by the Fock operator X_F is equal to the C^*-algebra generated by the shift operator S; that is, it is the Toeplitz algebra (cf. [Dv96, Theorem V.1.5]).

14. Suppose $q \in \mathbb{C}$, $|q| = 1$, $q^2 \neq 1$, and fix a square root $q^{1/2}$ of q. Let a, b be hermitian elements of a unital complex $*$-algebra A such that a is invertible. Define $c := (q^{-1/2}b + 1)a^{-1}$. Show that the following are equivalent:

  (i) $ab = qba$.
  (ii) $ac - qca = 1 - q$.
  (iii) $c = c^+$.

15. Suppose q is as in Exercise 14. Use the operator pairs $\{A, B\}$ from (11.52) and (11.53) to construct symmetric operators A, C satisfying the operator relation $AC - qCA = (1 - q)I$. Find a dense domain $\mathcal{D} \subseteq \mathcal{D}(AC) \cap \mathcal{D}(CA)$ such that $\mathcal{D}$ is invariant under A, C and $AC\varphi - qCA\varphi = (1 - q)\varphi$ for $\varphi \in \mathcal{D}$. (Further results on these operators can be found in [Sch94].)

11.10 Notes

The representation theory of finitely many operator relations with finitely many operators was developed by the Kiev School, notably by Yu.S. Samoilenko, V.L. Ostrovskyi, and E. Ye. Vaisleb. Important pioneering papers include [OS88, OS89, VS90, VS94, O96, STS96]. This theory is treated extensively in the monograph [OS99] by Ostrovskyi and Samoilenko, which contains detailed lists of representations for many relations and also an excellent bibliography. Further, for some basic relations operator-theoretic models are developed. We have given only a glimpse into this theory by treating a single but important relation.

For power partial isometries we refer to [HW63] and for centered operators to [MM74]. The hermitian q-plane and q-normal operators were studied in [O02, OS04, OS07, CSS14], the q-oscillator algebra in [CGP], and the real quantum plane in [Sch94, OS14].

An interesting related topic is the so-called q-deformed commutation relations. They are not treated in this book, but they are studied in many research papers; see, e.g., [BS94, JSW95, OPT08, Pr08, BLW17].

Chapter 12
Induced *-Representations

Induced representations are a fundamental tool in representation theory of groups and algebras. They were first defined and investigated for finite groups by G. Frobenius (1898), for algebras by D.G. Higman (1955), and for C^*-algebras by M. Rieffel and J.M.G. Fell. In this chapter, we develop this method for general *-algebras.

If B is a subalgebra of an algebra A and V is a left B-module, the induced module of V is the left A-module $A \otimes_B V$, where the left action is given by the formula $a\left(\sum_j a_j \otimes v_j\right) = \sum_j aa_j \otimes v_j$. If B is a *-subalgebra of a *-algebra A and the left module V stems from a *-representation of B, we shall try to make the induced module into a *-representation of A. This requires additional technical tools.

Conditional expectations are used to define inner products on induced modules, but they are also of interest in themselves. In Sect. 12.1, we investigate this concept and develop a number of examples, and all of them are coming from groups. In Sect. 12.2, we define induced representations. Section 12.3 is concerned with G-graded *-algebras $A = \sum_{g \in G} A_g$ for which $B := A_e$ is commutative. We study induced representations from characters of B that are nonnegative on $B \cap \sum A^2$.

In this chapter, B is a **unital *-subalgebra of a unital complex *-algebra** A. The unit elements of B and A are denoted by 1_B and 1_A, respectively.

To shorten formulas we often write a_j^+ instead of $(a_j)^+$ and a_g^+ instead of $(a_g)^+$.

12.1 Conditional Expectations

The following definition collects a number of basic notions for this chapter.

Definition 12.1 A B-*bimodule projection* of A on B is a linear map $\Phi : A \mapsto B$ such that for $a \in A$ and $b_1, b_2 \in B$,

© The Editor(s) (if applicable) and The Author(s), under exclusive license to Springer Nature Switzerland AG 2020

K. Schmüdgen, *An Invitation to Unbounded Representations of *-Algebras on Hilbert Space*, Graduate Texts in Mathematics 285, https://doi.org/10.1007/978-3-030-46366-3_12

$$\Phi(1_A) = 1_B, \quad \Phi(a^+) = \Phi(a)^+, \quad \Phi(b_1 a b_2) = b_1 \Phi(a) b_2. \tag{12.1}$$

A B-bimodule projection Φ is *faithful* if $\Phi(a^+ a) = 0$ for $a \in A$ implies $a = 0$.

A B-bimodule projection Φ of A on B is called a *conditional expectation* if

$$\Phi\left(\sum A^2\right) \subseteq \sum A^2 \tag{12.2}$$

and a *strong conditional expectation* if

$$\Phi\left(\sum A^2\right) \subseteq \sum B^2. \tag{12.3}$$

Let $\Phi : A \mapsto B$ be a linear map satisfying (12.1). Then, for $b \in B$, we have $\Phi(b) = \Phi(b 1_A 1_B) = b 1_B 1_B = b$; that is, Φ is indeed a *projection of A on B*.

The next lemma characterizes B-bimodule projections in terms of their kernels.

Lemma 12.2 *There exists a B-bimodule projection of A on B if and only if there exists a $*$-invariant subspace $\mathcal{T}$ of A such that $A = B \oplus \mathcal{T}$ and*

$$B \mathcal{T} B \subseteq \mathcal{T}. \tag{12.4}$$

If this is true, the B-bimodule projection Φ is uniquely defined by the requirement $\ker \Phi = \mathcal{T}$ and we have

$$\Phi\left(\sum A^2\right) = \sum B^2 + \Phi\left(\sum \mathcal{T}^2\right). \tag{12.5}$$

Proof Let Φ be a B-bimodule projection of A on B and put $\mathcal{T} = \ker \Phi$. For $t \in \mathcal{T}$ and $b_1, b_2 \in B$, we have $\Phi(b_1 t b_2) = b_1 \Phi(t) b_2 = 0$ and $\Phi(t^+) = \Phi(t)^+ = 0$. Hence the space $\mathcal{T}$ satisfies (12.4) and it is $*$-invariant. For $a \in A$, we have $\Phi(a) \in B$ and $\Phi(a - \Phi(a)) = \Phi(a) - \Phi(\Phi(a)) = 0$, so $a - \Phi(a) \in \mathcal{T}$. If $b \in B \cap \mathcal{T}$, then $0 = \Phi(b) = b$. This shows that $A = B \oplus \mathcal{T}$. Further, for $a = b + t$ with $b \in B$ and $t \in \mathcal{T}$, we obtain

$$\Phi(a^+ a) = \Phi((b + t)^+(b + t)) = \Phi(b^+ b) + \Phi(b^+ t) + \Phi(t^+ b) + \Phi(t^+ t)$$
$$= b^+ b + b^+ \Phi(t) + \Phi(t)^+ b + \Phi(t^+ t) = b^+ b + \Phi(t^+ t),$$

which proves (12.5).

Conversely, if $\mathcal{T}$ is given as above, one easily checks that the linear map Φ defined by $\Phi(b) = b$, $b \in B$, and $\Phi(t) = 0$, $t \in \mathcal{T}$, is a B-bimodule projection. $\qquad\square$

Example 12.3 (*Weyl algebra*)

Let $A := \mathbb{C}\langle p, q \,|\, p = p^+, q = q^+, pq - qp = -i \rangle$ be the Weyl algebra (Example 2.12). We show that *there is no B-bimodule projection of A on B* $:= \mathbb{C}[p]$.

Assume to the contrary that there is such a projection Φ and let $\mathcal{T}$ be its kernel. Then, since $A = B \oplus \mathcal{T}$, there exists a polynomial f such that $q + f(p) \in \mathcal{T}$. By

(12.4) we have $p(q + f(p))1_B \in \mathcal{T}$ and $1_B(q + f(p))p \in \mathcal{T}$ which implies that $pq - qp = -i\,1_A \in \mathcal{T}$. Hence $1_A \in \mathcal{T}$, so that $\Phi = 0$, which is a contradiction.

In Example 12.8 below we will see that there exists a unique B-bimodule projection of A on $B := \mathbb{C}[N]$, where $N = a^+a$. $\quad\bigcirc$

Example 12.4 (*States as conditional expectations*)
Let f be a hermitian functional on A such that $f(1) = 1$. Then $\Phi(a) := f(a) \cdot 1$, $a \in A$, is a B-bimodule projection of A on $B := \mathbb{C} \cdot 1$. Clearly, Φ is a conditional expectation if and only if $f(a^+a) \geq 0$ for $a \in A$; that is, f is a state on A. $\quad\bigcirc$

In the remaining part of this section we develop various methods for constructing conditional expectations.

Example 12.5 (*Group graded ∗-algebras*)
Suppose $A = \bigoplus_{g \in G} A_g$ is a unital G-graded ∗-algebra (Definition 2.19) and H is a subgroup of G. Then it follows from (2.20) that $A_H := \bigoplus_{h \in H} A_h$ is a unital ∗-subalgebra of A. Let Φ_H denote the canonical projection of A on A_H, that is, $\Phi_H(a) = \sum_{h \in H} a_h$ for $a = \sum_{g \in G} a_g \in A$, where $a_g \in A_g$ for $g \in G$.

For a subset X of G, let A_X denote the vector space $\sum_{g \in X} A_g$.

Proposition 12.6 *The map Φ_H is a conditional expectation of A on A_H.*

Proof The three conditions of (12.1) follow at once from (2.20). We prove (12.2).

We choose one element $k_i \in G$, $i \in G/H$, in each left coset of H in G. Let $a \in A$. We can write $a = \sum_{i \in G/H} b_i$, where $b_i \in A_{k_i H}$. If $i, j \in G/H$, then $b_j^+ b_i \in A_{Hk_j^{-1}k_i H}$, hence $\Phi_H(b_i^+ b_i) = b_i^+ b_i$ and $\Phi_H(b_j^+ b_i) = 0$ if $i \neq j$. Using these facts we derive

$$\Phi_H(a^+a) = \Phi_H\left(\sum_{j \in G/H}\sum_{i \in G/H} b_j^+ b_i\right) = \sum_{i \in G/H} b_i^+ b_i \in \sum A^2. \qquad (12.6)$$

Thus, $\Phi_H(\sum A^2) \subseteq \sum A^2$, so Φ_H is a conditional expectation. $\qquad\square$

Corollary 12.7 *An element $x \in A_H$ belongs to $\sum A^2$ if and only if x is a finite sum $\sum b_i^+ b_i$, where each b_i belongs to some vector space A_{gH} for $gH \in G/H$.*

Proof The if part is trivial. We prove the only if part. Suppose $x \in A_H \cap \sum A^2$. Then $x = \sum_k a_k^+ a_k$, where $a_k \in A$. Since $x = \Phi_H(x) = \sum_k \Phi(a_k^+ a_k)$, it follows from (12.6), applied with $a = a_k$, that x is of the described form. $\qquad\square$

From Eq. (12.6) it follows that Φ_H is faithful when $\sum_{i=1}^n b_i^+ b_i = 0$ for elements $b_1, \ldots, b_n \in A$ implies that $b_1 = \cdots = b_n = 0$. This holds if A has a faithful ∗-representation (Corollary 4.45) and in particular if A is an O^*-algebra. Thus, if A is an O^*-algebra, the conditional expectation Φ_H is faithful.

An important special case is $H = \{e\}$. Then the map $\Phi := \Phi_H$ is called the *canonical conditional expectation* of the G-graded ∗-algebra A on $B = A_e$.

In this case, Corollary 12.7 says that an element x is in $\mathsf{B} \cap \sum \mathsf{A}^2$ if and only if x is of the form $x = \sum_i b_i^+ b_i$, where $b_i \in \mathsf{A}_{g_i}$ for all i. In other words, if $x \in \sum \mathsf{A}^2$ belongs to B, then x can be represented as a finite sum of hermitian squares $b_i^+ b_i$ of *homogeneous* elements b_i. $\bigcirc$

We illustrate the preceding for our guiding example.

Example 12.8 (*Weyl algebra*)
In this example we use the Weyl algebra in the form $\mathsf{A} = \mathbb{C}\langle a, a^+ | aa^+ - a^+ a = 1 \rangle$. Recall from Example 2.23 that A is $\mathbb{Z}$-graded with $\mathbb{Z}$-grading determined by $a \in \mathsf{A}_1$, $a^+ \in \mathsf{A}_{-1}$ and we have $\mathsf{B} = \mathsf{A}_0 = \mathbb{C}[N]$, where $N = a^+ a$.

Then the canonical conditional expectation $\Phi : \mathsf{A} \mapsto \mathsf{B}$ acts by

$$\Phi\left(p_0(N) + \sum_{k \geq 1} \left(a^k p_k(N) + (a^k)^+ p_{-k}(N) \right) \right) = p_0(N), \qquad (12.7)$$

where $p_j(N) \in \mathbb{C}[N]$ for $j \in \mathbb{Z}$.

We want to apply Corollary 12.7 to describe the cone $\mathsf{B} \cap \sum \mathsf{A}^2$. Let $k \in \mathbb{N}_0$. Then $a_k \in \mathsf{A}_k$ is of the form $a_k = a^k p_k$, where $p_k \in \mathbb{C}[N]$, and by (8.13),

$$a_k^+ a_k = p_k^+ (a^+)^k a^k p_k = N(N-1) \cdots (N-k+1) p_k^+ p_k.$$

Each $a_{-k} \in \mathsf{A}_{-k}$ is of the form $a_{-k} = (a^+)^k p_{-k}$, with $p_{-k} \in \mathbb{C}[N]$, so by (8.14),

$$(a_{-k})^+ a_{-k} = (p_{-k})^+ a^k (a^+)^k p_{-k} = (N+1)(N+2) \cdots (N+k)(p_{-k})^+ p_{-k}.$$

By induction on k one can show that each product $(N+1)(N+2) \cdots (N+k)$ belongs to $\sum \mathsf{B}^2 + N \sum \mathsf{B}^2$, so that $(a_{-k})^+ a_{-k} \in \sum \mathsf{B}^2 + N \sum \mathsf{B}^2$. Therefore, it follows from the preceding and Corollary 12.7, applied in the case $H = \{e\}$, that

$$\mathsf{B} \cap \sum \mathsf{A}^2 = \sum \mathsf{B}^2 + N \sum \mathsf{B}^2 + N(N-1) \sum \mathsf{B}^2 + \cdots \qquad (12.8)$$

It is clear that the element $N = a^+ a \in \mathsf{B} \cap \sum \mathsf{A}^2$ does *not* belong to $\sum \mathsf{B}^2$. Hence, $\mathsf{B} \cap \sum \mathsf{A}^2 \neq \sum \mathsf{B}^2$ and the canonical conditional expectations $\Phi : \mathsf{A} \mapsto \mathsf{B}$ are not strong. $\bigcirc$

Example 12.9 (*Group algebras*)
Suppose G is a group and H is a subgroup of G. Let $\mathsf{A} = \mathbb{C}[G]$ and $\mathsf{B} = \mathbb{C}[H]$ be the corresponding group algebras (Definition 2.11), and let Φ be the canonical projection of $\mathbb{C}[G]$ on $\mathbb{C}[H]$ defined by $\Phi(g) = g$ if $g \in H$ and $\Phi(g) = 0$ if $g \notin H$.

Proposition 12.10 Φ *is a strong conditional expectation of* $\mathbb{C}[G]$ *on* $\mathbb{C}[H]$.

Proof It is clear that Φ is a $\mathbb{C}[H]$-bimodule projection.

We prove that $\Phi(\sum \mathbb{C}[G]^2) \subseteq \sum \mathbb{C}[H]^2$. We fix precisely one element $k_i \in G$ in each left coset $i \in G/H$. Let $a = \sum_{g \in G} \alpha_g g$, $\alpha_g \in \mathbb{C}$, be an element of the group algebra $\mathbb{C}[G]$. Then there exist elements $a_i \in \mathbb{C}[H]$, $i \in G/H$, such that $a = \sum_{g \in G} \alpha_g g = \sum_{i \in G/H} k_i a_i$. If $i, j \in G/H$ and $i \neq j$, then $k_i^{-1} k_j \notin H$ and hence $\Phi(k_i^{-1} k_j) = 0$. Obviously, $\Phi(k_i^{-1} k_i) = e$. Therefore,

$$\Phi(a^+ a) = \Phi\left(\left(\sum_{i \in G/H} k_i a_i\right)^+ \left(\sum_{j \in G/H} k_j a_j\right)\right) = \Phi\left(\sum_{i,j \in G/H} a_i^+ k_i^{-1} k_j a_j\right)$$

$$= \sum_{i,j \in G/H} \Phi(a_i^+ k_i^{-1} k_j a_j) = \sum_{i,j \in G/H} a_i^+ \Phi(k_i^{-1} k_j) a_j = \sum_{i \in G/H} a_i^+ a_i.$$

Thus, $\Phi(a^+ a) \in \sum \mathbb{C}[H]^2$ and Φ is a strong conditional expectation. □ ○

Example 12.11 (*Conditional expectations from groups of ∗-automorphisms*)
Suppose G is a compact group which acts as a group of ∗-automorphisms $\theta_g, g \in G$, on A. We assume in addition that the action is *locally finite-dimensional*; that is, for every $a \in$ A there exists a finite-dimensional linear subspace $V \subset$ A such that $a \in V$, $\theta_g(V) \subseteq V$ for all $g \in G$, and the map $g \mapsto \theta_g(a)$ of G into V is continuous. Let μ denote the Haar measure of G. Then the mapping Φ given by

$$\Phi(a) = \int_G \theta_g(a) \, d\mu(g), \quad a \in \mathsf{A}, \tag{12.9}$$

is well defined. One easily verifies that Φ is a B-bimodule projection of A on the ∗-subalgebra $\mathsf{B} := \{a \in \mathsf{A} : \theta_g(a) = a \text{ for all } g \in G\}$ of stable elements.

Using the representation theory of compact groups it can be shown [SS13] that $\Phi(\sum \mathsf{A}^2) \subseteq \sum \mathsf{A}^2$. Therefore, Φ is a conditional expectation of A on B. ○

Example 12.12 (*Crossed product algebras*)
Let B be a unital complex ∗-algebra and G a discrete group which acts as a group of ∗-automorphisms $g \mapsto \theta_g$ on B. Then, as developed in Sect. 2.2, the crossed product ∗-algebra $\mathsf{A} := \mathsf{B} \times_\theta G$ is defined. As a vector space, A is the complex tensor product $\mathsf{B} \otimes \mathbb{C}[G]$. The product and involution of A (see (2.16)) are given by

$$(a \otimes g)(b \otimes h) = a\theta_g(b) \otimes gh \quad \text{and} \quad (a \otimes g)^+ = \theta_{g^{-1}}(a^+) \otimes g^{-1}. \tag{12.10}$$

Then, A is a G-graded ∗-algebra with $\mathsf{A}_g := \mathsf{A} \otimes g$, $g \in G$, and the canonical conditional expectation Φ on $\mathsf{A}_e = \mathsf{B} \otimes e \cong \mathsf{B}$ is defined by $\Phi(a \otimes e) = a \otimes e$ and $\Phi(a \otimes g) = 0$ if $g \neq e$.

Proposition 12.13 *The conditional expectation* $\Phi : \mathsf{A} \mapsto \mathsf{B}$ *is strong.*

Proof Let $x = \sum_{g \in G} a_g \otimes g \in \mathsf{A}$. Using the formulas (12.10) we compute

$$\Phi(xx^+) = \Phi\left(\sum_{g,h\in G} (a_g \otimes g)(a_h \otimes h)^+\right) = \Phi\left(\sum_{g,h\in G} a_g\theta_{gh^{-1}}(a_h^+) \otimes gh^{-1}\right)$$

$$= \sum_{g\in G} a_g a_g^+ \otimes e = \sum_{g\in G} (a_g \otimes e)(a_g \otimes e)^+ \in \sum \mathsf{B}^2.$$

$\square\,\bigcirc$

12.2   Induced *-Representations

For the moment, let B be a subalgebra of an algebra A and V a left B-module. Then there is a left A-action on the tensor product $\mathsf{A} \otimes_{\mathsf{B}} V$ given by $a_0(a \otimes v) := a_0 a \otimes v$. The corresponding left A-module $\mathsf{A} \otimes_{\mathsf{B}} V$ is called the *induced module* of V. If in addition B is a *-subalgebra of a *-algebra A and V is a complex inner product space on which B acts as a *-representation, we want to define an inner product on $\mathsf{A} \otimes_{\mathsf{B}} V$, or on some quotient space, such that the left action of A becomes a *-representation. This is where a conditional expectation of A on B is used.

Throughout this section, we suppose that Φ is a **conditional expectation** of the unital complex *-algebra A on its unital *-subalgebra B.

For a *-representation ρ of B, we denote by $\mathsf{A} \otimes_{\mathsf{B}} \mathcal{D}(\rho)$ the quotient of the complex tensor product $\mathsf{A} \otimes \mathcal{D}(\rho)$ by the subspace

$$\mathcal{N}_\rho = \left\{ \sum_{k=1}^{r} x_k b_k \otimes \varphi_k - \sum_{k=1}^{r} x_k \otimes \rho(b_k)\varphi_k : x_k \in \mathsf{A}, \, b_k \in \mathsf{B}, \, \varphi_k \in \mathcal{D}(\rho), \, r \in \mathbb{N} \right\}.$$

Lemma 12.14 *Let ρ be an arbitrary *-representation of B on $(\mathcal{D}(\rho), \langle\cdot,\cdot\rangle)$. Then:*

(i) $\left\langle \sum_k x_k \otimes \varphi_k, \sum_l y_l \otimes \psi_l \right\rangle_0 := \sum_{k,l} \langle \rho(\Phi(y_l^+ x_k))\varphi_k, \psi_l \rangle,$ (12.11)

where $x_k, y_l \in \mathsf{A}$ and $\varphi_k, \psi_l \in \mathcal{D}(\rho)$, gives a well-defined hermitian sesquilinear form $\langle\cdot,\cdot\rangle_0$ on the complex vector space $\mathsf{A} \otimes_{\mathsf{B}} \mathcal{D}(\rho)$.

(ii) *There is a well-defined algebra homomorphism $\pi_0 : \mathsf{A} \to L(\mathsf{A} \otimes_{\mathsf{B}} \mathcal{D}(\rho))$,*

$$\pi_0(a)\left(\sum_k x_k \otimes \varphi_k\right) = \sum_k ax_k \otimes \varphi_k,$$ (12.12)

where $a \in \mathsf{A}, x_k \in \mathsf{A}, \varphi_k \in \mathcal{D}(\rho)$, satisfying

$$\langle \pi_0(a)\zeta, \eta \rangle_0 = \langle \zeta, \pi_0(a^+)\eta \rangle_0 \quad \text{for} \quad a \in \mathsf{A}, \, \zeta, \eta \in \mathsf{A} \otimes_{\mathsf{B}} \mathcal{D}(\rho).$$ (12.13)

Proof (i): Obviously, $\langle\cdot,\cdot\rangle_0$ defines a sesquilinear form on the complex tensor product $\mathsf{A} \otimes \mathcal{D}(\rho)$. To prove that $\langle\cdot,\cdot\rangle_0$ is also well defined on the quotient space $\mathsf{A} \otimes_{\mathsf{B}} \mathcal{D}(\rho) = (\mathsf{A} \otimes \mathcal{D}(\rho))/\mathcal{N}_\rho$ it suffices to show that $\langle \zeta, \eta \rangle_0 = 0$ and

$\langle \eta, \zeta \rangle_0 = 0$ for arbitrary vectors

$$\eta = \sum_j y_j \otimes \psi_j \in A \otimes \mathcal{D}(\rho), \quad \zeta = \sum_k x_k b_k \otimes \varphi_k - \sum_k x_k \otimes \rho(b_k)\varphi_k \in \mathcal{N}_\rho.$$

Using that $\Phi(y_l^+ x_k b_k) = \Phi(y_l^+ x_k)b_k$ by condition (12.1) we obtain

$$\sum_{k,l} \langle \rho(\Phi(y_l^+ x_k b_k))\varphi_k, \psi_l \rangle = \sum_{k,l} \langle \rho(\Phi(y_l^+ x_k))\rho(b_k)\varphi_k, \psi_l \rangle.$$

By (12.11), this implies $\langle \zeta, \eta \rangle_0 = 0$. Similarly, $\langle \eta, \zeta \rangle_0 = 0$.
Now let $\eta := \sum_k x_k \otimes \varphi_k$ and $\xi := \sum_l y_l \otimes \psi_l$ be vectors of $A \otimes_B \mathcal{D}(\rho)$. Using the relation $\Phi(a)^+ = \Phi(a^+)$, $a \in A$, we compute

$$\langle \eta, \xi \rangle_0 = \sum_{k,l} \langle \rho(\Phi(y_l^+ x_k))\varphi_k, \psi_l \rangle = \sum_{k,l} \langle \varphi_k, \rho(\Phi(y_l^+ x_k)^+)\psi_l \rangle$$
$$= \sum_{k,l} \langle \varphi_k, \rho(\Phi(x_k^+ y_l))\psi_l \rangle = \sum_{k,l} \overline{\langle \rho(\Phi(x_k^+ y_l))\psi_l, \varphi_k \rangle} = \overline{\langle \xi, \eta \rangle_0},$$

which proves that the sesquilinear form $\langle \cdot, \cdot \rangle_0$ on $A \otimes_B \mathcal{D}(\rho)$ is hermitian.

(ii): For $\sum_k x_k b_k \otimes \varphi_k \in A \otimes_B \mathcal{D}(\rho)$ and $a \in A$, we have

$$\pi_0(a)\Big(\sum_k x_k b_k \otimes \varphi_k\Big) = \sum_k a(x_k b_k) \otimes \varphi_k = \sum_k (ax_k)b_k \otimes \varphi_k$$
$$- \sum_k ax_k \otimes \rho(b_k)\varphi_k = \pi_0(a)\Big(\sum_k x_k \otimes \rho(b_k)\varphi_k\Big).$$

Therefore, the operator $\pi_0(a)$ is well defined on the quotient space $A \otimes_B \mathcal{D}(\rho)$. It is obvious that π_0 is a homomorphism of the algebra A into the algebra of linear operators on the vector space $A \otimes_B \mathcal{D}(\rho)$. For $a, x \in A$ and $\varphi, \psi \in \mathcal{D}(\rho)$,

$$\langle \pi_0(a)(x \otimes \varphi), y \otimes \psi \rangle_0 = \langle ax \otimes \varphi, y \otimes \psi \rangle_0 = \langle \rho(\Phi(y^+ ax))\varphi, \psi \rangle$$
$$= \langle \varphi, \rho(\Phi(y^+ ax)^+)\psi \rangle = \overline{\langle \rho(\Phi(x^+(a^+ y)))\psi, \varphi \rangle}$$
$$= \overline{\langle a^+ y \otimes \psi, x \otimes \varphi \rangle_0} = \langle x \otimes \varphi, \pi_0(a^+)(y \otimes \psi) \rangle_0.$$

By linearity this yields (12.13). □

Note that in Lemma 12.14 neither the positivity of the inner product of $\mathcal{D}(\rho)$ nor the positivity condition (12.2) was used. Now we turn to the case when the form $\langle \cdot, \cdot \rangle_0$ is positive semi-definite.

Lemma 12.15 *Suppose ρ is a ∗-representation of B such that the sesquilinear form $\langle \cdot, \cdot \rangle_0$ on $A \otimes_B \mathcal{D}(\rho)$, defined by Eq. (12.11), is positive semi-definite, that is, $\langle \eta, \eta \rangle_0 \geq 0$ for $\eta \in A \otimes_B \mathcal{D}(\rho)$. Let $\langle \cdot, \cdot \rangle$ denote the inner product on the quotient vector space $\mathcal{D}(\pi_0) := (A \otimes_B \mathcal{D}(\rho))/\mathcal{K}_\rho$ defined by $\langle [\eta], [\zeta] \rangle = \langle \eta, \zeta \rangle_0$, where $\mathcal{K}_\rho := \{\eta : \langle \eta, \eta \rangle_0 = 0\}$ and $[\eta]$ denotes the equivalence class $[\eta] := \eta + \mathcal{K}_\rho$. Then,*

$$\pi_0(a)[\eta] = [\pi_0(a)\eta], \ a \in \mathsf{A}, \ \eta \in \mathsf{A} \otimes_\mathsf{B} \mathcal{D}(\rho),$$

*defines a *-representation* π_0 *of* A *on the unitary space* $(\mathcal{D}(\pi_0), \langle \cdot, \cdot \rangle)$. *Let* π *denote the closure of* π_0.

Proof Because of Lemma 12.14 it suffices to check that $\pi_0(a)$ is well defined on $\mathcal{D}(\pi_0)$, that is, $\pi_0(a)\mathcal{K}_\rho \subseteq \mathcal{K}_\rho$. Let $\eta \in \mathcal{K}_\rho$. Using (12.13) and the Cauchy–Schwarz inequality for the positive semi-definite sesquilinear form $\langle \cdot, \cdot \rangle_0$ we obtain

$$\langle \pi_0(a)\eta, \pi_0(a)\eta \rangle_0 = \langle \eta, \pi_0(a^+)\pi_0(a)\eta \rangle_0 = \langle \eta, \pi_0(a^+a)\eta \rangle_0$$
$$\leq \langle \eta, \eta \rangle_0^{1/2} \langle \pi_0(a^+a)\eta, \pi_0(a^+a)\eta \rangle_0^{1/2} = 0,$$

so that $\pi_0(a)\eta \in \mathcal{K}_\rho$. $\qquad\qquad\qquad\qquad\qquad\qquad\qquad\qquad\qquad\qquad\qquad\square$

Definition 12.16 A *-representation ρ of B is called *inducible* (from B to A with respect to the conditional expectation Φ) if the form (12.11) is positive semi-definite. In this case, we say the closed *-representation π of A in Lemma 12.15 is *induced* from the *-representation ρ and denote π by $\mathrm{Ind}_{\mathsf{B}\uparrow\mathsf{A}}\,\rho$ or simply by $\mathrm{Ind}\,\rho$.

The next proposition describes a class of inducible representations.

Let $\mathrm{Rep}_c^+\,\mathsf{B}$ denote the family of *-representations of B that are direct sums of cyclic *-representations ρ_i with cyclic vectors ξ_i such that $\langle \rho_i(c)\xi_i, \xi_i \rangle \geq 0$ for $c \in \mathsf{B} \cap \sum \mathsf{A}^2$.

Proposition 12.17 *Each *-representation* ρ *of* $\mathrm{Rep}_c^+\,\mathsf{B}$ *is inducible.*

Proof By Lemma 12.15 and Definition 12.16 it suffices to prove that the form $\langle \cdot, \cdot \rangle_0$ is positive semi-definite for ρ in $\mathrm{Rep}_c^+\,\mathsf{B}$.

Assume first ρ is cyclic with cyclic vector ξ. Let $\eta = \sum_{k=1}^n x_k \otimes \varphi_k \in \mathsf{A} \otimes_\mathsf{B} \mathcal{D}(\rho)$ and fix $\varepsilon > 0$. That ξ is cyclic means that $\rho(\mathsf{B})\xi$ is dense in $\mathcal{D}(\rho)$ in the graph topology of ρ. Hence there are elements $b_1, \ldots, b_n \in \mathsf{B}$ such that $\|\rho(b_k)\xi - \varphi_k\| < \varepsilon$ and $\|\rho(\Phi(x_l^+ x_k))(\rho(b_k)\xi - \varphi_k)\| < \varepsilon$ for all $k, l = 1, \ldots, n$. Then

$$|\langle \rho(\Phi(x_l^+ x_k))\varphi_k, \varphi_l \rangle - \langle \rho(\Phi(x_l^+ x_k))\rho(b_k)\xi, \rho(b_l)\xi \rangle|$$
$$= |\langle \rho(\Phi(x_l^+ x_k))\varphi_k, \varphi_l - \rho(b_l)\xi \rangle - \langle \rho(\Phi(x_l^+ x_k))(\rho(b_k)\xi - \varphi_k), \rho(b_l)\xi \rangle|$$
$$\leq \|\rho(\Phi(x_l^+ x_k))\varphi_k\|\,\varepsilon + \|\rho(b_l)\xi\|\,\varepsilon \leq \|\rho(\Phi(x_l^+ x_k))\varphi_k\|\,\varepsilon + \|\varphi_l\|\,\varepsilon + \varepsilon^2.$$

Now we set $a := \sum_k x_k b_k \in \mathsf{A}$. From the preceding inequality we conclude that $\langle \eta, \eta \rangle_0 = \sum_{k,l=1}^n \langle \rho(\Phi(x_l^+ x_k))\varphi_k, \varphi_l \rangle$ can be approximated as small as we want by

$$\sum_{k,l=1}^n \langle \rho(\Phi(x_l^+ x_k))\rho(b_k)\xi, \rho(b_l)\xi \rangle = \sum_{k,l=1}^n \langle \rho(b_l^+ \Phi(x_l^+ x_k)b_k)\xi, \xi \rangle$$

$$= \sum_{k,l=1}^n \langle \rho(\Phi((x_l b_l)^+ x_k b_k))\xi, \xi \rangle = \langle \rho(\Phi(a^+ a))\xi, \xi \rangle.$$

Since $a^+a \in \sum A^2$ and hence $\Phi(a^+a) \in B \cap \sum A^2$ by condition (12.2), we have $\langle \rho(\Phi(a^+a))\xi, \xi \rangle \geq 0$ by the positivity assumption on ρ. Therefore, $\langle \eta, \eta \rangle_0 \geq 0$.

The case when ρ is a direct sum of cyclic representations ρ_i from $\mathrm{Rep}_c^+ B$ is easily reduced to the cyclic case, using that $A \otimes_B \mathcal{D}(\rho) \subseteq \oplus_i (A \otimes_B \mathcal{D}(\rho_i))$. $\quad\square$

Let us summarize the preceding considerations. Suppose ρ is a *-representation of the *-algebra B of the class $\mathrm{Rep}_c^+ B$.

Then Eq. (12.11) defines a positive semi-definite hermitian sesquilinear form $\langle \cdot, \cdot \rangle_0$ on $A \otimes_B \mathcal{D}(\rho)$. Let $\mathcal{K}_\rho := \{\eta \in A \otimes_B \mathcal{D}(\rho) : \langle \eta, \eta \rangle_0 = 0\}$ be its kernel, and let $\langle \cdot, \cdot \rangle$ denote the inner product on the quotient space $\mathcal{D}_0 := (A \otimes_B \mathcal{D}(\rho))/\mathcal{K}_\rho$ defined by $\langle [\eta], [\zeta] \rangle = \langle \eta, \zeta \rangle_0$, where $[\eta] := \eta + \mathcal{K}_\rho$.

The induced *-representation $\mathrm{Ind}\,\rho$ of A acts on the complex inner product space $(\mathcal{D}_0, \langle \cdot, \cdot \rangle)$ by

$$(\mathrm{Ind}\,\rho)(a)\left[\sum_k x_k \otimes \varphi_k\right] = \left[\sum_k ax_k \otimes \varphi_k\right], \quad a \in A, \tag{12.14}$$

where $x_k \in A$, $\varphi_k \in \mathcal{D}(\rho)$. The domain $\mathcal{D}_0$ is a core for $\mathrm{Ind}\,\rho$.

Remark 12.18 The following slight reformulation of this construction is sometimes convenient: We define first the positive semi-definite sesquilinear form $\langle \cdot, \cdot \rangle_0$ on the complex tensor product $A \otimes \mathcal{D}(\rho)$ by (12.11) and then directly the induced representation $\mathrm{Ind}\,\rho$ by (12.14) on the quotient inner product space $((A \otimes \mathcal{D}(\rho))/\mathcal{G}_\rho, \langle \cdot, \cdot \rangle)$ of $A \otimes \mathcal{D}(\rho)$ by the kernel $\mathcal{G}_\rho := \{\zeta \in A \otimes \mathcal{D}(\rho) : \langle \zeta, \zeta \rangle_0 = 0\}$ of the form $\langle \cdot, \cdot \rangle_0$. $\circ$

The next result is often useful to detect inducible and induced *-representations.

Proposition 12.19 *Let ρ be a *-representation of B. Suppose there exist a unitary space $(\mathcal{D}_1, \langle \cdot, \cdot \rangle_1)$ and a (well-defined!) linear map Ψ of $A \otimes \mathcal{D}(\rho)$ on $\mathcal{D}_1$ such that*

$$\langle \Psi(x \otimes \varphi), \Psi(y \otimes \psi) \rangle_1 = \langle \rho(\Phi(y^+x))\varphi, \psi \rangle, \quad x, y \in A, \varphi, \psi \in \mathcal{D}(\rho). \tag{12.15}$$

*Then ρ is inducible (from B to A with respect to the conditional expectation Φ) and $\mathrm{Ind}\,\rho$ is unitarily equivalent to the closure of the *-representation π_1 of A on $\mathcal{D}_1$:*

$$\pi_1(a)\left(\Psi\left(\sum_k x_k \otimes \varphi_k\right)\right) := \Psi\left(\sum_k ax_k \otimes \varphi_k\right), \quad a, x_k \in A, \varphi_k \in \mathcal{D}(\rho).$$

Proof In this proof we use the reformulation given in Remark 12.18. We define a linear mapping U of $A \otimes \mathcal{D}(\rho)$ on $\mathcal{D}_1$ by

$$U\left(\sum_k x_k \otimes \varphi_k\right) = \Psi\left(\sum_k x_k \otimes \varphi_k\right), \quad x_k \in A, \varphi_k \in \mathcal{D}(\rho).$$

Now we use twice the fact that the inner product $\langle \cdot, \cdot \rangle_1$ is positive definite. First, comparing (12.11) and (12.15) it follows that the sesquilinear form $\langle \cdot, \cdot \rangle_0$, defined on the complex tensor product $A \otimes \mathcal{D}(\rho)$ by (12.11), is positive semi-definite. Hence ρ

is inducible. Next, by assumption (12.15) it implies that $\sum_k x_k \otimes \varphi_k \in \mathcal{G}_\rho$ if and only if $\Psi(\sum_k x_k \otimes \varphi_k) = 0$. Therefore, U passes to an isometric linear mapping, denoted again U, of the quotient inner product space $\mathcal{D}_0 := ((\mathsf{A} \otimes \mathcal{D}(\rho))/\mathcal{G}_\rho, \langle \cdot, \cdot \rangle)$ on the inner product space $(\mathcal{D}_1, \langle \cdot, \cdot \rangle_1)$ and hence to a unitary operator of the corresponding Hilbert space completions. By construction, $\pi_1(a)\eta = U(\mathrm{Ind}\,\rho)(a)U^{-1}\eta$ for any $a \in \mathsf{A}$, $\eta \in \mathcal{D}_1$. Thus, since $(\mathrm{Ind}\,\rho) \upharpoonright \mathcal{D}_0$ is a *-representation of A, so is π_1. Since $\mathcal{D}_0$ is a core for $\mathrm{Ind}\,\rho$ and $\mathcal{D}_1$ is a core for $\overline{\pi}_1$, we conclude $\overline{\pi}_1 = U(\mathrm{Ind}\,\rho)U^{-1}$. $\square$

Example 12.20 (*States—Example* 12.4 *continued*)
Suppose f is a state on A and let Φ denote the conditional expectation of A on $\mathsf{B} := \mathbb{C} \cdot 1$ defined by $\Phi(a) = f(a) \cdot 1$, $a \in \mathsf{A}$; see Example 12.4. If ρ is the identity representation of B on $\mathbb{C}$ (i.e., $\rho(\lambda \cdot 1) = \lambda$ for $\lambda \in \mathbb{C}$), it follows from the preceding construction (and also from Proposition 12.19) that the *-representation $\mathrm{Ind}\,\rho$ of A is just the GNS representation $\overline{\pi}_f$ associated with the state f. Thus, the induction procedure can be viewed as a generalization of the GNS construction. $\bigcirc$

12.3 Induced Representations of Group Graded *-Algebras from Hermitian Characters

Throughout this section, $\mathsf{A} = \bigoplus_{g \in G} \mathsf{A}_g$ is a G-graded unital complex *-algebra and we assume that the *-subalgebra $\mathsf{B} := \mathsf{A}_e$ is **commutative**.

For the study of induced representations we need condition (12.20) below. First we derive this condition. For a *-algebra X, set $\mathcal{R}(\mathsf{X}) := \{x \in \mathsf{X} : x^+ x = 0\}$. Clearly, $\mathcal{R}(\mathsf{X})$ is a right ideal of X and contained in the *-radical $\mathrm{Rad}\,\mathsf{X}$ by Corollary 4.45. Hence $\mathcal{R}(\mathsf{X})$ is annihilated by all *-representations of X.

Lemma 12.21 *Let $g \in G$ and $c, d \in \mathsf{A}_g$. Then*

$$c^+ c d^+ d - c^+ d d^+ c \in \mathcal{R}(\mathsf{B}). \tag{12.16}$$

*For arbitrary *-representations ρ of B and π of A we have*

$$\rho(c^+ d)\rho(d^+ c) = \rho(c^+ c)\rho(d^+ d)\,, \quad \pi(c)\pi(d^+ d) = \pi(dd^+)\pi(c). \tag{12.17}$$

Proof In this proof we set $a := cd^+ d - dd^+ c$ and $b := c^+ a = c^+ c d^+ d - c^+ dd^+ c$.

Note that for arbitrary $x, y \in \{c, d\}$ the elements $x^+ y$ and xy^+ are in B, and hence they commute. Using this fact several times we compute

$$
\begin{aligned}
cd^+ dc^+ cd^+ d &= cd^+(dc^+)cd^+ d = (dc^+)cd^+ cd^+ d \\
&= dc^+ c(d^+ c)d^+ d = d(d^+ c)c^+ cd^+ d = dd^+ cc^+ cd^+ d, \tag{12.18} \\
c^+ cd^+ dc^+ dd^+ c &= (c^+ c)d^+ dc^+ dd^+ c = d^+ dc^+ dd^+ c(c^+ c) \\
&= d^+ d(c^+ d)d^+ cc^+ c = (c^+ d)d^+ dd^+ cc^+ c \\
&= c^+ dd^+ dd^+(cc^+)c = c^+ dd^+(cc^+)dd^+ c = c^+ dd^+ cc^+ dd^+ c. \tag{12.19}
\end{aligned}
$$

Since $c^+cd^+d = d^+dc^+c$ by the commutativity of B, we have $b^+ = b$. Therefore,

$$b^+b = b^2 = (c^+cd^+d - c^+dd^+c)(c^+cd^+d - c^+dd^+c)$$
$$= c^+[cd^+dc^+cd^+d - dd^+cc^+cd^+d] + [c^+dd^+cc^+dd^+c - c^+cd^+dc^+dd^+c] = 0.$$

Here the two expressions in squared brackets vanish by (12.18) and (12.19). Hence, since $b \in \mathsf{B}$, we have $b \in \mathcal{R}(\mathsf{B})$, which proves (12.16). Since ρ annihilates $\mathcal{R}(\mathsf{B})$, $\rho(b) = 0$. Inserting the definition of b this yields the first equality of (12.17).

Further, using once again the commutativity of B we derive

$$a^+a = (d^+dc^+ - c^+dd^+)(cd^+d - dd^+c)$$
$$= [d^+d(c^+c) - c^+dd^+c]d^+d + c^+d(d^+d)d^+c - d^+dc^+dd^+c$$
$$= [(c^+c)d^+d - c^+dd^+c]d^+d + (d^+d)c^+dd^+c - d^+dc^+dd^+c = bd^+d.$$

Since $b \in \mathcal{R}(\mathsf{B})$, we have $\pi(b) = 0$, so $\pi(a)^+\pi(a) = \pi(a^+a) = \pi(b)\pi(d^+d) = 0$. Hence $\pi(a) = 0$. By the definition of a this gives the second equality of (12.17). □

Recall that $\hat{\mathsf{B}}$ denotes the set of hermitian characters of B. Since hermitian characters are one-dimensional ∗-representations, the first equation of (12.17) yields

$$\chi(c^+d)\chi(d^+c) = \chi(c^+c)\chi(d^+d) \quad \text{for all} \quad \chi \in \hat{\mathsf{B}}, \ \ c,d \in \mathsf{A}_g, g \in G. \quad (12.20)$$

The following notions play a crucial role in what follows.

Definition 12.22 $\hat{\mathsf{B}}^+ := \left\{ \chi \in \hat{\mathsf{B}} : \chi(a) \geq 0 \ \ \text{for} \ \ a \in \mathsf{B} \cap \sum \mathsf{A}^2 \right\}.$

Definition 12.23 Let $\chi \in \hat{\mathsf{B}}^+$ and $g \in G$. We say that χ^g is *defined* if there exists an element $a_g \in \mathsf{A}_g$ such that $\chi(a_g^+a_g) \neq 0$. In this case, we set

$$\chi^g(b) := \frac{\chi(a_g^+ba_g)}{\chi(a_g^+a_g)} \quad \text{for} \ \ b \in \mathsf{B}. \quad (12.21)$$

For $\chi \in \hat{\mathsf{B}}^+$, let G_χ denote the set of elements $g \in G$ for which χ^g is defined.

The next proposition collects a number of basic properties of the map $\chi \mapsto \chi^g$.

Proposition 12.24 *Suppose that* $\chi \in \hat{\mathsf{B}}^+$ *and* $g, h \in G$.

(i) *If* χ^g *is defined, then* $\chi^g \in \hat{\mathsf{B}}^+$ *and* χ^g *does not depend on the choice of* a_g.
(ii) *If* χ^g *and* $(\chi^g)^h$ *are defined, then* χ^{hg} *is defined and equal to* $(\chi^g)^h$.
(iii) *If* χ^g *is defined, then* $(\chi^g)^{g^{-1}}$ *is defined and equal to* χ.
(iv) χ^e *is defined and equal to* χ.
(v) *If* $a_g \in \mathsf{A}_g$ *and* $\chi(a_g^+a_g) \neq 0$, *then* $\chi^g(a_ga_g^+) = \chi(a_g^+a_g)$.
(vi) *If* $a_g, c_g \in \mathsf{A}_g$ *and* $\chi(a_g^+c_g) \neq 0$, *then* χ^g *is defined and*

$$\chi^g(b) = \frac{\chi(a_g^+ b c_g)}{\chi(a_g^+ c_g)}, \quad b \in \mathsf{B}. \tag{12.22}$$

Proof (i): Let $c, d \in \mathsf{A}_g$ be such that $\chi(d^+d) \neq 0$ and $\chi(c^+c) \neq 0$. Since $cd^+ \in \mathsf{B}$ and B is commutative, $bcd^+ = cd^+b$ for $b \in \mathsf{B}$ and therefore

$$\chi(c^+bc)\chi(d^+d) = \chi(c^+bcd^+d) = \chi(c^+cd^+bd) = \chi(c^+c)\chi(d^+bd),$$

so that

$$\frac{\chi(c^+bc)}{\chi(c^+c)} = \frac{\chi(d^+bd)}{\chi(d^+d)}.$$

This shows that $\chi^g(b)$ in (12.21) does not depend on the particular choice of a_g. We prove that χ^g is a character of $\hat{\mathsf{B}}^+$. Let $b_1, b_2 \in \mathsf{B}$. Since B is commutative, $a_g a_g^+ b_1 = b_1 a_g a_g^+$. Using that χ is a character on B and $a_g^+ a_g \in \mathsf{B}$, we derive

$$\chi^g(b_1 b_2) = \frac{\chi(a_g^+ b_1 b_2 a_g)}{\chi(a_g^+ a_g)} = \frac{\chi(a_g^+ a_g)\,\chi(a_g^+ b_1 b_2 a_g)}{\chi(a_g^+ a_g)\,\chi(a_g^+ a_g)} = \frac{\chi(a_g^+ a_g a_g^+ b_1 b_2 a_g)}{\chi(a_g^+ a_g)\chi(a_g^+ a_g)}$$

$$= \frac{\chi(a_g^+ b_1 a_g a_g^+ b_2 a_g)}{\chi(a_g^+ a_g)\chi(a_g^+ a_g)} = \frac{\chi(a_g^+ b_1 a_g)\,\chi(a_g^+ b_2 a_g)}{\chi(a_g^+ a_g)\chi(a_g^+ a_g)} = \chi^g(b_1)\chi^g(b_2).$$

The character χ is hermitian and so is χ^g.

Let $b \in \mathsf{B} \cap \sum \mathsf{A}^2$. Then $a_g^+ b a_g$ and $a_g^+ a_g$ belong to $\mathsf{B} \cap \sum \mathsf{A}^2$. Hence, since $\chi \in \hat{\mathsf{B}}^+$, we have $\chi(a_g^+ b a_g) \geq 0$ and $\chi(a_g^+ a_g) \geq 0$, so $\chi^g(b) \geq 0$. Thus, $\chi^g \in \hat{\mathsf{B}}^+$.

(ii): That χ^g and $(\chi^g)^h$ are defined means that there exist elements $a_g \in \mathsf{A}_g$ and $a_h \in \mathsf{A}_h$ such that $\chi(a_g^+ a_g) \neq 0$ and $\chi^g(a_h^+ a_h) = \chi(a_g^+ a_h^+ a_h a_g)\chi(a_g^+ a_g)^{-1} \neq 0$. Thus, $\chi((a_h a_g)^+ a_h a_g) \neq 0$. Therefore, since $a_h a_g \in \mathsf{A}_{hg}$, χ^{hg} is defined. It is easily verified that $(\chi^g)^h = \chi^{hg}$.

(iii): There exists an $a_g \in \mathsf{A}_g$ such that $\chi(a_g^+ a_g) \neq 0$. Since $a_g^+ \in \mathsf{A}_{g^{-1}}$ and

$$\chi^g(a_g a_g^+) = \frac{\chi(a_g^+ a_g a_g^+ a_g)}{\chi(a_g^+ a_g)} = \frac{\chi(a_g^+ a_g)\,\chi(a_g^+ a_g)}{\chi(a_g^+ a_g)} = \chi(a_g^+ a_g) \neq 0, \quad (12.23)$$

$(\chi^g)^{g^{-1}}$ is defined. It is straightforward to check that $(\chi^g)^{g^{-1}} = \chi$.

(vi): Since $\chi(a_g^+ c_g) \neq 0$, we have $\chi(c_g^+ a_g) = \overline{\chi(a_g^+ c_g)} \neq 0$. Therefore, (12.20) implies that $\chi(a_g^+ a_g) \neq 0$, so χ^g is defined. Now (12.22) follows by combining (12.21) and the equality

$$\chi(a_g^+ b a_g)\chi(a_g^+ c_g) = \chi(a_g^+ b(a_g a_g^+) c_g) = \chi(a_g^+ (a_g a_g^+) b c_g) = \chi(a_g^+ a_g)\chi(a_g^+ b c_g).$$

(iv) is trivial, and (v) follows from (12.23). $\square$

For $g \in G$, let $\mathcal{D}_g$ denote the set of characters $\chi \in \hat{B}^+$ such that χ^g is defined. From Proposition 12.24 it follows that for $g \in G$ the mapping $\chi \mapsto \alpha_g(\chi) := \chi^g$ defines a bijection $\alpha_g : \mathcal{D}_g \mapsto \mathcal{D}_{g^{-1}}$ obeying the following properties:

(i) $\alpha_e(\chi) = \chi$ for $\chi \in \mathcal{D}_e = \hat{B}^+$.

(ii) $\alpha_h(\alpha_g(\chi)) = \alpha_{hg}(\chi)$ for $\chi \in \mathcal{D}_g$, $\chi^g \in \mathcal{D}_h$, and $g, h \in G$.

We call this family of mappings $g \mapsto \alpha_g$ a *partial action* of the group G on $\hat{B}^+$.

Example 12.26 below shows that χ^g is not always defined, so in general the map $g \mapsto \alpha_g$ is not an action of the group G on $\hat{B}^+$. We illustrate this with two examples.

Example 12.25 (*Crossed product *-algebras—Example 12.12 continued*)
Suppose B is a commutative unital *-algebra, and let $A = B \times_\theta G$ be the crossed product algebra considered in Example 12.12. By Proposition 12.13, the corresponding conditional expectation of A on B is strong. Hence $\sum A^2 \cap B = \sum B^2$, so that $\hat{B}^+ = \hat{B}$. Then $1 \otimes g \in A_g$ and $\chi((1 \otimes g)^+(1 \otimes g)) = \chi(1 \otimes e) = 1$ for $\chi \in \hat{B}^+$. Thus χ^g is defined for all $\chi \in \hat{B}^+$ and $g \in G$. Since $\theta_g(b) = gbg^{-1}$ (by (2.17)) in the algebra $A = B \times_\theta G$, we obtain $\chi^g(b) = \chi(\theta_{g^{-1}}(b))$. This formula implies that the map $g \mapsto \alpha_g$ is an *action of the group G* on the set $\hat{B}^+$.   ○

Example 12.26 (*Weyl algebra—Example 12.8 continued*)
Let $A = \mathbb{C}\langle a, a^+ | aa^+ - a^+a = 1 \rangle$ be the Weyl algebra. We retain the notation of Example 12.8. Recall that A is $\mathbb{Z}$-graded with $a \in A_1$ and $B = A_0 = \mathbb{C}[N]$. From (12.8) it follows easily that a character $\chi \in \hat{B}$ is nonnegative on $B \cap \sum A^2$ if and only if $\chi(N) \in \mathbb{N}_0$. For $k \in \mathbb{N}_0$, let χ_k denote the character of $\hat{B}^+$ defined by $\chi_k(N) = k$.

Let $n \in \mathbb{N}_0$. Any element of A_n is of the form $a^n p(N)$ with $p \in \mathbb{C}[N]$. Clearly, $\chi_k((a^n p(N))^+ a^n p(N)) \neq 0$ implies that $\chi_k(a^{+n} a^n) \neq 0$. Thus, $\alpha_n(\chi_k)$ is defined if and only if $\chi_k(a^{+n} a^n) \neq 0$. By (8.13) and (8.12), we have

$$\chi_k(a^{+n} a^n) = \chi_k(N(N-1) \cdots (N-n+1)) = k(k-1) \cdots (k-n+1),$$
$$\chi_k(a^{+n} N a^n) = \chi_k(a^{+n} a^n (N-n)) = \chi_k(a^{+n} a^n)(k-n).$$

Hence $\alpha_n(\chi_k)$ is defined if and only if $k \geq n$. In this case, $\alpha_n(\chi_k) = \chi_{k-n}$. Similarly, using (8.14) and (8.12) we obtain

$$\chi_k(a^n a^{+n}) = \chi_k((N+1)(N+2) \cdots (N+n)) = (k+1)(k+2) \cdots (k+n),$$
$$\chi_k(a^n N a^{+n}) = \chi_k(a^n a^{+n}(N+n)) = \chi_k(a^n a^{+n})(k+n),$$

so $\alpha_{-n}(\chi_k)$ is defined for all $n \in \mathbb{N}$ and we have $\alpha_{-n}(\chi_k) = \chi_{k+n}$.
Thus, $G_{\chi_k} = \{n \in \mathbb{Z} : n \leq k\}$ and $\mathcal{D}_n = \{\chi_j : n \leq j\}$ for $k \in \mathbb{N}_0$, $n \in \mathbb{Z}$.   ○

Now we are ready to describe the induced representations.

The characters of $\hat{B}^+$ are nonnegative on $B \cap \sum A^2$, so they are one-dimensional representations of $\text{Rep}_c^+ B$ and hence inducible by Proposition 12.17. The next proposition contains explicit formulas for the corresponding *-representations.

Proposition 12.27 *Suppose that* $\chi \in \hat{\mathsf{B}}^+$ *and let* $\pi = \operatorname{Ind} \chi$. *For each* $g \in G_\chi$ *we choose an element* $a_g \in \mathsf{A}_g$ *such that* $\chi(a_g^+ a_g) \neq 0$. *Then the vectors*

$$e_g := \frac{[a_g \otimes 1]}{\sqrt{\chi(a_g^+ a_g)}}, \; g \in G_\chi, \tag{12.24}$$

form an orthonormal basis of the Hilbert space $\mathcal{H}(\pi)$. *For* $h, g \in G$ *and* $c_h \in \mathsf{A}_h$,

$$\pi(c_h) e_g = \frac{\chi(a_{hg}^+ c_h a_g)}{\sqrt{\chi(a_{hg}^+ a_{hg}) \chi(a_g^+ a_g)}} \; e_{hg} \; \text{ if } \; hg \in G_\chi \tag{12.25}$$

and $\pi(c_h) e_g = 0$ *otherwise. In particular,* $\pi(b) e_g = \chi^g(b) e_g$ *for* $b \in \mathsf{B}$.

Proof We recall the construction of the representation $\pi = \operatorname{Ind} \chi$ from Lemmas 12.14 and 12.15. In the special case $\rho = \chi$ the Hilbert space $\mathcal{H}(\pi)$ is the closed span of vectors $[a \otimes 1]$, where $a \in \mathsf{A}$. By (12.11), the inner product is given by

$$\langle [a \otimes 1], [c \otimes 1] \rangle = \chi(\Phi(c^+ a)), \quad a, c \in \mathsf{A}.$$

If $a \in \mathsf{A}_g$ and $c \in \mathsf{A}_h$, then $\Phi(c^+ a) = 0$ if $g \neq h$ and $\Phi(c^+ a) = c^+ a$ if $g = h$.

First suppose $g \notin G_\chi$. Then $\|[c_g \otimes 1]\|^2 = \chi(c_g^+ c_g) = 0$. Hence the Hilbert space $\mathcal{H}(\pi)$ is the closed span of vectors $[c_g \otimes 1]$, where $c_g \in \mathsf{A}_g$ and $g \in G_\chi$.

Let $g \in G_\chi$. Then we have $\langle [a_g \otimes 1], [a_g \otimes 1] \rangle = \chi(a_g^+ a_g) > 0$, so the vector e_g has norm one. Now suppose that $g, h \in G_\chi$, $g \neq h$. Therefore, $\Phi(a_h^+ a_g) = 0$ and $\langle [a_g \otimes 1], [a_h \otimes 1] \rangle = \chi(\Phi(a_h^+ a_g)) = 0$.

Now let $c_g \in \mathsf{A}_g$. Using (12.20), with $d := a_g, c := c_g \in \mathsf{A}_g$, we obtain

$$\left| \langle [a_g \otimes 1], [c_g \otimes 1] \rangle \right|^2 = |\chi(\Phi(c_g^+ a_g))|^2 = \chi(c_g^+ a_g) \overline{\chi(c_g^+ a_g)}$$
$$= \chi(c_g^+ a_g) \chi(a_g^+ c_g) = \chi(a_g^+ a_g) \chi(c_g^+ c_g) = \|[a_g \otimes 1]\|^2 \|[c_g \otimes 1]\|^2.$$

That is, we have equality in the Cauchy–Schwarz inequality. Therefore, $[c_g \otimes 1]$ is a complex multiple of $[a_g \otimes 1]$ and hence of e_g. Putting the preceding together, we have proved that the vectors e_g, $g \in G_\chi$, form an orthonormal basis of $\mathcal{H}(\pi)$.

Next we prove formula (12.25). Suppose $hg \in G_\chi$. Note that $c_h a_g \in \mathsf{A}_{hg}$. Hence, as shown in the preceding paragraph, $[c_h a_g \otimes 1]$ is a multiple of $[a_{hg} \otimes 1]$ and so of the unit vector e_{hg}. Therefore, using (12.24) we derive

$$[c_h a_g \otimes 1] = \langle [c_h a_g \otimes 1], e_{hg} \rangle \, e_{hg} = \chi(a_{hg}^+ a_{hg})^{-1/2} \langle [c_h a_g \otimes 1], [a_{hg} \otimes 1] \rangle \, e_{hg}$$
$$= \chi(a_{hg}^+ a_{hg})^{-1/2} \chi(a_{hg}^+ c_h a_g) \, e_{hg}. \tag{12.26}$$

Further, by (12.24) and the definition of $\pi(c_h)$,

$$\pi(c_h)e_g = \chi(a_g^+ a_g)^{-1/2}\pi(c_h)[a_g \otimes 1] = \chi(a_g^+ a_g)^{-1/2}[c_h a_g \otimes 1]. \qquad (12.27)$$

Inserting (12.26) into (12.27) yields (12.25).

Let $c_h = b \in \mathsf{B} = \mathsf{A}_e$. We apply (12.25) with $h = e$ and get

$$\pi(b)e_g = \chi(a_g^+ b a_g)\chi(a_g^+ a_g)^{-1}\, e_g = \chi^g(b)\, e_g.$$

Finally, suppose $hg \notin G_\chi$. Then χ^{hg} is not defined, so it follows from Proposition 12.24(vi) that $0 = \chi(a_{hg}^+ c_h a_g) = \langle[c_h a_g \otimes 1], [a_{hg} \otimes 1]\rangle$. Hence, by (12.27) and (12.26), $\langle\pi(c_h)e_g, e_{hg}\rangle = 0$. If $k \in G, k \neq hg$, then $\Phi(e_k^+ c_h a_g) = 0$ and hence $\langle\pi(c_h)e_g, e_k\rangle = 0$. Thus, $\pi(c_h)e_g \perp e_k$ for all $k \in G$, so that $\pi(c_h)e_g = 0$. $\qquad\square$

Example 12.28 (*Weyl algebra—Example* 12.26 *continued*)
Fix $k \in \mathbb{N}_0$. Recall from Example 12.26 that χ_k is the character on $\mathsf{B} = \mathbb{C}[N]$ given by $\chi_k(p(N)) = p(k), p \in \mathbb{C}[N]$. Further, $G = \mathbb{Z}$ and $G_{\chi_k} = \{m \in \mathbb{Z} : m \leq k\}$.

We use Proposition 12.27 to describe the induced representation $\pi := \operatorname{Ind} \chi_0$ of the character χ_0. For $-n \in G_{\chi_0} = -\mathbb{N}_0$ we choose $a_{-n} := (a^+)^n \in \mathsf{A}_{-n}$. As shown in Example 12.26, we have $\chi_0((a_{-n})^+ a_{-n}) = \chi_0(a^n(a^+)^n) = n!$, so that $e_{-n} = n!^{-1/2}[(a^+)^n \otimes 1]$. The vectors $e_{-n}, n \in \mathbb{N}_0$, form an orthonormal basis of $\mathcal{H}(\pi)$ by Proposition 12.27.

Next we determine the action of $\pi(a)$. Set $h = 1, c_1 = a, g = -n$. If $n = 0$, then $hg = 1 + 0 \notin G_{\chi_0}$, so $\pi(a)e_0 = 0$. Suppose now $n \in \mathbb{N}$. Then $1 - n \in G_{\chi_0}$, $a_{hg} = a_{-(n-1)} = (a^+)^{n-1}$ and $c_h a_g = a(a^+)^n$. We insert all this in Eq. (12.25) and compute

$$\begin{aligned}
\pi(a)e_{-n} &= \frac{\chi_0(a^{n-1}a(a^+)^n)}{\sqrt{\chi_0(a^{n-1}(a^+)^{n-1})\chi_0(a^n(a^+)^n))}}\, e_{-(n-1)} \\
&= \frac{n!}{\sqrt{(n-1)!\, n!}}\, e_{-(n-1)} = \sqrt{n}\, e_{-(n-1)}.
\end{aligned}$$

Similarly, setting $c_{-1} = a^+$ in (12.25), we obtain $\pi(a^+)e_{-n} = \sqrt{n+1}\, e_{-(n+1)}$ for $n \in \mathbb{N}_0$. (This follows also from the relation $\langle\pi(a)e_{-k}, e_{-n}\rangle = \langle e_{-k}, \pi(a^+)e_{-n}\rangle$.) Comparing these actions with (8.22) we conclude that $\pi = \operatorname{Ind} \chi_0$ *is unitarily equivalent to the Bargmann–Fock representation of the Weyl algebra.* This shows the usefulness of unbounded induced representations for general ∗-algebras. $\qquad\bigcirc$

12.4 Exercises

1. Let $p_1, \ldots, p_n \in \mathsf{A}$ be a decomposition of the unit of the ∗-algebra A, that is, $p_1 + \cdots + p_n = 1$ and $p_j = p_j^2 = p_j^+$ for $j = 1, \ldots, n$. Show the following:

   a. $p_j p_k = 0$ for all $j \neq k$.

   b. The map $\Phi : a \mapsto p_1 a p_1 + \cdots + p_n a p_n$ is a conditional expectation of A
      on $\mathsf{B} = \{b \in \mathsf{A} : b = \Phi(b)\}$.
   c. If A is a unital O^*-algebra, then Φ is faithful.

The following Exercises 2–4 show that A satisfies the assumptions of Sect. 12.3;
details can be found in [SS13].

2. Let $f \in \mathbb{R}[x]$ and $\mathsf{A} := \mathbb{C}\langle x, x^+ \mid xx^+ = f(x^+x)\rangle$. Prove the following:

   a. The *-algebra A is $\mathbb{Z}$-graded with grading defined by $x \in \mathsf{A}_1, x^+ \in \mathsf{A}_{-1}$.
   b. $\mathsf{B} := \mathsf{A}_0$ is commutative.
   c. B spanned by $(x^+)^{k_1} x^{n_1} \cdots (x^+)^{k_r} x^{n_r}$, where $k_i, n_i \in \mathbb{N}_0, \sum_i k_i = \sum_i n_i$.
   d. $\mathsf{A}_n = \mathsf{B} \cdot x^n$ and $\mathsf{A}_{-n} = (x^+)^n \cdot \mathsf{B}$ for $n \in \mathbb{N}$.

3. (*Enveloping algebras $\mathcal{E}(su(2))$ and $\mathcal{E}(su(1, 1))$*)
   The enveloping algebras $\mathcal{E}(su(2))$ and $\mathcal{E}(su(1, 1))$ are the complex unital $*$-
   algebras with generators e, f, h, defining relations

$$he - eh = 2e, \quad hf - fh = -2f, \quad ef - fe = h,$$

   and involution $e^+ = f, h^+ = h$ for $su(2)$, $e^+ = -f, h^+ = h$ for $su(1, 1)$. Let
   A be $\mathcal{E}(su(2))$ or $\mathcal{E}(su(1, 1))$. Prove that:

   a. A is $\mathbb{Z}$-graded with grading determined by $e \in \mathsf{A}_1, f \in \mathsf{A}_{-1}, h \in \mathsf{A}_0$.
   b. $\mathsf{B} := \mathsf{A}_0$ is commutative and equal to $\mathbb{C}[ef, h]$.
   c. $\mathsf{A}_n = e^n \cdot \mathsf{B}$ and $\mathsf{A}_{-n} = f^n \cdot \mathsf{B}$ for $n \in \mathbb{N}_0$.

4. (*Coordinate algebra of the quantum group $SU_q(2)$, see, e.g., [KS97, 4.1.4]*)
   Suppose $q > 0$. Let A be the unital complex $*$-algebras with generators a, c and
   defining relations

$$ac = qca, \quad ac^+ = qc^+a, \quad c^+c = cc^+, \quad a^+a + c^+c = 1, \quad aa^+ + q^2c^+c = 1.$$

   Prove the following assertions:

   a. A is algebraically bounded.
   b. A is $\mathbb{Z}$-graded with grading given by $a \in \mathsf{A}_1, a^+ \in \mathsf{A}_{-1}, c \in \mathsf{A}_0$.
   c. $\mathsf{B} := \mathsf{A}_0$ is commutative and equal to $\mathbb{C}[c, c^+, N]$, where $N := a^+a$.
   d. $\mathsf{A}_n = a^n \cdot \mathsf{B}$ and $\mathsf{A}_{-n} = (a^+)^n \cdot \mathsf{B}$ for $n \in \mathbb{N}_0$.

5. Let $q > 0, q \neq 1$. Develop the theory of Sect. 12.3 for the $\mathbb{Z}$-graded $*$-algebra
   $\mathsf{A} = \mathbb{C}\langle x, x^+ \mid xx^+ = qx^+x\rangle$ with grading given by $x \in \mathsf{A}_1, x^+ \in \mathsf{A}_{-1}$. Compare
   the outcome with the $*$-representations of A obtained in Sect. 11.6.

In Exercises 6–10, A denotes the Weyl algebra $\mathbb{C}\langle a, a^+ \mid aa^+ - a^+a = 1\rangle$.

6. Show that for each $z \in \mathbb{T}$ there is a $*$-automorphism α_z of A such that $\alpha_z(a) = za$
   and $\alpha_z(a^+) = \bar{z} a^+$. Show that the action $z \mapsto \alpha_z$ of $\mathbb{T}$ defines a conditional
   expectation by Example 12.11. What is the $*$-algebra B of stable elements?

7. Decide whether the following polynomials of $N := a^+ a$ are in $\sum \mathsf{A}^2$.

   a. $(N - 1)(N + 1)$.
   b. $(N + 1)(N + \frac{3}{2})$.
   c. $(N - 1)(N - \frac{3}{2})$.
   d. $(N - k)(N - (k + 1))$, where $k \in \mathbb{Z}$.

8. Let $p(N) = (N - k_0) \cdots (N - k_n)$, where $n \in \mathbb{N}_0$, $k_0, k_1, \ldots, k_n \in \mathbb{Z}$, and $0 \le k_0 < k_1 < \cdots < k_n$. Show that $p(N) \in \sum \mathsf{A}^2$ if and only if we have $k_0 = 0, k_1 = 1, \ldots, k_n = n$.

9. Show that there exists a positive linear functional f on A that is *not* of the form $f(\cdot) = \mathrm{Tr}\, t\pi_{\mathbb{C}}(\cdot)$ for some trace class operator $t \in \mathbf{B}_1(\pi_{\mathbb{C}}(\mathsf{A}))_+$, where $\pi_{\mathbb{C}}$ is the Bargmann–Fock representation.
   Hint: Use $(N - 1)(N - 2) \notin \sum \mathsf{A}^2$ and a separation argument (Theorem 10.36).

10. Show that for each $k \in \mathbb{N}_0$ the induced representation Ind χ_k of A in Example 12.28 is unitarily equivalent to the Bargmann–Fock representation.

12.5 Notes

Induced representations of C^*-algebras were invented by M. Rieffel [Rf74a] and J.M.G. Fell [F72]. A detailed and very readable treatment of this theory is presented in the monograph [FD88]; see also [RW98].

Induced representations for general ∗-algebras were introduced and first studied in the joint paper [SS13] of the author with Y. Savchuk. In this chapter we have given a glimpse into this subject. The paper [SS13] contains an extensive treatment including various imprimitivity theorems and many examples. Induced representations for the ∗-algebras in Exercises 2 and 3 are developed in [SS13]. Further examples are elaborated in [Dy15] and [DS13]. Exercise 8 is taken from [FS89].

Chapter 13
Well-Behaved Representations

In contrast to single operators, the self-adjointness of a representation is not enough to rule out pathological behavior. For instance, as shown in Example 7.6, there exists a self-adjoint irreducible representation of the commutative *-algebra $\mathbb{C}[x_1, x_2]$ acting on an infinite-dimensional Hilbert space. It is natural to look for additional conditions to select classes of "well-behaved" representations that exclude pathological phenomena. For commutative *-algebras the integrable representations defined in Sect. 7.2 are such a class. For the Weyl algebra and enveloping algebras there are also self-adjoint representations with pathological behavior, but it is not difficult to "guess" how to define well-behaved representations. For the Weyl algebra one takes self-adjoint representations for which the closures of the images of the generators p, q are self-adjoint and satisfy the Weyl relation (8.44). Then the Stone–von Neumann theorem implies that, up to unitary equivalence, the well-behaved representations are precisely the direct sums of Schrödinger representations. For an enveloping algebra one requires that the representation is integrable with respect to the corresponding connected and simply connected Lie group.

No general or canonical method is known to select well-behaved representations of an arbitrary *-algebra A. In this chapter we develop three possible approaches to this problem. In Sect. 13.1, we consider group graded *-algebras A for which the base *-algebras A_e are commutative and finitely generated. In this case we define well-behaved representations by requiring that their restrictions to A_e are integrable. Section 13.2 deals with representations which are associated with torsion-free bounded representations of certain *-algebras of fractions (Theorem 13.12). In Sect. 13.4, we consider an auxiliary *-algebra X on which A acts and derive representations of A from those of X (Theorem 13.20). In Sects. 13.3 and 13.5, we apply the methods from Sects. 13.2 and 13.4 to the Weyl algebra and to enveloping algebras, respectively, and show that in both cases the natural classes of well-behaved representations are obtained (Theorems 13.17 and 13.22).

© The Editor(s) (if applicable) and The Author(s), under exclusive license
to Springer Nature Switzerland AG 2020
K. Schmüdgen, *An Invitation to Unbounded Representations of *-Algebras
on Hilbert Space*, Graduate Texts in Mathematics 285,
https://doi.org/10.1007/978-3-030-46366-3_13

Throughout this chapter, A denotes a **unital complex** ∗-algebra, and all ∗-algebras are **complex**.

13.1   Well-Behaved Representations of Some Group Graded ∗-Algebras

In this section, we suppose that $A = \oplus_{g \in G} A_g$ is a G-graded unital ∗-algebra such that the ∗-algebra $B := A_e$ is **commutative** and **finitely generated.**

Our aim is to propose a concept of well-behaved representations for such ∗-algebras. The corresponding definition is the following.

Definition 13.1 A nondegenerate ∗-representation π of A is called *well-behaved* if its restriction $\pi \restriction B$ to the commutative ∗-algebra B is integrable according to Definition 7.7.

The following proposition holds without any assumption on the ∗-subalgebra B.

Proposition 13.2 *Suppose that π is a ∗-representation of the G-graded ∗-algebra $A = \oplus_{g \in G} A_g$ and let $B := A_e$. Then the graph topologies of π and $\pi \restriction B$ coincide. In particular, π is closed if and only if $\pi \restriction B$ is closed.*

Proof Since B is a subalgebra of A, the graph topology of $\pi \restriction B$ is obviously weaker than that of π. For $a_g \in A_g$ and $\varphi \in \mathcal{D}(\pi)$,

$$\|\pi(a_g)\varphi\|^2 = \langle \pi((a_g)^+ a_g)\varphi, \varphi \rangle \leq (\|\pi((a_g)^+ a_g)\varphi\| + \|\varphi\|)^2.$$

Since $(a_g)^+ a_g \in B$, this implies that the graph topology of π is weaker than that of $\pi \restriction B$. Hence both graph topologies coincide.

Recall that a ∗-representation is closed if and only if its graph topology is complete. Therefore, π is closed if and only if $\pi \restriction B$ is closed.   □

Each well-behaved representation of A is self-adjoint, because integrable representations of B are self-adjoint by Proposition 7.9.

Proposition 13.3 *Let π be a well-behaved representation of A. Then any self-adjoint subrepresentation π_0 of π is well-behaved.*

Proof Since π_0 is self-adjoint, by Corollary 4.31 there exists a ∗-representation π_1 of A such that $\pi = \pi_0 \oplus \pi_1$. Since $\pi \restriction B$ is integrable, so is obviously $\pi_0 \restriction B$.   □

Example 13.4 (*Weyl algebra*)
Let $A = \mathbb{C}\langle a, a^+ | aa^+ - a^+a = 1 \rangle$ be the Weyl algebra, considered as $\mathbb{Z}$-graded ∗-algebra with $\mathbb{Z}$-grading given by $a \in A_1$; see Example 2.23. Then $B = A_0 = \mathbb{C}[N]$, where $N = a^+a$.

Let π be ∗-representation of A. If $\pi \restriction B$ is integrable, then $\pi \restriction B$, hence π, is self-adjoint and the operator $\pi(N)$ is essentially self-adjoint by Proposition 4.19(vi).

Hence, by Theorem 8.9, π is unitarily equivalent to some representation π_g and so to a direct sum of Bargmann–Fock representations. Conversely, let π be a direct sum of Bargmann–Fock representations. Then, π is self-adjoint and for any $b = b^+ \in \mathsf{B} = \mathbb{C}[N]$, the operator $\pi(b)$ acts as a diagonal operator, so it is essentially self-adjoint. Hence $\pi \upharpoonright \mathsf{B}$ is integrable by Theorem 7.11.

Summarizing, a ∗-representation of the Weyl algebra is well-behaved according to Definition 13.1 if and only if it is unitarily equivalent to a direct sum of Bargmann–Fock representations. $\bigcirc$

Integrable representations of finitely generated commutative ∗-algebras have been described in Theorem 7.23 by means of spectral measures. The next proposition restates this result in the present setting for the ∗-algebra B. Recall from Definition 12.22 that $\hat{\mathsf{B}}^+$ is the set of characters of B that are nonnegative on $\mathsf{B} \cap \sum \mathsf{A}^2$. For $b \in \mathsf{B}$, f_b denotes the function on $\hat{\mathsf{B}}^+$ defined by $f_b(\chi) = \chi(b)$, $\chi \in \hat{\mathsf{B}}^+$. Equipped with the corresponding weak topology, $\hat{\mathsf{B}}^+$ is a locally compact topological Hausdorff space and each function f_b is continuous.

Proposition 13.5 *Suppose π is a well-behaved ∗-representation of A. Then there exists a unique spectral measure E^π on the locally compact space $\hat{\mathsf{B}}^+$ such that*

$$\overline{\pi(b)} = \int_{\hat{\mathsf{B}}^+} f_b(\chi)\, dE^\pi(\chi) \equiv \int_{\hat{\mathsf{B}}^+} \chi(b)\, d\,E^\pi(\chi) \quad \text{for } b \in \mathsf{B}. \tag{13.1}$$

The spectral projections $E^\pi(M)$ leave the domain $\mathcal{D}(\pi)$ invariant.

Proof Since π is well-behaved, $\pi_0 := \pi \upharpoonright \mathsf{B}$ is integrable. Clearly, $Q := \mathsf{B} \cap \sum \mathsf{A}^2$ is a quadratic module of B and $\hat{\mathsf{B}}^+ = \mathsf{B}(Q)_+$. If $c = \sum_j a_j^+ a_j \in \mathsf{B} \cap \sum \mathsf{A}^2$, then $\langle \pi_0(c)\varphi, \varphi \rangle = \sum_j \|\pi(a_j)\varphi\|^2 \geq 0$ for all $\varphi \in \mathcal{D}(\pi)$, that is, $\pi_0(c) \geq 0$. Thus, Theorem 7.23 applies to π_0 and Q, so there exists a unique spectral measure $E^\pi := E_{\pi_0}$ supported on $\hat{\mathsf{B}}^+$ such that (7.7) holds. This gives (13.1). By Theorem 7.23(ii), $E^\pi(M) = E_{\pi_0}(M)$ leaves $\mathcal{D}(\pi_0) = \mathcal{D}(\pi)$ invariant. $\square$

Corollary 13.6 *Let π be a well-behaved ∗-representation of A and $b \in \mathsf{B}$. Then the operator $\overline{\pi(b)}$ is normal, its spectral measure $E_{\overline{\pi(b)}}$ is $E_{\overline{\pi(b)}}(\cdot) = E^\pi(f_b^{-1}(\cdot))$, that is, $E_{\overline{\pi(b)}}(M) = E^\pi(\{\chi \in \hat{\mathsf{B}}^+ : \chi(b) \in M\})$ for any Borel subset M of $\mathbb{C}$, and its spectral decomposition is*

$$\overline{\pi(b)} = \int_{\mathbb{C}} \lambda\, d\,E^\pi(f_b^{-1}(\lambda)). \tag{13.2}$$

Proof By Theorem 7.11, $\overline{\pi(b)}$ is normal. Transforming (13.1) under the mapping $\hat{\mathsf{B}}^+ \ni \chi \mapsto \lambda := f_b(\chi) \in \mathbb{C}$ yields (13.2). Hence, by the uniqueness of the spectral measure of a normal operator, $E^\pi(f_b^{-1}(\cdot))$ is the spectral measure of $\overline{\pi(b)}$. $\square$

General elements of A may behave badly in well-behaved ∗-representations. However, for *homogeneous* elements $a_g \in \mathsf{A}_g$ there are the following nice results.

Proposition 13.7 *Suppose π is a well-behaved $*$-representation of* A. *Let $a \in A_g$ and $c \in A_h$, where $g, h \in G$. Then:*

$$\left| \overline{\pi(a)} \right|^2 = \overline{\pi(a^+ a)} = \pi(a)^* \overline{\pi(a)}, \quad \overline{\pi(a)} = \pi(a^+)^*, \tag{13.3}$$

$$\overline{\pi(ac)} = \overline{\overline{\pi(a)} \cdot \overline{\pi(c)}}. \tag{13.4}$$

Proof Clearly, $\pi(a^+ a) \subseteq \pi(a)^* \pi(a) \subseteq \pi(a)^* \overline{\pi(a)}$, hence $\overline{\pi(a^+ a)} \subseteq \pi(a)^* \overline{\pi(a)}$. Since π is well-behaved, $\pi \restriction B$ is integrable. Therefore, since $a^+ a \in B_{her}$, $\overline{\pi(a^+ a)}$ is self-adjoint by Theorem 7.11, so that $\overline{\pi(a^+ a)} = \pi(a)^* \overline{\pi(a)} = |\overline{\pi(a)}|^2$. This proves the first half of (13.3).

Similarly, $\pi(a^+ a) \subseteq \overline{\pi(a^+)} \, \pi(a^+)^*$ implies that $\overline{\pi(a^+ a)} = \overline{\pi(a^+)} \pi(a^+)^*$, so $\overline{\pi(a^+ a)} = \pi(a^+)^{**} \pi(a^+)^* = |\pi(a^+)^*|^2$. Combining these relations with the fact that $\mathcal{D}(T) = \mathcal{D}(|T|)$ for any densely defined closed operator T we get

$$\mathcal{D}\left(\overline{\pi(a)}\right) = \mathcal{D}\left(|\overline{\pi(a)}|\right) = \mathcal{D}\left(\left(\overline{\pi(a^+ a)}\right)^{1/2}\right) = \mathcal{D}\left(|\pi(a^+)^*|\right) = \mathcal{D}(\pi(a^+)^*). \tag{13.5}$$

Since $\overline{\pi(a)} \subseteq \pi(a^+)^*$, Eq. (13.5) yields $\overline{\pi(a)} = \pi(a^+)^*$. This is the second half of (13.3).

Now we prove (13.4). Let us abbreviate $T := \overline{\pi(a)} \cdot \overline{\pi(c)}$. Clearly,

$$\pi(c^+ a^+ a c) = \pi(c^+)\pi(a^+)\pi(a)\pi(c) \subseteq \left(\overline{\pi(a)} \cdot \overline{\pi(c)}\right)^* \overline{\pi(a)} \cdot \overline{\pi(c)} = T^* T,$$

so that $\overline{\pi(c^+ a^+ a c)} \subseteq T^* T$. Since $\pi \restriction B$ is integrable and $c^+ a^+ a c \in B_{her}$, the operator $\overline{\pi(c^+ a^+ a c)}$ is self-adjoint. But $T^* T$ is also self-adjoint. Hence $\overline{\pi(c^+ a^+ a c)} = T^* T$ which implies $\mathcal{D}((\overline{\pi(c^+ a^+ a c)})^{1/2}) = \mathcal{D}(|T|) = \mathcal{D}(T)$. Since $ac \in A_{gh}$, Eq. (13.5) remains valid if a is replaced by ac and yields $\mathcal{D}(\overline{\pi(ac)}) = \mathcal{D}((\overline{\pi(c^+ a^+ a c)})^{1/2})$. From the two preceding domain equalities we conclude that $\mathcal{D}(\overline{\pi(ac)}) = \mathcal{D}(T)$. Combined with the obvious inclusion $\overline{\pi(ac)} \subseteq \overline{\pi(a)} \cdot \overline{\pi(c)} = T$, this gives $\overline{\pi(ac)} = T$, which is the equality (13.4). $\qquad\square$

13.2 Representations Associated with $*$-Algebras of Fractions

Throughout this section, we deal with the following setup:

(1) A *is a unital $*$-algebra and* S *is a countable submonoid of* $A \backslash \{0\}$ *which is invariant under the involution (i.e., $1 \in S$ and $st \in S$, $s^+ \in S$ for $s, t \in S$).*

(2) B *is a unital $*$-algebra such that* A *is a $*$-subalgebra of* B *containing the unit element of* B *and all elements of* S *are invertible in* B.

(3) X *is a unital* *-*subalgebra of* B *such that* S^{-1} *is a right Ore subset of* X *(i.e.,* $S^{-1} \subseteq X$ *and for* $s \in S$, $x \in X$ *there exist* $t \in S$, $y \in X$ *such that* $s^{-1}y = xt^{-1}$, *or equivalently,* $sx = yt$).

It should be emphasized that (3) requires that the set S^{-1} (!) of inverses of elements of S is an Ore set of X. Further, A and B have the same unit elements and

$$S \subseteq A \subseteq B, \quad S^{-1} \subseteq X \subseteq B.$$

Since S is *-invariant, so is S^{-1}. Hence the right Ore set S^{-1} of X is also a *left Ore set* (i.e., for $s \in S$, $x \in X$ there exist $t \in S$, $y \in Y$ such that $ys^{-1} = t^{-1}x$, or equivalently, $xs = ty$). Note that the *-algebras B and X contain "fractions" of the form as^{-1}, where $a \in A$, $s \in S$.

A simple example for the preceding setup is the following.

Example 13.8 Let $A = \mathbb{C}[x]$ and $S = \{s^n : n \in \mathbb{N}_0\}$, where $s := x^2 + 1$. Let B be the *-algebra of rational functions on $\mathbb{R}$ generated by A and $s^{-1} = (x^2 + 1)^{-1}$ and let X be the unital *-subalgebra of B generated by $a := s^{-1}$ and $b := xs^{-1}$. Since X is commutative, S^{-1} is obviously a right Ore subset of X. ○

Remark 13.9 We mention two interesting and important examples of Ore sets. These results will not be used in this book. Let A be the enveloping algebra $\mathcal{E}(\mathfrak{g})$ of a finite-dimensional Lie algebra $\mathfrak{g}$ (see Sect. 9.1) or the Weyl algebra $W(d)$ (see Example 2.10). Then the set $A \setminus \{0\}$ is a right Ore subset of A and also a left Ore set, because it is *-invariant. For $A = \mathcal{E}(\mathfrak{g})$ this is proved in [Di77a, 3.6.13], while for $A = W(d)$ this follows from [Di77a, 4.6.4], [Co95, Exercise 5.5], or [Lm99, p. 318, Exercise 12]. ○

Some simple facts are collected in the next lemma. Set

$$XS := \{xs : x \in X, s \in S\} \subseteq B, \quad SX := \{sx : x \in X, s \in S\} \subseteq B.$$

Lemma 13.10 *Suppose that the conditions of (3) are satisfied.*

(i) *Let* $\mathcal{F}$ *be a finite subset of* S. *There exists an element* $t_0 \in S$ *such that* $st^{-1} \in X$ *and* $t^{-1}s \in X$ *for all* $s \in \mathcal{F}$, *where* $t := t_0^+ t_0 \in S$.
(ii) $XS = SX$.
(iii) XS *is a unital* *-*subalgebra of* B.
(iv) *Let* A_0 *be a set of generators of the* *-*algebra* A. *If* $A_0 \subseteq XS$, *then* $A \subseteq XS$.

Proof (i): We first prove by induction on the cardinality that for each finite set $\mathcal{F} \subseteq S$ there exists a $t_1 \in S$ such that $st_1^{-1} \in X$ for all $s \in \mathcal{F}$. Suppose this is true for $\mathcal{F}$. Let $s_1 \in S$. Since $s_1^{-1} \in X$ and S^{-1} is a right Ore set, there are elements $t_2 \in S$ and $y \in X$ such that $s_1^{-1}t_2^{-1} = t_1^{-1}y$. Then $s(t_2s_1)^{-1} = (st_1^{-1})y \in X$ for $s \in \mathcal{F}$ and $s_1(t_2s_1)^{-1} = t_2^{-1} \in X$. This proves our claim with $t_2s_1 \in S$ for $\mathcal{F} \cup \{s_1\}$.

Again, let $\mathcal{F}$ be a finite subset of S. Applying the statement proved in the preceding paragraph to the set $\mathcal{F} \cup \mathcal{F}^+$, there exists an element $t_0 \in S$ such that

$st_0^{-1} \in X$ and $s^+ t_0^{-1} \in X$ for $s \in F$. Then we have $s(t_0^+ t_0)^{-1} = (st_0^{-1})(t_0^+)^{-1} \in X$ and $(t_0^+ t_0)^{-1} s = ((s^+ t_0^{-1})(t_0^+)^{-1})^+ \in X$ for $s \in \mathcal{F}$.

(ii): That S^{-1} is a right Ore set means that $SX \subseteq XS$. Hence, since X and S are $*$-invariant, $XS \subseteq SX$. Thus, $XS = SX$.

(iii): Let $x_1, x_2 \in X$ and $s_1, s_2 \in S$. By the right Ore property of S^{-1}, there are elements $t_1 \in S$, $y_1 \in X$ such that $s_1 x_2 = y_1 t_1$. Then $x_1 s_1 x_2 s_2 = x_1 y_1 t_1 s_2 \in XS$. By (i), we can find an element $t \in S$ such that $s_1 t^{-1} \in X$ and $s_2 t^{-1} \in X$. For $\lambda_1, \lambda_2 \in \mathbb{C}$, we obtain $\lambda_1 x_1 s_1 + \lambda_2 x_2 s_2 = (\lambda_1 x_1 s_1 t^{-1} + \lambda_2 x_2 s_2 t^{-1})t \in XS$. This proves that XS is a linear subspace of B which is closed under multiplication. Obviously, $1 = 1 \cdot 1 \in XS$. Since $XS = SX$ by (ii), XS is invariant under the involution. Summarizing, we have shown that XS is a unital $*$-subalgebra of B.

(iv): Follows at once from (iii). $\square$

Retain the conditions (1), (2), (3) and suppose $A \subseteq XS$. Recall that $S^{-1} \subseteq X$.

Definition 13.11 A $*$-representation ρ of the $*$-algebra X is called S^{-1}-*torsionfree* if $\ker \rho(s^{-1}) = \{0\}$ for all $s \in S$.

Suppose ρ is a nondegenerate S^{-1}-*torsionfree* $*$-representation of X acting by bounded operators on a Hilbert space $\mathcal{H}$ such that $\mathcal{D}(\rho) = \mathcal{H}$. Then $\rho(1) = I$, because ρ is nondegenerate. Our aim is to associate an (in general unbounded) $*$-representation π_ρ of the $*$-algebra A with ρ. The domain of π_ρ is

$$\mathcal{D}(\pi_\rho) := \bigcap_{s \in S} \rho(s^{-1})\mathcal{H}. \tag{13.6}$$

Let $a \in A$. Since $A \subseteq XS$, there exists an $s \in S$ such that $as^{-1} \in X$. Then the inverse operator $\rho(s^{-1})^{-1}$ exists (because ρ is S^{-1}-*torsionfree*)), and we can define

$$\pi_\rho(a)\varphi := \rho(as^{-1})\rho(s^{-1})^{-1}\varphi, \quad \varphi \in \mathcal{D}(\pi_\rho). \tag{13.7}$$

The main result of this section is the following theorem.

Theorem 13.12 *Suppose* A, S, B, X *satisfy the assumptions (1), (2), (3) and, in addition,* $A \subseteq XS$. *Let* ρ *be a bounded nondegenerate* S^{-1}-*torsionfree* $*$-*representation of the* $*$-*algebra* X *on a Hilbert space* $\mathcal{D}(\rho) = \mathcal{H}$.

Then π_ρ *is a well-defined closed* $*$-*representation of the* $*$-*algebra* A *on the dense domain* $\mathcal{D}(\pi_\rho)$ *of* $\mathcal{H}$. *It has a Frechet graph topology, and we have* $\pi_\rho(1) = I$,

$$\pi_\rho(s)\mathcal{D}(\pi_\rho) = \mathcal{D}(\pi_\rho) \quad and \quad \pi_\rho(s)\varphi = \rho(s^{-1})^{-1} \upharpoonright \mathcal{D}(\pi_\rho) \quad for \ s \in S. \tag{13.8}$$

Some technical steps of the proof are contained in the following lemma.

Lemma 13.13 (i) $\mathcal{D}(\pi_\rho)$ *is dense in the Hilbert space* $\mathcal{H}$.
(ii) $\rho(x)\mathcal{D}(\pi_\rho) \subseteq \mathcal{D}(\pi_\rho)$ *for* $x \in X$.
(iii) $\rho(s^{-1})\mathcal{D}(\pi_\rho) = \mathcal{D}(\pi_\rho)$ *for* $s \in S$.

Proof (i): The proof is based on the Mittag-Leffler lemma (Proposition 3.30).

We enumerate the countable set $S = \{r_j : j \in \mathbb{N}\}$ such that $r_1 = 1$ and put $S_n := \{r_1, \ldots, r_n\}$. From Lemma 13.10(i) it follows that for any $n \in \mathbb{N}$ there exists an element $t_n = t_n^+ \in S$ such that $st_n^{-1} \in X$ for all $s \in S_n$ and $t_n t_{n+1}^{-1} \in X$.

Let E_n denote the vector space $\rho(t_n^{-1})\mathcal{H}$, equipped with the inner product

$$(\varphi, \psi)_n := \langle \rho(t_n^{-1})^{-1}\varphi, \rho(t_n^{-1})^{-1}\psi \rangle, \quad \varphi, \psi \in E_n, \ n \in \mathbb{N}_0.$$

Since E_n is the range of the bounded injective operator $\rho(t_n^{-1})$, $(E_n, (\cdot, \cdot)_n)$ is a Hilbert space with norm $|| \cdot ||_n = ||\rho(t_n^{-1})^{-1} \cdot ||$.

We first show that E_{n+1} is a subspace of E_n and that $|| \cdot ||_n \leq c_n || \cdot ||_{n+1}$ for some constant $c_n > 0$. Let $\psi \in \mathcal{H}$ and set $\varphi := \rho(t_{n+1}^{-1})\psi$. Since $t_n t_{n+1}^{-1} \in X$ by the choice of elements t_k, we have $\varphi = \rho(t_{n+1}^{-1})\psi = \rho(t_n^{-1})\rho(t_n t_{n+1}^{-1})\psi$. This proves that $E_{n+1} \subseteq E_n$. Set $c_n = ||\rho(t_n t_{n+1}^{-1})||$. By definition, $||\varphi||_{n+1} = ||\psi||$ and hence

$$||\varphi||_n = ||\rho(t_n t_{n+1}^{-1})\psi|| \leq ||\rho(t_n t_{n+1}^{-1})|| \, ||\psi|| = c_n \, ||\varphi||_{n+1}.$$

Next we prove that E_{n+1} is dense in $(E_n, || \cdot ||_n)$. It suffices to show that the null vector is the only vector $\zeta \in E_n$ which is orthogonal to E_{n+1} in the Hilbert space $(E_n, (\cdot, \cdot)_n)$. Put $\xi := \rho(t_n^{-1})\zeta$. Then, since ζ is orthogonal to E_{n+1} in $(E_n, (\cdot, \cdot)_n)$,

$$
\begin{aligned}
0 = (\zeta, \rho(t_{n+1}^{-1})\varphi)_n &= \langle \rho(t_n^{-1})^{-1}\zeta, \rho(t_n^{-1})^{-1}\rho(t_{n+1}^{-1})\varphi \rangle \\
&= \langle \xi, \rho(t_n^{-1})^{-1}\rho(t_n^{-1})\rho(t_n t_{n+1}^{-1})\varphi \rangle = \langle \xi, \rho(t_n t_{n+1}^{-1})\varphi \rangle \\
&= \langle \rho(t_n t_{n+1}^{-1})^*\xi, \varphi \rangle = \langle \rho(t_{n+1}^{-1}t_n)\xi, \varphi \rangle \\
&= \langle \rho(t_{n+1}^{-1}t_n)\rho(t_n^{-1})\zeta, \varphi \rangle = \langle \rho(t_{n+1}^{-1})\zeta, \varphi \rangle
\end{aligned}
$$

for all $\varphi \in \mathcal{H}$. Thus, $\rho(t_{n+1}^{-1})\zeta = 0$. Since ρ is torsionfree, $\ker \rho(t_{n+1}^{-1}) = \{0\}$. Hence $\zeta = 0$. This proves that E_{n+1} is dense in E_n.

In the preceding paragraphs we have shown that the assumptions of Proposition 3.30 are fulfilled. Therefore, $E_\infty := \cap_{n \in \mathbb{N}_0} E_n$ is dense in the normed space $E_0 = \mathcal{H}$. Obviously, $\mathcal{D}(\pi_\rho) \subseteq E_\infty$. Let $s \in S$. Then $s \in S_n$ for some $n \in \mathbb{N}$ and hence $\rho(t_n^{-1})\mathcal{H} = \rho(s^{-1})\rho(st_n^{-1})\mathcal{H} \subseteq \rho(s^{-1})\mathcal{H}$. This yields $E_\infty \subseteq \mathcal{D}(\pi_\rho)$. Therefore, $\mathcal{D}(\pi_\rho) = E_\infty$ is dense in $\mathcal{H}$.

(ii): Suppose $\varphi \in \mathcal{D}(\pi_\rho)$ and $x \in X$. Let $s \in S$. By assumption (3) there exist $t \in S$, $y \in X$ such that $xt^{-1} = s^{-1}y$. By (13.6), there is a vector $\psi \in \mathcal{H}$ such that $\varphi = \rho(t^{-1})\psi$. Then $\rho(x)\varphi = \rho(xt^{-1})\psi = \rho(s^{-1})\rho(y)\psi \in \rho(s^{-1})\mathcal{H}$. Since $s \in S$ was arbitrary, $\rho(x)\varphi \in \cap_{s \in S} \rho(s^{-1})\mathcal{H} = \mathcal{D}(\pi_\rho)$.

(iii): Let $s \in S$. Since $\rho(s^{-1})\mathcal{D}(\pi_\rho) \subseteq \mathcal{D}(\pi_\rho)$ by (ii), it suffices to show that each vector $\varphi \in \mathcal{D}(\pi_\rho)$ is in $\rho(s^{-1})\mathcal{D}(\pi_\rho)$. By (13.6), we have $\varphi \in \rho(s^{-1})\mathcal{H}$ and $\varphi \in \rho((ts)^{-1})\mathcal{H}$ for all $t \in S$, so there exist vectors $\psi \in \mathcal{H}$ and $\eta_t \in \mathcal{H}$ such that $\varphi = \rho(s^{-1})\psi = \rho((ts)^{-1})\eta_t = \rho(s^{-1})\rho(t^{-1})\eta_t$. Since $\ker \rho(s^{-1}) = \{0\}$, the latter implies $\psi = \rho(t^{-1})\eta_t$, so that $\psi \in \cap_{t \in S} \rho(t^{-1})\mathcal{H} = \mathcal{D}(\pi_\rho)$ and $\varphi = \rho(s^{-1})\psi$.

Proof of Theorem 13.12 First we prove that the operator $\pi_\rho(a)$ is well defined, that is, $\pi_\rho(a)$ in (13.7) does not depend on the particular element s of S satisfying $as^{-1} \in$ X. Let $\tilde{s} \in$ S be another element such that $a\tilde{s}^{-1} \in$ X. By Lemma 13.10(i), there exists an element $t \in$ S such that $st^{-1} \in$ X and $\tilde{s}t^{-1} \in$ X. Then we have $at^{-1} = (as^{-1})(st^{-1}) \in$ X. Let r denote s or $\tilde{s}$. Suppose $\varphi \in \mathcal{D}(\pi_\rho)$. We write $\varphi = \rho(t^{-1})\psi$ with $\psi \in \mathcal{H}$ and compute

$$\rho(ar^{-1})\rho(r^{-1})^{-1}\varphi = \rho(ar^{-1})\rho(r^{-1})^{-1}\rho(t^{-1})\psi$$
$$= \rho(ar^{-1})\rho(r^{-1})^{-1}\rho(r^{-1}rt^{-1})\psi = \rho(ar^{-1})\rho(rt^{-1})\psi$$
$$= \rho(ar^{-1}rt^{-1})\rho(t^{-1})^{-1}\psi = \rho(at^{-1})\rho(t^{-1})^{-1}\varphi.$$

We apply this for $r = s$, $\tilde{s}$ and obtain $\rho(as^{-1})\rho(s^{-1})^{-1}\varphi = \rho(a\tilde{s}^{-1})\rho(\tilde{s}^{-1})^{-1}\varphi$. This shows that the operator $\pi_\rho(a)$ is well defined.

Since $\rho(s^{-1})^{-1}\varphi \in \mathcal{D}(\pi_\rho)$ and $\rho(as^{-1})\rho(s^{-1})^{-1}\varphi \in \mathcal{D}(\pi_\rho)$ by Lemma 13.13,(ii) and (iii), $\pi_\rho(a)\varphi \in \mathcal{D}(\pi_\rho)$. Hence $\pi_\rho(a)$ leaves $\mathcal{D}(\pi_\rho)$ invariant.

Let $a, b \in$ A. We prove that $\pi_\rho(a+b) = \pi_\rho(a) + \pi_\rho(b)$ and $\pi_\rho(ab) = \pi_\rho(a)\pi_\rho(b)$. For this we use essentially the fact, proved in the first paragraph, that in the definition (13.7) of $\pi_\rho(a)$ any element $s \in$ S such that $as^{-1} \in$ X can be taken.

Since $A \subseteq$ XS, there are $s_1, s_2 \in$ S such that $as_1^{-1} \in$ X and $bs_2^{-1} \in$ X. By Lemma 13.10(i) we can find $s \in$ S such that $s_1s^{-1} \in$ X and $s_2s^{-1} \in$ X. Since $as^{-1} \in$ X, $bs^{-1} \in$ X, and $(a+b)s^{-1} \in$ X, the relation

$$\rho(as^{-1})\rho(s^{-1})^{-1}\varphi + \rho(bs^{-1})\rho(s^{-1})^{-1}\varphi = \rho((a+b)s^{-1})\rho(s^{-1})^{-1}\varphi$$

for $\varphi \in \mathcal{D}(\pi_\rho)$ implies that $\pi_\rho(a+b)\varphi = \pi_\rho(a)\varphi + \pi_\rho(b)\varphi$.

Again by $A \subseteq$ XS, there exist $t_1, t_2, t_3, t_4 \in$ S such that $at_1^{-1}, bt_2^{-1}, abt_3^{-1} \in$ X and $t_1bt_4^{-1} \in$ X. By Lemma 13.10(i) there is an element $t \in$ S such that $t_jt^{-1} \in$ X, $j = 1, 2, 3, 4$. Then $abt^{-1} = (abt_3^{-1})(t_3t^{-1}) \in$ X, $t_1bt^{-1} = (t_1bt_4^{-1})(t_4t^{-1}) \in$ X and $bt^{-1} = (bt_2^{-1})(t_2t^{-1}) \in$ X. Let $\varphi \in \mathcal{D}(\pi_\rho)$. Inserting the preceding facts and the definitions of $\pi_\rho(ab)$, $\pi_\rho(a)$, $\pi_\rho(b)$ we derive

$$\pi_\rho(ab)\varphi = \rho(abt^{-1})\rho(t^{-1})^{-1}\varphi = \rho(at_1^{-1})\rho(t_1bt^{-1})\rho(t^{-1})^{-1}\varphi$$
$$= \rho(at_1^{-1})\rho(t_1^{-1})^{-1}\rho(t_1^{-1})\rho(t_1bt^{-1})\rho(t^{-1})^{-1}\varphi$$
$$= \pi_\rho(a)\rho(t_1^{-1}t_1bt^{-1})\rho(t^{-1})^{-1}\varphi = \pi_\rho(a)\rho(bt^{-1})\rho(t^{-1})^{-1}\varphi$$
$$= \pi_\rho(a)\pi_\rho(b)\varphi.$$

Next we show that $\langle \pi_\rho(a)\varphi, \psi \rangle = \langle \varphi, \pi_\rho(a^+)\psi \rangle$ for $a \in$ A and $\varphi, \psi \in \mathcal{D}(\pi_\rho)$. From Lemma 13.10(i) there are elements $t_1, t_2 \in$ S and $t = t^+ \in$ S such that $at_1^{-1}, a^+t_2^{-1} \in$ X and $t_1t^{-1}, t_2t^{-1} \in$ X. Then $at^{-1} = (at_1^{-1})(t_1t^{-1}) \in$ X and $a^+t^{-1} = (a^+t_1^{-1})(t_2t^{-1}) \in$ X. Using that ρ is *-representation defined on $\mathcal{H}$ and $\rho(t^{-1})$ is a bounded self-adjoint operator we compute

$$\langle \pi_\rho(a)\varphi, \psi \rangle = \langle \rho(at^{-1})\rho(t^{-1})^{-1}\varphi, \psi \rangle = \langle \rho(t^{-1})^{-1}\varphi, \rho((at^{-1})^+)\psi \rangle$$
$$= \langle \rho(t^{-1})^{-1}\varphi, \rho(t^{-1}a^+)\rho(t^{-1})\rho(t^{-1})^{-1}\psi \rangle$$
$$= \langle \rho(t^{-1})^{-1}\varphi, \rho(t^{-1}a^+t^{-1})\rho(t^{-1})^{-1}\psi \rangle$$
$$= \langle \rho(t^{-1})^{-1}\varphi, \rho(t^{-1})\rho(a^+t^{-1})\rho(t^{-1})^{-1}\psi \rangle$$
$$= \langle \rho(t^{-1})\rho(t^{-1})^{-1}\varphi, \pi_\rho(a^+)\psi \rangle = \langle \varphi, \pi_\rho(a^+)\psi \rangle,$$

where in the second line we can write $\psi \in \mathcal{D}(\pi_\rho)$ as $\psi = \rho(t^{-1})\rho(t^{-1})^{-1}\psi$ because $\rho(t^{-1})\mathcal{D}(\pi_\rho) = \mathcal{D}(\pi_\rho)$ by Lemma 13.13(iii).

Setting $a = s = 1$ in (13.7) yields $\pi_\rho(1)\varphi = \varphi$ for $\varphi \in \mathcal{D}(\pi_\rho)$, so $\pi_\rho(1) = I$.

Recall from Lemma 13.13(i) that $\mathcal{D}(\pi_\rho)$ is dense in $\mathcal{H}$. Thus, we have shown that π_ρ is a ∗-representation of A on the dense domain $\mathcal{D}(\pi_\rho)$ of the Hilbert space $\mathcal{H}$.

Let $s \in \mathsf{S}$. Then, $\pi_\rho(s)\varphi = \rho(ss^{-1})\rho(s^{-1})^{-1}\varphi = \rho(s^{-1})^{-1}\varphi$ for $\varphi \in \mathcal{D}(\pi_\rho)$, because $\rho(1) = I$. Since $\rho(s^{-1})\mathcal{D}(\pi_\rho) = \mathcal{D}(\pi_\rho)$ by Lemma 13.13(iii), we obtain $\pi_\rho(s)\mathcal{D}(\pi_\rho) = \mathcal{D}(\pi_\rho)$. This proves (13.8).

Finally we prove the assertion concerning the graph topology of π_ρ. Let us retain the notation of the proof of Lemma 13.13(i). Then, since $\pi_\rho(t_n)\varphi = \rho(t_n^{-1})^{-1}\varphi$ for $\varphi \in \mathcal{D}(\pi_\rho)$ by (13.8), $||\pi_\rho(t_n)\varphi||$ is just the norm $||\varphi||_n$. Let $a \in \mathsf{A}$. Applying again that $\mathsf{A} \subseteq \mathsf{XS}$, there exist $n \in \mathbb{N}$ and $t \in \mathsf{S}_n$ such that $at^{-1} \in \mathsf{X}$. Then $tt_n^{-1} \in \mathsf{X}$, so that $at_n^{-1} = (at^{-1})(tt_n^{-1}) \in \mathsf{X}$ and hence

$$||\pi_\rho(a)\varphi|| = ||\rho(at_n^{-1})\rho(t_n^{-1})^{-1}\varphi|| = ||\rho(at_n^{-1})\pi_\rho(t_n)\varphi|| \leq ||\rho(at_n^{-1})|| \, ||\varphi||_n.$$

The preceding shows that the graph topology of π_ρ is generated by the countable family of norms $|| \cdot ||_n, n \in \mathbb{N}$. Hence it is metrizable. It is complete and so a Frechet topology, since $\mathcal{D}(\pi_\rho) = \cap_n E_n$ is the intersection of Hilbert spaces $(E_n, || \cdot ||_n)$. In particular, π_ρ is closed. This completes the proof of Theorem 13.12.   □

Corollary 13.14 *Retain the assumptions and the notation of Theorem 13.12. Suppose there is a set X_0 of generators of the algebra X such that each $x \in \mathsf{X}_0$ is of the form $x = as^{-1}$ in B, with $a \in \mathsf{A}$, $s \in \mathsf{S}$. Then $\pi_\rho(\mathsf{A})'_s = \rho(\mathsf{X})'$. In particular, the ∗-representation π_ρ of A is irreducible if and only if the ∗-representation ρ of X is irreducible.*

Proof Suppose $T \in \pi_\rho(\mathsf{A})'_s$. Then, $T\varphi \in \mathcal{D}(\pi_\rho)$ and $T\pi_\rho(a)\varphi = \pi_\rho(a)T\varphi$ for $a \in \mathsf{A}$, $\varphi \in \mathcal{D}(\pi_\rho)$. Let $x = as^{-1} \in \mathsf{X}_0$, where $a \in \mathsf{A}$, $s \in \mathsf{S}$, and $\psi \in \mathcal{D}(\pi_\rho)$. By Lemma 13.13(ii), $\varphi := \rho(s^{-1})\psi$ is in $\mathcal{D}(\pi_\rho)$. Applying (13.7) twice we derive

$$T\rho(x)\psi = T\rho(as^{-1})\rho(s^{-1})^{-1}\varphi = T\pi_\rho(a)\varphi = \pi_\rho(a)T\varphi = \rho(as^{-1})\rho(s^{-1})^{-1}T\varphi$$
$$= \rho(as^{-1})\pi_\rho(s)T\varphi = \rho(as^{-1})T\pi_\rho(s)\varphi = \rho(x)T\rho(s^{-1})^{-1}\varphi = \rho(x)T\psi.$$

Since $\mathcal{D}(\pi_\rho)$ is dense in $\mathcal{H}$, this gives $T\rho(x) = \rho(x)T$ on $\mathcal{H}$ for $x \in \mathsf{X}_0$. Hence $T \in \rho(\mathsf{X})'$, because X_0 generates the algebra X.

Conversely, let $T \in \rho(\mathsf{X})'$. From (13.6) it follows at once that T leaves $\mathcal{D}(\pi_\rho)$ invariant. Therefore, since T commutes with $\rho(as^{-1})$ and $\rho(s^{-1})$, hence with $\rho(s^{-1})^{-1}$, (13.7) implies that T commutes with $\pi_\rho(a)$. Thus, $T \in \pi_\rho(\mathsf{A})'_s$.

The assertion concerning the irreducibility follows from the fact (Proposition 4.26) that π_ρ is irreducible if and only if $0, I$ are the only projections in $\pi_\rho(\mathsf{A})'_s$. $\square$

Example 13.15 (*Example* 13.8 *continued*)
Retain the notation of Example 13.8. Recall that X is the unital $*$-algebra of rational functions generated by $a := s^{-1} = (x^2 + 1)^{-1}$ and $b := xs^{-1} = x(x^2 + 1)^{-1}$. Then $(a - \frac{1}{2})^2 + b^2 = \frac{1}{4}$, so the $*$-algebra X is bounded by Corollary 10.8. Hence each closed $*$-representation ρ of X acts by bounded operators on $\mathcal{D}(\rho) = \mathcal{H}(\rho)$.

Let $\mathcal{C}$ be the circle in $\mathbb{R}^2$ defined by the equation $(\lambda - \frac{1}{2})^2 + \mu^2 = \frac{1}{4}$. Then, since $\rho(a)$ and $\rho(b)$ are commuting bounded self-adjoint operators on $\mathcal{H}(\rho)$ satisfying $(\rho(a) - \frac{1}{2}I)^2 + \rho(b)^2 = \frac{1}{4}I$, it follows from the multi-dimensional spectral theorem that there exists a spectral measure E supported on $\mathcal{C}$ such that

$$\rho(p(a, b)) = \int_{\mathcal{C}} p(\lambda, \mu) \, dE((\lambda, \mu)), \quad p \in \mathbb{C}[a, b].$$

Then $\ker \rho(a) = \ker \rho(s^{-1}) = E((0, 0))\mathcal{H}(\rho)$. Therefore, ρ is torsionfree if and only if $E((0, 0)) = 0$.

Assume that $E((0, 0)) = 0$. Note that $x = ba^{-1}$ in the larger algebra B. By (13.7) and (13.6), $\pi_\rho(x)\varphi = \rho(b)\rho(a)^{-1}\varphi$ for $\varphi \in \mathcal{D}(\pi_\rho) = \cap_{n=1}^{\infty}\rho(a^n)\mathcal{H}(\rho)$, so

$$\pi_\rho(q(x))\varphi = \int_{\mathcal{C}} q(\mu\lambda^{-1}) \, dE((\lambda, \mu))\varphi, \quad q \in \mathsf{A} = \mathbb{C}[x], \ \varphi \in \mathcal{D}(\pi_\rho). \qquad \bigcirc$$

13.3 Application to the Weyl Algebra

In this section, A is the Weyl algebra $\mathbb{C}\langle p, q \mid p = p^+, q = q^+, pq - qp = -\mathrm{i}\rangle$ and the submonoid S consists of 1 and all finite products of elements from the set

$$\{s_1 = p - \mathrm{i}, \ s_2 = p + \mathrm{i}, \ s_3 = q - \mathrm{i}, \ s_4 = q + \mathrm{i}\}.$$

Recall that the Schrödinger representation π_S of A on $L^2(\mathbb{R})$ acts by

$$(\pi_S(p)\varphi)(t) = -\mathrm{i}\varphi'(t), \quad (\pi_S(q)\varphi)(t) = t\varphi(t) \text{ for } \varphi \in \mathcal{D}(\pi_S) = \mathcal{S}(\mathbb{R}).$$

Since multiplication by the function $(t \pm \mathrm{i})^{-1}$ leaves the Schwartz space $\mathcal{S}(\mathbb{R})$ invariant, $\pi_S(s_3)$ and $\pi_S(s_4)$ are bijections of $\mathcal{S}(\mathbb{R})$. Applying the Fourier transform (Example 8.11) it follows that the same is true for $\pi_S(s_1)$ and $\pi_S(s_2)$. Hence $x := \pi_S(s_1)^{-1}$ and $y := \pi_S(s_3)^{-1}$ are bounded operators of $\mathcal{L}^+(\mathcal{S}(\mathbb{R}))$. Clearly, $x^+ = \pi_S(s_2)^{-1}$ and $y^+ = \pi_S(s_4)^{-1}$. Then the following relations are valid:

$$x - x^+ = 2i\, x^+x, \quad y - y^+ = 2i\, y^+y, \quad xx^+ = x^+x, \quad yy^+ = y^+y, \qquad (13.9)$$

$$xy - yx = -ixy^2x = -iyx^2y, \quad xy^+ - y^+x = -ix(y^+)^2x = -iy^+x^2y^+. \qquad (13.10)$$

Indeed, the relations (13.9) are obvious resolvent identities, while the relations (13.10) are just the resolvent Eqs. (8.53) and (8.54) which hold by Example 8.21.

Let B denote the $*$-subalgebra of $\mathcal{L}^+(\mathcal{S}(\mathbb{R}))$ generated by the operators x, y, and $\pi_S(A)$ and let X be the unital $*$-subalgebra of B generated by x, y. Since π_S is faithful, A becomes a $*$-subalgebra of B by identifying $\pi_S(f)$ and $f \in A$. From the relations (13.10) it follows immediately that $xX = Xx$ and $yX = Xy$. Hence $x^+X = Xx^+$ and $y^+X = Xy^+$. Since $x \cong s_1^{-1}$, $x^+ \cong s_2^{-1}$, $y \cong s_3^{-1}$, $y^+ \cong s_4^{-1}$, the latter means that the four generators s_j^{-1} of the monoid S^{-1} in X obey the right Ore condition and so does S^{-1}. Thus assumption (3) is fulfilled. Clearly, $p - i = s_1^{-1}s_1^2 \in XS$ and $q - i = s_3^{-1}s_3^2 \in XS$. Therefore, from Lemma 13.10(iv), applied with $A_0 = \{p - i, q - i\}$, it follows that $A \subseteq XS$. Summarizing, all assumptions of Theorem 13.12 and Corollary 13.14 (with $X_0 = \{x, x^+, y, y^+\}$) are satisfied.

Lemma 13.16 *Suppose z is a bounded normal operator on a Hilbert space $\mathcal{H}$ such that $\ker z = \{0\}$ and $z - z^* = 2\,iz^*z$. Then $b := z^{-1} + iI$ is self-adjoint.*

Proof Because z is normal, we obtain $\ker z^* = \{0\}$. Since $z^* = z(I - 2iz^*)$ and $z = z^*(I + 2\,iz)$, we have $z^*\mathcal{H} = z\mathcal{H}$ and so $\mathcal{D}((z^*)^{-1}) = \mathcal{D}(z^{-1})$. The identity $z^* = z(I - 2iz^*)$ implies that $z^{-1}z^* = I - 2iz^*$ on $\mathcal{H}$. Let $\psi \in \mathcal{H}$. Then, for $\varphi := z^*\psi \in \mathcal{D}((z^*)^{-1})$, we get $z^{-1}\varphi = z^{-1}z^*\psi = \psi - 2iz^*\psi = (z^*)^{-1}\varphi - 2\,i\varphi$. This means that $z^{-1} \supseteq (z^*)^{-1} - 2iI$. Since $\mathcal{D}((z^*)^{-1}) = \mathcal{D}(z^{-1})$ as noted above, it follows that $z^{-1} = (z^*)^{-1} - 2\,iI$. Using the latter identity we derive

$$b = z^{-1} + iI = (z^*)^{-1} - iI = (z^{-1})^* - iI = b^*. \qquad \square$$

Theorem 13.17 *Suppose ρ is a nondegenerate bounded S^{-1}-torsionfree $*$-representation of the $*$-algebra X on a Hilbert space $\mathcal{D}(\rho) = \mathcal{H}$. Then the $*$-representation $\pi_\rho^\sim$ of the Weyl algebra A from Theorem 13.12 is unitarily equivalent to a direct sum of Schrödinger representations.*

Proof Set $z := \rho(x) = \rho(s_1^{-1})$. Then $z^* = \rho(x^+)$. From the first and the third relations of (13.9) it follows that z is a normal operator satisfying $z - z^* = 2\,iz^*z$. Since ρ is torsionfree, $\ker z = \{0\}$. Therefore, Lemma 13.16 applies, so the operator $P := z^{-1} + iI = \rho(s_1^{-1})^{-1} + iI$ is self-adjoint. Replacing x by y the same reasoning shows that $Q := \rho(s_3^{-1})^{-1} + iI$ is a self-adjoint operator.

Let $\varphi \in \mathcal{D}(\pi_\rho)$. We show that $PQ\varphi - QP\varphi = -i\varphi$. The relation $pq - qp = -i$ implies $s_1s_3 - s_3s_1 = -i$. Since π_ρ is a representation of A by Theorem 13.12,

$$\pi_\rho(s_1)\pi_\rho(s_3)\varphi - \pi_\rho(s_3)\pi_\rho(s_1)\varphi = -i\varphi. \qquad (13.11)$$

From (13.8) and the definition of P we obtain $\pi_\rho(s_1)\varphi = \rho(s_1^{-1})^{-1}\varphi = (P - \mathrm{i})\varphi$.
Similarly, $\pi_\rho(s_3)\varphi = (Q - \mathrm{i})\varphi$. Inserting these expressions into Eq. (13.11) we get
$(P - \mathrm{i})(Q - \mathrm{i})\varphi - (Q - \mathrm{i})(P - \mathrm{i})\varphi = -\mathrm{i}\varphi$, so that $PQ\varphi - QP\varphi = -\mathrm{i}\varphi$.

By (13.8), $(P - \mathrm{i})(Q - \mathrm{i})\mathcal{D}(\pi_\rho) = \pi_\rho(s_1)\pi_\rho(s_3)\mathcal{D}(\pi_\rho) = \mathcal{D}(\pi_\rho)$. Therefore, the
self-adjoint operators P and Q satisfy the assumptions of Kato's Theorem 8.22. By
this theorem, $\{P, Q\}$ is unitarily equivalent to a direct sum of Schrödinger pairs.

We show that π_ρ is unitarily equivalent to a direct sum of Schrödinger representations. The map $\rho \mapsto \pi_\rho$ (according to Theorem 13.12) respects unitary equivalences and direct sums. Hence it suffices to prove this in the case when $\{P, Q\}$ is
the Schrödinger pair. By construction, $\pi_\rho(p) = \pi_\rho(s_1 + \mathrm{i}) \subseteq (P - \mathrm{i}) + \mathrm{i} = P$. Similarly, $\pi_\rho(q) \subseteq Q$. Since the Schrödinger representation is the largest $*$-representation
π of A on $L^2(\mathbb{R})$ such that $\pi(p) \subseteq P$ and $\pi(q) \subseteq Q$, this implies $\pi_\rho \subseteq \pi_S$.

We prove the converse inclusion. From the relation $P = \rho(s_1^{-1})^{-1} + \mathrm{i}I$ we get
$(P - \mathrm{i})^{-1} = \rho(s_1^{-1})$. Similarly, we derive $(P + \mathrm{i})^{-1} = \rho(s_2^{-1})$, $(Q - \mathrm{i})^{-1} = \rho(s_3^{-1})$,
and $(Q + \mathrm{i})^{-1} = \rho(s_4^{-1})$. The Schwartz space $\mathcal{S}(\mathbb{R})$ is invariant under $(P \pm \mathrm{i})^{-1}$
and $(Q \pm \mathrm{i})^{-1}$, so it is contained in the intersection of ranges of finite products
of operators $\rho(s_j^{-1})$ and hence in $\mathcal{D}(\pi_\rho)$ by (13.6). Thus, $\mathcal{D}(\pi_S) = \mathcal{S}(\mathbb{R}) \subseteq \mathcal{D}(\pi_\rho)$.
Combined with the result from the preceding paragraph, we obtain $\pi_\rho = \pi_S$. $\square$

13.4 Compatible Pairs of $*$-Algebras

In this section, A is a **complex unital** $*$-algebra and X is a (not necessarily unital)
complex $*$-algebra.

Definition 13.18 We shall say that (A, X) is a *compatible pair of $*$-algebras* if X
is a left A-module, with left action denoted by $\triangleright$, that is, there exists a linear map
$\theta : A \otimes X \mapsto X$, written as $\theta(a \otimes x) = a \triangleright x$, such that

$$(ab) \triangleright x = a \triangleright (b \triangleright x) \quad \text{and} \quad 1_A \triangleright x = x \quad \text{for} \quad a, b \in A, \ x \in X, \qquad (13.12)$$

satisfying the condition

$$(a \triangleright x)^+ y = x^+(a^+ \triangleright y) \quad \text{for} \quad a \in A, \ x, y \in X. \qquad (13.13)$$

We illustrate this notion with a simple example.

Example 13.19 Suppose that B is a $*$-algebra which contains A and X as
$*$-subalgebras such that $a \cdot x \in X$ and $1_A \cdot x = x$ for $a \in A$, $x \in X$. We define
$\theta(\sum_j a_j \otimes x_j) := \sum_j a_j \cdot x_j$, that is, $a \triangleright x := a \cdot x$, where "$\cdot$" always denotes the
product of the larger $*$-algebra B.

Then (A, X) is a compatible pair of $*$-algebras. Condition (13.13) and the first
equation of (13.12) follow from the $*$-algebra axioms for B. We verify (13.13) by

$$(a \triangleright x)^+ y = (a \cdot x)^+ y = (x^+ \cdot a^+) \cdot y = x^+ \cdot (a^+ \cdot y) = x^+(a^+ \triangleright y).   \quad \bigcirc$$

Any compatible pair (A, X) allows one to construct ∗-representations of A from nondegenerate ∗-representations of X, as shown by the following theorem.

Theorem 13.20 *Suppose* (A, X) *is a compatible pair of* ∗-*algebras and let* ρ *be a nondegenerate* ∗-*representation of the* ∗-*algebra* X. *Then there exists a unique closed* ∗-*representation* π^ρ *of* A *on the Hilbert space* $\mathcal{H}(\pi^\rho) = \mathcal{H}(\rho)$ *such that* $\pi^\rho(1_\mathsf{A}) = I$, $\rho(\mathsf{X})\mathcal{D}(\rho) := \mathrm{Lin}\,\{\rho(x)\varphi : x \in \mathsf{X}, \varphi \in \mathcal{D}(\rho)\}$ *is a core for* π^ρ, *and*

$$\pi^\rho(a)\rho(x)\varphi = \rho(a \triangleright x)\varphi \quad \text{for } a \in \mathsf{A}, \ x \in \mathsf{X}, \ \varphi \in \mathcal{D}(\rho). \tag{13.14}$$

Proof The uniqueness assertion is obvious, so it suffices to prove the existence.

Let $\zeta = \sum_j \rho(x_j)\varphi_j$ and $\eta = \sum_k \rho(y_k)\psi_k$ be arbitrary vectors from $\rho(\mathsf{X})\mathcal{D}(\rho)$, where $x_j, y_k \in \mathsf{X}$ and $\varphi_j, \psi_k \in \mathcal{D}(\rho)$, and let $a \in \mathsf{A}$. Using the compatibility condition (13.13) and the ∗-representation properties of ρ we compute

$$\left\langle \sum_j \rho(a \triangleright x_j)\varphi_j, \sum_k \rho(y_k)\psi_k \right\rangle = \sum_{j,k} \langle \varphi_j, \rho((a \triangleright x_j)^+)\rho(y_k)\psi_k \rangle$$

$$= \sum_{j,k} \langle \varphi_j, \rho((a \triangleright x_j)^+ y_k)\psi_k \rangle = \sum_{j,k} \langle \varphi_j, \rho(x_j^+(a^+ \triangleright y_k))\psi_k \rangle$$

$$= \sum_{j,k} \langle \varphi_j, \rho(x_j^+)\rho(a^+ \triangleright y_k)\psi_k \rangle = \left\langle \sum_j \rho(x_j)\varphi_j, \sum_k \rho(a^+ \triangleright y_k)\psi_k \right\rangle,$$

which means that

$$\left\langle \sum_j \rho(a \triangleright x_j)\varphi_j, \eta \right\rangle = \left\langle \zeta, \sum_k \rho(a^+ \triangleright y_k)\psi_k \right\rangle. \tag{13.15}$$

Now suppose $\zeta = 0$. Then Eq. (13.15) implies $\langle \sum_j \rho(a \triangleright x_j)\varphi_j, \eta \rangle = 0$ for all vectors $\eta \in \rho(\mathsf{X})\mathcal{D}(\rho)$. Since the representation ρ is nondegenerate, the set of such vectors η is dense in $\mathcal{H}(\rho)$. Therefore, $\sum_j \rho(a \triangleright x_j)\varphi_j = 0$. Hence there exists a well-defined (!) linear operator $\pi(a)$ on the domain $\rho(\mathsf{X})\mathcal{D}(\rho)$ given by

$$\pi(a)\left(\sum_j \rho(x_j)\varphi_j \right) := \sum_j \rho(a \triangleright x_j)\varphi_j, \quad a \in \mathsf{A}. \tag{13.16}$$

We insert this definition into Eq. (13.15). Then (13.15) reads as

$$\langle \pi(a)\zeta, \eta \rangle = \langle \zeta, \pi(a^+)\eta \rangle \quad \text{for } a \in \mathsf{A}, \ \zeta, \eta \in \rho(\mathsf{X})\mathcal{D}(\rho). \tag{13.17}$$

From (13.16) and the first equality of (13.12) it follows that π is an algebra homomorphism of A in the algebra $L(\rho(\mathsf{X})\mathcal{D}(\rho))$ of linear operators on $\rho(\mathsf{X})\mathcal{D}(\rho)$. Since $\pi(1_\mathsf{A})(\rho(x)\varphi) = \rho(1_\mathsf{A} \triangleright x)\varphi = \rho(x)\varphi$ by the second equality of (13.12), $\pi(1_\mathsf{A})$ is the identity map of $\rho(\mathsf{X})\mathcal{D}(\rho)$. Combined with (13.17) we have shown that π is a ∗-representation of the ∗-algebra A on the domain $\rho(\mathsf{X})\mathcal{D}(\rho)$ such that $\pi(1_\mathsf{A}) = I$. Let π^ρ be the closure of π. Then all assertions of Theorem 13.20 hold. $\square$

Representations of the form π^ρ, for appropriate compatible pairs and bounded ∗-representations ρ, can be used to define well-behaved representations of A. In the next section, we elaborate this idea with an important example. Another much simpler example is sketched in Exercise 3.

13.5 Application to Enveloping Algebras

In this section, we use the notation and some results on Lie groups and enveloping algebras from Sect. 9.1. Let $A := \mathcal{E}(\mathfrak{g})$ be the enveloping algebra of the Lie algebra $\mathfrak{g}$ of a Lie group G and X the $*$-subalgebra $C_0^\infty(G)$ of the $*$-algebra $L^1(G; \mu_l)$.

First we recall some basic facts. The multiplication of X is the convolution $f_1 * f_2$, see (9.4), and the involution of X is given by $f^+(g) := \Delta_G(g)^{-1}\, \overline{f(g^{-1})}$, see (9.7), where Δ_G is the modular function of G. Note that $\overline{\Delta_G(g)^{-1}} = \Delta_G(g)^{-1}$ and right and left Haar measures of G are related by $d\mu_r(g) = \Delta_G(g)^{-1} d\mu_l(g)$, $g \in G$.

Each element $x \in \mathfrak{g}$ acts on $C_0^\infty(G)$ by $(\tilde{x}f)(g) = \frac{d}{dt} f(\exp(-tx)g)_{|t=0}$; see (9.1). Set $\tilde{1}f = f$. The map $x \mapsto \tilde{x}$ extends uniquely to a unit preserving algebra homomorphism of $\mathcal{E}(\mathfrak{g})$. Then X becomes a left A-module satisfying (13.12) by

$$x \triangleright f := \tilde{x}f \quad \text{for} \quad x \in A = \mathcal{E}(\mathfrak{g}),\ f \in X = C_0^\infty(G). \tag{13.18}$$

Lemma 13.21 *With this left action $\triangleright$, (A, X) is a compatible pair of $*$-algebras.*

Proof By the preceding it only remains to prove that (13.13) is satisfied. This can be derived from Example 13.19; in Exercise 5 we sketch such a proof. Here we prefer to give a direct proof of the crucial condition (13.13) by an explicit computation.

It is easily seen (see Exercise 4) that if (13.13) holds for a set of algebra generators, it does for the whole algebra A. Therefore, since (13.13) is trivial for 1, it suffices to prove it for elements of $\mathfrak{g}$.

Let $x \in \mathfrak{g}$ and f_1, $f_2 \in X$. Using the above formulas and $x^+ = -x$ we derive

$$
\begin{aligned}
((x \triangleright f_1)^+ * f_2)(g) &= \int (x \triangleright f_1)^+(k)\, f_2(k^{-1}g)\, d\mu_l(k) \\
&= \int \Delta_G(k)^{-1} \overline{(\tilde{x}f_1)(k^{-1})}\, f_2(k^{-1}g)\, d\mu_l(k) \\
&= \int \overline{(\tilde{x}f_1)(k^{-1})}\, f_2(k^{-1}g)\, d\mu_r(k) \\
&= -\frac{d}{dt}\Big|_{t=0} \int \overline{f_1(\exp(tx)k^{-1})}\, f_2(k^{-1}g)\, d\mu_r(k) \\
&= -\frac{d}{dt}\Big|_{t=0} \int \overline{f_1(\exp(tx)k^{-1})}\, f_2(k^{-1}g)\, d\mu_r(k\exp(-tx)) \\
&= -\frac{d}{dt}\Big|_{t=0} \int \overline{f_1(h^{-1})}\, f_2(\exp(-tx)h^{-1}g)\, d\mu_r(h) \\
&= -\int \overline{f_1(h^{-1})}\, (\tilde{x}f_2)(h^{-1}g)\, d\mu_r(h) \\
&= -\int \Delta_G(h)^{-1} \overline{f_1(h^{-1})}\, (\tilde{x}f_2)(h^{-1}g)\, d\mu_l(h) \\
&= -\int f_1^+(h)\, (\tilde{x}f_2)(h^{-1}g)\, d\mu_l(h) \\
&= -(f_1^+ * (x \triangleright f_2))(g) = (f_1^+ * (x^+ \triangleright f_2))(g). \qquad \square
\end{aligned}
$$

The following theorem says that certain representations π^ρ from Theorem 13.20 are infinitesimal representations dU.

Theorem 13.22 *Let $\| \cdot \|_1$ denote the norm of $L^1(G; \mu_l)$. Suppose ρ is a bounded nondegenerate $*$-representation of the $*$-algebra X on a Hilbert space $\mathcal{D}(\rho) = \mathcal{H}$ that is continuous in the norm $\| \cdot \|_1$. Then there exists a unitary representation of the Lie group G on $\mathcal{H}$ such that $dU = \pi^\rho$.*

Proof Since $X = C_0^\infty(G)$ is dense in $(L^1(G; \mu_l), \| \cdot \|_1)$ and ρ is $\| \cdot \|_1$-continuous, ρ extends by continuity to a continuous nondegenerate $*$-representation, denoted again by ρ, of the Banach $*$-algebra $L^1(G; \mu_l)$. A basic result from group representation theory (see, e.g., [F95, Theorem 3.11]) states that ρ comes from a unitary representation U of the Lie group G, that is, we have

$$\rho(f)\varphi = \int_G f(g)U(g)\varphi \, d\mu_l(g) = U_f\varphi \quad \text{for } f \in L^1(G; \mu_l), \ \varphi \in \mathcal{D}(\rho). \quad (13.19)$$

In particular, (13.19) holds for $f \in X$. By definition, the domain $\rho(X)\mathcal{D}(\rho)$ is the span of vectors $U_f\varphi$, where $f \in C_0^\infty(G)$, $\varphi \in \mathcal{H}$. This is just the Gårding domain $\mathcal{D}_G(U)$ of U. Let $x \in \mathcal{E}(\mathfrak{g})$, $f \in C_0^\infty(G)$, and $\varphi \in \mathcal{H}$. Then, by formula (13.14) and Proposition 9.6(iii), we derive

$$\pi^\rho(x)U_f\varphi = \pi^\rho(x)\rho(f)\varphi = \rho(x \triangleright f)\varphi = \rho(\tilde{x}f)\varphi = U_{\tilde{x}f}\varphi = dU(x)U_f\varphi.$$

Thus, $\pi^\rho \lceil \mathcal{D}_G(U) = dU \lceil \mathcal{D}_G(U)$. By definition, $\rho(X)\mathcal{D}(\rho)$ is a core for π^ρ, and by Corollary 9.13, $\mathcal{D}_G(U)$ is a core for dU. Therefore, since both $*$-representations π^ρ and dU are closed, we obtain $\pi^\rho = dU$. □

For each unitary representation U of the Lie group G, there is a nondegenerate continuous $*$-representation ρ of $(L^1(G; \mu_l), \| \cdot \|_1)$, hence of $(X, \| \cdot \|_1)$, such that $\rho(f) = U_f$, $f \in X$ (see, e.g., [F95, Theorem 3.9]). Using this result one can reverse the reasoning of the preceding proof and show that the converse of Theorem 13.22 is also true, that is, any $*$-representation dU of $A = \mathcal{E}(\mathfrak{g})$ is of the form π^ρ; see Exercise 6. Thus, if the Lie group G is connected and simply connected, these $*$-representations π^ρ are precisely the integrable representations according to Definition 9.47.

Each $*$-representation of a Banach $*$-algebra on a Hilbert space is continuous in the operator norm [DB86, Theorem 26.13]; hence all $*$-representations of the $*$-algebra $L^1(G; \mu_l)$ are automatically continuous in the norm $\| \cdot \|_1$.

13.6   Exercises

1. Retain the notation and the assumptions of Theorem 13.12. Show that for $s \in \mathsf{S}$ we have $\overline{\pi_\rho(s)} = \rho(s^{-1})^{-1}$ and $\mathcal{D}(\pi_\rho)$ is a core for the operator $\rho(s^{-1})^{-1}$.
   Hint: In the proof of Lemma 13.13, use the fact that $E_\infty = \mathcal{D}(\pi_\rho)$ is dense in each Hilbert space $(E_n, \|\cdot\|_n)$, $n \in \mathbb{N}$, by the Mittag-Leffler lemma.

2. (*Integrable representations of* $\mathsf{A} := \mathbb{C}[x_1, \dots, x_d]$ *and fraction algebras*)
   Let S be the submonoid of A generated by $1, s_j := x_j^2 + 1$ for $j = 1, \dots, d$.
   Let B the $*$-algebra of all rational functions on $\mathbb{R}^d$ and X the $*$-subalgebra of B generated $1, s_j^{-1}, x_j s_j^{-1}$ for $j = 1, \dots, d$.

   a. Verify that $\mathsf{A}, \mathsf{S}, \mathsf{B}, \mathsf{X}$ satisfy the assumptions of Theorem 13.12.
   b. Prove that each $*$-representation π_ρ from Theorem 13.12 is an integrable representation of A in the sense of Definition 7.7.
   c. Show that each integrable representation of A is of the form π_ρ.

3. (*Integrable representations of* $\mathsf{A} := \mathbb{C}[x_1, \dots, x_d]$ *and compatible pairs*)
   Let X be the $*$-algebra $C_c(\mathbb{R}^d)$ of compactly supported continuous functions on $\mathbb{R}^d$ with pointwise multiplication and involution $f^+(t) = \overline{f(t)}$. We define $(p \triangleright \varphi)(t) = p(t)f(t)$ for $p \in \mathsf{A}, \varphi \in \mathsf{X}$.

   a. Show that (A, X) is a compatible pair.
   b. Prove that each $*$-representation π^ρ of A in Theorem 13.20 is integrable.
   c. Show that each integrable $*$-representation of A is of the form π^ρ.

4. Let A be a unital $*$-algebra and X a $*$-algebra. Let A_0 be a set of algebra generators of A. Suppose that X is a left A-module; see Definition 13.18. Prove that if condition (13.13) holds for $a \in \mathsf{A}_0$, it does for all $a \in \mathsf{A}$.

5. Consider the $*$-algebras $\mathsf{A} = \mathcal{E}(\mathfrak{g})$ and $\mathsf{X} = C_0^\infty(G)$ from Sect. 13.5.

   a. Show that the map $f \mapsto f * \varphi$, $f \in \mathsf{X}$, $\varphi \in C_0^\infty(G)$, define a nondegenerate faithful $*$-representation of X on the domain $C_0^\infty(G)$ in $L^2(G; \mu_l)$.
   b. Prove that $(\tilde{x} f) * \varphi = \tilde{x}(f * \varphi)$ for $x \in \mathcal{E}(\mathfrak{g})$ and $f, \varphi \in C_0^\infty(G)$.
   c. Use Example 13.19 to prove that (A, X) is a compatible pair, with action of A on X defined by (13.18).
      Hints: Consider A and X (by a.) as $*$-subalgebras of $\mathcal{L}^+(\mathcal{D})$, $\mathcal{D} = C_0^\infty(G)$. By b., the product of $\mathcal{L}^+(\mathcal{D})$ gives the corresponding action of A on X.

6. Let $\mathsf{A} = \mathcal{E}(\mathfrak{g})$ and $\mathsf{X} = C_0^\infty(G)$ be as in Sect. 13.5 and let U be a unitary representation of the Lie group G. Show that there is a continuous $*$-representation ρ of $(\mathsf{X}, \|\cdot\|_1)$ such that $dU = \pi^\rho$.

13.7 Notes

Well-behaved representations for group graded ∗-algebras A with commutative base algebras A_e were introduced in [SS13], with another definition, and studied therein in detail.

The method of fraction algebras elaborated in Sect. 13.2 is taken from [Sch10], where it was used to prove noncommutative Positivstellensätze. The algebra X from Sect. 13.3 is the one-dimensional version of the so-called resolvent algebra studied in [BG08]. Compatible pairs as in Sect. 13.4 were invented in [Sch02], which contains a number of further examples.

Another method for the construction of well-behaved representations was proposed by A. Inoue and his coworkers; it is based on unbounded C^*-seminorms. An *unbounded C^*-seminorm* on a ∗-algebra A is a C^*-seminorm (see Definition 10.1 and [Yo96]) on some ∗-subalgebra of A. This approach was developed in [BIO01, BIT01], and its relation to the method of compatible pairs was investigated in [BIK04]. A detailed treatment can be found in [AIT02, Chap. 8].

Chapter 14
Representations on Rigged Spaces and Hilbert C^*-Modules

This chapter gives a short digression into the theory of $*$-representations of $*$-algebras on rigged spaces. There, in contrast to Hilbert space representations, the complex-valued inner product is replaced by an algebra-valued sesquilinear map. In Sect. 14.1, we develop right or left rigged A-spaces and $*$-representations on such spaces. In Sect. 14.2, we study weak A–B-imprimitivity bimodules. These are A–B-bimodules which are right B-rigged and left A-rigged spaces satisfying various compatibility axioms. They allow us to convert algebraic properties of A into those of B and vice versa. In Sect. 14.3, we require the positive semi-definiteness of the riggings. If A acts by a $*$-representation on a positive semi-definite right B-rigged space, then each $*$-representation π of B on a complex inner product space induces a $*$-representation Ind π of A on a complex inner product space. Section 14.4 deals with A–B-imprimitivity bimodules. These are weak A–B-imprimitivity bimodules for which the riggings are positive semi-definite. An A–B-imprimitivity bimodule yields an equivalence between $*$-representations of A and B (Theorem 14.31). In Sects. 14.5 and 14.6, we introduce Hilbert C^*-modules for C^*-algebras and study $*$-representations on Hilbert C^*-modules.

Throughout this chapter, A, B, C are **unital complex** $*$-algebras and $\mathfrak{A}$ denotes a C^***-algebra**.

14.1 Rigged Spaces

Roughly speaking, a rigged space is a right or left module equipped with an algebra-valued sesquilinear mapping which is compatible with the module action.

© The Editor(s) (if applicable) and The Author(s), under exclusive license
to Springer Nature Switzerland AG 2020
K. Schmüdgen, *An Invitation to Unbounded Representations of *-Algebras on Hilbert Space*, Graduate Texts in Mathematics 285,
https://doi.org/10.1007/978-3-030-46366-3_14

Definition 14.1 A *right* A-*rigged space* is a right A-module $\mathcal{E}_A$, with right action written as $x \cdot a$, together with a map $[\cdot, \cdot]_A : \mathcal{E}_A \times \mathcal{E}_A \mapsto A$, which is linear in the second variable, such that:

(i) $([x, y]_A)^+ = [y, x]_A$ for $x, y \in \mathcal{E}_A$.
(ii) $[x, y \cdot a]_A = [x, y]_A\, a$ for $x, y \in \mathcal{E}_A, a \in A$.

A right A-rigged space $\mathcal{E}_A$ is called *full* if

$$\mathrm{Lin}\left\{ [x, y]_A : x, y \in \mathcal{E}_A \right\} = A$$

and *nondegenerate* if $[x, y]_A = 0$ for all $y \in \mathcal{E}_A$ implies that $x = 0$.

The symbol "$\cdot$" always refers to the action of elements of A on a left or right A-module. That $\mathcal{E}_A$ is a right A-module with right action "$x \cdot a$" means that

$$(x \cdot a) \cdot b = x \cdot (ab) \quad \text{and} \quad x \cdot 1_A = x \quad \text{for} \quad x \in \mathcal{E}_A, \ a, b \in A.$$

Note that $[x, y]_A\, a$ in (ii) denotes the product of the elements $[x, y]_A$ and a in A.

Let $\mathcal{E}_A$ be a right A-rigged space. Then, by condition (i), $[\cdot, \cdot]_A : \mathcal{E}_A \times \mathcal{E}_A \mapsto A$ is conjugate linear in the first variable. Further, using conditions (i) and (ii) we derive

$$a\,[x, y]_A = ([y, x]_A a^+)^+ = ([y, x \cdot a^+]_A)^+ = [x \cdot a^+, y]_A,$$

so that

$$a\,[x, y]_A = [x \cdot a^+, y]_A \quad \text{for} \quad x, y \in \mathcal{E}_A, \ a \in A. \tag{14.1}$$

A rigging $[\cdot, \cdot]_A$ can have a large kernel or it could be even identically zero. By passing to the quotient with respect to its kernel we obtain a nondegenerate rigged space, as stated in the next lemma. We omit the simple proof.

Lemma 14.2 *Suppose* $(\mathcal{E}_A, [\cdot, \cdot]_A)$ *is a right* A-*rigged space. Then the kernel space*

$$\mathcal{N}_A := \left\{x \in \mathcal{E}_A : [x, y]_A = 0, \ y \in \mathcal{E}_A \right\} = \{y \in \mathcal{E}_A : [x, y]_A = 0, \ x \in \mathcal{E}_A\}$$

is a right A-*module and the quotient right* A-*module* $\mathcal{E}_A / \mathcal{N}_A$ *is a nondegenerate right* A-*rigged space with rigging given by* $[x + \mathcal{N}_A, y + \mathcal{N}_A]_A := [x, y]_A, x, y \in \mathcal{E}_A.$

Definition 14.3 A *left* A-*rigged space* is a left A-module ${}_A\mathcal{E}$, equipped with a map ${}_A[\cdot, \cdot] : {}_A\mathcal{E} \times {}_A\mathcal{E} \mapsto A$, which is linear in the first variable and satisfies:

(i) $({}_A[x, y])^+ = {}_A[y, x]$ for $x, y \in {}_A\mathcal{E}$.
(ii) ${}_A[a \cdot x, y] = a\,{}_A[x, y]$ for $x, y \in {}_A\mathcal{E}, a \in A$.

For a left A-rigged space, the form ${}_A[\cdot, \cdot]$ is conjugate linear in the second variable. The left module action and notions such as fullness and nondegeneracy are defined similarly as for right rigged spaces. For a left A-rigged space ${}_A\mathcal{E}$, we obtain

$$_\mathsf{A}[x, y] \, a = {}_\mathsf{A}[x, a^+ \cdot y] \quad \text{for } x, y \in {}_\mathsf{A}\mathcal{E}, \, a \in \mathsf{A}. \tag{14.2}$$

Example 14.4 $(\mathcal{E}_\mathsf{A} = \mathsf{A} \text{ and } {}_\mathsf{A}\mathcal{E} = \mathsf{A})$
The $*$-algebra A itself is a *right* A-rigged space $\mathcal{E}_\mathsf{A} = \mathsf{A}$ with multiplication as right action and rigging $[a, b]_\mathsf{A} := a^+ b, a, b \in \mathsf{A}$. Likewise, A is a *left* A-rigged space ${}_\mathsf{A}\mathcal{E} = \mathsf{A}$ with multiplication as left action and rigging ${}_\mathsf{A}[a, b] := ab^+, a, b \in \mathsf{A}$. $\bigcirc$

The following are two slight modifications of the preceding example.

Example 14.5 Suppose $c = c^+$ is a central element of A. Then $\mathcal{E}_\mathsf{A} = \mathsf{A}$ is a right A-rigged space and ${}_\mathsf{A}\mathcal{E} = \mathsf{A}$ is a left A-rigged space with actions given by the multiplication and riggings $[a, b]_\mathsf{A} := c \, a^+ b$ and ${}_\mathsf{A}[a, b] := c \, ab^+$, respectively. $\bigcirc$

Example 14.6 Suppose $(\mathcal{D}, \langle \cdot, \cdot \rangle)$ is a complex inner product space. The complex tensor product $\mathcal{E}_\mathsf{A} := \mathcal{D} \otimes \mathsf{A}$ becomes a right A-rigged space with action given by $(\varphi \otimes a) \cdot b = \varphi \otimes ab$ and rigging

$$\left[\sum_i \varphi_i \otimes a_i, \sum_j \psi_j \otimes b_j \right]_\mathsf{A} := \sum_{i,j} \langle \psi_j, \varphi_i \rangle (a_i)^+ b_j. \qquad \bigcirc \tag{14.3}$$

Remark 14.7 In the theory of Hilbert modules of C^*-algebras (see, e.g., [La94], [RW98]) it is common to work with *right* modules and algebra-valued inner products that are *conjugate linear* in the first variable. We follow this convention. However, in this book, inner products of "ordinary" Hilbert spaces are always *linear* in the first variable. This requires a changing of the order of variables when we pass from *right* rigged spaces to Hilbert spaces and vice versa. Note that for a *left* rigged space the algebra-valued map is linear in the first variable. $\bigcirc$

For a vector space $\mathcal{E}$, let $\overline{\mathcal{E}}$ denote the complex conjugate vector space, that is, the sets and the additive groups of $\mathcal{E}$ and $\overline{\mathcal{E}}$ are the same, but the scalar multiplication is replaced by the complex conjugate. We write $\overline{x}$ to indicate that an element x of $\mathcal{E}$ is viewed as an element of $\overline{\mathcal{E}}$. Then, by definition, $\lambda \overline{x} = \overline{\lambda} x$ for $\lambda \in \mathbb{C}$ and $x \in \mathcal{E}$.

Clearly, if $\mathcal{E}_\mathsf{A}$ is a *right* A-rigged space, then the complex conjugate vector space $\overline{\mathcal{E}}$ becomes a *left* A-rigged space ${}_\mathsf{A}\overline{\mathcal{E}}$ with definitions

$$a \cdot \overline{x} := \overline{x \cdot a^+} \quad \text{and} \quad {}_\mathsf{A}[\overline{x}, \overline{y}] := [x, y]_\mathsf{A}, \quad x, y \in \mathcal{E}_\mathsf{A}, \, a \in \mathsf{A}.$$

Similarly, if ${}_\mathsf{A}\mathcal{E}$ is a *left* A-rigged space, then $\overline{\mathcal{E}}$ is a *right* A-rigged space $\overline{\mathcal{E}}_\mathsf{A}$ with

$$\overline{x} \cdot a := \overline{a^+ \cdot x} \quad \text{and} \quad [\overline{x}, \overline{y}]_\mathsf{A} := {}_\mathsf{A}[x, y], \quad x, y \in {}_\mathsf{A}\mathcal{E}, \, a \in \mathsf{A}. \tag{14.4}$$

Indeed, axiom (ii) in Definitions 14.1 and 14.3 follows from (14.1) and (14.2), respectively. Roughly speaking, passing to $\overline{\mathcal{E}}$ interchanges right and left structures.

Another important notion is introduced in the following definition.

Definition 14.8 An A−B-*bimodule* is a vector space $\mathcal{E}$ which is both a left A-module and a right B-module and satisfies

$$a \cdot (x \cdot b) = (a \cdot x) \cdot b \quad \text{for} \quad a \in A, b \in B, x \in \mathcal{E}. \tag{14.5}$$

To indicate the module actions one may write $_A\mathcal{E}_B$ for an A−B-bimodule $\mathcal{E}$.

For a right B-rigged space $\mathcal{E}_B$, let $\mathcal{L}(\mathcal{E}_B)$ denote the algebra of all $\mathbb{C}$-linear and B-linear mappings t of $\mathcal{E}_B$ into itself. Here a mapping t of $\mathcal{E}_B$ is called B-*linear* if $t(x \cdot b) = t(x) \cdot b$ for $b \in B$ and $x \in \mathcal{E}_B$.

Definition 14.9 A $*$-*representation* of a $*$-algebra A on a right B-rigged space $\mathcal{E}_B$ is an algebra homomorphism π of A in $\mathcal{L}(\mathcal{E}_B)$ satisfying

$$[\pi(a)x, y]_B = [x, \pi(a^+)y]_B \quad \text{for} \quad a \in A, x, y \in \mathcal{E}_B. \tag{14.6}$$

If π is a $*$-representation of A on $\mathcal{E}_B$, then by the definition of $\mathcal{L}(\mathcal{E}_B)$ we have

$$\pi(a)(x \cdot b) = (\pi(a)x) \cdot b \quad \text{for} \quad a \in A, b \in B, x \in \mathcal{E}_B. \tag{14.7}$$

Thus, $\mathcal{E}_B$ is an A−B-bimodule with left action of A given by $a \cdot x := \pi(a)x$.

If π is an algebra homomorphism of A in $L(\mathcal{E}_B)$ such that (14.6) and (14.7) hold, then π is a map into $\mathcal{L}(\mathcal{E}_B)$ (by (14.7)) and hence a $*$-representation of A on $\mathcal{E}_B$.

The following important construction is called the *internal tensor product* of rigged spaces.

Proposition 14.10 *Suppose $\mathcal{E}_A$ is a right A-rigged space and π is a $*$-representation of A on a right B-rigged space $\mathcal{F}_B$. Let $\mathcal{E} \otimes_A \mathcal{F}$ denote the quotient of the complex tensor product $\mathcal{E}_A \otimes \mathcal{F}_B$ by*

$$\mathcal{N} := \text{Lin}\left\{x \cdot a \otimes y - x \otimes \pi(a)y : x \in \mathcal{E}_A, \, y \in \mathcal{F}_B, \, a \in A\right\}. \tag{14.8}$$

This space is a right B-rigged space $(\mathcal{E} \otimes_A \mathcal{F})_B$ with right action and rigging

$$(x \otimes y) \cdot b := x \otimes y \cdot b, \quad x \in \mathcal{E}_A, y \in \mathcal{F}_B, b \in B, \tag{14.9}$$

$$[x \otimes y, x' \otimes y']_B := [y, \pi([x, x']_A) y']_B, \quad x, x' \in \mathcal{E}_A, y, y' \in \mathcal{F}_B. \tag{14.10}$$

Proof Clearly, formula (14.10) extends to a map $[\cdot, \cdot]_B$ of the *complex* tensor product $\mathcal{E}_A \otimes \mathcal{F}_B$ into B which is linear in the second and conjugate linear in the first variables. Then, by Definition 14.1(i) and (14.6),

$$([x \otimes y, x' \otimes y']_B)^+ = ([y, \pi([x, x']_A)y']_B)^+ = [\pi([x, x']_A)y', y]_B$$
$$= [y', \pi(([x, x']_A)^+)y]_B = [y', \pi([x', x]_A)y]_B = [x' \otimes y', x \otimes y]_B. \tag{14.11}$$

Further, using (14.1) and (14.6) we derive

$$\sum_{j,k} [x_j \cdot a \otimes y_j, x_k' \otimes y_k']_{\mathsf{B}} = \sum_{j,k} [y_j, \pi([x_j \cdot a, x_k']_{\mathsf{A}}) y_k']_{\mathsf{B}}$$

$$= \sum_{j,k} [y_j, \pi(a^+[x_j, x_k']_{\mathsf{A}}) y_k']_{\mathsf{B}} = \sum_{j,k} [\pi(a) y_j, \pi([x_j, x_k']_{\mathsf{A}}) y_k']_{\mathsf{B}}$$

$$= \sum_{j,k} [x_j \otimes \pi(a) y_j, x_k' \otimes y_k']_{\mathsf{B}}$$

and similarly, by (14.11),

$$\sum_{j,k} [x_k' \otimes y_k', x_j \cdot a \otimes y_j]_{\mathsf{B}} = \sum_{j,k} [x_k' \otimes y_k', x_j \otimes \pi(a) y_j]_{\mathsf{B}}.$$

These two equations imply that $[\cdot, \cdot]_{\mathsf{B}}$ passes to a well-defined sesquilinear map, denoted also $[\cdot, \cdot]_{\mathsf{B}}$, of the quotient $(\mathcal{E} \otimes_{\mathsf{A}} \mathcal{F})_{\mathsf{B}} = (\mathcal{E}_{\mathsf{A}} \otimes \mathcal{F}_{\mathsf{B}})/\mathcal{N}$ into B. By (14.11) this map satisfies condition (i) of Definition 14.1. Since π is a $*$-representation of A on $\mathcal{F}_{\mathsf{B}}$, we have $\pi(a)(y \cdot b) = (\pi(a)y) \cdot b$ for $y \in \mathcal{F}_{\mathsf{B}}$, $a \in \mathsf{A}$, $b \in \mathsf{B}$ by (14.7). Hence the action of B on the complex tensor product $\mathcal{E}_{\mathsf{A}} \otimes \mathcal{F}_{\mathsf{B}}$ defined by (14.9) leaves $\mathcal{N}$ invariant, so it gives a well-defined right action of B on the quotient $(\mathcal{E} \otimes_{\mathsf{A}} \mathcal{F})_{\mathsf{B}}$. We verify condition (ii) of Definition 14.1. Using again (14.7) we compute for $b \in \mathsf{B}$,

$$[x \otimes y, (x' \otimes y') \cdot b]_{\mathsf{B}} = [x \otimes y, x' \otimes (y' \cdot b)]_{\mathsf{B}} = [y, \pi([x, x']_{\mathsf{A}}) (y' \cdot b)]_{\mathsf{B}}$$
$$= [y, (\pi([x, x']_{\mathsf{A}}) y') \cdot b]_{\mathsf{B}} = [y, \pi([x, x']_{\mathsf{A}}) y']_{\mathsf{B}} b = [x \otimes y, (x' \otimes y')]_{\mathsf{B}} b.$$

Summarizing, we have shown that $(\mathcal{E} \otimes_{\mathsf{A}} \mathcal{F})_{\mathsf{B}}$ is a right B-rigged space. □

Proposition 14.11 *Let $\mathcal{E}_{\mathsf{A}}$, $\mathcal{F}_{\mathsf{B}}$, $(\mathcal{E} \otimes_{\mathsf{A}} \mathcal{F})_{\mathsf{B}}$ be as in Proposition 14.10. In addition, suppose $c \mapsto c\cdot$ is a $*$-representation of C on the right A-rigged space $\mathcal{E}_{\mathsf{A}}$. Then there is a $*$-representation ρ of C on the right B-rigged space $(\mathcal{E} \otimes_{\mathsf{A}} \mathcal{F})_{\mathsf{B}}$ such that*

$$\rho(c)(x \otimes y) = c \cdot x \otimes y, \quad x \in \mathcal{E}_{\mathsf{A}}, \ y \in \mathcal{F}_{\mathsf{B}}, \ c \in \mathsf{C}. \tag{14.12}$$

Proof Since C acts as a $*$-representation on $\mathcal{E}_{\mathsf{A}}$, $(c \cdot x) \cdot a = c \cdot (x \cdot a)$ for $x \in \mathcal{E}_{\mathsf{A}}$, $a \in \mathsf{A}$, $c \in \mathsf{C}$ by (14.7). Therefore, the left action $\rho(c)$ of C on the complex tensor product $\mathcal{E}_{\mathsf{A}} \otimes \mathcal{F}_{\mathsf{B}}$ defined by (14.12) leaves $\mathcal{N}$ (see (14.8)) invariant. Hence it passes to a well-defined left action of C on the quotient $(\mathcal{E} \otimes_{\mathsf{A}} \mathcal{F})_{\mathsf{B}}$. By the corresponding definitions, $\rho(c)((x \otimes y) \cdot b) = (c \cdot x) \otimes (y \cdot b) = (\rho(c)(x \otimes y)) \cdot b$, so (14.7) holds. Using condition (14.6) for the $*$-representation $c \mapsto c\cdot$ of C we derive

$$[\rho(c)(x \otimes y), x' \otimes y']_{\mathsf{B}} = [y, \pi([c \cdot x, x']_{\mathsf{A}}) y']_{\mathsf{B}} = [y, \pi([x, c^+ \cdot x']_{\mathsf{A}}) y']_{\mathsf{B}}$$
$$= [x \otimes y, (c^+ \cdot x') \otimes y']_{\mathsf{B}} = [x \otimes y, \rho(c^+)(x' \otimes y')]_{\mathsf{B}}.$$

Therefore, by linearity, ρ is a $*$-representation of C on $(\mathcal{E} \otimes_{\mathsf{A}} \mathcal{F})_{\mathsf{B}}$. □

14.2 Weak Imprimitivity Bimodules

The main notions of this section are contained in the following two definitions.

Definition 14.12 A *weak* A$-$B*-imprimitivity bimodule* is a triple $(\mathcal{E}, {}_A[\cdot,\cdot], [\cdot,\cdot]_B)$ such that $\mathcal{E}$ is an A$-$B-bimodule satisfying the following conditions:

(i) $(\mathcal{E}, [\cdot,\cdot]_B)$ is a full right B-rigged space such that $[a \cdot x, y]_B = [x, a^+ \cdot y]_B$ for $a \in A$ and $x, y \in \mathcal{E}$.

(ii) $(\mathcal{E}, {}_A[\cdot,\cdot])$ is a full left A-rigged space such that ${}_A[x \cdot b, y] = {}_A[x, y \cdot b^+]$ for $b \in B$ and $x, y \in \mathcal{E}$.

(iii) ${}_A[x, y] \cdot z = x \cdot [y, z]_B$ for $x, y, z \in \mathcal{E}$.

Definition 14.13 Two $*$-algebras A and B are called *weakly Morita equivalent* if there exists a weak A$-$B-imprimitivity bimodule.

Note that axiom (iii) in Definition 14.12 is a compatibility condition of the two riggings ${}_A[\cdot,\cdot]$ and $[\cdot,\cdot]_B$. It is crucial for many applications.

If $\mathcal{E}$ is a right B-rigged space and a left A-rigged space, then the A$-$B-bimodule property (14.5) follows already from the conditions (i)–(iii) in Definition 14.12. Indeed, let $a \in A$ and $x, y, z \in \mathcal{E}$. Using (iii), the left A-module property of $\mathcal{E}$ and Definition 14.3(ii) we derive

$$a \cdot (x \cdot [y, z]_B) = a \cdot ({}_A[x, y] \cdot z) = (a\,{}_A[x, y]) \cdot z = {}_A[a \cdot x, y] \cdot z = (a \cdot x) \cdot [y, z]_B.$$

Since the right B-rigged space $\mathcal{E}$ is full by (i), each $b \in B$ is a sum of terms $[y, z]_B$. Hence the preceding equality yields (14.5).

Note that (14.5) and the last condition in Definition 14.12(i) imply that the map $a \mapsto a\cdot$ is a $*$-representation of A on the right B-rigged space $\mathcal{E}$ according to Definition 14.9.

Example 14.14 If A and B are $*$-isomorphic, then A and B are weakly Morita equivalent. Indeed, let θ be a $*$-isomorphism of A on B and set $\mathcal{E} := B$. Then, for $a \in A, b \in B, x, y \in \mathcal{E}$, we define the operations

$$a \cdot x := \theta(a)x, \quad x \cdot b := xb, \quad [x, y]_B := x^+ y, \quad {}_A[x, y] := \theta^{-1}(xy^+).$$

One easily verifies that $\mathcal{E} = B$ is a weak imprimitivity module of A and B. $\bigcirc$

Example 14.15 (*Matrix* $*$-*algebras*)
Let C be a fixed unital complex $*$-algebra and $k, m \in \mathbb{N}$. Recall from Subsection 2.2 that $A := M_k(C)$ and $B := M_m(C)$ are also unital complex $*$-algebras. The vector space $\mathcal{E} := M_{k,m}(C)$ of (k, m)-matrices over C is an A$-$B-bimodule with left and right actions given by the multiplication of matrices. The bimodule axiom (14.5) holds by the associativity of matrix multiplication. Also, $\mathcal{E}$ is a right B-rigged space and a left A-rigged space with riggings defined by

$$[X, Y]_B := X^+Y \quad \text{and} \quad {}_A[X, Y] := XY^+, \quad X, Y \in \mathcal{E} = M_{k,m}(\mathbb{C}).$$

It is straightforward to verify that $\mathcal{E}$, with these definitions, is a weak A−B-imprimitivity bimodule. Hence A and B are weakly Morita equivalent.   ○

As noted above, passing to the complex conjugate $\overline{\mathcal{E}}$ interchanges left and right rigged structures of $\mathcal{E}$. Thus, if $\mathcal{E}$ is a weak A−B-imprimitivity bimodule, then $\overline{\mathcal{E}}$ becomes a weak B−A-imprimitivity bimodule with the corresponding operations.

The next proposition describes the tensor product of weak imprimitivity bimodules. This result is in fact the transitivity of the weak Morita equivalence.

Proposition 14.16 *Suppose $\mathcal{E}$ is a weak A−B-imprimitivity bimodule and $\mathcal{F}$ is a weak B−C-imprimitivity bimodule. Then the quotient $\mathcal{E} \otimes_B \mathcal{F}$ of the complex tensor product $\mathcal{E} \otimes \mathcal{F}$ by $\mathcal{N} := \text{Lin}\{x \cdot b \otimes y - x \otimes b \cdot y : x \in \mathcal{E}, y \in \mathcal{F}, b \in B\}$ is a weak A−C-imprimitivity module. The left action of A and the right action of C on $\mathcal{E} \otimes_B \mathcal{F}$ and the corresponding riggings are defined by*

$$a \cdot (x \otimes y) := (a \cdot x) \otimes y, \tag{14.13}$$

$$(x \otimes y) \cdot c := x \otimes (y \cdot c), \tag{14.14}$$

$$[x \otimes y, u \otimes v]_C := [[u, x]_B \cdot y, v]_C, \tag{14.15}$$

$${}_A[x \otimes y, u \otimes v] := {}_A[x, u \cdot {}_B[v, y]], \tag{14.16}$$

where $a \in A$, $c \in C$, $x, u \in \mathcal{E}$ and $y, v \in \mathcal{F}$.

Proof First we show that $\mathcal{E} \otimes_B \mathcal{F}$ satisfies condition (i) of Definition 14.12. Here we use essentially the conditions of the two weak imprimitivity bimodules $\mathcal{E}$ and $\mathcal{F}$. Then $\mathcal{E}$ is a right B-rigged space, $\mathcal{F}$ is a right C-rigged space and the map $b \mapsto b \cdot$ is a $*$-representation of B on $\mathcal{F}$. Therefore, by Proposition 14.10, $\mathcal{E} \otimes_B \mathcal{F}$ is a right C-rigged space and Eq. (14.9) gives just the action (14.14) of C. Then, since $[[u, x]_B \cdot y, v]_C = [y, [x, u]_B \cdot v]_C$, the rigging (14.10) coincides with (14.15). Further, since the map $a \mapsto a \cdot$ is a $*$-representation of A on $\mathcal{E}$, it follows from Proposition 14.11 that (14.13) defines a $*$-representation of A on $\mathcal{E} \otimes_B \mathcal{F}$.

To complete the proof of Definition 14.12(i) it remains to verify the fullness of the right C-rigged space $\mathcal{E} \otimes_B \mathcal{F}$. Let $c \in C$. The right B-rigged space $\mathcal{E}$ and the right C-rigged space $\mathcal{F}$ are full (by Definition 14.12(i)), so there exist elements $u_j, u'_j \in \mathcal{E}$ and $v_k, v'_k \in \mathcal{F}$ such that $1_B = \sum_j [u'_j, u_j]_B$ and $c = \sum_k [v_k, v'_k]_C$. Then

$$\sum_{j,k} [u_j \otimes v_k, u'_j \otimes v'_k]_C = \sum_k \left[\left(\sum_j [u'_j, u_j]_B\right) \cdot v_k, v'_k\right]_C$$

$$= \sum_k [1_B \cdot v_k, v'_k]_C = \sum_k [v_k, v'_k]_C = c.$$

Now the proof of condition (i) is complete. The proof of (ii) is similar; Propositions 14.10 and 14.11 are replaced by their counterparts which are obtained by interchanging left and right actions and riggings.

Finally, we prove condition (iii). For $x, u, u' \in \mathcal{E}$ and $y, v, v' \in \mathcal{F}$, we deduce

$$
\begin{aligned}
(x \otimes y) \cdot [u \otimes v, u' \otimes v']_C &= x \otimes (y \cdot [\,[u', u]_B \cdot v, v']_C) \\
&= x \otimes (_B[y, [u', u]_B \cdot v] \cdot v') = (x \cdot {}_B[y, [u', u]_B \cdot v]) \otimes v' \\
&= (x \cdot (_B[y, v] [u, u']_B)) \otimes v' = (x \cdot [u \cdot {}_B[v, y], u']_B) \otimes v' \\
&= (_A[x, u \cdot {}_B[v, y]] \cdot u') \otimes v' = {}_A[x \otimes y, u \otimes v] \cdot (u' \otimes v').
\end{aligned}
$$

Here the first and seventh equalities are the definitions (14.13) and (14.14) of the actions of C and A. The second and sixth equalities follow from Definition 14.12(iii) and the fourth and fifth from (14.1) and (14.2). The third equality holds because $\mathcal{N}$ is annihilated in the quotient space.

Thus, we have poved that $\mathcal{E} \otimes_B \mathcal{F}$ is a weak A–C-imprimitivity bimodule. □

Proposition 14.17 *Weak Morita equivalence is an equivalence relation for unital complex ∗-algebras.*

Proof Example 14.14, with A = B and θ the identity map, shows that weak Morita equivalence is reflexive. As noted above, if $\mathcal{E}$ is a weak A–B-imprimitivity bimodule, then $\overline{\mathcal{E}}$ is a weak B–A-imprimitivity bimodule, so weak Morita equivalence is symmetric. The transitivity follows from Proposition 14.16. □

In the rest of this section, suppose $\mathcal{E}$ *is a weak* A–B-*imprimitivity bimodule.* We will show how pre-quadratic modules can be transported from A to B and vice versa.

Lemma 14.18 *If Q is a subset of A_{her}, then the finite sums of elements $[a \cdot x, x]_B$, where $a \in Q$ and $x \in \mathcal{E}$, form a pre-quadratic module $Q_{\mathcal{E}}$ of B.*

Proof Since $([a \cdot x, x]_B)^+ = [x, a \cdot x]_B = [a \cdot x, x]_B$, $Q_{\mathcal{E}}$ is contained in B_{her}. Obviously, $\lambda c \in Q_{\mathcal{E}}$ for $\lambda > 0$ and $c \in Q_{\mathcal{E}}$. Let $b \in B$, $a \in Q$, and $x \in \mathcal{E}$. Then

$$
\begin{aligned}
b^+[a \cdot x, x]_B b &= b^+[a \cdot x, x \cdot b]_B = ([x \cdot b, a \cdot x]_B b)^+ \\
&= ([x \cdot b, (a \cdot x) \cdot b]_B)^+ = [x \cdot b, a \cdot (x \cdot b)]_B)^+ = [a \cdot (x \cdot b), (x \cdot b)]_B \in Q_{\mathcal{E}}.
\end{aligned}
$$

Thus, $Q_{\mathcal{E}}$ is a pre-quadratic module of B. □

For a subset P of B_{her}, let $_{\mathcal{E}}P$ denote the finite sums of elements $_A[y, y \cdot b]$, where $b \in P$ and $y \in \mathcal{E}$. Then, a similar reasoning as in the proof of Lemma 14.18 shows $_{\mathcal{E}}P$ *is a pre-quadratic module of* A.

Proposition 14.19 *If Q is a pre-quadratic module of A and P is a pre-quadratic module of B, then we have $_{\mathcal{E}}(Q_{\mathcal{E}}) \subseteq Q$ and $(_{\mathcal{E}}P)_{\mathcal{E}} \subseteq P$.*

Proof We prove only the first inclusion; the proof for the second is similar.

Since $_{\mathcal{E}}(Q_{\mathcal{E}})$ consists of sums of elements of the form $_A[y, y \cdot [a \cdot x, x]_B]$, where $a \in Q$ and $x, y \in \mathcal{E}$, it suffices to show that these elements are in Q. Using the axioms of a weak A–B-imprimitivity bimodule we compute

$$_A[y, y \cdot [a \cdot x, x]_B] = {}_A[y \cdot [x, a \cdot x]_B, y] = {}_A[{}_A[y, x] \cdot (a \cdot x), y]$$
$$= {}_A[y, x]_A[a \cdot x, y] = ({}_A[x, y])^+ \, a \, ({}_A[x, y]).$$

This element is in Q, since $a \in Q$ and Q is a pre-quadratic module of A. $\qquad\square$

We illustrate these general constructions with an interesting example.

Example 14.20 (*Matrix algebras- Example* 14.15 *continued*)
As in Example 14.15, C is a unital complex $*$-algebra, $A = M_k(C)$, $B = M_m(C)$, and $\mathcal{E} = M_{k,m}(C)$ is a weak A–B-imprimitivity bimodule.

If Q and P are pre-quadratic modules of $A = M_k(C)$ and $B = M_m(C)$, respectively, the pre-quadratic modules $Q_\mathcal{E}$ of B and $_\mathcal{E}P$ of A are

$$Q_{m,k} := Q_\mathcal{E} = \left\{ \sum_{i=1}^{s} X_i^+ A_i X_i \; : \; A_i \in Q, \; X_i \in M_{k,m}(C), \; s \in \mathbb{N} \right\}, \quad (14.17)$$

$$P_{k,m} := {}_\mathcal{E}P = \left\{ \sum_{i=1}^{s} Y_i B_i Y_i^+; \; B_i \in P, \; Y_i \in M_{k,m}(C), \; s \in \mathbb{N} \right\}.$$

We now specialize this to the case of the polynomial algebra $C := \mathbb{C}[t_1, \ldots, t_d]$. Let Q be the quadratic module of hermitian $k \times k$ matrices over $\mathbb{C}[t_1, \ldots, t_d]$ that are positive semi-definite for all $t \in \mathbb{R}^d$. It is easily checked that the identity matrix is in $Q_{m,k}$. Hence each set $Q_{m,k}$ defined by (14.17) is a quadratic module of B. Put $Q_{m,0} := \sum B^2$. Then we obtain from (14.17) an increasing chain (see Exercise 6) of quadratic modules of the $*$-algebra $B = M_m(\mathbb{C}[t_1, \ldots, t_d])$:

$$Q_{m,0} \subseteq Q_{m,1} \subseteq Q_{m,2} \subseteq \cdots \subseteq Q_{m,m}. \quad (14.18)$$

The matrices of $Q_{m,k}$ are called *k-positive*.

Let $d = 1$. Any matrix $A \in Q_{m,m}$ is positive semi-definite for all $t \in \mathbb{R}$. Therefore, by a classical result [Ja70], [Dj70] on matrix polynomials in one variable, A is of the form $A = B^+ B$ for some $B \in M_m(\mathbb{C}[t])$. Hence all quadratic modules in (14.18) coincide with $Q_{m,0} = \sum M_m(\mathbb{C}[t])^2$.

Now suppose $d \geq 2$. As shown in [FS89], the matrix

$$\begin{pmatrix} 1 + t_1^4 t_2^2 & t_1 t_2 \\ t_1 t_2 & 1 + t_1^2 t_2^4 \end{pmatrix} \quad (14.19)$$

is in $Q_{2,2}$, but it is not in $Q_{2,1}$. Therefore, $Q_{2,1} \neq Q_{2,2}$.

The quadratic modules $Q_{m,k}$ are not as artificial, as one might think at first glance. They can be used to characterize *k-positive* $*$-representations of the $*$-algebra $\mathbb{C}[t_1, \ldots, t_d]$ (see [Sch90, pp. 307–308] for precise statements). $\qquad\bigcirc$

14.3   Positive Semi-definite Riggings

All considerations in the preceding two sections were purely algebraic. In this section, we turn to Hilbert space representations and study positive semi-definite riggings. This requires some preliminaries.

Lemma 14.21 *For a matrix $A = (a_{ij})_{i,j=1}^n \in M_n(A)$ the following are equivalent:*

(i) *For any $*$-representation ρ of A and vectors $\varphi_1, \ldots, \varphi_n \in \mathcal{D}(\rho)$, we have*

$$\sum_{i,j=1}^n \langle \rho(a_{ij})\varphi_j, \varphi_i \rangle \geq 0. \tag{14.20}$$

(ii) *For each positive functional f on the $*$-algebra $M_n(A)$ we have $f(A) \geq 0$.*

The set of hermitian matrices of $M_n(A)$ satisfying these equivalent conditions is denoted by $M_n(A)_+$. Then $M_n(A)_+$ a quadratic module of $M_n(A)$.

Proof Using (14.20) it is straightforward to verify that $M_n(A)_+$ is a quadratic module; we leave this to the reader as Exercise 7.

(i)$\rightarrow$(ii): Let f be a positive functional on the $*$-algebra $M_n(A)$ and let π_f be its GNS representation. By Proposition 10.23, there is a $*$-representation ρ of A such that $\pi_f = \rho_n$, that is, $\mathcal{D}(\pi_f) = \oplus_{k=1}^n \mathcal{D}(\rho)$, $\varphi_f = (\varphi_1, \ldots, \varphi_n)$, $\varphi_j \in \mathcal{D}(\rho)$, and

$$(\pi_f(A)\varphi_f)_k = \sum_{j=1}^n \rho(a_{kj})\varphi_j \quad \text{for} \quad A = (a_{ij}) \in M_n(A), \ k = 1, \ldots, n. \tag{14.21}$$

Then, by (14.21) and (14.20),

$$f(A) = \langle \pi_f(A)\varphi_f, \varphi_f \rangle = \sum_{i,j=1}^n \langle \rho(a_{ij})\varphi_j, \varphi_i \rangle \geq 0.$$

(ii)$\rightarrow$(i): Let ρ and $\varphi_1, \ldots, \varphi_n$ be as in (i). Define a functional f on $M_n(A)$ by

$$f(B) = \sum_{i,j=1}^n \langle \rho(b_{ij})\varphi_j, \varphi_i \rangle \quad \text{for} \quad B = (b_{ij}) \in M_n(A). \tag{14.22}$$

Let $C = (c_{ij}) \in M_n(A)$. Then $(C^+C)_{ij} = \sum_k c_{ki}^+ c_{kj}$ and therefore

$$f(C^+C) = \sum_{i,j,k=1}^n \langle \rho(c_{ki}^+ c_{kj})\varphi_j, \varphi_i \rangle = \sum_{k=1}^n \Big\langle \sum_{j=1}^n \rho(c_{kj})\varphi_j, \sum_{i=1}^n \pi(c_{ki})\varphi_i \Big\rangle \geq 0.$$

This shows that f is a positive linear functional on $M_n(A)$. Hence $f(A) \geq 0$ by (ii), which gives (14.20) by (14.22). $\qquad\square$

The following simple lemma is based on a standard separation argument.

Lemma 14.22 *Let* X *be a complex* $*$*-algebra and* $a = a^+ \in X$. *If* $f(a) \geq 0$ *for all positive linear functionals of* X, *then* a *is in the closure of the cone* $\sum X^2$ *in the finest locally convex topology of* X *(see Appendix C).*

Proof Let C denote the closure of $\sum X^2$ in the finest locally convex topology. Assume to the contrary that $a \notin C$. Then, by Proposition C.2, there exists an $\mathbb{R}$-linear functional f on the real vector space X_{her} such that $f(a) < 0$ and $f(c) \geq 0$ for $c \in C$. Its extension, denoted also f, to a $\mathbb{C}$-linear functional on X is a positive functional such that $f(a) < 0$, a contradiction. $\square$

Definition 14.23 A right A-rigged space $(\mathcal{E}_A, [\cdot, \cdot]_A)$ is called *positive semi-definite* if for all $x_1, \ldots, x_n \in \mathcal{E}$, $n \in \mathbb{N}$, the matrix $([x_i, x_j]_A)_{i,j=1}^n$ is in $M_n(A)_+$.

Similarly, a left B-rigged space $(_B\mathcal{F}, _B[\cdot, \cdot])$ is called *positive semi-definite* if the matrix $(_B[x_i, x_j])_{i,j=1}^n$ belongs to $M_n(B)_+$ for all $x_1, \ldots, x_n \in \mathcal{E}$ and $n \in \mathbb{N}$.

Example 14.24 (*Examples* 14.4 *and* 14.5 *continued*)
Consider the right A-rigged module $\mathcal{E}_A = A$ with rigging $[a, b]_A = c\,a^+b$, where $c = c^+$ is a central element of A. Let $x_1, \ldots, x_n \in \mathcal{E}_A$. Then, for any $*$-representation ρ of A and vectors $\varphi_1, \ldots, \varphi_n \in \mathcal{D}(\rho)$,

$$\sum_{i,j=1}^n \langle \rho([x_i, x_j]_A)\varphi_j, \varphi_i \rangle = \sum_{i,j=1}^n \langle \rho(c)\rho(x_i^+)\rho(x_j)\varphi_j, \varphi_i \rangle$$

$$= \Big\langle \rho(c)\Big(\sum_{j=1}^n \rho(x_j)\varphi_j\Big), \sum_{i=1}^n \rho(x_i)\varphi_i \Big\rangle. \tag{14.23}$$

Hence, if $\rho(c) \geq 0$ for all $*$-representations ρ, then the matrix $([x_i, x_j]_A)_{i,j=1}^n$ is in $M_n(A)_+$ by Lemma 14.21 and the rigging $[\cdot, \cdot]_A$ is positive semi-definite. In particular, this holds for $c = 1$, so the "canonical" rigging $[a, b]_A = a^+b$ of $\mathcal{E}_A$ is always positive semi-definite. The same is true for the rigging $_A[a, b] = ab^+$ of $_A\mathcal{E}$.

Now suppose there exist a $*$-representation ρ of A and a vector $\varphi \in \mathcal{D}(\rho)$ such that $\langle \rho(c)\varphi, \varphi \rangle < 0$. (An example is the Motzkin polynomial for $A = \mathbb{C}[t_1, t_2]$; see Examples 2.30 and 7.5.) Then, for $x_1 = 1, \varphi_1 = \varphi, n = 1$, the expression in (14.23) is negative, so the rigging is not positive semi-definite. $\bigcirc$

The next proposition is the main result of this section. It says that the positive semi-definiteness of riggings carries over to the internal tensor product.

Proposition 14.25 *Suppose* $\mathcal{E}_A$ *is a right* A*-rigged space,* $\mathcal{F}_B$ *is a right* B*-rigged space, and* π *is a* $*$*-representation of* A *on* $\mathcal{F}_B$. *Then, if* $\mathcal{E}_A$ *and* $\mathcal{F}_B$ *are positive semi-definite, so is the right* B*-rigged space* $(\mathcal{E} \otimes_A \mathcal{F})_B$ *from Proposition 14.10 with rigging given by (14.10).*

Proof Let $z_1, \ldots, z_m \in \mathcal{E} \otimes_A \mathcal{F}$. We can write z_i as a finite sum $z_i = \sum_k x_{ik} \otimes y_{ik}$, where $x_{ik} \in \mathcal{E}_A$, $y_{ik} \in \mathcal{F}_B$. Then, by (14.10), we have

$$[z_i, z_j]_B = \sum_{k,l} [y_{ik}, \pi([x_{ik}, x_{jl}]_A) y_{jl}]_B. \tag{14.24}$$

Since $\mathcal{E}_A$ is positive semi-definite, Lemma 14.22, applied to a $*$-algebra of matrices over A, implies that the matrix $X = ([x_{ik}, x_{jl}]_A)_{ik,jl}$ is the limit of a net of matrices $A = (a_{ik,jl})_{ik,jl}$, which are finite sums $\sum_r (A_r)^+ A_r$ of hermitian squares of matrices. Then we have $a_{ik,jl} = \sum_{r,p,q} (a_{r;pq,ik})^+ a_{r;pq,jl}$. We abbreviate $b_{rpq;i} := \sum_l \pi(a_{r;pq,il}) y_{il}$. Let us fix i, j and compute

$$\sum_{k,l} [y_{ik}, \pi(a_{ik,jl}) y_{jl}]_B = \sum_{k,l} \left[y_{ik}, \pi\left(\sum_{r,p,q} (a_{r;pq,ik})^+ a_{r;pq,jl}\right) y_{jl}\right]_B$$

$$= \sum_{r,p,q} \left[\sum_k \pi(a_{r;pq,ik}) y_{ik}, \sum_l \pi(a_{r;pq,jl}) y_{jl}\right]_B = \sum_{r,p,q} [b_{rpq;i}, b_{rpq;j}]_B.$$

Since $\mathcal{F}_B$ is positive semi-definite, each matrix $([b_{rpq;i}, b_{rpq;j}]_B)_{i,j=1}^m$, hence also

$$\left(\sum_{r,p,q} [b_{rpq;i}, b_{rpq;j}]_B\right)_{i,j=1}^m = \left(\sum_{k,l} [y_{ik}, \pi(a_{ik,jl}) y_{jl}]_B\right)_{i,j=1}^m, \tag{14.25}$$

belongs to $M_m(B)_+$. Now we pass to the limit $A \to X$ in the finest locally convex topology. Then the matrix in (14.25) goes to the matrix in (14.24). Since $M_m(B)_+$ is obviously closed in the finest locally convex topology and the matrix in (14.25) is in $M_m(B)_+$, so is the matrix $([z_i, z_j]_B)_{i,j=1}^m$ by (14.24). This proves that the right B-rigged space $(\mathcal{E} \otimes_A \mathcal{F})_B$ is positive semi-definite.   □

Recall that induced representations were introduced and studied in Chap. 12. Now we present an approach to this concept in the context of rigged spaces. That is, we use Propositions 14.10, 14.11, and 14.25 to define *induced representations on Hilbert spaces*. In order to get representations of A that are induced from representations of B, we apply these results with A, B, C replaced by B, C, A, respectively, and set C $:= \mathbb{C}$.

For the following discussion we assume that $\mathcal{E}_B$ is a positive semi-definite right B-rigged space and $a \mapsto a\cdot$ is a $*$-representation of A on $\mathcal{E}_B$.

Suppose π is an "ordinary" $*$-representation of B on a complex inner product space $(\mathcal{D}, \langle \cdot, \cdot \rangle_0)$. Setting $[\varphi, \psi]_{\mathbb{C}} := \langle \psi, \varphi \rangle_0$ for $\varphi, \psi \in \mathcal{D}$, $\mathcal{F}_{\mathbb{C}} := \mathcal{D}$ becomes a right $\mathbb{C}$-rigged space and π is a $*$-representation on $\mathcal{F}_{\mathbb{C}}$. Further, by (14.8), the B-balanced tensor product $\mathcal{E} \otimes_B \mathcal{F}$ is the quotient of the complex tensor product by

$$\mathcal{N} := \text{Lin}\left\{x \cdot b \otimes \varphi - x \otimes \pi(b)\varphi : x \in \mathcal{E}_B, \, \varphi \in \mathcal{D}, \, b \in B\right\}.$$

By Proposition 14.25, the rigging (14.10) is positive semi-definite. The riggings of right rigged spaces are conjugate linear, and our inner products are linear in the first variables. Therefore, as noted in Remark 14.7, in order to get a positive semi-definite inner product $\langle \cdot, \cdot \rangle$ on $\mathcal{D}_0 := \mathcal{E} \otimes_B \mathcal{D}$ from the rigging (14.10), we have to interchange the order of variables twice, for $\langle \cdot, \cdot \rangle$ and $\langle \cdot, \cdot \rangle_0$. Then we obtain

$$\langle x \otimes \varphi, y \otimes \psi \rangle := \langle \pi([y, x]_B)\varphi, \psi \rangle_0, \quad x, y \in \mathcal{E}_B, \ \varphi, \psi \in \mathcal{D}. \tag{14.26}$$

Then, for $a \in A$, there is a linear mapping $\rho_0(a)$ on $\mathcal{D}_0$ such that

$$\rho_0(a)(x \otimes \varphi) = a \cdot x \otimes \varphi, \quad a \in A, \ \varphi \in \mathcal{D}_0, \ a \in A. \tag{14.27}$$

By Proposition 14.11, the map $a \mapsto \rho_0(a)$ is an algebra homomorphism of A in the algebra $L(\mathcal{D}_0)$ of linear operators on $\mathcal{D}_0$ satisfying

$$\langle \rho_0(a)\eta, \xi \rangle = \langle \eta, \rho_0(a^+)\xi \rangle, \quad \eta, \xi \in \mathcal{D}_0. \tag{14.28}$$

To obtain a positive *definite* inner product we consider the quotient $\mathcal{D}(\rho)$ of $\mathcal{D}_0$ by

$$\mathcal{N}_\pi := \left\{ \eta \in \mathcal{D}_0 : \langle \eta, \xi \rangle = 0, \xi \in \mathcal{D}_0 \right\} = \left\{ \eta \in \mathcal{D}_0 : \langle \xi, \eta \rangle = 0, \xi \in \mathcal{D}_0 \right\}. \tag{14.29}$$

The inner product of $\mathcal{D}_0$ yields an inner product, also denoted $\langle \cdot, \cdot \rangle$, on the quotient space $\mathcal{D}(\rho) = \mathcal{D}_0/\mathcal{N}_\pi$. For $\eta \in \mathcal{D}_0$, let $[\eta]$ denote the equivalence class $\eta + \mathcal{N}_\pi$ in $\mathcal{D}(\rho)$. From (14.28) it follows that the operators $\rho_0(a)$ leave $\mathcal{N}_\pi$ invariant, so $\rho(a)[\eta] = [\rho_0(a)\eta]$, $\eta \in \mathcal{D}_0$, is well defined, and that the map $a \mapsto \rho(a)$ is a $*$-representation ρ of A on the complex inner product space $(\mathcal{D}(\rho), \langle \cdot, \cdot \rangle)$.

Definition 14.26 The $*$-representation ρ of A on $\mathcal{D}(\rho)$, given by

$$\rho(a)[x \otimes \varphi] = [a \cdot x \otimes \varphi], \quad \varphi \in \mathcal{D}(\rho), \ a \in A, \tag{14.30}$$

is called the *induced representation* of π and denoted by $\mathcal{E}-\mathrm{Ind}\,\pi$.

Note that, in contrast to Definition 12.16 in Sect. 12.2, the induced representation $\mathcal{E}-\mathrm{Ind}\,\pi$ according to Definition 14.26 is not necessarily closed.

14.4 Imprimitivity Bimodules

Recall that weak $A-B$-imprimitivity bimodules were introduced in Definition 14.12.

Definition 14.27 A weak $A-B$-imprimitivity bimodule $(\mathcal{E}, {}_A[\cdot, \cdot], [\cdot, \cdot]_B)$ is called an $A-B$-*imprimitivity bimodule* if the right B-rigged space $(\mathcal{E}, [\cdot, \cdot]_B)$ and the left A-rigged space $(\mathcal{E}, {}_A[\cdot, \cdot])$ are positive semi-definite.

Two $*$-algebras A and B are called *Morita equivalent* if there exists an $\mathsf{A}-\mathsf{B}$-imprimitivity bimodule.

Proposition 14.28 *Morita equivalence is an equivalence relation for complex unital $*$-algebras.*

Proof The proof is almost the same as for the weak Morita equivalence (Proposition 14.17). From Example 14.24 it follows that $\mathcal{E} = \mathsf{A}$ is an $\mathsf{A}-\mathsf{A}$-imprimitivity bimodule, which gives the reflexivity. It is easily shown that if a weak $\mathsf{A}-\mathsf{B}$-imprimitivity bimodule $\mathcal{E}$ has positive semi-definite riggings, so has $\overline{\mathcal{E}}$. This yields the symmetry. The transitivity follows by combining Proposition 14.16 and its counterpart for left rigged spaces with Proposition 14.25.   □

Remark 14.29 Morita equivalence was invented in ring theory by K. Morita (1958) in order to study the equivalence of module categories of rings (see, e.g., [Lm99, §18]). The corresponding notion for C^*-algebras [RW98, Definition 3.1] is based on Hilbert C^*-modules which are complete and have *positive definite* riggings (see Definition 14.32(iv) below). Our Definitions 14.12 and 14.27 for general $*$-algebras are much weaker; we assume neither the nondegeneracy of the riggings nor the completeness in some topology.   ○

Example 14.30 (*Matrix $*$-algebras-Example* 14.15 *continued*)
We retain the notation of Example 14.15 and prove that the weak $\mathsf{A}-\mathsf{B}$-imprimitivity bimodule $\mathcal{E} = M_{k,m}(\mathbb{C})$ defined therein is an $\mathsf{A}-\mathsf{B}$-imprimitivity bimodule. Hence the $*$-algebras $\mathsf{A} = M_k(\mathbb{C})$ and $\mathsf{B} = M_m(\mathbb{C})$ are Morita equivalent.

It suffices to show that both riggings are positive semi-definite. We carry out the proof for $[\cdot, \cdot]_\mathsf{B}$. The reasoning is very similar to the one use in Example 14.24.

Let $X_1, \ldots, X_n \in \mathcal{E}, n \in \mathbb{N}$. By Definition 14.23 and Lemma 14.21, we have to show for any $*$-representation π of $\mathsf{B} = M_m(\mathbb{C})$ and vectors $\varphi_1, \ldots, \varphi_n \in \mathcal{D}(\pi)$ the matrix $([X_i, X_j]_\mathsf{B})_{i,j=1}^n$ satisfies (14.20). By Proposition 10.23, the $*$-representation π of $M_m(\mathbb{C})$ is of the form ρ_m for some $*$-representation ρ of $\mathbb{C}$. Let $\varphi_i = (\varphi_{i,1}, \ldots, \varphi_{i,m})$ and $X_i = (x_{i;p,q})_{p,q} \in M_{k,m}(\mathbb{C})$. The matrix $([X_i, X_j]_\mathsf{B})$ has the entries $([X_i, X_j]_\mathsf{B})_{p,q} = ((X_i)^+ X_j)_{p,q} = \sum_l (x_{i;l,p})^+ x_{j;l,q}$. Hence

$$\sum_{i,j} \langle \pi([X_i, X_j]_\mathsf{B}) \varphi_j, \varphi_i \rangle = \sum_{i,j,p,q,l} \langle \rho((x_{i;l,p})^+ x_{j;l,q}) \varphi_{j,q}, \varphi_{i,p} \rangle$$

$$= \sum_l \left\langle \sum_{j,q} \rho(x_{j;l,q}) \varphi_{j,q}, \sum_{i,p} \rho(x_{i;l,p}) \varphi_{i,p} \right\rangle \geq 0.$$

Thus, condition (14.20) holds for the rigging $[\cdot, \cdot]_\mathsf{B}$, which completes the proof. ○

Let $\mathcal{E}$ be an $\mathsf{A}-\mathsf{B}$-imprimitivity bimodule. By the induction procedure developed at the end of the preceding section, each $*$-representation of B induces a $*$-representation of A (see Definition 14.26). Since $\overline{\mathcal{E}}$ is a $\mathsf{B}-\mathsf{A}$-imprimitivity bimodule, any $*$-representation of A induces a $*$-representation of B as well. The following theorem says that for nondegenerate $*$-representations, up to unitary equivalence, these maps are inverse to each other.

Theorem 14.31 *Suppose $\mathcal{E}$ is an A–B-imprimitivity bimodule. Let π and π' be non-degenerate $*$-representations of B and A, respectively. Then the $*$-representations π and $\overline{\mathcal{E}}$–$\mathrm{Ind}(\mathcal{E}$–$\mathrm{Ind}\,\pi)$ of B are unitarily equivalent and so are the $*$-representations π' and $\mathcal{E}$–$\mathrm{Ind}(\overline{\mathcal{E}}$–$\mathrm{Ind}\,\pi')$ of A.*

Proof We prove the assertion for π; the proof for π' is similar. In the following proof we use freely the axioms of imprimitivity bimodules from Definition 14.12. Let ρ_0 and $\rho := \mathcal{E}$–$\mathrm{Ind}\,\pi$ be defined by (14.27) and (14.30), respectively.

First we note that there exists a well-defined linear (!) map U of the complex tensor product $\overline{\mathcal{E}} \otimes \mathcal{E} \otimes \mathcal{D}(\pi)$ into $\mathcal{D}(\pi)$ such that

$$U(\overline{x} \otimes (y \otimes \varphi)) = \pi([x, y]_\mathsf{B})\varphi, \quad x, y \in \mathcal{E}, \varphi \in \mathcal{D}(\pi). \tag{14.31}$$

The map U passes to the balanced tensor product $\overline{\mathcal{E}} \otimes_\mathsf{A} \mathcal{E} \otimes_\mathsf{B} \mathcal{D}(\pi)$. Indeed, for $a \in \mathsf{A}$ and $b \in \mathsf{B}$, we derive

$$U(\overline{x} \cdot a \otimes (y \cdot b \otimes \varphi)) = U(\overline{a^+ \cdot x} \otimes (y \cdot b \otimes \varphi)) = \pi([a^+ \cdot x, y \cdot b]_\mathsf{B})\varphi$$
$$= \pi([a^+ \cdot x, y]_\mathsf{B})\pi(b)\varphi = \pi([x, a \cdot y]_\mathsf{B})\pi(b)\varphi = U(\overline{x} \otimes (a \cdot y \otimes \pi(b)\varphi)).$$

Our next aims are to show that U passes to a map of $\overline{\mathcal{E}} \otimes \mathcal{D}(\rho)$ into $\mathcal{D}(\pi)$ and that it preserves the inner product. Let $\xi, \eta \in \mathcal{D}(\rho)$. Then ξ and η are of the form $\xi = \sum_i y_i \otimes \varphi_i, \eta = \sum_j v_j \otimes \psi_j$, where $y_i, v_j \in \mathcal{E}, \varphi_i, \psi_j \in \mathcal{D}(\pi)$. We compute

$$\langle \overline{x} \otimes \xi, \overline{u} \otimes \eta \rangle = \langle \rho_0([\overline{u}, \overline{x}]_\mathsf{A})\xi, \eta \rangle = \sum_{i,j} \langle \rho_0({}_\mathsf{A}[u, x])(y_i \otimes \varphi_i), v_j \otimes \psi_j \rangle$$

$$= \sum_{i,j} \langle ({}_\mathsf{A}[u, x] \cdot y_i) \otimes \varphi_i, v_j \otimes \psi_j \rangle = \sum_{i,j} \langle \pi([v_j, {}_\mathsf{A}[u, x] \cdot y_i]_\mathsf{B})\varphi_i, \psi_j \rangle. \tag{14.32}$$

Let $\xi = \sum_i y_i \otimes \varphi_i \in \mathcal{N}_\pi$, see (14.29). Then, for $a \in \mathsf{A}$, $a \cdot \xi = 0$ and hence

$$\langle a \cdot \xi, w \otimes \zeta \rangle = \sum_i \langle a \cdot y_i \otimes \varphi_i, w \otimes \zeta \rangle = \sum_i \langle \pi([w, a \cdot y_i]_\mathsf{B})\varphi_i, \zeta \rangle = 0.$$

Inserting the latter into (14.32), with $a = {}_\mathsf{A}[u, x]$, $w = v_j$, $\zeta = \psi_j$, it follows that $\langle \overline{x} \otimes \xi, \overline{u} \otimes \eta \rangle = 0$. Similarly, $\eta \in \mathcal{N}_\pi$ implies $\langle \overline{x} \otimes \xi, \overline{u} \otimes \eta \rangle = 0$. Therefore, we can pass to the equivalence classes $[\xi], [\eta]$ of ξ, η in $\mathcal{D}(\rho) = \mathcal{D}_0/\mathcal{N}_\pi$ and continue

$$\langle \overline{x} \otimes [\xi], \overline{u} \otimes [\eta] \rangle = \sum_{i,j} \langle \pi([v_j, {}_\mathsf{A}[u, x] \cdot y_i]_\mathsf{B})\varphi_i, \psi_j \rangle$$

$$= \sum_{i,j} \langle \pi([v_j, u \cdot [x, y_i]_\mathsf{B}]_\mathsf{B})\varphi_i, \psi_j \rangle = \sum_{i,j} \langle \pi([v_j, u]_\mathsf{B}[x, y_i]_\mathsf{B})\varphi_i, \psi_j \rangle$$

$$= \sum_{i,j} \langle \pi([v_j, u]_\mathsf{B})\pi([x, y_i]_\mathsf{B})\varphi_i, \psi_j \rangle = \sum_{i,j} \langle \pi([x, y_i]_\mathsf{B})\varphi_i, \pi([u, v_j]_\mathsf{B})\psi_j \rangle$$

$$= \langle U(\overline{x} \otimes \xi), U(\overline{u} \otimes \eta) \rangle.$$

Hence, U extends by linearity to a linear mapping, denoted again U, of the complex tensor product $\overline{\mathcal{E}} \otimes \mathcal{D}(\rho)$ into $\mathcal{D}(\pi)$, which preserves the inner product.

Now we consider the representation $\sigma := \overline{\mathcal{E}}-\text{Ind } \rho$ of B which is induced from the representation ρ of A. Let σ_0 denote the corresponding homomorphism defined by (14.27) with ρ replaced by σ. Let $b \in \mathsf{B}$. Then, for $\xi = \sum_i y_i \otimes \varphi_i \in \mathcal{E} \otimes \mathcal{D}(\rho)$,

$$U(\sigma_0(b)(\overline{x} \otimes \xi)) = U((b \cdot \overline{x}) \otimes \xi) = U\left(\overline{x \cdot b^+} \otimes \left(\sum_i y_i \otimes \varphi_i\right)\right)$$

$$= \sum_i \pi([x \cdot b^+, y_i]_\mathsf{B})\varphi_i = \sum_i \pi(b\,[x, y_i]_\mathsf{B})\varphi_i$$

$$= \sum_i \pi(b)\pi([x, y_i]_\mathsf{B})\varphi_i = \pi(b)U\left(\overline{x} \otimes \left(\sum_i y_i \otimes \varphi_i\right)\right) = \pi(b)U(\overline{x} \otimes \xi),$$

which implies that

$$\pi(b)U\zeta = U\sigma_0(b)\zeta \quad \text{for} \quad \zeta \in \overline{\mathcal{E}} \otimes \mathcal{D}(\rho). \tag{14.33}$$

Since $\text{Lin}\{[x, y]_\mathsf{B}; x, y \in \mathcal{E}\} = \mathsf{B}$ (because $\mathcal{E}$ is full) and $\pi(\mathsf{B})\mathcal{D}(\pi) = \mathcal{D}(\pi)$ (because π is nondegenerate), it follows from (14.31) that U maps $\overline{\mathcal{E}} \otimes \mathcal{D}(\rho)$ onto $\mathcal{D}(\pi)$. Since $U : \overline{\mathcal{E}} \otimes \mathcal{D}(\rho) \mapsto \mathcal{D}(\pi)$ preserves the inner product, $\ker U$ is just the kernel $\mathcal{N}_\rho$ defined by (14.29), applied to ρ. Hence U passes to an isometric linear mapping U of $\mathcal{D}(\sigma) = (\mathcal{E} \otimes \mathcal{D}(\rho))/\mathcal{N}_\rho$ on $\mathcal{D}(\pi)$ and the equality (14.33) yields $\pi(b)U = U\sigma(b)$. Thus, $\pi(b) = U\sigma(b)U^{-1}$ for $b \in \mathsf{B}$. Since $\sigma = \overline{\mathcal{E}}-\text{Ind } (\mathcal{E}-\text{Ind}\pi)$ by the above definitions, this is the assertion. $\qquad\square$

14.5 Hilbert C^*-modules

In this section, $\mathfrak{A}$ is a (not necessarily unital) C^*-**algebra** with norm $\|\cdot\|_\mathfrak{A} = \|\cdot\|$. For $a, b \in \mathfrak{A}_{\text{her}}$, "$a \geq b$" refers to the order relation in $\mathfrak{A}$, that is, $\sigma_\mathfrak{A}(a - b) \subseteq \mathbb{R}_+$. We also use the fact that $0 \leq a \leq b$ implies that $\|a\| \leq \|b\|$.

A Hilbert C^*-module for $\mathfrak{A}$ is a very special case of a right $\mathfrak{A}$-rigged space $\mathcal{X}$: The rigging is an $\mathfrak{A}$-valued inner product, denoted now by $\langle \cdot, \cdot \rangle_\mathcal{X}$, and the module is complete in the corresponding norm. First we define pre-Hilbert C^*-modules.

Definition 14.32 A *pre-Hilbert C^*-module over $\mathfrak{A}$*, or a *pre-Hilbert $\mathfrak{A}$-module*, is a right $\mathfrak{A}$-module $\mathcal{X}$, together with a map $\langle \cdot, \cdot \rangle_\mathcal{X} : \mathcal{X} \times \mathcal{X} \mapsto \mathfrak{A}$, such that for $x, y, z \in \mathcal{X}, \alpha, \beta \in \mathbb{C}$, and $a \in \mathfrak{A}$:

  (i) $\langle x, \alpha y + \beta z \rangle_\mathcal{X} = \alpha\langle x, y \rangle_\mathcal{X} + \beta\langle x, z \rangle_\mathcal{X}$,
  (ii) $(\langle x, y \rangle_\mathcal{X})^+ = \langle y, x \rangle_\mathcal{X}$,
  (iii) $\langle x, y \cdot a \rangle_\mathcal{X} = \langle x, y \rangle_\mathcal{X}\, a$,
  (iv) $\langle x, x \rangle_\mathcal{X} \geq 0$, and $\langle x, x \rangle_\mathcal{X} = 0$ implies $x = 0$.

Before we define Hilbert C^*-modules we derive an auxiliary lemma.

Lemma 14.33 *Suppose $\mathcal{X}$ is a pre-Hilbert C^*-module over $\mathfrak{A}$. Then*

$$\|x\|_{\mathcal{X}} := \left\|\langle x, x\rangle_{\mathcal{X}}\right\|_{\mathfrak{A}}^{1/2}, \quad x \in \mathcal{X},$$

defines a norm $\|\cdot\|_{\mathcal{X}}$ on $\mathcal{X}$. For $x, y \in \mathcal{X}$ and $a \in \mathfrak{A}$, we have

$$\|\langle x, y\rangle_{\mathcal{X}}\|_{\mathfrak{A}} = \|x\|_{\mathcal{X}}\|y\|_{\mathcal{X}}, \tag{14.34}$$

$$\|x \cdot a\|_{\mathcal{X}} \le \|x\|_{\mathcal{X}}\|a\|_{\mathfrak{A}}. \tag{14.35}$$

Proof First we prove (14.34). We abbreviate $a := \langle y, x\rangle_{\mathcal{X}} \in \mathfrak{A}$, $b := \langle x, x\rangle_{\mathcal{X}} \in \mathfrak{A}$ and $c := \langle y, y\rangle_{\mathcal{X}} \in \mathfrak{A}$. If $a = 0$, then $\langle x, y\rangle_{\mathcal{X}} = 0$ and (14.34) holds trivially.

Now we assume that $a \ne 0$. Let $\lambda \in \mathbb{R}$. Using the axioms (i)–(iii) of Definition 14.32 we derive

$$\begin{aligned}
0 &\le \langle x - y \cdot (\lambda a), x - y \cdot (\lambda a)\rangle_{\mathcal{X}} \\
&= \langle x, x\rangle_{\mathcal{X}} - \lambda\langle x, y\rangle_{\mathcal{X}} a - \lambda a^+\langle y, x\rangle_{\mathcal{X}} + \lambda^2 a^+\langle y, y\rangle_{\mathcal{X}} a \\
&= b - 2\lambda a^+ a + \lambda^2 a^+ c a.
\end{aligned}$$

Therefore, $2\lambda a^+ a \le b + \lambda^2 a^+ c a$ and

$$2\lambda\|a^+ a\| = 2\lambda\|a\|^2 \le \|b + \lambda^2 a^+ c a\| \le \|b\| + \lambda^2\|c\|\,\|a\|^2$$

for all $\lambda \in \mathbb{R}$. Hence it follows that $\|a\|^4 \le \|b\|\,\|c\|\,\|a\|^2$. Since $a \ne 0$, this yields

$$\|\langle x, y\rangle_{\mathcal{X}}\|^2 = \|a^+\|^2 = \|a\|^2 \le \|b\|\,\|c\| = \|\langle x, x\rangle_{\mathcal{X}}\|\,\|\langle y, y\rangle_{\mathcal{X}}\|,$$

which gives (14.34).

As in the scalar case, (14.34) implies the triangle inequality for $\|\cdot\|_{\mathcal{X}}$. It is obvious that $\|\lambda x\|_{\mathcal{X}} = |\lambda|\,\|x\|_{\mathcal{X}}$ for $\lambda \in \mathbb{C}$, $x \in \mathcal{X}$. Hence $\|\cdot\|_{\mathcal{X}}$ is a seminorm. From Definition 14.32(iv) it follows that it is a norm.

To prove (14.35) we derive

$$\|x \cdot a\|_{\mathcal{X}}^2 = \|\langle x \cdot a, x \cdot a\rangle_{\mathcal{X}}\|_{\mathfrak{A}} = \|a^+\langle x, x\rangle_{\mathcal{X}} a\|_{\mathfrak{A}} \le \|x\|_{\mathcal{X}}^2\|a\|_{\mathfrak{A}}^2. \qquad \square$$

Definition 14.34 A *Hilbert C^*-module over $\mathfrak{A}$*, or a *Hilbert $\mathfrak{A}$-module*, is a pre-Hilbert $\mathfrak{A}$-module $\mathcal{X}$ that is complete in the norm $\|\cdot\|_{\mathcal{X}}$.

An "ordinary" complex Hilbert space $(\mathcal{H}, \langle\cdot, \cdot\rangle)$ is a Hilbert C^*-module for the C^*-algebra $\mathfrak{A} = \mathbb{C}$ if we define $\langle\varphi, \psi\rangle_{\mathcal{X}} = \langle\psi, \varphi\rangle$. The interchange of variables is necessary, because in our convention Hilbert space inner products are linear and inner products of Hilbert C^*-modules are conjugate linear in the first variables.

The simplest example of a Hilbert $\mathfrak{A}$-module is the C^*-algebra $\mathfrak{A}$ itself.

Example 14.35 ($\mathcal{X} = \mathfrak{A}$)

The C^*-algebra $\mathfrak{A}$ is a Hilbert $\mathfrak{A}$-module $\mathcal{X}$ with multiplication as right action and $\mathfrak{A}$-valued inner product $\langle x, y \rangle_{\mathcal{X}} := x^+ y$, where $x, y \in \mathfrak{A}$. Then

$$\|x\|_{\mathcal{X}}^2 = \|\langle x, x \rangle_{\mathcal{X}}\|_{\mathfrak{A}} = \|x^+ x\|_{\mathfrak{A}}^2 = \|x\|_{\mathfrak{A}}^2$$

by the C^*-property of the norm $\|\cdot\|_{\mathfrak{A}}$, so we have $\|x\|_{\mathcal{X}} = \|x\|_{\mathfrak{A}}$ for $x \in \mathfrak{A}$. Hence, since $\mathfrak{A}$ is complete, so is $\mathcal{X}$. Thus, $\mathcal{X} = \mathfrak{A}$ is a Hilbert $\mathfrak{A}$-module. $\bigcirc$

Example 14.36   (*Direct sum of a finite set of Hilbert $\mathfrak{A}$-modules*)

Let $(\mathcal{X}_k, \langle \cdot, \cdot \rangle_{\mathcal{X}_k})$, $k = 1, \ldots, n$, be Hilbert $\mathfrak{A}$-modules. Then the direct sum $\mathcal{X} = \sum_{k=1}^{n} \oplus \mathcal{X}_k$ of right $\mathfrak{A}$-modules becomes a Hilbert $\mathfrak{A}$-module with inner product

$$\langle (x_k), (y_k) \rangle_{\mathcal{X}} := \sum_{k=1}^{n} \langle x_k, y_k \rangle_{\mathcal{X}_k}, \quad (x_k), (y_k) \in \mathcal{X}.$$ $\bigcirc$

The simplest infinite-dimensional Hilbert space is $l^2(\mathbb{N})$. Its counterpart for Hilbert C^*-modules is more subtle, as the following example shows.

Example 14.37 ($\mathcal{X} = l_2(\mathfrak{A})$)

The *Hilbert $\mathfrak{A}$-module* $\mathcal{X} = l_2(\mathfrak{A})$ is defined by

$$l_2(\mathfrak{A}) := \left\{ (x_k)_{k \in \mathbb{N}} : x_k \in \mathfrak{A}, \ \sum_{k=1}^{\infty} (x_k)^+ x_k \ \text{converges in } \mathfrak{A} \right\}.$$

In this example the following two facts will be essentially used.

I. A sequence (x_k), with $x_k \in \mathfrak{A}$, belongs to $l^2(\mathfrak{A})$ if and only if for each $\varepsilon > 0$ there exists an $n(\varepsilon)$ such that $\|\sum_{k=r}^{s}(x_k)^+ x_k\| < \varepsilon$ for all $s \geq r \geq n(\varepsilon)$.

II. $\|\sum_{k=r}^{s}(x_k)^+ x_k\| \leq \|\sum_{k=n}^{m}(x_k)^+ x_k\|$ for elements $x_k \in \mathfrak{A}$, $r \geq n, m \geq s$, and $\|\sum_{k=r}^{s}(y_k)^+ y_k\| \leq \|\sum_{k=1}^{\infty}(y_k)^+ y_k\|$ for $(y_k) \in l^2(\mathfrak{A})$.

(Note that II. holds, since $0 \leq a \leq b$ for $a, b \in \mathfrak{A}$ implies that $\|a\| \leq \|b\|$.)

Now we begin to prove that $l^2(\mathfrak{A})$ is a Hilbert $\mathfrak{A}$-module.

Let $(x_k), (y_k) \in l^2(\mathfrak{A})$, $\lambda \in \mathbb{C}$, and $a \in \mathfrak{A}$. From I. and the relations

$$(x + y)^+(x + y) \leq (x + y)^+(x + y)^+ + (x - y)^+(x - y) = 2x^+ x + 2y^+ y$$

we conclude that $(x_k + y_k) \in l^2(\mathfrak{A})$. It is clear that $(\lambda x_k) \in l^2(\mathfrak{A})$. Combining I. and the inequality $\|\sum_{k=r}^{s}(x_k a)^+ x_k a\| \leq \|a\|^2 \|\sum_{k=r}^{s}(x_k)^+ x_k\|$ it follows that $(x_k a) \in l^2(\mathfrak{A})$. Thus, $l^2(\mathfrak{A})$ is a right $\mathfrak{A}$-module.

From the polarization identity (2.27) it follows that $\sum_k (x_k)^+ y_k$ converges in $\mathfrak{A}$ if $(x_k), (y_k) \in l^2(\mathfrak{A})$. Therefore,

$$\langle (x_k), (y_k) \rangle_{\mathcal{X}} := \sum_{k=1}^{\infty} (x_k)^+ y_k \quad \text{for } (x_k), (y_k) \in l_2(\mathfrak{A}),$$

defines an $\mathfrak{A}$-valued inner product on $l^2(\mathfrak{A})$.

It remains to prove the completeness. For suppose $x^{(n)} = (x_k^{(n)}), n \in \mathbb{N}$, is a Cauchy sequence in $l^2(\mathfrak{A})$. Fix $k \in \mathbb{N}$. Then, by II., we have for $n, m \in \mathbb{N}$,

$$\left\| (x_k^{(n)} - x_k^{(m)})^+ (x_k^{(n)} - x_k^{(m)}) \right\| \leq \left\| \sum_{j=1}^{\infty} (x_j^{(n)} - x_j^{(m)})^+ (x_j^{(n)} - x_j^{(m)}) \right\|.$$

Hence the kth coordinate sequence $(x_k^{(n)})_{n \in \mathbb{N}}$ is a Cauchy sequence in $\mathfrak{A}$, so it converges to some $x_k \in \mathfrak{A}$. Let x denote the sequence $(x_k)_{k \in \mathbb{N}}$. We have to show that $x \in l^2(\mathfrak{A})$ and $\lim_n \|x - x^{(n)}\|_{\mathcal{X}} = 0$. Let $n, r, s \in \mathbb{N}, s \geq r$. Again by II.,

$$\left\| \sum_{j=r}^{s} (x_j - x_j^{(n)})^+ (x_j - x_j^{(n)}) \right\| \tag{14.36}$$

$$= \lim_m \left\| \sum_{j=r}^{s} (x_j^{(m)} - x_j^{(n)})^+ (x_j^{(m)} - x_j^{(n)}) \right\|$$

$$\leq \overline{\lim_m} \left\| \sum_{j=1}^{\infty} (x_j^{(m)} - x_j^{(n)})^+ (x_j^{(m)} - x_j^{(n)}) \right\| = \overline{\lim_m} \|x^{(m)} - x^{(n)}\|_{\mathcal{X}}^2. \tag{14.37}$$

Let $\varepsilon > 0$. Since $(x^{(n)})$ is a Cauchy sequence, $\|x^{(m)} - x^{(n)}\|^2 < \varepsilon/2$ for large n, m. Fix such an n. Then (14.37) is $< \varepsilon$ for large m and so is (14.36) for all $r, s, s \geq r$. By I., this implies that the series $\sum_{j=1}^{\infty} (x_j - x_j^{(n)})^+ (x_j - x_j^{(n)})$ converges in the norm of $\mathfrak{A}$. Therefore, $x \quad x^{(n)} \in l^2(\mathfrak{A})$ and hence $x \in l^2(\mathfrak{A})$.

Further, setting $r = 1$ and letting $s \to \infty$ in (14.36)–(14.37), we obtain

$$\|x - x^{(n)}\|_{\mathcal{X}}^2 = \left\| \sum_{j=1}^{\infty} (x_j - x_j^{(n)})^+ (x_j - x_j^{(n)}) \right\| < \varepsilon.$$

Thus, $\lim_n \|x - x^{(n)}\|_{\mathcal{X}} = 0$. This proves that $l^2(\mathfrak{A})$ is a Hilbert $\mathfrak{A}$-module.

It might be instructive to compare $\mathcal{X} = l^2(\mathfrak{A})$ with

$$\mathcal{X}_1 := \left\{ (x_k)_{k \in \mathbb{N}} : x_k \in \mathfrak{A}, \ \sum_{k=1}^{\infty} \|x_k\|^2 < \infty \right\}.$$

It is not difficult to see that $\mathcal{X}_1 \subseteq \mathcal{X}$. However, $\mathcal{X}_1 = \mathcal{X}$ if and only if the C^*-algebra $\mathfrak{A}$ is finite-dimensional [Fk90]. ○

The Hilbert $\mathfrak{A}$-module $l^2(\mathfrak{A})$ is also denoted $\mathcal{H}_\mathfrak{A}$ and called the *standard Hilbert C^*-module* over $\mathfrak{A}$. It plays an important role in the theory of Hilbert C^*-modules. The Kasparov stabilization theorem [La94, Theorem 6.2] says for each countably generated Hilbert $\mathfrak{A}$-module $\mathcal{X}$ the direct sum $\mathcal{X} \oplus \mathcal{H}_\mathfrak{A}$ is unitarily equivalent to $\mathcal{H}_\mathfrak{A}$. (A Hilbert $\mathfrak{A}$-module $\mathcal{X}$ is called countably generated if it has a countable subset such that $\mathcal{X}$ is the smallest Hilbert $\mathfrak{A}$-submodule which contains this set.)

Definition 14.38 A subset M of $\mathcal{X}$ is called *essential* if $M^\perp = \{0\}$, where

$$M^\perp := \{x \in \mathcal{X} : \langle x, y \rangle_\mathcal{X} = 0 \text{ for } y \in M \}. \tag{14.38}$$

Obviously, if M is dense in $(\mathcal{X}, \|\cdot\|_\mathcal{X})$, it is essential. The converse is not true, as the following simple example shows.

Example 14.39 (*A proper closed submodule which is essential*)
Let $\mathcal{X}$ be the Hilbert $\mathfrak{A}$-module from Example 14.35 for the C^*-algebra $\mathfrak{A} := C([0,1])$. Then $M := \{f \in \mathcal{X} : f(0) = 0\}$ is a closed $\mathfrak{A}$-submodule of $\mathcal{X}$ such that $M \neq \mathcal{X}$ and $M^\perp = \{0\}$. In particular, $M \oplus M^\perp \neq \mathcal{X}$. $\qquad\qquad\bigcirc$

This example shows that the Riesz theorem about orthogonal projections does not hold for Hilbert C^*-modules! From the technical side, this failure is one of the main reasons for most of the difficulties in operator theory on Hilbert C^*-modules.

14.6 Representations on Hilbert C^*-modules

In this section, $(\mathcal{X}, \langle \cdot, \cdot \rangle_\mathcal{X})$ is a **Hilbert $\mathfrak{A}$-module** and $\mathfrak{B}$ is a **$*$-subalgebra** of $\mathfrak{A}$.

By $\mathfrak{B}$-*submodule* of $\mathcal{X}$ we mean a linear subspace $\mathcal{D}$ of $\mathcal{X}$ which is invariant under the right action of $\mathfrak{B}$.

First we develop some elementary definitions and facts on operators.

Definition 14.40 A $\mathfrak{B}$-*operator* on $\mathcal{X}$ is a $\mathbb{C}$-linear and $\mathfrak{B}$-linear map t of a $\mathfrak{B}$-submodule $\mathcal{D}(t)$ of $\mathcal{X}$, called the *domain* of t, into $\mathcal{X}$, that is,

$$t(\lambda x) = \lambda t(x) \quad \text{and} \quad t(x \cdot b) = t(x) \cdot b \text{ for } x \in \mathcal{D}(t), \lambda \in \mathbb{C}, b \in \mathfrak{B}.$$

Note that the $\mathfrak{B}$-linearity is a very strong requirement in general.

Example 14.41 Let $\mathcal{X} = \mathfrak{A}$ be the Hilbert $\mathfrak{A}$-module for the C^*-algebra $\mathfrak{A} = C([0,1])$ and $\mathfrak{B} = \mathbb{C}[x]$. The multiplication operator t by the variable x with domain $\mathcal{D}(t) = \mathbb{C}[x]$ is a $\mathfrak{B}$-operator. The domain $\mathcal{D}(t)$ is not an $\mathfrak{A}$-submodule and t is not an $\mathfrak{A}$-operator. The multiplication operator by the variable x with domain $\mathcal{X}$ is the continuous extension of t and an $\mathfrak{A}$-operator.

Suppose s is another $\mathfrak{B}$-operator such that $\mathcal{D}(s) = \mathbb{C}[x]$ and $t1 = s1$. Then $s(b) = s(1 \cdot b) = s(1)b = t(1)b = t(1 \cdot b) = t(b)$ for $b \in \mathbb{C}[x]$, so that $s = t$. $\qquad\bigcirc$

Definition 14.42 Suppose t is a $\mathfrak{B}$-operator on $\mathcal{X}$ such that $\mathcal{D}(t)$ is essential. Define

$$\mathcal{D}(t^*) = \{y \in \mathcal{X} : \text{There exists a } z \in \mathcal{X} \text{ such that } \langle tx, y \rangle_{\mathcal{X}} = \langle x, z \rangle_{\mathcal{X}}, \ x \in \mathcal{D}(t)\}.$$

Since $\mathcal{D}(t)$ is essential, the element z in Definition 14.42 is uniquely determined by y. (Indeed, if z and z' are two such elements, it follows that $z - z' \in \mathcal{D}(t)^\perp = \{0\}$, so $z = z'$.) Hence $t^*y := z$ gives a well-defined map t^* of $\mathcal{D}(t^*)$ into $\mathcal{X}$. It is easily verified (and follows from Lemma 14.44 below) that t^* is a $\mathfrak{B}$-operator. It is called the *adjoint operator* of t and by definition we have

$$\langle tx, y \rangle_{\mathcal{X}} = \langle x, t^*y \rangle_{\mathcal{X}} \quad \text{for} \quad x \in \mathcal{D}(t), \ y \in \mathcal{D}(t^*).$$

It should be emphasized that the adjoint operator t^* is already well defined if the domain $\mathcal{D}(t)$ is essential; it is not needed that $\mathcal{D}(t)$ is dense in $\mathcal{X}$.

The counterpart of the $*$-algebra $\mathcal{L}^+(\mathcal{D})$ (see Definition 3.1) is the following.

Definition 14.43 Suppose $\mathcal{D}$ is a $\mathfrak{B}$-submodule of $\mathcal{X}$. Let $\mathcal{L}^+_{\mathfrak{B}}(\mathcal{D})$ denote the set of maps $t : \mathcal{D} \mapsto \mathcal{D}$ for which there exists a map $s : \mathcal{D} \mapsto \mathcal{D}$ such that

$$\langle tx, y \rangle_{\mathcal{X}} = \langle x, sy \rangle_{\mathcal{X}} \text{ for } x, y \in \mathcal{D}. \tag{14.39}$$

From now on we suppose that $\mathcal{D}$ is a $\mathfrak{B}$-**submodule** of $\mathcal{X}$.

Lemma 14.44 *Suppose $t \in \mathcal{L}^+_{\mathfrak{B}}(\mathcal{D})$ and let s be as in Definition 14.43. Then both maps t and s are $\mathfrak{B}$-operators with domain $\mathcal{D}$. Further, s is uniquely determined by t and will be denoted by t^+.*

Proof In order to prove that t and s are $\mathfrak{B}$-operators we have to show that these maps are $\mathbb{C}$-linear and $\mathfrak{B}$-linear. As a sample we verify that t is $\mathfrak{B}$-linear. Let $b \in \mathfrak{B}$ and $x, y \in \mathcal{D}$. Then, $x \cdot b \in \mathcal{D}$, so using (14.39) we deduce

$$\langle t(x \cdot b), y \rangle_{\mathcal{X}} = \langle x \cdot b, sy \rangle_{\mathcal{X}} = b^+ \langle y, sy \rangle_{\mathcal{X}} = b^+ \langle tx, y \rangle_{\mathcal{X}} = \langle (tx) \cdot b, y \rangle_{\mathcal{X}}.$$

Therefore, $\langle t(x \cdot b) - (tx) \cdot b, y \rangle_{\mathcal{X}} = 0$ for all $y \in \mathcal{D}$. Since $(t(x \cdot b) - (tx) \cdot b) \in \mathcal{D}$, we can set $y := t(x \cdot b) - (tx) \cdot b$. Then $\langle y, y \rangle_{\mathcal{X}} = 0$. Hence $y = 0$ by Definition 14.32(iv) and so $t(x \cdot b) = (tx) \cdot b$, which proves that t is $\mathfrak{B}$-linear.

We show that s is uniquely determined by t. Let $\tilde{s} : \mathcal{D} \mapsto \mathcal{D}$ be another map satisfying (14.39) and let $y \in \mathcal{D}$. Then we have $\langle tx, y \rangle_{\mathcal{X}} = \langle x, sy \rangle_{\mathcal{X}} = \langle x, \tilde{s}y \rangle_{\mathcal{X}}$ and so $\langle x, sy - \tilde{s}y \rangle_{\mathcal{X}} = 0$ for all $x \in \mathcal{D}$. Setting $x = sy - \tilde{s}y$ we obtain $x = 0$ and hence $sy = \tilde{s}y$. Thus, $s = \tilde{s}$. $\square$

Lemma 14.45 *$\mathcal{L}^+_{\mathfrak{B}}(\mathcal{D})$ is a unital complex $*$-algebra with operator multiplication and involution $t \mapsto t^+$.*

Proof The proof is verbatim the same as in the Hilbert space case (Lemma 3.2). $\square$

If $\mathcal{D}$ is essential in $\mathcal{X}$, it follows from (14.39), compared with the definitions of $\mathcal{D}(t^*)$ and t^*, that t^+ is the restriction to $\mathcal{D}$ of the adjoint operator t^*.

Now we can give the main definition of this section.

Definition 14.46 A *-*representation* of a *-algebra A on a $\mathfrak{B}$-submodule $\mathcal{D}$ of $\mathcal{X}$ is a *-homomorphism π of A in the *-algebra $\mathcal{L}_{\mathfrak{B}}^+(\mathcal{D})$. We write $\mathcal{D}(\pi) := \mathcal{D}$.

A large number of basic notions and facts for Hilbert space representations carry over almost verbatim to *-representations on $\mathfrak{B}$-submodules.

Let π be a *-representation of A on a $\mathfrak{B}$-submodule $\mathcal{D}$. The *graph topology* of π is the locally convex topology on $\mathcal{D}$ defined by the seminorms

$$\|x\|_a := \|\pi(a)x\|_\mathcal{X}, \quad a \in \mathsf{A}, \ x \in \mathcal{D}.$$

As in the Hilbert space case (Lemma 3.5), this family of seminorms is directed.

Proposition 14.47 *Suppose that $\mathfrak{B}$ is dense in the C^*-algebra $\mathfrak{A}$. Let π be a *-representation of A on a $\mathfrak{B}$-submodule $\mathcal{D}$. Then the $\mathfrak{B}$-operator $\pi(a)$ on $\mathcal{X}$ is closable for $a \in \mathsf{A}$. The completion $\hat{\mathcal{D}}$ of $\mathcal{D}$ in the graph topology of π is a right $\mathfrak{A}$-submodule such that*

$$\hat{\mathcal{D}} = \bigcap_{a \in \mathsf{A}} \mathcal{D}(\overline{\pi(a)}). \tag{14.40}$$

*There is a *-representation $\bar{\pi}$ on the $\mathfrak{A}$-submodule $\mathcal{D}(\bar{\pi}) := \hat{\mathcal{D}}$ given by*

$$\bar{\pi}(a) := \overline{\pi(a)} \restriction \hat{\mathcal{D}}, \quad a \in \mathsf{A}.$$

Proof We show that the $\mathfrak{B}$-operator $\pi(a)$ is closable. Suppose (x_n) is a sequence of $\mathcal{D}$ such that $x_n \to 0$ and $\pi(a)x_n \to y$ in $\mathcal{X}$ for some $y \in \mathcal{X}$. Then, for $z \in \mathcal{D}$,

$$\langle \pi(a)x_n, z \rangle_\mathcal{X} = \langle x_n, \pi(a^+)z \rangle_\mathcal{X} \to \langle y, z \rangle_\mathcal{X} = 0.$$

Hence $y \in \mathcal{D}^\perp$. Since $\pi(a)x_n \in \mathcal{D}$, we obtain $0 = \langle y, \pi(a)x_n \rangle_\mathcal{X} \to 0 = \langle y, y \rangle_\mathcal{X}$, so that $y = 0$. This proves that $\pi(a)$ is closable.

Next we prove that the closure $\overline{\pi(a)}$ of $\pi(a)$ is an $\mathfrak{A}$-operator. Let $x \in \mathcal{D}$ and $b \in \mathfrak{A}$. Since $\mathfrak{B}$ is dense in $\mathfrak{A}$, there is a sequence (b_n) of elements of $\mathfrak{B}$ converging to b. Using Eq. (14.35) and the fact that $\pi(a)$ is a $\mathfrak{B}$-operator we deduce

$$\pi(a)(x \cdot b_n) = (\pi(a)x) \cdot b_n \to (\pi(a)x) \cdot b \quad \text{and} \quad x \cdot b_n \to x \cdot b \text{ in } \mathcal{X}.$$

Therefore, from the definition of the closure it follows that $x \cdot b \in \mathcal{D}(\overline{\pi(a)})$ and $\overline{\pi(a)}\,(x \cdot b) = (\pi(a)x) \cdot b = (\overline{\pi(a)}\,x) \cdot b$. Thus, $\overline{\pi(a)}$ is an $\mathfrak{A}$-operator.

Using essentially the fact that the family of seminorm $\| \cdot \|_a, a \in \mathsf{A}$, is directed, the remaining part of the proof follows the same pattern as in the Hilbert space case; see the proof of Proposition 3.8. We do not carry out the details. $\qquad\square$

Many examples of $*$-representations on Hilbert C^*-modules can be derived from the following simple lemma.

Lemma 14.48 *Let $\mathcal{X} = \mathfrak{A}$ be the Hilbert $\mathfrak{A}$-module from Example 14.35, with $\mathfrak{A}$-valued inner product $\langle x, y \rangle_{\mathcal{X}} = x^+ y$, and let $\mathfrak{B}$ be a $*$-subalgebra of $\mathfrak{A}$. Let $\mathcal{D}$ be a complex inner product space. Suppose that $\mathfrak{B}$ is also a $*$-subalgebra of $\mathcal{L}^+(\mathcal{D})$ and there is a $*$-representation ρ of A on the complex inner product space $\mathcal{D}$ such that*

$$\rho(a) \cdot b \in \mathfrak{B} \quad \text{for} \quad a \in \mathsf{A}, b \in \mathfrak{B}. \tag{14.41}$$

Then there is a $*$-representation π of A (according to Definition 14.46) on the $\mathfrak{B}$-submodule $\mathfrak{B}$ of $\mathcal{X}$ defined by $\pi(a)b := \rho(a) \cdot b$ for $a \in \mathsf{A}, b \in \mathfrak{B}$.

The dot "$\cdot$" in (14.41) refers to the product of operators in $\mathcal{L}^+(\mathcal{D})$. That is, $\rho(a) \cdot b$ is the product of $\rho(a)$ and b in the algebra $\mathcal{L}^+(\mathcal{D})$, while $\pi(a)b$ means the action of $a \in \mathsf{A}$ on $b \in \mathfrak{B}$ by the representation π.

Proof From the assumptions it follows easily that π is an algebra homomorphism of A into $L(\mathfrak{B})$. Let $a \in \mathsf{A}$ and $b, c \in \mathfrak{B}$. Again we denote the product in $\mathcal{L}^+(\mathcal{D})$ by "$\cdot$" in this proof. Using the $*$-algebra properties of ρ and $\mathcal{L}^+(\mathcal{D})$ we derive

$$\langle \pi(a)b, c \rangle_{\mathcal{X}} = (\rho(a) \cdot b)^+ \cdot c = (b^+ \cdot \rho(a)^+) \cdot c = b^+ \cdot (\rho(a^+) \cdot c) = \langle b, \pi(a^+)c \rangle_{\mathcal{X}}.$$

Here the first equality holds, because for elements of $\mathfrak{B}$ the $\mathfrak{A}$-valued inner product $\langle \cdot, \cdot \rangle_{\mathcal{X}}$ is the product in $\mathfrak{B}$ and hence the product in $\mathcal{L}^+(\mathcal{D})$ by the assumption. The third equality is just the associativity of the multiplication in the algebra $\mathcal{L}^+(\mathcal{D})$. Hence $\pi(a) \in \mathcal{L}^+_{\mathfrak{B}}(\mathcal{D})$ by Definition 14.43 and π is a $*$-homomorphism of A into $\mathcal{L}^+_{\mathfrak{B}}(\mathcal{D})$. $\qquad\square$

The following example shows that each "ordinary" Hilbert space representation gives rise to a $*$-representation on the Hilbert C^*-module of compact operators.

Example 14.49 *(Hilbert space representations and C^*-algebras of compacts)* Suppose ρ is a $*$-representation of A on a Hilbert space $\mathcal{H}(\rho)$. Let $\mathfrak{A}$ be the C^*-algebra of compact operators on the Hilbert space $\mathcal{H}(\rho)$ and let $\mathcal{X} = \mathfrak{A}$ be the corresponding C^*-module from Example 14.35. Then the $*$-subalgebra $\mathcal{F}$ of all finite rank operators of $\mathcal{L}^+(\mathcal{D}(\rho))$ is also a $*$-subalgebra of $\mathfrak{A}$. Clearly, $\rho(a) \cdot x \in \mathcal{F}$ for $a \in \mathsf{A}, x \in \mathcal{F}$. Therefore, by Lemma 14.48, *there is a $*$-representation π of A on the (dense) $\mathcal{F}$-submodule $\mathcal{F}$ of the Hilbert C^*-module $\mathcal{X} = \mathfrak{A}$ defined by $\pi(a)x = \rho(a) \cdot x$ for $a \in \mathsf{A}$ and $x \in \mathcal{F}$.* $\qquad\bigcirc$

Since Hilbert C^*-modules have positive definite riggings (by axiom (iv) in Definition 14.32), all constructions and results of Sects. 14.3 and 14.4 remain valid here. In particular, if A acts by a $*$-representation on a $\mathfrak{A}$-submodule of a Hilbert $\mathfrak{A}$-module, then each $*$-representation π of the C^*-algebra $\mathfrak{A}$ on a Hilbert space yields an induced $*$-representation $\mathcal{E}-\text{Ind}\,\pi$ of A according to Definition 14.26.

In the rest of this section we develop three interesting examples of representations on Hilbert C^*-modules.

Example 14.50 (*Representations on $\mathfrak{A} = C_0(\mathbb{R}^d)$*)
Suppose $\mathfrak{A}$ is the C^*-algebra $C_0(\mathbb{R}^d)$ of continuous functions on $\mathbb{R}^d$ vanishing at infinity and $\mathcal{X} = \mathfrak{A}$ is the Hilbert $\mathfrak{A}$-module from Example 14.35. It is not difficult to verify that there is a $*$-representation π of the $*$-algebra $\mathsf{A} := \mathbb{C}_d[\underline{x}]$ acting on the $\mathfrak{A}$-submodule $\mathcal{D}(\pi)$ of $\mathcal{X}$ given by

$$\pi(p)f := p \cdot f, \quad \mathcal{D}(\pi) := \big\{ f \in C_0(\mathbb{R}^d) : p \cdot f \in C_0(\mathbb{R}^d), \ p \in \mathsf{A} \big\}. \quad (14.42)$$

Since $\mathsf{A} = \mathbb{C}_d[\underline{x}]$, the domain $\mathcal{D}(\pi)$ contains $C_c(\mathbb{R}^d)$, so $\mathcal{D}(\pi)$ is dense in $\mathcal{X}$.

Now let $d = 1$ and let K be a nonempty nowhere dense subset of $\mathbb{R}$. Let A be the $*$-algebra of rational functions with poles (if there are any) in K. Then (14.42) defines also a $*$-representation of A on the $\mathfrak{A}$-submodule $\mathcal{D}(\pi)$ of $\mathcal{X}$. Since the functions of the domain $\mathcal{D}(\pi)$ vanish at K, $\mathcal{D}(\pi)$ is not dense in $\mathcal{X}$. But $\mathcal{D}(\pi)$ is essential in $\mathcal{X}$, because K is nowhere dense.   ○

In Definition 14.46 we required only the $\mathfrak{B}$-linearity for the representation operators $\pi(a)$ rather than the $\mathfrak{A}$-linearity. The reason is that the $\mathfrak{B}$-linearity for some appropriate dense $*$-subalgebra $\mathfrak{B}$ is often easily obtained, as shown by the following Examples 14.51 and 14.52. Then, by Proposition 14.47, in both examples the $*$-representation $\overline{\pi}$ of A acts on an $\mathfrak{A}$-submodule by $\mathfrak{A}$-linear operators $\overline{\pi}(a)$.

Example 14.51 (*Representation of the Hermitian quantum plane on a C^*-algebra*)
In this example we use some facts from Sect. 11.6. Fix a number $q > 0$, $q \neq 1$. Let $\mathsf{A} := \mathbb{C}\langle x, x^+ | xx^+ = qx^+x \rangle$ be the $*$-algebra of the Hermitian quantum plane.

Suppose μ is a Radon measure on $(0, +\infty)$ such that $\mu(M) = \mu(q^{1/2}M)$ for all Borel sets M and let $\mathcal{H} := L^2((0, +\infty); \mu)$. Then the operator U on $\mathcal{H}$ defined by $(U\varphi)(t) = \varphi(q^{1/2}t)$ is unitary. Let C_c denote the set of continuous functions f on $(0, +\infty)$ such that $\operatorname{supp} f \subseteq [a, b]$ for some numbers a, b, where $0 < a < b < +\infty$. Let M_f be the multiplication operator by $f \in C_c$ on $\mathcal{H}$.

Let $\mathfrak{B}$ be the linear span of operators $U^n M_f$, where $f \in C_c$, $n \in \mathbb{Z}$. Since $UM_f = M_{f(q^{1/2}.)}U$, $\mathfrak{B}$ is a $*$-algebra of bounded operators on $\mathcal{H}$. The completion of $\mathfrak{B}$ in the operator norm is a C^*-algebra $\mathfrak{A}$. Again let $\mathcal{X} := \mathfrak{A}$ the corresponding Hilbert $\mathfrak{A}$-module from Example 14.35.

As noted in Sect. 11.6, the operator X on the Hilbert space $\mathcal{H}$, defined by

$$(X\varphi)(t) := q^{1/2}t\varphi(q^{1/2}t), \quad \mathcal{D}(X) = \big\{ \varphi(t) \in \mathcal{H} : t\varphi(t) \in \mathcal{H} \big\}, \quad (14.43)$$

is q-normal with adjoint $(X^*\varphi)(t) = t\varphi(q^{-1/2}t)$. Since the dense domain $\mathcal{D} := C_c$ of $\mathcal{H}$ is invariant under X and X^*, there is a $*$-representation ρ of A on $\mathcal{D}$ given by $\rho(x) = X \restriction \mathcal{D}$. Clearly, $\mathcal{D}$ is invariant under $\mathfrak{B}$ and $\mathfrak{B}$ is a $*$-subalgebra of $\mathcal{L}^+(\mathcal{D})$. For $f \in C_c$, $n \in \mathbb{Z}$, we have $XU^n M_f = U^{n+1}M_g$ and $X^*U^n M_f = U^{n-1}M_h$ with $g, h \in C_c$. This implies that $\rho(a) \cdot b \in \mathfrak{B}$ for $a \in \mathsf{A}$, $b \in \mathfrak{B}$. Thus, the assumptions of Lemma 14.48 are fulfilled. Hence *there is a $*$-representation π of A on the $\mathfrak{B}$-submodule $\mathfrak{B}$ of $\mathcal{X}$ such that $\pi(a)b = \rho(a) \cdot b$ for $a \in \mathsf{A}$, $b \in \mathfrak{B}$.*   ○

Example 14.52 (*Representations of enveloping algebras on group C^*-algebras*)
In this example, we freely use the notation and some facts on Lie groups from
Sects. 9.1 and 9.2. Let G be a Lie group and let $\mathsf{A} := \mathcal{E}(\mathfrak{g})$ be the enveloping algebra of the Lie algebra $\mathfrak{g}$ of G. Recall that $C_0^\infty(G)$ is a $*$-algebra with convolution
multiplication (9.4) and involution (9.7). Let $f \in C_0^\infty(G)$. Then, for each unitary
representation U of the Lie group G, there is a bounded operator U_f on $\mathcal{H}(U)$
defined by $U_f = \int U(g)f(g)\,d\mu_l(g)$. By Proposition 9.6(vi), the map $f \mapsto U_f$ is
a (nondegenerate) $*$-representation of the $*$-algebra $C_0^\infty(G)$. Let $\|f\|_*$ denote the
supremum of operator norms $\|U_f\|$ over all unitary representations U of the Lie
group G.

Statement: $\|\cdot\|_*$ *is a C^*-norm (see* (B.2)) *on the $*$-algebra $C_0^\infty(G)$.*

Proof Let $f, f_1, f_2 \in C_0^\infty(G)$. Since $\|U_f\| \leq \int |f(g)|d\mu_l(g)$, the supremum $\|f\|_*$
is always finite. Using that $f \mapsto U_f$ is a $*$-representation we derive

$$\|U_{f_1 f_2}\| = \|U_{f_1}U_{f_2}\| \leq \|U_{f_1}\|\,\|U_{f_2}\|,$$
$$\|U_{f^+}\| = \|(U_f)^*\| = \|U_f\|,$$
$$\|U_{f^+f}\| = \|U_{f^+}U_f\| = \|(U_f)^*U_f\| = \|U_f\|^2.$$

Taking the supremum over U we conclude that $\|\cdot\|_*$ is a C^*-seminorm.

Suppose $\|f\|_* = 0$ for some function $f \in C_0^\infty(G)$. Let U be the left regular repre-
sentation of G (see Example 9.9). Then, for $\varphi \in C_0^\infty(G)$, $U_f\varphi$ is just the convolution
$f * \varphi$, as shown by (9.19). Taking an approximate identity (that is, a sequence (φ_n)
of functions $\varphi_n \in C_0^\infty(G)$ such that $\int \varphi_n d\mu_l = 1$ and $\cap_n \mathrm{supp}\,\varphi_n = \{e\}$), we obtain
$0 = \lim_n f * \varphi_n = f$. Thus, $f = 0$, so $\|\cdot\|_*$ is a C^*-norm.   □

From this Statement it follows that the completion of $(C_0^\infty(G), \|\cdot\|_*)$ is a C^*-
algebra, called the *group C^*-algebra* of G and denoted by $C^*(G)$. (The same C^*-
algebra is obtained by completing $L^1(G; \mu_l)$ in the C^*-norm $\|\cdot\|_*$. This is the usual
way to define the group C^*-algebra in the literature; see, e.g., [F89].) Again let $\mathcal{X}$ be
the Hilbert C^*-module for the C^*-algebra $\mathfrak{A} := C^*(G)$ from Example 14.35.

Each element $a \in \mathsf{A} = \mathcal{E}(\mathfrak{g})$ acts as a right-invariant differential operator $\tilde{a}$ on G,
see (9.1), and there is a $*$-representation ρ of A on the dense domain $\mathcal{D} := C_0^\infty(G)$ of
the Hilbert space $L^2(G; \mu_l)$ given by $\rho(a)\varphi := \tilde{a}\varphi$, $\varphi \in C_0^\infty(G)$, see Example 9.9.
It is not difficult to verify that $\mathfrak{B} = C_0^\infty(G)$ is a $*$-subalgebra of $\mathcal{L}^+(\mathcal{D})$ and we have
$\rho(a) \cdot f = \tilde{a}f \in \mathfrak{B}$ for $a \in \mathsf{A}$, $f \in \mathfrak{B}$. Hence Lemma 14.48 applies, so *there is a
$*$-representation π of A on the $\mathfrak{B}$-submodule $\mathfrak{B}$ of $\mathcal{X}$ defined by $\pi(a)f = \rho(a) \cdot f$*
for $a \in \mathsf{A}$, $f \in \mathfrak{B}$.

The $*$-representations π on $\mathfrak{B}$-submodules of Hilbert $\mathfrak{A}$-modules in Examples
14.51 and 14.52 are closely related to the compatible pairs studied in Sect. 13.4. For
instance, Lemma 13.21 states that the $*$-algebras $\mathsf{A} = \mathcal{E}(\mathfrak{g})$ and $\mathsf{X} = C_0^\infty(G)$ form a
compatible pair. This means the action of A on X, defined by $a \triangleright f := \tilde{a}f$, satisfies
the compatibility condition (13.13):

$$(a \triangleright f_1)^+ f_2 = f_1^+(a^+ \triangleright f_2) \quad \text{for } a \in \mathsf{A}, \ f_1, f_2 \in C_0^\infty(G). \tag{14.44}$$

Set $\pi(a)f := a \triangleright f$. Then, in terms of the $\mathfrak{A}$-valued inner product on $C_0^\infty(G)$, Eq. (14.44) reads as $\langle \pi(a)f_1, f_2 \rangle_{\mathcal{X}} = \langle f_1, \pi(a^+)f_2 \rangle_{\mathcal{X}}$. Therefore, condition (13.13) implies that π is a $*$-representation of A on $\mathfrak{B} = C_0^\infty(G)$.   ○

14.7   Exercises

1. Let $\mathcal{E}_A$ be a right A-rigged space. Show that $\mathrm{Lin}\{[x, y]_A : x, y \in \mathcal{E}_A\}$ is a two-sided $*$-ideal of A.
2. Suppose $\mathcal{E}$ is a weak A–B-imprimitivity bimodule. Carry out the proof that $\overline{\mathcal{E}}$ is a weak B–A-imprimitivity bimodule.
3. Show that the rigging in Example 14.6 defined by (14.3) is positive semi-definite.
4. Let A be the $*$-algebra $\mathbb{C}[t_1, \ldots, t_d]$ of polynomials or the $*$-algebra of rational functions on $\mathbb{R}^d$, equipped with the rigging $[a, b]_A = c\, a^+ b$ as in Example 14.5. Are these right A-rigged spaces nondegenerate? When are the rigged spaces full?
5. Carry out the proofs that the A–B-bimodules $\mathcal{E}$ in Examples 14.14 and 14.15 are weak A–B-imprimitivity bimodules.
6. Consider the quadratic modules $Q_{m,k}$ in Example 14.20 defined by (14.17).

   a. Prove that $Q_{m,j} \subseteq Q_{m,j+1}$ for $j = 0, \ldots, m - 1$.
   b. Prove that $Q_{m,0} \neq Q_{m,1}$ if $C = \mathbb{C}[t_1, \ldots, t_d], d \geq 2$.

7. Prove that the set $M_n(A)_+$ of hermitian matrices defined in Lemma 14.21 is a quadratic module of $M_n(A)$.
8. Show that the completion of a pre-Hilbert $\mathfrak{A}$-module is a Hilbert $\mathfrak{A}$-module.
9. Suppose $\mathcal{Z}$ is a compact Hausdorff space and $\mathcal{H}$ is a Hilbert space. Show that the space $\mathcal{X} = C(\mathcal{Z}; \mathcal{H})$ of continuous functions $z \mapsto \varphi(z)$ from $\mathcal{Z}$ to $\mathcal{H}$ is a Hilbert C^*-module over the C^*-algebra $\mathfrak{A} = C(\mathcal{Z})$ with $\mathfrak{A}$-valued inner product $\langle \varphi, \psi \rangle_{\mathcal{X}}(z) = \langle \varphi(z), \psi(z) \rangle_{\mathcal{H}}$ and pointwise multiplication as right $\mathfrak{A}$-action.
10. Let $\mathfrak{A}$ be a C^*-algebra and $(x_n)_{n \in \mathbb{N}}$ a sequence of elements $x_n \in \mathfrak{A}$ such that $\sum_n \|x_n\|^2 < \infty$. Show that $(x_n) \in l_2(\mathfrak{A})$.
11. Let $\mathfrak{A}$ be the C^*-algebra of compact operators on an infinite-dimensional Hilbert space. Find a sequence $(x_n) \in l_2(\mathfrak{A})$ such that $\sum_n \|x_n\|^2 = +\infty$.
    Hint: Set $x_n = \lambda_n \langle \cdot, \varphi_n \rangle \varphi_n$, where (φ_n) is an orthonormal sequence and (λ_n) is a null sequence which is not in $l^2(\mathbb{N})$.
12. Develop Example 14.52 for $G = \mathbb{R}^d$. Prove in this case directly that π is a $*$-representation on the Hilbert C^*-module $C^*(\mathbb{R}^d)$.

14.8   Notes

The pioneering work on induced representations, Morita equivalence and rigged spaces for C^*-algebras is due to M. Rieffel [Rf74a], [Rf74b], see also [RW98]. General rigged spaces without positivity conditions, as in the text, are developed in the monograph [FD88]. Morita equivalence and rigged modules for general $*$-algebras were studied extensively in [Ar99], [BW01], [BW05]. Standard texts on Hilbert C^*-modules are [La94] and [RW98]. The transport of quadratic modules as in Sect. 14.2 is taken from [Sch09]. Representations of $*$-algebras on Hilbert C^*-modules are used in [My16].

A well-studied class of unbounded operators on Hilbert C^*-modules are the regular operators introduced by S. Baaj [B81] and investigated in [Wo91], [WN92]; see [La94, Chap. 9 and 10] for a nice treatment. Let $\mathcal{X}$ and $\mathcal{Y}$ be Hilbert C^*-modules over $\mathfrak{A}$. An $\mathfrak{A}$-operator of $\mathcal{X}$ into $\mathcal{Y}$ is called *regular* if t is closed, $\mathcal{D}(t)$ is dense in $\mathcal{X}$, $\mathcal{D}(t^*)$ is dense in $\mathcal{Y}$ and $(I + t^*t)\mathcal{X}$ is dense in $\mathcal{X}$. A more general class of operators are the graph regular operators introduced in [GS15]; their domains are essential, but not necessarily dense. To illustrate the difference between regular and graph regular operators we consider the special case $\mathcal{X} = \mathcal{Y} = \mathfrak{A} := C_0(\mathcal{Z})$, where $\mathcal{Z}$ is a locally compact Hausdorff space. Then regular operators are multiplication operators by functions of $C(\mathcal{Z})$, while graph regular operators are multiplication operators by functions which are continuous up to nowhere dense sets and for which the modulus goes to infinity in a neighborhood of each discontinuity.

An interesting and important problem is how to associate a C^*-algebra to a class of unbounded representations of a $*$-algebra. There are various attempts and approaches, the affiliated operators of Woronowicz [Wo95], the C^*-envelope of Dowerk and Savchuk [DS13], and the C^*-hull of representations on Hilbert C^*-modules of Meyer [My16].

Appendix A
Unbounded Operators on Hilbert Space

The following basic operator-theoretic notions and facts can be found in most books on Hilbert space operators such as [RS72, RS75, BS87], or [Sch12].

Suppose $\mathcal{H}$, $\mathcal{H}_1$, $\mathcal{H}_2$ are complex Hilbert spaces with inner products $\langle \cdot, \cdot \rangle$, $\langle \cdot, \cdot \rangle_1$, $\langle \cdot, \cdot \rangle_2$, respectively. The bounded operators of $\mathcal{H}_1$ into $\mathcal{H}_2$ are denoted by $\mathbf{B}(\mathcal{H}_1, \mathcal{H}_2)$, and we set $\mathbf{B}(\mathcal{H}) := \mathbf{B}(\mathcal{H}, \mathcal{H})$. The Hilbert–Schmidt operators on $\mathcal{H}$ are denoted by $\mathbf{B}_2(\mathcal{H})$, the trace class operators by $\mathbf{B}_1(\mathcal{H})$, and the trace by Tr.

By an *operator* of $\mathcal{H}_1$ into $\mathcal{H}_2$ we mean a linear mapping T of a linear subspace $\mathcal{D}(T)$ of $\mathcal{H}_1$, the *domain* of T, into $\mathcal{H}_2$. The *kernel* of the operator T is defined by $\ker T := \{\varphi \in \mathcal{D}(T) : T\varphi = 0\}$ and its *range* is $\operatorname{ran} T := \{T\varphi : \varphi \in \mathcal{D}(T)\}$.

If T and S are operators of Hilbert spaces $\mathcal{H}$ and $\mathcal{H}_1$, respectively, into $\mathcal{H}_2$, we say S is an *extension* of T or T is a *restriction* of S and write $T \subseteq S$ if $\mathcal{H}$ is a subspace of $\mathcal{H}_1$, $\mathcal{D}(T) \subseteq \mathcal{D}(S)$ and $T\varphi = S\varphi$ for $\varphi \in \mathcal{D}(T)$. The restriction of T to a linear subspace $\mathcal{D}$ of $\mathcal{D}(T)$ is denoted by $T \restriction \mathcal{D}$.

Let T be an operator of $\mathcal{H}_1$ into $\mathcal{H}_2$. Then T is called *closed* if for each sequence (φ_n) from $\mathcal{D}(T)$ such that $\lim_n \varphi_n = \varphi$ in $\mathcal{H}_1$ and $\lim_n T\varphi_n = \psi$ in $\mathcal{H}_2$ we have $\varphi \in \mathcal{D}(T)$ and $\psi = T\varphi$. Further, T is called *closable* if it has a closed extension. In this case, T has a smallest closed extension, called the *closure* of T and denoted by $\overline{T}$. Note that T is closable if and only if for each sequence (φ_n) from $\mathcal{D}(T)$ such that $\lim_n \varphi_n = 0$ in $\mathcal{H}_1$ and $(T\varphi_n)$ converges in $\mathcal{H}_2$ we have $\lim_n T\varphi_n = 0$.

A linear subspace $\mathcal{D}$ of the domain $\mathcal{D}(T)$ is called a *core* for an operator T if for each $\varphi \in \mathcal{D}(T)$ there exists a sequence (φ_n) of vectors $\varphi_n \in \mathcal{D}$ such that $\varphi = \lim_n \varphi_n$ in $\mathcal{H}_1$ and $T\varphi = \lim_n T\varphi_n$ in $\mathcal{H}_2$. If T is closable, this holds if and only if the closures of $T \restriction \mathcal{D}$ and T coincide.

Suppose $\mathcal{D}(T)$ is dense in $\mathcal{H}_1$. The *adjoint operator* T^* is defined as follows: Its domain $\mathcal{D}(T^*)$ is the set of vectors $\psi \in \mathcal{H}_2$ for which there exists an $\eta \in \mathcal{H}_1$ such that $\langle T\varphi, \psi \rangle_2 = \langle \varphi, \eta \rangle_1$ for all $\varphi \in \mathcal{D}(T)$; in this case $T^*\psi := \eta$. Then T^* is a closed operator of $\mathcal{H}_2$ into $\mathcal{H}_1$ and $(\overline{T})^* = T^*$. If $T \subseteq S$ on $\mathcal{H}_1$, then $S^* \subseteq T^*$.

© The Editor(s) (if applicable) and The Author(s), under exclusive license
to Springer Nature Switzerland AG 2020
K. Schmüdgen, *An Invitation to Unbounded Representations of *-Algebras on Hilbert Space*, Graduate Texts in Mathematics 285,
https://doi.org/10.1007/978-3-030-46366-3

The operator T is closable if and only if the domain $\mathcal{D}(T^*)$ of its adjoint is dense in $\mathcal{H}_2$; in this case, $T \subseteq T^{**}$ and $\overline{T} = T^{**}$. In particular, T is closed if and only if $T = T^{**}$. Further, $\ker T^* = (\operatorname{ran} T)^{\perp}$ and $\ker \overline{T} = (\operatorname{ran} T^*)^{\perp}$.

If an operator T maps $\mathcal{H}$ into $\mathcal{H}$, we say that T is an operator *on* $\mathcal{H}$.

For a linear subspace $\mathcal{D}$ of $\mathcal{H}$, the identity map of $\mathcal{D}$ is denoted by $I_{\mathcal{D}}$.

If T_1 and T_2 are operators of $\mathcal{H}_1$ and $\mathcal{H}_2$, respectively, into $\mathcal{H}$, the *direct sum* $T := T_1 \oplus T_2$ is the operator T of $\mathcal{H}_1 \oplus \mathcal{H}_2$ into $\mathcal{H}$ defined by

$$T(\varphi_1, \varphi_2) = (T_1\varphi_1, T_2\varphi_2), \quad \varphi_1 \in \mathcal{D}(T_1), \varphi_2 \in \mathcal{D}(T_2).$$

A closed subspace $\mathcal{K}$ of $\mathcal{H}$ is called *reducing* for an operator T on $\mathcal{H}$ if T is a direct sum of operators acting on $\mathcal{K}$ and $\mathcal{K}^{\perp}$.

We shall say that an operator T on $\mathcal{H}$ is *irreducible* if $\{0\}$ and $\mathcal{H}$ are the only reducing subspaces for T, or equivalently, if T cannot be written as a direct sum of two operators acting on nonzero Hilbert spaces. This holds if and only if 0 and I are the only orthogonal projections P acting on $\mathcal{H}$ and satisfying $PT \subseteq TP$.

Now we suppose that T and S are operators on $\mathcal{H}$.

Suppose $\mathcal{D}(S)$ is dense. If $\mathcal{D}(ST)$ is dense, then $T^*S^* \subseteq (ST)^*$. If $\mathcal{D}(T + S)$ is dense, then $T^* + S^* \subseteq (T + S)^*$. Further, if $\mathcal{D}(S^n)$ is dense, then $(S^*)^n \subseteq (S^n)^*$. In all three cases we do not have equality in general. But, if S is bounded and $\mathcal{D}(S) = \mathcal{H}$, then $T^*S^* = (ST)^*$ and $T^* + S^* = (T + S)^*$.

For a closed operator T the *resolvent set* $\rho(T)$ is the set of numbers $z \in \mathbb{C}$ for which the operator $T - zI$ has a bounded inverse $(T - zI)^{-1}$ that is defined on the whole Hilbert space $\mathcal{H}$. The set $\sigma(T) := \mathbb{C}\backslash\rho(T)$ is the *spectrum* of T. The subset of all eigenvalues of the operator T is denoted by $\sigma_p(T)$.

An operator T is called *symmetric* if $\langle T\varphi, \psi \rangle = \langle \varphi, T\psi \rangle$ for $\varphi, \psi \in \mathcal{D}(T)$, or equivalently, if $\langle T\varphi, \varphi \rangle$ is real for all $\varphi \in \mathcal{D}(T)$. A symmetric operator is called *positive* if $\langle T\varphi, \varphi \rangle \geq 0$ for $\varphi \in \mathcal{D}(T)$; in this case write $T \geq 0$.

Suppose T is an operator on $\mathcal{H}$ with dense domain $\mathcal{D}(T)$. Then T is called *formally normal* if $\mathcal{D}(T) \subseteq \mathcal{D}(T^*)$ and $\|T\varphi\| = \|T^*\varphi\|$ for $\varphi \in \mathcal{D}(x)$ and *normal* if T is formally normal and $\mathcal{D}(T) = \mathcal{D}(T^*)$. Note that T is normal if and only if T is closed and $T^*T = TT^*$. Further, a normal operator T is maximal formally normal, that is, if S is formally normal on $\mathcal{H}$ and $T \subseteq S$, then $T = S$.

An operator T with dense domain $\mathcal{D}(T)$ is symmetric if and only if $T \subseteq T^*$. Any densely defined symmetric operator is closable and its closure is also symmetric.

A densely defined operator T is called *self-adjoint* if $T = T^*$ and *essentially self-adjoint* if its closure $\overline{T}$ is self-adjoint, or equivalently, if $\overline{T} = T^*$.

Each self-adjoint operator T is maximal symmetric, that is, if S is a symmetric operator acting on the same Hilbert space such that $T \subseteq S$, then $T = S$.

Some important self-adjointness criteria are given in the following proposition.

Proposition A.1 *Let $z_+, z_- \in \mathbb{C}$ be fixed such that $\operatorname{Im} z_+ > 0$, $\operatorname{Im} z_- < 0$. For any densely defined symmetric operator T on $\mathcal{H}$ the following are equivalent:*

   (i) *T is essentially self-adjoint.*
   (ii) *T has a unique self-adjoint extension on $\mathcal{H}$.*
   (iii) *$(T - z_+ I)\mathcal{D}(T)$ and $(T - z_- I)\mathcal{D}(T)$ are dense in $\mathcal{H}$.*
   (iv) *$\mathcal{D}_-(T) := \ker(T^* - z_- I) = \{0\}$ and $\mathcal{D}_+(T) := \ker(T^* - z_+ I) = \{0\}$.*

If the symmetric operator T is positive, these conditions are equivalent to

   (v) *$(T + zI)\mathcal{D}(T)$ is dense in $\mathcal{H}$ for one (then for all) $z > 0$.*

Proof [Sch12, Propositions 3.8 and 3.15 and Theorem 13.10]. $\qquad\qquad$ □

The vector spaces $\mathcal{D}_\pm(T)$ in Proposition A.1(iv) are the *deficiency spaces* of T. Their dimensions $d_\pm = \dim \mathcal{D}_\pm$ are called the *deficiency indices* of the symmetric operator T; they are independent of the particular choice of numbers $z_\pm$ satisfying $\operatorname{Im} z_+ > 0$ and $\operatorname{Im} z_- < 0$.

Suppose T is a symmetric operator and $T\mathcal{D}(T) \subseteq \mathcal{D}(T)$. Then

$$\|T^k \varphi\|^2 \le \|T^n \varphi\|^2 + \|\varphi\|^2 \quad \text{for } \varphi \in \mathcal{D}(T),\ k, n \in \mathbb{N}_0, k \le n. \tag{A.1}$$

From (A.1) it can be derived that $\overline{T^n} \subseteq (\overline{T})^n$. Further, if $\mathcal{D} \subseteq \mathcal{D}(T)$ is a core for some power T^n, it is also a core for each power T^k, $k \le n$.

Corollary A.2 *Suppose T is a densely defined symmetric operator and $n \in \mathbb{N}$. If T^n is essentially self-adjoint, so is T. In this case, $\overline{T^n} = (\overline{T})^n$.*

Proof Since T^n is essentially self-adjoint, $\overline{T^n} = (T^n)^* \supseteq (T^*)^n$. We choose $z \in \mathbb{C}$ such that $z^n = i$ and let $\xi \subset \ker(T^* - zI)$. Then $T^*\xi = z\xi$, so $(T^n)^*\xi = (T^*)^n\xi = z^n\xi = i\xi$. Since $\overline{T^n} = (T^n)^*$ is self-adjoint, Proposition A.1,(i)→(iv), gives $\xi = 0$. Thus, $\ker(T^* - zI) = \{0\}$ and $\ker(T^* - \bar{z}I) = \{0\}$, upon replacing z by $\bar{z}$. Hence T is essentially self-adjoint by Proposition A.1,(iv)→(i).

Further, since $\overline{T^n} \subseteq (\overline{T})^n$ as noted above, $(\overline{T})^n$ is a symmetric extension of the self-adjoint operators $\overline{T^n}$. Therefore, $\overline{T^n} = (\overline{T})^n$. $\qquad\qquad$ □

Let $\mathcal{K}_j$ be a closed linear subspace of $\mathcal{H}_j$, $j = 1, 2$. A *partial isometry* with *initial space* $\mathcal{K}_1$ and *final space* $\mathcal{K}_2$ is a linear operator T of $\mathcal{H}_1$ into $\mathcal{H}_2$ with domain $\mathcal{H}_1$ such that T is an isometric mapping of $\mathcal{K}_1$ onto $\mathcal{K}_2$ and $T\varphi = 0$ for $\varphi \in \mathcal{K}_1^\perp$.

Proposition A.3 *Suppose X is a densely defined closed operator of $\mathcal{H}_1$ into $\mathcal{H}_2$. Then $|X| := (X^* X)^{1/2}$ is a positive self-adjoint operator on $\mathcal{H}_1$, and there is a partial isometry U of $\mathcal{H}_1$ into $\mathcal{H}_2$ with initial space $\mathcal{K}_1$ and final space $\mathcal{K}_2$, where*

$$\mathcal{K}_1 := (\ker X)^\perp = \overline{\operatorname{ran} X^*} = (\ker |X|)^\perp = \overline{\operatorname{ran} |X|} = U^*(\overline{\operatorname{ran} X}), \tag{A.2}$$

$$\mathcal{K}_2 := \overline{\operatorname{ran} X} = (\ker X^*)^\perp = (\ker |X^*|)^\perp = \overline{\operatorname{ran} |X^*|} = U(\overline{\operatorname{ran} X^*}), \tag{A.3}$$

such that $X = U|X|$. *Moreover,*

$$U^*U = P_{\mathcal{K}_1}, \quad UU^* = P_{\mathcal{K}_2}, \quad |X^*| = UX^* = XU^* = U|X|U^*, \qquad (A.4)$$

where $P_{\mathcal{K}_1}$ and $P_{\mathcal{K}_2}$ are the projections of $\mathcal{H}_1$ and $\mathcal{H}_2$ onto $\mathcal{K}_1$ and $\mathcal{K}_2$, respectively.
If $X = VC$, where C is a positive self-adjoint operator on $\mathcal{H}_1$ and V is a partial isometry of $\mathcal{H}_1$ into $\mathcal{H}_2$ with initial space $\overline{\operatorname{ran}|X|}$, then $C = |X|$ and $V = U$.

Proof [Sch12, Theorem 7.2]. $\square$

The representation $X = U|X|$ is called the *polar decomposition* of X and U is the *phase operator* of X. Then $X^* = U^*|X^*|$ is the polar decomposition of X^*.

Next we recall some basics on spectral integrals (see [Sch12, Chap. 4] for precise definitions and further results).

Let $\mathfrak{A}$ be a σ-algebra of subsets of a set Ω. A *spectral measure* on $\mathfrak{A}$ is a mapping E of $\mathfrak{A}$ into the orthogonal projections of a Hilbert space $\mathcal{H}$ such that $E(\Omega) = I$ and $E(\cup_{n=1}^{\infty} M_n)\varphi = \sum_{n=1}^{\infty} E(M_n)\varphi$, $\varphi \in \mathcal{H}$, for any sequence (M_n) of pairwise disjoint sets $M_n \in \mathfrak{A}$. Suppose E is a spectral measure on $\mathfrak{A}$. Then, for any $\mathfrak{A}$-measurable complex function f on Ω, there exists the *spectral integral*

$$\mathbb{I}(f) := \int_{\Omega} f(\lambda)\, dE(\lambda).$$

This operator-valued integral is a closed normal operator on $\mathcal{H}$ with domain

$$\mathcal{D}(\mathbb{I}(f)) = \left\{ \varphi \in \mathcal{H} : \int_{\Omega} |f(\lambda)|^2\, d\langle E(\lambda)\varphi, \varphi \rangle < \infty \right\}.$$

If the function f is bounded, the operator $\mathbb{I}(f)$ is bounded. If $\overline{f}$ is the complex conjugate function of f, then $\mathbb{I}(f)^* = \mathbb{I}(\overline{f})$. For $\mathfrak{A}$-measurable functions f, g,

$$\mathbb{I}(f) + \mathbb{I}(g) \subseteq \mathbb{I}(f + g) \quad \text{and} \quad \mathbb{I}(f)\mathbb{I}(g) \subseteq \mathbb{I}(f \cdot g). \qquad (A.5)$$

In general, we do not have equalities in (A.5), but for any polynomial $p \in \mathbb{C}[x]$ we have $p(\mathbb{I}(f)) = \mathbb{I}(p(f))$; see [Sch12, Theorem 4.16 and Proposition 4.22].

We say that two self-adjoint operators T and S on $\mathcal{H}$ *strongly commute* if their resolvents $(T - zI)^{-1}$ and $(S - wI)^{-1}$ commute for some (then for all) $z, w \in \mathbb{C}\backslash\mathbb{R}$.

The following result is the *multi-dimensional spectral theorem*.

Proposition A.4 *Let $d \in \mathbb{N}$. Suppose $T = \{T_1, \ldots, T_d\}$ is a d-tuple of pairwise strongly commuting self-adjoint operators T_j on a Hilbert space $\mathcal{H}$. Then there exists a unique spectral measure E_T on the Borel σ-algebra of $\mathbb{R}^d$ such that*

$$T_j = \int_{\mathbb{R}^d} \lambda_j\, dE_T(\lambda_1, \ldots, \lambda_d), \quad j = 1, \ldots, d.$$

Proof [Sch12, Theorem 5.23].   □

Then, for any Borel function f on $\mathbb{R}^d$ there exists the spectral integral

$$f(T) := \mathbb{I}(f) \equiv \int_{\mathbb{R}^d} f(\lambda)\, dE_T(\lambda).$$

Let T be a single self-adjoint operator. By the *spectral theorem* (Proposition A.4), there exists a unique spectral measure E_T on the Borel σ-algebra of $\mathbb{R}$ such that

$$T = \int_{\mathbb{R}} \lambda\, dE_T(\lambda).$$

The assignment $f \mapsto f(T) := \int_{\mathbb{R}} f(\lambda)\, dE_T(\lambda)$ is the *functional calculus* of T. The support of E_T is equal to the spectrum of T. Two self-adjoint operators T and S on $\mathcal{H}$ strongly commute if and only if their spectral measures E_T and E_S commute.

A *one-parameter unitary group* on a Hilbert space $\mathcal{H}$ is a homomorphism U of $\mathbb{R}$ into the unitary operators on $\mathcal{H}$ such that $\lim_{t \to 0} U(t)\varphi = \varphi$ for all $\varphi \in \mathcal{H}$.

If T is a self-adjoint operator, then $U(t) := e^{\mathrm{i}tT}$, $t \in \mathbb{R}$, is a one-parameter unitary group. The following converse of this fact is *Stone's theorem*.

Proposition A.5 *For each one-parameter unitary group U on a Hilbert space $\mathcal{H}$, there exists a unique self-adjoint operator T on $\mathcal{H}$ such that $U(t) = e^{\mathrm{i}tT}$ for $t \in \mathbb{R}$. The operator $\mathrm{i}T$ is called the* infinitesimal generator *of U. In fact,*

$$\mathcal{D}(T) = \left\{\varphi \in \mathcal{H} : \lim_{t \to 0} t^{-1}(U(t) - I)\varphi \text{ exists in } \mathcal{H}\right\},$$
$$T\varphi = -\mathrm{i}\big(\lim_{t \to 0} t^{-1}(U(t) - I)\varphi\big), \quad \varphi \in \mathcal{D}(T).$$

Proof [Sch12, Theorem 6.2].   ⊔

In the rest of this appendix we develop the **complexification** of a real Hilbert space. Suppose $(\mathcal{K}, \langle \cdot, \cdot \rangle)$ is a *real* Hilbert space. This means that $\langle \cdot, \cdot \rangle$ is a real-valued inner product on the real vector space $\mathcal{K}$ such that $\mathcal{K}$ is complete in the norm defined by $\|\varphi\| := \sqrt{\langle \varphi, \varphi \rangle}$, $\varphi \in \mathcal{K}$. We define a *complex* Hilbert space $\mathcal{K}_{\mathbb{C}}$.

The elements of $\mathcal{K}_{\mathbb{C}}$ are pairs $\varphi + \mathrm{i}\psi := (\varphi, \psi) \in \mathcal{K} \times \mathcal{K}$. It is easily verified that $\mathcal{K}_{\mathbb{C}}$ becomes a $\mathbb{C}$-vector space with algebraic operations

$$(\varphi + \mathrm{i}\psi) + (\xi + \mathrm{i}\eta) := \varphi + \xi + \mathrm{i}(\psi + \eta),$$
$$(\alpha + \mathrm{i}\beta)(a + \mathrm{i}b) := \alpha a - \beta b + \mathrm{i}(\alpha b + \beta a),$$

and there is an inner product $\langle \cdot, \cdot \rangle_{\mathbb{C}}$ on the complex vector space $\mathcal{K}_{\mathbb{C}}$ defined by

$$\langle \varphi + \mathrm{i}\psi, \xi + \mathrm{i}\eta \rangle_{\mathbb{C}} := \langle \varphi, \xi \rangle + \langle \psi, \eta \rangle + \mathrm{i}\langle \psi, \xi \rangle - \mathrm{i}\langle \varphi, \eta \rangle,$$

where $\varphi, \psi, \xi, \eta \in \mathcal{K}$ and $\alpha, \beta \in \mathbb{R}$. For the corresponding norm we have

$$\|\varphi + \mathrm{i}\,\psi\|_{\mathbb{C}}^2 = \|\varphi\|^2 + \|\psi\|^2, \quad \varphi, \psi \in \mathcal{K}.$$

Therefore, $(\mathcal{K}_{\mathbb{C}}, \|\cdot\|_{\mathbb{C}})$ is complete and hence a complex Hilbert space.

Let T be an $\mathbb{R}$-linear operator on $\mathcal{K}$. We define a $\mathbb{C}$-linear operator $T_{\mathbb{C}}$ on $\mathcal{K}_{\mathbb{C}}$ by

$$\mathcal{D}(T_{\mathbb{C}}) = \mathcal{D}(T) + \mathrm{i}\mathcal{D}(T), \; T_{\mathbb{C}}(\varphi + \mathrm{i}\psi) = T\varphi + \mathrm{i}T\psi, \; \varphi, \psi \in \mathcal{D}(T). \qquad (A.6)$$

Basic properties of T carry over almost verbatim to $T_{\mathbb{C}}$. For instance, if T is bounded, so is $T_{\mathbb{C}}$ and $\|T\| = \|T_{\mathbb{C}}\|$. Further, if T is densely defined, then $T_{\mathbb{C}}$ is also densely defined on $\mathcal{K}_{\mathbb{C}}$ and for the corresponding adjoints we have

$$\mathcal{D}((T_{\mathbb{C}})^*) = \mathcal{D}(T^*) + \mathrm{i}\mathcal{D}(T^*), \; (T_{\mathbb{C}})^*(\varphi + \mathrm{i}\psi) = T^*\varphi + \mathrm{i}T^*\psi, \; \varphi, \psi \in \mathcal{D}(T^*).$$

Hence T is self-adjoint on $\mathcal{K}$ if and only if $T_{\mathbb{C}}$ is self-adjoint on $\mathcal{K}_{\mathbb{C}}$. In this case, the spectra $\sigma(T)$ and $\sigma(T_{\mathbb{C}})$ coincide. For the self-adjoint operators T and $T_{\mathbb{C}}$ the spectral theorem holds and the spectral measures E_T and $E_{T_{\mathbb{C}}}$ are related by $E_T(M) = E_{T_{\mathbb{C}}}(M)\!\upharpoonright\!\mathcal{K}$. (Proofs of these facts and further results about operators on $\mathcal{K}$ and $\mathcal{K}_{\mathbb{C}}$ can be found, for instance, in [MV97, Remark 20.18].)

Appendix B
C^*-Algebras and Representations

Standard references for (complex) C^*-algebras and their representations are [Di77b, KR83, Dv96]. Real C^*-algebras do not occur in most books on operator algebras. They are treated extensively in the books [Gol82] and [Li03].

A *Banach algebra* is a complex algebra A equipped with a norm with respect to which A is complete and which satisfies

$$\|ab\| \leq \|a\|\,\|b\| \quad \text{for} \quad a, b \in \mathsf{A}. \tag{B.1}$$

A *Banach $*$-algebra* is a Banach algebra which is simultaneously a $*$-algebra.

Definition B.1 A (complex) C^*-algebra is a Banach $*$-algebra with norm $\| \cdot \|$ and involution $a \mapsto a^+$ such that the norm satisfies the "C^*-condition"

$$\|a^+ a\| = \|a\|^2 \quad \text{for} \quad a \in \mathsf{A}. \tag{B.2}$$

A norm satisfying (B.1) and (B.2) is called a C^*-*norm*.

By (B.1) and (B.2), $\|a\|^2 = \|a^+ a\| \leq \|a^+\|\,\|a\|$, so that $\|a\| \leq \|a^+\|$. Replacing a by a^+ yields $\|a^+\| = \|a\|$. Thus, $\|a^+\| = \|a\|$, $a \in \mathsf{A}$, for each C^*-algebra A.

Suppose A is a C^*-algebra. Let $\mathsf{A}_{\mathrm{her}} := \{a \in \mathsf{A} : a^+ = a\}$. An element $a \in \mathsf{A}_{\mathrm{her}}$ is called *positive*, denoted $a \geq 0$, if $a \in \sum \mathsf{A}^2$, or equivalently, if the spectrum of a is contained in $\mathbb{R}_+$. For $a, b \in \mathsf{A}_{\mathrm{her}}$ we write $a \leq b$ or $b \geq a$ if $b - a \geq 0$. Note that $0 \leq a \leq b$ implies $\|a\| \leq \|b\|$.

Two C^*-algebras are called *isomorphic* if there exists a $*$-isomorphism of the corresponding $*$-algebras which preserves the norm.

If $\mathcal{X}$ is a locally compact Hausdorff space, then the algebra $C_0(\mathcal{X})$ of continuous functions on $\mathcal{X}$ vanishing at infinity is a commutative C^*-algebra with the supremum norm and complex conjugation as involution. Note that $C_0(\mathcal{X})$ is unital if and only if the space $\mathcal{X}$ is compact. Each commutative C^*-algebra is isomorphic to some C^*-algebra $C_0(\mathcal{X})$.

© The Editor(s) (if applicable) and The Author(s), under exclusive license   353
to Springer Nature Switzerland AG 2020
K. Schmüdgen, *An Invitation to Unbounded Representations of *-Algebras on Hilbert Space*, Graduate Texts in Mathematics 285,
https://doi.org/10.1007/978-3-030-46366-3

Let $\mathcal{H}$ be a Hilbert space. For a subset $\mathcal{N}$ of $\mathbf{B}(\mathcal{H})$ its *commutant* is defined by

$$\mathcal{N}' := \{T \in \mathbf{B}(\mathcal{H}) : TS = ST \text{ for } S \in \mathcal{N}\}.$$

Now we introduce two locally convex topologies on $\mathbf{B}(\mathcal{H})$: The *weak operator topology* is defined by the family of seminorms $T \mapsto |\langle T\varphi, \psi \rangle|$, where $\varphi, \psi \in \mathcal{H}$, and the *strong operator topology* is defined by the seminorms $T \mapsto \|T\varphi\|$, where $\varphi \in \mathcal{H}$. A net $(T_j)_{j \in J}$ of operators $T_j \in \mathbf{B}(\mathcal{H})$ converges to $T \in \mathbf{B}(\mathcal{H})$ in the weak operator topology if and only if $\lim_j \langle (T_j - T)\varphi, \psi \rangle = 0$ for $\varphi, \psi \in \mathcal{H}$ and in the strong operator topology if and only if $\lim_j \|(T_j - T)\varphi\| = 0$ for $\varphi \in \mathcal{H}$.

Definition B.2 A $*$-subalgebra $\mathcal{N}$ of $\mathbf{B}(\mathcal{H})$ which contains the identity operator is called a *von Neumann algebra* if it is closed in the weak operator topology, or equivalently, in the strong operator topology.

By the double commutant theorem [KR83, 5.3.1], a $*$-subalgebra $\mathcal{N}$ of $\mathbf{B}(\mathcal{H})$ with $I \in \mathcal{N}$ is a von Neumann algebra if and only if $\mathcal{N}$ is equal to $(\mathcal{N}')'$.

Let $\mathcal{N}$ be a von Neumann algebra on $\mathcal{H}$. A densely defined closed operator A on $\mathcal{H}$ is *affiliated* with $\mathcal{N}$ if $TA \subseteq AT$ for $T \in \mathcal{N}'$, or equivalently, for all unitaries $T \in \mathcal{N}'$ [KR83, p. 342]. If $A = U|A|$ is the polar decomposition of A, this holds if and only if $U \in \mathcal{N}$ and the spectral projections of $|A|$ are in $\mathcal{N}$. A self-adjoint operator is affiliated with $\mathcal{N}$ if and only if its spectral projections are in $\mathcal{N}$.

Now we suppose that $\mathbf{A}$ is a **complex C^*-algebra**. We collect some basic notions and results on $*$-representations of $\mathbf{A}$.

A $*$-*representation* of $\mathbf{A}$ on a (complex) Hilbert space $\mathcal{H}$ is a $*$-homomorphism π of $\mathbf{A}$ into the (complex) $*$-algebra $\mathbf{B}(\mathcal{H})$. Then we write $\mathcal{H}(\pi) := \mathcal{H}$.

Let π_i, $i \in I$, be a family of $*$-representations of $\mathbf{A}$. The direct sum $\pi = \oplus_{i \in I} \pi_i$ is the $*$-representation π on the direct sum Hilbert space $\mathcal{H}(\pi) := \oplus_{i \in I} \mathcal{H}(\pi_i)$ defined by $\pi(a)(\varphi_i) = (\pi_i(a)\varphi_i)$ for $a \in \mathbf{A}$ and $(\varphi_i) \in \mathcal{H}(\pi)$.

Two $*$-representations π_1 and π_2 of $\mathbf{A}$ are called *unitarily equivalent* if there is a unitary operator U of $\mathcal{H}(\pi_1)$ onto $\mathcal{H}(\pi_2)$ such that $\pi_2(a) = U\pi_1(a)U^{-1}$ for $a \in \mathbf{A}$.

Suppose π is a $*$-representation of $\mathbf{A}$.

A closed linear subspace $\mathcal{K}$ of $\mathcal{H}(\pi)$ is called *invariant* under π if $\pi(a)\mathcal{K} \subseteq \mathcal{K}$ for all $a \in \mathbf{A}$. In this case, $\pi_{\mathcal{K}}(a) := \pi(a){\restriction}\mathcal{K}$, $a \in \mathbf{A}$, defines a $*$-representation $\pi_{\mathcal{K}}$ of $\mathbf{A}$ on $\mathcal{K}$. The orthogonal complement $\mathcal{K}^\perp = \{\psi \in \mathcal{H}(\pi) : \langle \psi, \varphi \rangle = 0, \varphi \in \mathcal{K}\}$ is also invariant under π, and we have $\pi = \pi_{\mathcal{K}} \oplus \pi_{\mathcal{K}^\perp}$.

Further, π is called *irreducible* if $\{0\}$ and $\mathcal{H}(\pi)$ are the only closed linear subspaces of $\mathcal{H}(\pi)$ that invariant under π, or equivalently, if π is a direct sum $\pi_1 \oplus \pi_2$ of $*$-representations, then $\mathcal{H}(\pi_1) = \{0\}$ or $\mathcal{H}(\pi_1) = \mathcal{H}(\pi)$. By Schur's lemma, π is irreducible if and only if the commutant $\pi(\mathbf{A})'$ of $\pi(\mathbf{A})$ is $\mathbb{C} \cdot I$.

A $*$-representation π is called *cyclic* if there exists a vector $\varphi \in \mathcal{H}$, called a *cyclic vector* for π, such that $\pi(\mathbf{A})\varphi$ is dense in $\mathcal{H}$. Each $*$-representation of $\mathbf{A}$ is a direct sum of cyclic $*$-representations.

We say π is *nondegenerate* if $\pi(\mathbf{A})\mathcal{H}(\pi) := \text{Lin}\{\pi(a)\varphi : a \in \mathbf{A}, \varphi \in \mathcal{H}(\pi)\}$ is dense in $\mathcal{H}(\pi)$ and π is *faithful* if $\pi(a) = 0$ implies $a = 0$.

The famous Gelfand–Naimark theorem [KR83, 4.5.6] states that every C^*-algebra has a faithful $*$-representation. Thus, each C^*-algebra is isomorphic to a norm-closed $*$-subalgebra of $\mathbf{B}(\mathcal{H})$ for some Hilbert space $\mathcal{H}$.

A linear functional f on A is *positive* if $f(a^+a) \geq 0$ for all $a \in A$. Each positive functional f on A is *hermitian* (that is, $f(a^+) = \overline{f(a)}$ for $a \in A$) and continuous, and we have $\|f\| = f(1)$ if A is unital. A *state* on a C^*-algebra is a positive linear functional of norm 1. A state f on A is called *pure* if it is an extreme point of the convex set of all states of A.

By the GNS construction [Dv96, Theorem I.9.6], for each positive functional on a C^*-algebra A there exists a, unique up to unitary equivalence, cyclic $*$-representation ρ_f of A with cyclic vector φ such that $f(a) = \langle \pi(a)\varphi, \varphi \rangle$, $a \in A$. If f is a state, this $*$-representation ρ_f is irreducible if and only if f is pure.

In the rest of this appendix we discuss **real C^*-algebras**. The definitions of real Banach algebras and Banach $*$-algebras are the same as in the complex case.

Definition B.3 A *real C^*-algebra* is a real Banach $*$-algebra A such that

$$\|a\|^2 \leq \|a^+a + b^+b\| \quad \text{for} \quad a, b \in A. \tag{B.3}$$

The norm $\| \cdot \|$ of each real C^*-algebra satisfies the C^*-condition (B.2). Indeed, from (B.3), applied with $b = 0$, and (B.1) we obtain

$$\|a\|^2 \leq \|a^+a\| \leq \|a^+\| \, \|a\| \tag{B.4}$$

so that $\|a\| \leq \|a^+\|$ and hence $\|a\| = \|a^+\|$. Inserting this into (B.4) gives (B.2).

In contrast to complex C^*-algebras, (B.3) does not follow from the C^*-condition. (For the *real* $*$-algebra $A := \mathbb{C}$ with involution $z^+ = z$ and absolute value as norm, (B.2) is valid, but (B.3) is not for $a = 1$, $b = $ i, since $1^+1 + i^+i = 0$.)

A real Banach $*$-algebra A is a real C^*-algebra if and only if the C^*-condition (B.2) holds, and the element $1 + a^+a$ of the unitization A^1 is invertible in A^1 for all $a \in A$. (This follows by combining Corollary 5.2.1 and Proposition 7.3.4 in [Li03].) This characterization is often taken as the definition of a real C^*-algebra in the literature (for instance, in [Gol82]) and [DB86]).

Let A be a real C^*-algebra. A $*$-representation of A is a $*$-homomorphism into the $*$-algebra $\mathbf{B}(\mathcal{H})$ for a *real* Hilbert space $\mathcal{H}$. Many notions and results are similar to the complex case, but there are also fine distinctions; we will discuss one in the next paragraph. First we note that, by Schur's lemma [Li03, Proposition 5.3.7], a $*$-representation π of A is irreducible if and only if $\pi(A)'_{\text{her}} = \mathbb{R} \cdot I$.

Suppose π is irreducible. Let $a \in \pi(A)'$, $a \neq 0$. Then, since aa^*, $a^*a \in \pi(A)'_{\text{her}}$, there are real numbers α, β such that $aa^* = \alpha I$ and $a^*a = \beta I$. By $a \neq 0$, we have $\alpha \neq 0$ and $\beta \neq 0$. Hence a has a left inverse and a right inverse, so a is invertible and $\pi(A)'$ is a division algebra. Thus $\pi(A)'$ is a real normed division algebra. Therefore, by Mazur's theorem [Ri47, Theorem 1.7.6], $\pi(A)'$ is isomorphic to $\mathbb{R}$ or $\mathbb{C}$ or $\mathbb{H}$, the quaternions. All three cases can occur. Recall that, for an irreducible $*$-representation π of a complex C^*-algebra, $\pi(A)'$ is isomorphic to $\mathbb{C}$.

Let $\mathcal{X}$ be a locally compact Hausdorff space and $\tau : \mathcal{X} \mapsto \mathcal{X}$ a homeomorphism such that $\tau^2 = \mathrm{Id}$. Then the real $*$-subalgebra

$$C_0(\mathcal{X}; \tau) := \left\{ f \in C_0(\mathcal{X}) : f(\tau(x)) = \overline{f(x)},\ x \in \mathcal{X} \right\}$$

of $C_0(\mathcal{X})$ is a commutative real C^*-algebra. Note that if $\tau = \mathrm{Id}$, then $C_0(\mathcal{X}; \tau)$ is just the real C^*-algebra of real-valued functions of $C_0(\mathcal{X})$. By a theorem of Arens and Kaplansky [AK48], [Gol82, Theorem 10.7], each commutative real C^*-algebra is isomorphic to some $C_0(\mathcal{X}; \tau)$.

The counterpart of the Gelfand–Naimark theorem for real C^*-algebras is the following result of Ingelstam [I64], [Gol82, Theorem 15.3]: Any real C^*-algebra is isomorphic to a norm-closed real $*$-subalgebra of $\mathbf{B}(\mathcal{H})$ for some *real* Hilbert space $\mathcal{H}$.

Suppose $\mathcal{H}$ is a real Hilbert space and A is a norm-closed real $*$-subalgebra of $\mathbf{B}(\mathcal{H})$. From the formula $\|T\| = \sup \{|\langle T\varphi, \varphi\rangle| : \|\varphi\| = 1\}$ for $T = T^* \in \mathbf{B}(\mathcal{H})$ it follows easily that condition (B.3) is fulfilled, so A is a real C^*-algebra according to Definition B.3.

Further, let $\mathcal{H}_{\mathbb{C}}$ be the complexification of $\mathcal{H}$. For $T \in \mathbf{B}(\mathcal{H})$ let $T_{\mathbb{C}}$ denote the operator of $\mathbf{B}(\mathcal{H}_{\mathbb{C}})$ defined by (A.6). Recall that $\|T\| = \|T_{\mathbb{C}}\|$. Then the map $T \mapsto T_{\mathbb{C}}$ is an isometric (real) $*$-isomorphism of A on a real $*$-subalgebra of $\mathbf{B}(\mathcal{H}_{\mathbb{C}})$, and there is a (complex) $*$-homomorphim Ψ of the complexification $\mathsf{A}_{\mathbb{C}} = \mathsf{A} + i\mathsf{A}$ of A on a complex $*$-subalgebra $\Psi(\mathsf{A}_{\mathbb{C}})$ of $\mathbf{B}(\mathcal{H}_{\mathbb{C}})$ given by $\Psi(T + iS) = T_{\mathbb{C}} + iS_{\mathbb{C}}$, $T, S \in \mathsf{A}$. Clearly, $T_{\mathbb{C}} + iS_{\mathbb{C}} = 0$ implies $T = S = 0$, so Ψ is a $*$-isomorphism. Since A is norm-closed, $\Psi(\mathsf{A}_{\mathbb{C}})$ is norm-closed in $\mathbf{B}(\mathcal{H}_{\mathbb{C}})$ and hence a complex C^*-algebra. We define a C^*-norm on $\mathsf{A}_{\mathbb{C}}$ by $\|a\|_{\mathbb{C}} := \|\Psi(a)\|$ for $a \in \mathsf{A}_{\mathbb{C}}$. Since $\|T\| = \|T_{\mathbb{C}}\|$ for $T \in \mathbf{B}(\mathcal{H})$, we have $\|a\|_{\mathbb{C}} = \|a\|$ for $a \in \mathsf{A}$. By the preceding sketch of arguments we have shown that the complexification of the real C^*-algebra $(\mathsf{A}, \|\cdot\|)$ is the complex C^*-algebra $(\mathsf{A}_{\mathbb{C}}, \|\cdot\|_{\mathbb{C}})$.

Combining the results of the two last paragraphs it follows that the complexification $\mathsf{A}_{\mathbb{C}}$ of *each* real C^*-algebra A is a complex C^*-algebra. Thus, a real Banach $*$-algebra $(A, \|\cdot\|)$ is a real C^*-algebra if and only if there is a norm $\|\cdot\|_{\mathbb{C}}$ on the complexification $\mathsf{A}_{\mathbb{C}}$ which coincide with $\|\cdot\|$ on A such that $(\mathsf{A}_{\mathbb{C}}, \|\cdot\|_{\mathbb{C}})$ is a complex C^*-algebra. This result can be also taken as the definition of a real C^*-algebra.

Given a complex C^*-algebra B, it is natural to ask whether it is the complexification $\mathsf{A}_{\mathbb{C}}$ of a real C^*-algebra A. In this case, A is called a *real form* of B. For instance, the complex C^*-algebra $M_2(\mathbb{C})$ has the nonisomorphic real forms $M_2(\mathbb{R})$ and $\mathbb{H}$ (see Exercise 2.6). As shown by A. Connes (1977), there exists a von Neumann algebra that has no real form. In contrast, other von Neumann algebras (for instance, the hyperfinite II_1 factor) have a unique real form.

Appendix C
Locally Convex Spaces and Separation of Convex Sets

First we define **locally convex topologies** and **locally convex spaces**.

A brief introduction into the theory of locally convex spaces is given in [Cw90, Chap. IV]. An advanced treatment is the monograph [Sh71].

In this appendix, E and F are $\mathbb{K}$-vector spaces, where $\mathbb{K} = \mathbb{R}$ or $\mathbb{K} = \mathbb{C}$.

A *seminorm* on E is a mapping $p : E \mapsto [0, +\infty)$ such that $p(\lambda x) = |\lambda| p(x)$ and $p(x + y) \leq p(x) + p(y)$ for all $\lambda \in \mathbb{K}$ and $x, y \in E$.

Suppose Γ_E is a family of seminorms on E such that $p(x) = 0$ for all $p \in \Gamma_E$ implies $x = 0$. The *locally convex topology* defined by Γ_E is the topology τ on E for which the sets

$$\left\{ y \in E : p_1(x - y) \leq \varepsilon, \ldots, p_k(x - y) \leq \varepsilon \right\},$$

where $p_1, \ldots, p_k \in \Gamma_E, k \in \mathbb{N}, \varepsilon > 0$, form a base of neighborhoods of $x \in E$. By the separation assumption on Γ_E (i.e., "$p(x) = 0$ for all $p \in \Gamma_E$ implies $x = 0$"), this topology τ is Hausdorff.

The family Γ_E is called *directed* if, given $p_1, p_2 \in \Gamma_E$, there exists a $p \in \Gamma_E$ such that $p_j(x) \leq p(x)$ for $x \in E$, $j = 1, 2$. The family of seminorms $p_1 + \cdots + p_k$, where $p_j \in \Gamma_E$ and $k \in \mathbb{N}$, is directed and defines the same topology as Γ_E.

A *locally convex space* is a vector space E equipped with some locally convex topology τ; in this case we also write $E[\tau]$.

A net $(x_j)_{j \in J}$ of elements $x_j \in E$ converges to an element $x \in E$ in the locally convex space $E[\tau]$ if and only if $\lim_j p(x_j - x) = 0$ for all $p \in \Gamma_E$.

A locally convex topology is *metrizable*, that is, it is given by a metric, if and only if the topology can be defined by a *countable* family of seminorms.

A *Frechet space* is a complete metrizable locally convex space.

Let E and F be locally convex spaces with defining families of seminorms Γ_E and Γ_F, respectively. A linear mapping $T : E \mapsto F$ is *continuous* if and only if for each $q \in \Gamma_F$ there exist seminorms $p_1, \ldots, p_k \in \Gamma_E$ and a constant $c > 0$ such that

© The Editor(s) (if applicable) and The Author(s), under exclusive license
to Springer Nature Switzerland AG 2020
K. Schmüdgen, *An Invitation to Unbounded Representations of *-Algebras on Hilbert Space*, Graduate Texts in Mathematics 285,
https://doi.org/10.1007/978-3-030-46366-3

$$q(T(x)) \leq c(p_1(x) + \cdots + p_k(x)) \quad \text{for} \quad x \in E.$$

If Γ_E is directed, this holds if and only if for any $q \in \Gamma_F$ there are $p \in \Gamma_E$ and $c > 0$ such that $q(T(x)) \leq cp(x)$, $x \in E$.

The vector space of all continuous linear functionals on a locally convex space $E[\tau]$ is called the *dual space* of $E[\tau]$ and denoted by E'.

The family of *all* seminorms on a vector space E defines the *finest locally convex topology* τ_{st} on E. Each linear mapping of $E[\tau_{st}]$ into another locally convex space is continuous. In particular, each linear functional on E is τ_{st}-continuous, so the dual space of $E[\tau_{st}]$ is the vector space E^* of all linear functionals on E.

A *dual pairing* of vector spaces E, F is a bilinear mapping $(\cdot, \cdot) : E \times F \mapsto \mathbb{K}$. The *weak topology* $\sigma(E, F)$ is the locally convex topology on E defined by the family of seminorms $e \mapsto |(e, f)|$, where $f \in F$. The dual space of the locally convex space $E[\sigma(E, F)]$ is given by the linear functionals $e \mapsto (e, f)$, $f \in F$.

In the second part of the appendix we deal with the **separation of convex sets**.

In what follows, E denotes a real vector space. We denote by E^* the vector space of all linear functionals $f : E \to \mathbb{R}$.

A subset C of E is called *convex* if $\lambda a + (1 - \lambda)b \in C$ for $a, b \in C, \lambda \in [0, 1]$.

A point $x \in C$ is said to be an *extreme point* of a convex set C if $x = \lambda y + (1 - \lambda)z$ with $y, z \in C$ and $\lambda \in (0, 1)$ implies $x = y = z$.

A point $e \in E$ is called an *algebraically interior point* of a set C if, given $a \in E$, there exists a number $\lambda_a > 0$ such that $e + \lambda a \in C$ for $\lambda \in [-\lambda_a, \lambda_a]$.

The following two propositions are basic separation results for convex sets.

Proposition C.1 *Let A and B be nonempty disjoint convex sets in E such that A has an algebraically interior point. Then there exists a linear functional $f \in E^*$, $f \neq 0$, such that*

$$\sup\{f(b) : b \in B\} \leq \inf\{f(a) : a \in A\}. \tag{C.1}$$

Proof [Kö60, §17, 1, (3), p. 187] or [H75, Chap. I, §4, B, Corollary, p.15]. □

Proposition C.2 *Suppose E is a locally convex space and A and B are nonempty disjoint convex subsets of E. If B is compact and A is closed in E, then there exists a continuous linear functional $f : E \mapsto \mathbb{R}$ such that*

$$\sup\{f(b) : b \in B\} < \inf\{f(a) : a \in A\}. \tag{C.2}$$

Proof [Sh71, Chap. II, 9.2]. □

Definition C.3 A subset P of E is called a *cone* if

$$a + b \in P \quad \text{and} \quad \lambda a \in P \quad \text{for} \quad a, b \in P, \ \lambda \geq 0.$$

For a cone P in E, the *dual cone* is the cone $P^\wedge$ in E^* defined by

$$P^{\wedge} = \{f \in E^* : f(a) \geq 0 \quad \text{for} \quad a \in P\}. \tag{C.3}$$

If A is a cone P in E, it is obvious that the infima in (C.1) and (C.2) are zero, so the functionals f in Propositions C.1 and C.2 belong to the dual cone $P^{\wedge}$.

From now on we suppose P is a cone in E. The cone P gives rise to an ordering on the vector space E by defining "$a \leq b$ if and only if $b - a \in P$" for $a, b \in E$.

A linear functional $f \in P^{\wedge}$ is called *extremal* if $g \in P^{\wedge}$ and $f - g \in P^{\wedge}$ imply that $g = \alpha f$ for some number $\alpha \in [0, 1]$.

We say an element $e \in E$ is an *order unit* for P if, given $a \in E$, there exists a $\lambda > 0$ such that $-\lambda e \leq a \leq \lambda e$. A subspace X of E is called *cofinal* for P if for any $b \in E$ there exists an $x \in X$ such that $x \geq b$.

Suppose e is an order unit for P. Clearly, then $E = P - P$ and $\mathbb{R} \cdot e$ is cofinal. The relations $-\lambda e \leq a \leq \lambda e$ for $\lambda > 0$ imply that $e \geq 0$, that is, $e \in P$. Further, it follows easily that e is an algebraically interior point of P.

Except from the extremality the following propositions are standard facts; the first is [Cw90, Theorem 9.8] and the second follows from Propositions C.1, applied to $A = P$, $B = \{c\}$. We carry out the proofs to obtain the *extremality* of f.

Proposition C.4 *Suppose P is a cone in E and E_0 is a linear subspace of E which is cofinal in E for P. Let $P_0 = P \cap E_0$. Each extremal functional $f_0 \in (P_0)^{\wedge}$ on E_0 has an extension to an extremal functional $f \in P^{\wedge}$ on E.*

Proof Suppose f_1 is a linear functional on a linear subspace $E_1 \supseteq E_0$ of E such that f_1 is an extension of f_0, nonnegative on $P_1 := P \cap E_1$ and extremal in $(P_1)^{\wedge}$.

If $E_1 = E$ we are done, so we can assume that $E_1 \neq E$. We fix an element $b \in E$ such that $b \notin E_1$ and set $E_2 = E_1 + \mathbb{R} \cdot b$. Put

$$\gamma := \sup \{f_1(a) : a \in E_1, \, a \leq b\}, \quad \delta := \inf \{f_1(a) : a \in E_1, \, b \leq a\}. \tag{C.4}$$

Since $E_1 \supseteq E_0$ is cofinal, the sets in (C.4) are not empty and $\gamma \in \mathbb{R}$ and $\delta \in \mathbb{R}$ are defined. By $b \notin E_1$, there is a well-defined linear functional f_2 on E_2 given by

$$f_2(a + \lambda b) := f_1(a) + \lambda \gamma, \quad a \in E_1, \, \lambda \in \mathbb{R}.$$

We prove that f_2 is nonnegative on $P_2 := P \cap E_2$. Let $x = a + \lambda b \in P_2$, where $a \in E_1$ and $\lambda \in \mathbb{R}$. If $\lambda = 0$, then $x = a \in P_1$ and so $f_2(x) = f_1(a) \geq 0$.

If $\lambda > 0$, then $b \geq -\lambda^{-1}a$, hence $\gamma \geq f_1(-\lambda^{-1}a)$ by the definition of γ. Therefore, $\lambda\gamma \geq -f_1(a)$ and $f_2(x) = f_1(a) + \lambda\gamma \geq 0$.

Now suppose $\lambda < 0$. Then $a \geq -\lambda b$ and $(-\lambda)^{-1}a \geq b$, so $f_1((-\lambda)^{-1}a) \geq \delta$ by the definition of δ and $f_1(a) \geq -\lambda\delta$. Since f_1 is nonnegative on P_1 by assumption, $\gamma \leq \delta$. Therefore, since $\lambda < 0$, we obtain $f_2(x) = f_1(a) + \lambda\gamma \geq f_1(a) + \lambda\delta \geq 0$. This completes the proof of the assertion that f_2 is nonnegative on P_2.

Next we show that f_2 is extremal in $(P_2)^{\wedge}$. Take a functional $g \in (P_2)^{\wedge}$ such that $f_2 - g \in (P_2)^{\wedge}$. Then, in particular, $g(a) \leq f_2(a) = f_1(a)$ for $a \in P_1$. Since f_1 is extremal in $(P_1)^{\wedge}$, there exists an $\alpha \in [0, 1]$ such that $g(a) = \alpha f_1(a)$ for all $a \in E_1$. We prove that $g(b) = \alpha\gamma$.

Suppose $a \in E_1$ and $b \geq a$. Since $g \in (P_2)^{\wedge}$, $g(b) - \alpha f_1(a) = g(b - a) \geq 0$, so that $g(b) \geq \alpha f_1(a)$. Taking the supremum over a and using that $\alpha \geq 0$, we get $g(b) \geq \alpha\gamma$. On the other side, since $f_1(a) = f_2(a)$ and $f_2(b) = \gamma$, we derive

$$\gamma - g(b) - (1-\alpha)f_1(a) = \gamma - f_1(a) - g(b) + \alpha f_1(a) = f_2(b-a) - g(b-a) \geq 0,$$

so that $\gamma - g(b) \geq (1 - \alpha)f_1(a)$. Taking again the supremum over a, we obtain $\gamma - g(b) \geq (1 - \alpha)\gamma$ by $1 - \alpha \geq 0$. Hence $g(b) \leq \alpha\gamma$. Therefore, combining both inequalities proved in the preceding, we obtain $g(b) = \alpha\gamma$. Since $f_2(b) = \gamma$, we have $g = \alpha f_2$, which proves that f_2 is extremal.

The proof is completed by a standard Zorn's lemma argument. Consider the set of pairs (f_1, E_1) as above with the ordering $(f_1, E_1) \preceq (\tilde{f}_1, \tilde{E}_1)$ if $E_1 \subseteq \tilde{E}_1$ and $f_1 = \tilde{f}_1 \lceil E_1$. Zorn's lemma implies that there is a maximal pair (f_1, E_1). Then, by the preceding proof, we must have $E_1 = E$, $P_1 = P$, so $f := f_1 \in P^{\wedge}$ is an extremal extension of f_0. $\square$

Proposition C.5 *Suppose P is a cone in a real vector space E which has an order unit e. Then, for any element $c \in E$ such that $c \notin P$, there exists an extremal linear functional $f \in P^{\wedge}$ such that $f(e) = 1$ and $f(c) \leq 0$.*

Proof Let $E_0 = \mathrm{Lin}\{c, e\}$ and $P_0 := P \cap E_0$. We show that there is an extremal functional $f_0 \in (P_0)^{\wedge}$ such that $f_0(e) = 1$ and $f_0(c) \leq 0$. Then, since e is an order unit, $E_0 \supseteq \mathbb{R} \cdot e$ is cofinal in E, so the desired extension exists by Proposition C.4.

First suppose c and e are linearly dependent. Then $c = \alpha e$ with $\alpha \in \mathbb{R}$. Since $e \in P$ as noted above, $\alpha \leq 0$, so the functional f_0 defined by $f_0(\lambda e) = \lambda$ has the desired properties. From now on we assume that c and e are linearly independent.

Next note that $-e \notin P$. (Otherwise, since $e \in P$ is an order unit, we then obtain $c \in P$, which contradicts the assumption $c \notin P$.) Set

$$\gamma_0 := \sup\{\gamma : \gamma e \leq c\}, \quad \delta_0 := \inf\{\delta : c \leq \delta e\}. \tag{C.5}$$

Since e is an order unit, there are such numbers γ, δ. Further, $\gamma e \leq c \leq \delta e$ implies $(\delta - \gamma)e \geq 0$. Thus, since $-e \notin P_0$, it follows that $\delta \geq \gamma$ and hence $\delta_0 \geq \gamma_0$.

Since e and c are linearly independent, we can define a linear functional f_0 on E_0 by $f_0(c) = \gamma_0$ and $f_0(e) = 1$. By $c \notin P_0$ and $e \in P_0$, $\gamma e \leq c$ implies $\gamma < 0$, so that $f_0(c) = \gamma_0 \leq 0$.

We show that f_0 is P_0-positive. Let $x = \alpha e + \beta c \in P_0$ with $\alpha, \beta \in \mathbb{R}$.

First let $\beta = 0$. Then $x = \alpha e \geq 0$. By $-e \notin P_0$, we have $\alpha \geq 0$ and so $f_0(x) = \alpha \geq 0$.

Next let $\beta > 0$. Upon scaling we can assume that $\beta = 1$. Then, $x = \alpha e + c \geq 0$ and $c \geq -\alpha e$. Hence $-\alpha \leq \gamma_0$ by (C.5), so that $f_0(x) = \alpha + \gamma_0 \geq 0$.

Finally, let $\beta < 0$. Again, upon scaling by $-\beta^{-1}$, we assume $x = \alpha e - c \geq 0$, so that $\alpha e \geq c$. Therefore, $\alpha \geq \delta_0$ by (C.5) and $f_0(x) = \alpha - \gamma_0 \geq \alpha - \delta_0 \geq 0$.

We prove that f_0 is extremal in $(P_0)^{\wedge}$. Let $g \in (P_0)^{\wedge}$ be such that $f_0 - g \in (P_0)^{\wedge}$. If $\gamma e \leq c$, then $g(\gamma e) = \gamma g(e) \leq g(c)$. Since $e \in P_0$, we have $g(e) \geq 0$. Taking the supremum over γ we get $\gamma_0 g(e) \leq g(c)$. On the other side,

$$(f_0 - g)(\gamma e) = \gamma_0 - \gamma g(e) \le (f_0 - g)(c) = \gamma_0 - g(c)$$

gives $g(c) \le \gamma g(e) \le \gamma_0 g(e)$. Thus $g(c) = \gamma_0 g(e)$. Therefore, $g = g(e) f_0$. Since $e \in P_0$, we have $(f_0 - g)(e) = 1 - g(e) \ge 0$, so that $g(e) \in [0, 1]$. This completes the proof of the extremality of f_0. $\square$

References

[Ag15]   Agranovich, M.S., *Sobolev Spaces, Their Generalizations and Elliptic Problems in Smooth and Lipschitz Domains*, Springer, New York, 2015.

[Al83]   Alberti, P.M., A note on the transition probability over C^*-algebras, Lett. Math. Phys. **7**(1983), 107–112.

[Al03]   Alberti, P.M., Playing with fidelities, Rep. Math. Phys. **51**(2003), 87–125.

[Ac82]   Alcantara, J., Order properties of a class of tensor algebras, Publ. RIMS Kyoto Univ. **18**(1982), 539–550.

[An67]   Allan, G.R., On a class of locally convex algebras, Proc. London Math. Soc. (3) **17**(1967), 91–114.

[AIT02] Antoine, J.-P., A. Inoue, and C. Trapani, *Partial *-Algebras and Their Operator Realizations*, Kluwer Acad. Publ., Dordrecht, 2002.

[AT02]   Antoine, J.-P. and C. Trapani, Bounded elements in topological *-algebras, Studia Math. **203**(2011), 223–251.

[Ar99]   Ara, P., Morita equivalence for rings with involution, Alg. Repres. Theory **2**(1999), 227–247.

[Ar46]   Arens, R., The space L^ω and convex topological rings, Bull. Amer. Math. Soc. **52**(1946), 931–935.

[AK48]   Arens, R. and I. Kaplansky, Topological representation of algebras, Trans. Amer. Math. Soc. **63**(1948), 457–481.

[Ak72]   Araki, H., Bures distance function and a generalization of Sakai's non commutative Radon-Nikodym theorem, Publ. RIMS Kyoto Univ. **8**(1972), 335–362.

[Ak09]   Araki, H., *Mathematical Theory of Quantum Fields*, Oxford University Press, Oxford, 2009.

[B81]   Baaj, S., Multiplicateurs non bornes, Thesis, Universite Pierre et Marie Curie **37**(1981), 1–44.

[BIT01] Bagarello, F., A. Inoue, and C. Trapani, Unbounded C^*-seminorms and *-representations on partial *-algebras, Z. Anal. Anw. **20**(2001), 295–314.

[Ba61]   Bargmann, V., On a Hilbert space of analytic functions and an associated integral transform I, Commun. Pure Appl. Math. **14**(1961), 187–214.

[BV13]   Barrreira, L. and C. Valls, *Dynamical systems*, Universitexts, Springer-Verlag, London, 2013.

[BR77]   Barut, A.O. and R. Raczka, *Theory of Group Representations and Applications*, PWN, Warsaw, 1977.

[BCR84] Berg, C., J.P.R. Christensen, and P. Ressel, *Harmonic Analysis on Semigroups*, Springer-Verlag, New York, 1984.

© The Editor(s) (if applicable) and The Author(s), under exclusive license   363
to Springer Nature Switzerland AG 2020
K. Schmüdgen, *An Invitation to Unbounded Representations of *-Algebras on Hilbert Space*, Graduate Texts in Mathematics 285,
https://doi.org/10.1007/978-3-030-46366-3

[BGV04]   Berline, N., E. Getzler, and M. Vergne, *Heat Kernels and Dirac Operators*, Springer-Verlag, Berlin, 2004.

[BJS66]   Bers, L., P. John, and M. Schechter, *Partial Differential Equations*, Interscience, New York, 1966.

[Bh84]   Bhatt, S.J., An irreducible representation of a symmetric star algebra is bounded, Trans. Amer. Math. Soc. **292**(1985), 645–652.

[BIO01]   Bhatt, S.J., A. Inoue, and H. Ogi, Unbounded C^*-seminorms and unbounded C^*-spectral algebras, J. Operator Theory **45**(2001), 53–80.

[BIK04]   Bhatt, S.J., A. Inoue, and K.-D. Kürsten, Well-behaved unbounded operator representations and unbounded C^*-seminorms, J. Math. Soc. Japan **56**(2004), 417–445.

[BS87]   Birman, M.S. and M.Z. Solomyak, *Spectral Theory of Self-adjoint Operators in Hilbert Space*, Kluwer, Dordrecht, 1987.

[BdL59]   Bishop, E. and K. de Leeuw, The representation of linear functioals by measures on sets of extreme points, Ann. Inst. Fourier (Grenoble) **9**(1959), 305–331.

[B62]   Borchers, H.J., On the structure of the algebra of field operators, Nuovo Cimento **24**(1962), 214–236.

[B72]   Borchers, H.J., Algebraic aspects of Wightman field theory, in: *Statistical Mechanics and Field Theory*, R.N. Sen and C. Weil (Eds), Halsted Press, New York, 1972, pp. 31–79.

[BY75a]   Borchers, H.J. and J. Yngvason, On the algebra of field operators. The weak commutant and integral decomposition of states, Commun. Math. Phys. **42**(1975), 231–252.

[BY75b]   Borchers, H.J. and J. Yngvason, Integral representation for Schwinger functionals and the moment problem over nuclear spaces, Commun. Math. Phys. **43**(1975), 255–271.

[BJ26]   Born, M. and P. Jordan, Zur Quantenmechanik, Zeitschr. für Physik **34**(1926), 858–888.

[B26]   Born, M., Zur Quantenmechanik der Stossvorgänge, Zeitschr. für Physik **34**(1926), 863–867.

[BS94]   Bozejko, M. and R. Speicher, Completely positive maps on Coxeter groups, deformed commutation relations and operator spaces, Math. Ann **300**(1994), 97–120.

[BLW17]   Bozejko, M., E. Lytvynov, and J. Wysoczanski, Fock representations of Q-deformed commutation relations, J. Math. Phys. **57**(2017), 073501.

[BGJR88]   Bratteli, O., F. Goodman, P.E.T. Jorgensen, and D.W. Robinson, The heat semigroup and integrability of Lie algebras, J. Funct. Analysis **79**(1988), 351–397.

[BR87]   Bratteli, O. and D.W. Robinson, *Operator Algebras and Quantum Statistical Mechanics*, vol. 1, Second Edition, Springer, Berlin, 1987.

[BR97]   Bratteli, O. and D.W. Robinson, *Operator Algebras and Quantum Statistical Mechanics*, vol. 2, Springer, Berlin, 1997.

[BG08]   Buchholz, D. and H. Grundling, The resolvent algebra: A new approach to canonical quantum systems, J. Funct. Analysis **254**(2008), 2725–2779.

[BK91]   Burban, I.M. and A.U. Klimyk, On spectral properties of q-oscillator operators, Lett. Math. Phys. **29**(1991), 13–18.

[B69]   Bures, D.J.C., An extension of Kakutani's theorem on infinite product measure to the tensor product of semifinite W^*-algebras, Trans. Amer. Math. Soc. **35**(1969), 199–212.

[BH11]   Burns, K. and B. Hasselblatt, The Sharkowsky theorem: a natural direct proof, Amer. Math. Monthly **118**(2011), 229–244.

[BW01]   Bursztyn, H. and S. Waldmann, Algebraic Rieffel induction, formal Morita equivalence, and applications to deformation quantization, J. Geom. Phys. **77**(2001), 307–364.

[BW05]   Bursztyn, H. and S. Waldmann, Completely positive inner products and strong Morita equivalence, Pacific J. Math. **222**(2005), 201–236.

[CD58]   Cartier, P. and J. Dixmier, Vecteurs analytiques dans les representations de groups de Lie, Amer. J. Math. **80**(1958), 131–145.

[CGP]   Chaichian, M., H. Grosse, and P. Presnajder, Unitary representations of the q-oscillator algebra, J. Phys. A: Math. Gen. **27**(1994), 2045–2051.

[Ch81] Chernoff, P.R., Mathematical obstructions to quantization, Hadronic J. **4**(1981), 879–898.

[CV59] Civin, P. and B. Yood, Involutions on Banach algebras, Pacific J. Math. **9**(1959), 415–436.

[C08] Cimprič, J., Maximal quadratic modules on ∗-rings, Algebras Represent. Theory **11**(2008), 83–91.

[C09] Cimprič, J., A representation theorem for quadratic modules on ∗-rings, Canad. Math. Bulletin **52**(2009), 39–52.

[CSS14] Cimprič, J., Y. Savchuk, and K. Schmüdgen, On q-normal operators and the quantum complex plane, Trans. Amer. Math. Soc. **27**(2014), 2045–2051.

[Cw90] Conway, J.B., *A Course in Functional Analysis*, Springer-Verlag, New York, 1990.

[Co95] Coutinho, S.C., *A Primer on Algebraic D-Modules*, Cambridge Univ. Press, Cambridge, 1995.

[Dv96] Davidson, K.R., *C*-Algebras by Example*, Fields Inst. Monographs, Amer. Math. Soc., Providence, R.I., 1996.

[Ds71] Davies, E.B., Hilbert space representations of Lie algebras, Commun. Math. Phys. **23**(1971), 159–168.

[D25] Dirac, P.A.M., The fundamental equations of quantum mechanics, Proc. Royal Soc. London, Ser. A **109**(1925), 642–653.

[Di58] Dixmier, J., Sur la relation i$(PQ - QP) = 1$, Comp. Math. **13**(1958), 263–270.

[Di68] Dixmier, J., Sur la algebras de Weyl, Bull. Soc. Math. France **96**(1968), 209–242.

[Di77a] Dixmier, J., *Enveloping Algebras*, North-Holland Publ., Amsterdam, 1977.

[Di77b] Dixmier, J., *C*-Algebras*, North-Holland Publ., Amsterdam, 1977.

[DM78] Dixmier, J. and P. Malliavin, Factorisations de functions et de vecteurs indefiniment differentiables, Bull. Sci. Math. **102**(1978), 305–330.

[Dj70] Djoković, D.Z., Hermitian matrices over polynomial rings, J. Algebra **43**(1976), 359–374.

[DB86] Doran, R.S. and V.A. Belfi, *Characterizations of C*-Algebras*, Marcel Dekker, New York, 1986.

[DS13] Dowerk, P. and Y. Savchuk, Induced representations and C^*-envelopes of some quantum ∗-algebras, J. Lie Theory **23**(2013), 229–250.

[DH89] Dubin, D.A. and M.A. Hennings, Regular tensor algebras, Publ. RIMS Kyoto Univ. **25**(1989), 1001–1020.

[Dy15] Dubray, D., Hilbert space representations of the quantum ∗-algebra $U_q(su_{1,1})$, Banach J. Math. Anal. **9**(2015), 261–277.

[Em72] Emch, G. G., *Algebraic Methods in Statistical Mechanics and Quantum Field Theory*, Wiley, New York, 1972.

[F72] Fell, J.M.G., A new look at Mackey's imprimitivity theorem, Lecture Notes Math. **266**, Springer-Verlag, Berlin, 1972, pp. 43–58.

[FD88] Fell, J.M.G. and R.S. Doran, *Representations of ∗-Algebras, Locally Compact Groups, and Banach ∗-Bundles*, vol. 2, Academic Press, Boston, MA, 1988.

[FS73] Flato, M. and D. Sternheimer, Separate and joint analyticity in Lie group representations, J. Funct. Analysis **13**(1973), 268–276.

[FSSS72] Flato, M., J. Simon, H. Snellman and D. Sternheimer, Simple facts about analytic vectors and integrability, Ann. Scient. de l'Ecole Norm. Sup. **5**(1972), 323–434.

[F32] Fock, V., Konfigurationsraum und zweite Quantelung, Zeitschr. für Physik **75**(1932), 622–647.

[FGN60] Foias, C., L. Gehér, and B. Sz.-Nagy, On the permutability condition of quantum mechanics, Acta Sci. Math. (Szeged) **21**(1960), 78–89.

[FG63] Foias, C. and L. Gehér, Über die Weylschen Vertauschungsrelationen, Acta Sci. Math. (Szeged) **24**(1963), 97–102.

[F89] Folland, G.B., *Harmonic Analysis in Phase Space*, Princeton Univ. Press, Princeton, 1989.

[F95]   Folland, G.B., *A Course in Abstract Harmonic Analysis*, CRC Press, Boca Raton, 1995.

[Fr05]   Fragoulopoulou, M., *Topological Algebras with Involution*, North-Holland, Amsterdam, 2005.

[Fk90]   Frank, M., Self-duality and C^*-reflexity of Hilbert C^*-moduli, Z. Anal. Anwendungen **9**(1990), 165–176.

[FS89]   Friedrich, J. and K. Schmüdgen, n-Positivity of unbounded $*$-representations, Math. Nachr. **41**(1989), 233–250.

[Fr53]   Friedrichs, K.O., *Mathematical Aspects of Quantum Theory of Fields*, Interscience, New York, 1953.

[Fu82]   Fuglede, B., Conditions for two self-adjoint operators to commute or to satisfy the Weyl relation, Math. Scand. **51**(1982), 163–178.

[Ga47]   Gårding, L., Note on continuous representations of Lie groups, Proc. Nat. Acad. Sci. U.S.A. **33**(1947), 331–332.

[GS15]   Gebhardt, R. and K. Schmüdgen, Unbounded operators on Hilbert C^*-modules, Intern. J. Math. **26**(2015), 1550094, 1–48.

[GN43]   Gelfand, I.M. and M.A. Naimark, On the embedding of normed rings into rings of operators in Hilbert space, Mat. Sbornik **54**(1943), 197–243.

[GJ60]   Gillman, L. and M. Jerison, *Rings of Continuous Functions*, van Nostrand, Princeton, 1960.

[Gl67]   Gleason, A.M., A characterization of maximal ideals, J. Analyse Math. **19**(1967), 171–172.

[Gol82]   Goodearl, K.R., *Notes on Real and Complex C^*-Algebras*, Birkhäuser, Boston, 1982.

[GM90]   Goodearl, K.R. and P. Menal, Free and residually finite-dimensional C^*-algebras, J. Funct. Analysis **90**(1990), 391–410.

[G69a]   Goodman, R., Analytic and entire vectors for representations of Lie groups, Trans. Amer. Math. Soc. **143**(1969), 55–76.

[G69b]   Goodman, R., Analytic domination by fractional powers of a positive operators, J. Funct. Analysis **3**(1969), 246–264.

[GGH95]   Gotay, M.J., H.R. Grundling, and C.H. Hurst, A Groenewold-van Hove theorem for S^2, Trans. Amer. Math. Soc. **348**(1996), 1579–1597.

[GGT96]   Gotay, M.J., H.R. Grundling, and G.M. Tuynman, Obstruction results in quantization theory, J. Nonlinear Sci. **6**(1996), 469–498.

[Gt99]   Gotay, M.J., On the Groenewold-van Hove problem for $\mathbb{R}^{2n}$, J. Math. Phys. **40**(1999), 2107–2116.

[Gr46]   Groenewold, H.J., On the principles of elementary quantum mechanics, Physica **12**(1946), 405–460.

[Gk73]   Grothendieck, A., *Topological Vector Spaces*, Gordon and Breach, New York, 1973.

[Gu79]   Gudder, S.P., A Radon-Nikodym theorem for $*$-algebras, Pacific J. Math. **80**(1979), 141–149.

[GH78]   Gudder, S.P. and R.L. Hudson, A noncommutative probability theory, Trans. Amer. Math Soc. **245**(1978), 1–41.

[GS84]   Guillemin, V. and S. Sternberg, *Symplectic Techniques in Physics*, Cambridge Univ. Press, Cambridge, 1984.

[Hg55]   Haag, R., On quantum field theories, Dan. Mat. Fys. Medd. **29**(1955), Nr. 12.

[Hg92]   Haag, R., *Local Quantum Physics. Fields, Particles, Algebras*, Springer-Verlag, Berlin, 1992.

[Ha13]   Hall, B.C., *Quantum Theory for Mathematicians*, Springer, New York, 2013.

[H67]   Halmos, P.R., *A Hilbert Space Problem Book*, van Nostrand, Princeton, 1967.

[HW63]   Halmos, P.R. and L. J. Wallen, Powers of partial isometries, Indiana U. Math. **19**(1970), 657–663.

[Hs53]   Harish-Chandra, Representations of a semi-simple Lie group on a Banach space, Trans. Amer. Math. Soc. **75**(1953), 185–243.

[HK03] Hasselblatt, B. and A. Katok, *A First Course in Dynamics*, Cambridge Univ. Press, Cambrigde, 2003.

[He85] Hegerfeldt, G.C., Extremal decomposition of Wightman functions and of states on nuclear *-algebras by Choquet theory, Commun. Math. Phys. **45**(1975), 133–135.

[H02] Helton, J. W., Positive noncommutative polynomials are sums of squares, Ann. Math. **156**(2002), 675–694.

[Hg27] Heisenberg, W., Über den anschaulichen Inhalt der quantentheoretischen Kinematik und Mechanik, Zeitschr. für Physik **43**(1927), 172–198.

[HN12] Hilgert, J. and K.-H. Neeb, *Lie Groups*, Springer, New York, 2012.

[Ho90] Hofmann, G., On topological tensor algebras, Wiss. Z. KMU Leipzig, Math.-Naturw. R. **39**(1990), 598–622.

[H75] Holmes, R.B., *Geometric Functional Analysis*, Springer, New York, 1975.

[vH51] van Hove, L., Sur certaines representations unitaires d'un groupe infini de transformations, Proc. Roy. Acad. Sci. Belgium **26**(1951), 1–102.

[vH52] van Hove, L., Les difficulties de divergences pour un modele particulier de champ quantifie, Physics **18**(1952), 145–149.

[Ho88] Howe, R., The oscillator semigroup, Proc. Symp. Pure Math. **48**(1988), 61–132.

[HT92] Howe, R. and E.C. Tan, *Non-Abelian Harmonic Analysis*, Universitext, Springer-Verlag, New York, 1992.

[I64] Ingelstam, L., Real Banach algebras, Arkiv för Mat. **5**(1964), 239–270.

[In77] Inoue, A., Unbounded representations of symmetric *-algebras, J. Math. Soc. Japan **29**(1977), 219–232.

[In83] Inoue, A., A Radon-Nikodym theorem for positive linear functionals on *-algebras, J. Operator Theory **10**(1983), 77–86.

[In98] Inoue, A., *Tomita-Takesaki Theory on Algebras of Unbounded Operators*, Lecture Notes Math. **1699**, Springer-Verlag, Berlin, 1998.

[Ja70] Jakubovich, V.A., Factorization of symmetric matrix polynomials, Dokl. Acad. Nauk SSSR **194**(1970), 532–535.

[Jo94] Josza, R., Fidelity for mixed quantum states, J. Mod. Optics **4**(1994), 2315–2323.

[JSW95] Jorgensen, P.E.T., L.M. Schmitt, and R.F. Werner, Positive representations of general relations allowing Wick ordering, J. Funct. Analysis **143**(1995), 33–99.

[J42] Jung, H.W.E., Über ganze birationale Transformationen der Ebene, J. reine Angew. Math. **184**(1942), 161–174.

[K65] Kadison, R.V., Transformation of states in operator theory, Topology **2**(1965), Suppl. 2, 177–198.

[KR83] Kadison, R.V. and J.R. Ringrose, *Fundamentals of the Theory of Operator Algebras*, vol. I, Academic Press, New York, 1983.

[KZ68] Kahane, J.-P. and W. Zelazko, A charaterization of maximal ideals in commutative Banach algebras, Studia Math. **29**(1968), 339–343.

[Kp47] Kaplansky, I., Topological rings, Amer. J. Math. **59**(1947), 153–183.

[Ka63] Kato, T., On the commutation relation $AB - BA = C$, Arch. Rat. Mech. Anal. **10**(1963), 273–275.

[Ka67] Kato, T., *Perturbation Theory for Linear Operators*, Springer, Berlin, 1967.

[Kz68] Katznelson, Y., *An Introduction to Harmonic Analysis*, Dover, New York, 1968.

[Ke39] Keller, O.H., Ganze Cremona-Transformationen, Monatsh. Math Phys. **47**(1939), 299–306.

[KS97] Klimyk, A.U. and K. Schmüdgen, *Quantum Groups and Their Representations*, Springer, Berlin, 1997.

[Kn96] Knapp, A.W., *Lie Groups Beyond an Introduction*, Birkhäuser, Boston, 1996.

[KMRT98] Knus, M.-A., A. Merkurjew, M. Rost, and J.-P. Tignol, *The Book of Involutions*, Amer. Math. Soc., Coll. Publ. vol. **44**, Providence, R.I., 1998.

[Kö60] Köthe, G.,*Topological Vector Spaces I.*, Springer, Berlin, 1960.

[Lm99] Lam, T.Y., *Lectures on Modules and Rings*, Springer, New York, 1999.

[La94]   Lance, E.C., *Hilbert C*-Modules: A toolkit for operator algebraists*, Cambridge Univ. Press, Cambridge, 1994.

[La97]   Lance, E.C., Finitely presented C^*-algebras, in: *Operator Algebras and Applications*, A. Katavolous (Ed.), Kluwer Acad. Publ., Dordrecht, 1997, pp. 255–266.

[Ln17]   Landsman, K., *Foundations of Quantum Theory*, Springer Open, 2017.

[Lg85]   Lang, S., $SL(2, \mathbb{R})$, Springer, New York, 1985.

[Ls60a]   Langlands, R.P., Semi-groups and representations of Lie groups, Thesis, Yale University, 1960.

[Ls60b]   Langlands, R.P., Some holomorphic semigroups. Proc. Nat. Acad. Sci. **46**(1960), 361–363.

[Lr74]   Lassner, G., Über die Realisierung topologischer Tensoralgebren, Math. Nachr. **62**(1974), 89–101.

[Li03]   Bingren Li, *Real Operator Algebras*, World Scientific, Singapore, 2003.

[Lt33]   Littlewood D.E., On the classification of algebras, Proc. London Math. Soc. **35**(1933), 200–240.

[Mk99]   Markus, A., *Representation Theory of Group Graded Algebras*, Nova Science Publ. Inc, Commack, N.Y., 1999.

[Ms08]   Marshall, M., *Positive Polynomials and Sums of Squares*, Amer. Math. Soc., Providence, R.I., 2008.

[MV97]   Meise, R. and D. Voigt, *Functional Analysis*, Oxford Univ. Press, Oxford, 1997.

[My16]   Meyer, R., Representations by unbounded operators: C^*-hulls, local-global principle, and induction, Doc. Math. **22**(2017), 1375–1466.

[Mo13]   Moretti, V., *Spectral Theory and Quantum Mechanics*, Springer, Cham, 2013.

[MM74]   Morrel, B. and P. Muhly, Centered operators, Studia Math. **51**(1974), 251–263.

[Nb00]   Neeb, K.-H., *Holomorphy and Convexity in Lie Theory*, de Gryuter, Berlin, 2000.

[N59]   Nelson, E., Analytic vectors, Ann. Math. **70**(1959), 572–612.

[NS59]   Nelson, E. and W.F. Stinespring, Representations of elliptic operators in an enveloping algebra, Amer. J. Math. **81**(1959), 547–560.

[vN31]   von Neumann, J., Die Eindeutigkeit der Schrödingerschen Operatoren, Math. Ann. **104**(1931), 570–578.

[vN32]   von Neumann, J., *Mathematische Grundlagen der Quantenmechanik*, Springer-Verlag, Berlin, 1932.

[O96]   Ostrovskyi, V.L., On operator relations, centered operators, and nonbijective dynamical systems, Methods Funct. Anal. Topol. **2**(1996), 114–121.

[OPT08]   Ostrovskyi, V.L., D. Proskurin, and L. Turowska, Unbounded representations of q-deformation of Cuntz algebras, Lett. Math. Phys. **85**(2008), 147–162.

[OS88]   Ostrovskyi, V.L. and Yu. S. Samoilenko, Application of the projection spectral theorem to noncommuting families of operators, Ukr. Math. Zh. **40**(1988), 469–481.

[OS89]   Ostrovskyi, V.L. and Yu. S. Samoilenko, Unbounded operators satisfying non-Lie commutation relations, Rep. Math. Phys. **28**(1989), 91–103.

[OS99]   Ostrovskyi, V.L. and and Yu. S. Samoilenko, *Introduction to the Theory of Representations of Finitely Presented Algebras. I. Representations by bounded operators*, Cambridge Sci. Publishers, Cambridge, 2014.

[OS14]   Ostrovskyi, V.L. and K. Schmüdgen, A resolvent approach to the real quantum plane, Integral Equ. Oper. Theory **79**(2014), 451–476.

[O02]   Ôta, S., Some classes of q-normal operators, J. Operator Theory **48**(2002), 151–186.

[OS04]   Ôta, S. and F.H. Szafraniec, Notes on q-deformed operators, Studia Math. **165** (2004), 295–301.

[OS07]   Ôta, S. and F.H. Szafraniec, q-positive definiteness and related topics, J. Math. Anal. Appl. **329** (2007), 987–997.

[Pl01]   Palmer, Th.W., *Banach Algebras and the General Theory of *-Algebras*, vol. II, Cambridge Univ. Press, Cambridge, 2001.

[Pa79]   Pauli, W. , *Wissenschaftlicher Briefwechsel mit Bohr, Einstein, Heisenberg u.a.*, A. Hermann, K.v. Meyenn and V.S. Weiskopf (Eds),vol. I: 1918–1929, Springer, New York, 1979.

[Ph01]   Phelps, R.R., *Lectures on Choquet's theorem*, Lecture Notes Math. **1757**, Springer, Berlin, 2001.

[Po02]   Popovych, S., Monomial *-algebras and Tapper's conjecture, Methods Funct. Anal. Topol. **8**(2002), 70–75.

[Po08]   Popovych, S., On O^*-representability and C^*-representability of *-algebras, Preprint, Chalmers Univ., Göteburg, 2008.

[Pu72]   Poulsen, N.S., On C^∞-vectors and intertwining bilinear forms for representations of Lie groups, J. Funct. Analysis **9**(1972), 87–120.

[Pr98]   Proskurin, D., Homogenous ideals in Wick *-algebras, Proc. Amer. Math. Soc. **126**(1998), 3371–3376.

[Pr08]   Proskurin, D., Unbounded representations of q-deformation of Cuntz algebra, Lett. Math. Phys. **85**(2008), 147–162.

[Pw71]   Powers, R.T., Self-adjoint algebras of unbounded operators, Commun. Math. Phys. **21**(1971), 85–124.

[Pw74]   Powers, R.T., Self-adjoint algebras of unbounded operators II., Trans. Amer. Math. Soc. **187**(1974), 261–293.

[Pu67]   Putnam, C.R., *Commutation Properties of Hilbert Space Operators and Related Topics*, Springer, Berlin, 1967.

[PW89]   Pusz, W. and S.L. Woronowicz, Twisted second quantization, Rep. Math. Phys. **27**(1989), 591–431.

[RW91]   Rade, L. and B. Westergren, *Springer's Mathematische Formeln*, Springer, Berlin, 1991.

[RW98]   Raeburn I. and D.P. Williams, *Morita Equivalence and Continuous-Trace C^*-Algebras*, Amer. Math. Soc., Providence, R.I., 1998.

[Ra46]   Raikov, D.A., To the theory of normed rings with involution, Dokl. Akad. Nauk S.S.S.R. **54**(1946), 387–390.

[RS72]   Reed, M. and B. Simon, *Methods of Modern Mathematical Physics I. Functional Analysis*, Academic Press, New York, 1972.

[RS75]   Reed, M. and B. Simon, *Methods of Modern Mathematical Physics II. Fourier Analysis, Self-Adjointness*, Academic Press, New York, 1975.

[RS79]   Reed, M. and B. Simon, *Methods of Modern Mathematical Physics III. Scattering Theory*, Academic Press, New York, 1979.

[RS78]   Reed, M. and B. Simon, *Methods of Modern Mathematical Physics IV. Analysis of Operators*, Academic Press, New York, 1978.

[Re46]   Rellich, F., Der Eindeutigkeitssatz für die Lösungen der quantenmechanischen Vertauschungsrelationen, Nachr. Akad. Wiss. Gött., Math.-Phys. Klasse **1946**, 107–115.

[Ri84]   Richter, P., Zur Zerlegung von Darstellungen nuklearer Algebren, Wiss. Z. KMU Leipzig, Math.-Naturw. R. **33**(1984), 63–65.

[Ri47]   Rickart, C.E., The singular elements of a Banach algebra, Duke Math. J. **14**(1947), 1063–1077.

[Ri60]   Rickart, C.E., *General Theory of Banach Algebras*, van Nostrand, Princeton, 1960.

[Rf74a] Rieffel, M.A., Induced representations of C^*-algebras, Adv. Math. **13**(1974), 176–257.

[Rf74b] Rieffel, M.A., Morita equivalence for C^*-algebras and W^*-algebras, J. Pure Appl. Algebra **5**(1974), 51–94.

[Rn91]   Robinson, D. W., *Elliptic Operators and Lie Groups*, Clarendon Press, Oxford, 1991.

[Rg03]   Rosenberg, J., A selective history of the Stone-von Neumann theorem, in: *Operator Algebras, Quantization, and Noncommutative Geometry: A Centennial Celebration Honoring John von Neumann and Marshall H. Stone*, R.S. Doran and R.V. Kadison (Eds), Contemp. Mathematics vol. **365**, Amer. Math. Soc., Providence, R.I., 2003.

[RST78] Rubel, L. A., W.A. Squires, and B.A. Taylor, Irreducibility of certain entire functions with applications to harmonic analysis, Ann. Math. **108**(1978), 553–567.

[Ru74]   Rudin, W., *Real and Complex Analysis*, Second Edition, McGraw-Hill Inc., New York, 1974.

[Rk87]   Rusinek, J., Analytic vectors and integrability of Lie algebra representations, J. Funct. Analysis **74**(1987), 10–23.

[S91]   Samoilenko, Yu.S., *Spectral Theory of Families of Self-adjoint Operators*, Kluwer Acad. Publ., Dordrecht, 1991.

[T99]   Tapper, P., Embedding of ∗-algebras into C^*-algebras and C^*-ideals generated by words, J. Operator Theory **41**(1999), 351–364.

[SN17]   Sakurai, J.J. and J. Napolitano, *Modern Quantum Mechanics*, Cambridge Univ. Press, Cambridge, 2017.

[STS96]   Samoilenko, Yu. S., L.B. Turowska, and V.S. Shulman, Semilinear relations and their ∗-representations, Methods Funct. Anal. Topol. **2**(1996), 55–111.

[SS13]   Savchuk, Y. and K. Schmüdgen, Unbounded induced representations of ∗-algebras, Algebras Represent. Theory **16**(2013), 309–376.

[Sh71]   Schäfer, H.H., *Topological Vector Spaces*, Springer, New York, 1971.

[Sch78]   Schmüdgen, K., On trace representation of linear functionals on unbounded operator algebras, Commun. Math. Phys. **63**(1978), 113–130.

[Sch79]   Schmüdgen, K., A proof of a theorem on trace representation of strongly positive linear functionals on Op∗-algebras, J. Operator Theory **2**(1979), 39–47.

[Sch83a]   Schmüdgen, K., On domains of powers of closed symmetric operators, J. Operator Theory **9**(1983), 53–75.

[Sch83b]   Schmüdgen, K., On the Heisenberg commutation relation I, J. Funct. Analysis **50**(1983), 8–49.

[Sch83c]   Schmüdgen, K., On the Heisenberg commutation relation II, Publ. RIMS Kyoto Univ. **19**(1983), 601–671.

[Sch84]   Schmüdgen, K., Graded and filtrated topological ∗-algebras II, The closure of the positive cone, Rev. Roumaine Math. Pures at Appl. **29**(1984), 89–96.

[Sch86]   Schmüdgen, K., A note on commuting self-adjoint operators affiliated to properly infinite von Neumann algebras, Bull. London Math. Soc **16**(1986), 287–292.

[Sch90]   Schmüdgen, K., *Unbounded Operator Algebras and Representation Theory*, Akademie-Verlag, Berlin, and Birkhäuser, Basel, 1990.

[Sch91]   Schmüdgen, K., Non-commutative moment problems, Math Z. **206**(1991), 623–649.

[Sch94]   Schmüdgen, K., Integrable representations of $\mathbb{R}_q^2$, $X_{q,\gamma}$ and $SL_q(2, \mathbb{R})$, Commun. Math. Phys. **159**(1994), 217–237.

[Sch02]   Schmüdgen, K., On well-behaved unbounded representations of ∗-algebras, J. Operator Theory **48**(2002), 487–502.

[Sch05]   Schmüdgen, K., A strict Positivstellensatz for the Weyl algebra, Math. Ann. **331**(2005), 779–794.

[Sch09]   Schmüdgen, K., Noncommutative real algebraic geometry– some basic concepts and first ideas, in: *Emerging Applications of Algebraic Geometry*, M. Putinar and S. Sullivant (Eds), Springer, New York, 2009, pp. 325–350.

[Sch10]   Schmüdgen, K., Algebras of fractions and strict Positivstellensätze for ∗-algebras, J. reine Angew. Math. **647**(2010), 57–86.

[Sch12]   Schmüdgen, K., *Unbounded Self-adjoint Operators on Hilbert Space*, Springer, New York, 2012.

[Sch15]   Schmüdgen, K., Transition probabilities of positive functionals on ∗-algebras, J. Operator Theory **73**(2015), 443–463.

[Sch17]   Schmüdgen, K., *The Moment Problem*, Springer, Cham, 2017.

[Sz18]   Schötz, M., On characters and pure states of ∗-algebras, arXiv: 1811.04882v1.

[Sz19]   Schötz, M., Gelfand-Naimark theorems for ordered ∗-algebras, arXiv: 1906.08752v2.

[Schr26]   Schrödinger, E., Quantisierung als Eigenwertproblem. III, Ann. Phys. **79**(1926), 734–756.

[Sb79]   Sebestyen, Z., Every C^*-seminorm is automatically submultiplicative, Period. Math. Hungar. **10**(1979), 1–8.

[Sb84]   Sebestyen, Z., On representability of linear functionals on ∗-algebras, Period. Math. Hungar. **15**(1984), 233–239.

[Sb86]   Sebestyen, Z., States and ∗-representations I., Period. Math. Hungar. **17**(1986), 163–171.

[Se47a]   Segal, I.E., Postulates of General Quantum Mechanics, Ann. Math. **48**(1947), 930–948.

[Se47b]   Segal, I.E., Irreducible representations of operator algebras, Bull. Amer. Math. Soc. **53**(1947), 73–88.

[Se51]   Segal, I.E., A class of operator algebras which are determined by groups, Duke Math. J. **18**(1951), 221–265.

[Se52]   Segal, I.E., Hypermaximality of certain operators on Lie groups, Proc. Amer. Math. Soc. **3**(1952), 13–15.

[Se62]   Segal, I.E., Mathematical characterization of the physical vacuum for a linear Bose-Einstein field, Illinois J. Math. **6**(1962), 500–523.

[Sh62]   Shale, D., Linear symmetries of free boson fields, Trans. Amer. Math. Soc. **103**(1962), 149–167.

[Sh64]   Sharkovsky, A.N., Co-existence of cycles of a continuous mapping of the line into itself, Ukrain. Math. J. **16**(1964), 61–71.

[Sh65]   Sharkovsky, A.N., On cycles and structure of a continuous mapping, Ukrain. Math. J. **17**(1965), 104–111.

[SKSF]   Sharkovsky, A.N., S.F. Kolyada, A.G. Sivak and V.V. Federenko, *Dynamical Systems of One-dimensional Maps*, Kluwer Acad. Publ., Dordrecht, 1997.

[SF70]   Shirali, S. and J.W.M. Ford, Symmetry in complex Banach algebras II., Duke Math. J. **37**(1980), 275–280.

[S72]   Simon, J., On the integrability of finite-dimensional Lie algebra representations, Commun. Math. Phys. **28**(1972), 39–42.

[SS02]   Stochel, J. and F.H. Szafraniec, A pecularity of the creation operator, Glasgow Math. J. **44**(2002), 137–147.

[St30]   Stone, M.H., Linear transformations in Hilbert space, III. Operational methods and group theory, Prod. Nat. Acad. Sci. USA **16**(1930), 172–175.

[SW00]   Streater, R.F. and A.S. Wightman, *PCT, Spin and Statistics, and All That*, Princeton Univ. Press, Princeton, 2000.

[S01]   Summers, S.J., On the Stone-von Neumann uniqueness theorem and its ramifications, in: *John von Neumann and the Foundations of Quantum Physics*, M. Redei and M. Stöltzner (Eds), Kluwer, Dordrecht, 2001, pp. 135–152.

[Ty86]   Taylor, M.E., *Noncommutative Harmonic Analysis*, Amer. Math. Soc., Providence, R.I., 1986.

[Ti63]   Tillmann, H.G., Zur Eindeutigkeit der Lösungen der quantenmechanischen Vertauschungsrelationen, Acta Sci. Math. (Szeged) **24**(1963), 258–270.

[Tr67]   Treves, F., *Topological Vector Spaces, Distributions and Kernels*, Academic Press, New York, 1967.

[Ts05]   Tsuchimoto, Y., Endomorphisms of Weyl algebras and p-curvatures, Osaka J. Math. **42**(2005), 435–452.

[T02]   Turowska, L. B., On the complexity of the description of ∗-algebra representations, Proc. Amer. Math. Soc. **130**(2002), 3051–3065.

[U62]   Uhlmann, A., Über die Definition der Quantenfelder nach Wightman und Haag, Wiss. Z. KMU Leipzig, Math. Naturw. Reihe **11**(1962), 213–217.

[U76]   Uhlmann, A., The "transition probability" in the state space of a ∗-algebra, Rep. Math. Phys. **9**(1976), 273–279.

[VE00]   Van den Essen, A., *Polynomial Automorphisms*, Birkhäuser, Basel, 2000.

[VS90]   Vaysleb, E.Ye. and Yu. S. Samoilenko, Representations of operator relations by unbounded operators and multidimensional dynamical systems, Ukrain. Math. J. **42**(1990), 1011–1019.

[VS94]   Vaysleb, E.Ye. and Yu. S. Samoilenko, Representations of the relations $UF(A) = UFA$) by unbounded self-adjoint and unitary operators, Selecta Math. Sov. **13**(1994), 35–54.

[Va64]   Varadarajan, V.S., *Lie Groups, Lie Algebras, and Their Representations*, Springer, New York, 1964.

[Wa05]   Waldmann, S., States and representation theory in deformation quantization, Rev. Math. Phys. **17**(2005), 15–75.

[Wa07]   Waldmann, S., *Poisson–Geometrie und Deformationsquantisierung*, Universitexts, Springer, Berlin, 2007.

[Wi64]   Weil, A., Sur certains groupes d'opérateurs unitaires, Acta Math. **111**(1964), 143–211.

[Wr72]   Warner, G., *Harmonic Analysis in Semi-Simple Lie Groups*, Springer-Verlag, Berlin, 1972.

[Wi74]   Wichmann, J., Hermitian ∗-algebras which are not symmetric, London Math. Soc. **8**(1974), 102–112.

[Wi76]   Wichmann, J., On the symmetry of matrix algebras, Proc. Amer. Math. Soc. **54**(1976), 237–240.

[Wi78]   Wichmann, J., The symmetric radical of algebras with involution, Arch. Math. (Basel) **30**(1978), 83–88.

[Wie49]   Wielandt, H., Über die Unbeschränktheit der Schrödingerschen Operatoren der Quantenmechanik, Math. Ann. **121**(1949), 21.

[Wi47]   Wintner, A., The unboundedness of quanten-mechanical matrices, Phys. Rev. **71**(1947), 738–739.

[Wo70]   Woronowicz, S.L., The quantum problem of moments I, Rep. Math. Phys. **1**(1970), 135–145.

[Wo91]   Woronowicz, S.L., Unbounded operators affiliated with C^*-algebras and non-compact quantum groups, Commun. Math. Phys. **136**(1991), 399–432.

[WN92]   Woronowicz, S.L. and K. Napiórkowski, Operator theory in the C^*-algebra framework, Reports Math. Phys. **31**(1992), 353–371.

[Wo95]   Woronowicz, S.L., C^*-algebras generated by unbounded elements, Rev. Math. Phys. **7**(1995), 481–521.

[Yn73]   Yngvason J., On the algebra of test functions for field operators, Commun. Math. Phys. **34**(1973), 315–333.

[Yo96]   Yood B., C^*-seminorms, Studia Math. **118**(1996), 19-26.

[Z13]   Zimmermann, K., Positivstellensätze for the Weyl algebra, Ph.D. Thesis, Leipzig University, 2013.

Index

© The Editor(s) (if applicable) and The Author(s), under exclusive license
to Springer Nature Switzerland AG 2020
K. Schmüdgen, *An Invitation to Unbounded Representations of *-Algebras
on Hilbert Space*, Graduate Texts in Mathematics 285,
https://doi.org/10.1007/978-3-030-46366-3

Symbol Index

© The Editor(s) (if applicable) and The Author(s), under exclusive license
to Springer Nature Switzerland AG 2020
K. Schmüdgen, *An Invitation to Unbounded Representations of *-Algebras on Hilbert Space*, Graduate Texts in Mathematics 285,
https://doi.org/10.1007/978-3-030-46366-3

379

Printed in the United States
by Baker & Taylor Publisher Services

Printed in the United States
by Baker & Taylor Publisher Services